# LATIN AMERICAN INSECTS AND ENTOMOLOGY

**Lantern bugs and cicadas, from Madam Merian's** *Metamorphoses insectorum surinamensium* (1705). Note the imagined specimen at the bottom of the plate which combines features of the two insects.

# LATIN AMERICAN INSECTS AND ENTOMOLOGY

Charles L. Hogue

UNIVERSITY OF CALIFORNIA PRESS • Berkeley • Los Angeles • Oxford

University of California Press
Berkeley and Los Angeles, California

University of California Press
Oxford, England

Library of Congress Cataloging-in-Publication Data

Hogue, Charles Leonard.
    Latin American insects and entomology / by
Charles L. Hogue.
        p.    cm.
    Includes bibliographical references (p.        )
and index.
    ISBN 0-520-07849-7 (cloth)
    1. Insects—Latin America.   I. Title.
QL476.5.H64   1993
595.7098—dc20                                91-48184
                                                CIP

Printed in the United States of America

1   2   3   4   5   6   7   8   9


The paper used in this publication meets the minimum requirements of American National Standard
for Information Sciences—Permanence of Paper for
Printed Library Materials, ANSI Z39.48-1984 ⊗

IN MEMORIUM

To the great regret of family, his friends, and his colleagues, Charles Hogue died suddenly in mid-1992, while the manuscript for this book was being typeset. The Press would like to thank his son, James Hogue, for skillfully and meticulously seeing the book through the final stages of proofreading and printing. His care ensured that *Latin American Insects and Entomology*, Charles Hogue's last major work, would appear when and as his father would have wished.

The ubiquity of spirits and the impossibility of killing them seem to personify a feeling of helplessness in the face of an environment so beautiful and so cruel. On the river or working in a garden the sun hurts, "It is eating," the Sharanahua say, and heads ache for the rest of the day. The incessant gnats feed all day, and, as one lies in a hammock, someone leans over and slaps hard and says, "sandfly," and a black fly, fat with human blood, falls dead. Sundown is a moment of relief which even a hundred mosquitoes cannot mar.

In the forest someone shouts to warn of an *isula*, the huge stinging ants that make one drunk with pain, and, reaching for a handhold on a tree, one must avoid a swarm of red fire ants. Returning, one looks for ticks, huge tapir ticks, gray and voracious, or worse, the almost invisible tiny red ticks that burrow into the skin and hurt for a week. The women dig the egg sacs of chiggers out of toes skillfully so that the sac does not break to leave a budding worm to swell the foot, and they break each and every tiny egg with a needle so that it does not lie in wait for another bare foot. An infected gnat drops a worm's egg into the leg while sucking blood, and two weeks later the pain of the worm turning in the leg is excruciating, and it must be removed by daubing an old, foul, drop of tobacco juice on the skin and slowly winding the worm out on a stick. Women and girls pick lice out of men's hair and their own, break them in their teeth and eat them. When faced by a new animal or insect, I learned to ask both, "Do we eat it?" and "Does it eat us?"

Janet Siskind,
*To Hunt in the Morning*

# Contents

# Preface

The idea for this book germinated in my mind for many years after my christening in Latin American insect research. As a result of travels to many countries, it became acutely apparent to me that a comprehensive entomological work was sorely needed to serve the many people, both visitors and residents, interested in insects and their other terrestrial arthropod relatives. The curiosity of tourists and general natural historians needed satisfaction, and the more serious minded student and practicing professional needed an up-to-date review of the subject. After a long period of note taking and preliminary organization of my then chaotic knowledge of the subject, I resolved to fashion such a volume.

Some of my colleagues were incredulous that I could cover such a vast territory. But my experience writing general insect guides told me that with cautious choosing, I could make something really useful, though, of course, far from complete. In fact, writing this book has been an exercise in selectivity, especially with respect to the choice of taxonomic groups to include. I relied on my own experience and the experience of others in this process, and I have tried to give information on the most common, conspicuous, or otherwise notable (economically or historically important) units, whether species, genera, or higher groups. The other topics of discussion, such as ecology and the study of insects,

have also been presented with an eye to the reader's need for an overall understanding of what has transpired and is transpiring in Latin American entomology and to providing a framework for review and citation of pertinent literature.

Some who read this book will feel slighted because of the lack of coverage of topics of particular interest to them, or they may consider that important facts, taxa, or publications have been omitted. I can only ask these readers to remember the vastness of the subject and the necessity for extreme conservatism in choices of matter for inclusion. Of course, I welcome suggestions for additions or changes in emphasis for future editions.

I have designed the book to answer questions. In my language and in the selection of taxa/phenomena, and points of information about each, I have been guided by my perception of what most readers want to know rather than a desire to produce an encyclopedia of all the facts that might be recorded. The technical literature, to which I have so freely referred, will serve the latter purpose. Indeed, to present in-depth data, keys to identification, and exhaustive treatments of even the major categories of Latin American insect life would require many volumes and years of effort to produce and would not produce the ready, readable, and portable text that I think is most needed now.

# Acknowledgments

For reviewing sections and offering criticisms and information on many topics, I have been fortunate to have the expert assistance of many specialists. I am greatly indebted to the individuals named below for information, identifications, and countless improvements in the manuscript (the appearance of their names does not necessarily imply agreement with my final interpretations and statements): J. Richard Andrews (Nahuatl names), Arthur G. Appel (cockroaches), Phillip A. Adams (Neuroptera), Richard W. Baumann (Plecoptera), Vitor Becker (Lepidoptera), Jackie Belwood (Orthoptera), Harry Brailovsky (bugs), A. Brindle (Dermaptera), Jacob Brodzinsky (fossils), Lincoln P. Brower (butterflies), John C. Brown (Lepidoptera), Richard C. Brusca (Crustacea, arthropod anatomy), Gary R. Buckingham (useful insects), James L. Castner (Orthoptera), Gilbert L. Challet (aquatic beetles), John A. Chemsak (Cerambycidae), James C. Cokendolpher (harvestmen), Julian P. Donahue (Lepidoptera), John T. Doyen (darkling beetles), Richard D. Durtsche (study), W. D. Edmonds (dung scarabs), George F. Edmunds, Jr. (mayflies), K. C. Emerson (Mallophaga), Marc E. Epstein (Lepidoptera), Terry L. Erwin (beetles), Arthur V. Evans (beetles), Eric M. Fischer (Diptera), Oliver S. Flint, Jr. (Trichoptera), Will Flowers (aquatics), Manfredo A. Fritz (Sphecoidea), Richard C. Froeschner (Hemiptera), David G. Furth (ecology), Rosser W. Garrison (Odonata), P. Genty (palm pollinators), Edmund Giesbert (Cerambycidae), Eric E. Grissel (Chalcidoidea), Robert J. Gustafson (botany), Alan R. Hardy (Scarabaeidae), Brian R. Harris (butterflies), Steven Hartman (mantids), Henry A. Hespenheide (beetles), Frank T. Hovore (Cerambycidae), Chistopher A. Ishida (practical entomology), D. K. McE. Kevan, deceased (orthopteroids), James E. Keirans (ticks), John E. Lattke (general), Claude Lemaire (Saturniidae), Herbert W. Levi (spiders), James E. Lloyd (Lampyridae), Wilson R. Lourenço (scorpions), Richard B. Loomis, deceased (chiggers), Robert J. Lyon (gall wasps), Volker Mahnert (pseudoscorpions), Sergio Martínez (fossils), Mildred E. Mathias (botany), Eustorgio Méndez (medical entomology, parasites), Arnold S. Menke (Cynipidae and other Hymenoptera), Edward L. Mockford (psocids), Jacqueline Y. Miller (Castniidae), Michael J. Nelson (medical entomology), David A. Nickel (orthopteroids), M. W. Nielson (leafhoppers), Lois B. O'Brien (Homoptera), David L. Pearson (tiger beetles), Stewart B. Peck (cave insects), Norman D. Penny (Neuroptera and general), Don R. Perry (canopy insects), Manuel L. Pescador (mayflies), John T. Polhemus (aquatic Hemiptera), Diomedes Quintero Arias, Jr. (whip scorpions), Shivaji Ramalingam (mosquitoes), Edward S. Ross (Embiidina), H. F. Rowell (Acrididae), Raymond E. Ryckman (ectoparasitic bugs), Ann L. Rypstra (spiders), Jorge A. Santiago-Blay (scorpions), Jack Schuster (Passalidae), Terry N. Seeno (beetles), Rowland M. Shelley (myriapods), David R. Smith (Symphyta), Roy R. Snelling (Hymenoptera), Omelio Sosa, Jr. (history), Paul J. Spangler (aquatics), Lionel A. Stange (Neuroptera), Orley R. Taylor (honeybee), Donald B. Thomas (heteropterans and veterinary entomology), Carlos Trenary (history), Fred S. Truxal (Notonectidae), Alan Watson (Lepidoptera), Thomas K. Wood (Membracidae), A. Willink (history and study), Stephen L. Wood (Scolytidae), and Thomas J. Zavortink (mosquitoes).

For suggestions and information on diverse or multiple topics, I also have many other entomologists to thank, including, José Artigas, Tomás Cekalovic, Ana Lia Estévez, Carlos H. W. Flechtmann, Luis F. Jirón, Alberto and Beatriz Larraín, Carlos Machado Allison, and Irene Rut-Wais.

There are several colleagues who are responsible for more pervasive involvement in the work as a whole, who have reviewed and contributed to the entire manuscript, to whom I owe a special debt of gratitude: Terry L. Erwin, Arthur V. Evans, Gerardo Lamas, Jack Longino, Scott E. Miller, José Palacios-Vargas, Nelson Papavero, and Allen M. Young. For tolerance of my many probings of their knowledge and points of view, I am particularly appreciative of the contributions of Julian P. Donahue, Chris Nagano, Roy R. Snelling, and Fred S. Truxal.

I am also indebted to the many local friends and contacts I have made on my Latin American travels—guides, foresters, farmers, Native Americans, and many others, too numerous to name, who freely shared their firsthand field knowledge of insects and thus contributed to the originality of this book in countless ways.

To all who worked with the manuscript primarily in an editorial capacity, I wish to express gratitude, especially to Ernest Callenbach. Several librarians contributed in no small way by finding and interpreting many difficult areas of the literature: Nicole Bouché, Katharine E. S. Donahue, Jennifer Edwards, Donald McNamee, Kathy Showers (Los Angeles County Museum of Natural History) and Nancy Axelrod (University of California, Berkeley, Entomology Library). The preparation of the ink drawings were greatly assisted by Leland E. Dietz and his Xerox machine. My appreciation is also due to Don Meyers for his careful and considerate handling of my black-and-white photographic needs, as well as to Tracy Robertson and James Robertson for other technical assistance with the figures.

I thank James L. Castner, George Dodge, and James N. Hogue for allowing me to use those color photographs bearing their names in the captions.

Finally, I acknowledge my wife, Barbara, for her critical role in helping this task to completion, primarily her enormous patience with my needs for time and other of our mutual resources. She also typed most of my original drafts and did a great deal of editing. For these contributions, not only extended here but lovingly provided in support of most of my entomological career, I dedicate this book to her.

# Introduction

This work provides an introduction to the common and notable insects of Latin America and a foundation for their study. All the countries of the Western Hemisphere south of the U.S.–Mexican border are covered, including the West Indies, the continental islands, and the oceanic islands within 2,500 kilometers of and usually associated with isthmian America and South America (but not Bermuda or Isla de Pascua or Easter Island).

In discussing broad geographic regions, I use the terms "Latin America," "New World tropics," and "Neotropics" (and their adjectival forms) as precisely as possible. The first refers to the most inclusive political region within the bounds stated above; the second refers to lowland, moist to wet, forested biotypes only; and the third refers to the biogeographic region that includes South America, the West Indies, and tropical North America.

In this volume, I discuss true insects and other kinds of terrestrial arthropods and related creatures (arachnids, millipedes, centipedes, onychophorans, etc.) commonly thought of as "insectlike." In many places, it is overly cumbersome to be exact when referring to these groups, although I have tried to avoid misleading the reader by adding some explanatory phrase, such as "and other terrestrial arthropods," but I have found it difficult to be perfectly consistent in doing so. The meaning of a broader grouping is sometimes implied by the term "insect."

For each kind of insect or other arthropod discussed in a separate section in the systematic chapters, I open with a list of the applicable scientific names of the category, including the most commonly used synonyms, and its place in the nomenclatural hierarchy. This is followed by established vernacular names in Spanish, Portuguese, and other regional languages. I use the widely familiar "Quechua" in place of the more correct "Runasimi." No attempt is made to give a complete synonymy, as this would require a lengthy treatment of its own. The determination of plurals in some antique or indigenous languages is a problem, and some of those given may not find acceptance by purists. I have replaced a few common names, for example, "wax bugs" for the order Homoptera, "dragon-headed bugs" for *Fulgora* spp., and "big-legged bugs" for the family Coreidae. I feel that these are more appropriate and correct than previously used names and that they are useful to enhance common parlance about these often-mentioned taxa. There are a few new common names (e.g., "viper worms" for *Hemeroplanes* sp., Sphingidae; "flag moths" for the subfamily Pericopinae of the Arctiidae; and "shiny scarabs" for the subfamily Rutelinae of the Scarabaeidae.

The chapter on study of insects is included as information for novices and to enhance the use of the book for teaching and for ready reference for professionals. Much of the information has not been compiled elsewhere.

Citations to the literature were chosen according to certain constraints. First, they are always given as authority to, and source of further information on, topics of special interest, statistical statements, historical remarks, and such. I also include basic references to general subjects or sections or broad categories but only the most modern, comprehensive, and well-referenced, presuming that the reader will find in its own bibliography all the earlier pertinent literature. Minor taxonomic and mostly regional papers are excluded; classic or standard papers are included, even if short or outdated. References are given at the end of the sections to which they apply, rather than assembled together at the end of the book, to facilitate compiling literature by subject. Because I use this format, a few references have been repeated, obviating the need for cross listings and making it easier for the reader to find them. I have seen all references except those rare ones followed by "[Not seen.]." Some new observations are included and denoted by "(orig. obs.)." I cover the literature published to December 31, 1991.

This book is intended for the widest possible audience: students, lay people, travelers, teachers, and professional entomologists alike. The latter will forgive my use of nontechnical terms whenever possible; but generally, I follow vernacular ("wing covers") with technical synonyms ("elytra"). Nevertheless, the use of many technical terms has been unavoidable. They arc explained in chapter 1 or are available in most any general entomology textbook.

## Illustrations

I chose the species for the ink drawings in the systematic portion of the book as representative generally of those found in the text. All the drawings are my own, and most are based on museum specimens entirely or in part to confirm details present in existing illustrations. I have used the latter only casually as an aid to composition.

The majority of insects and other types illustrated in the line drawings are placed in stylized natural settings and are depicted as living animals. I have not drawn figures to scale, except that an attempt was made to indicate comparative size, that is, larger species are larger than smaller, although not in proportion. The need occasionally arose to greatly magnify microscopic forms. With the exception of those with the names of other photographers or sources noted in the captions, all the color and black-and-white photographs of insects and phenomena are mine also, converted from 35 mm transparencies taken of live specimens in the field.

## Abbreviations

Throughout the book, only the following three abbreviations are used: BL (body length, i.e., front of head to tip of abdomen); BWL (length of insect from front of head to tips of wings when folded in repose); WS (wingspan).

# 1 GENERAL ENTOMOLOGY

Entomology (Greek: *entomon* insect + *logos* discourse) is the scientific study of insects. A general knowledge of insects and their arthropod relatives is basic to an understanding of their classification and biology in Latin America. There are many good reference works (Parker 1982, Kükenthal 1923, Grassé 1949) and entomology textbooks (Barth 1972, Borror and DeLong 1969, Borror et al. 1981, Coronado et al. 1976, Daly et al. 1978, Carrera 1963, Hayward 1971, Lara 1977, Richards and Davies 1977, 1983) as well as comprehensive treatments of anatomy, physiology, classification, phylogeny, development, and behavior. Because these works are thorough and widely applicable, it would be redundant and inefficient to repeat their contents in a specialized book such as this. The remainder of this chapter offers a synopsis of fundamental topics and an explanation of terminology to make the book more usable and instructive for general readers.

## References

BARTH, R. 1972. Entomología geral. Fund. Insto. Oswaldo Cruz, Rio de Janeiro.

BORROR, D. J., AND D. M. DeLONG. 1969. Introdução ao estudo dos insetos. Ed. Edgard Bluecher, São Paulo. Brazilian edition translated and edited by D. D. Correa et al.

BORROR, D. J., D. M. DeLONG, AND C. H. TRIPLEHORN. 1981. An introduction to the study of insects. 5th ed. Saunders College, Philadelphia.

CARRERA, M. 1963. Entomologia para você. Ed. Univ. São Paulo, São Paulo.

CORONADO, P., R. A. MÁRQUEZ, AND A. MÁRQUEZ. 1976. Introducción a la entomología. Ed. Limusa, Mexico City.

DALY, H. V., J. T. DOYEN, AND P. R. EHRLICH. 1978. Introduction to insect biology and diversity. McGraw Hill, New York.

GRASSÉ, P., ed. 1949-. Traité de zoologie. Vols. 6–10. Masson, Paris.

HAYWARD, K. J. 1971. Guía para el entomólogo principiante. 2d ed. Univ. Nac. Tucumán, Insto. Miguel Lillo, Misc. no. 37: 1–159.

KÜKENTHAL, W. 1923-. Handbuch der zoologie. Vol. 4. Arthropoda. de Gruyter, Berlin.

LARA, F. 1977. Principios de entomología. Fac. Cien. Agr. Vet., Univ. Estad. Paulista "Julio de Mesquita Filho," São Paulo.

PARKER, S. B., ed. 1982. Synopsis and classification of living organisms. Vol. 2. McGraw-Hill, New York.

RICHARDS, O. W., AND R. G. DAVIES. 1977. Imms' general textbook of entomology, 10th ed. Vols. 1–2. Chapman and Hall, London.

RICHARDS, O. W., AND R. G. DAVIES. 1983. Tradado de entomología Imms. Vol. 1, Estructura, fisiología y desarrollo; Vol. 2, Clasificación y biología. Ed. Omega, Barcelona.

## HISTORY OF LATIN AMERICAN ENTOMOLOGY

Studies on Latin American insects and related arthropods began late in the history of biology because of the belated discovery of the New World by Europeans and its academic isolation for almost two centuries thereafter. The history of entomology in the region is best traced as a series of overlapping accomplishments by different categories of searchers and formats of investigation, rather than as traditional chronological periods that are here not readily identifiable. These categories are largely determined by the kinds of education available or customary at the time. The earliest disciples were broadly trained in philosophy, theology, or medicine; later, the narrower disciplines of natural history, biology, and zoology evolved. Not until very late in the nineteenth century did courses in entomology exist and still later full curricula leading to a degree in the subject.

No general discussion of Latin American entomology is available, although there are some regional treatments: Lizer y Trelles (1947) recounts the overall scene from South America, as do Lamas (1981) from Peru, Willink (1969) from Argentina, Kevan (1977) from the West Indies, Barrera (1955) from Mexico, Fernández (1978) from Venezuela, and Jirón and Vargas (1986) from Costa Rica. See also Bodenheimer (1929) and Chardon (1949) for many basic notes and Howard (1930) for events in the origins and growth of practical entomology in most parts of Latin America. Gilbert (1977) provides an index to published biographies of deceased entomologists. (In the following sketch, dates of birth and death of major figures are given in brackets []. The titles and publication dates of historically important works are woven into the text; because they are well known, they are not cited in the references.)

## Antiquity

There are many evidences of pre-Columbian appreciation for insects, arachnids, and myriapods among the classic civilizations of Mexico, Central America, and northwestern South America. Most references are to species that affected human health and welfare. Surviving Mayan (Stempell 1908) and other ancient Mexican murals and codices depict various species of economic and religious importance, including stingless bees, scorpions, and butterflies (Teotihuacán). Early Mexican pottery, also from the Teotihuacán period

(A.D. 200–800), are adorned with insect designs, and early Mochica pottery from the northern coast of Peru shows human figures clearly engaged in delousing and infested with the chigoe. Other representations of insect forms appear in sculptures, petroglyphs, and textiles from various cultures (Morge 1973, Tozzer and Allen 1910). Ancient languages and myths contain many entomological allusions, especially to noxious or ubiquitous species, for example, in Nahuatl, Xochiquetzal, butterfly flower goddess (Beutelspacher 1976).

## Chroniclers

With the arrival of Columbus, the insects of the New World became known to Western civilization. One might speculate that the lights seen on the shores of Hispaniola, that night of October 11, 1492, were not native camp fires but glowing *Pyrophorus* beetles and thus that an insect was the first thing sighted in America: "After the Admiral had spoken he saw the light once or twice and it was like a wax candle rising and falling" (*J. First Voyage*).

Among the conquistadors and colonists who followed were scribes and reporters appointed by the Spanish crown to chronicle the discoveries and bring the influence of Western thinking to the new settlements. Many of the sixteenth-century technical reports of the natural wonders of the newfound lands contained references to insects. One of the earliest, the *Historia General y Natural de las Indias, Islas y Tierrafirme del Mar Oceano* (first 20 volumes), was written by Gonzalo Oviedo in 1535. It described for the first time such American curiosities as the cucuyo (headlight beetle), chigger, chigoe, cochineal insect, and stingless bees. Mentions of the same conspicuous species appeared in other, similar treatments of the period, such as José de Acosta's *Historia Natural y Moral de las Indias* (1590) and Bernal Diaz's *Historia Verdadera de la Conquista de la Nueva España* (1568–

1632). Bernabé Cobo [1572–1659] wrote about white butterflies (*Ascia monuste*) that attacked crops in Lima in his *Historia del Nuevo Mundo* (1653), the most important work of the period on the natural history of Peru.

Fray Bernardino de Sahagún [?–1590] (fig. 1.1) completed his *Historia General de las Cosas de Nueva España* in 1560, but it was not published until the early nineteenth century. It described many insects, arachnids, and myriapods and was later accompanied by illustrations originally intended for it, but from which the text was long separated (*Codex Florentino*, fig. 1.2.). The work explained how the Aztecs treated black widow spider bites and scorpion stings and made special mention of useful

**Figure 1.2** Figure from Codex Florentino. The stinging arthropod is described by Sahagún in the early sixteenth century as a "scorpion," but in the figure, it is more similar to the larva of *Corydalus cornutus* (Megaloptera), called the "water dog" (perro del agua), an insect widely feared as venomous even today in Mexico.

insects such as the maguey worm and the cochineal bug (Curran 1937).

Francisco García Hernández [1514–1578] collected natural objects of medical significance in early colonial Mexico, including thirty insects and "worms." His manuscripts were published in various illustrated, annotated editions after his death, the best known being *Rerum Medicarum Novae Hispaniae Thesaurus sev Plantarum Animalium Mineralium Mexicanorum Historia* (published 1648–1651) (d'Ardois 1959–60, Barrera 1981) in which *Tractatus Quartus, Historia Insectorum Novae Hispaniae*, was ostensibly the first unified work on Latin American insects.

The chroniclers were savants not schooled in biology or in the methods of scientific investigation. Consequently, their statements sometimes contained considerable errors, these often the result of believing too literally the accounts of the Indians. But the firsthand recording of natural history by courtiers, travelers, explorers,

**Figure 1.1** Fray Bernardino de Sahagún, post-conquest chronicler of insect life in the New World. (Frontispiece from *Historia General de la Cosas de Nueva España*, Edition Pedro Robredo, Mexico, 1938)

traders, soldiers, missionaries, historians, and adventurers continued in subsequent centuries (e.g., A. de Herrera, *Historia General de las Indias Occidentalis* [1728]) and remains a tradition even today (e.g., Jacques Cousteau's *Amazon Journey* [1984]). The creatures described tend to be the same as those described by earlier authors, although stinging ants, large centipedes, Mexican wild silkworms, and tarantulas also appear.

## Naturalists

Following the chroniclers onto the scene were the naturalists, distinguished from the former by possessing some education in the biological sciences. An early example was George Marcgrave [1610–1644], who, during the Dutch invasion of Brazil in 1638–1644, traveled widely and studied insects in that country. An important edition of his works, citing many indigenous insects, was published in 1648 by one of his traveling companions, Guilielmus Piso, in *De Indiae Utriusque re Naturali et Medica Libri XIV*.

Following her ten-year stay (1690–1701) in Surinam, where she collected information on insect life histories, Madame Maria Sybilla Merian [1647–1717] (fig. 1.3) produced her famous *Metamorphoses Insectorum Surinamensium* (1705), with superbly done color plates (Erlanger 1976). The work contained some errors, including a confusion of the headlight beetles (*Pyrophorus*), cicadas, and dragon-headed bugs (*Fulgora*), that engendered misconceptions of the latter's ability to luminesce and stridulate which persist even today (one plate actually shows a mongrel insect, a cicada, bearing the head of *Fulgora*) (Frontispiece).

Later naturalists, following in this tradition and notable for significant observations on Latin American insects, were Hans Sloane, *A Voyage to the Islands Madera, Barbadoes, Nieves, St. Christophers and Jamaica* (1707, 1725); G. Gardner, *Travels in*

**Figure 1.3** Maria Sybilla Merian, famous for her observations of insect natural history in the Guianas in the seventeenth century. (Frontispiece from her botanical work, *Erucarum hortensis . . .* , Amsterdam, 1718)

*the Interior of Brazil* (1849); Thomas Belt, *A Naturalist in Nicaragua* (1874); Theodore Roosevelt, *Through the Brazilian Wilderness* (1914); Konrad Guenther, *A Naturalist in Brazil* (1931); R. Hingston, *A Naturalist in the Guiana Forest* (1932); and others.

## Renaissance Scholars

The first works on Latin American insects by those fully qualified as scientists were carried out by established European renaissance scholars in absentia. They received specimens and reports from collectors and correspondents on the scene but never set foot in the new lands themselves. Stingless bees were described in Konrad Gesner's encyclopedic *Historia Animalium* (1607) but were referenced therein to a work by one Andre Thevet, who "amonst other matters [in the New-found World] reporteth that he did see a company of flies of Honey-bees about a tree . . . : of which trees there were a great number in a hole that was in a tree, wherein they made Honey and Wax" (Topsel 1967). In *De Animalibus Insectis* (1602–1618), considered to be the world's first book on entomology, Ulisse Aldrovandi [1522–1605] wrote and figured some Latin American insects, including the cucuyo. This insect, by now famous, also appeared in Thomas Mouffet's *Theatrum Insectorum* (1634) alongside a rhinoceros beetle (*Megasoma*). Réné de Réaumur figures and describes in fine detail a dragon-headed bug (*Fulgora*) in his *Memoir pour servir à l'histoire des insectes* (1734–1742).

New World specimens were incorporated into the rapidly growing European collections of the time. Nehemiah Grew figures many from the cabinets of the Royal Society in England (*Museum regalis societatis*, 1685).

Culminating this phase of historical development were the great taxonomists, Carolus Linnaeus [1707–1778] and J. C. Fabricius [1745–1808], who were able to include a large number of species from the American tropics in their landmark editions of *Systema Naturae* (1st ed., 1735; 10th ed., 1758) and various *Systemas*, respectively.

## Collectors

Many of these descriptions were based on specimens provided by a new breed of naturalists to the region, the collectors. Some of the first, who made insect-catching trips to South America in the mid-1700s, were Pehr Loefling [1729–1756], Carl Dahlberg [1721–1781], Daniel Rolander [1726–1793], and Daniel Solander [1733–1782].

As travel to and conditions in the colonies improved, the number of collectors quickly increased, as did their range (Lamas 1979, 1981; Papavero 1971, 1973). The majority of these individuals worked independently, and many paid their expenses through the sale of their collections to museums and private collectors in the United States and Europe.

A famous duo of pioneer collectors was Karl von Martius [1794–1868] and Johann von Spix [1781–1826] (Fittkau 1983). In 1817–1820, they traversed much of eastern and Amazonian Brazil, collecting thousands of insects that were studied subsequently by European specialists. A later pair were Osbert Salvin [1835–1898] and Frederick Godman [1834–1919] who traveled widely and amassed specimens in Central America and Mexico in the late 1800s. Their extensive collections were assembled in London, systematically worked up by a number of entomologists, and published in a lengthy series of volumes under the title *Biologia Centrali-Americana* (1879–1915), a classic faunal report. Some of the libérés of the infamous French penal colonies in French Guiana from the late 1800s to 1938 made a living by supplying foreign markets with butterflies from the local jungles (Le Moult 1955). Today, many collectors, both commercial and voluntary, continue to provide material to specialists in their own and other countries.

## Scientific Expeditions

Other collectors and naturalists participated in or led the organized scientific or biological expeditions that are a very important part of the growth of entomological science in Latin America. These were often sponsored by governments or agencies and included multiple investigators, each with specialized assignments, and were much more elaborate than the simple forays of individuals. Primary examples are numerous and date from the early eighteenth century.

Antonio de Ulloa [1716–1795] was a military man appointed as Spanish crown representative to the French Académie de Science Expedition to South America in

1735–1746 with La Condamine to measure the length of an arc of the meridian at the equator. His *Noticias Americanas* (1772) contained specific mentions of equatorial insect life, including an account of a locust plague. The monumental expedition of the times, however, must be that of Baron Alexander von Humboldt [1769–1859] and Aimé Bonpland [1773–1858] to explore northern South America and Mexico in 1799–1804. Their extensive insect collections were researched by Latreille in Europe (Papavero 1971, chap. 4).

Other exemplary expeditions that furthered entomology in Latin America were several sea voyages with frequent land stops for collecting, such as the expeditions of the French vessel *La Coquille* (1822–1823), the Swedish *Eugenies* (1851–1852), and the Austrian *Novara* (1857–1859). Of special interest also were the Hamburger Südperu Expedition in 1936 (Titschack 1951–1954) and the Machris Brazilian Expedition of 1956 (entomologist F. Truxal; Delacour 1957). A modern example is the report of entomological results of the 1978–79 Danish Expedition to Patagonia and Tierra del Fuego (Madsen et al. 1980).

Later came those expeditions undertaken by investigators trained in biology or zoology who emphasized work with insects and who conducted studies on their own collections. Three categories of investigators may be recognized: visiting, expatriate, and native.

### Visiting Biologists and Zoologists

Perhaps the first biologist to produce significant entomological results from his own excursions in Latin America was Claudio Gay [1800–1873] (fig. 1.4), an ambitious French traveler who began to work with Chilean insects as early as 1836. Later, he published many research papers, his most important being *Historia Física y Política de Chile* (arthropod portions, 1849–1852).

About the same time, Charles Darwin

**Figure 1.4** Claudio Gay, born in France but first trained biologist to make a major contribution to Latin American entomology through his work in Chile. (From portrait in Universidad de Chile; courtesy of José Valencia)

[1809–1882] made his epic global voyage that included major sojourns in South America. He was inclined toward entomology and gained some insights into his revolutionary theory of natural selection from observations of South American insects. In the initial sentences of his introduction to the *Origin of Species* (1859), he states, "When on board H.M.S. 'Beagle,' as naturalist, I was much struck with certain facts in the distribution of the organic beings inhabiting South America . . . [which] throw some light on the origin of species." Some of these facts concerned the distribution of insects and insect examples of sexual selection (e.g., the Chilean stag beetle, *Chiasognathus*) much elaborated in his later *Descent of Man* (1871).

Also celebrated among itinerant biologists of this period was Henry Walter Bates [1825–1892] (fig. 1.5). He spent eleven years on the Amazon and its tributaries

**Figure 1.5**  Henry Walter Bates, first entomologist explorer of the Amazon Valley in the mid-1800s. (Frontispiece from *The Naturalist on the River Amazons*, John Murray, London, 1892)

(1848–1859) and collected some 14,000 specimens, including 8,000 species new to science. For the first five years of his travels, he was accompanied by Alfred Russel Wallace [1823–1913], who was also an avid collector but who chose to continue his studies in the Malay Archipelago where he produced his own theory of natural selection paralleling Darwin's. Wallace recounts his South American experiences in *A Narrative of Travels on the Amazon and Río Negro* (1853); Bates recounts his in a later parallel work, *The Naturalist on the River Amazons* (1892). Bates collected, but he also observed and analyzed, producing many papers on Neotropical Coleoptera. The work that distinguished him as an entomologist was his *Contributions to an Insect Fauna of the Amazon Valley, Lepidoptera: Heliconidae*, published in 1862 (Moon 1976).

The period of the early to mid-1800s was

a time of independence for most of the countries of Latin America and establishment of national universities, museums, and other learned institutions, with departments paying attention to terrestrial arthropod life forms. Many scholars from Europe emigrated to Latin America. Biology came of age, and considerable progress was made in the academic phases of entomology, primarily insect systematics. But agricultural and medical entomology, knowledge of pesticides, and the role of insects as vectors of disease awaited the threshold of the twentieth century.

Other visiting biologists of note were William Beebe [1877–1962], who was gifted with an extraordinary ability of expression and published on many aspects of Neotropical insect biology, for example, *High Jungle* (1949) (Berra 1977), and A. S. Calvert, who produced works on Costa Rican insects, including the book, written with his wife, *A Year of Costa Rican Natural History* (1917).

### Expatriate Biologists

A special group of early biologists who worked on insects were expatriates. They were trained in Europe or North America but were drawn to the Neotropics by its exotic and poorly known insect life. They brought with them their education from Western schools and did not merely travel to the New World but spent the rest of their days in their adopted homes. Deserving special mention in this category is Fritz Müller [1822–1897], born in Germany, who settled in Blumenau, Santa Catarina, Brazil, in 1852. He was a correspondent of Darwin and is best known for his discovery of the type of mimicry named after him.

Another outstanding expatriate biologist was German-born Hermann Burmeister [1807–1892]. After a sojourn in Brazil (1850), he settled in Argentina and became the director of the natural history museum in Buenos Aires and published many important entomological papers. A more modern example is Felix Woytkowski [1892–1966], who migrated from his native Poland to Peru in 1929. There he collected more than a thousand insect species new to science (Woytkowski 1978). Other notable expatriate entomologists were Hermann von Ihering [1850–1930], who emigrated from Germany to Brazil, Emilio Goeldi [1859–1917], Switzerland to Brazil, Adolfo Lutz [1855–1940], Germany to Brazil, Paul Biolley [1862–1909], Switzerland to Costa Rica, and Henri Pittier [1857–1950], Switzerland to Costa Rica.

### Native Zoologists

The first full-fledged biologist specializing in insects who was born in Latin America was Cuban Felipe Poey [1799–1891] (fig. 1.6). He left his birthplace, La Habana, to study in France but returned to produce works in entomology in his own land, especially on Lepidoptera (*Centuria de Lepi-*

**Figure 1.6**  Felipe Poey, first native-born Latin American entomologist. (From *Memorias de la Sociedad Cubana de Historia Natural,* 1, facing p. 8, 1915)

*dópteros de Cuba*, 1847). Clodomiro Picado [1887–1944] also left for study in France, completing his doctoral dissertation on Neotropical bromeliad communities. He returned to his native Costa Rica to become its most famous biologist. The Argentinian Arribálzaga brothers, Felix [1854–1894] and Enrique [1856–1935], were educated in their homeland where they carried out extensive studies on insect biology and taxonomy, especially on Diptera.

### The Entomologists

By reason of their generalized training, the biologists and zoologists could not be considered entomologists in the strict sense. But because of their scientific abilities, interest, and emphasis on investigation and publication, the title could be logically bestowed on them.

Full curricula in the discipline of entomology were not offered in universities until the very late nineteenth century, so professionals in the study of insect biology are virtually all twentieth-century products. Their numbers now range in the thousands. Who they are and the nature of their accomplishments are best appreciated by reference to the modern literature and bibliographies such as the *Zoological Record,* Parts Insecta, Arachnida, and Myriapoda.

The amateur entomologist deserves some special recognition here. An active cadre of educated and often highly sophisticated individuals exists who find pleasure in the study of insects. Most are collectors, perhaps the majority working with showy insects like butterflies and beetles, but not always merely for the sake of accumulating specimens. Many take advantage of financial security acquired in other enterprises to pursue serious questions in entomology. They may even find time to carry out investigations for which the professional finds no support and make valuable contributions directly in their own publications or in collaboration with professionals. They are therefore distinct from the dealers, whose primary aim in collecting is to profit financially from the sale of their catches.

## Practical Entomology

Mention of diverse pestiferous Latin American insects is common in the earliest chronicles and later works. The first reference to control was made by Francisco Hernández in his *Historia de los Insectos de Nueva España*, written in manuscript in about the mid-sixteenth century and stating that the Mexicans used a concoction of tobacco that they spread over the walls to kill fleas in a house (Hoeppli 1969:177). It may have been Henry Hawks, a Vera Cruz (Mexico) merchant, who provided one of the earliest clues to the connection between mosquitoes and human disease, when he wrote in 1572, "This towne is inclined to many kinde of diseases, by reason of the great heat, and a certeine gnat or flie which they call a musquito, which biteth both men and women in their sleepe. . . . Many there are that die of this annoyance" (Keevil 1957).

While evolutionary and taxonomic studies of insects continued following the birth of scientific entomology and expanded into the early twentieth century, the discovery of arthropod vectors of animal and human disease and the development of chemical control of crop pests fostered increased work in the applied phases of entomology in Latin America. In medical entomology, major strides were made in the battle against yellow fever and malaria because of the newfound knowledge that mosquitoes were the critical link in the spread of these diseases. It was the application of entomological principles by physicians Carlos Finlay [1833–1915] (fig. 1.7), Walter Reed [1851–1902], and William Gorgas [1854–1920] which rid La Habana of yellow fever in 1901 and made possible the construction

**Figure 1.7** Carlos Finlay, whose ideas led to the mosquito's role in transmission of yellow fever. (From a portrait formerly hung in the Regional Office of the Pan American Health Organization, Washington, D.C.; presently owned by Dr. J. Fermoselle-Bacardi, Coral Gables, Florida. Reproduced with owner's permission)

of the Panama Canal shortly thereafter (Le Prince et al. 1916, McCullough 1977). In 1909, Carlos Chagas [1879–1934] demonstrated that a lethal trypanosome parasite of humans (*Trypanosoma cruzi*) was transmitted by a kissing bug (*Panstrongylus megistus*).

Modern knowledge of agricultural entomology (Doreste et al. 1981, Howard 1930), the identification and control of crop pests, was primarily imported, workers and technology in Europe and North America largely determining the course of events. Although numerous references to pest insects are scattered throughout the literature of the sixteenth, seventeenth, and eighteenth centuries (Kevan 1977), possibly the earliest scientific investigation into what could really be called economic entomology was made in 1801 when a special commission composed of members of the General Assembly of the Bahamas was sent to the West Indies to look into damage done to cotton by red bugs (*Dysdercus*) and the chenille (*Alabama argillacea*).

In 1870, B. Pickman Mann, of Cambridge, was sent to Brazil with a commission authorized by Dom Pedro to study the country's natural history. He specialized in coffee and maize insects and prepared reports on each. In 1897, the Comisión para la Extinción la Langosta (antilocust commission) was established in Argentina, the first of many similar agencies, with appointed entomologists, to be formed in most countries during the early part of the twentieth century. Among the pioneers were W. H. T. Townsend and Johannes Wille in Peru, George Wolcott in Puerto Rico, and G. E. Bodkin in British Guiana. The earliest book on agricultural entomology was *Las Epidemias de las Plantas en la Costa del Perú* by Manuel García y Merino (1878).

Parasitoids and predators of several pests were introduced into problem areas with varying results by the 1930s (Myers 1931), and several sites became the scene of important experimental trials in biological control (Hagen and Franz 1973). Hopes were realized in the Brazilian Amazon fly (*Metagonistylum minense*, Tachinidae) for control of sugarcane moths. The sterile male technique for the control of screwworm was first tested successfully on the island of Curaçao in 1954.

Notes on the history of the various insects of commercial value in Latin America are to be found in the systematic portion of this book (see cochineal insects, silk moth, stingless bees, honeybee, etc.).

## References

BARRERA, A. 1955. Ensayo sobre el desarrollo histórico de la entomología en México. Rev. Soc. Mex. Entomol. 1: 23–38.

BARRERA, A. 1981. Notas sobre la interpretación de los artrópodos citados en el tratado cuarto, Historia de los insectos de Nueva España, de Francisco Hernández. Fol. Entomol. Mex. 49: 27–34.

BERRA, T. M. 1977. William Beebe, an annotated bibliography. Archon, Hamden, Conn.

BEUTELSPACHER, C. R. 1976. La diosa Xochiquetzal. Soc. Mex. Lepidop. Bol. Inf. 2: 1–3.

BODENHEIMER, F. S. 1929. Materialien zür Geschichte der Entomologie bis Linné. 1: 1–498; 2: 1–486. Junk, Berlin.

CHARDON, C. E. 1949. Los naturalistas en la América Latina. Sec. Agric., Pec. Col., Rep. Dominicana, Ciudad Trujillo.

CURRAN, C. H. 1937. Insect lore of the Aztecs. Nat. Hist. 39: 196–203.

D'ARDOIS, G. S. ed. 1959–60. Francisco Hernández, Obras completas. Univ. Nac. México, Mexico City. Vol. 1. 1960, Vida y obra de Francisco Hernández, España y Nueva España en la época de Felipe II by José Miranda.

DELACOUR, J. 1957. The Machris Brazilian Expedition: general account. Los Angeles Co. Mus. Contrib. Sci. 1: 1–11.

DORESTE, E., F. FERNÁNDEZ, AND P. P. PAREDES. 1981. Contribución a la historia de la entomología agrícola en Venezuela. 5th Cong. Venezolano Entomol. (Maracay), Mem. Pp. 29–50.

ERLANGER, L. 1976. Maria Sybilla Merian, seventeenth-century entomologist, artist and traveler. Ins. World Dig. 3(1): 12–21.

FERNÁNDEZ, F. 1978. Contribución a la historia de la entomología en Venezuela. Red. Fac. Agron. (Maracay) 26: 11–27.

FITTKAU, E. J. 1983. Johann Baptist Ritter von Spix, sein Leben und sein wissenschaftliches Werk. Spixiana suppl. 9: 11–18.

GILBERT, P. 1977. A compendium of the biographical literature on deceased entomologists. Brit. Mus. Nat. Hist., London.

HAGEN, K. S., AND J. M. FRANZ. 1973. A history of biological control. *In* R. F. Smith, T. E. Mittler, and C. N. Smith, eds., History of entomology. Annual Reviews, Palo Alto. Pp. 433–476.

HOEPPLI, R. 1969. Parasitic diseases in Africa and the Western Hemisphere, early documentation and transmission by the slave trade. Acta Trop. suppl. 10: 1–240.

HOWARD, L. O. 1930. A history of applied entomology. Smithsonian Misc. Coll. 84: 1–564. (See Pt. VI, South and Central America and the West Indies, 417–462.)

JIRÓN, L. F., AND R. G. VARGAS. 1986. La entomología en Costa Rica: Una reseña histórica. Rev. "Quipu" de Historia de la Ciencia (México) 3(1): 67–77.

KEEVIL, J. J. 1957. Medicine and the Navy 1200–1900. Vol. 1. E. and S. Livingstone, Edinburgh.

KEVAN, D. K. McE. 1977. Mid-eighteenth-century entomology and helminthology in the West Indies: Dr. James Grainger. Soc. Bibliog. Nat. Hist. J. 8: 193–222.

LAMAS, G. 1979. Otto Michael (1859–1934), el cazador de mariposas del Amazonas. Col. Suiza Perú Bol. 1979(2): 35–38.

LAMAS, G. 1981 [1980]. Introducción a la historia de la entomología en el Perú. Rev. Peruana Entomol. 23: 17–37.

LE MOULT, E. 1955. Mes chasses aux papillons. Ed. Pierre Horay, Paris.

LE PRINCE, J. A., A. J. ORENSTEIN, AND L. O. HOWARD. 1916. Mosquito control in Panama. Putnams, New York.

LIZER Y TRELLES, C. A. 1947. Introducción e historia de la entomología. Argentino Cien. Nat. "Bernardino Rivadavia," Publ. Ext. Cul. Didac. 1 (Curso de entomología) (Buenos Aires), 1–52. (See p. 20ff., Historia de la entomología sudamericana.)

McCULLOUGH, D. 1977. The imperturbable Dr. Gorgas. In D. McCullough, The path between the seas: The creation of the Panama Canal 1870–1914. Simon & Schuster, New York. Pp. 405–426.

MADSEN, H. B., E. S. NIELSEN, AND S. ODUM, eds. 1980. The Danish scientific expedition to Patagonia and Tierra del Fuego 1978–1979. Geogr. Tidsskr. 80: 1–28.

MOON, H. P. 1976. Henry Walter Bates F.R.S. 1825–1892, Explorer, scientist and Darwinian. Leicestershire Museums, Leicestershire.

MORGE, G. 1973. Entomology in the Western world in antiquity and in medieval times. In R. F. Smith, T. E. Mittler, and C. N. Smith, eds., History of entomology. Annual Reviews, Palo Alto. Pp. 37–80.

MYERS, J. G. 1931. A preliminary report on an investigation into the biological control of West Indian insect pests. Empire Mrkt. Bd. Publ. 42:1–173.

PAPAVERO, N. 1971, 1973. Essays on the history of Neotropical dipterology, with special reference to collectors (1750–1905) 2 vols. Mus. Zool. Univ. São Paulo, São Paulo.

STEMPELL, W. 1908. Die Tierbilder der Maya-handschriften, Zeit. Ethnol. 1908: 704–743.

TITSCHACK, E., ed. 1951–1954. Beiträge zür Fauna Perus: Nach der Hamburger Südperu-Expedition 1936, anderer Sammlungen, wie auch auf Grund von Literaturangaben. Vols. 1–4. Fischer, Jena.

TOPSEL, E. 1967 [1658]. The history of four-footed beasts and serpents and insects. Vol. 2. Da Capo, New York.

TOZZER, A. M., AND G. M. ALLEN. 1910. Animal figures in the Maya Codices. Harvard Univ., Peabody Mus. Pap. Amer. Archaeol. Ethnol. 4: 273–372, pls. 1–39.

WILLINK, A. 1969. Contribución a la historia de la entomología Argentina. Univ. Nac. Tucumán, Fund. Insto. Miguel Lillo Misc. 28: 1–30.

WOYTKOWSKI, F. 1978. Peru, my unpromised land. Smithsonian Inst. and Natl. Sci. Found., Washington, D.C.

## LATIN AMERICAN ENTOMOLOGY TODAY

We have seen how the foundations of entomology in Latin America were laid, through four centuries of effort by many types of investigators: chroniclers, general observers, renaissance scholars, collectors, and trained entomologists. By the middle of the twentieth century, there was a firm establishment of the trend toward specialization, begun first with the choice of systematists to study certain limited taxa, followed by the separation of the applied agricultural and medicoveterinary fields and maturation of the principles of insect ecology and genetics.

An important modern specialization has been the strong interest in tropical biology by a large number of local students and young entomologists from North America and Europe. The Organization for Tropical Studies has been a prime mover in this area, principally through training at field stations in Costa Rica. Fundamental discoveries are now being made about the ecological and evolutionary strategies of insects in the humid lowland environments, for example, the theory of Pleistocene relictual centers of distribution (Brown 1982), the theory of island biogeography (MacArthur and Wilson 1967), and the organization of insect societies (Wilson 1971).

The vindication of the idea of continental drift (largely from data collected during the International Geophysical Year, 1957–1958) has opened the door to understanding many enigmatic distributions of Latin American insect species. These had been explained by improbable oversea dispersals and land bridges. Amphinotic patterns of austral forms are now best explained by their origination on the composite continent, Gondwanaland. A landmark study in this area is Lars Brundin's, *Transantarctic Relationships and Their Significance, as Evidenced by Chironomid Midges* (Brundin 1967).

Operationalism (analysis according to a prescribed and consistent procedure) has now become a canon of phylogenetic studies on Neotropical groups, thanks to the logic forced into the procedures of taxonomists by the approaches of phenetics, promoted in R. Sokal and P. H. A. Sneath's, *Principles of Numerical Taxonomy* (1963), and cladistics, found in W. Hennig's, *Grundzüge einer Theorie der Phylogenetischen Systematik* (1950). "Trees of descent," or phyletic dendrograms, now are superimposed on patterns of origin and dispersion ("area diagrams"). Also of great promise in the study of insect systematics, physiology, and genetics are newly developed molecular and biochemical techniques for obtaining comparative data from insect tissues for nucleic acid sequences, metabolic enzymes and other proteins, immunological compatibility, and gene probing (Hillis and Moritz 1990). The modern miniaturization and improved availability of computers and mathematical techniques for analysis of morphology and distributions has helped immensely in making the use of these logical procedures feasible. Specific entomological expeditions to remote and unexplored areas and fieldwork by both local and foreign specialists are making possible discoveries in all geographic and study areas (e.g., Madsen et al. 1980).

Today, entomology continues to be a rapidly developing field of science, seeking to understand the diversity of insect life in Latin America and its significance to humankind. In most every country, governments and private enterprise are recognizing the importance of insect forms and employing entomologists. There has been an increase in the numbers of positions filled by local graduates, although workers trained or imported from other countries still fill many posts. Educated amateurs also remain important contributors. Gradually, with the help of new technologies for acquiring, recording, and dispensing knowledge, the fauna is becoming known, local biological phenomena are being revealed and integrated into universal schemes, losses from destruction of food and fiber and disease are being reduced, and an appreciation for the value of insects and their arthropod relatives is being realized. The need remains for more facilities and fuller staffing of research institutions and greater local activity, including the popularization of insect natural history to the general public.

## References

BROWN, JR., K. S. 1982. Paleoecology and regional patterns of evolution in Neotropical forest butterflies. *In* G. T. Prance, ed., Biological diversification in the tropics. Columbia Univ. Press, New York. Pp. 255–308.

BRUNDIN, L. 1967. Insects and the problem of austral disjunctive distribution. Ann. Rev. Entomol. 12: 149–168.

HENNIG, W. 1950. Grundzüge einer Theorie der Phylogenetischen Systematik. Deutcher Zentralverlag, Berlin.

HILLIS, D. M., AND C. MORITZ. 1990. Molecular systematics. Sinauer Associates, Sunderland, Mass.

MACARTHUR, R. H., AND E. O. WILSON. 1967. The theory of island biogeography. Princeton Univ. Press, Princeton.

MADSEN, H. B., E. S. NIELSEN, AND S. ODUM, eds. 1980. The Danish scientific expedition to Patagonia and Tierra del Fuego 1978–1979. Geogr. Tidssk. 80: 1–28.

SOKAL, R. R., AND P. H. A. SNEATH. 1963. Principles of numerical taxonomy. Freeman, San Francisco.

WILSON, E. O. 1971. The insect societies. Harvard Univ. Press, Cambridge.

# INSECT STRUCTURE AND FUNCTION

The physical form and body workings (Snodgrass 1935, 1952; Wigglesworth 1984; Chapman 1982; Kerkut and Gilbert 1984; King and Akai 1982–1984; Manton 1977; Rockstein 1964–1974, 1978; Smith 1968; Treherne et al. 1963–1982) of insects and their terrestrial arthropod relatives are as remarkable and complex as those of any animal type. Numerous structural and functional systems will be used in the text following as organizational topics for a basic review.

## References

CHAPMAN, R. F. 1982. The insects: structure and function. 3d ed. American Elsevier, New York.

KERKUT, G. A., AND L. I. GILBERT. 1984. Comprehensive insect physiology, biochemistry and pharmacology. Vols. 1–13. Pergamon, Elmsford, N.Y.

KING, R. C., AND H. AKAI, eds. 1982–1984. Insect ultrastructure. Vols. 1–2. Plenum, New York.

MANTON, S. M. 1977. The Arthropoda, habits, functional morphology, and evolution. Clarendon, Oxford.

ROCKSTEIN, M., ed. 1964–1974. The physiology of insects. Vols. 1–6. Academic, New York.

ROCKSTEIN, M., ed. 1978. Biochemistry of insects. Academic, New York.

SMITH, D. S. 1968. Insect cells: Their structure and function. Oliver and Boyd, Edinburgh.

SNODGRASS, R. E. 1935. Principles of insect morphology. McGraw-Hill, New York.

SNODGRASS, R. E. 1952. A textbook of arthropod anatomy. Comstock, Ithaca.

TREHERNE, J. E., M. J. BERRIDGE, AND V. B. WIGGLESWORTH. 1963–1982. Advances in insect physiology. Vols. 1–16. Academic, New York.

WIGGLESWORTH, V. B. 1984. Insect physiology. 8th ed. Chapman and Hall, London.

## Integument

The outer, living epidermis in insects is a single layer of generally simple, cuboidal cells that secrete an external nonliving cuticle. The cuticle (Neville 1975) is very durable and resistant because of its composition of waterproof waxes and complex molecules of such substances as chitin (a nitrogenous polysaccharide) and sclerotin (protein). In combination with the epidermis, it forms the insect's integument (Hepburn 1976).

The integument may be generally thin and flexible, as in insect larvae, or thick and rigid, as in most adults and larval structures such as the head. Rigidity is the result of the abundance of sclerotin, and hard areas, or "sclerites," are said to be well sclerotized. Flexibility is allowed by membranous joints or articulations between the rigid portions. Thus, the integument gives the insect its basic form and is its primary protective system, forming a barrier to water loss and entry of pathogenic microorganisms as well as providing resistance to physical trauma.

Sclerites may also be separated by infoldings, known as apodemes (if linear, called sutures; if pitlike, apophyses). It is to the internal portions of apodemes that the main muscles of motion are attached, giving the integument a secondary function, that of a skeleton (exoskeleton).

The cuticle derives its color not only from its structural components but from infusions of pigments (Crowmartie 1959) and microstructural developments (lamellae, gratings, etc.) that cause scattering, refraction, and defraction of light waves striking them, resulting in spectral phenomena. Among the pigments are common colored compounds such as melanin (black), pterines (white, red, yellow), carotenoids (red, brown), carminic acid (carmine), and flavones (red, yellow). Physical colors are often metallic or iridescent blues, greens, and reddish hues. Many Neotropical butterflies are beautifully colored from combinations of both pigmentary and physical colors localized in the wing scales (Ghiradella 1984). Gold and silver are interference colors also, but unlike the other metallics, which are produced by pure, narrow wavelengths, these are broad-band reflective mixes of radiation (Neville 1975, 1977). By providing a surface for display of color patterns, the integument serves additional functions—protection by crypsis and mimicry, sexual recognition, and so forth.

## References

CROWMARTIE, R. I. T. 1959. Insect pigments. Ann. Rev. Entomol. 4: 59–76.

GHIRADELLA, H. 1984. Structure of iridescent lepidopteran scales: Variations on several themes. Entomol. Soc. Amer. Ann. 77: 637–645.

HEPBURN, H. R., ed. 1976. The insect integument. Elsevier, New York.

NEVILLE, A. C. 1975. Biology of the arthropod cuticle. Springer, New York.

NEVILLE, A. C. 1977. Metallic gold and silver colours in some insect cuticles. Ins. Physiol. 23: 1267–1274.

## Body Cavity

The body cavity of all arthropods is not considered a true coelom, as it lacks a complete mesodermal lining. Morphologists call it a "mixocoel" because of its formation embryologically from the fusion of the blastocoel with parts of the secondary body cavity. Because it is filled with blood, emptying into it from an open-ended circulatory system, it is also known as a hemocoel.

## Segmentation

From their annelid and marine arthropod ancestors, insects and their terrestrial relatives have inherited a segmented body. Between an anteriormost (acron) and posteriormost segment (telson), a varying number of segments are interposed, depending on the group. There were originally 18 (or 19, if a second antennal segment is recognized) in insects, 19 in chelicerates, and as many as 100 in myriapods. These were more or less equal in form and in the possession of a pair of walking appendages in the first terrestrial arthropods, much like modern-day centipedes. Evolution eventually favored the fusion of adjoining segments (a process called tagmosis) for various functional purposes (e.g., flight in higher insects), and body regions were formed. Of these, insects display a triple set, including the head (composed of the acron plus four or five highly fused original segments), a thorax (of three segments, pro-, meso-, and metathorax) and abdomen (with eleven segments, the posteriormost being highly modified into genitalia).

Arachnids and myriapods show different patterns of fusion. In the former, the head is undefined and its segments totally incorporated into the thorax (cephalothorax), which itself may also join into the abdomen, as in mites and ticks. Segmentation in these anterior two regions is concealed by a shield or carapace and is evident ventrally only by the serial set of appendages. The abdomen is either undefined or formed from several segments. Myriapods display only a well-developed head and uniformly segmented thorax-abdomen, with each segment bearing similar legs.

Uniquely, the insect thorax may bear a pair of wings on the meso- or metathorax, or both, but never on the prothorax. Thus, according to the appendages they possess, the three body regions are specialized for separate functions, the head for ingestion and perception, the thorax for locomotion, and the abdomen for metabolic processes and reproduction.

### The Head and Its Appendages

The head (Matsuda 1965) is the most highly modified body region, being a separate organ (except in arachnids), in which the primitive segmentation is almost obliterated. It is normally a rigid capsule, containing the main perceptive and integrative neural elements of the animal as well as ingestive organs.

The many sensory appendages of the

head include the antennae in insects, centipedes, and millipedes, all with one pair. Arachnids lack antennae, their place usually being assumed by the pedipalps that have become antennalike. However, in some arachnids, the pedipalps take other forms and functions, as the claws of scorpions or walking legs in sun spiders.

Around the mouth, modified segmented appendages serve as jaws or stylets for chewing or imbibing liquids and bear food-tasting and smelling organs called palpi. In arachnids, these organs are the chelicerae, with the basic scissor form, but they are used directly in feeding by tearing or stabbing the food, not chewing. The chelicerae may lose the movable element and become a piercing needle in mites, especially parasitic ones, and in spiders they are modified into fangs. Arachnids use the inner portion of the leg coxae to scoop liquid nutrient into the simple mouth. Among insects and millipedes, there are two pairs of jaws, the anterior mandibles and behind them the maxillae; centipedes have two sets of maxillae. In insects, the mandibles and maxillae may retain a primitive toothed or molar form for biting and chewing solid foods, or they may be greatly elongated and bladelike or hypodermiclike for piercing and siphoning. The labium may form only a supportive sheath around the latter or be itself spongy and absorptive and act directly in food collection. Flexibility in adaptation of mouth parts has been a major factor in the success of insects as a group, the variety of morphological types making possible an enormous diversity of food niches and feeding strategies.

Although not of appendicular origin, the eyes are of major sensory importance to the head capsule (Horridge 1975). There is a pair of larger multifaceted compound eyes in adult insects laterally and usually one to three smaller, single-faceted simple eyes medially on top of the head. In other terrestrial arthropod groups and immature insects, only simple eyes (ocelli) are present, either in lateral clumps on the sides of the head (millipedes) or on the back of the cephalothorax (arachnids). Eyes are also often absent altogether (many centipedes).

The structure and function of the compound eyes are complex. They bulge out on either side of the head to give a wide range of vision in all directions. Each is an aggregation of similar rod-shaped facets called ommatidia, the number of which varies from one per eye in some ants to over 10,000 in dragonflies. Each ommatidium is composed of elongate sensory cells containing light-sensitive pigments, these concentrated toward the center (thus seen as a dark rod, called the rhabdome) and exposed on the exterior through a capping, duplex lens that gathers and focuses light. There are also cells with diffuse pigment around the lens. The sensory cells are nerve cells and are connected directly to the brain, there being no optic nerve in insects.

There are many variations in the detailed structure of the ommatidium, such as the "apposition" versus the "superposition" types. In the former, the rhabdome is long, and the diffuse pigment cells isolate each ommatidium. In the latter, the rhabdome is short, and the screening pigment moves depending on the amount of light in the environment. Image formation is believed to be basically different in the two types, but little is certain about this aspect of eye function. It is known that insects generally have good visual acuity and light level accommodation. Wavelength discrimination varies considerably, with a tendency toward the ultraviolet in many species (Silberglied 1979). Many insects, such as bees and butterflies, have good color vision and can orient by polarized light.

## References

Horridge, G. A. 1975. The compound eye and vision of insects. Clarendon, Oxford.

Matsuda, R. 1965. Morphology and evolution

of the insect head. Amer. Entomol. Inst., Mem. 4: 1–334.

SILBERGLIED, R. E. 1979. Communication in the ultraviolet. Ann. Rev. Ecol. Syst. 10: 373–398.

### The Thorax and Its Appendages

The thorax (Matsuda 1970), when present as in insects, is a boxlike unit, primarily concerned with locomotion (Herreid and Fourtner 1981). It is the site of the largest muscles in the body, those that move legs and, in insects, the wings.

In all insects and arachnids and some myriapods, the number of ambulatory legs is reduced from one pair (primitive) per segment, and the legs arise only from the thoracic region. This arrangement affords more efficient and more rapid mobility than the original, dispersed condition retained by centipedes and millipedes. Adult insects all have three pairs, one pair from each of the three thoracic segments. In arachnids, the typical number of pairs is four, although the pedipalps preceding these may sometimes be involved with locomotion, and the first true leg pair may substitute as feelers (Amblypygi).

The insect leg itself is multisegmented and typically composed, from base to tip, of a coxa ("hip"), trochanter, femur ("thigh"), tibia ("shin"), and tarsus ("foot"). The last section has one to five segments and is tipped with grasping claws, pads, or both. The number of leg articles in the Myriapoda and Chelicerata differs in the various orders but is always six or more. Simple, generalized legs are for walking and running (Wilson 1966). Modifications of form occur in the legs of all groups but are the most diverse in the insects, among which are molelike digging legs, jumping legs with greatly enlarged, muscle-filled femora, hairy legs for carrying pollen, grasping and clasping legs in ectoparasites, and flattened, oarlike swimming legs in aquatic insects. In centipedes, the first pair are sharp fangs associated with poison glands.

### References

HERREID, C. F., II, AND C. R. FOURTNER. 1981. Locomotion and energetics in arthropods. Plenum, New York.

MATSUDA, R. 1970. Morphology and evolution of the insect thorax. Entomol. Soc. Can., Mem. 76: 1–431.

WILSON, D. M. 1966. Insect walking. Ann. Rev. Entomol. 11: 103–122.

### Wings and Flight

Insects are the only invertebrates with wings. These unique structures are normally present only in the adult stage (mayflies with winged subimagos being the only exception) and always arise from the meso- and metathoracic segments. Their historic origin is still a debated question, there being evidence of independent derivation from the body wall and their serial homology to the legs (Kukalova-Peck 1983) and abdominal gill plates of aquatic ancestors (Kukalova-Peck 1978, Matsuda 1981).

In the most primitive insect orders, the Apterygota, wings never evolved. Some groups, in particular, those adapted to a parasitic way of life, have lost their wings through retrograde evolution on one or both segments. The Diptera are characterized by the replacement of the metathoracic pair by nonflight, sensory organs, the halteres.

Anatomically, wings are flattened, expansive outgrowths of the dorsolateral integument of the thorax. They have an upper and lower cuticle, between which run nerves and blood channels, the epidermis, muscle, and other tissues having been largely obliterated. The cuticle is membranous and usually transparent, although it may be pigmented or covered by dense coverings of hairs or scales. It is also thickened in a linear pattern to form veinlike struts for strength. The pattern of the latter is not random but determined phylogenetically and is relatively constant among taxa (Comstock 1918), thus providing useful criteria for identification and study of relationships. Each vein root is named, and its homologue is recognizable in all insects. So are the branches of some adventitious veins (such as crossveins) and cells formed by closed sets of veins. Slightly different venational plans are recognized by various authors (see Wootton 1979).

The primary purpose of wings, of course, is flight, although other ends may be served. They may be modified as protective shields (such as the elytra of beetles and hemelytra of bugs) or possess color patterns that provide protection or recognition signals to other individuals.

The flight process in insects has been studied extensively and found aerodynamically unique (Ellington 1984, Goldsworthy and Wheeler 1989, Rainey 1976). All movement is imparted by muscles located within the thorax, not in the wing itself, through a kind of fulcrum formed by the lateral body wall and elastic membranes and sclerotic articulations in the wing base. The shape and timing of the stroke is also determined by these structures. It ranges from a slow, simple, up-and-down flapping action as in large butterflies, with a wing beat frequency of only 4 to 5 per second, to a complex rotational or twisting movement to and fro as well as up and down, with as many as 1,000 beats per second (*Forcipomyia*, Diptera). Insects in the latter category have the capacity for extremely agile aerobatics, and some can attain flight speeds of nearly 20 kilometers per hour (Hocking 1953). Most flight is trivial and of short duration, taking insects through their regular life routines. Some insects such as locusts and dragonflies, however, are capable of sustained flight over long distances for migration and dispersal. Small insects found high in the atmosphere are moved primarily by wind and air drafts and form a kind of aerial "plankton."

The land-bound myriapods and chelicerates seldom move long distances on their own, although young spiders may be wafted hundreds of kilometers, like kites on air currents, by letting out long silk threads grasped by the winds (a process called "ballooning").

## References

COMSTOCK, J. H. 1918. The wings of insects. Priv. pub., Ithaca.

ELLINGTON, C. P. 1984. The aerodynamics of flapping animal flight. Amer. Zool. 24: 95–105.

GOLDSWORTHY, G. J., AND C. H. WHEELER, eds. 1989. Insect flight. CRC, Boca Raton.

HOCKING, B. 1953. The intrinsic range and speed of flight of insects. Royal Entomol. Soc. London, Trans. 104: 223–345, pl. I–VI.

KUKALOVA-PECK, J. 1978. Origin and evolution of insect wings and their relation to metamorphosis, as documented by the fossil record. J. Morph. 156: 53–126.

KUKALOVA-PECK, J. 1983. Origin of the insect wing and wing articulation from the arthropod leg. Can. J. Zool. 61: 1618–1669.

MATSUDA, R. 1981. The origin of insect wings (Arthropoda: Insecta). Intl. J. Ins. Morph. Embryol. 10: 387–398.

RAINEY, R. C., ed. 1976. Insect flight. Blackwell, Oxford.

WOOTTON, R. J. 1979. Function, homology and terminology in insect wings. Syst. Entomol. 4: 81–93.

### The Abdomen and Its Appendages

The abdomen is generally the least modified of the three body regions (Matsuda 1976). It retains its primitive homomerous segmentation in insects, although in many, the number of segments is reduced through fusion. In spiders and acarids, only traces of segmentation remain, and it is evanescent in other groups. Myriapods have a large number of equal segments, but diplopods exhibit a fusion of adjoining segment pairs to form duplex, secondary segments (hence the double-paired legs that give the group its name). Fully developed abdominal walking legs persist only in this order and the Chilopoda. Apterygote insects also possess ventral appendages on some basal abdominal segments, which represent vestigial legs.

Also found at the base of the abdomen in varied groups are special modifications such as hearing organs (grasshoppers, geometrid moths), copulatory organs (Odonata), or book lungs (spiders). This region is even intimately fused with and functionally a part of the thorax in higher Hymenoptera.

The terminus of the abdomen also has acquired many modifications. In insects, because the sex opening (gonopore) exits here, there are complex structures for copulation and oviposition, collectively termed genitalia (Scudder 1971). In males, these are for sexual purposes only and are composed basically of paired lateral grasping or sensory or manipulative appendages and medial intromittent organs and in many cases, pheromone dispersing structures. Corresponding female structures respond to or receive the male elements. Eggs are placed by special appendages (ovipositors). Because many genitalic structures are often directly involved with the mechanical determination of sexual isolation, they present useful characters for species discrimination in taxonomy (Tuxen 1970).

Modifications of the posterior extremity of the abdomen in Chelicerata include single long antennalike tactile structures (Uropygi), the scorpion sting, and the spinnerets of spiders. Centipedes and insects also often have paired, segmented, terminal sensory appendages (cerci) that operate like a kind of rear set of antennae.

In noninsect groups, the gonopore is located ventrally near the base of the abdomen, elaborate genitalia are absent, and copulation is usually remote from the primary exits of the reproductive system. For example, in spiders, the male fills tubules in the pedipalps with seminal fluid from its gonopore and transfers the liquid to the female "manually" with these appendages. Males of other chelicerates, apterygote insects, and chilopods deposit sperm packets (spermatophores) on the substratum to be picked up by the female. Male millipedes transfer the sperm with modified legs (gonopods), by contacting the female gonopore directly or by leaving spermatophores to be picked up by the female.

Miscellaneous terminal appendages occur, including large forceps for prey capture and defense, which earwigs have, spinnerets for manipulating silk in spiders, or even snorkellike breathing apparatuses in many immature aquatic insects. The sting of scorpions is the highly modified last abdominal segment.

## References

MATSUDA, R. 1976. Morphology and evolution of the insect abdomen: With special reference to developmental patterns and their bearings upon systematics. Pergamon, New York.

SCUDDER, G. G. E. 1971. Comparative morphology of insect genitalia. Ann. Rev. Entomol. 16: 379–406.

TUXEN, S. L., ed. 1970. Taxonomist's glossary of genitalia in insects. 2d ed. Munksgaard, Copenhagen.

## Muscular System

Closely tied functionally to the exoskeleton is the main muscular system. All muscles attach to the integument internally and provide motion to the arthropod body in all its varied actions. They never form a body wall plexus but lie in bundles running between insertions. The latter may be broad or attenuated, cover extensive areas on sclerites, or fasten to invaginated extensions of the latter, the apodemes. The latter, when long and slender, are tendonlike but are histologically unlike vertebrate connective tissue, which is virtually absent in insects and their arthropod relatives. Visceral muscles, as circular, oblique, or longitudinal bands, are confined to the walls of the digestive tract and ducts of the reproductive system. In insects and their relatives, all muscle tissue is striated, whether skeletal or visceral.

Physiologically, insect muscle tissue (Usherwood 1975) is basically the same as that of vertebrates, although it (apparently) is capable of slightly more rapid twitches. Very rapid movements of insects, such as wing beat frequencies of 200 to 300 per second, are made possible by vibratory action of elastic portions of the cuticle, the muscles themselves contracting no more rapidly per stimulus than those of a bird. The tremendous power per body weight and size of many insects (such as the giant horned scarabs, *Dynastes*) is also an illusion. Strength results from the exertion of short fibers arranged along the entire length of leg joint surfaces so that the load is evenly and widely distributed. Power output and metabolic rates of insects, however, are much higher than in vertebrates, the result of a direct and continuous oxygen supply via the tracheal system.

Gross anatomy of the musculature is highly complex, and there may be hundreds of discrete muscles in even a small insect. An early anatomist described over 4,000 in the goat moth caterpillar, as compared to a mere 529 in humans. This richness of muscles combined with mechanically diverse articulations permits a diverse repertoire of intricate movements by these animals.

## Reference

Usherwood, P. N. R. 1975. Insect muscles. Academic, London.

## Digestive System

In all groups, the alimentary canal is a continuous, fairly straight tube, with openings anteriorly via the mouth and posteriorly via the anus. There are three regions of the gut, defined by their embryonic origins: a foregut and hindgut, both formed by invagination of the blastocoel and lined with epidermis and cuticle, and a mesodermal midgut, lacking a cuticle. There are various diverticula depending on the group, most commonly a crop (temporary storage sac leading off the esophagus) and blind gastric ceca arising from the midgut or stomach. Salivary glands empty into the mouth cavity or from the tip of a proboscis via long ducts associated with the hypopharynx.

Most of the digestive enzymes are produced by the cells in the walls of the midgut and ceca. The nature of these enzymes varies according to dietary adaptations, proteases and lipases predominating in carnivores, cellulases and related compounds in wood feeders, keratinase and collagenase in scavengers of vertebrate connective tissues and hair, and so on.

### Nutrition and Metabolism

Food enters the gut by the mouth located on the front or underside of the head in insects and myriapods or the head region in arachnids. There it is mixed with predigestive enzymes from the salivary glands, fangs, or regurgitations. Most digestive processes are reserved for the interior of the stomach and intestine but in spiders begin prior to swallowing. The latter regurgitate on their prey, causing enzymatic liquification externally. Some early changes in food in other types of arthropods may be wrought by secretions of the salivary glands.

On its way to the stomach via the esophagus, food may be diverted into a crop for storage or ground up by a region of the gut set with spines or teeth moved by extra heavy muscles (proventriculus).

The nutritional requirements of insects, arachnids, and so on, and their metabolic processes (Gilmour 1961) also vary enormously (Dadd 1973). The same essential elements important to most animals for energy (Downer 1981) and growth seem to be needed by all, either supplied in the diet or synthesized metabolically or by intestinal symbionts.

Nucleic acids are synthesized by all insects, as are some vitamins. The ten essential amino acids, however, all must be ingested in appropriate proportions to sustain growth. Carbohydrates serve as a major energy source, and although they are often present in the diet, they are not always essential and can be converted from protein or fats. There are considerable differences in the ability of different insects to utilize polysaccharides. Wood roaches and termites, for example, rely on protozoans or bacteria in the gut to break down cellulose to assimilable sugars.

Fats are the chief form in which energy is stored, and the ability to synthesize them is widespread. Other lipids such as cholesterol must be acquired from foodstuffs. Vitamin needs vary considerably, although generally only water soluble B vitamins must be present in food. Vertebrate blood is notably lacking in the latter, and many hematophagous insects rely on symbionts for these compounds. Parasites store B vitamins, ingested by the scavenging larvae that feed on bacteria-rich food, and these are passed on to the adult. Inorganic salts and trace minerals are needed much as they are in vertebrate animals.

### References

DADD, R. H. 1973. Insect nutrition: Current developments and metabolic implications. Ann. Rev. Entomol. 18: 381–420.

DOWNER, R. G. H., ed. 1981. Energy metabolism in insects. Plenum, New York.

GILMOUR, D. 1961. The metabolism of insects. Freeman, San Francisco.

### Growth

Food serves not only to provide energy for activity but also to build up stores for long dormant periods, and, of course, it is the basis for growth. As arthropods, with a confining, almost nonexpandable, nonliving exterior cuticle, insects and their kin achieve size increases and maturity only by periodic spurts of growth following molting. This process takes place from a few to many times during the animal's life, although it ceases after adulthood in insects. Molting (or ecdysis) is preceded by a cessation of activity and catalysis of the lower cuticle layers when muscles and sense or-

gans assume new attachments. Only the old outer cuticular layer is shed, including the linings of the larger tracheae, foregut, and hindgut.

### Rhythms and Seasonality

Terrestrial arthropods display rhythms in their activity, correlated with and adapting them generally to changes in their environment. Some of these rhythms are independent of signals from their surroundings (endogenous or circadian rhythms; Brady 1974). Examples of such functions are daily periods of sleep alternating with active locomotion or feeding and periods of singing or courting. Events in long-term life cycles are also cyclic and have internal controls interacting with changes in ambient stimuli, day length being a very strong one (Beck 1980). Although the physiological basis for such functions is still not understood, an underlying "biological clock" mechanism is postulated (Saunders 1982).

Long periods of quiescence commonly occur in insects and relatives, often to carry the animal through adverse seasons. This is called diapause and is characteristic of high latitude or high elevation species in the wintertime (hibernation) or of desert species during dry periods (aestivation). At these times, growth, development, and activity is attenuated. Finally, diapause is broken with the return of favorable conditions, and emergence occurs. Sometimes large numbers may return to action simultaneously, resulting in population explosions. Periods of dormancy are less profound in tropical than temperate insects because of more equable environmental conditions in the lower latitudes (Denlinger 1986).

### References

BECK, S. D. 1980. Insect photoperiodism. 2d ed. Academic, New York.

BRADY, J. 1974. The physiology of insect circadian rhythms. Adv. Ins. Physiol. 10: 1–115.

DENLINGER, D. L. 1986. Dormancy in tropical insects. Ann. Rev. Entomol. 31: 239–264.

SAUNDERS, D. S. 1982. Insect clocks. 2d ed. Pergamon, Oxford.

## Luminescence

Another specialized metabolic job to which certain body chemicals are put is bioluminescence (Harvey 1957). Quite a number of insects, primarily beetles (glowworms, fireflies, headlight beetles, railroad worms) and millipedes, have evolved light-producing organs (McElroy et al. 1974). The mechanism of light production is complex (Case and Strause 1978) but basically involves the oxidation of luciferin in the presence of the enzyme luciferase. Luciferin is first activated by ATP in the presence of magnesium, then oxidized to an excited form (adenyl-oxy-luciferin) that decays to a lower energy form with the liberation of light. The reaction is cool and very efficient, some 98 percent of the energy involved being released as light.

### References

CASE, J. F., AND L. G. STRAUSE. 1978. Neurally controlled luminescent systems. *In* P. J. Herring, ed., Bioluminescence in action. Academic, New York. Pp. 331–366.

HARVEY, E. N. 1957. A history of luminescence, from the earliest times until 1900. Amer. Phil. Soc., Philadelphia.

McELROY, W. D., H. H. SELIGER, AND M. DELUCA. 1974. Insect bioluminescence. Physiol. Ins. 2: 411–460.

## Blood and Circulation

All the arthropods that are the subject of this book possess an open circulatory system (Jones 1977). That is, the blood moves for the most part over and around the tissues and organs, bathing them and exchanging molecules with them directly, in a continuous body cavity, the hemocoel. In insects, there are no blood vessels save the main aorta that leads anteriorly, directly from the heart (McCann 1970), and empties into sinuses surrounding the brain. In centipedes, there are short lateral arteries leading to the gut and other minor vessels. There is a "pulmonary artery" to the book lungs in spiders as well as secondary vessels to the legs, tail, and so on, in other arachnids. The heart, which lies dorsally in the hemocoel, just beneath the abdominal roof, propels the blood forward with peristaltic contractions. After passing through the body cavity, including the legs, antennae, wings, and other appendages, and often aided by auxiliary, pulsatile organs at their bases, the blood reenters the heart through lateral pores (ostia).

The blood itself (or hemolymph) consists of a fluid plasma (Florkin and Jeuniaux 1974) in which nucleated cells are suspended (Crossley 1975). The latter are of many types but normally do not possess hemoglobin like vertebrate corpuscles, the oxygen/carbon dioxide transport function being assumed by the tracheal system (see below). Insect blood is not red (except in a few specialized types that have hemoglobin, such as blood worms, *Chironomus*) but green, a mixture of carotenoid and bile pigment ("insectoverdin"), bluish, or almost clear. The functions of the blood cells include phagocytosis, wound healing, coagulation, storage, and regulation of intermediate metabolism. The plasma serves primarily as the carrier of substances to tissues and also provides a store of nutritive compounds such as sugars and proteins. Its water acts as a reservoir for the maintenance of cellular fluids.

### References

CROSSLEY, A. C. 1975. The cytophysiology of insect blood. Adv. Ins. Physiol. 11: 117–221.

FLORKIN, M., AND C. JEUNIAUX. 1974. Hemolymph composition. Physiol. Ins. 5: 255–307.

JONES, J. C. 1977. The circulatory system of insects. C. T. Thomas, Springfield, Ill.

McCANN, F. V. 1970. Physiology of insect hearts. Ann. Rev. Entomol. 15: 173–200.

## Hormones

An important class of chemicals transported by the blood are hormones (Novák 1975, Sláma et al. 1974). There are many types, and they vary in their effects, even those from a single endocrine organ. A few activities mediated by hormones are molting, metamorphosis, egg production, color changes, daily activity rhythms, dormancy, and caste determination in social insects. Many of these processes are controlled by the balance and timely production of only a few basic hormones such as the "molting hormone" (ecdysone) and the "juvenile hormone" (neotenin), often under the overriding command of neurosecretory hormones.

Insect hormones are complex biochemicals produced by two types of endocrine organs. These are the neurosecretory cells in the central nervous system and the endocrine glands which are separate masses of tissue specialized for hormone production. Well-known examples of the latter are the corpora cardiaca, forming part of the wall of the aorta, the corpora allata, situated on either side or surrounding the esophagus, and the prothoracic glands, diffuse tissue aggregations at the back of the head or in the floor of the prothorax. The neurosecretory cells send their hormonal products to the target organs (often glands of the second type) along the axons of nerve cells. Secretions from the endocrine glands are released into the blood.

### References

Novák, V. J. A. 1975. Insect hormones. Chapman & Hall, London.
Sláma, K., M. Romanuk, and F. Sorm. 1974. Insect hormones and bioanalogues. Springer, New York.

## Pheromones

Much like hormones (sometimes called "ectohormones"), pheromones (Jacobson 1972) are special kinds of biologically active substances released by one individual which cause other individuals of the same species to act in a specific way. These substances are extremely numerous in kind and influence among insects and their relatives. In fact, entomologists have realized in recent years that the dominant means of communication between these creatures is via these messenger substances (Shorey 1976), perceived by olfactory sense organs, especially on the antennae, mouthparts, and tarsi (Lewis 1984). They are produced by ectodermal (exocrine) glands on the abdomen, wings, or other parts of the body.

Some pheromonal systems that have been particularly well studied are the aphrodisiacal scents from the wings of male butterflies and moths or eversible abdominal glands of the females. These chemicals serve to draw the sexes together and elicit courtship and copulatory behavior. The trail-marking substances and alarm chemicals of ants and bees that foster aggregation are also pheromonal, as are the caste and activity controlling regulators in social insect colonies.

### References

Jacobson, M. 1972. Insect pheromones. 2d ed. Physiol. Ins. 3: 229–276.
Lewis, T., ed. 1984. Insect communication. Academic, New York.
Shorey, H. H. 1976. Animal communication by pheromones. Academic, New York.

## Other External Secretions

Allomones are compounds produced by insects and their relatives that elicit antagonistic reactions between individuals (Bell and Carde 1984). They benefit the sender only, usually protecting it by warding off an attack by the receiver (Blum 1981). The pain-giving (not prey-seducing) venoms of female aculeate Hymenoptera, repugnant odors of many true bugs and beetles, and emetic body chemicals (cardiac glycocides and the like) in a few butterflies are of this category. Such also is the function of cantharidin (Young 1984a, 1984b), a terpenoid produced by "blister beetles" (Meloidae). When provoked, these beetles exude blood containing this substance from the tibio-tarsal articulations, and they are strongly avoided by insectivorous vertebrates and carnivorous insects.

Other secretions are external but cause no interactive response in other or the same species. These are utilitarian substances involved in the life processes of the producer. Examples are silk (Denny 1980) for cocoons and webs, adhesives to bind eggs in place, and materials such as wax or gums for building structures. Venom used by spiders, centipedes, scorpions, and others to obtain food also belong in this category. Regardless of function, arthropod venoms are usually compared from chemical or pharmacological standpoints (Bettini 1978).

## References

Bell, W. J., and R. T. Carde, eds. 1984. Chemical ecology of insects. Sinauer, Sunderland, Mass.

Bettini, S., ed. 1978. Arthropod venoms. Springer, Berlin.

Blum, M. S. 1981. Chemical defenses of arthropods. Academic, New York.

Denny, M. W. 1980. Silks—their properties and functions. Soc. Exper. Biol., Symp. 34: 247–272.

Young, D. K. 1984a. Cantharidin and insects: An historical review. Great Lakes Entomol. 17: 187–194.

Young, D. K. 1984b. Field records and observations of insects associated with cantharidin. Great Lakes Entomol. 17: 195–199.

## Nervous System

In insects, as with other animals, the nervous tissue is composed of nerve cells (neurons), which are grouped into linear nerves and gangliar masses to form a central nervous system (Treherne 1974, Miller 1979), an autonomic (or stomatogastric) system, and a peripheral or sensory nerve system. The first is ventral, lying in the floor of the hemocoel, and is characterized by a succession of ganglia interspersed along a paired, ventral nerve cord. The nerve cell bodies are located peripherally in the ganglia, the center of which are occupied by a complex of nerve fibers (the neuropile) that connect the ganglia as the nerve cord.

The largest and most complex ganglion is the anteriormost. It is dorsal, above the pharynx, in the head. This is the brain (Howse 1970), which may actually be composed of two or more fused primary ganglia. It is the overriding center of neural integration to which the other ventral ganglia are ultimately subjugated, although each of the latter may have some degree of autonomy. A beheaded insect may continue to live and exhibit locomotory and sensory activity for some time before it eventually dies from such injury.

The major sensory organs of the head, the eyes, antennae, and palpi, are connected by large nerves directly to the brain. The brain also contains neurosecretory cells and functions partly as an endocrine organ as explained above.

The first ventral ganglion is also located in the head region and is associated with ingestive processes. There follows a varying number of segmented ganglia, primitively, one per segment, up to eleven in insects, and many more in myriapods, but the number is often less, due to fusion of segments, especially in the thorax.

The autonomic system is closely associated with the digestive tract and consists of a small number of small ganglia and short fine nerves. Its function is to control visceral activity. It is also involved with parts of the endocrine system.

Efferent nerves run from the central nervous system to the muscles in all parts of the body. Afferent nerves lead from the sensory system, mainly the integumentary sense organs, to the central nervous system. The cell bodies of sensory neurons are located near the sensilla themselves, and their axons connect them directly to the ganglia without intervening synapses.

## References

Howse, P. E. 1970. Brain structure and behavior in insects. Ann. Rev. Entomol. 20: 359–379.

Miller, T. A. 1979. Insect neurophysiological techniques. Springer, New York.

Treherne, J. E. 1974. Insect neurobiology. North Holland, Amsterdam.

### Integumentary Sense Organs

The arthropod would be isolated from its environment by the nonliving, encapsulating cuticle, but thousands of structures sensitive to external stimuli (Dethier 1963), collectively called sensilla, cover the surface. They are especially numerous on the antennae, tarsal pads, and palpi and communicate with the nervous system via nerves of the peripheral system. The anatomy of sensilla is extremely varied, each type specifically adapted to the perception of a certain subset of stimuli important to the animal's safety and other life processes.

The most common and often most abundant sensilla are hairlike extensions (setae, chaetae) responsible for mechanoreception (McIver 1975). These may respond to touch, stretching, or bending, directly from an outward force or by pressure from another part of the body (often they are found in articulations). The hair base may be simple and level with the surface or recessed and quite complex, as is the trichobothria of arachnids.

Portions of the body wall can be innervated so that deformations transmit information about mechanical stresses. Even slight vibrations from air currents or compression waves are perceived. For example, masses of stretch receptors in the swollen, subbasal segment of the antennae of mosquitoes and other flies (Johnston's organ) respond to deflexions of the flagellum by air movements, giving these insects an acute sense of hearing. When the integument is especially sensitive in this way, and there are structural modifications for reception such as thin, vibrating membranes, these portions are considered auditory organs (Michelsen 1979). Such are the thoracic and abdominal tympana of many moths and grasshoppers, fore tibial hearing pits of katydids, and acoustical windows in the cicada thorax.

Sensilla are often structured for the reception of chemicals in air or liquids. Such chemoreceptors (Slifer 1970) usually have thin or porous walls so the molecules may pass through the outer part of the organ and stimulate inner receptive surfaces. They may be extremely sensitive. Calculations for the sex attractant of the domestic silk moth indicate that a single molecule may elicit a response.

Certain sensilla also react to ambient temperature changes, radiant heat, pressure, humidity, and surface moisture (Altner and Loftus 1985). Perception of related factors internally are by direct cellular sensitivity.

## References

Altner, H., and R. Loftus. 1985. Ultrastructure and function of insect thermo- and hygroreceptors. Ann. Rev. Entomol. 30: 273–295.

Dethier, V. G. 1963. The physiology of insect senses. Methuen, London.

McIver, S. B. 1975. Structure of cuticular mechanoreceptors of arthropods. Ann. Rev. Entomol. 20: 381–397.

Michelsen, A. 1979. Insect ears as mechanical systems. Amer. Sci. 67: 696–706.

Slifer, E. H. 1970. The structure of arthropod chemoreceptors. Ann. Rev. Entomol. 15: 121–142.

### Sound Production

Correlated with hearing in many insects and a few arachnids is sound production, another means of communication (Haskell 1961). There are a great variety of mechanisms for making sounds that are audible to other insects and to the human ear. Some, such as those resulting from the vibration of wings in flight, may be adventitious and apparently have no value to the animal, but most have a specific function and originate from unique, sometimes elaborate structures. Extraspecific uses usually are to startle and are protective (Masters 1979); intraspecific functions include the calling and courtship stimulations between the sexes, aggregation, spreading alarm, and giving the location of other colony members in social and semisocial forms.

Sounds may be produced as a byproduct of some activity such as feeding or wing movement, tapping the substrate, and ejections of air, but the major and most effective means of sonification involve frictional mechanisms and vibrating membranes (tymbals). The former, called stridulation, involves two facing surfaces that are roughened and that, when moved against each other, produce a sound. Such are the narrow scraper and file in the base of the fore wings of crickets and katydids. Many other insects, beetles, lepidopterous larvae and pupae, and so on, have broad corrugated or ridged areas that when rubbed together, give a variety of grinding, hissing, squeaking, and clicking sounds.

Sounds produced by the vibration of a membrane driven by muscles are common in Homoptera, Heteroptera, and some moths but are best developed in male cicadas. This sound-producing organ is located in the dorsolateral part of the first abdominal segment. Sound is made when the tymbal muscle contracts, pulling it back rapidly. Release allows it to return to the starting position suddenly against the air, and the resulting vibrations set up high-intensity air waves that may sound to the human ear like a deafening screech or harsh scream.

## References

Haskell, P. T. 1961. Insect sounds. Quadrangle Books, Chicago.

Masters, W. M. 1979. Insect disturbance stridulation: Its defensive role. Behav. Ecol. Sociobiol. 5: 187–200.

## Excretion

The typical insect nephritic organs (Malpighian tubules) are long, thin, blindly ending tubes arising from the gut near the junction of midgut and hindgut and extending freely in the body cavity. Their numbers vary among different groups from a few to hundreds. The wall of the tubule is one cell thick, encircling a lumen. These cells extract waste products of metabolism from the blood, nitrogenous by-products usually in the form of uric acid but also as urea and ammonia. Potassium, sodium, and other inorganic ions are also eliminated, along with a quantity of water.

The maintenance of constant salt levels, water, osmotic pressure in the hemolymph, and elimination of nitrogenous wastes are the main excretory tasks (Maddrell 1971) of the Malpighian tubules in insects. Those organs are present in the other groups, although they may be replaced by nephridial glands in arachnids and some chilopods.

## Reference

Maddrell, S. H. P. 1971. The mechanisms of insect excretory systems. Adv. Ins. Physiol. 8: 200–331.

### *Water Relations*

Terrestrial arthropods are subject to water loss (Barton-Browne 1964, Stobbart and Shaw 1974) from excretion, in the feces, and through the cuticle, including that lining the respiratory system. The loss is especially intense in species living in arid environments. Water is gained primarily in the food but also by drinking and general absorption from humid air. Special organs of conservation are also present in association with the hindgut, whose normal functions include reabsorption of water from the feces. One of these, the cryptonephridium, incorporates the distal ends of Malpighian tubules which loop back onto or into a thickened portion of the rectum. Water is recycled from the latter back into the tubules and reused; feces from these insects emerge in a very dry state.

Aquatic insects have salt and water control problems different from but no less severe than those faced by terrestrial types. Since the hemolymph is hypertonic to the outside medium, there is a constant tendency for water to pass into the insect through the cuticle. This uptake is counterbalanced by a copious liquid outpouring, which, however, results in a loss of salts. This is corrected by reabsorption by the rectum.

## References

BARTON-BROWNE, L. B. 1964. Water regulation in insects. Ann. Rev. Entomol. 9: 63–82.

STOBBART, R. H., AND J. SHAW. 1974. Salt and water balance; excretion. Physiol. Ins. 5: 361–446.

## Respiration

Exchange of respiratory gases in insects and allied terrestrial arthropods takes a very different form from that found in other animals. The anatomy of most contains a system of tubules (tracheal system) dedicated directly to the tasks of bringing oxygen to the tissues and carrying off carbon dioxide and other waste gases. The blood plays no significant role in this process except in very small, immature forms that live in damp conditions and aquatics with blood-filled gills. The tracheae open to the outside through segmentally arranged pores, the spiracles, which generally have a closing device to keep water loss to a minimum. Large tubes run inward from the spiracles and branch profusely, often interconnecting with sacs or other tubules and terminating finally in minute blind endings (tracheoles) directly on the cells. Derived from integumentary epidermal cells, the entire system, except the tracheoles, is lined with cuticle that has a circular ringed structure for strength against collapse.

The rates of diffusion of oxygen and carbon dioxide are sufficient to allow these gases to passively reach all tissues, but ventilatory movements are necessary in large and very active forms. This is accomplished by abdominal compression, contraction, and other muscular movements. The length of the diffusion path, however, is a factor limiting the size of insects, in particular, those with the bulky muscle masses needed for flight.

Spiders have a tracheal system in the abdomen only, including a variety of modifications, among them "sieve trachea," which are large trunks from the ends of which originate numerous individual fine tracheae. Many also possess unique respiratory structures called "book lungs," which are lamellate, trachealike plates extending into the body cavity. Blood flows between the plates, exchanging molecules with the chambers of the tracheoid tubules, thus functioning much like a vertebrate lung.

The mechanisms of external respiratory adaptations in aquatic insects (Miller 1974) go in a great variety of directions. They rely on tapping atmospheric air, or extraction of dissolved oxygen from the surrounding liquid, or combinations of both. Among the former, most are often associated air stores of one kind or another. The tracheal system itself may have sacs or enlargements to accommodate air supplies, or bubbles may be carried beneath the wings or held onto the general body surface by hairs or other extensions of the integument. Prevented from collapse by these extensions, these air bubbles act as "physical gills," oxygen and carbon dioxide passing in and out of them through their surface, which acts like a membrane ("plastron respiration"). Spiracles communicating with the bubbles tap the air store and can also function normally should the water dry up or the animal emerge to assume a terrestrial phase of existence.

Species utilizing atmospheric air must come to the surface from time to time to restore their gaseous provisions, although some, such as certain mosquito larvae, may stay below for very long periods of time, tapping air carried in the vessels of aquatic plants.

Small aquatic insects may employ the general cuticle as a gill. Large types have other forms of gill structures, expansive plates or fingerlike extensions filled with blood, or a rich tracheal network to carry on gas exchange. Respiration in many endoparasitic types relies on similar mechanisms, their lives being spent in a liquid ambience for long periods.

## Reference

MILLER, P. L. 1974. Respiration: Aquatic insects. 2d ed. Physiol. Ins. 6: 403–467.

## Reproduction

Insects and like arthropods are normally bisexual and require sexual communion or mating (Blum and Blum 1979, Thornhill and Alcock 1983), with subsequent gamete fusion, for reproduction (Davey 1965, Englemann 1970). Only in a few cases has parthenogenesis—and in still fewer cases, hermaphroditism—evolved. The production of normal young by unfertilized females is part of the regular reproductive process in many Homoptera, alternating with the sexual process. Unfertilized eggs may be the means of sex determination in others, such as the honeybee, which produces drones by this method. In the cottony cushion scale (*Icerya purchasi*), both male and female gonads develop in the female, and self-fertilization takes place.

The gonads and their immediate ducts are almost always paired. The generative organ may be single or multiple in myriapods, derived from mesodermal embryonic tissue. The gonoducts join paired or single ectodermal invaginations that lead to the outside via the gonopore. This may be located either terminally as in most insects or near the base of the abdomen in arachnids.

Male insects and myriapods usually have a complex set of genitalia surrounding the gonopore, an extension of which terminates in an intromittent organ or penis (often called the aedeagus). These genitalia, especially the claspers of one sort or another, are important in locking the pair securely and precisely together while the penis is inserted, forming a physical connection that is normally species specific (Eberhard 1985). They may also play a part in physical or chemical stimulation necessary for successful copulation (their inner surfaces often bear sensillar patches) (Alexander 1964). The gonopore is unelaborated in spiders, the function of the genitalia being assumed by the pedipalps.

The external female genitalia are relatively simple compared to the male's, but some special structures (ovipositors) may be present for egg placement.

## References

ALEXANDER, R. D. 1964. The evolution of mating behaviour in arthropods. Royal Entomol. Soc., Symp. 2: 78–94.

BLUM, M. S., AND N. A. BLUM. 1979. Sexual selection and reproductive competition in insects. Academic, New York.

DAVEY, K. G. 1965. Reproduction in the insects. Freeman, San Francisco.

EBERHARD, W. G. 1985. Sexual selection and animal genitalia. Harvard Univ. Press, Cambridge.

ENGLEMANN, F. 1970. The physiology of insect reproduction. Pergamon, Oxford.

THORNHILL, R., AND J. ALCOCK. 1983. The evolution of insect mating systems. Harvard Univ. Press, Cambridge.

### *Fertilization*

The sperm cells produced by the testes are introduced internally into the female in most forms, that is, fertilization is internal. They may be first kept in storage in diverticulae of the common oviduct, however, and released to fuse with the eggs only as they pass, the female thus controlling the time of fertilization.

Introduction of sperm is not always directly via the gonopore. Secondary genitalia are developed most notably in Odonata and spiders. The former transfer the sperm from the gonopore to the accessory copulatory organs on the venter of the third abdominal segment; male spiders use syringes in the bulbous apex of the pedipalps for this purpose. Sperm is carried in a liquid medium, or more commonly, compressed into packets (spermatophores) that may be inserted into, or formed, in the common oviduct or its outpocketings (spermathecae), or are placed on the substratum to be picked up by the female.

## Size

Terrestrial arthropods are subject to size limitations because of the combined restrictions of rigidity, lack of permeability, and weight of the cuticle, which becomes too much of an encumbrance to movement in very large forms. Also, the diffusion rates of respiratory gases is insufficient to traverse the distances necessary through prolonged tracheal systems, although this is overcome to some extent by breathing movements. Environmental determinants, such as moisture and food availability, are also important (Schoener and Janzen 1968).

In spite of these restrictions, some extremely large insects are found in Latin America, all long lived, herbivorous, forest types. In terms of bulk, the record must be adult males of the large horned scarab, like *Megasoma elephas*, which may weigh 40 grams or more. Wing expanse is another measure of size and finds its greatest expression in the birdwing moth (*Thysania agrippina*), with a spread from wing tip to wing tip of up to 30 centimeters. Those with the longest, although slender, bodies are the Neotropical centipede *Scolopendra gigantea*, which extends 27 centimeters, and walkingsticks, some 26 centimeters (*Philbalosoma phyllinum*) from the head to the tip of the abdomen. Indeed, the wet forests of the Neotropics are traditionally thought to harbor many insect goliaths. While not the largest overall, some that are

the biggest of their category or impressive in any sense are many horned beetles such as *Dynastes hercules* (17 cm, including horn), morpho butterflies, *Morpho hecuba* (wingspan 18 cm), tarantulas, *Theraphosa lablondi* (20 cm leg span), and lubber grasshoppers, *Tropidacris* (wingspan 25 cm, length to folded wing tips, 13 cm). The largest flies in the world are the Neotropical *Pantophthalmus* (Pantophthalmidae) that measure 4 centimeters in length and weigh over 2.5 grams.

At the low end of the size scale are the smallest known insects, parasitic wasps of the genus *Alaptus* (Myrmaridae) with body lengths of only 0.2 millimeters.

Insects and their terrestrial relatives, by and large, are small, the vast majority 6 to 10 millimeters long and 25 to 50 milligrams in weight. This is their single most important structural characteristic, enabling the exploitation of the infinite number of small niches of nature. Insects need little space and minimal sustenance to live and hide from predators.

## Reference

SCHOENER, T. W., AND D. H. JANZEN. 1968. Notes on environmental determinants of tropical versus temperate insect size patterns. Amer. Nat. 102: 207–224.

## Genetics and Cytology

Insect genetics has been a fruitful field and has contributed a great deal to this field of general science, particularly through studies on *Drosophila*. Much of this success is attributable to the ease with which many insects are maintained in the laboratory, their rapid turnover of generations, diversity of phenotypic expressions of gene effects, and in many cases, giant, well-marked chromosomes.

The genetic control of a large number of particular insect characteristics has been elucidated, such as the distribution of different types of hairs, color patterns, resistance to insecticides, and wing venation. Gross changes in Lepidoptera wing color patterns are known to be determined by simple gene differences (Robinson 1971).

Sex in insects is basically determined by the production of different gametes, although epigenetic factors, such as hormones, are also important (Langé 1970). Sex chromosomes may be involved, a variety of combinations being found. Males heterozygous XY and XO and females homozygous XX is the usual situation. The reverse is true of Lepidoptera and Trichoptera. In Hymenoptera, fertilized eggs develop into females, unfertilized eggs into males, the latter therefore being haploid individuals.

Genotype and gene frequencies are properties of populations rather than of individual insects. Their behavior is important to the understanding of evolutionary processes when it is realized that it is shifts in their frequency, either randomly (genetic drift), by mutation, selection, or external events, that lead to speciation and higher order phylogenetic changes. A classic case of the latter is the increase to normalcy of melanism in populations of European moths living in industrial environments where heavy soot pollution darkens their resting substrates (Kettlewell 1973). No melanics of this type are yet known in Latin America.

Mutations are easily induced in insects by means of radiation and chemicals. The former is even used routinely to create sterile individuals for mass release in genetic control schemes (Pal and Whitten 1974).

The mode of gene operation is also becoming known in insects. In the giant chromosomes of fly larvae, characteristic swellings, forming after natural hormones contact the cell, appear to indicate activity of specific genes.

Genetic work with other terrestrial arthropod groups aside from insects has lagged behind work with insects.

## References

Kettlewell, H. B. D. 1973. The evolution of melanism. Clarendon, Oxford.

Langé, G. 1970. Relations entre le déterminisme génétique du sexe et la contrôle hormonal de sa différentiation chez les arthropodes: Comparaison avec les vertèbrés. Ann. Biol. 9: 189–230.

Pal, R., and M. J. Whitten. 1974. The use of genetics in insect control. Elsevier/North-Holland, Amsterdam.

Robinson, R. 1971. Lepidoptera genetics. Pergamon, Oxford.

## INSECT BEHAVIOR

Insect behavior (Matthews and Matthews 1978) is a rapidly developing field of study that attempts to explain both the complex anatomical and physiological bases and higher, integrative mechanisms for activity. Only short-term, decisively determined actions are recognized in this framework. Long-lasting, slowly induced actions, such as diapause or maturation, are considered physiologic or developmental phenomena (see other parts of this chapter).

Physiochemically and anatomically, insects possess the same elements that control behavior in all animals. Foremost of these is the nervous system (Roeder 1963), including its sensory component, but the muscular and hormonal components play an essential, if secondary, part. It is the degree of complexity of the first that determines the levels on which lines of action lie.

A key element of the nervous system in determining behavior is the associative (adjustor, internuncial) neuron, which intercedes between receptor (efferent) and effector (afferent) neurons and has the capacity to redirect and modify otherwise simple reflex reactions. Large numbers of these form masses (neuropiles) in the brain and ventral ganglia and serve as centers of neural integration. These are something like the cortex or gray matter of the human brain and define the overall function of a ganglion. They represent the main areas where activities are generated and organized. A major such center is the corpus pedunculatum ("mushroom-shaped body"), believed to be the site of summation of simultaneous excitation from all sources. It tends to be small in arthropods with simple behavior, large in those with complicated lives, such as the social Hymenoptera. These cells both stimulate and inhibit.

Endocrine secretions are not only caused to flow in response to nervous command but are actually part of the nervous system in the form of neurosecretory cells. These cells produce hormones that move along the axons and direct other nerve and endocrine tissues to emote.

Of course, activity is finally the result of muscular contraction. Insects and their relatives may have very large numbers of discrete muscle bundles that predicate a likewise elaborate system of efferent nerves. It is fortunate that a lack of obstructive connective tissue in these animals makes it possible to dissect and experiment to determine pathways relatively easily. The largest nerves lead to the most active locomotor organs, the wings and legs. Other major efferents control the mouthparts, antennae, cerci, genitalia, and numerous other muscularized structures.

The insect behaviorist looks for chains or pathways of stimulation-integration-action to explain activities (Browne 1974). The latter can be considered to be composed of bits or units that meld together into sequences first, then complexes or systems. The simplest movements have the simplest nerve control and fewest muscles involved. The most complex systems have very large numbers of pathways and processes and are so complicated that it is possible to analyze them only in general. An understanding of the way the whole insect acts requires an extension of the rudimentary functioning of the neural, hormonal, and muscular elements. This extension progresses along a scale of increasing complexity, beginning with so-called automatic or instinctive behavior and terminating with learned activity.

The simplest instinctive actions are reflex arcs, so-called knee-jerk responses, where a part of the body reacts directly to a stimulus without the intercession of an association nerve. An example is the retraction of the tarsus from a hot surface. A step up from this level occurs when the whole body is coordinated but by nonmodifiable reactions. Where only a single action is identifiable, such as movement away from or toward light or touching or shunning other individuals or objects, the behavior is called a taxis or tropism. Such behavior may be positive or negative. The attraction of moths to artificial light, the catatonic freezing or "death feigning" display many species use to escape harm, and the following of odor trails by dung beetles to find food for their young are specific examples.

A series of these tropistic elements may be strung together, one triggering the next to form a fixed action pattern. These may take up a sizable part of the behavioral repertoires of most insects. Pupation in giant silk moth larvae offers an appropriate example: changes in photoperiod or some internal stimulus causes them to cease feeding. This initiates defecation and a wandering, searching activity, leading to the discovery of a suitable pupation site. Even if the latter is not found, the larvae will begin to spin silk and form a cocoon of a specific shape in which it finally settles and pupates. This sequence follows the same steps regardless of changes in external stimuli (unless acute) and does not vary according to any information learned by the individual.

Insects and other terrestrial arthropods are capable of limited learning (Alloway 1972), defined as any relatively permanent change in behavior that results from practice. Such learning is of a low order and often short lived, but it is often essential to the animal's existence. At least two types have been seen, classical Pavlovian conditioning and, much more commonly, instrumental conditioning, where reinforcement stimuli direct the performance of the insect. The latter is a characteristic especially of social insects, like the honeybee, which can be trained artifically to fly to a colored surface by food offerings. Under natural conditions, this ability is important in recruiting foragers and in efficient utilization of a flower nectar food source. Some forms, such as cockroaches and ants, facilitate to mazes. The vast majority of these arthropods, however, probably are capable of virtually no learning whatever.

The complexity of some behavior in insects, particularly social insects, most especially ants, whose lives parallel our own in some ways, has suggested to some the possibility of the existence of intelligence. As possessed by higher vertebrates, including ourselves, no such high degree of learning and reasoning can be truly ascribed to these creatures. All activity, regardless of how cunning and comprehending it seems, can be explained on the basis of fixed action sequences, with very limited learning. The nesting of digger wasps (*Ammophila*) is a classic example: the female wasp first digs a burrow in sandy soil which it then closes over at the mouth. It then leaves to search for prey, captures it, and returns to the location of the burrow. To do this, it has had to learn a few landmarks by which it navigates. Their misplacement, however, may lead the digger wasp to conclude wrongly on the exact location. The nest, when found, is opened and the prey packed within, an egg is laid on it, and the female exits, closes the nest permanently, and leaves to repeat the process elsewhere. All of these are innate, unmodifiable acts.

The remarkable thing about insect behavior is that it may be highly complicated, comparable in this respect alone to vertebrates, yet it is nearly all controlled by instinctive mechanisms. Fundamental life processes are thus served efficiently, although automatically and unswervingly, and have contributed to their success as a group.

It is useful to segregate and classify the kinds of motivation driving the insect body because it is often found that single action sequences operate within them. The following are only representative, as many examples fit into the categories given; additional types will appear in the main text of this book.

1. *Alimentation.* Finding food and feeding involve specific movements, often elaborate. Mosquitoes respond to visual and odor cues to find warm-blooded hosts and then follow tactile stimuli to select a proper station and find a capillary. Internal pressure from expansion of the stomach causes cessation of feeding and induces flight.

2. *Survival.* Its host, discovering a mosquito in the act of feeding, will attempt to destroy or remove it. The insect displays flight as a survival act, an extremely common one with winged types. Other survival-related behavior is shelter seeking, catalepsis, and biting. Most protective coloration is accompanied by postures that enhance deception or warning patterns.

3. *Aggression.* Both intra- and interspecific agonistic (fighting) behavior occurs in insects, including male-male competition for females, as in the horned scarabs. Bees may grapple for a nectary or over territory and females. Raiding for food, such as found in many ants, should not be confused with aggression, although the results are the same. The vanquished colony is perceived as food, not as a rival faction.

4. *Sex.* This essential, overriding drive in all organisms has led to some of the most incredibly complex and even bizarre activities in all groups of terrestrial arthropods. These are divided into

mate finding, courtship, copulation, and insemination (Thornhill and Alcock 1983).

5. *Brood care.* Parental behavior occurs in relatively few insects and other terrestrial arthropods and is a precursor to social organization in general. It greatly increases survivability and is necessary for the maintenance of colonies.

6. *Intraspecific communication.* The ways in which information is transmitted between individuals of the same species are tremendously varied, employing visual, chemical, auditory, tactile, and other methods. The use of airborne pheromones seems to dominate, although nutritive chemicals, ingested by the receiver (trophallaxis), are transmitted among members of social insect colonies. Sound also ties many nonsocial types together.

7. *Tool using.* It is an amazing fact that a few insects actually use tools—in an instinctive way, of course. The prime example is the pebble employed by digger wasps to tamp the soil plug of their burrow nests.

8. *Construction.* Many types form structures from a variety of building materials, both extraneous (mud, paper, wood) and intrinsic (silk). Architecture may be elaborate and the size and strength of many edifices prodigious. A high level of cooperation can be required between members of social forms to put up nests. Individual efforts are also intricate and consistent with regard to geometry and engineering.

9. *Migration.* A large number of species regularly move from one territory to another, some even on long-established and precise migratory routes. Unidirectional flight is a conspicuous manifestation of this behavior, and it is most conspicuous in larger, active forms such as butterflies and day-flying moths.

Some behavioral traits apparently not found in insects and their relatives are play, expression of grief or sorrow, and humor. These are characteristics of a vertebrate cerebrum and set these higher creatures apart from insects and other arthropods, which function as virtual automatons.

## References

ALLOWAY, T. M. 1972. Learning and memory in insects. Ann. Rev. Entomol. 17: 43–56.

BROWNE, L. B. 1974. Experimental analysis of insect behavior. Springer, New York.

MATTHEWS, R. M., AND J. R. MATTHEWS. 1978. Insect behavior. Wiley, New York.

ROEDER, K. D. 1963. Nerve cells and insect behavior. Harvard Univ. Press, Cambridge.

THORNHILL, R., AND J. ALCOCK. 1983. The evolution of insect mating systems. Harvard Univ. Press, Cambridge.

## DEVELOPMENT AND LIFE CYCLES

### Eggs

Whether external to the female parent's body (oviparity) or temporarily within (viviparity), all insects, spiders, and allied terrestrial arthropods start their lives as eggs (Hinton 1981). Eggs come in an amazing variety of shapes and sizes. They are usually placed singly or in groups in proximity to the juvenile's food source but may be scattered indiscriminately only in the general habitat where development occurs. Many have elaborate cuticular sculpturing, and some possess devices for attachment to the substratum or caps (opercula) that open to allow egress of the young. A number of species protect their eggs from moisture loss and trauma by covering them with froth or encasing them in other substances that harden around them (oothecae).

## Reference

HINTON, H. E., ed. 1981. Biology of insect eggs. Vols. 1–3. Pergamon, Oxford.

## Embryology

Just prior to fertilization, insect eggs are composed mostly of yolk and small islands of cytoplasm surrounding the female nucleus on one edge. When the egg is laid, the nucleus is usually in the metaphase of the first meiotic division, in which state it receives the sperm, one of which unites with the oocyte after meiosis is complete. The nucleus then migrates to the center of the egg and begins to divide mitotically. The resulting cells move to the periphery and form the blastoderm, or early embryo, which later lodges on one side of the egg. The germ layers and embryonic membranes soon develop, and determination of segmentation and the primary organs and tissues ensues. The appendages appear, and after a time, the perfect body of the first juvenile stage is complete. This stage takes different forms depending on the evolutionary level of the group. Fairly similar embryological steps are followed by other terrestrial arthropods (Johannsen and Butt 1941). A major exception are the springtails (Collembola), whose eggs undergo holoblastic cleavage.

## Reference

Johannsen, O. A., and F. H. Butt. 1941. Embryology of insects and myriapods. McGraw-Hill, New York.

## Development

Insects and related arthropods must pass through a series of developmental stages on their way to becoming sexually mature adults (Agrell and Lundquist 1973). These stages are all the more discrete because of the necessity of molting and growth in stepwise phases. The animal itself between molts is referred to as an "instar," the time period, "stadium." In virtually all insects, the first instar possesses the complete number of segments after hatching; in other groups, segments are added as development proceeds.

As the animal progresses toward maturity, it increases in size, and changes in internal and external form and proportions occur to a greater or lesser degree (Sehnal 1985). In most noninsects and primitive apterous insects, the immatures are fairly similar to the adults. Juvenile insects of the higher orders that possess wings, however, undergo a fair amount of body modification, called metamorphosis, primarily associated with the growth of the wings and exploitation of habitats different from the adult. Metamorphosis is said to be "gradual" (incomplete) in lower winged insects with externally developing wing buds; the single juvenile type is called a nymph (or sometimes naiad in aquatics). Nymphs generally have feeding and other habits similar to the adult; naiads live rather different lives because of their water habitats. Metamorphosis is "complete" in the higher winged insects. In these, there are two fundamental juvenile stages: a larva, which has several instars; it finally molts into a pupa, which eventually yields the adult. These early stages look totally unlike and live in ways very different from the adult and indeed diverge from them in almost every way. This has contributed to the evolutionary success of these insects through the dichotomous specialization of life functions (feeding and growth by immatures, dispersal and reproduction by adults). Divergence of body form and function has even taken a further step in many species with varying types of larvae (hypermetamorphosis) such as found in the blister beetles (Meloidae), chalcidoid wasps, and others. Immatures of different insect groups are called by various names. For example, larvae of Lepidoptera are caterpillars; pupae of butterflies, chrysalids; larvae of muscoid flies, maggots; and beetle larvae, grubs. Pupae generally are protected by their location, underground in cells or in wood or other material or encased in a cocoon of silk spun by the prepupal instar.

## References

AGRELL, I. P. S., AND A. M. LUNDQUIST. 1973. Physiology and biochemical changes during insect development. Physiol. Ins. 1: 159–247.

SEHNAL, F. 1985. Morphology of insect development. Ann. Rev. Entomol. 30: 89–109.

## Maturation

The adult is the sexually capable instar, whose responsibility is to find a mate and reproduce, thus perpetuating the species. Development of the internal and external sexual organs completes growth and usually molting, although in some noninsect terrestrial arthropods, molting may continue throughout life. Wings in the insects also become fully grown and functional at this time, the one exception being the mayflies, which have a winged instar (subimago) preceding the full imago.

Some aberrant conditions occur in insects, such as neoteny, in which the adult retains its outward larval body form but completes development of the internal reproductive organs. Neotenic adults mate and parent offspring while continuing to feed and live as immatures. This is a common condition in railroad worms and other beetles and in some primitive flies.

## Life Cycles

The way an insect or allied arthropod develops in relation to its seasonal environment constitutes its life cycle (Tauber et al. 1985). Life cycles are as varied as the kinds of animals living them. Perhaps a majority of species in arctic or temperate life zones have annual generations, that is, one complete turnover, egg to egg per year. Others have biannual or multiannual cycles. The latter implies the existence of prolonged feeding periods, often on food that is poor in nutrition (e.g., wood-boring beetle larvae), or the intercession of a period of diapause. Still others are semiannual (bivoltine) or multivoltine, with two to several generations per year. The latter are more typical of tropical or other stable environments where unfavorable drought or cold does not force temporary arrests in development. Some insects, such as pomace flies (*Drosophila*), develop very rapidly and repeatedly and may have almost continuous reproduction throughout the year. Some mosquitoes mature very quickly in transient water following infrequent rains but remain dormant in the egg stage for most of the remainder of the year.

Certain insect types regularly incorporate asexual reproduction in their life cycles in addition to sexual reproduction. This alternation of generations is typical of aphids, for example. When conditions are best for plant growth and therefore feeding, as in the beginning of the rainy season, emphasis is on multiplication of numbers. This is accomplished by the "stem mothers" that bear live, sterile, wingless females parthenogenetically and as rapidly as possible. As the season favorable for dispersal approaches, when there is less or no rain and winds may increase, sexually active, winged males and females appear, to mate, mix genes, and disperse to new localities. The females lay eggs that hatch into the asexual forms once again. Production of sexual forms is controlled by changes in temperature and photoperiod; under constant tropical conditions, cyclical alternation of generations may not occur.

## Reference

TAUBER, M. J., C. A. TAUBER, AND S. MASAKI. 1985. Seasonal adaptations of insects. Oxford Univ. Press, New York.

## EVOLUTION AND CLASSIFICATION

The reconstruction of the historical evolution and determination of the interrelationships of the presently extant orders of insects and other terrestrial invertebrates has not been settled by any means. There remain many controversies, even over major theses, such as the monophyly (descent from a single ancestral line) of the Arthropoda or of the apterygote hexapods. There is extensive literature on these disagreements and relevant argumentation (Anderson 1973; Boudreaux 1979; Gupta 1979; Manton 1977; Sharov 1966).

The arthropod groups included in this book are all basically terrestrial, probably by way of several independent, parallel evolutionary pathways, from varied precursors among the Onycophora, Crustacea (Isopoda), Uniramia (myriapods and insects), and Chelicerata (arachnids), and are thus only distantly related (Manton 1977: 257–258).

The onycophoran line seems to attach most closely to the myriapodan, and these animals can no longer be considered intermediate phylogenetically between annelids and arthropods, the latter now being recognized as a polyphyletic group. They are not ancestral to either that group or Hexapoda, nor is the latter descended from the Myriapoda. Embryological evidence indicates that all three have diverged independently from common ancestors with uniramous (lobate) legs.

Although the Crustacea are distant, with fundamentally different, biramous appendages, they share many features with insects. They are almost entirely marine, with only a few types secondarily adapted for life in fresh waters and on land.

The chelicerates, distinguished fundamentally by their chelicerate mouthparts, are virtually all terrestrial, although probably derived from originally marine ancestors. Evolution within the subphylum is not clear. All efforts to subdivide the orders have remained inconclusive, as have associated phylogenetic speculations. Those with book lungs (Scorpionida, Uropygi, Amblypygi, spiders) presumably can be grouped; scorpions, with their complete segmentation, are the most primitive. In body shape and external genitalia, the Opiliones resemble some primitive mites, with which they seem to form a close branch. The other groups are all isolated.

The phylogeny of the primitively mandibulate Uniramia (appendages with single stem) is fairly well understood, at least for the insects in general (Kristensen 1981). Myriapods retain homomerism, having only a distinct head, but seem to possess the basic body structure likeliest to precede that of insects. The ancestors of the insects (Hennig 1981) evolved a three-somite thorax and three pairs of legs at an early time, reducing the many equal body parts of the myriapods. This arrangement is preserved in all the true insects and three primitive orders (the Parainsecta) that differ from the insects in several basic ways, including the mouthparts, which are not exposed as usual (ectognathy) but overgrown by cranial folds (entognathy). There are also wingless (Apterygota) predecessors in body design to the dominant insects that evolved wings early in their history (Pterygota) (Kukalová-Peck 1987).

At first (Paleoptera), wings were clumsy, outwardly projecting, fixed, flight organs, as seen in many extinct groups of the Paleozoic (e.g., Palaeodictyoptera) and extant mayflies and Odonata, but soon acquired improvements, among them the ability to be flexed over the body which all the higher orders have (Neoptera). Even those that have secondarily lost wings altogether, often in association with ectoparasitism (fleas, lice, bedbugs, etc.), retain the thoracic structure of their fully winged ancestors.

Three major lines emerged within the higher winged insects. The first, most primitive assemblage (Polyneoptera), which some workers question as monophyletic, includes the "orthopteroid" groups, the Orthoptera, Grylloptera, Dermaptera, and other orders that display an enlarged, fanlike hind wing with many longitudinal veins, multisegmented tarsi, and many Malpighian tubules and ganglia internally.

The second, "hemipteroid" group (Paraneoptera) is largely characterized by regressive traits, loss or reduction in the number of tarsal segments, few Malpighian tubules, and ganglia in the central nervous system. The main orders found here are the Hemiptera, Psocoptera, Mallophaga, Anoplura, and Thysanoptera. Both of these major groups display incomplete metamorphosis.

The third and the most recently derived and highly successful group within the Neoptera, are those with complete metamorphosis (Holometabola). It makes up more than 80 percent of the species of living insects. Within the group, various subdivisions can be recognized, but their definition and the orders to be included are not all supported by incontrovertible evidence. At least "neuropteroid" (Megaloptera, Neuroptera) and "panorpoid" (Mecoptera, Trichoptera, Lepidoptera, Diptera, Siphonaptera) lines are readily identifiable, but the Coleoptera and Hymenoptera are independent and stand apart.

The foregoing can be summarized in a linear classification such as that following. It should be emphasized that such schemes are matters of personal preference and are always controversial. This classification is conservative and intended to reflect relationships of broad groups as well as to provide convenient categories for the presentation of the systematic material forming the major portion of this book. (See also taxon table.) The reader interested in alternative ideas should consult the references cited at the end of this section.

CLASSIFICATION OF INSECTS AND
MAJOR GROUPS OF RELATED
TERRESTRIAL ARTHROPODS
(* = *groups not covered in this book; included for reference only.*)

Phylum Onychophora-onychophorans
Phylum Arthropoda-arthropods
    Subphylum Biramia
        Class Crustacea—crustaceans
            Subclass Peracarida
                Order Isopoda—isopods
                Order Amphipoda—amphipods
    Subphylum Chelicerata
        Class Arachnida
                Order *Ricinulei—ricinulids
                Order Araneae—spiders
                Order Opiliones—harvestmen
                Superorder Acari—mites and ticks
                Order *Palpigradi—palpigrades
                Order *Schizomida—micro whip scorpions
                Order Uropygi—whip scorpions
                Order Amblypygi—whipless whip scorpions
                Order Pseudoscorpionida—pseudoscorpions
                Order Scorpionida—scorpions
                Order Solpugida—sunspiders
    Subphylum Uniramia
        Class Myriapoda—myriapods
            Subclass Chilopoda—centipedes
            Subclass Diplopoda—millipedes
            Subclass *Symphyla—symphylans
            Subclass *Pauropoda—pauropods
        Class Hexapoda—insects (broad sense)
            Subclass Parainsecta
                Order *Protura—proturans
                Order *Diplura—diplurans
                Order Collembola—springtails
            Subclass Insecta—true insects
                Infraclass Apterygota
                    Order Thysanura—silverfish and bristletails
                Infraclass Pterygota—winged insects
                    Superorder Paleoptera-ancient—winged insects
                        Order Ephemeroptera—mayflies

Order Odonata–dragonflies
and damselflies
Superorder Neoptera–
modern-winged insects

"ORTHOPTEROIDS"

Order Plecoptera–stoneflies
Order Grylloptera–katydids
and crickets
Order Orthoptera–
grasshoppers and allies
Order Blattodea–
cockroaches
Order Mantodea–mantids
Order Phasmatodea–
walkingsticks
Order Dermaptera–earwigs
Order Isoptera–termites
Order Embiidina–web
spinners

"HEMIPTEROIDS"

Order Psocoptera–psocids
Order *Zoraptera–
zorapterans
Order Mallophaga–biting
lice
Order Anoplura–sucking
lice
Order Hemiptera–true bugs
(heteropterans and
homopterans)
Order Thysanoptera–thrips

"NEUROPTEROIDS"

Order *Mecoptera–
scorpionflies
Order Megaloptera–
dobsonflies
Order Neuroptera–nerve-
winged insects

"PANORPOIDS"

Order Diptera–flies and
gnats
Order Siphonaptera–fleas
Order Trichoptera–
caddisflies
Order Lepidoptera–
butterflies and moths

ORDERS OF UNCERTAIN AFFINITIES

Order Coleoptera–beetles
Order Hymenoptera–ants,
bees, and wasps

## References

ANDERSON, D. T. 1973. Embryology and phylog-
eny in annelids and arthropods. Pergamon,
Oxford.
BOUDREAUX, H. B. 1979. Arthropod phylogeny,
with special reference to insects. Wiley, New
York.
GUPTA, A. P., ed. 1979. Arthropod phylogeny.
Van Nostrand Reinhold, New York.
HENNIG, W. 1981. Insect phylogeny. Wiley,
Chinchester, Eng.
KRISTENSEN, N. P. 1981. Phylogeny of insect
orders. Ann. Rev. Entomol. 26: 135–157.
KUKALOVA-PECK, J. 1987. New Carboniferous
Diplura, Monura, and Thysanura, the hexa-
pod ground plan, and the role of thoracic
side lobes in the origin of wings. (Insecta).
Can. J. Zool. 65: 2327–2345.
MANTON, S. M. 1977. The Arthropoda, habits,
functional morphology and evolution. Claren-
don, Oxford.
SHAROV, A. G. 1966. Basic arthropodan stock
with special reference to insects. Pergamon,
Oxford.

## FOSSIL INSECTS

Known Latin American fossil insect sites
are few, but they have produced considerable material representing several types of
fossilization. Most represent relatively recent strata (Cenozoic).

Impressions in sedimentary rock from
the Eocene in South America are most
significant (Martínez 1982). One of the best
known beds is found at Sunchal, Argentina,
in the province of Jujuy. Many specimens of
weevils and other insects were excavated
there by entomologist T. D. A. Cockerell
early in this century. The oldest insects from
the region are of an unidentified order
(*Eugeropteron* and *Geropteron*) from middle
Carboniferous beds in the Sierra de los
Llanos of the province of Rioja, also in
Argentina. Other important sites of insect

fossils preserved in sedimentary rocks are located in Rio Grande do Sul, Brazil, and at Bajo de Veliz, in the province of San Luis in Argentina (Upper Carboniferous) (Martínez pers. comm.). An enormous theraphosine spider (*Megarachne servinei*), measuring 34 centimeters from the chelicerae to the tip of the abdomen and having a leg span of more than 50 centimeters, was discovered in the Bajo de Veliz Formation (Hünicken 1980). The Santana Formation near Ceará, Brazil, has yielded fossils of Lower Cretaceous age (Grimaldi 1990).

Two amber deposits have yielded abundant specimens, although of relatively recent times. Preservation is exceedingly good. The insects were caught in resinous exudates from various trees. The viscous substance later hardened with burial and time and remained clear within so that the most minute structures (hairs, genitalia, mouthparts) can be observed today in perfect detail, even on the smallest mites and midges. Preservation is so perfect in some cases that attempts are being made to recover and replicate fossilized DNA from cells of fungus gnats with large chromosomes (Poinar and Hess 1982).

A famous site of late Oligocene to early Miocene Age is near Simojovel, in the state of Chiapas, Mexico (Hurd et al. 1962). It has been particularly well studied and has produced a wide variety of taxa (various authors 1963, 1971). These deposits were known to early peoples of Central America, and amber pieces with insect inclusions were fashioned into ornaments. The original resin is believed to have been secreted by a leguminous tree of the genus *Hymenaea*.

Localities in the Dominican Republic have commanded attention in recent years (Baroni-Urbani and Saunders 1982) and are also the source of a thriving "gemstone" industry (Rice 1979, Rice and Rice 1980). The deposits are generally accepted as of about the same age as the Chiapas amber and likewise derived from *Hymenaea*. The many mines scattered about the country have produced amber fragments with thousands of specimens of more than one hundred families of insects plus spiders and other arachnida (Cokendolpher 1986) and invertebrates (Sanderson and Farr 1960), including the first fossils of gardening ants (Attini, *Trachymyrmex*, Baroni-Urbani 1980). Wilson (1985) found 37 genera of ants, of which 34 survive in neighboring areas of Latin America.

In the Americas, there are also a number of other known but unexplored amber deposits, for example, in Colombia (Cockerell 1923), Brazil (Froes Abreu 1937), and surely other countries (Poinar and Agudelo 1980).

Good preservation is also characteristic of the Quaternary Age remains found in asphalt deposits. Some sites in this category are located in Trinidad (Blair 1927) and at Talara on the northern Peruvian coast (Churcher 1966). Here, because of the stickiness of the tarlike medium and the attractiveness of the surface, which looks like water, asphalt seeps form very efficient small animal traps. The most numerous kinds of insects found as fossils in these deposits are hard-bodied ground beetles, aquatics, and carrion feeders. Aquatic insects are sometimes indicative of the presence of freshwater pools near the asphalt or overlying it. They were entrapped when the water dried up during drought periods. Carrion-feeding species were caught along with the carcasses of vertebrates that died in the black quagmires.

Because of their small size and delicateness, insects and their relatives produce good fossils only in fine-grained or homogeneous matrices. The foregoing are of this type. Other modes of fossilization that may be important in Latin America, and which have been scarcely investigated, are permineralization (such as in mineral-charged waters), peat and soft coal encapsulation, cave sediments (Miller 1986), and silicification, especially evident in calcareous nodules. Evidences of feeding, boring, coprolites, and trails should also be common in deposits of plant fossils. Insect remains in association with ancient human remains may also be of considerable archaeological significance (e.g., Warner and Smith 1968).

## References

BARONI-URBANI, C. B. 1980. First description of fossil gardening ants. Amber collection Stuttgart and Natural History Museum Basel: Hymenoptera: Formicidae. I: Attini. Stutt. Beitr. Naturk. Ser. B (Geol. Paleon.) 54: 1–13.

BARONI-URBANI, C. B., AND J. B. SAUNDERS. 1982. The fauna of the Dominican amber: The present state of knowledge. 9th Carib. Geol. Conf. (Santo Domingo, 1980) Trans. 1: 213–223.

BLAIR, K. G. 1927. Insect remains from oil sands in Trinidad. Entomol. Soc. London Trans. 75: 137–141.

CHURCHER, C. S. 1966. The insect fauna from the Talara tar-seeps, Peru. Can. J. Zool. 44: 985–993.

COCKERELL, T. D. A. 1923. Insects in amber from South America. Amer. J. Sci. 5: 331–333.

COKENDOLPHER, J. C. 1986 (1987). A new species of fossil *Pellobunus* from Dominican Republic amber (Arachnida: Opiliones: Phalangodidae). Carib. J. Sci. 22: 205–211.

FROES ABREU, S. 1937. Sobre a ocorrência de ambur nos arenitos da serie Bahia: Brasil. Inst. Nac. Tech. (Rio de Janeiro) Bol. Inf. 2(4): 8.

GRIMALDI, D. A., ed. 1990. Insects from the Santana Formation, Lower Cretaceous, of Brazil. Amer. Mus. Nat. Hist. Bull. 195: 1–191.

HÜNICKEN, M. A. 1980. A giant fossil spider (*Megarachne servinei*) from Bajo de Veliz. Acad. Nac. Cien. Córdoba, Bol. 53: 317–325.

HURD, JR., P. D., R. F. SMITH, AND J. W. DURHAM. 1962. The fossiliferous amber of Chiapas, Mexico. Ciencia 21(3): 107–118, Pl. I–II.

MARTÍNEZ, S. 1982. Catálogo sistemático de los insectos fósiles de América del Sur. Fac. Hum. Cien. (Univ. Rep., Montevideo) Ser. Cien. Tierra, Rev. 1(2): 29–83.

MILLER, S. E. 1986. Phylum Arthropoda, Class Insecta. *In* D. W. Steadman, Holocene vertebrate fossils from Isla Floreana, Galápagos. Smithsonian Contrib. Zool. 413: 1–103.

POINAR, JR., G. O., AND F. AGUDELO. 1980. El ambar: Oro fósil del nuevo mundo. Americas 32(10): 33–40.

POINAR, JR., G. O., AND R. HESS. 1982. Ultra-structure of 40-million-year-old insect tissue. Science 215: 1241–1242.

RICE, P. C. 1979. Amber of Santo Domingo—mining in the Dominican Republic. Lapidary J. (Nov. 1979): 1804–1810.

RICE, H. E., AND P. C. RICE. 1980. Pepitas de sol antillano. Americas 32 (10): 37–41.

SANDERSON, M. W., AND T. H. FARR. 1960. Amber with insect and plant inclusions from the Dominican Republic. Science 131: 1313–1314.

VARIOUS AUTHORS. 1963, 1971. Studies of fossiliferous amber arthropods of Chiapas, Mexico, Pts. I, II. Univ. Calif. Publ. Entomol. 31: 1–53, pls. 1–3; 63: i–vi, 1–106, pls. 1–3.

WARNER, R. E., AND G. E. SMITH, JR. 1968. Boll weevil found in pre-Columbian cotton from Mexico. Science 162(3856): 911–912.

WILSON, E. O. 1985. Invasion and extinction in the West Indian ant fauna: Evidence from the Dominican amber. Science 229(4710): 265–267.

## INSECT NAMES

All known organisms, including insects and their relatives, have a scientific name, and many also have a common name (Goto 1982).

Scientific names are applied according to rigorous procedures (Ride et al. 1985), with consistency, universality, and stability as primary considerations. Each species must bear a unique Latinized two-part epithet (for subspecies, three-part), consisting of a genus and specific name, for example, for the common housefly, *Musca domestica*. Species are grouped into a hierarchy of categories, the most usual being (in ascending order) the tribe (name ends in suffix, *-ini*, e.g., Muscini), subfamily (*-inae*, Muscinae), family (*-idae*, Muscidae), and order (no standard suffix, e.g., Diptera).

Names are given to species or higher taxa as they are discovered and established by publication. The first describer is entitled to authorship, and all others are obliged to use that name. The term "new species" refers to one that has been so found for the first time, not to freshly evolved ones; none of the latter has yet been observed in nature.

Scientific names are properly pronounced according to the rules of Latin, but their way of being spoken usually varies according to the native accent of the speaker. This should bother no one except Latin scholars, as long as the name is understood.

Common names, or vulgates, are applied to the insects and their relatives in all Latin American countries. Lexica have been published for Chile (Brücher 1942, Perez D'Angello 1966), Peru (Dourojeanni 1965), Brazil (Baucke 1961, Biezanko and Link 1972, Monte 1928, da Silva 1930–1934) and Haiti (Audant 1941). Many vulgates are adopted directly from indigenous languages, some tribes and local cultures being prolific nomenclaturists, especially in Brazil (Monte 1928). These suffer from frequent spelling and pronunciation variations, particularly in Brazil (where, in general, I follow von Ihering's [1968] orthography). At least partial entomological glossaries exist for the following native tongues: Mayan (Welling 1958), Aztec (= Nahuatl, etc.; Ordoño 1982), Kunza (= Atacameño; Munizaga and Herrera 1957), Jívaro (Guallart 1968), Tupí-Guaraní (Tastevin 1923), and Quechua (García 1976). Vernacular names appear according to no consistent set of standards, varying from place to place or time to time with different origins and related to the nature of the society employing them (Stoetzel 1989). Phonetic variations in spelling are common.

Scientists and educated people often form simple transliterations of technical names (muscids or muscideos, from Muscidae) or accommodate names of classic origin (scarabs or escarabajos, from Greek *karabos*). Laymen and country folk are likely to invent quaint, often descriptive appellations that frequently apply to an insect's behavior (saltamonte = "hill jumper") or

stinging abilities (lagarta de fogo = "fire worm"), anatomy (tijeretas = "scissor bearers"), or that are onamatapoetic (cricket, chicharras), or that may be without obvious derivation (gallinipper). Sometimes these are literal translations from modern languages (scorpions, escorpiones) or usages (tarantulas) not common to the region. Mixtures of sympatric languages also occur (sede [Spanish] + ocuilin [Nahuatl] = sedeocuilin = silkworm). The only attempt to standardize common names has been made with pest species in English (Stoetzel 1989).

Most languages have a broad term for insects and like animals, roughly equivalent to the English, for example, "bug" ("worm" or "grub"): bicho (Spanish and Portuguese) and ocuilin (Nahuatl).

## References

AUDANT, A. 1941. Identification des insectes d'Haiti par leur nom créole. Soc. Hist. Geogr. Haiti, Rev. 12(42): 51–55. [Not seen.]

BAUCKE, O. 1961. Os nomes comuns dos insectos no Rio Grande do Sul. Sec. Agric., Porto Alegre.

BIEZANKO, C. M., AND D. LINK. 1972. Nomes populares dos Lepidópteros no Rio Grande do Sul (Segundo Catalogo). Univ. Fed. Santa Maria, Bol. Tec. 4: 3–15.

BRÜCHER, G. 1942. Lista de algunos nombres vulgares de insectos. Dept. San. Veg. (Min. Agric., Santiago) Bol. 2(2): 120–125.

DA SILVA, B. R. 1930–1934. Nomenclatura popular dos Lepidópteros do Distrito Federal and seus arredores. Vols. 1–5. O Campo, Rio de Janeiro.

DOUROJEANNI, M. J. 1965. Denominaciones vernaculares de insectos y algunos otros invertebrados en la selva del Perú. Rev. Peruana Entomol. 8: 131–137.

GARCÍA, R. J. 1976. Nombre de algunos insectos y otros invertebrados en "Quechua." Rev. Peruana Entomol. 19: 13–16.

GOTO, H. E. 1982. Animal taxonomy. Arnold (Inst. Biol., Stud. Biol. no. 143), London.

GUALLART, J. M. 1968. Nomenclature Jíbaro-Aguaruna de la fauna del Alto Marañón (Invertebrados). Biota 7: 195–209.

IHERING, R. VON. 1968. Dicionário dos animais do Brasil. Ed. Univ. Brasília, São Paulo.

Monte, O. 1928. Os nomes vulgares dos in-sectos de Brasil. Almanak Agric. Brasil. 1928: 228–289.

Munizaga, C., and J. Herrera. 1985. Notas entomológicas de Socaire (Obtenidas durante la Expedición Chileno-Alemana a Socaire, en mayo de 1957). Notas Centr. Est. Antropol., Univ. Chile, 1: 3–13.

Ordoño, C. M. 1982. Diccionario de zoología Náhuatl. Ed. Innovación, Mexico.

Pérez D'Angello, V. 1966. Concordancia entre los nombres vulgares y científicos de los insectos chilenos. Mus. Nac. Hist. Nat. Not. Mens. 10(119): 2–7.

Ride, W. D. L., C. W. Sabrosky, G. Bernardi, and R. V. Melville, eds. 1985. International code of zoological nomenclature. 3d ed. Intl. Trust Zool. Nomen., London.

Stoetzel, M. B. 1989. Common names of insects and related organisms. Entmol. Soc. Amer. Lanham, Md.

Tastevin, C. 1923. Nomes de plantas e animaes em Lingua Tupy. Rev. Mus. Paulista 13: 687–763.

Welling, E. C. 1958. Some Mayan names for certain Lepidoptera in the Yucatán penin-sula. J. Lepidop. Soc. 12: 118.

## INSECTS AND HUMAN CULTURE

Aside from their importance as pests and our academic interest in insects, these crea-tures, spiders, and related arthropods have considerable influence in that portion of human activity that may be called the humanities—music, art, literature, lan-guage, religion, and folklore (fig. 1.8). The study of these influences is a general area of insect study called cultural entomology (Hogue 1987). Examples appear among historical, modern, and indigenous peo-ples. (Some of the more general are cited below; many other specific cases are scat-tered through the remainder of this book in the sections on the various insects in-volved; for Mexico, see MacGregor 1969.)

Insects, spiders, centipedes, and scorpi-ons appear in the Mayan Codices (Dresden, Tro-Cortesianus, and Peresianus), indicat-

**Figure 1.8** Decorative plates from modern Peru prominently featuring images of the fly (*chuspi*), revered in Incan times and a design motif in Andean art today. (Original, author's collection)

ing an appreciation of their existence and their inclusion in cultural events, such as rituals, ceremonies, and dances. The fa-mous Nasca figures include an immense spider (fig. 1.9). Portions of same are also stylized as glyphs having linguistic signifi-cance (Tozzer and Allen 1910). In the eighteenth century, it was believed that a small, red insect (still unidentified but called "coya" in the Orinoco region) caused severe skin eruptions; its effects could only be remedied by ceremoniously passing the body through a fire made from a specific grass ("guayacán") (Kamen-Kaye 1979). Many such curious accounts of insects fill the accounts of early visitors and colonists in the New World (Cowan 1865).

Insects have lent their names to many places in Latin America. Among the better known are Chapultepec, the "hill of the grasshoppers" (*chapulin* = grasshopper + *tepec* = hill) where the Aztec Emperor Montezuma's castle was built in what is now part of Mexico City, and Urubamba, "plain of the insect" (*uru* = spider or

**Figure 1.9** The spider was an eminent symbol in Peruvian cultures of prehistory. It is displayed on a grand scale among the Nasca figures in the southern desert.

**Figure 1.10** Political graffiti on wall in San José, Costa Rica, by group using a social insect, the ant, as a symbol of Socialist doctrine; 1978. (Original, author's collection)

caterpillar + *pampa* = plain), the sacred valley of the Incas near Cuzco in Peru.

In modern times, insects symbolize numerous ideas (fig. 1.10), especially in literature and folklore (Lenko and Papavero 1979). Science fiction and fantasy novels often use the dangerous qualities of many types to instill horror or malevolence. Superstitions and fanciful stories attributing good or bad fortune to many insects, spiders, or the like, are believed by sectors of the population, especially those in remote or primitive areas (Hogue 1985).

The cultural use of insects is perhaps best developed among Indian tribes still surviving in many parts of Latin America (Berlin and Prance 1978; Hitchcock 1962; Kevan 1983; Posey 1978, 1983). The study of this aspect of cultural entomology is referred to as "ethnoentomology" (and includes some of the odd practical uses of

**Figure 1.12** In a variation of the "toucandira ritual" in which giant hunting ants of the genera *Dinoponera* and *Paraponera* are used, a mat tied with paper wasps is applied to the chest of this Roucouyenne Indian (French Guiana) to test his courage. (From H. Davis, *The Jungle and the Damned* 1952, Duell, Sloan and Pearce, New York; reproduced with permission)

**Figure 1.11** Image from the Codex Telleriano Remensis of the Aztec deity, Itzpapálotl, in nature represented by wild silk moths of the genus *Rothschildia*. (Hand copy by Carlos Beutelspacher in *Mariposas entre los Antiguos Mexicanos,* 1989; reproduced with author's permission)

insects, such as for food or medicine; see valuable insects, chap. 3).

In many Amazonian Indian groups, insects were in the past and are still today venerated religiously, and they play central roles as deities or mythic figures (fig. 1.11). The four guardians of the cardinal points in Warao cosmology are social insects—two wasps, a bee, and a termite (Wilbert 1985). Ritual also incorporates insects, for example, the giant hunting ants (*Dinoponera*) in puberty ceremonies practiced by various Amazonian tribes (Liebrecht 1886; fig. 1.12). Similarly, pain is endured from the stings of wasps whose nests are purposely molested as a part of rites of passage among the Gorotire-Kayapó in Brazil (Posey 1981).

**Figure 1.13**   Modern Peruvian Indian (Yagua) necklace using beetle parts as main decorative element (Original, author's collection)

**Figure 1.14**   A bottle of mezcal containing a maguey worm (*Comadia redtenbacheri,* Cossidae) as an extra treat for the drinker. The beverage was important in ancient and modern Mexican culture. The insect retains today its natural association with the plant and its product. (Los Angeles County Museum of Natural History collection)

Metallic beetle parts and even galls (Berlin and Prance 1978) are used in body ornamentation by Indians in all parts of the region (fig. 1.13). Insects, especially musical species (crickets and katydids), luminescent forms (headlight beetles), and large beetles, and orthopterans are kept as pets or curiosities. Many species are eaten, both for sustenance and as delicacies (fig. 1.14).

## References

BERLIN, B., AND G. T. PRANCE. 1978. Insect galls and human ornamentation: The ethnobotanical significance of a new species of *Licania* from Amazonas, Peru. Biotropica 10: 81–86.

COWAN, F. 1865. Curious facts in the history of insects. Lippincott, Philadelphia.

HITCHCOCK, S. W. 1962. Insects and Indians of the Americas. Entomol. Soc. Amer. Bull. 8(4): 181–187.

HOGUE, C. L. 1985. Amazonian insect myths. Terra 23(6): 10–15.

HOGUE, C. L. 1987. Cultural entomology. Ann. Rev. Entomol. 32: 181–199.

KAMEN-KAYE, D. 1979. A bug and a bonfire. J. Ethnopharm. 1: 103–110.

KEVAN, D. K. McE. 1983. The place of grasshoppers and crickets in Amerindian cultures. 2d Trien. Meet. Pan American Acrid. Soc. (Bozeman, Mont., 1979) Proc. P. 8–74c.

LENKO, K., AND N. PAPAVERO. 1979. Insetos no folclore. Conselho Estad. Artes Cien. Human., São Paulo.

LIEBRECHT, F. 1886. Tocandyrafestes. Zeit. Ethnol. 18: 350–352.

MACGREGOR, R. 1969. La représentation des insectes dans l'ancien Mexique. L'Entomologiste 25: 1–8.

POSEY, D. A. 1978. Ethnoentomological survey of Amerind groups in lowland Latin America. Fla. Entomol. 61: 225–228.

POSEY, D. A. 1981. Wasps, warriors and fearless

men: Ethnoentomology of the Kayapó Indians of central Brazil. J. Ethnobiol. 1: 165–174.

Posey, D. A. 1983. Ethnomethodology as an *emic* guide to cultural systems: The case of the insects and the Kayapó Indians of Amazonia. Rev. Brasil. Zool. 1: 135–144.

Tozzer, A. M., and G. M. Allen. 1910. Animal figures in the Maya Codices. Harvard Univ., Peabody Mus., Pap., Amer. Archaeol. Ethnol. 4: 273–372, pls. 1–39.

Wilbert J. 1985. The house of the swallow-tailed kite: Warao myth and the art of thinking in images. *In* G. Urton, ed., Animal myths and metaphors. Univ. Utah, Salt Lake City.

# 2 ECOLOGY

As elsewhere, a wide variety of complex, interacting ecological factors determine the distributions and survival adaptations of insects in Latin America (Huffaker and Rabb 1984, Price 1984). These may be categorized either as geographic (physical factors) or biotic (living agents) (Walter 1979). Both are viewed from historical as well as contemporary perspectives.

### References

HUFFAKER, C. B., AND R. L. RABB, eds. 1984. Ecological entomology. Wiley, New York.
PRICE, P. W. 1984. Insect ecology. 2d ed. Wiley, New York.
WALTER, H. 1979. Vegetation of the earth and ecological systems of the geo-sphere. 2d ed. Springer, New York. Translated from 3d rev. German edition.

## GEOGRAPHY

Of fundamental importance to the occurrence of insect life is the surface character of the land, its configuration, chemistry, and relief, or its physiography, which provides footholds for the very existence and continuity of individuals. Equally basic are the conditions of climate, that is, temperature, moisture, sunlight, as well as the conditions of the medium (soil, water, or atmosphere), which give suitable substance and support for life. These components of geography work together to create broad habitats for insects called life zones. From a consideration of the historical geography of the flora and fauna of such areas,

biogeographic provinces may also be defined which contain characteristic groups of species (faunistics).

### Physiography

Past and present-day shapes and positions of landmasses, their elevations, connections, and surface textures, delineate broad areas within Latin America which determine in a most elemental way the distribution of insects. Physiographic subdivisions in Latin America have been outlined according to various schemes. A simplified version is presented below (Fig. 2.1); it is modified from Sauer (1950) and Sick (1969).

### References

SAUER, C. O. 1950. Geography of South America, Handbk. So. Amer. Indians 6: 319–344.
SICK, W. D. 1969. Geographic substance. In E. J. Fittkau, J. Illies, H. Klinge, G. H. Schwabe, and H. Sioli, eds., Biogeography and ecology in South America. 2: 449–474. Junk, The Hague.

### Climate and Medium

Each climatic factor exerts its critical effects in a variety of ways: temperature as freezing point, highs, lows, means, ranges, heat, cold, daily fluctuations; moisture as rainfall, dew, fog, clouds (and, of course, by determining the fundamental life-supporting media, aquatic versus terrestrial); sunlight as day and night, shade, illumination, radiation, and photoperiod. All act over long time periods as weather

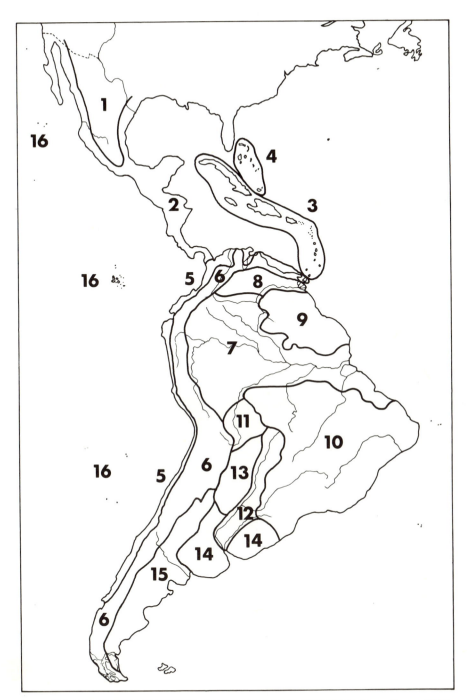

**Figure 2.1   MAJOR PHYSIOGRAPHIC AREAS OF LATIN AMERICA** (from Sauer 1950 and Sick 1969).    MIDDLE AMERICA: 1. Mexican Highlands; 2. Isthmian America (lowland Mexico, Central America); 3. West Indies (Greater and Lesser Antillean Islands); 4. Bahamas;    SOUTH AMER-ICA: 5. Pacific Coastal plain; 6. Andes (including Caribbean Borderlands and Bolivian Altiplano); 7. Amazon Basin; 8. Orinoco Basin; 9. Guiana Highlands; 10. Brazilian Highlands (including Brazilian Coastal Mountains) and Mato Grosso, Plateau of Paraná; 11. Llanos de Mamoré; 12. Paraná-Paraguay Depression; 13. Gran Chaco; 14. Pampas; 15. Patagonia;    INSULAR AMERICA: 16. Oceanic islands (Pacific: Galapagos and Revillagigedo Archipelagos, Cocos Island, Easter Island, Juan Fernandez etc. Atlantic: Fernando de Noronha Archipelago, Ascensión Island); 17. Continental islands (Falklands-Malvinas, Tres Marias, Isla de Coiba, Pearl Islands, Magellanic Archipelago).

climate, and seasonality and, of course, interact with physiography. The interplay of physical agents in the atmosphere creates a multitude of environments that are also circumscribable and can even be mapped. Many schemes have been devised to classify these directly as climatic areas (Köppen and Geiger 1931–1934).

Major shifts in the earth's climatic pattern occur at irregular intervals, changing the rainfall, temperature, and other aspects of the weather, sometimes drastically over wide areas. In Latin America, the best-known manifestation of such planetary-scale atmospheric fluctuations is the El Niño phenomenon, itself part of the so-called, larger Southern Oscillation that involves much of the Pacific and Indian Ocean basins of the Southern Hemisphere (Philander 1990). At roughly seven- to ten-year intervals, the warm equatorial counter-current in the Panama bight shifts strongly to the south, unduly heating and displacing the normally cool, northward-flowing Humboldt Current. As a result, heavy rains come to the Peruvian coastal deserts, and vegetation flourishes (especially on the lomas; see Special Habitats, below). The effects of El Niño on coastal insect populations have been found to be significant in some cases (Beingolea 1987a, 1987b). Associated weather anomalies may be felt even far inland on the continent, for example, prolonged dryness in the western Amazonian rain forest or cold snaps in sub-Andean valleys.

Brief but severe periods of cold weather fronts, originating in Antarctica, pass across the Amazon Basin in odd years, usually during early July (Días Frios de San Juan, *friagem*) (Ratisbona 1976: 226, 237). Daily minimum temperature may drop from a normal 20° to 8° C. The sky becomes heavily clouded, a strong wind blows, and it fails to rain.

Responses of insects to these stresses may be dramatic or subtle but are as yet virtually unstudied in Latin America. The impact of such sudden weather changes on these small, ectothermic creatures surely must be intense, especially on temperature-sensitive species. Interference with reproduction and population "die-offs" should be expected.

The effects of climate are felt most immediately in the atmosphere. They are manifest equally but more slowly in the aquatic medium, in standing (lentic) waters (ponds, lakes, etc.), in moving (lotic) waters (streams, rivers) (Fittkau 1964; Macan 1962, 1974), and in the soil (Kühnelt 1961).

## References

BEINGOLEA, O. D. 1987a. El fenómeno "El Niño" 1982–83 y algunos insectos-plaga en la costa peruana. Rev. Peruana Entomol. 28: 55–57.

BEINGOLEA, O. D. 1987b. La langosta *Schistocerca interrita* en la costa norte del Perú, durante 1983. Rev. Peruana Entomol. 28: 35–40.

KÖPPEN, W., AND R. GEIGER. 1931–1934. Handbuch der Klimatologie. Vols. 1–4. Borntraeger, Berlin.

KÜHNELT, W. 1961. Soil biology with special reference to the animal kingdom. Faber & Faber, London.

FITTKAU, E. J. 1964. Remarks on limnology of central-Amazon rain-forest streams. Int. Verh. Limnol., Verh. 15: 1092–1096.

MACAN, T. T. 1962. The ecology of aquatic insects. Ann. Rev. Entomol. 7: 261–288.

MACAN, T. T. 1974. Freshwater ecology. 2d ed. Wiley, New York.

PHILANDER, S. G. H. 1990. El Niño, La Niña and the Southern Oscillation. Academic, San Diego.

RATISBONA, L. R. 1976. The climate of Brazil. In W. Schwerdtfeger, ed., Climates of Central and South America, world survey of climatology. 12: 219–293. Elsevier, Amsterdam.

## Vegetation Zones

Physicochemical environmental factors not only act directly on insect life forms but influence them indirectly through the kinds of plant growth they allow (biotic factors). Groupings of plants adapted to a particular set of soil and climatic conditions delimit broad environments for insects.

Many types of vegetation are present in Latin America, and these are classified according to various systems (e.g., Beard 1944, Graham 1973, Hueck and Seibert 1972, Sauer 1950, Weber 1969). These systems are only regionally applicable because they often define units in terms of specific plant taxa present. A universal scheme, globally applicable, is Holdridge's (1967, 1982), which combines the effects of elevation, latitude, rainfall, and temperature to define vegetational formations, independent of floristic elements.

It is very apparent that the nature of vegetation plays a primary role in determining the Neotropical insect fauna in each physiographic area. The special richness of the Amazon Basin is a good case in point. Contrary to former ideas, the region's vegetation has not existed continuously unchanged for tens of millions of years but has varied considerably from near desert to lush forest in recent geologic periods, particularly during the Pleistocene Age, during alternating arid and humid conditions. In the drier phases, moisture-requiring vegetation shrunk greatly and fragmented into forest patches where rainfall persisted which was adequate for their survival ("refuge theory," Haffer 1982; but see Endler 1982).

This disjunction into forest islands isolated from each other by grassland or even desertlike plant cover divided many formerly continuous populations and led to their evolution into new species. Wet phases, such as the world is now experiencing, allowed the patches to expand again and the gaps between them to close. But evidences of the former islands, or *refugia*, are still present as concentrations of endemics. Among insects, this is shown especially well by butterflies (Brown 1982) and among arachnids by scorpions (Lourenço 1986).

This heterogeneity in Amazonian forest and wet forests in other areas partly explains the latest and one of the most extreme forces shaping the modern entomofauna of the basin (Simpson and Haffer 1978).

## References

BEARD, J. S. 1944. Climax vegetation in tropical America. Ecology 25: 127–158.

BROWN, JR., K. S. 1982. Paleoecology and regional patterns of evolution in Neotropical forest butterflies. *In* G. T. Prance, ed., Biological diversification in the tropics. Columbia Univ. Press, New York. Pp. 255–308.

ENDLER, J. A. 1982. Pleistocene forest refuges: Fact or fancy. *In* G. T. Prance, ed., Biological diversification in the tropics. Columbia Univ. Press, New York. Pp. 641–657.

GRAHAM, A., ed. 1973. Vegetation and vegetational history of northern Latin America. Elsevier, Amsterdam.

HAFFER, J. 1982. General aspects of the Refuge Theory. *In* G. T. Prance, ed., Biological diversification in the tropics. Columbia Univ. Press, New York. Pp. 6–24.

HOLDRIDGE, L. R. 1967. Life Zone ecology. Trop. Sci. Ctr., San José, Costa Rica.

HOLDRIDGE, L. R. 1982. Ecología basada en zonas de vida. Insto. Interamer. Coop. Agric. San José, Costa Rica.

HUECK, K., AND P. SEIBERT. 1972. Vegetationskarte von Südamerika (Mapa de la vegetación de America del sur). Fischer, Stuttgart.

LOURENÇO, W. R. 1986. Diversité de la faune scorpionique de la région amazonienne; centres d'endémisme; nouvel appui à la théorie des refuges forestiers du Pléistocène. Amazoniana 9: 559–580.

SAUER, C. O. 1950. Geography of South America. Handbk. So. Amer. Indians 6: 319–344.

SIMPSON, B. B., AND J. HAFFER. 1978. Speciation patterns in the Amazonian forest biota. Ann. Rev. Ecol. Syst. 9: 497–518.

WEBER, H. 1969. Zur natürlichen Vegetationsgliederung von Südamerika. *In* E. J. Fittkau, J. Illies, H. Klinge, G. H. Schwabe, and H. Sioli, eds., Biogeography and ecology in South America. 2:475–518. Junk, The Hague.

## Artificial Environments

The foregoing discussion has been concerned with natural or original conditions and patterns of native flora and fauna. Since coming to the southern lands of the

New World 20,000 to 50,000 years ago, humans have modified the original life zones to varying degrees (Künkel 1963), even creating large tracts of essentially new, "artificial life zones." Such are the cities, farms, and vast grasslands for cattle grazing that have come into being since the Conquest. Even before Columbus, the indigenous population cut and burned sizable tracts of forest to support shifting agriculture, turning them into plots of cultivated species that only slowly returned to climax status through successional stages of different vegetation. The Incan civilization fashioned a new Andean landscape by their extensive terracing, grazing, irrigation, and road building.

In addition to changes contrived for living space and agriculture, there have also been impacts on natural populations of animals and plants through hunting and gathering, unintentional pollution, and erosion. Such modifications force adjustments by the insect inhabitants. This is most intense in the urban setting where new complexes of species adapted to civilization come into being. The entomology of cultivated fields of single or mixed crops also departs widely from the norm in natural environments.

### Reference

KÜNKEL, G. 1963. Vegetationszerstörung und Bodenerosion in Lateinamerika. Arch. Naturschutz Landschaft. 3(1): 59–80.

## ECOSYSTEMS

Not only the vegetation but the community of all the plants, insects, and other animals must be considered together to describe the conditions of a particular habitat. The organisms (biocenoses) located in a particular place (biotope) constitute an ecosystem. Within the ecosystem, each organism has a functional role (niche). The niches of insects in Latin America have been little studied, other than in agricultural monocultures.

Habitats are recognizable on different levels. The most immediate environment where the determinants of life specifically affect an organism is its microhabitat. Examples are infinite in number; the interior of a decaying log or water-filled cup of a bromeliad are examples. The vertical distribution of different species of mosquitoes (Bates 1944) and butterflies (Papageorgis 1975) in forests is evidence of the effects of subtle environmental determinants. The existence of so many microhabitats partly accounts for the vast diversity of the insects, creatures so adept at calling home the smallest and most demanding of living spaces.

Larger, more inclusive areas containing microhabitats, such as the vegetation zones (e.g., forest canopy) and small physiographic features (caves, lakes), are macrohabitats. Still greater groupings and larger expanses of earth forming grossly recognizable macrohabitats are called life zones or biomes (e.g., tropical forest, deserts). The statuses of Latin American insects in several special habitats are discussed below.

### Special Habitats

The great number of natural Neotropical insect macro- and microhabitats makes it impossible to discuss more than a few of the most distinctive, peculiar, and important. They form particular places for insects whose structure and activities may be very different from those of the other insects in the surrounding general environment.

Most insects studied in artificial habitats are injurious species and are discussed in Chapter 3, Practical Entomology. A few investigations have focused on other faunal elements of plantations (Young 1986a, 1986b), habitations, and like areas under human management.

## References

BATES, M. 1944. Observations on the distribution of diurnal mosquitoes in a tropical forest. Ecology 25: 159–170.

PAPAGEORGIS, C. 1975. Mimicry in Neotropical butterflies. Amer. Sci. 63: 522–532.

YOUNG, A. M. 1986a. Notes on the distribution and abundance of Dermaptera and Staphylinidae (Coleoptera) in some Costa Rican cacao plantations. Entomol. Soc. Wash. Proc. 88: 328–343.

YOUNG, A. M. 1986b. Notes on the distribution and abundance of ground and arboreal nesting ants (Hymenoptera: Formicidae) in some Costa Rican cacao habitats. Entomol. Soc. Wash. Proc. 88: 550–571.

### *Forest Canopy*

Until very recently, the forest canopy has been penetrated by insect researchers only with great difficulty (Hingston 1932). The earliest workers had to be content with fortuitous tree falls even to gain a glimpse of upper-story life. The canopy could be reached by skilled native climbers, but they carried no scientific expertise aloft, and their activities had to be directed ineffectively by their earthbound employers. Other approaches, still remote ones, have been to elevate various kinds of traps to catch some of the fauna or knock it down with quick-acting, biodegradable insecticides (Erwin 1983).

Greater improvement in access came with the construction of arboreal ladders and platforms (Porter and DeFoliart 1981), towers (a famous one on Barro Colorado Island, Panama), or elevated causeways, from which observations and collections could be made. However, these offer very limited mobility and are costly to construct and maintain.

Lately, some practical means have been found not only to move about and obtain specimens from this complex realm but even to carry out extended studies within it (Mitchell 1982). Even scientists themselves are now able to go into the canopy, using mountaineering techniques (Perry 1980,

1984; Perry and Williams 1981). Unfortunately, few entomologists are trained in both the academic and athletic aspects of this demanding, although highly rewarding, approach.

The results of these efforts, although fragmented and imperfect, demonstrate the startling fact that the canopy has an extremely rich insect complement. This stratum of forests is considered to be the "last frontier" in tropical entomology and is attracting study from a number of viewpoints: faunistics, pollination ecology, and even populational biology (see below).

## References

ERWIN, T. L. 1983. Beetles and other insects of tropical forest canopies at Manaus, Brazil, sampled by insecticidal fogging techniques. *In* S. L. Sutton, T. C. Whitmore, and A. C. Chadwick, eds., Tropical rain forest: Ecology and management. Blackwell, Oxford. Pp. 59–75.

HINGSTON, R. W. 1932. A naturalist in the Guiana forest. Longmans Green, New York.

MITCHELL, A. W. 1982. Reaching the rain forest roof (A handbook on techniques of access and study in the canopy). Leeds Phil. Lit. Soc. Leeds, Eng.

PERRY, D. R. 1980. I probe the jungle's last frontier. Int. Wildlife 10: 5–11.

PERRY, D. R. 1984. The canopy of the tropical rain forest. Sci. Amer. 251: 138–147.

PERRY, D. R., AND J. WILLIAMS. 1981. The tropical rain forest canopy: A method providing total access. Biotropica 13: 283–285.

PORTER, C. H., AND G. R. DEFOLIART. 1981. The man-biting activity of phlebotomine sand flies (Diptera: Psychodidae) in a tropical wet forest environment in Colombia. Arq. Zool. São Paulo 30: 81–158.

### *Amazon Inundation Forests*

Vast expanses bordering the Amazon and its major tributaries are annually flooded to a depth of several meters for periods of five to six months or more. These forests (called *igapó* locally) harbor an insect fauna especially adapted to the stresses of alternately rising and receding waters.

The terrestrial species that live in these

forests are numerous and very diverse. They have evolved strategies to compensate for the periodic loss of their terrestrial habitat by exercising mobility to escape into the canopy or to other dry areas or acquiring survival adaptations for coping with inundation (Adis 1977, Adis et al. 1988). Spiders and many flightless species of ground beetles, for examples, are known to ascend into the upper forest level (Erwin and Adis 1982) during flood periods. Members of the cockroach genus *Epilampra,* however, have acquired the capacity to swim and have the hindmost spiracles situated on short stalks to aid breathing while they are in the water. Only minor elements of the ground fauna are present here (e.g., pseudoscorpions; Adis and Mahnert 1985, Adis et al. 1988), there being no persistent ground litter or stable soil.

In response to the drying phase, aquatic insect inhabitants of these areas return to the main channels from which the floodwaters arose. Some may become trapped in small, dammed depressions, but normally they either take to the air and fly to water or burrow into the mud where they may remain inactive for the duration of the dry season.

## References

ADIS, J. 1977. Programa mínimo para analises de ecosistemas: Artrópodos terrestres em florestas inundáveis da Amazonia central. Acta Amazonica 7: 223–229.

ADIS, J., AND V. MAHNERT. 1985. On the natural history and ecology of Pseudoscorpions (Arachnida) from an Amazonian blackwater inundation forest. Amazoniana 9: 297–314.

ADIS, J., V. MAHNERT, J. W. DE MORAIS, AND J. M. GOMES. 1988. Adaptation of an Amazonian pseudoscorpion (Arachnida) from dryland forests to inundation forests. Ecology 69: 287–291.

ERWIN, T. L., AND J. ADIS. 1982. Amazonian inundation forests: Their role as short-term refuges and generators of species richness and taxon pulses. *In* G. T. Prance, ed., Biological diversification in the tropics. Columbia Univ. Press, New York. Pp. 358–371.

### Soil and Ground Litter

The soil and its overlying layer of organic litter constitute the habitat of a great many small to very small arthropods (Kühnelt 1961). Dominant among these are ants, springtails, spiders, psocids, thrips, cryptostigmatic mites, and microbeetles (Penny et al. 1978), but larger insects are sometimes also found here (Young 1983). It is incredible that such arthropods may comprise almost 50 percent of the total soil and litter biota in the central Amazon forest, ants and termites alone forming about 60 percent of this (Fittkau and Klinge 1973). Most occupy the upper 5 to 8 centimeters of the soil.

The composition of this fauna and its ecology have been the subject of many studies in the Neotropics, especially in forests (e.g., Williams 1941, Deboutteville and Rapoport 1962–1968, Lieberman and Dock 1982) where moisture seems to be the major factor determining the seasonal distribution of species and population fluctuations. Augmentations in diversity and abundance generally follow increases in water content (Levings and Windsor 1984), usually from rainfall during the wet season in all life zones (Dantas and Schubart 1980, Lieberman and Dock 1982, Willis 1976). Strickland (1945) concluded that the general environment of an area is more important than the soil type in determining the sizes of ground and litter populations. That a variety of conditions would be determinants was confirmed by studies of forest litter arthropods in southeastern Peru (Pearson and Derr 1986).

## References

DANTAS, M., AND H. O. R. SCHUBART. 1980. Correlação dos índices de agregação de Acari e Collembola com 4 fatores ambientais numa pastagem de terra firme da Amazonia. Acta Amazonica 10: 771–774.

DEBOUTTEVILLE, C. D., AND E. RAPOPORT, eds. 1962–1968. Biologie de l'Amérique Australe. Etudes sur la faune de sol; 2: Etudes sur la faune du sol; 3: Etudes sur la faune du sol +

Documents biogéographiques; 4, Documents biog. et ecologiques. Ed. Cen. Nat. Rech. Sci., Paris.

FITTKAU, E. J., AND H. KLINGE. 1973. On biomass and trophic structure of the central Amazonian rain forest ecosystem. Biotropica 5: 2–14.

KÜHNELT, W. 1961. Soil biology with special reference to the animal kingdom. Faber & Faber, London.

LEVINGS, S. C., AND D. M. WINDSOR. 1984. Litter moisture content as a determinant of litter arthropod distribution and abundance during the dry season on Barro Colorado Island, Panama. Biotropica 16: 125–131.

LIEBERMAN, S., AND C. F. DOCK. 1982. Analysis of the leaf litter arthropod fauna of a lowland tropical evergreen forest site (La Selva, Costa Rica). Rev. Biol. Trop. 30: 27–34.

PEARSON, D. L., AND J. A. DERR. 1986. Seasonal patterns of lowland forest floor arthropod abundance in southeastern Perú. Biotropica 18: 244–256.

PENNY, N. D., J. R. ARIAS, AND H. O. R. SCHUBART. 1978. Tendências populacionais de fauna de Coleópteros do solo sob floresta de terra firme na Amazonia. Acta Amazonica 8: 259–265.

STRICKLAND, A. H. 1945. A survey of the arthropod soil and litter fauna of some forest reserves and cacao estates in Trinidad, British West Indies. J. Anim. Ecol. 14: 1–11.

WILLIAMS, E. C. 1941. An ecological study of the floor fauna of the Panamanian rain forest. Chicago Acad. Sci. Bull. 6: 63–124.

WILLIS, E. D. 1976. Seasonal changes in the invertebrate litter fauna on Barro Colorado Island, Panamá. Rev. Brasil. Biol. 36: 643–657.

YOUNG, A. M. 1983. Patterns of distribution and abundance in small samples of litter-inhabiting Orthoptera in some Costa Rican cacao plantations. New York Entomol. Soc. J. 91: 312–327.

### Black Water Lakes and Rivers

Some lowland tropical river basins contain tributaries and landlocked basins (oxbow lakes, *cochas*) with tea-colored water that in the depths appears black. These are distinguished from so-called white waters not only by their color but by physical and chemical properties. White waters (actually a milky chocolate color) are nutrient rich, neutral to slightly alkaline, and turbid.

Black waters usually flow from nutrient-poor, sandy soils and thus are low in minerals but are acidic and may contain high concentrations of organic compounds (tannic acids, phenolics) leached from vegetation and toxic to insects. The water and surrounding land thus afford unfavorable conditions for insects, and such black water basins generally have depauperate entomofaunas (Janzen 1974). Some specially adapted aquatic types, however, such as certain chironomid midges (Fittkau 1971) and water mites (Tundisi et al. 1979), can be very numerous, even in the most heavily charged water.

The Guiana Shield of northern South America is a large black water area and is notorious for the poorness of its productivity. Other similar such regions are found in the Brazilian Highlands and on the Yucatán Peninsula.

Clear waters (greenish to clear) are also recognized but are biologically similar to black water.

### References

FITTKAU, E. J. 1971. Distribution and ecology of Amazonian chironomids (Diptera). Can. Entomol. 103: 407–413.

JANZEN, D. H. 1974. Tropical blackwater rivers, animals and mast fruiting by the Dipterocarpaceae. Biotropica 6: 69–103.

TUNDISI, J. G., A. M. P. MARTINS, AND T. MATSUMURA. 1979. Estudos ecológicos preliminares em sistemas aquáticos em Aripuanã. Acta Amazonica 9: 311–315.

### Caves

Insects and many related terrestrial arthropods of diverse groups are true troglobites (obligate cavernicoles, i.e., animals narrowly and specifically adapted for life deep in caves; Culver 1982, Hoffmann et al. 1986). However, tropical caves are usually dominated by species classed as troglophiles, which also live in noncave habitats. Some are omnivores, but they are more normally specialized for particular foods and are of two basic types: scavengers and predators. The former feed on the droppings of bats (Gnaspini 1989), oil birds, and other higher fauna (guanophages) or take nourishment from organic debris such as insect and vertebrate carrion and plant matter that washes or falls into the caves or is brought in and dropped by other animals (detritivores). Those among the predators survive by catching and consuming other live cave dwellers. A widespread example are the long-legged cave crickets of the genus *Amphiacusta*.

The scavengers are more numerous than predators in kinds and numbers of individuals. Among these are the terrestrial isopods (*Trichorhina*), millipedes (*Eurhinocricus*), cockroaches (*Periplaneta, Blaberus*), tineid moths, dung beetles (*Ataenius*), and many others. They are often extremely numerous; darkling beetles (Tenebrionidae) or mites (Acari) may carpet portions of the floor of caves to a depth of a centimeter or more.

Predators are conspicuous for their large size and aggressiveness. Spectacular in these respects are the giant tailless whip scorpions (Amblypygi, *Phrynus*) that prey on crickets and cockroaches. Others in the category are centipedes, many spiders, ants, and various beetles, principally rove beetles (Staphylinidae) and ground beetles (Carabidae). Many soil mites may occupy these niches as well.

Cave insects often exhibit morphological adaptations to life in the dark (Dessen et al. 1980), including eyes reduced or absent, lack of integumentary pigmentation, reduction and loss of wings, and greatly elongated, highly sensitive appendages (especially the antennae).

The entomology of Neotropical caves has been the subject of much study (Peck 1977, Strinati 1971), but more remains to be done. Best known are the cave faunas of Mexico (Reddell 1971), Cuba (Orghidan et al. 1973–1983), Jamaica (Peck 1975), Puerto Rico (Peck 1974, 1981*a*), Barbados (Peck 1981*b*), and Venezuela (Chapman 1980). Recent work, yet unpublished, has found over thirty species of eyeless cave and soil arthropods in volcanic caves in the Galápagos Islands (Peck pers. comm.).

## References

CHAPMAN, P. 1980. The invertebrate fauna of caves of the Serrania de San Luis, Edo. Falcon, Venezuela. Brit. Cave Res. Assoc. Trans. 7: 179–199.

CULVER, D. C. 1982. Cave life, evolution and ecology. Harvard Univ. Press, Cambridge.

DESSEN, E. M. B., V. R. ESTON, M. S. SILVA, M. T. TEMPERINI-BECK, AND E. TRAJANO. 1980. Levantamento preliminar de fauna de cavernas de algumas regiões do Brasil. Cien. Cult. 32: 414–725.

GNASPINI, N. P. 1989. Análise comparativa da fauna associada a depósitos de guano de morcegos cavernícolas no Brasil: Primeira Approximação. Rev. Brasil. Entomol. 32: 183–192.

HOFFMANN, A., J. G. PALACIOS-VARGAS, AND J. B. MORALES-MALACARA. 1986. Manual de biospeleología. Univ. Nac. Aut. México, México.

ORGHIDAN, T., A. NÚÑEZ JIMÉNEZ, V. DECON, S. NEGREA, AND N. V. BAYES. 1973–1983. Résultats des expéditions biospéleologique cubano-Roumaines à Cuba. Vols. 1–4. Ed. Academiei, Bucharest, Republicii Socialiste România.

PECK, S. B. 1974. The invertebrate fauna of tropical American caves. Pt. II: Puerto Rico, an ecological and zoogeographic approach. Biotropica 6: 14–31.

PECK, S. B. 1975. The invertebrate fauna of tropical American caves. Pt. III: Jamaica, an introduction. Int. J. Speleol. 7: 303–326.

PECK, S. B. 1977. Recent studies on the invertebrate fauna and ecology of sub-tropical and tropical American caves. 6th Intl. Cong. Speleol. Proc. 5: 185–193.

PECK, S. B. 1981*a*. Zoogeography of invertebrate cave faunas in southwestern Puerto Rico. Natl. Speleol. Soc. Bull. 43: 70–79.

PECK, S. B. 1981*b*. Community composition and zoogeography of the invertebrate cave fauna of Barbados. Fla. Entomol. 64: 519–527.

REDDELL, J. R. 1971. A review of the cavernicole fauna of Mexico, Guatemala, and Belize. Texas Mem. Mus. Austin Bull. 27: 1–327.

STRINATI, P. 1971. Recherches biospéleologiques en Amerique du Sud. Ann. Spéleol. 26: 439–450.

## Lomas

The low coastal desert hills of central Peru provide one of the most unusual habitats for insects in the Neotropic (Dogger and Risco 1970). It is speculated that, historically, the lomas were part of a larger chaparral biome that once extended along the entire western slopes of the Andes (Péfaur 1978). In this zone of extreme general aridity, wetness in the form of rain comes only at multiyear intervals, decades or more. At these times, explosions of plant life occur on the otherwise parched hills, and they become green islands in the bleak desert. Normally, only the regular annual fogs (*garúas*, May–October) bring moisture to these slopes to maintain a less plush but more reliable vegetation.

Also adapted to these climatic and vegetative cycles and living on the lomas vegetation is a complex and diverse community of insects and other terrestrial arthropods, some species of which are known from nowhere else (Aguilar 1964). The more important groups are spiders (Aguilar, Pacheco, and Silva 1987), mites, springtails, wax insects, beetles (especially Tenebrionidae), flies, and ants (Aguilar 1981), which are most abundant and numerous in kinds in the damper upper elevations. Some special forms are wingless sticklike forms ("palitos vivientes de Lima"), including two walkingsticks, a jumping stick (Proscopiidae), and assassin bugs (Reduviidae). One walkingstick (*Libethra minuscula*) is omnivorous but dies with the failing plants at the end of the damp season. The other (*Bostra scabrinota*) feeds on one plant but can change its color to match seasonal changes and is present year-round (Aguilar 1970). An unexpected element is the water measurer (*Bacillometra woytkowskii*, Hydrometridae), a heteropterous insect normally associated with bodies of fresh water (Aguilar, Oyeyama, and Aguilar 1987). Its presence is associated with the high humidity of the lomas in the winter.

## References

AGUILAR, P. G. 1964. Especies de artrópodos registrados en las lomas de los alredededores de Lima. Rev. Peruana Entomol. 7: 93–95.

AGUILAR, P. G. 1970. Los "palitos vivientes de Lima." I: Phasmatidae de las lomas. Rev. Peruana Entomol. 13: 1–8.

AGUILAR, P. G. 1981. Fauna desértico-costera Peruana. VII: Apreciaciones sobre diversidad de invertebrados en la costa central. Rev. Peruana Entomol. 24: 127–132.

AGUILAR, P. G., F. OYEYAMA, AND Z. P. AGUILAR. 1987. Los "palitos vivientes de Lima." III: Un Hydrometridae de las lomas costeras. Rev. Peruana Entomol. 28: 89–92.

AGUILAR, P. G., V. R. PACHECO, AND R. SILVA. 1987. Fauna desértico-costera peruana. VIII: Arañas de las lomas Zapallal, Lima (nota preliminar). Rev. Peruana Entomol. 29: 99–103.

DOGGER, J., AND S. H. RISCO. 1970. La fauna insectil de las lomas de Trujillo, Estudio del cerro "Campana." Bol. Tec. Circ. Entomol. Norte (Lambayeque) 1(2): 1–5.

PÉFAUR, J. E. 1978. Composition and structure of communities in the lomas of southern Peru. Ph.D. diss., Univ. Kansas, Lawrence.

## The Ocean

At first thought, it seems incongruous that the insects, so tremendously successful in dominating the land and the inland waters of the world, have failed to conquer the seas. Intolerance of hypersaline water is certainly not to blame, because many aquatic insects live in highly salinated lakes and ponds inland and at the ocean's margin. The reasons apparently lie in the fact that they arrived on the scene long after their other invertebrate predecessors had locked up all the ecological niches. There are a few examples, however, of marine-adapted groups (Cheng 1976).

Insects truly adapted to life far out to sea are the fewest and consist only of the marine water striders or "ocean skaters" (several genera of Gerridae, see water striders, chap. 7). Closer to shore, the number and kinds of marine insects greatly increases. There one finds the marine midges (Chironomidae). These live on rocky shores and have larvae that are completely at home in the salt spray and are even submerged by seawater during high tide. Other intertidal insects and arachnids are found among the springtails and mites (especially Halacaridae).

Several kinds of lice are parasitic on seagoing mammals, seals and sea lions particularly. When the host submerges, they escape osmotic dessication by seawater and drowning by hiding close to the skin in air pockets under the host's fur.

The seashores of the continents and islands of Latin America also support an even more extensive group of marine littoral insects (Evans 1968). Many of these are freshwater groups, including inland hypersaline lake types, that have shifted to the similar habitats at the sea margins, especially salt marshes and mangroves. A majority prey on invertebrates, feeding on other shore life, such as tiger beetles, rove beetles, pseudoscorpions, spiders, and shore bugs (Saldidae).

Many of these are associated with seaweed and kelp that accumulates on beaches, feeding either directly on this material (kelp fly larvae, *Fucellia;* see chap. 11; Coelopidae, shore flies [Ephydridae]) or catching and devouring the wrack scavengers (spiders, ground beetles, rove beetles). The adults of some, such as salt marsh mosquitoes and tabanids and mudflat-breeding punkies, are feeders on vertebrate blood. Their population explosions can make human life impossible in seaside areas. Several water boatmen species (especially *Trichocorixa reticulata*) live in highly saline shore pools (Davis 1966) and have even been collected in plankton tows in the Gulf of California.

## References

CHENG, L., ed. 1976. Marine insects. North-Holland, Amsterdam.

DAVIS, C. C. 1966. Notes on the ecology and reproduction of *Trichocorixa reticulata* in a Jamaican salt-water pool. Ecology 47: 850–852.

EVANS, W. G. 1968. Some intertidal insects from western Mexico. Pan-Pacific Entomol. 44: 236–241.

### Torrential Waters

Stream waters flowing in the range of 60 to 200 centimeters per second or greater are considered torrential and are common in the upland drainages of mountainous regions. Originating in countless springs and snowfields on the heights of the Cordillera, bounding over boulders and crashing into foam-covered pools, torrential streams descend through gorges and narrow canyons, over hard, rocky beds, before reaching the lower, gentle slopes.

Fast, cold water offers a refugial niche to many insects, comparatively free from vertebrate predation because few such large animals can live in such an inhospitable medium. Exceptions are a few hardy insectivorous types, such as torrent ducks, dippers, and a few fish (introduced trout and possibly some astroblephid catfishes). As a result, some very ancient representatives of several aquatic groups have persisted for geologic ages as torrenticolous (rheophilous) relicts. These include entire families, most prominently the net-winged midges (Blephariceridae) and various beetles (Elmidae, Dryopidae, Psephenidae). Other similarly adapted taxa are the larval stages of lance-winged moth flies (*Maruina*, Psychodidae) and blackflies (Simuliidae) and many species of Trichoptera (especially in the family Hydrobiosidae), Hemiptera (*Cryphocricos*, Naucoridae), Plecoptera (*Araucanioperla*, Gripopterygidae), water midges (Chironomidae), and Ephemeroptera (*Epeorus*, Heptageniidae).

Extreme structural and physiological adaptations have evolved in these forms in response to the strong selective pressures of fast current and smooth substrates. These include holdfast structures (ventral suckers, claws), streamlining, and plastron respiration. The latter makes use of the function of air films and pockets as "physical gills." A requirement for proper function of this system is cold, clean, oxygen-rich water. The presence of these insects therefore indicates healthy stream conditions.

Food habits for the relatively passive grazers and predators do not require rapid movement. Adult emergence is often "explosive." To avoid drowning, the imago rises to the surface in a bubble and takes wing immediately after contacting the air.

Torrenticolous insects are not well known in Latin America. Some famous early studies were made on Diptera in southeastern Brazil by Fritz Müller (1879, 1895) and Adolfo Lutz (1930). New species are commonly discovered in the habitat, especially in remote mountains.

## References

LUTZ, A. 1930. Biologia das águas torrenciais e encachoeiradas. Soc. Biol. Montevideo, Arch. suppl. 1: 114–120.

MÜLLER, F. 1879. A metamorphose de um insecto diptero. Mus. Nac. Rio de Janeiro, Arch. 5/6: 47–85, pls. 4–7.

MÜLLER, F. 1895. Contributions towards the history of a new form of larvae of Psychodidae (Diptera) from Brazil. Entomol. Soc. London, Trans. 1895: 479–482, pls. 10–11.

### Tank Plants

Some Neotropical plants have parts of their anatomy developed into cup-shaped water-holding structures (phytotelmata) (Frank and Lounibos 1983) and are referred to as reservoir or tank plants. They are of several types. The best known are the bromeliads (Bromeliaceae) (Gómez 1977) whose water accumulations provide a home for small aquatic animals. This microcosm was first studied comprehensively as an ecosystem by Picado (1913), who distinguished between the aquatic milieu ("aquarium") and terrestrial portion ("terrarium").

The aquarium consists of a spacious central cup and expanded lateral leaf bases that collect rainwater. Large plants may store a liter or more and may have a cup 5 to 10 centimeters, in diameter and equally deep. These reservoirs of water provide habitats suitable for the development of many aquatic insects, including mosquitoes, especially *Wyeomyia* and other sabethines and *Aedes*. An abundance of bromeliads in a particular area can foster a substantial mosquito population, a good example being *Anopheles darlingi*, the main malaria vector in Trinidad in the 1940s (Downs and Pittendrigh 1946). Other representative aquatic inhabitants include water midges (*Chironomus*), punkies (*Bezzia*), damselflies (Pseudostigmatidae), beetles (Helodidae), and water mites. The detailed ecology of this fraction has been studied by many entomologists, including Laessle (1961). (See the bibliography of bromeliad and pitcher plant reservoir plant entomology by Fish and Beaver [1978].)

Detritus collects also in the lateral bracts, and a special kind of arboreal soil is created which is like a "terrarium" to another group of insects. Here are found spiders, carabid beetles, ants, isopods, millipedes, mites, springtails, and other terrestrial forms (Murillo et al. 1983). A few insects actually feed on the leaves of bromeliads and form yet another guild in association with these plants (Beutelspacher 1972).

Another category of Neotropical tank plants are the insectivorous "pitcher plants" of the family Sarraceniaceae. In these, deep, urn-shaped leaves have evolved to hold fluids that normally kill and digest hapless insects that fall into them. However, immatures of some insect species, mostly mosquitoes of the tribe Sabethini, are immune to the corrosive action of these chemicals and actually develop normally in this very peculiar aquatic microhabit. In the Neotropics, pitcher plants of only the genus *Heliamphora* are known, found in the Venezuelan-Guianan region. They are known to be occupied by *Wyeomyia* mosquitoes of the subgenus *Zinzala* (Zavortink 1985), but their other occupants are not studied.

Flower bracts, especially those of the

genus *Heliconia* (Musaceae), hold watery fluids, secreted by the plant itself, as well as rainwater and host a specific micro-community of insects similar to that of bromeliads (Seifert 1982). The inflorescences harbor one of two kinds of insect groupings depending on size and shape, amount of water, and age. The first is composed of fly larvae (nonmosquito) and beetles in older, less voluminous bracts with small amounts of water. The most common fly larvae are of pomace flies (Drosophilidae), hover flies (Syrphidae, *Copestylum*), and soldier flies (Stratiomyiidae, *Merosargus*). The beetles are primarily scavenging water beetles (Hydrophilidae, *Gillisius*), leaf beetles (Chrysomelidae, Hispinae, *Cephaloleia*), and rove beetles (Staphylinidae, *Odontolinus* and *Quichuana*). There are also earwigs (Dermaptera, *Carcinophora*) and cockroaches (Blattidae, *Litopeltis*).

In the second group, mosquito larvae are dominant (Seifert 1980), in younger, larger inflorescences with larger amounts of water. At least five genera are represented: *Wyeomyia, Trichoprosopon, Toxorhynchites, Culex,* and *Sabethes.*

Various interactions among the insects in these communities have been observed, including predation, commensalism, colonization, and competition, although the last seems to be weak. The clearly defined nature of the communities, the commonness of *Heliconia* plants, and the ease with which the insects may be manipulated experimentally make them ideal for investigations of species interactions and other ecological phenomena (Seifert 1975, 1981; Seifert and Seifert 1976a, 1976b, 1979).

## References

BEUTELSPACHER, C. R. 1972. Some observations on the Lepidoptera of bromeliads. J. Lepidop. Soc. 26: 133–137.

DOWNS, W. G., AND C. S. PITTENDRIGH. 1946. Bromeliad malaria in Trinidad, British West Indies. Amer. J. Trop. Med. Hyg. 26: 46–66.

FISH, D., AND R. A. BEAVER. 1978. A bibliography of the aquatic fauna inhabiting bromeliads (Bromeliaceae) and pitcher plants (Nepenthaceae and Sarraceniaceae). Florida Antimosq. Assoc., Proc. 49: 11–19.

FRANK, J. H., AND L. P. LOUNIBOS, eds. 1983. Phytotelmata, terrestrial plants as hosts for aquatic insect communities. Plexus, Medford, N.J.

GÓMEZ, L. D. 1977. La biota bromelícola excepto anfíbios y reptiles. Hist. Nat. Costa Rica 1: 45–62.

LAESSLE, A. M. 1961. A micro-limnological study of Jamaican bromeliads. Ecology 42: 499–517.

MURILLO, R. M., J. G. PALACIOS, J. M. LABOUGLE, E. M. HENTSCHEL, J. E. LLORENTE, K. LUNA, P. ROJAS, AND S. ZAMUDIO. 1983. Variación estacional de la entomofauna asociada a *Tillandsia* spp. en una zona de transición biótica. Southwest. Entomol. 8: 292–302.

PICADO, C. 1913. Les broméliacées épiphytes considérée comme milieu biologique. Bull. Scient. France Belgique 47: 215–360.

SEIFERT, R. P. 1975. Clumps of *Heliconia* inflorescences as ecological islands. Ecology 56: 1416–1422.

SEIFERT, R. P. 1980. Mosquito fauna of *Heliconia aurea*. J. Anim. Ecol. 49: 687–697.

SEIFERT, R. P. 1981. Principal components analysis of biogeographic patterns among *Heliconia* insect communities. New York Entomol. Soc., J. 89: 109–122.

SEIFERT, R. P. 1982. Neotropical *Heliconia* insect communities. Quart. Rev. Biol. 57: 1–28.

SEIFERT, R. P., AND F. H. SEIFERT. 1976a. Natural History of insects living in inflorescences of two species of *Heliconia*. New York Entomol. Soc., J. 84: 233–242.

SEIFERT, R. P., AND F. H. SEIFERT. 1976b. A community matrix analysis of *Heliconia* insect communities. Amer. Nat. 110: 461–483.

SEIFERT, R. P., AND F. H. SEIFERT. 1979. A *Heliconia* insect community in a Venezuelan cloud forest. Ecology 60: 462–467.

ZAVORTINK, T. J. 1985. *Zinzala*, a new subgenus of *Wyeomyia* with two new species from pitcher-plants in Venezuela (Diptera, Culicidae, Sabethini). Wasmann J. Biol. 43: 46–59.

### Miscellaneous Special Habitats

A few additional important special habitats that have been ignored or received minimal attention with regard to their entomofaunal aspects in Latin America are the following.

1. *Hot, mineral springs.* De Oliveira (1954) noted ephydrid flies in the hot effluent of a geyser at El Tatío in Chile. Some general collections from Termas de Chillán in the same country have been identified by Ruiz and Stuardo (1935).

2. *Inland salt lakes and playas,* such as the immense salares of the Bolivian altiplano (Salar de Coipasa, Lago de Poopó, Salar de Uyuni). A few insects of hypersaline waters have been described, notably mosquitoes (Bachmann and Casal 1963) and Corixidae of the genus *Trichocorixa* from Argentina (Bachmann 1979: 326f.) and the Mexican lakes, Texcoco and Ahuatle (see water boatmen, chap. 8). Ephydrid shore flies of the genus *Dimecoenia* (ecologically analogous to the Mexican and North American *Ephydra*) are abundant in these waters as well (de Oliveira 1958).

3. *Streams.* Entomological studies of lotic waters are numerous but not comprehensive and deal mostly with highland, hard-bottom streams (Illies 1964; Campos et al. 1984; Patrick and collabs. 1966; Turcotte and Harper 1982*a*, 1982*b*). Adaptive strategies were studied on stream invertebrates, mostly insects in the Argentine Río Negro, by Wais (1985).

4. *Alpine ponds and lakes.* Available are only the reports of Robeck et al. (1980) for miscellaneous Andean habitats and Gilson (1939) for Lake Titicaca.

5. *Bogs.* No published observations.

6. *Deserts,* both local and large, as the great Sonoran or Atacama. No apparent comprehensive observations.

7. *Sand dunes,* both coastal and inland desert. No apparent observations.

8. *Inland freshwater swamps.* No apparent observations.

9. *Volcanoes.* Old volcanoes may provide information on the effects of isolation on mountainous faunas (Palacios-Vargas 1985); dispersal and recolonization of new terrain will be demonstrated by insects and terrestrial arthropods found on new volcanoes, both at sea and on the continents. In both cases, the principles of island biogeography are applicable.

10. *Guianan tepuis.* The flat-topped jungle mesas of the Guiana Shield are acknowledged islands supporting many endemic biotic elements. They have been studied mostly by botanists, but some unique insects have been collected atop Mount Roraima and Cerro Neblina (Spangler 1981, 1985; Waterhouse 1900).

11. *Lakes and rivers.* Limnological studies, particularly of bodies of fresh water larger than streams, are numerous, but in these, more attention is usually paid to the macrofauna than to insects. Some works have concentrated on the latter (Wais 1984).

12. *Land crab burrows.* The burrows of tropical coastal land crabs support a complex ecosystem, including many insects and related arthropods (Bright and Hogue 1972), most of which have aquatic life stages in the water that accumulates within them from rainfall and groundwater seepage.

13. *Lava tubes.* No apparent observations.

Also remaining to be elucidated are the entomological characteristics of the great number of vegetation types throughout the region, for example, cerrado, palm forests, caatinga, cloud forest, mangroves, grassland, puna, páramo (Bernal 1985), alpine, and savanna.

The foregoing are all fruitful fields of investigation for future students of ecological and biogeographic entomology.

## References

BACHMANN, A. O. 1979. Notas para una monografía de las Corixidae Argentinas (Insecta, Heteroptera). Acta Zool. Lilloana 35: 305–350.

BACHMANN, A. O., AND O. H. CASAL. 1963. Mosquitos Argentinos que crian en aguas salobres y saladas. Soc. Entomol. Argentina, Rev. 25: 21–27.

BERNAL, A. 1985. Estudio comparativo de la artropofauna de un bosque alto-andino y de un pajonal paramuno en la región de Monserrate (Cund.). *In* H. Sturm and O. Rangel, Ecología de los paramos andinos: Una visión preliminar integrada. Univ. Nac. Colombia (Inst. Cien. Nat., Mus. Nat. Hist., Bibl. José Jeronimo Triana no. 9). Bogotá. Pp. 225–260.

BRIGHT, D. B., AND C. L. HOGUE. 1972. A synopsis of the burrowing land crabs of the world and list of their arthropod symbionts and burrow associates. Los Angeles Co. Mus. Nat. Hist., Contrib. Sci. 220: 1–58.

CAMPOS, H., J. ARENAS, C. JARA, T. GONSER, AND R. PRINS. 1984. Macrozoobentos y fauna lótica de las aguas limnéticas de Chiloé y Aysén continentales (Chile). Medio Ambiente 7: 52–64.

DE OLIVEIRA, S. J. 1954. Contribuição para o conhecimento do gênero "*Dimecoenia*" Cresson, 1916. I: "*Dimecoenia lenti*" sp. n. encontrado no Chile (Diptera, Ephydridae). Rev. Brasil. Biol. 14: 187–194.

DE OLIVEIRA, S. J. 1958. Contribuição para o conhecimento do gênero "*Dimenia*" Cresson, 1916. IV: Descrição da larva e do pupário de "*Dimecoenia grumanni*" Oliveira, 1954 (Diptera, Ephydridae). Rev. Brasil. Biol. 18: 167–169.

GILSON, H. C. 1939. The Percy Sladen Trust Expedition to Lake Titicaca in 1937. I: Description of the expedition. Linnean Soc. London, Trans. 1: 1–20.

ILLIES, J. 1964. The invertebrate fauna of the Huallaga, a Peruvian River, from the sources down to Tingo María. Intl. Verein Theor. Angew. Limnol., Verh. 15: 1077–1083.

PALACIOS-VARGAS, J. 1985. Microartrópodos del Popocatépetl (aspectos ecológicos y biogeográficos de los ácaros oribátidos e insectos colémbolos). Ph.D. diss., Univ. Nac. Aut. México, Mexico City.

PATRICK, R., AND COLLABORATORS. 1966. The Catherwood Foundation Peruvian-Amazon Expedition: Limnological and systematic studies. Acad. Nat. Sci. Philadelphia, Monogr. 14: 1–495.

ROBECK, S. S., L. BERNER, O. S. FLINT, JR., N. NIESER, AND P. J. SPANGLER. 1980. Results of the Catherwood Bolivian-Peruvian alti-plano expedition. Pt. I, Aquatic insects except Diptera. Acad. Nat. Sci. Philadelphia Proc. 132: 176–217.

RUIZ, F., AND C. STUARDO O. 1935. Insectos colectados en las Termas de Chillán. Rev. Chilena Hist. Nat. 39: 313–322.

SPANGLER, P. J. 1981. New and interesting water beetles from Mt. Roraima and Ptari-tepui, Venezuela (Coleoptera: Dytiscidae and Hydrophilidae). Aquat. Ins. 3: 1–11.

SPANGLER, P. J. 1985. A new genus and species of riffle beetle, *Neblinagena prima*, from the Venezuelan tepui, Cerro de la Neblina (Coleoptera, Elmidae, Larinae). Entomol. Soc. Wash., Proc. 87: 538–544.

TURCOTTE, P., AND P. P. HARPER. 1982a. The macroinvertebrate fauna of a small Andean stream. Freshwater Biol. 12: 411–419.

TURCOTTE, P., AND P. P. HARPER. 1982b. Drift patterns in a high andean stream. Hydrobiologia 89: 141–151.

WAIS, I. R. 1984. Two Patagonian basins— Negro (Argentina) and Valdivia (Chile)—as habitats for Plecoptera. Ann. Limnol. 20: 115–122.

WAIS, I. R. 1985. Stratégies adaptatives aux eaux courantes des invertébrés du bassin du fleuve Negro, Patagonia, Argentina. Int. Verein. Limnol., Verh. 22: 2167–2172.

WATERHOUSE, C. O. 1900. Report on a collection made by Messrs. F. V. McConnell and J. J. Quelch at Mt. Roraima in British Guyana. Linnean Soc. London, Trans. 8: 74–76.

## HISTORICAL BIOGEOGRAPHY

The historical perspective in zoogeography provides insights into the origins of species and faunas and their changes through time relative to geologic events (Fittkau et al. 1969). Insects make excellent material for such studies because of the very long history of many taxa and the variety of types available to test hypotheses of all kinds (Gressitt 1974, Munroe 1965). A group may be found which will fulfill almost any given set of conditions for any geologic period. For example, upland, sedentary types with limited vagility (such as narrowly adapted, torrenticolous Diptera) are particularly useful for tracing past mountainous land connections as far back as the late Paleozoic.

The existence of two, contrasting means of population disruption, that is, dispersal to isolated areas versus fragmention in place, is the basis for the "dispersalist" and "vicariance" schools of biogeographic thought (Ferris 1980) to explain the speciation process in evolving organisms. Actually, both mechanisms may cause the branching of phyletic lines and are part of a modern, unified theory of biogeography (Pielou 1979, Brown and Gibson 1983).

Because insects have been on earth for a very long time, at least since the middle of the Paleozoic era, continental connections and disjunctions (tectonics) (Dietz and Holden 1970, Marvin 1973, Smith et al. 1981) are major vicariant events that have affected their evolution and dispersal (Carbonell 1977: 155–161). In Latin America, the tectonic development of the Caribbean and isthmian regions seems to be much more complicated (Bonini et al. 1984, Durham 1985, Rosen 1985) than that of the South American portions (Jenks 1956, Harrington 1962) with consequent problems in explaining the origins of organisms there (Liebherr 1988, Woods 1989).

The origin of many groups on Gondwanaland, the great southern continent that was composed of what is now South America, Africa, Antarctica, Australia, and India, is still evident in the restricted occurrences of their descendants in those areas and in the southernmost portions of South America today (e.g., water midges [Brundin 1966, 1967]; see Keast 1973 for other insect examples of these so-called amphinotic or austral disjunctive distributions). Close affinities of some eastern Brazilian insects with West African species, such as among *Schistocerca* grasshoppers (Carbonell 1977:169), the amblypygid genus *Phrynus* (= Tarantula) (Quintero 1983), the psocid genera *Belaphapsocus* and *Notiopsocus* (New 1987), and the termite genus *Mimeutermes* (Emerson 1955), are current evidence of past union of the two continents at midlatitudes.

The formation of the Amazon Basin is a direct result of events brought into play by the break in the South America–Africa connection about 90 million years ago. According to one theory (Putzer 1984), the Amazon River system was probably continuous during the early Mesozoic with the Niger River, and the main flow was westward to the Pacific Ocean. But with the relatively recent (Miocene) uplift of the Andes, the flow was dammed and an enormous lake formed at their foot. As the plates separated, the western portion of the river reversed its flow and came to empty into the Atlantic. An extensive coastal plain was also created along the west side of the Cordillera from ejection of volcanic material and pluvial outwash. These vast physiographic changes have created a mixed heritage for the basin's present-day insect populations. Some are derived from the Old World, others are derived from highlands to the south and north, and still others have evolved in situ after long periods of isolation by river and climatic barriers. This variety of causes is one reason for the basin's incredible species richness.

On a much more limited scale, local natural disasters, such as volcanic eruptions and hurricanes, take a toll on insect life and may actually cause the extinction of small populations or even very regional species and constitute vicariant events, although such consequences have yet to be documented, especially those of low-density forest species (Elton 1975). Changes in the course of rivers, an especially common occurrence in meandering lowland drainages, such as the Amazon, can break up continuous populations and halt gene flow sufficiently to create new entities.

Dispersalist examples of the effects of geology on the history of insect life come from the oceanic islands scattered on the fringes of the main landmasses, all derived from marine volcanic eruptions at various times in the past and none of which have ever been connected by dry land to the mainland (Kuschel 1963). The length of time that these islands have remained continuously above the sea surface and their distance from the nearest stable large landmass are the prime variables determining opportunities for oversea colonization and subsequent speciational events (MacArthur and Wilson 1967). The Galápagos Islands are the best studied in this respect (Linsley 1977, Linsley and Usinger 1966). From the fairly high percentage of cryptozoic insects with reduced body pigmentation and vestigial eyes among the entomofauna, at least portions of the islands are considered to have been dry for at least a million years (Leleup 1976). In a few cases (e.g., *Schistocerca*), differentiation of endemic species complexes have also radiated like the Galápagos finches and groups on other remote archipelagos, such as the Hawaiian Islands (Thornton 1971: 82–87). Continental mountains, such as the Sierra Nevada de Santa Marta in Colombia or the Venezuelan tepuis, may also exhibit biogeographic features paralleling those of oceanic islands because of isolation (Vuilleumier 1970).

Dispersal across wide areas of inhospitable territory (Sparks et al. 1986), even between widely separated continents (Johnson and Bowden 1973), may take place through active, long-range flight by larger, powerful insects; drifting on wind currents by smaller, weaker species; flotation on the ocean surface by forms resistant to sea water and by rafting types (Heatwole and Levins 1972), particularly wood borers, leaf miners, and soil dwellers that can survive on or in driftwood, and on so-called floating islands discharged by major rivers. Very long distance dispersal across the Atlantic from Africa, aided by winds, is a possible way in which the Neotropics may have been populated by some species, including some pests (midges, thrips,

aphids) (Hespenheide 1977, Holzapfel 1978).

Human activity often artificially fosters dispersion or, more rarely, vicariance. Examples of the spread of pest species through commerce are numerous. Frequent transport of insects between islands in modern times, particularly on ships, promises to mix populations in the future and will probably destroy many of the natural segregates in the Galápagos Islands (Silberglied 1978). The construction of highways and canals, such as the proposed sea-level canal in Panama, facilitate the movement of many species along the pathway (Weber 1972) but disrupt interchanges across it.

## References

BONINI, W. E., R. B. HARGRAVES, AND R. SHAGAM, eds. 1984. The Caribbean–South American plate boundary and regional tectonics. Geol. Soc. Amer., Mem. 162: 1–421.

BROWN, J. H., AND A. C. GIBSON. 1983. Biogeography. Mosby, St. Louis.

BRUNDIN, L. 1966. Transantarctic relationships and their significance, as evidenced by chironomid midges. K. Swenska Vetenskakad. Avh. Naturskydd 11(1): 1–472.

BRUNDIN, L. 1967. Insects and the problem of austral disjunctive distribution. Ann. Rev. Entomol. 12: 149–168.

CARBONELL, C. S. 1977. Origin, evolution, and distribution of the Neotropical acridomorph fauna (Orthoptera): A preliminary hypothesis. Soc. Entomol. Argentina, Rev. 36: 153–175.

DIETZ, R. S., AND J. C. HOLDEN. 1970. Reconstruction of Pangea: Breakup and dispersion of continents, Permian to present. J. Geophys. Res. 75: 4939–4956.

DURHAM, J. W. 1985. Movement of the Caribbean Plate and its importance for biogeography in the Caribbean. Geology 13: 123–125.

ELTON, C. S. 1975. Conservation and the low population density of invertebrates inside Neotropical rain forest. Biol. Conserv. 7: 3–15.

EMERSON, A. E. 1955. Geographical origins and dispersion of termite genera. Fieldiana, Zool. 37: 465–521.

FERRIS, V. R. 1980. A science in search of a paradigm? Review of the symposium "Vicariance biogeography: A critique." Syst. Zool. 29: 67–76.

FITTKAU, E. J., J. ILLIES, H. KLINGE, G. H. SCHWABE, AND H. SIOLI, eds. 1969. Biogeography and ecology in South America. Vol. 2. Junk, the Hague.

GRESSITT, J. L. 1974. Insect biogeography. Ann. Rev. Entomol. 19: 293–321.

HARRINGTON, H. J. 1962. Paleogeographic development of South America. Amer. Assoc. Petrol. Geol. 46: 1773–1814.

HEATWOLE, H., AND R. LEVINS. 1972. Biogeography of the Puerto Rican Bank: Flotsam Transport of terrestrial animals. Ecology 53: 112–117.

HESPENHEIDE, H. A. 1977. Dispersion and the size composition of the aerial insect fauna. Ecol. Entomol. 2: 139–141.

HOLZAPFEL, E. P. 1978. Transoceanic airplane sampling for organisms and particles. Pacific Ins. 18: 169–189.

JENKS, W. F., ed. 1956. Handbook of South American geology. Geol. Soc. Amer. Mem. 65: 1–378.

JOHNSON, C. G., AND J. BOWDEN. 1973. Problems related to the transoceanic transport of insects, especially between the Amazon and Congo areas. In B. J. Meggers, E. S. Ayensu, and W. D. Duckworth, Tropical forest ecosystems in Africa and South America: A comparative review. Smithsonian Inst., Washington, D. C. Pp. 207–222.

KEAST, A. 1973. Contemporary biotas and the separation sequence of the southern continents. In D. H. Tarling and S. K. Runcorn, eds., Implication of continental drift to the earth sciences. 1:309–343. Academic, London.

KUSCHEL, G. 1963. Composition and relationship of the terrestrial faunas of Easter, Juan Fernandez, Desventuradas and Galápagos. Calif. Acad. Sci., Occ. Pap. 44: 79–95.

LELEUP, N. 1976. Les implications de l'existence d'elements relictuels parmi la faune entomologique cryptique des Iles Galapagos. Soc. Royal Belgique Entomol., Bull. Ann. 112: 90–100.

LIEBHERR, J. K., ed. 1988. Zoogeography of Caribbean insects. Cornell Univ. Press, Ithaca.

LINSLEY, E. G. 1977. Insects of the Galápagos (Supplement). Calif. Acad. Sci. Occ. Pap. 125: 1–50.

LINSLEY, E. G., AND R. L. USINGER. 1966. Insects of the Galápagos Islands. Calif. Acad. Sci., Proc. (ser. 4) 33: 113–196.

MACARTHUR, R. H., AND E. O. WILSON. 1967. The theory of island biogeography. Princeton Univ. Press, Princeton.

MARVIN, U. B. 1973. Continental drift, the evolution of a concept. Smithsonian Inst., Washington, D.C.

MUNROE, E. 1965. Zoogeography of insects and allied groups. Ann. Rev. Entomol. 10: 325–344.

NEW, T. R. 1987. Biology of the Psocoptera. Oriental Ins. 21: 1–109.

PIELOU, E. C. 1979. Biogeography. Wiley-Interscience, New York.

PUTZER, H. 1984. The geological evolution of the Amazon basin and its mineral resources In H. Sioli, The Amazon. Junk, Dordrecht. Pp. 15–46.

QUINTERO, JR., D. 1983. Revision of the amblypygid spiders of Cuba and their relationships with the Caribbean and Continental American amblypygid fauna. Stud. Fauna Curaçâo Caribbean Is. 196: 1–54.

ROSEN, D. E. 1985. Geological hierarchies and biogeographic congruences in the Caribbean. Missouri Bot. Gardens Ann. 72: 636–659.

SILBERGLIED, R. E. 1978. Inter-island transport of insects aboard ships in the Galápagos Islands. Biol. Conserv. 13: 273–278.

SMITH, A. G., A. M. HURLEY, AND J. C. BRIDEN. 1981. Phanerozoic paleocontinental world maps. Cambridge Univ. Press, Cambridge, Mass.

SPARKS, A. N., R. D. JACKSON, J. E. CARPENTER, AND R. A. MILLER. 1986. Insects captured at light traps in the Gulf of Mexico. Entomol. Soc. Amer. Ann. 79: 132–139.

THORNTON, I. 1971. Darwin's islands, a natural history of the Galápagos. Natural Hist., New York.

VUILLEUMIER, F. 1970. Insular biogeography in continental regions. I. The northern Andes of South America. Amer. Nat. 104: 373–388.

WEBER, N. A. 1972. The entomology of Panamá (Contributions to The Panamic Biota: Some observations prior to a sea-level canal). Biol. Soc. Wash. Bull. 2: 187–197.

WOODS, C. A., ed. 1989. Biogeography of the West Indies: Past, present, future. Sandhill Crane, Gainesville.

# BIOGEOGRAPHIC PROVINCES

Ecosystems found in certain physiographic areas vary in response to historical as well as physical and biotic factors. All tropical deserts, for example, will not have the same species or genera of darkling beetles or all high mountains the same butterfly groups. Past dispersals and evolution of organisms will result in different assemblages of taxa in geographic areas regardless of similarities they may have in their environment (Mooney 1977). These assemblages, particularly the existence of endemics, give taxonomic character to any region from which it may be recognized as a distinct biogeographic unit with biotic uniformity.

The Neotropical Region itself is one of the earth's broadest and most distinctive of such categories. The reason for this seems to be due primarily to the fact that South America and Africa, before their disjunction, constituted the largest block of land (western Gondwanaland) in the tropics and was therefore a separate evolutionary center (Raven and Axelrod 1975). Subsequent long isolation permitted undisturbed and independent development of an already rich biota (Fittkau 1969).

The composition and spatial arrangement of the world's biogeographic regions have changed through time. Past evolutionary patterns have been postulated for some other organisms (Hallam 1973) but are not known for insects in Latin America, mainly because of their poor fossil record there. The origins of these regions and contemporary insect geography are determinable, nevertheless, by combining knowledge of a group's morphology, mobility, geographic and ecological dispersal opportunities in the past, and historical disruptions in its range. There are relatively few such studies on Neotropical insects (Halffter 1975, 1976, 1987), but it is common for taxonomists to treat the zoogeography of their groups routinely in systematic papers (e.g., Nielsen and Robinson 1983).

In the Neotropics, different systems of classification of today's biogeographic subregions have been applied by authors choosing varied organisms as indicators and more or less inclusive amounts of land area. For all or most of Latin America, few are based on the entire flora (Gentry 1982, Cabrera and Willink 1973) or fauna (Gilmore 1950, Goldman and Moore 1946, Fittkau 1969), fewer on both (Darlington 1965, Udvardy 1975). Some are derived from an analysis of diverse entomofaunal elements (Halffter 1964, 1974, 1975, 1987; O'Brien 1971) or specific taxa (Porter 1980, Ichneumonidae; Kuschel 1969, beetles; Halffter 1975, dung beetles). Most, however, are limited to much narrower taxonomic groups and restrictive areas (Peña 1966, Chilean darkling beetles; Scott 1972, Antillean butterflies; de Armas 1982, Antillean scorpions; Lane 1943, sabethine mosquitoes). Some of the latter have produced complex results, such as Lamas's (1982) recognition of forty-eight biotic regions for Peru, based on the country's butterfly fauna alone.

The aquatic medium has been treated to a limited degree. Stream insects fall into two main groups, the cool-adapted (oligostenothermic) types found in certain mountainous areas (Andes, especially southern Andes, Guianas, southeastern Brazil), and warm-adapted (polystenothermic) types spread almost everywhere else in the lowlands (Illies 1965, 1969).

For entomological purposes, in Latin America, the simple classification of biogeographic regions shown in figure 2.2 is useful (although a much more detailed system is proposed by Udvardy [1975]).

## References

Cabrera, A. L., and A. Willink. 1973. Biogeografía de América Latina. Org. Est. Americanos, Ser. Biol., Monogr. 13: 1–122.

Darlington, Jr., P. J. 1965. Biogeography of the southern end of the world. McGraw-Hill, New York.

de Armas, L. F. 1982. Algunos aspectos zoogeográficos de la escorpionfauna antillana. Poeyana 238: 1–17.

Fittkau, E. J. 1969. The fauna of South Amer-

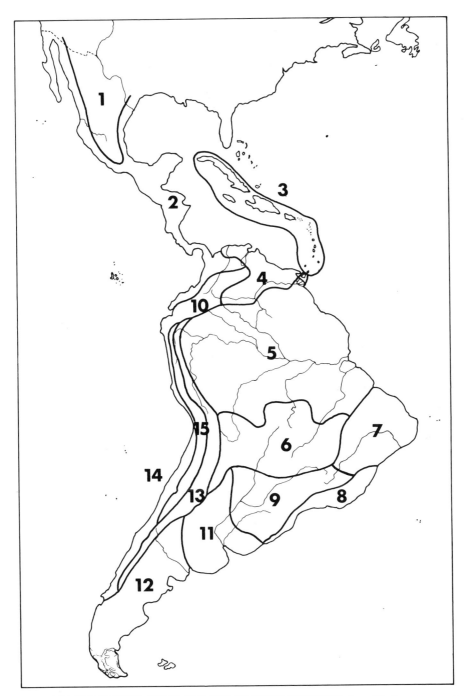

**Figure 2.2.     MAJOR NEOTROPICAL BIOGEOGRAPHIC REGIONS**  (modified and expanded from Fittkau 1969).     MIDDLE AMERICA:  1. Mexican Highlands (part of Nearctic Region); 2. Mexican Lowlands and Central America; 3. West Indies (Antilles);     SOUTH AMERICA  *Guiana-Brazil;* 4. Caquetio (Orinoco and south); 5. Hylaea (Amazonia); 6. Bororó (Brazilian plateau); 7. Cariri (northeastern Brazil); 8. Tupi; 9. Guaraní (southern Brazil); 10. Incasia (northern Andes) *Andes-Patagonia;* 11. Pampas; 12. Patagonia (includes Juan Fernández, Falklands, etc.); 13. Subandes (eastern Andean foot ranges); 14. Chile; 15. Andes (central and southern).

ica. *In* E. J. Fittkau, J. Illies, H. Klinge, G. H. Schwabe, and H. Sioli, eds., Biogeography and ecology in South America. 2:624–658. Junk, The Hague.

GENTRY, A. H. 1982. Neotropical floristic diversity: Phytogeographical connections between Central and South America, Pleistocene climatic fluctuations, or an accident of the Andean orogeny? Missouri Bot. Gardens Ann. 69: 557–593.

GILMORE, R. M. 1950. Fauna and ethnozoology of South America. Handbk. So. Amer. Indians 6: 345–464.

GOLDMAN, E. A., AND R. T. MOORE. 1946. The biotic provinces of Mexico. J. Mammal. 26: 347–360.

HALFFTER, G. 1964. La entomofauna Americana, ideas acerca de su origen y distribución. Fol. Entomol. Mex. 6: 1–108.

HALFFTER, G. 1974. Elements anciens de l'entomofaune néotropicale: Ses implications biogéographiques. Quaest. Entomol. 10: 223–262.

HALFFTER, G. 1975. Elements anciens de l'entomofaune néotropicale: Ses implications biogéographiques. Mus. Nat. Hist. Natur. (Paris) Mem. Ser. A. (Zool.) 2: 114–145.

HALFFTER, G. 1976. Distribución de los insectos en la zona de transición Mexicana: Relaciones con la entomofauna de norteamérica. Fol. Entomol. Mex. 35: 1–64.

HALFFTER, G. 1987. Biogeography of the montane entomofauna of Mexico and Central America. Ann. Rev. Entomol. 32: 95–114.

HALLAM, A. 1973. Atlas of palaeobiogeography. Elsevier, Amsterdam.

ILLIES, J. 1965. Phylogeny and zoogeography of the Plecoptera. Ann. Rev. Entomol. 10: 117–140.

ILLIES, J. 1969. Biogeography and ecology of Neotropical freshwater insects, especially those from running waters. *In* E. J. Fittkau, J. Illies, H. Klinge, G. H. Schwabe, and H. Sioli, eds., Biogeography and ecology in South America. 2: 685–708. Junk, The Hague.

KUSCHEL, G. 1969. Biogeography and ecology of South American Coleoptera. *In* E. J. Fittkau, J. Illies, H. Klinge, G. H. Schwabe, and H. Sioli, eds., Biogeography and ecology in South America. Vol. 2. Junk, The Hague.

LAMAS, G. 1982. A preliminary zoogeographical division of Peru, based on butterfly distributions (Lepidoptera, Papilionoidea). *In* G. T. Prance, ed., Biological diversification in the tropics. Columbia Univ. Press, New York. Pp. 336–357.

LANE, J. 1943. The geographic distribution of Sabethini (Dipt., Culicidae). Rev. Entomol. São Paulo 14: 409–429.

MOONEY, H. A., ed. 1977. Convergent evolution in Chile and California, Mediterranean climate ecosystems. Dowden, Hutchinson and Ross, Stroudsburg, Penn.

NIELSEN, E. S., AND G. S. ROBINSON. 1983. Ghost moths of southern South America (Lepidoptera: Hepialidae). Entomonograph 4: 1–192.

O'BRIEN, C. W. 1971. The biogeography of Chile through entomofaunal regions. Entomol. News 82: 197–207.

PEÑA, L. E. 1966. A preliminary attempt to divide Chile into entomofaunal regions based on the Tenebrionidae (Coleoptera). Postilla 97: 1–17.

PORTER, C. 1980. Zoogeografía de las Ichneumonidae Latino-Americanas (Hymenoptera). Acta Zool. Lilloana 36: 1–52.

RAVEN, P. H., AND D. I. AXELROD. 1975. History of the flora and fauna of Latin America. Amer. Sci. 63: 420–429.

SCOTT, J. A. 1972. Biogeography of Antillean butterflies. Biotropica 4: 32–45.

UDVARDY, M. D. F. 1975. A classification of the biogeographical provinces of the world. IUCN, Occ. Pap. 18: 1–48.

## Faunistics

The present-day composition of species in a particular geographic area comprises its fauna. As shown above, the nature of faunas, the relative number of different ecological types, and the taxonomic spectrum offer clues to the geologic history of the area and are important to ecologists as indicators of communities and ecosystems.

The extent of the entire Latin American entomofauna is not yet known. Estimates go as high as 20 million species. Attempts have been made to identify the insects and related arthropods of many portions of the region, but none can be considered complete. It involves a great deal of effort and taxonomic expertise even to compile a general list of all the speices in a small area. Only a few general lists and cursory faunal reviews have been published; consult the Faunal Surveys in chapter 13 for a listing of these.

# POPULATION BIOLOGY

The size, structure, energetics, composition, and mobility of insect populations in various life zones, macrohabitats, and microhabitats are highly significant, and a very large body of literature on both theoretical and applied insect demography exists (Young 1982). Such investigations always require quantitative analysis of population parameters, and particular sampling techniques are needed in each specialty area (Southwood 1980, Wolda 1984).

Most works deal with specific insects (Ehrlich and Gilbert 1973), often of economically important types, limited taxa (Torres 1984), or guilds (unrelated species with similar niches) (Heithaus 1979) within particular environments. Some of this research may include comparisons between different habitat types. Investigations of whole communities are very few owing to their complexity and their multifold interactions with the environment.

The lower layers of tropical forests (lowland, midlatitude moist to rain forests) have attracted the most attention from insect population biologists in Latin America. Sampling is direct; specimens can be observed directly or captured fairly easily with nets, the "sweeping" technique often being used for soft foliage, herbs, or grasses (Allan et al. 1973, Janzen 1973). An intensive thirteen-month survey of Amazonian forest was undertaken in 1977–78 in the Ducke Forest Preserve near Manaus (Penny and Arias 1982), for which passive Malaise traps caught most of the samples.

Long-term studies are rare (Wolda 1983). Over a period of several years, Elton (1973) sampled the forest "field layer" (15 cm to 1.8 m above the ground) in a variety of localities and with different methods and concluded that insect life generally exists in low numbers and that the sizes of most are small but that the species richness is very high. The most common forms found were ants, spiders, and orthopteroids. The very diverse nocturnal fauna of flying insects is composed of much larger insects.

Using sweep net samples to study diversity and distribution, various authors (e.g., Janzen 1973) have discovered profound differences along elevational transects, for example, in Costa Rica and the Venezuelan Andes (Janzen et al. 1976). Latitudinal effects (Janzen and Pond 1975) as well as seasonal effects (Janzen and Schoener 1968, Tanaka and Tanaka 1982) have also been included.

Light trapping has revealed some populational traits of Neotropical insects (Ricklefs 1975; Wolda 1978a, 1978b). In lowland forest, the seasonality of flight activity of nocturnal forms is pronounced. Generally, numbers are depressed in the mid-dry period and maximally expanded in the early weeks of wetness in both mainland (Wolda 1980) and island (Tanaka and Tanaka 1982) areas. The suppressive effect of moonlight on the activity of night-flying insects is also often noted (Wolda and Flowers 1985).

The insects of the forest canopy, although now accessible by special techniques, resist quantitative study because of the difficulties remaining in taking sizable samples and covering a significant area of habitat. Some advances have been made in the study of canopy bee biology by climbing entomologists (Perry 1983). Traps (Malaise, "photocollectors") have also been raised to upper levels for remote assessment. In this way, three responses by arboreal insects in Amazon inundation forests have been detected during the flood season: temporary immigration, survival in place, and dying out (Adis 1977).

A fruitful approach for obtaining statistically adequate data from the canopy has come from the insecticide fogging technique. Rapid-acting pyrethroids are blown into the trees. (These mildly toxic chemicals rapidly degrade in the environment and do not harm nontarget organisms.) All the exposed insects are stunned and fall to the ground, where they are collected, identified, and counted. In this way, almost complete samples can be taken. Cockroaches (Fisk 1982), wax bugs (Wolda 1980), grasshoppers (Roberts 1973), and carabid beetles (Erwin 1983) have been analyzed.

Some comparisons of species richness of temperate rheophilic (Stout and Vandermeer 1975) forms to those in Costa Rican and Andean (Patrick 1966) streams have been made. Results generally show that midlatitude stream faunas are significantly more diverse than those in comparable, high-latitude streams.

Soil and litter types are studied by means of the Berlese (Tullgren) funnel (Beebe 1916). Samples may also be taken with a core auger and separation by flotation in liquid suspensions. Substrata may be experimentally manipulated, which allows the investigator to measure minute habitat characteristics with great accuracy and compare them to natural plots (Stanton 1979).

From these various population studies, it is generally conceded that in tropical forests, insect abundance is low but the diversity very high. It seems also that the dispersion of insect populations over a particular area is seldom even but usually very patchy. This may be caused by the irregular distribution of habitats (Wiens 1976), but it can be so even if environmental needs appear continuously and abundantly available. Moisture content of litter and soil may be very unevenly distributed, for example, and a major factor affecting the presence or absence of ants, springtails, mites, and other arthropods (Levings and Windsor 1984). However, in spite of the enormous numbers of leaves available on a single tree or in a forest, only a few are normally utilized by herbivores at one time (notwithstanding population explosions when defoliation may occur).

## References

Adis, J. 1977. Programa mínimo para análises de ecosistemas: Artrópodos terrestres em florestas inundáveis da Amazonia central. Acta Amazonica 7: 223–229.

Allan, J. D., L. W. Barnhouse, R. H. Prestbye, and D. R. Strong. 1973. On foliage arthropod communities of Puerto Rican second growth vegetation. Ecology 54: 628–632.

Beebe, W. 1916. Fauna of four square feet of jungle debris. Zoologica 2: 107–119.

Ehrlich, P. R., and L. E. Gilbert. 1973. Population structure and dynamics of the tropical butterfly Heliconius ethilla. Biotropica 5: 69–82.

Elton, C. S. 1973. The structure of invertebrate populations inside Neotropical rain forest. J. Anim. Ecol. 42: 55–104.

Erwin, T. L. 1983. Beetles and other insects of tropical forest canopies at Manaus, Brazil, sampled by insecticidal fogging techniques. In S. L. Sutton, T. C. Whitmore, and A. C. Chadwick, eds., Tropical rain forests: Ecology and management. Blackwell, Oxford. Pp. 59–75.

Fisk, F. W. 1982. Abundance and diversity of arboreal Blattaria in moist tropical forest of the Panama Canal area and Costa Rica. Amer. Entomol. Soc., Trans. 108: 479–489.

Heithaus, E. R. 1979. Community structure of Neotropical flower visiting bees and wasps: Diversity and phenology. Ecology 60: 190–202.

Janzen, D. H. 1973. Sweep samples of tropical foliage insects: Effects of seasons, vegetation types, elevation, time of day and insularity. Ecology 54: 687–708.

Janzen, D. H., M. Attaroff, M. Fariñas, S. Reyes, N. Rincón, A. Soler, P. Soriano, and M. Vera. 1976. Changes in the arthropod community along an elevational transect in the Venezuelan Andes. Biotropica 8: 193–203.

Janzen, D. H., and C. M. Pond. 1975. A comparison, by sweep sampling, of the arthropod fauna of vegetation in Michigan, England and Costa Rica. Royal Entomol. Soc. London, Trans. 127: 30–50.

Janzen, D. H., and T. W. Schoener. 1968. Differences in insect abundance and diversity between wetter and drier sites during a tropical dry season. Ecology 49: 96–110.

Levings, S. C., and D. M. Windsor. 1984. Litter moisture content as a determinant of litter arthropod distribution and abundance during the dry season on Barro Colorado Island, Panama. Biotropica 16: 125–131.

Patrick, R. 1966. The Catherwood Foundation

Peruvian Amazon Expedition: Limnological and systematic studies. Monogr. Acad. Nat. Sci. Phil. 14: 1–495.

PENNY, N. D., AND J. R. ARIAS. 1982. Insects of an Amazon forest. Columbia Univ. Press, New York.

PERRY, D. R. 1983. Access methods, observations, pollination biology, bee foraging behavior, and bee community structure within a Neotropical wet forest canopy. Ph.D. diss., Univ. Calif., Los Angeles.

RICKLEFS, R. E. 1975. Seasonal occurrence of night-flying insects on Barro Colorado Island, Panama Canal Zone. New York Entomol. Soc., J. 83: 19–32.

ROBERTS, H. R. 1973. Arboreal Orthoptera in the rain forests of Costa Rica collected with insecticide: A report on the grasshoppers (Acrididae), including new species. Acad. Nat. Sci. Philadelphia Proc. 125: 49–66.

SOUTHWOOD, T. R. E. 1980. Ecological methods, with particular reference to the study of populations. 2d ed. Chapman & Hall, London.

STANTON, N. L. 1979. Abundance and diversity of Homoptera in the canopy of a tropical forest. Ecol. Entomol. 4: 181–190.

STOUT, J., AND J. VENDERMEER. 1975. Comparison of species richness for stream-inhabiting insects in tropical and mid-latitude streams. Amer. Nat. 109: 263–280.

TANAKA, L. K., AND S. K. TANAKA. 1982. Rainfall and seasonal changes in arthropod abundance on a tropical oceanic island. Biotropica 14: 114–123.

TORRES, J. A. 1984. Niches and coexistence of ant communities in Puerto Rico: Repeated patterns. Biotropica 16: 284–295.

WIENS, J. A. 1976. Population responses to patchy environments. Ann. Rev. Ecol. Syst. 7: 81–120.

WOLDA, H. 1978a. Fluctuations in abundance of tropical insects. Amer. Nat. 112: 1017–1045.

WOLDA, H. 1978b. Seasonal fluctuations in rainfall, food and abundance of tropical insects. J. Anim. Ecol. 47: 369–381.

WOLDA, H. 1980. Seasonality of tropical insects. I. Leafhoppers (Homoptera) in Las Cumbres, Panama. J. Anim. Ecol. 49: 277–290.

WOLDA, H. 1983. "Long-term" stability of tropical insect populations. Res. Pop. Ecol. Suppl. 3: 112–126.

WOLDA, H. 1984. Diversidad de la entomofauna y como medirla. 9th Cong. Latinoamericano Zool. (Arequipa), Inf. Final. Pp. 181–186.

WOLDA, H., AND R. W. FLOWERS. 1985. Seasonality and diversity of mayfly adults (Ephemeroptera) in a "nonseasonal" environment. Biotropica 17: 330–335.

YOUNG, A. M. 1982. Population biology of tropical insects. Plenum, New York.

## FOOD RELATIONS

Important to their ecological roles are the kinds of food and feeding habits exhibited by insects (Brues 1946, Cummins 1973). A majority, herbivores (or phytophages), act as primary reducers of plant life (d'Araújo y Silva et al. 1967–68, Guagliumi 1966, Martell 1974, Martorell 1976), including a relatively few saprophages that live only on dead plant material such as dead wood. Necrophages specialize on animal carrion, and coprophages utilize animal feces. Higher on the pyramid is another large and highly competitive group, the predators, which catch, kill, and consume other living animals, most often other insects (Clausen 1962) but sometimes even healthy vertebrates (Formanowicz et al. 1981, Hayes 1983). (Body toxins sometimes protect anurans from predation by spiders; the latter may even react negatively to aposematic coloration of this potential prey [Szelistowski 1985].) Modifying the strategy of predation are the parasites, which rob food (including blood) from hosts on or within whose bodies they intimately live without causing their death, and parasitoids, which live like parasites but eventually kill the host. The latter two types are not always clearly separated from each other (Askew 1971, Price 1975). There are also a variety of other narrow feeding specialists on odd nutrient sources, such as hair follicle secretions (hair follicle mites), beeswax (wax moths), cultured fungi (gardening ants), and many others. Whole guilds may be tied to a particular vertebrate host, a good example being the arthropod associates of sloths (Waage and Best 1985). Kleptoparasites and social parasites steal unguarded prey from spider webs or wasp food caches. Many insects combine such habits and are considered omnivores.

Herbivory and saprophagy are vastly complex feeding practices because of the variety of plant types and anatomical parts available as food for insects, including leaves (Ernest 1989), stems, buds, seeds, and fruits (Strong et al. 1984). Species range from highly restricted in their choice of species or part (host specific) to catholic in their tastes (generalists). There are insects that graze on diatoms and algae from films on rocks in streams, lichen browsers, and fern feeders (Forno and Bourne 1984, Hendrix 1980), in addition to those that eat angiosperms generally, both in aquatic (Cummins 1973) and terrestrial (Zimmerman et al. 1979) environments.

Living, dead, and even decomposing (Winder 1977) matter is utilized as food. The greatest availability to herbivores is found in forests, especially tropical rain forests, where the number of plant species is prodigious. Some general studies in this venue are those comparing herbivore damage in riparian versus dry forests (Stanton 1975), the effects of seed predation in determining tree distribution (Janzen 1970), and host specificity (Pipkin et al. 1966). The importance of insect reducers of dead wood and its conversion to soil in forests cannot be overstated (Morón 1985).

The superabundance of plants for herbivores would seem to allow enormous explosions of insect numbers, but the latter are rare events that are often associated with human perturbations, such as misuse of insecticides and monoculture. Natural factors limiting expansions of plant-eating insects are parasitism, predation, and unpalatability (from both physical and chemical deterrents), leading to intraspecific competition. There is evidence that the latter may be increased by the plant in direct response to insect attack (Wratten et al. 1988).

A special category of herbivores are the gall makers. Many flies, wasps, and mites, including entire families in these orders, the Cecidomyiidae, Cynipidae, and Eriophyidae, respectively, stimulate their plant hosts to form abnormal neoplastic growths called galls (Ananthakrishnan 1984). These take many forms and occur on all parts of the plant and are often harmful (Fernandes 1987). They serve the developing stage of the insect as a nutrient-rich, abundant, and reliable food supply. The mechanism of gall formation is not fully understood and varies among the groups involved. Very little is known of the galls of Latin America (Occhioni 1979).

Leaf miners (Hering 1951) form another specialized group of herbivores. It is the larvae of several families of Diptera (Agromyzidae, Tephritidae, Anthomyiidae) and Lepidoptera (Nepticulidae, Tischeriidae, Lyonetiidae, etc.), in particular, that have adapted to the confined microhabitat between the upper and lower epidermal tissues of leaves. Understandably, their chief morphological characteristics are small size and a compressed shape. Appendages are reduced or absent, although the mouthparts are well developed. The same is true of borers, in wood and other tissues, found among the Coleoptera (Cerambycidae, Buprestidae) and Lepidoptera (Pyralidae, Castniidae, Cossidae).

Necrophagous insects occur chiefly in the orders Coleoptera and Diptera (Jirón and Cartín 1981). Several beetle families feed entirely on dead vertebrate animals or their skins and hair, such as the carrion beetles (Silphidae) and dermestids (Dermestidae). Important carrion fly groups are the blowflies (Calliphoridae) and flesh flies (Sarcophagidae). The succession of communities of such insects in human cadavers is the key to their use in determining time of death as sometimes employed in forensic medicine (see Valuable Insects, chap. 3). Dead insect carcasses also provide a form of carrion for necrophagous species (Young 1986). These insects are important in nutrient turnover and, from the standpoint of environmental hygiene, fortunately, are abundant and widespread in nature (Morón and Terrón 1984).

A major group of the ecologically important coprophages, insects that feed in one stage or another on the feces of other animals, are the Phanaeine and Coprine dung scarabs (Scarabaeidae) (Peck and Howden 1984). Some groups are highly specific, for instance, larval sloth moths (Pyralidae, see Sloth Moths, chap. 10) and the scarab beetle, *Uroxys gorgon,* on sloth dung (Young 1981).

A great many insects with sucking mouthparts survive entirely on, or frequently supplement their diets with, liquid foods. The variety and quality of nutrients available in solution is great. Carbohydrates, amino acids, minerals, vitamins, and even fats are dissolved in such diverse and unlikely fluids as blood and lymph from animals, sap, rainwater solutions, nectar, animal and plant secretions (sweat, milk, tears), honeydew, decomposition products, weeping wounds, decay juices, plant juices, fecal matter, and urine.

Whatever the food source, its nutrient content must be appropriate for each species (Dadd 1973). Insects generally require the same basic classes of dietary substances—minerals, calories, protein, carbohydrates, fats, vitamins—as vertebrates, but many peculiar or restricted factors are needed to fulfill the metabolic quirks of idiosyncratic species. Some are incapable even of digesting their food without the intervention of symbiotic microorganisms in their guts (e.g., primitive termites and some cockroaches), while others must cultivate their food, the best example of which are the gardening ants (Attinae).

## References

ANANTHAKRISHNAN, T. N. 1984. Biology of gall insects. Edward Arnold, London.

D'ARAÚJO Y SILVA, A. G., C. R. GONÇALVES, D. M. GALVÃO, A. J. LOBO GONÇALVES, J. GOMES, M. DO NASCIMENTO SILVA, AND L. DE SIMONI. 1967–68. Quarto catálogo dos insetos que vivem nas plantas do Brasil. Pts. 1–2. Min. Agric., Dept. Def. Insp. Agropec., Rio de Janeiro.

ASKEW, R. R. 1971. Parasitic insects. American Elsevier, New York.

BRUES, C. T. 1946. Insect dietary: An account of the food habits of insects. Harvard Univ. Press, Cambridge.

CLAUSEN, C. P. 1962. Entomophagous insects. Hafner, New York.

CUMMINS, K. W. 1973. Trophic relations of aquatic insects. Ann. Rev. Entomol. 18: 183–206.

DADD, R. H. 1973. Insect nutrition: Current developments and metabolic implications. Ann. Rev. Entomol. 18: 381–420.

ERNEST, K. A. 1989. Insect herbivory on a tropical understory tree: Effects of leaf age and habitat. Biotropica 21: 194–199.

FERNANDES, G. W. 1987. Gall forming insects: Their economic importance and control. Rev. Brasil. Entomol. 31: 379–398.

FORMANOWICZ, JR., D. R., M. M. STEWART, K. TOWNSEND, F. H. POUGH, AND P. F. BRUSSARD. 1981. Predation by giant crab spiders on the Puerto Rican frog *Eleutherodactylus coqui.* Herpetologica 37: 125–129.

FORNO, I. W., AND A. S. BOURNE. 1984. Studies in South America of arthropods on the *Salvinia auriculata* complex of floating ferns and their effects on *S. molesta.* Bull. Entomol. Res. 74: 609–621.

GUAGLIUMI, P. 1966. Insetti e aracnidi delle pianti comuni del Venezuela segnalati nel periodo 1938–1963. Rel. Mono. Agrar. Subtrop. Trop. (Nov. Ser. 86): 1–391.

HAYES, M. P. 1983. Predation on the adults and prehatching stages of glass frogs (Centrolenidae). Biotropica 15: 74–76.

HENDRIX, S. D. 1980. An evolutionary and ecological perspective of the insect fauna of ferns. Amer. Nat. 115: 171–196.

HERING, M. 1951. Biology of the leaf miners. Junk, The Hague.

JANZEN, D. H. 1970. Herbivores and the number of tree species in tropical forests, Amer. Nat. 104: 501–528.

JIRÓN, L. F., AND V. M. CARTÍN. 1981. Insect succession in the decomposition of a mammal in Costa Rica. New York Entomol. Soc. J. 89: 158–165.

MARTELL, C. 1974. Primer catálogo de los insectos fitófagos de México. Fitófilo 27: 1–176.

MARTORELL, L. F. 1976. Annotated food plant

catalog of the insects of Puerto Rico. Agric. Expt. Sta., Univ. Puerto Rico, Puerto Rico.

MORÓN, M. A. 1985. Los insectos degradadores, un factor poco estudiado en los bosques de México. Fol. Entomol. Mexicana 65: 131–137.

MORÓN, M. A., AND R. A. TERRÓN. 1984. Distribución altitudinal y estacional de los insectos necrófilos en la Sierra Norte de Hildago, México. Acta Zool. Mexicana (n.s.) 3: 1–47.

OCCHIONI, P. 1979. "Galhas," "cecídeas" ou "tumores vegetais" em plantas nativas da flora do Brasil. Leandra 8–9: 5–35.

PECK, S. B., AND H. F. HOWDEN. 1984. Response of a dung beetle guild to different sizes of dung bait in a Panamanian rainforest. Biotropica 16: 235–238.

PIPKIN, S. B., R. L. RODRIGUEZ, AND J. LEON. 1966. Plant host specificity among flower feeding Neotropical *Drosophila* (Diptera: Drosophilidae). Amer. Nat. 100: 135–156.

PRICE, P. W., ed. 1975. Evolutionary strategies of parasitic insects and mites. Plenum, New York.

STANTON, N. 1975. Herbivore pressure on two types of tropical forests. Biotropica 7: 8–11.

STRONG, D. R., J. H. LAWTON, AND R. SOUTHWOOD. 1984. Insects on plants, community patterns and mechanisms. Harvard Univ. Press, Cambridge.

SZELISTOWSKI, W. A. 1985. Unpalatability of the poison arrow frog *Dendrobates pumilio* to the ctenid spider *Cupiennius coccineus*. Biotropica 17: 345-346.

WAAGE, J. K., AND R. C. BEST. 1985. Arthropod associates of sloths. *In* G. G. Montgomery, ed., The evolution and ecology of armadillos, sloths, and vermilinguas. Smithsonian Inst., Washington, D.C. Pp. 297–311.

WINDER, J. A. 1977. Some organic substrates which serve as insect breeding sites in Bahian cocoa plantations. Rev. Brasil. Biol. 37: 351–356.

WRATTEN, S. D., P. J. EDWARDS, AND L. WINDER. 1988. Insect herbivory in relation to dynamic changes in host plant quality. Biol. J. Linnean Soc. 35: 339–350.

YOUNG, A. M. 1986. Carcass-scavenging by *Taeniopoda reticulata* (Orthoptera: Acrididae) in Costa Rica. Entomol. News 97: 175–176.

YOUNG, O. P. 1981. The utilization of sloth dung in a Neotropical forest. Coleop. Bull. 35: 427–430.

ZIMMERMAN, H. G., H. E. ERB, AND R. E. MCFADYEN. 1979. Annotated list of some cactus-feeding insects of South America. Acta Zool. Lilloana 33: 101–112.

## DIVERSITY

The best recent count of the total number of named insect species in the world is 1,111,225 (Arnett 1983), although many textbooks still place the figure at 650,000 to 700,000. Plausible arguments are offered by authors for many more, 3 to 5 million or 10 million (May 1988) existing species. By liberal extrapolation of some data, there are indications of even as many as 30 million actual living species of tropical forest arthropods alone (Erwin 1982) or possibly several times that amount (Stork 1988). Whatever figure is accurate, most of the total could be expected to occupy Latin America, because of the region's very rich flora, large area, and complexity of physical and biotic habitats. An educated guess of 40 percent of the total world's biota could be made for the number of species ultimately discoverable in the region, based on available insect host diversity and land habitat area present compared to that of the other zoogeographic regions. Most of these species occur in lowland forests.

The reasons for the especially high diversity in the tropical portions are not fully understood, although various theories have been proposed (Dobzhansky 1950, Pianka 1966). No doubt, the inclusion of vast areas of wet lowland forest that provide abundant niches and a stable climate, at least in refuges, has much to do with it. There is little doubt that there are more species of insects of most major groups in the tropics than in temperate areas, with the possible exception of some parasitic Hymenoptera and bees (Wolda 1983), aphids (Dixon et al. 1987), the Plecoptera, and a few others. According to Janzen (1976), the explanation lies chiefly in the amount here of "harvestable productivity" arranged in a sufficiently heterogeneous manner. This allows utilization by a vast complex of small and very adaptable reducing animals, most fittingly, the insects.

Heterogeneity is provided by climatic and geographic variation over wide areas for long periods of time. The best example is provided by the Amazon Basin, where insect diversity is very high as a result of habitat disturbances—from a constantly changing vegetation, dissection of the landscape by rivers, alternation of inundated–dry forest types, and possibly other phenomena—for at least the duration of the Pliocene–Pleistocene era (Erwin and Adis 1982).

The stability of the wet lowland tropics gives the insects that live there a head start on speciation and allows them to accumulate, but because of competition, no single area becomes greatly enriched. The fact that there are a great number of species but relatively few individuals is thought by some to be largely due to the high productivity and availability of resources in such forests and the reduced seasonality that makes it possible for insects to occupy marginal niches (MacArthur 1969).

## References

ARNETT, JR., R. H. 1983. Status of the taxonomy of the insects of America north of Mexico: A preliminary report prepared for the subcommittee for the insect fauna of North America Project. For Stand. Comm. System Res., Entomol. Soc. America.

DIXON, A. F. G., P. KINDLMANN, L. LEPS, AND J. HOLMAN. 1987. Why there are so few species of aphids, especially in the tropics. Amer. Nat. 129: 580–592.

DOBZHANSKY, T. 1950. Evolution in the tropics. Amer. Sci. 38: 209–221.

ERWIN, T. L. 1982. Tropical forests: Their richness in Coleoptera and other arthropod species. Coleop. Bull. 36: 74–75.

ERWIN, T. L., AND J. ADIS. 1982. Amazonian inundation forests: Their role as short-term refuges and generators of species richness and taxon pulses. In G. T. Prance, ed., Biological diversification in the tropics. Columbia Univ. Press, New York. Pp. 358–371.

JANZEN, D. H. 1976. Why are there so many species of insects? 15th Int. Congr. Entomol. Proc. Pp. 84–94.

MACARTHUR, R. H. 1969. Patterns of communities in the tropics. Biol. J. Linnean Soc. 1: 19–30.

MAY, R. M. 1988. How many species are there on earth? Science 341: 1441–1449.

PIANKA, E. R. 1966. Latitudinal gradients in species diversity: A review of concepts. Amer. Nat. 100: 33–44.

STORK, N. E. 1988. Insect diversity: Facts, fiction and speculation, Biol. J. Linnean Soc. 35: 321–337.

WOLDA, H. 1983. Diversity, diversity indices and tropical cockroaches. Oecologia 58: 290–298.

## ENEMIES

Insects suffer diseases from microbial pathogens, as do higher organisms (Cantwell 1974, Steinhaus 1963). No species is immune from infections by a host of viruses, rickettsias, bacteria, nematodes (Nickle and Welch 1984), and protozoans. Some of these have practical application in the control of insect pests (Roberts and Castillo 1980). Inocula of several types of *Bacillus thuringiensis* already are produced commercially and are very successful in the battle against many caterpillar pests and blackfly larvae (see Control of Insect Pests, chap. 3).

Insects are also afflicted with many fungal infections (Madelin 1966). These are especially well developed in the more moist portions of the Latin American tropics. Species of the genus *Cordyceps* (Ascomycotina: Clavicipitales) are the most common. Dead specimens of many kinds of insects are often found with the fruiting bodies of these fungi projecting from them. In parts of the Peruvian rain forest, they are well known to the natives as *tamshi* and are thought by the naive to be metamorphosing plants (see Giant Solitary Hunting Ant, chap. 12). Such entomophagous fungi are undoubtedly important in the regulation of arthropod populations and help to maintain stability in rain forest ecosystems (Evans 1982).

Insects bear the burden of being the

principal food of a great many types of vertebrate animals. Undoubtedly, the largest single group of insectivores are birds, the majority of which require insects in their diet at some time of their lives. Many eat insects exclusively, such as flycatchers (Sherry 1984), swifts and swallows (Collins 1968, Hespenheide 1975), and woodpeckers. Some show specific foraging adaptations, even selecting upper versus lower leaf surfaces (Greenberg and Gradwohl 1980). Ant birds specialize in eating insects forced to expose themselves while escaping army ant feeding swarms.

There are many insect-feeding terrestrial mammals, primarily opossums, edentates (anteaters and armadillos), and primates. Anteaters consume great numbers of termites and ants in tropical areas and are strong determinants of the latter's distribution and adaptations for survival (Lubin et al. 1977).

Enormous quantities of night-flying insects, mostly moths, beetles, and gnats, are consumed by bats. Approximately 70 percent of living species are insectivores, eating on the average a quarter to a half of their body weight in insects nightly (Hill and Smith 1984: 63).

Reptiles and amphibians are heavy insect predators (Lieberman 1986). Most lizards (Lescure and Fretey 1977) and a few kinds of snakes and small crocodilians (Seijas and Ramos 1980) rely on an insect diet and may be strong selective agents in insect adaptation (Hunt 1983). Legless lizards living in rotting wood appear to prey almost entirely on termites. Amphibians, particularly frogs, are also entomophages. Arboreal species tend to specialize in ants and mites (Formanowicz et al. 1981), while litter-dwelling forms have a more varied insect diet (Toft 1981). Salamanders are significant only from the northern Andes northward. Fossorial caecilians may also take some soil insects in addition to their primary earthworm diet, although the aquatic species may be more insectivorous.

Carnivorous plants are highly adapted for trapping and digesting insect prey (Lloyd 1931). There are several categories represented in Latin America, differing in method of capture and form of the trapping device. Those with pitfalls, the so-called pitcher plants, all belong to the genus *Heliamphora* (Sarraceniaceae) and are found only in the Guiana area and Venezuela. They grow in dense clusters in bogs. The leaves are modified into urn-shaped structures, filled with a secretion that is corrosive to most insects that fall into it. Some aquatic insects have evolved special mechanisms to protect them from this chemical action and populate this generally unsuitable microhabitat (see Tank Plants, above).

The sundews (*Drosera*, Droseraceae) employ a totally different means of entrapment. The plant consists of a basal rosette of leaves, covered with glands raised on elongated tentaclelike stalks, each covered with a sticky, extremely viscid mucilage. Unwary insects are first snared by the glands; then the stalks and the leaves bend inward, completely imprisoning the prey, which is then digested by secretions within. Sundews grow where the soil is poor in nutrients, in swamps, in bogs, and on rotting logs and decomposing vegetation.

Bladderworts (the widespread *Utricularia* and Cuban *Biovularia*, Lentibulariaceae) are mostly rootless, floating or epiphytic plants. The most common species live freely suspended in ponds and slow watercourses, often backwater lakes and sluggish jungle streams. The traps are attached to the plant by a stalk and are borne on the roots and have the form of a small, flattened, hollow, saclike body. The opening or door is set with numerous glandular hairs. Small aquatic insects, such as mosquito larvae, water fleas, and midge larvae, become caught by these hairs and engulfed by the bladder, in which they are finally digested.

Some bromeliads apparently have

evolved the insectivorous habit (Frank and O'Meara 1984).

Competition exists between insects and other organisms. Sometimes this is active, as in the case of hummingbirds driving butterflies away from common nectar sources (Thomas et al. 1986), but more often it is passive, for example, usurpation of nectar sources by Africanized bees leading to loss of food to other honeybee strains and native bees (Roubik 1979).

## References

CANTWELL, G. E., ed. 1974. Insect diseases. Vol. 1. Marcel Dekker, New York.

COLLINS, C. T. 1968. The comparative biology of two species of swifts in Trinidad, West Indies. Fla. St. Mus. Bull. 11: 257–320.

EVANS, H. C. 1982. Entomogenous fungi in tropical forest ecosystems: An appraisal. Ecol. Entomol. 7: 47–60.

FORMANOWICZ, JR., D. R., M. M. STEWART, K. TOWNSEND, F. H. POUGH, AND P. F. BRUSSARD. 1981. Predation by giant crab spiders on the Puerto Rican frog *Eleutherodactylus coqui*. Herpetologica 37: 125–129.

FRANK, J. H., AND G. F. O'MEARA. 1984. The bromeliad *Catopsis berteroniana* traps terrestrial arthropods but harbors *Wyeomyia* larvae (Diptera: Culicidae). Fla. Entomol. 67: 418–424.

GREENBERG, R., AND J. GRADWOHL. 1980. Leaf surface specializations of birds and arthropods in a Panamanian forest. Oecologia 46: 115–124.

HESPENHEIDE, H. A. 1975. Selective predation by two swifts and a swallow in Central America. Ibis 117: 82–99.

HILL, J. E., AND J. D. SMITH. 1984. Bats, a natural history. Univ. Texas Press, Austin.

HUNT, J. H. 1983. Foraging and morphology in ants: The role of vertebrate predators as agents of natural selection. *In* P. Jaisson, Social insects in the tropics. 1st Int. Symp., Int. Union Stud. Soc. Ins. and Soc. Mexicana Entomol. (Cocoyoc 1980) Proc. Pp. 83–104.

LESCURE, J., AND J. FRETEY. 1977. Alimentation du lézard *Anolis marmoratus speciosus* Garman (Iguanidae) en Guyane française. Mus. Nat. Hist. Natur. (Paris) Bull. (Ecol. Gen.) 35: 45–52.

LIEBERMAN, S. S. 1986. Ecology of the leaf litter herpetofauna of a Neotropical rain forest: La Selva, Costa Rica. Acta Zool. Mexicana (n.s.) 15: 1–72.

LLOYD, F. E. 1931. The carnivorous plants. Chronica Botanica, Waltham, Mass.

LUBIN, Y. D., G. G. MONTGOMERY, AND O. P. YOUNG. 1977. Food resources of anteaters (Edentata: Myrmecophagidae). I. A year's census of arboreal nest of ants and termites on Barro Colorado Island, Panama Canal Zone. Biotropica 9: 26–34.

MADELIN, M. F. 1966. Fungal parasites of insects. Ann. Rev. Entomol. 11: 423–448.

NICKLE, W. R., AND H. E. WELCH. 1984. History, development, and importance of insect nematology. *In* W. R. Nickle, ed., Plant and insect nematodes. Marcel Dekker, New York. Pp. 627–653.

ROBERTS, D. W., AND J. M. CASTILLO. 1980. Bibliography on pathogens of medically important arthropods: 1980. World Health Org. Bull. 50 (suppl.): 1–197.

ROUBIK, D. W. 1979. Africanized honeybees, stingless bees and the structure of tropical plant-pollinator communities. Proc. 4th Int. Symp. Pollination, Maryland Agric. Exper. Sta., Spec. Misc. Publ. 2: 403–417.

SEIJAS, A. E., AND S. RAMOS. 1980. Características de la dieta de la baba (*Caiman crocodilus*) durante la estación seca en las sabanas moduladas del Estado Apure, Venezuela. Acta Biol. Venezolana 10: 373–389.

SHERRY, T. W. 1984. Comparative dietary ecology of sympatric, insectivorous Neotropical flycatchers (Tyrannidae). Ecol. Monogr. 54: 313–338.

STEINHAUS, E. A., ed. 1963. Insect pathology. 2 vols. Academic, New York.

THOMAS, C. D., P. M. LACKIE, M. J. BRISCO, AND D. N. HEPPER. 1986. Interactions between hummingbirds and butterflies at a *Hamelia patens* bush. Biotropica 18: 161–165.

TOFT, C. A. 1981. Feeding ecology of Panamanian litter anurans: Patterns in diet and foraging mode. J. Herpetol. 15: 139–144.

## PROTECTION FROM ENEMIES

Insects employ a vast array of direct devices to protect them from their enemies. The most elemental is flight to escape danger. Species with wings use them often for this purpose, some with incredible swiftness and agility, such as skipper butterflies. Others may run to safety, like ground beetles that scurry into soil cracks or burrow among ground litter when threatened. Insects resting on leaves or tree trunks often drop to the ground (weevils), jump or hop (leafhoppers), or snap (click beetles) from their perches to disappear among debris or vegetation on the soil surface. There they remain motionless, playing dead (catalepsy) until the threat passes.

Large size and an armored, often spined, body is another way many insects, especially some of the gigantic horned scarabs (*Dynastes*, *Megasoma*), avoid harm. They are simply too much of a mouthful for insectivorous predators.

A great many insects are biochemically unpalatable, possessing substances in their blood and tissues rendering them bitter or otherwise obnoxious to the palates of predators. The most common compounds in this category are cardenolides, terpenoids, alkaloids, and amines, which in many cases, tint the body fluids yellow (leaf beetles, fireflies). Some of these compounds are produced by the insects' own metabolic processes, but many are sequestered from the plants on which insects feed and are stored in body tissues and blood (Rothschild 1973).

Such chemicals are also ejected from the body as sprays (winged phasmids, millipedes), injected in stings or bites (vespoid wasps), secreted onto the body surface (wax bugs), or exposed in other ways to hurt, frighten, or disgust enemies. The purely physical effects of bites can also discourage, as anyone can attest who has felt the closing jaws of a large long-horned beetle or giant lubber grasshopper.

Highly complex designs of color and form, as well as behavior, provide the insect with a resemblance to some object in its environment which either possesses one of its own direct protective mechanisms or is somehow uninteresting to a potential predator (Bernardi 1985). Such deceptions are extremely well developed among insects, especially in the rich Neotropical entomofauna, and are examples of the well-known phenomenon of mimicry (Pasteur 1982, Rettenmeyer 1970), directed toward the visual sense of vertebrate predators (Robinson 1982). The concept of mimicry historically originated in classical studies on Neotropical butterflies by Henry Walter Bates (1862) (Stearn 1981), Alfred Russell Wallace (1870), and Fritz Müller (1879) and is still studied actively in this group of insects (Williamson and Nelson 1972).

All mimetic systems have three components: a model, or object having some direct protective capability; a mimic, the species obtaining protection by resembling the model; and a dupe, the operator or enemy that is fooled by the resemblance (Turner 1977). Gaudy, noticeable colors and patterns often provide ways of enhancing the efficacy of these direct protective devices and of otherwise deceiving or confusing enemies (protective coloration). Bright color and/or conspicuous markings can advertise unpalatability, reminding a predator of a past unpleasant experience and keeping it from advancing toward or injuring the insect even with an exploratory bite (Rothschild 1973). The bright reds and yellows found on tiger moths and lubber grasshoppers are for this purpose. Designs may be startling, such as the bright bars or dazzlingly bright color fields on the upper wing surfaces of many butterflies (e.g., *Morpho*). These confuse the senses of a pursuer, which sees conspicuous color in contrast to inconspicuous and fails to recognize or to be able to follow the movements of the insect.

Two distinct types of mimicry are recognized, one in which the model is some innocuous, inanimate object like a stone, a leaf, a twig, or wood. This is crypsis and is exemplified by the walkingstick's similarity to a piece of wood or the katydid's imitation of a leaf.

When the model is an animal actively avoided by the predator as noxious, unpalatable, or dangerous, or even just difficult to catch (Hespenheide 1973), true mimicry is in effect, a large number of variations of which are possible and are practiced by insects in nature (Vane-Wright 1976). Numerous similarly appearing species exhibit a common defense signal (pattern) that advertises their universal unacceptability. This is the classic situation referred to as Müllerian mimicry. More simply, one or more edible mimics may simulate another organism in some way repellent to the dupe, so escaping attention, an arrangement known as Batesian mimicry. Rarely, the dupe may itself also be the model (Hogue 1984: 149).

Mimicry is an important area of study in the Neotropics where spectacular examples abound (e.g., Chai 1988, Poole 1970, Young 1971). The best-known Müllerian mimicry series (pl. 3f) are those involving common patterns among various lepidopteran families (see Butterflies, chap. 10). In some instances, it is not yet clear whether a mimetic group is Müllerian or Batesian, as in those Diptera (Blephariceridae, Bibionidae, Tipulidae) and Lepidoptera (Zygaenidae, Pyralidae, Ctenuchinae) with a black body and legs, darkened wing membrane, and golden yellow scutum (Hogue 1981). Research includes experiments to measure the actual survival value of presumed mimicry systems (Brower et al. 1963, 1967).

Mimicry may have aggressive as well as purely defensive purposes. Here the mimic simulates an organism with which the dupe normally interacts. The model takes advantage of the dupe to prey on or parasitize it. Examples are female *Photuris* fireflies that lure males of different species to their death by sending signal flashes like those of their prey species females (see fireflies, chap. 9).

The type of mimicry employed by a species may vary according to the life stage involved (developmental mimicry). Thus, the nymphs of certain mantids may resemble stinging ants, while the adults may practice simple cryptic resemblance of leaves or sticks.

It is important to recognize also that mimetic similarities extend to behavior and structure as well as color and pattern. The attine jumping spiders (Salticidae, Attinae) would be only poor ant simulators by their dark colors alone, if their attitudes (forelegs held forward like antennae) and morphology (narrowed base of abdomen) did not also follow their model's physical appearance and comportment. A single species usually relies on a variety of protective abilities, not only one (Kettlewell 1959).

## References

BATES, H. W. 1862. Contributions to an insect fauna of the Amazon valley, Lepidoptera: Heliconidae. Linnean Soc. London Trans. 23: 495–566, Pl. LV–LVI.

BERNARDI, G., ed. 1985. Camouflage et mimetisme. Soc. Entomol. France Bull. 90: 1004–1103.

BROWER, L. P., J. V. Z. BROWER, AND C. T. COLLINS. 1963. Experimental studies of mimicry, 7. Relative palatability and Müllerian mimicry among Neotropical butterflies of the subfamily Heliconiinae. Zoologica 48: 65–84.

BROWER, L. P., L. M. COOK, AND H. J. CROZE. 1967. Predator responses to artificial Batesian mimics released in a Neotropical environment. Evolution 21: 11–23.

CHAI, P. 1988. Wing coloration of free-flying Neotropical butterflies as a signal learned by a specialized avian predator. Biotropica 20: 20–30.

HESPENHEIDE, H. A. 1973. A novel mimicry complex: Beetles and flies. J. Entomol. 48: 49–56.

HOGUE, C. L. 1981. Blephariceridae. *In* S. H. Hurlbert, G. Rodríguez, and N. Dias de Santos, eds., Aquatic biota of tropical South America. Pt. 1. Arthropoda. San Diego State Univ., San Diego. Pp. 285–287.

HOGUE, C. L. 1984. Observations on the plant hosts and possible mimicry models of "lantern bugs" (*Fulgora* spp.) (Homoptera: Fulgoridae). Rev. Biol. Trop. 32: 145–150.

KETTLEWELL, H. B. D. 1959. Brazilian insect adaptations. Endeavour 18: 200–210.

MÜLLER, F. 1879. *Ituna* and *Thyridia;* a remarkable case of mimicry in butterflies. Translated by R. Meldola. Entomol. Soc. London Proc. 1879: 20–29.

Pasteur, G. 1982. A classificatory review of mimicry systems. Ann. Rev. Ecol. Syst. 13: 169–199.

Poole, R. W. 1970. Habitat preferences of some species of a Müllerian-mimicry complex in northern Venezuela, and their effects on evolution of mimic wing pattern. New York Entomol. Soc. J. 78: 121–129.

Rettenmeyer, C. W. 1970. Insect mimicry. Ann. Rev. Entomol. 15: 43–74.

Robinson, M. H. 1982. Defensa contra depredadores que cazan por medios visuales. In G. A. de Alba and R. W. Rubinoff, eds., Evolución en los trópicos. Smithsonian Trop. Res. Inst., Panama. Pp. 57–76.

Rothschild, M. 1973. Secondary plant substances and warning coloration in insects. Royal Entomol. Soc. London Symp. 6: 59–83.

Stearn, W. T. 1981. Henry Walter Bates (1825–1892), discoverer of Batesian mimicry. Biol. J. Linnean Soc. 6: 5–7.

Turner, J. R. G. 1977. Mimicry: A study in behaviour, genetics, ecology and biochemistry. Ann. Rev. Ecol. Syst. 13: 169–199.

Vane-Wright, R. I. 1976. A unified classification of mimetic resemblances. Biol. J. Linnean Soc. 8:25–56.

Wallace, A. R. 1870. Mimicry, and other protective resemblances among animals. In A. R. Wallace, Contributions to the theory of natural selection: A series of essays. Macmillan, London. Pp. 45–129.

Williamson, G. B., and C. E. Nelson. 1972. Fitness set analysis of mimetic adaptive strategies. Amer. Nat. 106: 525–537.

Young, A. M. 1971. Mimetic associations in natural populations of tropical papilionid butterflies (Lepidoptera: Papilionidae). New York Entomol. Soc. J. 79: 210–224.

## SOCIAL INSECTS

Evolution has carried several groups of insects to a remarkably high level of group organization. These are the social insects, the termites, ants, and certain categories of wasps and bees (Jaisson 1983, Richards 1953, Wilson 1971). Each represents an independent lineage of social development from different, nonsocial ancestors.

True sociality (eusociality) is characterized by several essential elements: (1) members of the colony are all siblings directly related to the founding parents, not individuals of diverse parentage that have come to live together secondarily; (2) members cooperate in the care of young and defense of the colony (Hermann 1984); (3) a division of labor exists between members, often correlated with differences in body form; (4) generations overlap so that the offspring aid the parents in the work of the colony; and (5) colonies usually make and maintain some sort of nest structure. Exchange of social messenger chemicals also often occurs (trophallaxis), which acts to bind and control group behavior. Castes are determined in several ways, nutrition and pheromones being primarily important (Lüscher 1977, Oster and Wilson 1979, Watson et al. 1985). It is interesting that no group of insects with sucking mouthparts has evolved sociality; mandibles seem to be necessary in manipulating nest material and nursing duties. So-called altruism (kin selection) is present also among social insects, that is, no reproduction in favor of another individual, and may also involve self-sacrifice in defense of the colony.

The path to the origin of sociality seems to be in increasing interaction of adults with their offspring, coupled with increased longevity of the parents (Andersson 1984) and acquisition of group defensive specializations such as the sting (Starr 1985). Several logical stages in the progression from completely solitary lives to eusocial insects may be recognized, although they have not necessarily occurred in the history of all groups (Alexander 1974, Haro 1982, Lin and Michener 1972, Michener 1958, Wilson 1975).

Simple parental care is the first step. The mother remains with and protects the young, sometimes for a considerable time, keeping them from harm and sometimes providing occasional food and shelter. This is found in such disparate groups as the earwigs, treehoppers, and cockroaches. When the mother constructs nest chambers in which she places provisions of food especially for the young, called mass provisioning, a more complex level of interaction has been reached, such as that practiced by thread-waisted wasps (*Ammophila*). This advances to a still higher level of organization when provisioning is progressive, that is, fresh food made available as development continues, the habit of sand wasps (*Bembix*). Other subsocial or semisocial stages are demonstrated by communal nesting of female bees or wasps, even by groups of spiders occupying a single web, although truly social spiders have not been discovered.

Because of their separate origins, the various groups of truly social insects display some fundamental differences in biology. Termites, because they retain gradual metamorphosis, require less care of the young, which are mobile nymphs quite capable of feeding themselves. They do not have a dormant pupal period requiring special protection and care, and they have more flexible control over the redirection of development to determine castes than social insects with complete metamorphosis. The reproductive male remains with the "queen" mother, continuously inseminating her during her long life, in contrast to his usual early demise after one mating in other social species. Alate reproductives leave the nest, pair off and mate, and found a new colony together. Among the offspring, worker and soldier castes may develop. The former have normal chewing mandibles.

The ants are another major group, like termites, to have evolved social behavior. New colonies are normally founded much like termites, large swarms of flying males attracting females, but the female begins the brood alone, her mate having died soon after mating with her. Castes are similar also to those of termites, the workers having unmodified mandibular tools and soldiers having outsized heads and jaws used as weapons for colony defense.

The eusocial wasps are the hornets and paper wasps, all in the family Vespidae. One social genus (*Microstigmus*) of Sphecidae is also known. In social wasps and bees, the worker caste is little differentiated anatomically from the queen, usually only smaller. Behaviorally, however, they are quite distinct, being sterile and responsible primarily for the gathering of food, nest construction, and nursing the larvae. New colonies are formed by splitting of old colonies, some nonsterile female offspring leaving the nest and mating with males of other colonies. As with ants, the males die, and the females build a new home and start a family whose members quickly take on tasks, releasing the mother for reproductive activities only. She may live several years, becoming dormant during cold seasons in temperate areas or ovipositing continuously in tropical regions.

There are several groups of eusocial bees—honeybees, stingless bees, bumblebees, and certain types of carpenter bees and sweat bees. They are much like social wasps in their biology but are never wholly carnivorous. Their food is normally of plant origin, consisting of nectar and pollen. They are, therefore, much more intimately associated with flowering plants than the wasps, which mainly take animal prey.

The study of social insects has attracted a great many entomologists and behaviorists. The popularity of the subject partly derives from the similarities of insect societies to the human condition, and many scientists believe that knowledge gained from the workings of the insect may have direct bearing on some aspects of our own lives (Wilson 1975). This is particularly true in the area of communication and neural integration. The brains of the social groups are the largest and most complex of insects and can serve as simplified models of integrative systems involving primitive learning as well as instinctive activity.

## References

ALEXANDER, R. D. 1974. The evolution of social behavior. Ann. Rev. Ecol. Syst. 5: 325–383.

ANDERSSON, H. 1984. The evolution of eusociality. Ann. Rev. Ecol. Sys. 15: 165–189.

HARO, A. DE. 1982. Algunas consideraciones sobre el origen y evolución de las sociedades de insectos. In P. Jaisson, Social insects in the tropics. 1st Int. Symp. (Cocoyoc 1980), Int. Union Stud. Soc. Ins. and Soc. Mexicana Entomol. Proc. 1: 65–71.

HERMANN, H. R., ed. 1984. Defensive mechanisms in social insects. Praeger, New York.

JAISSON, P. 1983. Social insects in the tropics. 1st Int. Symp. (Cocoyoc 1980), Int. Union Stud. Soc. Ins. and Soc. Mexicana Entomol. Proc. Vols. 1–2.

LIN, N., AND C. D. MICHENER. 1972. Evolution of sociality in insects. Quart. Rev. Biol. 46: 131–159.

LÜSCHER, M. 1977. Phase and caste determination in insects. Pergamon, Oxford.

MICHENER, C. D. 1958. The evolution of social behavior in bees. 10th Int. Cong. Entomol. (Montreal) Proc. 2: 441–447.

OSTER, G. F., AND E. O. WILSON. 1979. Caste and ecology in social insects. Princeton Univ. Press, Princeton.

RICHARDS, O. W. 1953. The social insects. Macdonald, London.

STARR, C. K. 1985. Enabling mechanisms in the origin of sociality in the Hymenoptera—The sting's the thing. Entomol. Soc. Amer. Ann. 78: 836–840.

WATSON, J. A. L., B. M. OKOT-KOTBER, AND C. NOIROT, eds. 1985. Caste differentiation in social insects. Pergamon, Oxford.

WILSON, E. O. 1971. The insect societies. Harvard Univ. Press, Cambridge.

WILSON, E. O. 1975. Sociobiology. Harvard Univ. Press, Cambridge.

## SYMBIOSIS

Two or more kinds of organisms living in close association to the benefit of one (commensalism) or all partners (mutualism) (Boucher 1982) are said to exhibit symbiosis. ("Symbiosis" is used in the broad sense here to include all forms of intimate associations, not just those of mutual benefit for which the term is used in a narrow sense.) The Neotropics provide countless, fascinating examples of this phenomenon involving insects.

Some of the best-known examples occur between insects and higher plants (Bernays 1989). Most often, the former are ants, and major examples are discussed in chapter 12, on Hymenoptera (see ants and plants). Many symbiotic relationships exist between insects and lower plants, especially fungi (Lichtwardt 1986, Wheeler and Blackwell 1984).

Nest sharing, or inquilinism, is a type of symbiosis wherein both partners are insects (Kistner 1982). Examples are certain silverfish, millipedes, beetles, mites, and cockroaches, called myrmecophiles, that share a common dwelling with ants. Other social insects are likewise visited by termitophiles, melittophiles, and sphecophiles, depending on whether they are termites, bees, or wasps, respectively. There are many variations in the degree of closeness in the association and its specific nature (Kistner 1979). In the Neotropics, army ants and leaf cutter ants are frequent hosts (see chap. 12). The visitors live on the debris, scraps of food, and even corpses of ants. This is mutualistic coexistence, the ants profiting by the nest debris removal, the guests by a reliable food source.

The invasive species finds its way into the host's colony and is accepted into its society by a variety of physical and chemical deceptions (Hölldobler 1971). Staphylinid beetles may actually resemble and behave like their hosts (Wasmannian mimicry) (Akre and Rettenmeyer 1966). The trail-marking substances secreted by the ants may serve as attractants to inquilines (Moser 1964). The latter may also produce "tranquilizing" chemicals that pacify the ants or mimic the food offerings of worker ants.

The strategies of inquilines vary; some prey directly on the host, others on the immatures of the latter or on accumulated nest refuse. Others are apparently totally neutral, the purposes of their invasions remaining unknown.

A related phenomenon is exhibited by ants and other insects that have a particular fondness for the honeydew secretions produced by several groups of sap-sucking insects. Honeydew is a sugar-rich solution excreted from the digestive tract (aphids) or from special integumentary glands (scale insects, metalmark and hairstreak butterfly larvae). The solution is a food in return for which the honeydew producer receives protective, dispersal, cleaning, and other tending favors from the feeding insects. (See wax bugs, aphids, chap. 8; metalmarks, hairstreaks, chap. 10.)

Phoresy is yet another symbiotic relationship between insects. The term is used for the temporary attachment of a much smaller, relatively sedentary form to a much larger, more vagile form. The former receives transportation and is able to disperse far more widely than it could on its own. Many of the mites often found living on beetles are practicing this habit. This is also the case with the pseudoscorpions so common under the elytra of large long-horned beetles, especially the harlequin beetle (see harlequin beetle, chap. 9). Phoresy is often a means for parasites and predators to find access to their prey (Clausen 1976).

The above examples are designated "ectosymbiosis" (Hartzill 1967) in contrast to "endosymbiosis" (Koch 1967). The latter refers to the habitation of many microorganisms in the gut, other body cavities, and tissues of insects (Boush and Coppel 1974). The best-known cases are the flagellate protozoans of primitive termites and trichomycete fungi associated with numerous insects (Lichtwardt 1986).

## References

AKRE, R. D., AND C. W. RETTENMEYER. 1966. Behavior of Staphylinidae associated with army ants (Formicidae: Ecitonini). Kans. Entomol. Soc. J. 39: 745–796.

BERNAYS, E. A., ed. 1989. Insect-plant interactions. Vol. 1. CRC, Boca Raton, Fla.

BOUCHER, D. H. 1982. The ecology of mutualism. Ann. Rev. Ecol. Syst. 13: 315–347.

BOUSH, G. M., AND H. C. COPPEL. 1974. Symbiology: Mutualism between arthropods and microorganisms. In G. E Cantwell, ed., Insect diseases. 1: 301–326. Marcel Dekker, New York.

CLAUSEN, C. P. 1976. Phoresy among entomophagous insects. Ann. Rev. Entomol. 21: 343–368.

HARTZILL, A. 1967. Insect ectosymbiosis. In S. M. Henry, ed., Symbiosis. 2: 107–140. Academic, New York.

HÖLLDOBLER, B. 1971. Communication between ants and their guests. Sci. Amer. 224(3): 86–93.

KISTNER, D. H. 1979. Social and evolutionary significance of social insect symbionts. In H. R. Hermann, ed., Social insects. 1:339–413. Academic, New York.

KISTNER, D. H. 1982. The social insects bestiary. In H. R. Hermann, ed., Social insects. 3: 1–24. Academic, New York.

KOCH, A. 1967. Insects and their endosymbionts. In S. M. Henry, ed., Symbiosis. 2: 1–10. Academic, New York.

LICHTWARDT, R. W. 1986. The Trichomycetes (fungal associates of arthropoda). Springer, Berlin.

MOSER, J. C. 1964. Inquiline roach responds to trail-making substance of leaf-cutting ants. Science 143: 1048–1049.

WHEELER, Q., AND M. BLACKWELL, eds. 1984. Fungus-insect relationships: Perspectives in ecology and evolution. Columbia Univ. Press, New York.

## POLLINATION

The majority of flowering plants are dependent on insects as vehicles for the transfer of their pollen from anther to stigma (Meeuse and Morris 1984, Faegri and Van der Pijl 1979, Real 1983, Richards 1978). The flower itself, by color and odor attractants (Yeo 1973) and nectar and pollen rewards, is the prime mechanism for involving insects in pollination (Bacior 1974). Plants also may entice insects to visit and pollinate them by providing wax and resins or false rewards such as the promise of sex or food, the latter usually in the form of carrion or feces.

Many insects, in turn, require plants, their leaves, stems, wood, and so on, or floral products to sustain their lives and have a stake in pollination equal to that of the plant. Fruit setting in figs (*Ficus*), for example, is entirely dependent on minute wasps whose whole lives are likewise dependent on the tree (see fig wasps, chap. 12). Although grasses are generally regarded as wind-pollinated plants, in the still atmosphere of tropical forests, insects may still be necessary to effect pollen transfer (Soderstrom and Calderón 1971).

Pollination biology of insects and plants in the tropical portions of Latin America is an active field of study (Heinrich and Raven 1972, Jones and Little 1983). Most work is concerned with the mechanisms of pollen acquisition and deposition (see examples below); few deal with the important topics of transport during foraging activity and with interflower relationships (Frankie and Baker 1974). Pollination of trees in isolated remnants of tropical forests has also been studied (Raw 1989).

Pollinating insects are of two general types: (1) large, powerful, long-distance fliers, including sphinx moths (Cruden et al. 1976, Haber and Frankie 1989, Linhart and Mendenhall 1977), large bees (Sazima and Sazima 1989), and butterflies, and (2) comparatively weak-flying, short-distance fliers, such as small flower and bee flies (Syrphidae and Bombyliidae), moth flies (Psychodidae), punkies (Ceratopogonidae), chalcidoid wasps, minute bees, and some beetles (Meloidae, Oedemeridae, Scarabaeidae).

In addition to flight, many have elongate, sucking mouthparts and possess specialized anatomical structures for carrying or storing pollen and nectar. Most also have good color vision for locating day-blooming flowers and excellent olfactory senses to find plants with nocturnal blossoms.

Flowers assume a variety of shapes and special adaptations to enhance or ensure visits and their pollination by coadapted insects. Many, for example, have spots or color marking near the nectaries ("nectar guides") to direct the insect's attentions toward them. These may be visible only to pollinators with ultraviolet vision, as many insects have. Other flowers have very long corollas (such as *Posoqueria, Calliandra*) with deep nectaries that only the long tongues of sphinx moths can reach. Some plants produce flowers without nectaries but which so resemble others that do give nectar reward that they are visited by pollinators nevertheless, a form of floral mimicry (Haber 1984).

Some woody tropical forest plants practice "mass flowering." Individuals bloom profusely and synchronously during a short period (often during the dry season) and attract large numbers and diverse kinds of insect pollinators, especially bees. These insects tend to range freely between individuals of such plants, ensuring cross-pollination. Gentry (1978) suggested that insects may be induced to move between plants by insectivorous birds chasing them. Other trees are slightly asynchronous bloomers, having different individuals with short overlapping flowering periods. This is thought to be a strategy to foster pollinator movement between trees (Perry 1980).

Many orchid flowers exhibit floral adaptations for pollination to an extreme degree (Darwin 1890, Van der Pijl and Dodson 1966, Williams 1982). The lip petal of the Colombian orchid *Masdevallia bella* has a gill-like, fleshy appearance exactly like the underside of a mushroom and is pollinated by fungus gnats. The latter ostensibly visit the flower as they would in search of an oviposition site in their normal mushroom hosts. Several orchid species attract male euglossine bees, providing them with oily, fragrant substances used in the bee's courtship, in return for securing their services as pollinators (see orchid bees, chap. 12).

*Coryanthes,* or "bucket orchid" flowers,

all found in tropical Central and South America, are complexly formed to ensure pollination by bees. The lower part is a cuplike container that holds a sweet fluid accumulated from fluid-producing glands above on the tip of the flower's stalk. Male euglossine bees visit the flowers to collect substances from the petals and in doing so often lose their footing and fall into the wet, sticky pool below. Unable to use their wings and fly out, they are forced to escape the bucket by crawling through an opening in its side, in one wall of which are the pollen packets. Because of the tight squeeze, the bee brushes past the packets, which detach and adhere to it. The next flower visited receives the packets, which are scraped off onto the stigma and pollinate it.

Scents emitted by flowers are not always sweet. Many members of the orchid subtribe Pleurothallideae (Orchidaceae) and the genera *Dracontium* (Araceae), *Sterculia,* and *Herrania* (Sterculiaceae) attract carrion flies with putrescent odors.

A curious collection of plants utilize "trap flowers" to ensure pollination. The principle is to detain the pollinator for a short time, to ensure that it will pick up and deposit pollen. It is released hours or days later to repeat the process in the flowers of other individuals of the same species.

The South American bladder-flower plant (*Araujia sericofera*) is an example. Moths may visit their first bladder flower without being detained, merely picking up pollen packets (pollinia) on their tongues and carrying them off. When the moth feeds from a second flower, these pollinia become wedged into a slitlike structure on the petals where they come in contact with the receptive female organs. Larger, stronger moths, like sphingids, can pull away from this snare, but smaller, weaker species remain caught permanently. Phenylacetaldehyde emitted from the flowers is the active chemical agent that these plants

use initially to attract the moths (Cantelo and Jacobson 1979).

Another trap flower is that of the aristolochia vine (*Aristolochia* spp.). The complex chambered blossom attracts insects with its bright outer lip and carrion-mimicking odors. Visiting insects push their way down the dark tubular part into the base of the flower, or "prison," where the receptive young stigma is located. Their advance is prompted by a lighter colored or semitransparent area in the chamber, and their exit is prevented by backward-pointing hairs in the narrowed funnel region. By this mechanism they are trapped for some days until the stamens mature and shed their pollen, which dusts the insects. Then the hairs wither and the insects can leave, to enter another young flower and go through the trial again but pollinating the young stigma with the grains they have picked up on their previous captive experience.

The flower buds of the giant Amazon water lily (*Victoria amazonica*) open at night and emit a strong fruity odor that attracts scarab beetles of the genus *Cyclocephala* (Lovejoy 1978; pl. 3h). They close by the next morning, trapping the beetles in a deep chamber inside for a day where they feed on sterile inner anthers, depositing pollen on the style. They are released in the afternoon when the flowers open again, picking up pollen from the outer ripe anthers. *Cyclocephala* also pollinate *Cyclanthus* (Cyclanthaceae) in a similar manner: they spend twenty-four hours in the flower spathe but are not trapped (Beach 1982). *Cyclocephala* are known to be pollinators of aroid's, such as *Dieffenbachia* (H. Young 1986), and beetles generally are important in reproduction of a variety of Neotropical plant families. Flowering in many beetle-pollinated plants is accompanied by an increase in metabolic rate that leads to the production of heat. This causes the volatilization of various odors from the flower that attract the beetles (Meeuse 1975).

Insect pollination is very important in agriculture (Free 1970). To set fruit adequately for a profitable crop, many cultivated plants require pollination by insects, whose presence in plantations is encouraged by growers. Honeybees have long been recognized as valuable in this way, and hives are purposely placed in fields and orchards to increase seed and fruit set (Martin and McGregor 1973).

The cacao tree is pollinated chiefly by punkie flies of the genus *Forcipomyia* (A. Young 1986; see punkies, chap. 11) and adult gall midges (A. Young 1985). Because many of these are bromeliad tank breeders, the proximity of these plants is essential to successful cacao fruiting (Privat 1979). Although common on the plants, ants and Homoptera and other insects are probably not important (Winder 1978).

The introduced oil palm (*Elaeis guineensis*) has been found to depend on certain fruit beetles (Nitidulidae—*Mystrops, Haptoncus*) for pollination. Formerly thought to be pestiferous, they are now welcomed in areas where this valuable tree is cultivated (Genty et al. 1986). Nitidulid and scarab beetles, weevils, and other insect pollinators are equally important to native palms (Barfod et al. 1987, Beach 1984), which are assuming use as oil producers in some areas.

The Brazil nut tree (*Bertholletia excelsa*) apparently depends on euglossine bees for pollination. Only large species are capable of uncurling the floral androecium (protective hood around the anthers) (Nelson et al. 1985).

## References

ARMBRUSTER, W. S., AND G. L. WEBSTER. 1979. Pollination of two species of *Dalechampia* (Euphorbiaceae) in Mexico by euglossine bees. Biotropica 11: 278–283.

BACIOR, L. W. 1974. Behavioral aspects of coadaptations between flowers and insect pollinators. Missouri Bot. Garden Ann. 61: 760–769.

BARFOD, A., A. HENDERSON, AND H. BALSLEV.

1987. A note on the pollination of *Phytelephas microcarpa* (Palmae). Biotropica 19: 191–192.

BEACH, J. H. 1982. Beetle pollination of *Cyclanthus bipartitus* (Cyclanthaceae). Amer. J. Bot. 69: 1074–1081.

BEACH, J. H. 1984. The reproductive biology of the peach of "Pejibayé" palm (*Bactris gasipaes*) and a wild cogener (*B. porschiana*) in the Atlantic lowlands of Costa Rica. Principes 28: 107–119.

CANTELO, W. W., AND M. JACOBSON. 1979. Phenylacetaldehyde attracts moths to bladder flower and to blacklight traps. Envir. Entomol. 8: 444–447.

CRUDEN, R. W., S. KINSMAN, R. E. STOCKHOUSE II, AND Y. B. LINHART. 1976. Pollination, fecundity, and the distribution of moth-flowered plants. Biotropica 8: 204–210.

DARWIN, C. 1890. The various contrivances by which orchids are fertilized by insects. 2d ed. John Murray, London.

FAEGRI, K., AND L. VAN DER PIJL. 1979. The principles of pollination ecology. 3d ed. Pergamon, Elmsford, N.Y.

FRANKIE, G. W., AND H. G. BAKER. 1974. The importance of pollinator behavior in the reproductive biology of tropical trees. Inst. Biol. Univ. Nat. Autón. México, Ser. Bot., Ann. 45(1): 1–10.

FREE, J. B. 1970. Insect pollination of crop plants. Academic, London.

GENTRY, A. H. 1978. Anti-pollinators for mass-flowering plants? Biotropica 10: 668–669.

GENTY, P., A. GARZÓN, AND F. LUCCHINI. 1986. Polinización entomófila de la palma Africana en América tropical. Oléagineaux 41: 99–112.

HABER, W. A. 1984. Pollination by deceit in a mass-flowering tropical tree *Plumeria rubra* L. (Apocynaceae). Biotropica 16: 269–275.

HABER, W. A., AND G. W. FRANKIE. 1989. A tropical hawkmoth community: Costa Rican dry forest Sphingidae. Biotropica 21: 155–172.

HEINRICH, B., AND P. H. RAVEN. 1972. Energetics and pollination ecology. Science 176: 597–602.

JONES, C. E., AND R. T. LITTLE, eds. 1983. Handbook of experimental pollination biology. Van Nostrand Reinhold, New York.

LINHART, Y. B., AND J. A. MENDENHALL. 1977. Pollen dispersal by hawk moths in a *Lindenia rivalis* Benth. population in Belize. Biotropica 9: 143.

LOVEJOY, T. E. 1978. Royal water lilies: Truly Amazonian. Smithsonian 7: 77–82.

MARTIN, E. C., AND S. E. McGREGOR. 1973.

Changing trends in insect pollination of commercial crops. Ann. Rev. Entomol. 18: 207–226.

MEEUSE, B. J. D. 1975. Thermogenic respiration in aroids. Ann. Rev. Plant. Physiol. 26: 117–126.

MEEUSE, B., AND S. MORRIS. 1984. The sex life of flowers. Facts on File, New York.

NELSON, B. W., M. L. ABSY, E. M. BARBOSA, AND G. T. PRANCE. 1985. Observations on flower visitors to *Bertholletia excelsa* H.B.K. and *Couratari tenuicarpa* A.C. Sm. (Lecythidaceae). Acta Amazonica Suppl. 15: 225–234.

PERRY, D. R. 1980. The pollination ecology and blooming strategy of a Neotropical emergent tree *Dipteryx panamensis*. Biotropica 12: 307–313.

PRIVAT, F. 1979. Les bromeliacees, lieu de developpment de quelques insectes pollinisateurs de fleurs de cacao. Brenesia 16: 197–212.

RAW, A. 1989. The dispersal of euglossine bees between isolated patches of eastern Brazilian wet forest. Rev. Brasil. Entomol. 33: 103–107.

REAL, L., ed. 1983. Pollination biology. Academic, New York.

RICHARDS, A. J., ed. 1978. The pollination of flowers by insects. Linnean Soc. Symp. 6: 1–214.

SAZINA, I., AND M. SAZIMA. 1989. Mamangavas e irapuás (Hymenoptera, Apoidea): Visitas, interações e conseqüêcias para polinização do maracujá (Passifloraceae). Rev. Brasil. Entomol. 33: 109–118.

SODERSTROM, T. R., AND C. E. CALDERÓN. 1971. Insect pollination in tropical rain forest grasses. Biotropica 3: 1–16.

VAN DER PIJL, L., AND C. H. DODSON. 1966. Orchid flowers: Their pollination and evolution. Univ. Miami, Coral Gables.

WILLIAMS, N. G. 1982. The biology of orchids and euglossine bees. *In* J. Arditti, ed., Orchid biology. II: 119–171. Cornell Univ. Press, Ithaca.

WINDER, J. A. 1978. The role of non-dipterous insects in the pollination of cocoa in Brazil. Bull. Entomol. Res. 68: 559–574.

YEO, P. F. 1973. Floral allurements for pollinating insects. Royal Entomol. Soc. London Symp. 6: 151–157.

YOUNG, A. M. 1985. Studies of cecidomyiid midges (Diptera: Cecidomyiidae) as cocoa pollinators (*Theobroma cacao* L.) in Central America. Entomol. Soc. Wash. Proc. 87: 49–79.

YOUNG, A. M. 1986. Habitat differences in cocoa tree flowering, fruit set, and pollinator availability in Costa Rica. J. Trop. Ecol. 2: 163–186.

YOUNG, H. J. 1986. Beetle pollination of *Dieffenbachia longispatha* (Araceae). Amer. J. Bot. 73: 931–944.

# INSECT CONSERVATION

The unnatural despoliation of wildlife in Latin America, as in most parts of the world today, is proceeding at an alarmingly rapid pace. Habitat destruction is the primary cause of insect extinction through deforestation, erosion, and land alteration for agriculture, animal husbandry, hydroelectric energy and raw material production, mining, and urbanization. Insect life suffers also from misplaced insecticides, importation of alien organisms, pollution, fires, and direct commercial exploitation (pets and decorative uses) (Faria 1940), but these are far less significant than loss of habitat. There is some effect from scientific and hobby collectors, but this is also of little harm and may actually have beneficial results through advances in the knowledge gained through these activities.

Unfortunately, very little direct effort is being expended to counter these negative trends. Economic development everywhere takes precedence over protection of nature, especially insects, a life form against which there is almost universal enmity. The idea of purposeful concern over the fate of insects in the environment is a new and little appreciated concept, in spite of the fact that insects are a form of wildlife ecologically essential to mankind in so many ways, including their direct value as pollinators, reducers of organic matter, maintenance of environmental cleanliness, and food for other "desirable" animals (Pyle et al. 1981).

The signing by most Latin American countries of the Convention on International Trade in Endangered Species of Fauna and Flora (CITES) establishes controls on and monitors importation and export of certain species (Fuller and Smith 1984); presently, the only specific terrestrial arthropod from these countries which is listed is the Mexican red-kneed tarantula (Hemley 1986). Most countries have laws regulating the collecting of insects for scientific and commercial purposes, but these have virtually no effect on ameliorating the greater causes of harm. Only Brazil (Otero and Brown 1986) and Mexico have enacted legislation prohibiting the taking of specific insects: a swallowtail butterfly (*Parides ascanius*) is given special protection in Brazil, and the monarch butterfly is protected in its very localized overwintering sites in the mountains of central Mexico. The *Invertebrate Red Data Book* (Wells et al. 1983) of the International Union for the Conservation of Nature and Natural Resources lists many Neotropical insect species as vunerable, rare, threatened, and endangered, but no legal regulations yet control their removal from the environment.

One solution to overexploitation of some species by commercial collectors may be insect "farming" or "ranching," a technique found successful in Papua New Guinea with some of the large bird-wing butterflies (*Ornithoptera*). A few such enterprises are now in operation in Latin America (e.g., Brazil; Kesselring pers. comm.). The idea has been suggested generally (Crane and Fleming 1953, National Research Council 1983).

Indirectly, considerable progress has been accomplished by those countries that have set aside habitat as national parks and nature preserves. Many of the latter have been made possible by private individuals, consortia, and agencies such as the Xerces Society, The Nature Conservancy, and World Wildlife Fund-U.S. International programs also foster conservation measures that help in the preservation and conservation of insects through the U.S. Agency for International Devel-

opment (USAID), the United Nations Educational and Social Cooperation Organization (UNESCO), and the U.S. Peace Corps.

A critical consideration in the establishment of preserves is the "minimal size factor." Little is known regarding the amount of habitat that needs to be protected in order for the insect fauna to continue to thrive. A major baseline study directed at this question, and including some insects, for Amazonian forest is the cooperative World Wildlife Fund-U.S. and Brazilian National Research Institute project north of Manaus on various-sized, measured plots of isolated forest ("Minimum Critical Size of Ecosystems Project," Lovejoy 1980).

What the future holds for the insects, arachnids, and other similar terrestrial arthropods of Latin America is unforeseeable, but it seems realistic to look forward with pessimism. But for a very few exceptions (Lamas 1974), the disappearance of species goes almost entirely undocumented. In the forests alone, species are probably becoming extinct before they are discovered. This reduction of insect diversity is certainly occurring, however, and will accelerate at a rapid pace because of deforestation. The trend is exacerbated by the low population densities of invertebrates in this habitat type (Elton 1975). Unquestionably, a greater appreciation of the benefits of protecting these life forms is needed, and more action to prevent wholesale ruination of their living space is called for. Laws alone are not the answer. Public awareness and understanding of the issue is critical to all progress toward ensuring the survival of this part of the Earth's heritage (Lamas 1978).

## References

CRANE, J., AND H. FLEMING. 1953. Construction and operation of butterfly insectaries in the tropics. Zoologica 38: 161–172.
ELTON, C. S. 1975. Conservation and the low

population density of invertebrates inside Neotropical rain forest. Biol. Conserv. 7: 3–15.

FARIA, A. 1940. Caça e commercio de borboletas e outros insetos ornamentaes. Rev. Entomol. 11: 607–608.

FULLER, K. S., AND G. SMITH. 1984. Latin American wildlife trade laws. World Wildlife Fund, Washington, D.C.

HEMLEY, G. 1986. Spotlight on the red-kneed tarantula trade. Traffic (U.S.A.) 6(4): 16–17.

LAMAS, G. 1974. Supuesta extinción de una mariposa en Lima, Perú (Lepidoptera, Rhopalocera). Rev. Peruana Entomol. 17: 119–120.

LAMAS, G. 1978. Mariposas y conservación de la naturaleza en el Perú. Col. Suiza Perú Bol. (July): 61–66.

LOVEJOY, T. E. 1980. Discontinuous wilderness: Minimum areas for conservation. Parks 5(3): 13–15.

NATIONAL RESEARCH COUNCIL (U.S.A). 1983. Butterfly farming in Papua New Guinea. Managing Animal Resources Series. National Academy Press, Washington, D.C.

OTERO, L. S., AND K. S. BROWN, JR. 1986. Biology and ecology of *Parides ascanius* (Cramer, 1775) (Lep., Papilionidae), a primitive butterfly threatened with extinction. Atala 10–12: 2–16.

PYLE, R. M., M. BENTZIEN, AND P. OPLER. 1981. Insect conservation. Ann. Rev. Entomol. 26: 233–258.

WELLS, S. M., R. M. PYLE, AND N. M. COLLINS. 1983. The IUCN invertebrate red data book. Int. Union Conserv. Nat. Res., Gland, Switzerland.

# 3 PRACTICAL ENTOMOLOGY

For the human population of Latin America, as elsewhere in the world, competition from insects for food and fiber and interference with health and welfare are intense. Insects are injurious in several ways. Each is the subject of formal fields of study, representing compartments of the vast and complex subject of "practical" ("economic" or "applied") entomology. Insects are harmful by damaging growing crops and other useful plants (agricultural and forest entomology), by annoying and inflicting disease on humans (medical entomology) and domesticated animals (veterinary entomology), and by destroying useful products (stored product and structural entomology). The relationships of insects to man in population centers form another special topic (urban entomology).

The economic damage caused in these ways is incalculable in monetary terms and human suffering. Most entomologists in Latin America are employed to combat harmful species. The research literature in the field is immense. Because of the great size of the subject, only a brief review is possible here.

Although most publications in practical entomology deal with the temperate parts of the world (North America and Europe [Metcalf et al. 1962, Pfadt 1978]), comprehensive research on tropical pests has developed in recent years (Lamb 1974). The latter includes large parts of Latin America, yet references giving detailed information on regional economic entomology remain sparse and in many cases, outdated. Some

lists and cursory reviews of problem species are available for specific areas. These include Bolivia (Squire 1972), Colombia (Gallego 1967, Posada 1976), El Salvador (Berry 1959), French Guiana (Remillet 1988), Guatemala (Alvarado 1939), Honduras (Koone and Bañegas 1958), Uruguay (Peluffo 1942), the West Indies (Wolcott 1933), and the Lesser Antilles (Ballou 1912).

Insects cause direct damage to humans through their feeding or by biting and stinging. Their mere presence may also be harmful, but they are more serious as vectors of organisms pathological to plants as well as to man and his domesticated animals.

## References

ALVARADO, J. A. 1939. Los insectos dañinos y los insectos auxiliares de la agricultura en Guatemala. Published by author, Guatemala City.

BALLOU, H. A. 1912. Insect pests of the Lesser Antilles. Comm. Agric., Imp. Dept. Agric. West. Ind. Pamph. Ser. 71: 1–210.

BERRY, P. 1959. Entomología económica de El Salvador. Min. Agric. Gan. (Santa Tecla, El Salvador) Bol. Tec. 24: 1–255.

GALLEGO, F. L. 1967. Lista preliminar de insectos de importancia económica y secundarios, que afectan los principales cultivos, animales domésticos y al hombre, en Colombia. Fac. Nac. Agron. (Medellín) Rev. 26(65): 32–66.

KOONE, H. D., AND A. D. BAÑEGAS. 1958. Entomología económica hondureña. Min. Recurs. Nat. (Tegucigalpa) Bol. Tecn. 6: 1–139. [Not seen.]

LAMB, K. P. 1974. Economic entomology in the tropics. Academic, London.

METCALF, C. L., W. P. FLINT, AND R. L. METCALF.

1962. Destructive and useful insects. 4th ed. McGraw-Hill, New York.

PELUFFO, A. T. 1942. Insectos y otros parásitos de la agricultura y sus productos en el Uruguay. Univ. Rep., Fac. Agron., Montevideo.

PFADT, R. D., ed. 1978. Fundamentals of applied entomology. 3d ed. Macmillan, New York.

POSADA, L. 1976. Lista de insectos dañinos y otras plagas en Colombia. Insto. Colombiano Agropec., Bogotá.

REMILLET, M. 1988. Catalogue des insectes ravageurs des cultures en Guyane française. ORSTRM, Cayenne.

SQUIRE, F. A. 1972. Entomological problems in Bolivia. Pest Art. News Sum. 18: 239–268.

WOLCOTT, G. N. 1933. An economic entomology of the West Indies. Entomol. Soc. Puerto Rico, San Juan.

## AGRICULTURAL ENTOMOLOGY

Insects damage crop plants and reduce the value of agricultural produce by eating the vegetative parts and fruits, by acting as vectors of diseases (Carter 1973), or by contamination with their presence. Damage may be reflected in a variety of ways: wilted leaves, dead stems, discolored or spoiled fruit, gall formation (Fernandes 1987), and often the death of the plant. Loss of food, textile fiber, and ornamental plant produce in Latin America cannot be calculated with accuracy but surely runs into equivalents of billions of dollars annually.

No country or any plant of the great spectrum of Neotropical cultigens is immune from the depredations of injurious insects and mites; the problems are universal, as evident from several general publications (Caswell 1962; Fröhlich and Rodewald 1970; Gallo 1988; Hill 1983; Kranz et al. 1979; Ebeling 1959; Flechtmann 1983). A variety of reviews or catalogs of agricultural pests describe local situations: Argentina (Rizzo 1977, Molinari 1948), Brazil (Bondar 1913), Central America (King and Saunders 1984, Saunders et al. 1983), Colombia (Anonymous 1968), Cuba (Bruner et al. 1975), Dominican Republic (Sontoro 1960), Guatemala and El Salvador (Bates 1932), Honduras (Passoa 1983), the Lesser Antilles (Fennah 1947), Mexico (MacGregor and Gutiérrez 1983, Morón and Terrón 1988), Peru (Wille 1952), Puerto Rico (Chiesa Molinari 1942), Surinam (van Dinther 1960), and Uruguay (Ruffinelli and Carbonell 1954).

Most injurious species are adapted to a particular host, but a few attack almost any plant, the best examples being migratory locusts (*Schistocerca*) and leaf cutter ants (*Acromyrmex, Atta;* Cherrett and Peregrine 1976). The following are some of the more important pests of widely grown crop species encountered by Latin American agronomists.

The most serious enemies of the cacao tree (Leston 1970, Entwhistle 1972) are scale insects (*Pseudococcus, Planococcus, Dysmicoccus*) that act as vectors of viruses and fungal diseases, causing dieback and significant fruit reduction on affected plants. The cacao thrips (*Selemnothrips rubrocinctus*) and aphids (e.g., *Toxoptera aurantii*) cause similar damage.

Coffee trees serve as hosts to over two hundred insect species (Le Pelley 1968, 1973). The trunks and stems are susceptible to the larvae of a variety of boring beetles in Latin America, especially the coffee berry borer (*Hypothenemus hampei,* Scolytidae), which causes the dropping and decaying of berries, and "black borers" (*Apate*, Bostrichidae) that hollow the stems. The coffee leaf miner (*Leucoptera coffeella,* Lyonetiidae) damages leaves to such a degree that they fall off the plant. Leaf curl, burning, and stunting are commonly caused by various mealybugs (*Pseudococcus*) and scale insects.

The attacks of several kinds of lepidopterous stem borers can cause the virtual loss of entire crops of rice (Cheaney and Jennings 1975, Grist and Lever 1969). In Latin America, the chief offenders in this category are *Diatraea* spp. and other Pyralidae (Kapur 1964).

Corn (maize) is native to the New World, but this gives it no immunity to many introduced pests such as earworms (*Helicoverpa* spp., Noctuidae) that destroy the kernals and stem borers, mainly pyralid moths of the genus *Diatraea*, that may cause the entire plant to droop and die. Other pests include leaf destroyers, many leaf beetles (Chrysomelidae), and a host of other noctuid leaf worms and cutworms.

Although manioc (mandioca, cassava, yuca, tapioca) suffers from relatively few arthropod pests, they can cause extensive damage to this important food crop (Bellotti and van Schoonhoven 1978, Samways and Ciociola 1980). The cassava shoot-fly, *Neosilva perezi* (Lonchaeidae; sometimes erroneously cited as *Silba* or *Lonchaea chalybea*), is widespread and does major damage. Young larvae mine in growing shoots and may cause the entire shrub to die. The manioc gall midge (*Iatrophobia* [= *Autodiplosis, Eudiplosis*] *brasiliensis*, Cecidomyiidae) is well known because it causes an obvious deformation, red galls, on the leaves of plants it attacks. These galls reduce photosynthesis but rarely bring about loss of vitality of the whole plant. In parts of Brazil, the galls are known as *mamica de rama* or *veruga da mandioca*. Spider mites should also be counted among the more serious pests, especially *Tetranychus* and *Mononychellus* spp. The larvae of the ashy sphinx, *Erinnyis ello* (Winder 1976), and pyralid moth *Chilozela* (Becker 1986) feed directly on manioc leaves.

The vegetative parts of the banana plant are attacked by the mealybug *Pseudococcus comstocki* (Ostmark 1974). Adult banana fruit-scarring beetles (*Colaspis hypochlora*) eat the young, unfurled leaves and stems plus the fruit, causing scars that make the latter unsalable and allowing the entry of decay microbes. The banana thrips (*Hercinothrips bicinctus*) feeds on the fruit, discoloring it and reducing its market value, but it does not cause systemic effects on the plant. Plants are killed in many areas from the borings of banana weevil (*Cosmopolites sordidus*) larvae in the rhizome and pseudostem at ground level.

The pink bollworm (*Pectinophora gossypiella*) is the most important pest of cotton and occurs in nearly all growing areas. Heavy feeding on the bolls by the larvae leads to fiber loss and seed destruction. The cotton bollworm (*Helicoverpa zea*) and the famous cotton boll weevil (*Anthonomus grandis*) (Burke et al. 1986) both do similar damage. Red spider mites, especially (*Tetranychus cinnabarinus*), cause the leaves of cotton plants to wither and drop off. The cotton aphid (*Aphis gossypii*) also often infests the foliage.

Pineapple is damaged directly by the root-feeding pineapple mealybug (*Dysmicoccus brevipes*, Pseudococcidae) but also suffers from an associated fungus wilt disease (*Phytophthora*) that causes leaf degeneration and small fruit.

Citrus is impossible to grow profitably in many areas because of the attacks of numerous insects (Ebeling 1959). Most serious are the citrus spider mites, including *Metatetranychus citri*, and the six-spotted mite (*Eotetranychus sexmaculatus*). Considerable damage also results from overwhelming populations of citrus white flies (*Trialeurodes, Dialeurodes*, Aleyrodidae) and scale insects (California red scale, *Aonidiella aurantii*; West Indian red scale, *Sclenaspidus articulatus*; and citrus mussel scale or purple scale, *Lepidosaphes beckii*). The maggots of the fruit flies (*Rhagoletis, Anastrepha*), especially the Mediterranean fruit fly (*Ceratitis capitata*, Tephritidae), may account for almost total fruit losses in heavily infested areas. Mealybugs (e.g., *Planococcus citri*) are also a major problem.

Sugarcane also comes under attack from a large number of insects and mites (Long and Hensley 1972, Guagliumi 1972–73, Williams et al. 1969). Froghoppers, particularly the sugarcane froghopper (*Aeneolamia varia saccharina*, Cercopidae), are this crop's most serious pest in many countries. Sap feeding by the nymphs on the roots results in withering of the leaves and stunted stalks. The cane leafhopper (*Saccharosydne saccharivora*) is a troublesome species in Jamaica and elsewhere. Many subterranean, root-feeding scale insects, mealybugs, cicadas, and scarab beetle larvae ("white grubs") are also prevalent in various areas. Several stalk-boring larvae of the pyralid moth genus *Diatraea* (especially the sugarcane borer, *D. saccharalis*) are recognized as primary pests as well. Aphids (*Aphis sacchari, Sipha flava*) serve as vectors of virus and fungal diseases and injure the plants directly by sap removal.

Other crops and their major pests in Latin America are avocado (fruit flies, Tephritidae); alfalfa (spotted alfalfa aphid, *Therioaphis maculata*); beans (bean aphid, *Aphis fabae*); coconut palm (Lever 1969) (coconut scale, *Aspidiotus destructor;* coconut mite, *Eriophyes guerreronis;* planthopper, *Myndus crudus*, vector of lethal yellowing disease) (Howard et al. 1983); guayaba (fruit flies) (Espinoza 1972); maguey (fruit flies, *Euxesta*); mango (fruit flies); potatoes (potato tuberworm, *Phthorimaea operculella*); papaya (fruit flies); sorghum (sugarcane borer, *Diatraea saccharalis*) (Young and Teetes 1977); tobacco (tobacco and tomato hornworms, *Manduca sexta* and *M. quinquemaculata*); and wheat (wheat thrips, *Frankliniella tritici*).

## References

ANONYMOUS. 1968. Catálogo de insectos en cultivos de importancia económica en Colombia. Assoc. Latinoamer. Entomol. Publ. 1: 1–156.

BATES, M. 1932. Insectos nocivos: Estudio de las principales plagas guatemaltecas con algunos datos de Honduras y El Salvador. Serv. Tec. Coop. Agric., Unit. Fruit Co., Guatemala City.

BECKER, V. O. 1986. Correct name for the species of *Chilozela* (Lepidoptera: Pyralidae) whose caterpillars damage cassava in South America. Bull. Entomol. Res. 76: 195–198.

BELLOTTI, K. A., AND A. VAN SCHOONHOVEN.

1978. Mite and insect pests of cassava. Ann. Rev. Entomol. 23: 39–67.

BONDAR, G. 1913. Os insectos daninhos na agricultura. Sec. Agric. Indus. Comm., São Paulo.

BRUNER, S. C., L. C. SCARAMUZZA, AND A. R. OTERO. 1975. Catálogo de los insectos que atacan a las plantas económicas de Cuba. 2d ed. Acad. Cien. Cuba, Insto. Zool., La Habana.

BURKE, H. R., W. E. CLARK, J. R. CATE, AND P. A. FRYXELL. 1986. Origin and dispersal of the boll weevil. Entomol. Soc. Amer. Bull. 32: 228–238.

CARTER, W. 1973. Insects in relation to plant disease. 2d ed. Wiley, New York.

CASWELL, G. H. 1962. Agricultural entomology in the tropics. Arnold, London.

CHEANEY, R. L., AND P. R. JENNINGS. 1975. Problemas en cultivos de arroz en América Latina. Centr. Int. Agric. Trop. (Ser. GS-15), Cali.

CHERRETT, J. M., AND D. J. PEREGRINE. 1976. A review of the status of leaf-cutting ants and their control. Ann. Appl. Biol. 84: 128–133.

CHIESA MOLINARI, O. 1942. Entomología agrícola, identificación y control de insectos y otros animales dañinos o útiles a las plantas. Talleres Gráficos Accurziol, San Juan.

CHIESA MOLINARI, O. 1948. Las plagas de la huerta y el jardín y modo de combatirlas. Ed. Bell (Bibl. Pampa Argentina, Rev. Min. Agric. Gan. Int. Gen. Serv. Pais), Buenos Aires.

EBELING, W. 1959. Subtropical fruit pests. Univ. Calif. Press, Berkeley.

ENTWHISTLE, P. F. 1972. Pests of cocoa. Longmans, London.

ESPINOZA, W. O. 1972. Control fitosanitario en plantaciones de guayaba. Univ. Indus. Santander, Bucaramanga, Colombia.

FENNAH, R. G. 1947. The insect pests of food crops in the Lesser Antilles. Dept. Agric. Antigua, British West Indies. [Not seen.]

FERNANDES, G. W. 1987. Gall forming insects: Their economic importance and control. Rev. Brasil. Entomol. 31: 379–398.

FLECHTMANN, C. H. W. 1983. Acaros de importância agrícola. Liv. Nobel, São Paulo.

FRÖHLICH, G., AND W. RODEWALD. 1970. Pests and diseases of tropical crops and their control. Pergamon, Oxford.

GALLO, D. 1988. Manual de entomología agrícola. 2d ed. Ed. Agron. Ceres Ltda., São Paulo. [Not seen.]

GRIST, D. H., AND R. J. A. W. LEVER. 1969. Pests of rice. Longmans, London.

GUAGLIUMI, P. 1972–73. Pragas de cana de

açúcar no nordeste do Brasil. MIC, Insto., Açúcar e do álcool, Dr. Ad., Ser. Doc. Coleção Canavieira 10, Rio de Janeiro.

HILL, D. S. 1983. Agricultural insect pests of the tropics and their control. 2d ed. Cambridge Univ. Press, Cambridge.

HOWARD, F. W., R. C. NORRIS, AND D. L. THOMAS. 1983. Evidence of transmission of palm lethal yellowing agent by a planthopper, *Myndus crudus* (Homoptera: Cixiidae). Trop. Agric. 60: 168–171.

KAPUR, A. P. 1964. Taxonomy of the rice stem borers. *In* Int. Rice Res. Inst., The major insect pests of the rice plant. Johns Hopkins Univ. Press, Baltimore. Pp. 3–43.

KING, A. B. S., AND J. L. SAUNDERS. 1984. The invertebrate pests of annual food crops in Central America. Trop. Devel. Res. Inst., Overseas Devel. Adm., London.

KRANZ, J., H. SCHMÜTTERER, AND W. KOCH. 1979. Diseases, pests, and weeds in tropical crops. Wiley, New York.

LE PELLEY, R. H. 1968. Pests of coffee. Longmans, London.

LE PELLEY, R. H. 1973. Coffee insects. Ann. Rev. Entomol. 18: 121–142.

LESTON, D. 1970. Entomology of the cocoa farm. Ann. Rev. Entomol. 15: 273–294.

LEVER, R. J. A. W. 1969. Pests of the coconut palm. United Nations Food Agric. Org. Agric. Ser. 77: 1–190.

LONG, W. H., AND S. D. HENSLEY. 1972. Insect pests of sugarcane. Ann. Rev. Entomol. 17: 149–176.

MACGREGOR, R., AND O. GUTIÉRREZ. 1983. Guía de insectos nocivos para la agricultura en México. Ed. Alhambra Mexicana, Mexico City.

MORÓN, M. A., AND R. A. TERRÓN. 1988. Entomología practica. Insto. Ecol., México.

OSTMARK, H. E. 1974. Economic insect pests of bananas. Ann. Rev. Entomol. 19: 161–176.

PASSOA, S. 1983. Lista de los insectos asociados con los granos básicos y otros cultivos selectos en Honduras. Ceiba 25: 1–97.

RIZZO, H. F. E. 1977. Catálogo de insectos perjudiciales en cultivos de la Argentina. 4th ed. Hemisferio Sur, Buenos Aires.

RUFFINELLI, A., AND C. S. CARBONELL M. 1954. Segunda lista de insectos y otros artrópodos de importancia económica en el Uruguay. Assoc. Ing. Agron. Montevideo Rev. 94: 33–82.

SAMWAYS, M. J., AND A. I. CIOCIOLA. 1980. O complexo de artrópodos da mandioca (*Manihot esculenta*) Crantz em Lavras, Minas Gerais, Brasil. Soc. Entomol. Brasil An. 9: 3–10.

SAUNDERS, J. L., A. B. S. KING, AND C. L. VARGAS. 1983. Plagas de cultivos en América Central. Centro. Agron. Trop. Inves. Ens., Turrialba, Costa Rica.

SONTORO, R. 1960. Notas de entomología agrícola dominicana. Sec. Estad. Agric. Comer., Rep. Dominicana, Santo Domingo. [Not seen.]

VAN DINTHER, J. B. M. 1960. Insect pests of cultivated plants in Surinam. Landbouwproefstation Suriname Bull. 76: 1–159.

WILLE, J. E. 1952. Entomología agrícola del Perú. 2d ed. Min. Agric., Lima.

WILLIAMS, J. R., J. R. METCALF, R. W. MONTGOMERY, AND R. MATHES, eds. 1969. Pests of sugarcane. Elsevier, Amsterdam.

WINDER, J. A. 1976. Ecology and control of *Erinnyis ello* and *E. alope*, important insects in the New World. Pest Art. News Sum. 22: 449–466.

YOUNG, W. R., AND G. L. TEETES. 1977. Sorghum entomology. Ann. Rev. Entomol. 22: 193–218.

## FOREST ENTOMOLOGY

A special branch of agricultural entomology deals with forest pests (Dourojeanni Ricordi 1963, Gray 1972). The tremendous value of wood products makes this one of the most important fields economically but one somewhat neglected in Latin America. For this reason and because there are a large number of commercial timber species and forest types, it is difficult to generalize about forest pests in this part of the world. Only a few area studies or surveys have been conducted (Martorell 1945). Investigations on eucalyptus, snapdragon tree (*Gmelina arborea*), and pine (*Pinus caribea*) pests may become more appropriate as these exotic timber types replace native Neotropical hardwood species.

There has been a tendency to regard insect communities in mixed tropical forests as relatively stable, that is, subject to only small population fluctuations, compared to temperate forests. Thus, the likelihood of severe outbreaks are thought to be remote. However, as various authors have reported a number of localized population explosions in Old World tropical forests similar to those encountered in temperate regions, the possibility remains for similar occurrences in Neotropical forests.

Wood-boring beetles are the most common and serious timber pests. Their larvae molest all parts of the young, mature, and harvested tree. Most belong to the families Cerambycidae, Scolytidae, Curculionidae, Platypodidae, and Buprestidae. Damage to standing timber is almost wholly due to termites of the family Termitidae, in particular, members of the genus *Coptotermes* (Harris 1966). In their resin-gathering activities, stingless bees may damage nursery seedlings by boring into and gouging the stems of new plantings (Gara 1970).

Many lepidopterous species are no doubt injurious to timber trees in Latin America as they are to those of temperate forests, but little is known of the economic impact of the numerous leaf-feeding and wood-boring types. The mahogany webworm (*Macalla thyrsisalis*, Pyralidae) is one recognized pest species of mahogany (Howard and Solis, 1989).

## References

DOUROJEANNI RICORDI, M. 1963. Introducción al estudio de los insectos que afectan la explotación forestal en la selva Peruana. Rev. Peruana Entomol. Agric. 6(1): 27–38.

GARA, R. I. 1970. Report of forest entomology consultant. Inter-American Inst. Agr. Sci., Org. Amer. States, Turrialba, Costa Rica, UNDP Project 80: 1–21.

GRAY, B. 1972. Economic tropical forest entomology. Ann. Rev. Entomol. 17: 313–354.

HARRIS, W. V. 1966. The role of termites in tropical forestry. Ins. Soc. 13: 255–266.

HOWARD, F. W., AND M. A. SOLIS. 1989. Distribution, life history, and host plant relationships of mahogany webworm, *Macalla thyrsisalis* (Lepidoptera: Pyralidae). Florida Entomol. 72: 469–479.

MARTORELL, L. F. 1945. A survey of the forest insects of Puerto Rico. Univ. Puerto Rico J. Agric. 29: 69–608.

# MEDICAL ENTOMOLOGY

Many insects, spiders, mites, myriapods, and other arthropods are medically important, acting either as agents of harm to humans or as vectors of pathogenic microorganisms. This is such an important aspect of our existence that several textbooks treat the subject from an overall perspective in considerable detail (Faust et al. 1962; Flechtmann 1973; Horsfall 1962; James and Harwood 1969; Kettle 1984; Smith 1973). Regional discussions are also available for Argentina (del Ponte 1958), Brazil (Pinto 1930), Central America (Baerg 1929), Panama (Méndez and Chaniotis 1987), and South America (Bücherl 1969).

Because they inject or dispense venoms, members of many groups (Bücherl and Buckley 1971) are serious agents of medical problems throughout Latin America. The most important offenders are scorpions (*Tityus, Centruroides*), spiders (*Latrodectus, Loxosceles,* and *Phoneutria*), and stinging Hymenoptera (*Apis*, ants, and wasps) (Akre and Davis 1978). Poisons that act topically (vesicants) are produced by millipedes, blister beetles, fire beetles (*Paederus*), and others (Hoffman 1927). Nettling hairs or spines, such as adorn many caterpillars (Saturniidae, Limacodidae, Megalopygidae, etc.), also implant toxins (urtication, erucism).

Reactions to toxic substances (Tu 1984) may be slight to severe, even fatal in rare cases. Such effects occur either through direct toxification or by eliciting allergic responses through antigens (Frazier 1969), superficially (Orkin and Maibach 1985) or systemically. Any protein derived from the insect's body may cause a harmful reaction if it comes in contact with tissue topically or by injection. A common means of injection is through the bite of blood-feeding forms, that is, mosquitoes, ticks, mites, and sand flies (Feingold et al. 1968). The antigen is contained in the saliva and enters the bloodstream directly or via the lymphatics. Hypersensitive individuals exhibit varied symptoms, ranging from mild dermatosis to anaphylactic shock, which can be fatal.

Many people have extreme fears or phobias of insects and similar creatures, leading to psychoneuroses. A common form is delusory parasitosis (Waldron 1963), the unshakable belief that one's skin and orifices are infested with minute, barely visible insects, mites, or other vermin. This condition appears to be a symptom of a variety of organic and mental disorders. Extreme phobias also affect many persons, especially against large, very hairy, dark, or noisy species. These emotional complaints are not as well documented in Latin America as in other parts of the world but are surely as widespread.

Human myiasis is another major problem caused directly by larval Diptera, especially of blowflies and a few special types like the human botfly (*Dermatobia*). The body may be invaded by maggots, leading to a variety of symptoms, many highly repugnant psychologically as well as physically (Beesley 1974; and see Myiasis, chap. 11).

Insects and their relatives are highly efficient and diverse as transmitters of other pathogenic organisms. Disease microorganisms may be carried by the insect passively (mechanically) or may pass through certain of its developmental stages in the arthropod host, which is then considered an obligatory or "biological" vector. Several major groups are vectors of human and animal pathogens, including biting flies, fleas (Bibikova 1977), kissing bugs, blood-feeding mites, and ticks.

Many viruses, bacteria, and amoebic or worm cysts are mechanically transmitted. They are carried on the bodies, on the mouthparts, and in the intestines of filth flies, cockroaches, and other insects that frequent contaminated matter and food eaten later by humans. Many dysenteries, tapeworm, and nematode diseases are spread in this way, particularly under very unsanitary conditions when poverty or social disruption, such as war or natural disasters, prevails in a human population. Poliomyelitis, typhoid fevers, cholera, leprosy, and other diseases may also find new human hosts in this manner. Research indicates that the AIDS virus is not transmitted by insects, blood feeding or otherwise.

Biological vectors are found entirely among blood-feeding types, especially mosquitoes and other biting midges and flies, although ticks and mites are also significant. Their ecology is a principal factor determining the effectiveness of these insects as vectors (Muirhead-Thomson 1968). The organisms of over a dozen major types of human diseases are transmitted by arthropods in Latin America. The most notorious and widespread of these is malaria, which is caused by four species of plasmodial protozoans. These organisms invade various organs and destroy red blood cells, releasing toxins into the circulation which cause racking chills and fever. Vectors are several species of mosquitoes in the genus *Anopheles*.

Leishmaniasis is an affliction also caused by protozoans, at least three species of flagellates in the genus *Leishmania*. Sand flies (*Lutzomyia*, Psychodidae) carry these agents that invade and chemically destroy both dermal and internal tissues of vital organs. Another flagellate protozoan, *Trypanosoma cruzi*, that develops in kissing bugs (Reduviidae, Triatominae) brings on a serious ailment called Chagas's disease in many parts of Latin America. Visceral organs suffer chronic damage, which leads ultimately to death in many untreated cases.

Various parasitic nematode worms introduced from the Old World have become established in certain areas and cause a variety of filarial infections. These include *Wuchereria bancrofti* (Bancroftian filariasis), *Onchocerca volvulus* (onchocerciasis), and *Dirofilaria imitis* (dog heartworm). While seldom fatal, they wreak considerable damage by invading the tissues, producing inflammation, enlargement, and destruction. When essential organs such as the eye or brain are involved, critical functions of the senses may be impaired. Vectors are mostly mosquitoes, but blackflies, punkies, and tabanids also serve as carriers.

A large and growing number of viruses ("arboviruses") are being discovered which require biting fly, mite, and tick vectors. The worst of these historically has been the yellow fever virus, transmitted by mosquitoes, especially the yellow fever mosquito, *Aedes aegypti*. Several kinds of encephalitides, hemorrhagic fever, and dengue fever are also in this category. The viral genus *Phlebovirus*, transmitted by phlebotomine sand flies (*Lutzomyia*) and mosquitoes, contains many species of human pathogens causing intense flulike diseases (Tesh 1988).

Epidemic (*Rickettsia prowazekii*) and endemic (*R. mooseri*) typhus organisms pass to humans from the bodies of lice and fleas. A third rickettsia (*R. rickettsii*), that of Rocky Mountain spotted fever, is borne by hard ticks. These microbes invade and destroy the inner lining of small blood vessels. High fevers, often followed by death, occur. Similar symptoms follow infection by the spirochetes of relapsing fevers (*Borrelia*) transmitted through the bite or body secretions of ticks and lice.

The infamous plague bacillus (*Yersinia pestis*) still resides in animal reservoirs in parts of Latin America and sometimes makes its way to the human population via flea bites. Fortunately, epidemics like those of the past in Europe and elsewhere have not occurred in recent times.

Most of these diseases have been controlled by modern insecticides applied against the carriers and by drugs that kill the pathogens. However, as a result of relaxation of abatement campaigns and development of chemical resistance by both insects and microorganisms, some diseases are experiencing a resurgence and are again causing major problems in areas formerly thought free of them.

## References

AKRE, R. D., AND H. G. DAVIS. 1978. Biology and pest status of venomous wasps. Ann. Rev. Entomol. 23: 215–238.

BAERG, W. J. 1929. Some poisonous arthropods of North and Central America. 4th Int. Cong. Entomol. (Ithaca, 1928) Trans. 2: 418–438.

BEESLEY, W. N. 1974. Arthropods—Oestridae, myiases and acarines. *In* E. J. L. Southby, Parasitic zoonoses. Academic, New York. Pp. 349–368.

BIBIKOVA, V. A. 1977. Contemporary views on the interrelationships between fleas and the pathogens of human and animal diseases. Ann. Rev. Entomol. 22: 23–32.

BÜCHERL, W. 1969. Giftige arthropoden. *In* E. J. Fittkau, J. Illies, H. Klinge, G. H. Schwabe, and H. Sioli, Biogeography and ecology in South America. 2: 764–793. Junk, The Hague.

BÜCHERL, W., AND E. E. BUCKLEY. 1971. Venomous animals and their venoms. 3. Venomous invertebrates. Academic, New York.

DEL PONTE, E. 1958. Manual de entomología médica y veterinaria Argentinas. Ed. Lib. Colegio, Buenos Aires.

FAUST, E. C., P. C. BEAVER, AND R. C. JUNG. 1962. Animal agents and vectors of human disease. Lea and Febiger, Philadelphia.

FEINGOLD, B. F., E. BENJAMINI, AND D. MICHAELI. 1968. The allergic responses to insect bites. Ann. Rev. Entomol. 13: 137–158.

FLECHTMANN, C. H. W. 1973. Acaros de importância médico veterinária. Liv. Nobel, São Paulo.

FRAZIER, C. A. 1969. Insect allergy, allergic and toxic reactions to insects and other arthropods. Green, St. Louis.

HOFFMAN, W. A. 1927. Irritation due to insect secretion. Amer. Med. Assoc. J. 88: 145–146.

HORSFALL, W. R. 1962. Medical entomology, arthropods and human disease. Ronald, New York.

JAMES, M. T., AND R. F. HARWOOD. 1969. Herm's medical entomology. 6th ed. MacMillan, London.

KETTLE, D. S. 1984. Medical and veterinary entomology. Wiley, Somerset, N.J.

MÉNDEZ, E., AND B. CHANIOTIS. 1987. Reseña de las principales enfermedades transmitidas por garrapatas en Panamá. Rev. Med. Panamá 12: 217–223.

MÉNDEZ, E., AND B. CHANIOTIS. 1987. Reseña

de las principales enfermedades transmitadas por insectos en Panamá. Rev. Med. Panamá 12: 205–216.

MUIRHEAD-THOMSON, R. C. 1968. Ecology of insect vector populations. Academic, London.

ORKIN, M., AND H. I. MAIBACH. 1985. Cutaneous infestations and insect bites. Marcel Dekker, New York.

PINTO, C. 1930. Arthrópodes parasitos e transmissores de doenças. Vols. 1–2. Pimento de Mello, Rio de Janeiro.

SMITH, K. G. V., ed. 1973. Insects and other arthropods of medical importance. Trus. Brit. Mus. Nat. Hist., London.

TESH, R. B. 1988. The genus *Phlebovius* and its vectors. Ann. Rev. Entomol. 33: 169–181.

TU, A. T. 1984. Handbook of natural toxins: Insect poisons, allergens, and other invertebrate venoms. Vol. 2. Marcel Dekker, New York.

WALDRON, W. 1963. Psychiatric and entomological aspects of delusory parasitosis. Amer. Med. Assoc. J. 186: 213–214.

## VETERINARY ENTOMOLOGY

The study of arthropod agents and vectors of diseases of domesticated animals, pets, game, and wildlife constitutes the field of veterinary entomology (Kettle 1984, Southby 1982, Williams 1985).

One of the most serious afflictions suffered by cattle and other livestock in Latin America is babesiosis or cattle fever (caused by *Babesia bigemina*). It is present throughout the region and is transmitted by ticks, mainly *Boophilus microplus*.

Numerous viruses spread by mosquito vectors infect horses, other quadrupeds, and poultry. These are mainly the encephalitides, such as Venezuelan equine and eastern equine types that can rapidly decimate herds or flocks and are also transmissible to humans.

The heartworm (*Dirofilaria imitis*) is fairly common in the more tropical portions of Latin America. Its insect hosts are punkies and mosquitoes. Worms living in the heart impair its function in dogs, which are particularly susceptible to this filarial parasite.

A great number of external parasites, including ticks of all types, mange mites, keds (Hippoboscidae), fleas, and biting and sucking lice, infest every kind of domestic animal (Steelman 1976). The feeding and allergenic effects of these pests cause considerable annoyance to their hosts and greatly diminish their growth and vitality. Similarly, the attacks of biting flies, mostly blackflies, mosquitoes, horseflies and deerflies, and hematophagous Muscidae, keep animals on edge and negatively affect their general health. The sites of bites may become septic, and death may ensue from loss of blood. Particularly harmful are the horn fly (*Haematobia irritans*), whose constant pestering can cause significant weight loss in cattle, and the stable fly (*Stomoxys calcitrans*), which can drive quadrupeds to fits. The former species has invaded South America as far south as northern Brazil but seems not to be a major problem because of its poor adaptation to tropical climates. It may become a serious pest if it reaches temperate regions farther south (Thomas pers. comm.).

As with humans, myiasis is a problem—but to a much greater degree on account of the greater vulnerability of livestock. Foremost in this category are the deep tissue invaders such as the screwworm (*Cochliomyia hominivorax*), whose attacks are a major menace to sheep and cattle ranching over wide areas, and the human botfly (*Dermatobia hominis*). The more superficial effects of warbles and tissue bots (Hypodermatidae, Oestridae), while less serious physiologically, reduce the value of hides and pelts. Stomach bots (*Gasterophilus*) are also widespread and markedly affect the health of horses and cattle.

Wildlife diseases caused by insects and other arthropods are poorly understood in Latin America. Perhaps the best studied are the so-called sylvatic forms of plasmodia ("bird malarias"), trypanosomes, and viruses ("jungle yellow fever"), because of their relationship to outbreaks in humans, causing such syndromes as malaria, Chagas's disease, and yellow fever.

## References

KETTLE, D. S. 1984. Medical and veterinary entomology. Wiley, Somerset, N.J.

SOUTHBY, E. J. L. 1982. Helminths, arthropods and protozoa of domesticated animals. 7th ed. Lea and Febiger, Philadelphia.

STEELMAN, C. D. 1976. Effects of external and internal arthropod parasites on domestic livestock production. Ann. Rev. Entomol. 21: 155–178.

WILLIAMS, R. E. 1985. Livestock entomology. Wiley, New York.

# STORED PRODUCT AND STRUCTURAL PESTS

Harvesting, storage, and packaging provide no guarantee of safety to food and useful items from the ravages of insects and mites. Many species adapted for feeding on seeds, cellulose, and animal tissues and hair are naturally attracted to items composed of these materials which are brought in from the field (Baur 1984). The majority of these pests (Cotton 1960) are now cosmopolitan as a result of their association with commercial products that are distributed throughout the world through trade. They survive well in warehouses, storerooms, the holds of ships, and in the marketplace, where their presence and feeding degrades or destroys cereals and grains, paper, wood, fur and hides, fabrics, and other organic materials. Stored product pests are not well investigated as a group in Latin America, with some exceptions (Granovsky 1976, Passoa 1983).

Damage to grain in elevators and silos constitutes the largest losses to stored products. Grain, meal, and flour are attacked by a variety of beetles (Hinton 1945), moths (Corbet and Tams 1943), and mites (Flechtmann 1983). Beetle adults and larvae eat the kernels of rice, wheat, corn, and so on. These include the rice and granary weevils (*Sitophilus*) and the various grain beetles (*Tribolium* and *Tenebrio, Oryzaephilus*, etc.). Fortunately, the dreaded khapra beetle (*Trogoderma granarium*), which prefers dried vegetable products but also attacks animal products, is not now known to exist anywhere in Latin America. Its potential introduction is a constant menace, however. A number of grain beetles serve as intermediate hosts for human tapeworms (*Hymenolepis*), thus assuming medical importance (Cáceres and Guillén de Tantaleán 1972).

The larvae of flour and meal moths eat milled seeds and contaminate provisions with their webbing and feces. The principal offenders in this category are the Mediterranean flour moths (*Ephestia, Anagasta*), Angoumois grain moth (*Sitotroga cerealella*), and Indian meal moth (*Plodia interpunctella*). A considerable variety of mites (Cáceres et al. 1989, Flechtmann 1983:145–160) infest stored grains, some of which actually feed on fungi growing there and not on the produce itself. Also, several types bite and cause dermatitis ("itch mites") in granary workers and bakers.

In Latin America, termites are the chief destroyers of finished wood products. Many species are important, especially *Coptotermes*, which feeds not only on houses and lumber but on forest trees as well. Other major wood pests include powderpost beetles (especially *Lyctus*). Stored paper, including books, frequently is damaged not only by these insects but by silverfish, psocids, ants, and bostrichid beetles.

The leather industry is plagued by hide beetles (*Dermestes*) that riddle cowhides during the tanning process and storage. The cigarette beetle (*Lasioderma serricorne*) and drugstore beetle (*Stegobium paniceum*) lay waste to dried tobacco. They also destroy all manner of dry animal and plant products in the home and shops (stored nuts, cereals, spices, candy, etc.).

Wool garments and furs are subject to destruction by Webbing clothes moths

(*Tineola bisselliella*) and case-bearing clothes moths (*Tinea pellionella*) in Latin America as elsewhere.

## References

BAUR, F. J., ed. 1984. Insect management for food storage and processing. Amer. Assoc. Cereal Chem., St. Paul.

CÁCERES, I. E., AND Z. GUILLÉN DE TANTALEÁN. 1972. Insectos de Lima relacionados con el cistercoide de *Hymenolepis diminuta* (Rudolphi, 1819), (Cestoda: Hymenolepididae). Rev. Peruana Entomol. 15: 142–147.

CÁCERES, I. E., A. ELLIOT, AND I. NAKASHIMA. 1989. Ácari Prostigmata y Mesostigmata en alimentos almacenados de Lima, Huaraz e Iquitos. Rev. Peruana Entomol. 31: 13–17.

CORBET, A. S., AND W. H. T. TAMS. 1943. Keys for the identification of the Lepidoptera infecting stored food products. Zool. Soc. London Proc. B 113: 55–148.

COTTON, R. T. 1960. Pests of stored grain and grain products. Burgess, Minneapolis.

FLECHTMANN, C. H. W. 1983. Acaros de importância agrícola. Liv. Nobel, São Paulo.

GRANOVSKY, T. A. 1976. Insects associated with stored grain in Paraguay, South America. Kans. Entomol. Soc. J. 49: 508.

HINTON, H. E. 1945. Monograph of beetles affecting stored products. Brit. Mus. Nat. Hist., London.

PASSOA, S. 1983. Lista de los insectos asociados con los granos básicos y otros cultivos selectos en Honduras. Ceiba (Tegucigalpa) 25(1): 1–97.

## URBAN ENTOMOLOGY

Urbanization is proceding at a rapid pace throughout the world. Large cities are becoming even larger, new land is being taken over, and densities within old metropolitan centers are ever-increasing. Such growth forces contact between certain kinds of insects that inhabit homes and buildings and those whose natural habitats are being invaded. The study of this phenomenon is the relatively recently established field of urban entomology (Ebeling 1975, Frankie and Koehler 1983).

Negative effects of urban insects are many and depend on the types of environment they occupy, including private dwellings, restaurants and other food handling establishments, warehouses, manufacturing plants, and buildings dedicated to business, medical care, and recreation. The principal problems are health related, not only from direct contact but through contamination of food, bedding, and circulating products. Some curious psychological syndromes also are exacerbated, among them delusory parasitosis and mass hysterias associated with real or imagined microscopic insects believed to infest the human skin. Wooden structures and their furnishings are destroyed by insects, as are stored products. Use of outdoor recreational areas often exposes humans to arthropod vectors of pathogens. Hotels, bathhouses, and like establishments foster the transmission of body parasites like lice and bedbugs.

The major offenders to human peace of mind and welfare in urban areas are semidomesticated species, often of tropical origin, seeking the warmth and high humidity that prevails in our abodes and working places (Frankie and Ehler 1978). The best examples of these are several species of cockroaches, termites, and silverfish that live in the walls and furniture and other wooden components of houses. Flies enter through doors and windows and both bite and annoy us. Ants of many varieties do likewise. Clothes moths destroy woolen fabrics, and grain, flour, and meal moths and beetles invade the pantry.

As virgin land is converted to brick, mortar, and asphalt, persisting populations of native insects may bring grief to the new tenants. Housing situated near freshwater marshes, from which mosquitoes and punkies emerge, may make life miserable for people in their gardens and even indoors. Kissing bugs (Triatominae) living in rodent nests may choose humans as hosts on their nocturnal wanderings.

Control of domestic and urban insect pests has special requirements (Mallis et al. 1982, Osmun and Butts 1966). Paramount among these is the need to be considerate of human sensitivities. Property owners and businessmen concerned with economic damage to themselves and commerce must deal with their fears and dislikes of insects and the dangers of using insecticides near places of frequent and prolonged human occupation. Consideration must also be given to insect species that should be protected from urban sprawl and preserved in nature preserves or parks. Everything natural need not be destroyed in the name of progress or economic gain.

A few kinds of desirable insects, such as butterflies, may even be favored by urban conditions and special faunas created (Ruszczyk 1987).

## References

EBELING, W. 1975. Urban entomology. Univ. Calif. Press, Berkeley and Los Angeles.

FRANKIE, G. W., AND L. E. EHLER. 1978. Ecology of insects in urban environments. Ann. Rev. Entomol. 23: 367–387.

FRANKIE, G. W., AND C. S. KOEHLER, eds. 1983. Urban entomology: Interdisciplinary perspectives. Praeger, New York.

MALLIS, A. 1982. Handbook of pest control: The behavior, life history, and control of household pests. 6th ed. Franzak and Foster, Cleveland.

OSMUN, J. V., AND W. L. BUTTS. 1966. Pest control. Ann. Rev. Entomol. 11: 515–548.

RUSZCZYK, A. 1987. Distribution and abundance of butterflies in the urbanization zones of Porto Alegre, Brazil. J. Res. Lepidop. 25: 157–178.

## CONTROL OF INSECT PESTS

Under natural conditions, insect numbers are controlled by various means, principally by climatic strictures and by predators, parasites, parasitoids, and disease (Aguilar 1989, Strong 1984). Insects living in unwanted proximity to humanity and competing with people for food and fiber require artificial control (Martin and Wood-cock 1983). Practical (applied or economic) entomologists have devised a great many strategies to deal with pest species. Today, eradication is not the goal, as in the past, so much as reducing damage to acceptable tolerance levels, an approach referred to as "pest management" (Metcalf and Luckman 1982). Methods of pest management are successful to different degrees, depending on local conditions, damage levels, and availability of funds. In recent years, they have been combined in appropriate ways to capitalize on the best aspects of each, in the technique of "integrated control" (Apple and Smith 1976, van Huis 1981). This is usually the most logical and productive approach, rather than relying entirely on just one method, primarily chemicals, to achieve quick and cheap results. The use of chemicals alone has not been totally successful because of the ability of insects to develop resistance to most poisons (Georghiou and Saito 1983) and the deleterious side effects that accompany pesticide use (environmental pollution and destruction of nontarget organisms) (Green 1976). In fact, heavy and exclusive use of insecticides may actually lead to a decrease in crop production. A classic case of this in Latin America took place in the Cañete Valley in Peru (Barducci 1971). In the 1920s, agricultural emphasis here shifted from sugarcane to cotton. In the following two decades, cotton pest control was accomplished unevenly with chemicals and some ecological methods, and yields varied. In the 1950s, however, treatments with organic insecticides increased greatly and became pervasive. Yields dropped dramatically, and pests increased in kinds and intensity of damage. Finally, integrated methods were introduced, and after a few years, the crisis abated.

A key aspect in successful integrated control is vigilance and monitoring, using various kinds of trapping and sampling techniques to assess pest and damage levels before control measures are instigated (e.g., Silveira Neto 1972). Integrated control takes advantage of the following diverse methods of keeping checks on injurious insects.

1. *Chemical control.* Insecticides (pesticides) kill insects by their chemical action, usually by interfering with some essential metabolic function, such as transmission of electrical impulses across nerve synapses or blocking nutritive pathways (Corbett et al. 1984, Wilkinson 1976). Poisonous compounds reach their target tissues by ingestion (stomach poisons), by passage through the integument and sense organs (contact poisons), or by entering the tracheal system (fumigants).

   Insecticides come in almost infinite variety and chemical structure (Martin and Worthing 1976, Wiswesser 1976), and their mode of action, application, safety, and effectiveness comprise the complex subject of insect toxicology (Matsumura 1975). The major categories of insecticides based on chemistry (Buchel 1983) are the inorganics, for example, arsenic compounds, cyanide gas, and the botanicals ("first-generation insecticides"), the naturally occurring types of which (rotenone, pyrethrum, nicotine) have been in use for centuries in many parts of the world. Then there are the synthetic organics ("second generation"), products of modern chemistry, including such well-known chlorinated hydrocarbons as DDT, benzene hexachloride, and Chlordane. To these have been added recently the organophosphates (Malathion, Parathion) and the carbamates, both classes noted for their great potency.

   Insecticides are formulated or applied in various ways, as sprays, aerosols, or gases (fumigants), in pellets or granules, oils, injectates, dusts, and so on.

   Other chemicals are useful in combating insects by actions other than killing them. Such are repellents that prevent the pest from doing damage in the first place (Davis 1985). These are most often used against biting flies and evolved from the use of natural substances for centuries by many peoples to spare them the ravages of mosquitoes, punkies, and blackflies. Indians and rural people everywhere build smoky fires to repel insects. The Peruvian Amazon Indians rub their skin with the fruit of the *Siparuna* (Monimiaceae) shrub, which produces a citronella-like odor that is intended to ward off mosquitoes. Very effective artificially made repellents are dimethyl phthalate and DEET (N,N-diethyl-3-methylbensamide).

   Another form of chemical control is the release of laboratory-produced pheromones to confuse the mating behavior of crop pests and suppress their numbers by interfering with reproduction (Jacobson 1965). This seems to have the best potential with lepidopterous pests (Roelofs and Cardé 1977). A substance mimicking the courtship pheromone of the pink bollworm ("gossyplure") has been fairly successful in this way in cotton fields. Sex-attractant pheromones also are used in traps to catch Mediterranean ("medlure") and Mexican Fruit Flies so that their presence can be detected and monitored in infested areas. Hormone analogues, mostly growth regulators, are now known which disrupt normal growth and kill. Such sophisticated chemical attacks, engineered to spurn only the offensive species and operating on fundamental life processes so that resistance is unlikely, constitute the modern front of pest control ("third generation insecticides" or semiochemicals).

2. *Physical control.* These methods often involve the use of special equipment, machinery, and electrical or radiation-producing devices. They are generally

costly in time and labor and rarely give general control. Some, however, are simple, such as the use of screens or barriers. A very direct approach is hand picking and destruction by gathering and disposing of the pest. This was done against larval Lepidoptera, such as sphinx caterpillars and earworms, by ancient Americans but is not practiced on a commercial scale today. However, not very long ago, bounties were offered for scorpions in Durango, Mexico. It is reported that from April 1785 to October 1787, prizes were paid on 506,644 scorpions in that city (Baerg 1929: 422).

Pests may be caught in machines where they are killed by exposure to fumigants or excessively high or low temperatures. Most recently, electromagnetic radiation has been applied in different forms for control. Light draws many insects to their death in traps set at the edges of fields or around habitations. Ionizing radiation is a very effective device in the war against the screwworm and various fruit flies. It is used to sterilize males in great numbers so that they may be released to flood local populations and eliminate effective reproduction ("sterile male technique"). Ultrasonic waves have been tried for insect control but are totally ineffective (Ballard and Gold 1983, Lewis et al. 1982). Other physical forces, such as lasers and radiowaves, are now undergoing trials as potential control agents.

3. *Biological control.* An attractive approach because it concentrates on the target organism and causes minimal harm to the environment is biological control (Cock 1985, Hoy and Hertzog 1985). Classically, this has depended on the introduction of the pest's natural enemies (Caltagirone 1957, Sweetman 1958), which include predators, parasites, parasitoids, and disease microorganisms (Maramorosch and Sherman 1985). Many of the first three are other insects but may include insectivorous vertebrates such as birds, fish, toads, or lizards. This method is usually also self-sustaining and can maintain pest populations at low levels indefinitely as long as some hosts are left to sustain predator-parasite populations.

There are many examples of biological control in Latin America. The area serves both as the recipient of control agents and as the source of them. In the former category are five parasitic wasp species introduced into Cuba and Mexico from the Middle East which finally controlled the citrus blackfly (*Aleurocanthus woglumi*) in the first half of this century after many attempts (De Bach 1974: 139f., 167f.). An example of the latter case is an internal wasp parasitoid from Brazil (*Tetracnemus peregrinus*) which helped stop the mealybug, *Pseudococcus adonidum,* in parts of the United States early in this century. The Amazon fly (*Metagonistylum minense,* Tachinidae) has been transported from its native South America to several areas of the Caribbean where it now helps to keep the sugarcane borers (*Diatraea*) in check (De Bach 1974: 143 f.).

Other forms of biological control are being devised in endlessly ingenious ways by contemporary researchers in practical entomology. *Bacillus thuringiensis* is a bacterium lethal to caterpillars and other larval forms. Commercial preparations of the protein crystalline inclusions in the spores may be disseminated and act like a specific insecticide (Thuricide). Genetic control (Kirschbaum 1985, Pal and Whitten 1974) takes advantage of lethal or repressive genes that entomologists artificially introduce into wild populations from laboratory-reared individuals carrying these genes.

4. *Cultural control.* This method of control uses ordinary farm or management prac-

tices to reduce damage from pests as much as possible. It is the cheapest of all control measures but must be planned far in advance of the season of potential damage. It is also necessary to understand in detail the life history and habits of the insect pests involved.

The most common application of cultural control is crop rotation and timed tilling or soil cultivation. The objective is to remove the insect's food source and modify its environment at critical times of its life cycle. By varying the time of planting and harvesting, infestations may be avoided or much reduced. The use of resistant crop or animal varieties also keeps pest problems to a minimum (Maxwell and Jennings 1980). Some strains or varieties of cultivars are more or less resistant to insect attack and can be selected for propagation in pest-prone areas.

5. *Legal control.* The law can be applied against insect infestations and considerably ameliorate major problems. A powerful weapon against both medical and agricultural pests is quarantine. Spread of the offending species outside of the primary area is prevented, and control efforts can be concentrated on eradication. Legal measures are also important in making sure that pests are kept out of a country or region. Historically, the most serious injurious insects have been imported from other places. Free of their natural enemies in their home territories, their populations explode in the new lands. Laws are necessary to enforce safe use of dangerous insecticides and movement of materials that might spread problem species. They also establish agencies to control and study injurious insects and related arthropods, such as agricultural schools and experiment stations, pest control commissions and boards, institutes, and abatement districts.

## References

AGUILAR, P. G. 1989. Las arañas como controladoras de plagas insectiles en la agricultura peruana. Rev. Peruana Entomol. 331: 1–8.

APPLE, J. L., AND R. F. SMITH. 1976. Integrated pest management. Plenum, New York.

BAERG, W. J. 1929. Some poisonous arthropods of North and Central America. 4th Int. Cong. Entomol. (Ithaca, 1928) Trans. 2: 418–438.

BALLARD, J. B., AND R. E. GOLD. 1983. The response of male German cockroaches to sonic and ultrasonic sound. Kans. Entomol. Soc. J. 56: 93–96.

BARDUCCI, T. B. 1971. Ecological consequences of pesticides used to control cotton insects in the Cañete Valley, Perú. *In* M. T. Farvar and J. P. Milton, eds., The careless technology—ecology and international development. [Not seen.]

BUCHEL, K. H. 1983. Chemistry of pesticides. Wiley, New York.

CALTAGIRONE, L. 1957. Insectos entomófagos y sus huéspedes anotados para Chile. Dir. Gen. Prod. Agrar. Pesq., Santiago, Agric. Téc. Min. Agr. 17: 16–48.

COCK, M. J. W., ed. 1985. A review of biological control of pests in the Commonwealth Caribbean and Bermuda up to 1982. Commwealth. Inst. Biol. Contr. Tech. Comm. 9: 1–218.

CORBETT, J. R., K. WRIGHT, AND A. C. BAILLIE. 1984. The biochemical mode of action of pesticides. Academic, New York.

DAVIS, E. E. 1985. Insect repellents: Concepts of their mode of action relative to potential sensory mechanisms in mosquitoes (Diptera: Culicidae). J. Med. Entomol. 22: 237–243.

DE BACH, P. 1974. Biological control by natural enemies. Cambridge Univ. Press, London.

GEORGHIOU, G. P., AND T. SAITO. 1983. Pest resistance to pesticides. Plenum, New York.

GREEN, M. B. 1976. Pesticides—Boon or bane. Westview, Boulder, Colo.

HOY, M. A., AND D. C. HERTZOG, eds. 1985. Biological control in agricultural IPM systems. Academic, Orlando.

JACOBSON, M. 1965. Insect sex attractants. Wiley Interscience, New York.

KIRSCHBAUM, J. B. 1985. Potential implication of genetic engineering and other biotechniques to insect control. Ann. Rev. Entomol. 30: 51–70.

LEWIS, D. O., W. L. FAIRCHILD, AND D. J. LEPRINCE. 1982. Evaluation of an electronic mosquito repeller. Can. Entomol. 114: 699–702.

MARAMOROSCH, K., AND K. E. SHERMAN, eds.

1985. Viral insecticides for biological control. Academic, Orlando.

MARTIN, H., AND D. WOODCOCK. 1983. The scientific principles of crop protection. 7th ed. Arnold, London.

MARTIN, H., AND C. R. WORTHING. 1976. Insecticide and fungicide handbook. Blackwell, Oxford.

MATSUMURA, F. 1975. Toxicology of insecticides. Plenum, New York.

MAXWELL, F. G., AND P. R. JENNINGS. 1980. Breeding plants resistant to insects. Wiley, New York.

METCALF, R. L., AND W. H. LUCKMAN, eds. 1982. Introduction to insect pest management. 2d ed. Wiley, New York.

PAL, R., AND M. J. WHITTEN. 1974. The use of genetics in insect control. Elsevier/North-Holland, Amsterdam.

ROELOFS, W. L., AND R. T. CARDÉ. 1977. Responses of Lepidoptera to synthetic sex pheromone chemicals and their analogues. Ann. Rev. Entomol. 22: 377–405.

SILVEIRA NETO, S. 1972. Levantamento de insectos e flutação da poblação de pragas da ordem Lepidoptera, con o uso de armadilhas luminosas en diversas regiões do Estado de São Paulo. Lib. Doc., São Paulo.

STRONG, D. R. 1984. Banana's best friend. Nat. Hist. 93(12): 50–57.

SWEETMAN, H. L. 1958. The principles of biological control. Brown, Dubuque.

VAN HUIS, A. 1981. Integrated pest management in the small farmer's maize crop in Nicaragua. Mededelingen Landbouwhogeschool 81–86, Wageningen, Netherlands.

WILKINSON, C. F., ed. 1976. Insecticide biochemistry and physiology. Plenum, New York.

WISWESSER, W. L., ed. 1976. Pesticide index. 5th ed. Entomol. Soc. Amer., College Park, Md.

# VALUABLE INSECTS

The economics of insect life should not be viewed only negatively. Insects are valuable, even essential, to mankind in a variety of ways; these may be categorized as (1) economic, (2) ecological, (3) scientific, and (4) aesthetic.

## Economic Value

Because of their products and services, insects have in the past possessed, and still possess, considerable worth in the market; the potential of many more is unrealized. It is not just primitive or extinct native cultures that profit from this resource; the modern world does so as well.

Undoubtedly, the most valuable and heavily used insect products are honey and wax from the honeybee and silk from the domestic silk moth. Both insects are the basis for large industries (see Stingless Bees and Honeybee, chap. 12) in Latin America, although the direct production of raw silk has waned from former days (see Domestic Silk Moth, chap. 10). Fine silk textiles are now woven in factories, but they rely on raw material from the Orient and elsewhere.

The honeybee (*Apis mellifera*) is maintained by agriculturists not only for its products but also for its service as a pollinator of crops. The monetary value of this service in all Latin American countries exceeds millions of dollars annually in increased production of fruit and seeds.

Many insects, especially showy butterflies, moths, and beetles, are collected as the raw material to make artistic, decorative, or curiosity items for sale around the world. Collectors buy many of these, but far more are bought by tourists and artisans who may use only a portion of the insect's body to incorporate into salable pieces. In some countries, cottage industries of major economic importance have developed around the insect trade (see Butterflies, chap. 10).

Phytophagous insects are used as control agents against weeds and other undesirable plants. The larvae of cactus moths (see Cactus Moths, chap. 10) have been exported to many parts of the world from their native Argentina to destroy *Opuntia* cactus stands that spoil vast acreages of pasture- and cropland. Other insects help to subdue explosive growths of water weeds, such as alligator weed (Buckingham et al. 1983) and the water hyacinth (Center 1982). A lymantriid moth (*Elnoria noyesi*), called *malunya* (or *malumbia*) in Peru, has been considered as a possible but probably ineffective control for illegal coca cultivation (Berenbaum 1991).

Insects continue to be a source of food for native Americans in all areas (Bodenheimer 1951). Large moth and beetle larvae and, to a lesser extent, ants, wasp larvae, and some true bugs (see agave worms, chap. 10; rhinoceros beetle, chap. 9; leaf cutter ant, chap. 12) are the types most often eaten. This was noted by the early observers of life in the New World (Wallace 1853) and continues to be documented as an active cultural practice in contemporary times (Ruddel 1973). Studies have been sanctioned by the Mexican government to determine the feasibility of expanding local entomophagy to supplement the diet of a growing rural population directly or for animal feed (Elorduy de Conconi 1982).

Perhaps the greatest potential for profit from insects in the future lies in their abilities to synthesize biochemicals for medical use. Shamans and village doctors employ insects in remedial concoctions or apply them directly to diseased or damaged organs. The Aztecs used the oil of black widow spiders (*vitztle*) to stop pain (Curran 1937). Insects may not provide the raw material for production of quantities of these substances directly but guide pharmacologists to unusual compounds with healing qualities in a fashion paralleling the already proven method with medicinal plants ("drug prospecting"). Aposematic species that advertise the presence of substances manufactured in their bodies which are capable of altering vertebrate metabolisms will be prime candidates in this process; their warning colors may also help us to find yet unknown pharmacologically potent plants from which they have sequestered powerful chemical agents (Brown 1979).

The succession of communities of necrophilous insects in human cadavers is the key to their use in determining time of death as sometimes employed in forensic medicine (Vargas-Alvarado 1983, Keh 1985, Smith 1986).

## Ecological Value

Insects are unseen and usually unappreciated benefactors of mankind through the roles they play in ecosystems. They pollinate most wild species of flowering plants, they help reduce tremendous quantities of animal and plant waste and thereby add nutrients to and improve quality of soils (Seastedt 1984), they help regulate the numbers of other organisms, and they act as a primary food source for all sorts of insectivorous animals, most conspicuously, vertebrates. They are the greatest converters of plant matter to animal protein.

## Scientific Value

Science makes frequent use of insects as experimental organisms for the study of basic biological and physical processes. The pomace fly, *Drosophila melanogaster*, has been responsible for providing insight into our basic understanding of genetics. Because of their small size, ease of maintenance, and hardiness, insects are convenient organisms to place aboard experimental rockets to study the effects of outer space on life. Many insects, especially aquatic species, are useful as bioindicators. Their presence or absence or body modifications are characteristics sensitive to changes in the atmosphere, to toxic substances in air or water, and even to radioactivity.

## Aesthetic Value

Finally, it may be argued also that insects are part of the world to be valued for their own sake, apart from any direct application to which their existence may be put. They should be aesthetically appreciated for their beauty, their intricacy of form, and the lessons they teach about their ways of life.

# References

BERENBAUM, M. 1991. Just say "Notodontid?" Amer. Entomol. 37: 196–197.

BODENHEIMER, F. S. 1951. Insects as human food. Junk, The Hague.

BROWN, JR., K. S. 1979. Insetos aposemáticos: Indicadores naturais de plantas medicinais. Cien. Cult. 32 (suppl.): 189–200.

BUCKINGHAM, G. R., D. BOUCIAS, AND R. F. THERIOT. 1983. Reintroduction of the alligatorweed flea beetle (*Agasicles hygrophila* Selman and Vogt) into the United States from Argentina. J. Aquat. Plant Man. 21: 101–102.

CENTER, T. D. 1982. The water hyacinth weevils. Aquatics 4(2): 8, 16, 18–19.

CURRAN, C. H. 1937. Insect lore of the Aztecs. Nat. Hist. 39: 196–203.

ELORDUY DE CONCONI, J. R. 1982. Los insectos como fuente de proteínas en el futuro. Ed. Limusa, Mexico City.

KEH, B. 1985. Scope and applications of forensic entomology. Ann. Rev. Entomol. 30: 137–154.

RUDDEL, K. 1973. The human use of insects: Examples from the Yukpa. Biotropica 5: 94–101.

SEASTEDT, T. R. 1984. The role of microarthropods in decomposition and mineralization processes. Ann. Rev. Entomol. 29: 25–46.

SMITH, K. G. V. 1986. A manual of forensic entomology. Brit. Mus. Nat. Hist., London.

VARGAS-ALVARADO, E. 1983. Medicina legal. 3d ed. Ed. Lehmann, San José, Costa Rica.

WALLACE, A. R. 1853. On the insects used as food by the Indians of the Amazon. Entomol. Soc. London Trans. (n.s.) 2: 241–244.

# 4 TERRESTRIAL ARTHROPODS AND PRIMITIVE INSECTS

Other similar-appearing lower animals are often confused with insects. A jointed body, legs, and a "crawly" countenance are all that are required for many people to lump a wide variety of terrestrial arthropods with the true insects. Furthermore, these groups, such as spiders and mites, are of popular interest and considerable economic significance. It is appropriate, therefore, to include these insectlike groups (Clarke 1973, Cloudsley-Thompson 1958, Kaestner 1968, Parker 1982) in this book. The fact that they merge evolutionarily with the primitive Hexapoda also justifies addressing them. They are discussed here, prior to the chapters on the insects themselves.

## References

BRUSCA, R. C., AND G. J. BRUSCA. 1990. Invertebrates. Sinauer, Sunderland, Mass.

CLARKE, K. U. 1973. The biology of the Arthropoda. American Elsevier, New York.

CLOUDSLEY-THOMPSON, J. L. 1958. Spiders, scorpions, centipedes and mites. Pergamon, London.

KAESTNER, A. 1968. Invertebrate zoology, arthropod relatives, Chelicerata, Myriapoda. Vol. 2. Wiley Interscience, New York.

PARKER, S. P., ed. 1982. Synopsis and classification of living organisms. Vol. 2. McGraw Hill, New York.

## ONYCHOPHORANS

Onychophora. *Spanish* and *Portuguese:* Onicoforos.

These moderate-sized (BL 2–5 cm), caterpillarlike terrestrial animals, neither Annelida nor Arthropoda, form a separate phylum but combine qualities of both groups (Marcus 1937). Their annelid characteristics include internally repetitious body segmentation, an eye with a simple lens, the presence of nephridia (kidneylike organs) in most body segments, and a soft, flexible, wormlike shape, lacking a hardened exoskeleton. Some of their arthropod features are an open body cavity and circulatory system, modification of a pair of appendages into mandibles, claws on the appendages, a breathing system of tracheae, and an elongate dorsal heart. They also grow by shedding their skins like arthropods.

Onychophorans have their own special structures, including a transversely wrinkled and well-pigmented integument, each fold with many regularly placed papillae. They also have a pair of annulate antennae and special glands in the mouth cavity used to shoot streams of slime to capture prey and fend off enemies.

Knowledge of the biology of these creatures is scant. They require moisture and survive only in humid tropical environments or damp microhabitats in temperate regions. Here they inhabit leaf litter, rotten wood, and other moist retreats, such as banana stems (Young 1980) and cavities under bark. If agitated, they face their antagonist and forcefully spurt streams of sticky mucus from the slime glands in the mouth. These solidify into sticky threads that entangle anything they touch, producing a noxious mess. Silk shooting is also used to immobilize prey.

These are nocturnal animals. Their food

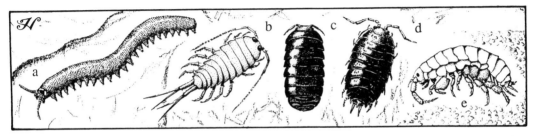

**Figure 4.1 ARTHROPODS.** (a) Onychophoran (*Macroperipatus torquatus*, Peripatidae). (b) Sea "roach" (*Ligia exotica*, Ligiidae). (c) Pillbug (*Armadillidium vulgare*, Armadillidiidae). (d) Sow bug (*Porcellio laevis*, Porcellionidae). (e) Sand "flea" (*Orchestia platensis*, Talitridae).

probably consists mainly of other small invertebrates or of partially decomposed leaf and wood tissue. One species is known to invade termite galleries in rotting wood to prey on their owners (Janvier 1975).

In the New World tropics, there are eight genera of Onychophora, including fifty-seven species in two families (Peck 1975). *Metaperipatus* (Peripatopsidae) live in damp forests in southern Chile (Claude-Joseph 1928). The various genera of Peripatidae are much more widespread in the rain forests of Amazonia and Central America (to southern Mexico) and the Caribbean. Here they crawl over and among the litter in search of other small invertebrates (termites, caterpillars, snails, etc.) on which they prey. *Peripatus heloisae* has been found in large numbers in ground-nesting termite mounds in Brazil by Carvalho (1942), who believes them to be termitophagous. *Speleoperipatus speloeus* is a blind and pale species found in caves in Jamaica (Peck 1975). The largest species, which measures up to 15 centimeters in length and lives in Trinidad, is the collared peripatus (*Macroperipatus torquatus;* fig. 4.1a) so named because of bright yellow markings around the bases of the antennae (Ghiselin 1985).

### References

CLAUDE-JOSEPH, F. 1928. Observations sur un Péripate du Chili. Ann. Sci. Nat. Zool. Ser. 10-11: 285–298.

GHISELIN, M. T. 1985. A movable feaster. Nat. Hist. 94(9): 54–61.

JANVIER, H. 1975. Un péripate du Chili chasseur de termites. Entomologiste 31: 63–68.

LEITÃO DE CARVALHO, A. 1942. Sobre "*Peripatus heloisae*," do Brasil Central. Mus. Nac. Rio de Janeiro (n.s.) Zool. Bol. 2: 57–73.

MARCUS, E. 1937. Sobre os Onychophoros. Insto. Biol. Sec. Agric. São Paulo Arch. 8: 255–266.

PECK, S. B. 1975. A review of the New World Onychophora with the description of a new cavernicolous genus and species from Jamaica. Psyche 82: 341–358.

YOUNG, A. M. 1980. On the patchy distribution of onychophorans in two cacao plantations in northeastern Costa Rica. Brenesia 17: 143–148.

## CRUSTACEANS

The Crustacea (Bliss 1982–, Schram 1986) are essentially marine, but many have become adapted to fresh water, and some live in moist places on land. Structurally, they are characterized primarily by the presence of biramous appendages, including mandibles whose grinding surface is served by the inner surface of the second primitive segment (gnathobasic jaws). The first-stage nauplius larva, a simple swimming form with three pairs of appendages and a single median eye, is a unique stage found only in this group of arthropods.

Crustaceans are extremely diverse and widespread in the sea. Only a few types are habitually terrestrial and appropriately included here.

## References

BLISS, D. E., ed. 1982–. The biology of the crustacea. Vols. 1–9. Academic, New York.

SCHRAM, F. R. 1986. Crustacea. Oxford, New York.

## TERRESTRIAL ISOPODS

Crustacea, Isopoda. *Spanish:* Cochinillas de la humedad, correderas (General). *Portuguese:* Tatuzinhos, baratinhas, bichos de conta (Brazil).

Many members of this large and diverse crustacean group (Mulaik 1960, Van Name 1936) are insectlike. They live in humid locations, often near water, including in tank plants such as bromeliads. Although they usually stay near moisture, they are capable of reproducing without depositing their eggs in free water. They may be particularly abundant in leaf litter and decaying vegetation, apparently feeding on the organic debris and fungi associated with such matter. Wood lice, which are found in decomposing leaf and wood litter, are the most familiar representatives. The group is also well represented in caves (Schultz 1981).

Terrestrial isopods are mostly small (BL 5–20 mm) and fairly uniform (onisciform): oval in outline and somewhat flattened; body segments distinct and more or less equal, except the posterior whose lateral portions are strongly curved to the rear; anterior segments may be expanded winglike to the sides. The thoracic region makes up most of the body length and consists of seven segments, each with a pair of simple legs. Both pairs of antennae are short, but one pair is much shorter than the other. They have stout mandibles and well-developed eyes.

There are terrestrial members of only a few families of Isopoda in the Neotropics. Some of the more common are the somewhat amphibious Tylidae and Ligiidae (*Ligia,* fig. 4.1b; *Ligidium*), which live along the seashore within or just above the tidal zone and inland by watercourses—"sea roaches." These have extra long antennae. The near-blind Trichoniscidae are very small (BL up to 5 mm) and inhabit dense organic litter in caves.

Pillbugs (fig. 4.1c), so called because they can roll themselves into a ball, belong to the Armadillidiidae and Armadillidae. These are dull colored, often solid gray or gray-brown with lighter mottling. To resist moisture loss, the cuticle is usually tough and leathery; in some, it is quite rigid and also a good protective armor for the internal organs.

In nature, the true woodlice or Porcellionidae (sowbugs) (fig. 4.1d) and Oniscidae live in accumulations of dead decomposing plant matter. They are also abundant, sometimes very abundant, in gardens, greenhouses, and domestic situations where they are considered pests.

The literature is scant on these terrestrial arthropods; few faunal papers exist (Vandel 1972).

## References

MULAIK, S. B. 1960. Contribución al conocimiento de los isópodos terestres de México (Isopoda, Oniscoidea). Soc. Mex. Hist. Nat. Rev. 21(1): 79–294.

SCHULTZ, G. A. 1981. Isopods (Oniscoidae) from caves in North America and northern South America. 8th Int. Cong. Speleol. Proc. 1: 551–552.

VANDEL, A. 1972. Les Isopodes terrestres de la Colombie. Stud. Neotrop. Fauna 7: 147–172.

VAN NAME, W. G. 1936. The American land and fresh-water isopod Crustacea. Amer. Mus. Nat. Hist. Bull. 71: 1–535.

## TERRESTRIAL AMPHIPODA

Crustacea, Amphipoda, Talytroidea, Talitridae, and other families. Beach hoppers, beach fleas, sand fleas, scuds.

Amphipods somewhat resemble isopods in basic structure and biology but are typically more slender and laterally compressed. Both pairs of antennae are usually well developed and long. The legs are variously modified, some of the anterior often as grasping devices (raptorial) and always with three pairs specialized for walking or hopping (uropods). The exoskeleton may be thin, mineralized, or occasionally heavily sclerotized. It is generally well pigmented and often pitted or otherwise microsculptured. The dorsal plates often have lateral, winglike flanges.

Like the isopods, amphipods require a moist or watery environment, not only for survival but for reproduction as well. They always return to water to deposit their eggs. There are several aquatic larval stages, the animal moving to land only with the last molts.

The Talitridae are known everywhere and normally encountered burrowing in damp sand, particularly that beneath beach-stranded seaweed and other debris. The dominant Central and South American genus is *Hyale,* with many species from all shores (Barnard 1979). The widespread sand flea *Orchestia platensis* (fig. 4.1e) is only common on Caribbean seashores, there being no members of the family on mainland South America in spite of the fact that the family is otherwise cosmopolitan.

This is a neglected group in the region. Much more is to be learned regarding species present and their biology in Latin America.

## Reference

BARNARD, J. L. 1979. Litorral gammaridean Amphipoda from the Gulf of California and the Galápagos Islands. Smithsonian Contrib. Zool. 271: 1–149.

# ARACHNIDS

## Arachnida

Arachnids comprise the majority of the chelicerates and are predominantly terrestrial (Besch 1969, Cloudsley-Thompson 1958, Savory 1977).

Chelicerates are defined by the structure of the anteriormost appendages (chelicerae), which are made up of a bulky basal segment with an apical movable finger. The chelicerae are modified to form varied organs, such as the venom apparatus (fangs) in spiders, piercing stylets in parasitic mites, or masticating jaws of scorpions.

The several orders are quite diverse. In all, the head is undefined, being fused with the thorax (cephalothorax); in some, the abdomen is also fused into a single body complex (mites).

Arachnids are an ancient stock, and they are diverse today, though poorly studied in Latin America. A great many species, especially among the spiders and mites, are yet to be discovered.

Arachnids exhibit a strongly climate-dependent distribution, mainly in two directions. There are those in moist habitats (Uropygi, Amblypygi) and deserticolous forms (Solpugida). Some transcend these divisions and are widespread and broadly adapted (Araneae, Scorpionida); other minor groups have specialized niches, such as ectoparasitism.

Several anatomical characteristics distinguish the arachnids. The head and thorax (leg-bearing portion) are fused into a single unit, the cephalothorax (prosoma). The abdomen (opisthosoma) may or may not be distinct. Mouthpart appendages are composed of a single pair of pinching or piercing chelicerae, with a stationary base and movable finger. A pair of segmented sensory appendages (pedipalps) precede the four pairs of legs. Some spiders and scorpions are capable of audible stridulation (Lucas and Bücherl 1972).

Arachnids are protected mainly by their secretive, commonly nocturnal habits and camouflage. Besch (1969) noted the predominance of green in the coloration of spiders and other Arachnida in South America. Many are capable of dropping appendages to escape capture (appendoto-my) (Roth and Roth 1984). Quite a few also produce noxious chemicals, some of which are very potent in repelling enemies.

Most arachnids are predaceous and feed by ejecting enzymes onto the prey from a preoral cavity and siphoning the resultant liquids, or by piercing the prey's skin and sucking blood or lymph. A few are plant feeders, using elongate sucking mouthparts to take sap or other liquids from the host.

## References

BESCH, W. 1969. South American Arachnida. *In* E. J. Fittkau, J. Illies, H. Klinge, G. H. Schwabe, and H. Sioli, eds., Biogeography and ecology in South America. 2: 723–740. Junk, The Hague.

CLOUDSLEY-THOMPSON, J. L. 1958. Spiders, scorpions, centipedes and mites. Pergamon, New York.

LUCAS, S., AND W. BÜCHERL. 1972. Aparelhos estriduladores do escorpião, *Rhopalurus iglesiasi dorsomaculatus* (Prado 1938), e da aranha caranguejeira, *Theraphosa blondi* (Latreille) 1804. Cien. Cult. 23: 635–637.

ROTH, V. D., AND B. M. ROTH. 1984. A review of appendotomy in spiders and other arachnids. Bull. Brit. Arachnol. Soc. 6: 137–146.

SAVORY, T. 1977. Arachnida. 2d ed. Academic, London.

## SPIDERS

Arachnida, Araneae (= Araneida).
*Spanish:* Arañas. *Portuguese:* Aranhas.
*Quechua:* Vilca-kuna. *Tupi-Guarani:*
Nhandui. *Nahuatl:* Tocameh, sing.
tócatl (Mexico).

Spiders (Foelix 1982, Gertsch 1979, Preston-Mafham and Preston-Mafham 1984) comprise a large and very familiar group and are fascinating and diverse in Latin America (Robinson 1984). The abdomen is almost always unsegmented and narrowly attached to the cephalothorax by a stalk. The chelicerae are modified into fangs connected to internal poison glands.

The pedipalps are leglike, and there are short terminal abdominal appendages developed in association with silk glands that open between them (spinnerets).

These creatures are found in all natural situations, often in close association with humans (Turnbull 1973). Most are reclusive and select dark retreats as living quarters. They are often found in caves (Gertsch 1973, Brignoli 1972). Others, however, are sun loving and free ranging. Several kinds even inhabit the marine intertidal zone (Roth and Brown 1975). In seasonal climates, they may be most abundant in wet periods when more prey is available than during dry periods (Lubin 1978).

Although spiders are typically solitary, a number of species show extended parental care and even relatively permanent social gatherings (Burgess 1978, Buskirk 1981). In a few species, colony members even cooperate in web building and nest maintenance, displaying mutual tolerance and communication (Buskirk 1981, Lubin 1980).

Since pre-Columbian times, people in Central Mexico have brought branches with the webs and spiders of a small (BL 5 mm) social species, called *el mosquero* (*Mallos gregalis*), into their homes to reduce the number of flies that invade during the rainy season (Simon 1909). Many individuals cooperate to build a dense sheet web with many chambers and retreats. They not only exhibit group tolerance but practice communal prey capture as well (Burgess 1979, but see Jackson 1979 and Tietjen 1986). The nest is occupied and expanded by successive generations (Burgess 1976). The theridiid *Anelosimus eximius* is the most widespread of several communal species in the genus in South America (Rypstra and Tirey 1989). It builds giant webs (1 m by nearly 5 m or more) containing hundreds, even thousands, of individuals who work, prey (Krafft and Pasquet 1991), and spin together (Levi 1963).

The perception and production of

sound is developed in many spiders (Legendre 1963) but is only one of their many means of communication (Witt and Rovner 1982). They also keep in touch by tugging and vibrating the web and through pheromones and, more rarely, vision.

Many spiders are cryptic in form, color, and behavior, and a considerable number exhibit mimicry, primarily of ants (Peckham 1889). They possess structural (elongate body and legs, constricted abdomen), behavioral (forelegs held in elevated position like antennae), and color modifications, all giving them a close resemblance to their formicine models. In the Neotropics, these mimics mostly belong to the families Clubionidae, subfamily Castianeirinae (Reiskind 1969), such as *Castianeira rica* (Reiskind 1970, 1977), and Salticidae (see Jumping Spiders, below). Some species even resemble several distinct types of ants through variation in colors, sexual dimorphism, and changes during development, so-called transformational mimicry. Velvet ants (Mutillidae) are also mimicked.

Several kinds of small flies are associated with spiders, as commensals (Robinson and Robinson 1977) or by using the web as protected perches (Lahmann and Zúñiga 1981). Symbioses involving spiders seem rare, although a few unstudied cases of interactions with ants are known (Noonan 1982). There are many kleptoparasitic types that live in the webs of other species from which they steal unattended prey (*arañas ladronas*, Restrepo 1948).

All spiders are obligate carnivores and possess venom produced by large internal glands that empty their products through ducts opening at the tips of the fangs. Spiders use poison to subdue prey or in defense. The quantity of venom and toxicity of most is slight, so that spider bites are usually of little or no medical importance. A few species, however, do have the capacity of harming (Bettini and Brignoli 1978, Schenone 1953) or even killing humans.

Some of these are found in Latin America, mainly in the genera *Trechona*, *Phoneutria*, and *Latrodectus*, whose venoms are neurotoxic. *Loxosceles* bites, containing hemolytic toxins, while occasionally severely disfiguring, are not lethal (Bücherl 1971).

Although other arthropods spin silk, nowhere is the process so well developed and web making so elaborate as in the spiders (Savory 1952). Spider silk has the highest tensile strength of any natural fiber and is ideal for constructing snares and traps for insects, as well as nests, retreats, and other structures. Webs may be relatively small, amorphous, and loosely formed or highly complex, symmetrical, and large, as in the orb weavers. The relative abundance of the different types depends on the taxonomic composition and ecological characteristics of an area (Lieberman-Jaffe 1981).

The number of species forming the Neotropical spider fauna is unknown but must be very large. There are few general works for the region (Pikelin and Schiapelli 1963). Two major groups are recognized, the Labidognatha (most), in which the chelicerae, or fangs, work laterally, like pliers against each other, and the Orthognatha, in which the fangs are parallel and fold back along the long axis of the body (tarantulas and relatives).

Most spiders have eight "eyes" located on the back of the anterior portion of the cephalothorax. The number and arrangement relative to these eyes are often used as identifying characteristics.

## References

BETTINI, S., AND S. M. BRIGNOLI. 1978. Review of the spider families, with notes on the lesser-known poisonous forms. *In* S. Bettini, ed., Arthropod venoms. Springer, Berlin. Pp. 101–120.

BRIGNOLI, P. M. 1972. Sur quelques araignés cavernicoles d'Argentine, Uruguay, Brésil et Vénézuela récoltées par le Dr. P. Strinate (Arachnida, Araneae). Rev. Suisse Zool. 79: 361–385.

BÜCHERL, W. 1971. Spiders. *In* W. Bücherl and E. E. Buckley, eds., Venomous animals and their venoms. 3. Venomous invertebrates. Academic, New York. Pp. 197–277.

BURGESS, J. W. 1976. Social spiders. Nat. Hist. 234: 101–106.

BURGESS, J. W. 1978. Social behavior in group-living spider species. Zool. Soc. London Symp. 42: 69–78.

BURGESS, J. W. 1979. Web-signal processing for tolerance and group predation in the social spider *Mallos gregalis*. Simon. J. Anim. Behav. 27: 157–164.

BUSKIRK, R. E. 1981. Sociality in the Arachnida. *In* H. R. Hermann, ed., Social insects. 2: 281–367. Academic, New York.

FOELIX, R. F. 1982. Biology of spiders. Harvard Univ. Press, Cambridge.

GERTSCH, W. J. 1973. A report on cave spiders from Mexico and Central America. Assoc. Mex. Cave Stud. Bull. 5: 141–163.

GERTSCH, W. J. 1979. American spiders. 2d ed. Van Nostrand Reinhold, New York.

JACKSON, R. R. 1979. Predatory behavior of the social spider *Mallos gregalis:* Is it cooperative? Ins. Soc. 26: 300–312.

KRAFFT, B., AND A. PASQUET. 1991. Synchronized and rhythmical activity during the prey capture in the social spider *Anelosimus eximius* (Araneae, Theridiidae). Ins. Soc. 38: 83–90.

LAHMANN, E. J., AND C. M. ZÚÑIGA. 1981. Use of spider threads as resting places by tropical insects. J. Arachnol. 9: 339–341.

LEGENDRE, R. 1963. L'audition et l'emission de sons chez les aranéides. Ann. Biol. Ser. 4(2): 371–390.

LEVI, H. W. 1963. The American spiders of the genus *Anelosimus* (Araneae: Theridiidae). Amer. Micro. Soc. Trans. 82: 30–48.

LIEBERMAN-JAFFE, S. 1981. Ecology of web-building spiders at Corcovado National Park, Costa Rica: A preliminary study. Stud. Neotrop. Fauna Environ. 16: 99–106.

LUBIN, Y. D. 1978. Seasonal abundance and diversity of web-building spiders in relation to habitat structure on Barro Colorado Island, Panama. J. Arach. 6: 31–51.

LUBIN, Y. D. 1980. Population studies of two colonial orb-weaving spiders. Zool. J. Linnaean Soc. 70: 265–287.

NOONAN, G. R. 1982. Notes on interactions between the spider *Eilica puno* (Gnaphosidea) and the ant *Camponotus inca* in the Peruvian Andes. Biotropica 14: 145–148.

PECKHAM, G. W. 1889. Protective resemblances in spiders. Nat. Hist. Soc. Wisc. Occ. Pap. 1: 61–113.

PIKELIN, B. S. G., AND R. D. SCHIAPELLI. 1963. Llave para la determinación de familias de arañas argentinas. Physis 24: 43–72.

PRESTON-MAFHAM, R., AND K. PRESTON-MAFHAM. 1984. Spiders of the world. Facts on File, New York.

REISKIND, J. 1969. The spider subfamily Castianeirinae of North and Central America. Mus. Comp. Zool. (Harvard Univ.) Bull. 138: 163–325.

REISKIND, J. 1970. Multiple mimetic forms in an ant-mimicking clubionid spider. Science 169: 587–588.

REISKIND, J. 1977. Ant-mimicry in Panamanian clubionid and salticid spiders (Araneae: Clubionidae, Salticidae). Biotropica 9: 1–8.

RESTREPO, A. 1948. La araña ladrona. Fac. Nac. Agron. (Medellín) Rev. 17: 157–168.

ROBINSON, M. H. 1984. Neotropical arachnology: Historic, ecological and evolutionary aspects. 9th Cong. Latinoamer. Zool. (Arequipa) Inf. Final. Pp. 89–100.

ROBINSON, M. H., AND B. ROBINSON. 1977. Association between flies and spiders: Bibiocommensalism and dipsoparasitism. Psyche 84: 150–157.

ROTH, V. D., AND W. L. BROWN. 1975. A new genus of Mexican intertidal zone spider (Desidae) with biological and behavioral notes. Amer. Mus. Nov. 2568: 1–7.

RYPSTRA, A. L., AND R. S. TIREY. 1989. Observations on the social spider, *Anelosimus domingo* (Araneae, Theridiidae), in southwestern Peru. J. Arachnol. 17: 368–370.

SAVORY, T. H. 1952. The spider's web. Warne, London.

SCHENONE, H. 1953. Mordeduras de arañas. Bol. Chileno Parasit. 8: 35–37.

SIMON, E. 1909. Sur l'araignée Mosquero. Acad. Sci. (Paris) Compt. Rend. Séance 148: 736–737.

TIETJEN, W. J. 1986. Social spider webs, with special reference to the web of *Mallos gregalis*. *In* W. A. Shear, ed., Spiders: Webs, behavior, and evolution. Stanford Univ. Press, Stanford. Pp. 172–206.

TURNBULL, A. L. 1973. Ecology of the true spiders (Araneomorphae). Ann. Rev. Entomol. 18: 305–348.

WITT, P. M., AND J. S. ROVNER, eds. 1982. Spider communication: Mechanisms and ecological significance. Princeton Univ. Press, Princeton.

## Tarantulas

Theraphosidae. *Spanish:* Tarántulas, arañas peludas (General), matacaballos (Mexico, Central America). *Portuguese:* Tarántulas, caranguejeiras (Brazil). *Tupi-Guarani:* Nhanduguaçá. Mygalomorphs, bird spiders.

This family is renowned for the enormous size of many of its members. The largest are males of *Theraphosa lablondi* (fig. 4.2a), which have a leg span as great as 25 centimeters. Specimens are known also with body lengths of 12 centimeters and with anterior femoral diameters up to 8 millimeters (Gerschman de Pikelin and Schiapelli 1966). The bulkier females are less expansive but may weigh 3 ounces (Gertsch 1979). Most species are smaller but are still large by spider standards (leg span 7.5 to 9.5 cm); others are small (less than 3 cm leg span).

Tarantulas are distinguished from other large hairy spiders by their large vertical fangs and legs that have only two claws, instead of the usual three. The equally sized, small eyes are closely grouped on a small tubercle. There are two pairs of large spinnerets.

Because of their formidable size and hairiness, tarantulas are widely regarded with great fear and are believed to be deadly. In spite of these attitudes, these spiders, especially the Mexican red-legged tarantula (*Brachypelma smithi*) from northwestern Mexico, are imported in large numbers to the United States to please pet fanciers (Hemley 1986). *Acanthoscurria,* and others, are legitimately feared by Matto Grosso Indians because of their powerful bite (Bücherl 1971: 229), and *Hapalopus* are known to carry potent venoms (Espinoza 1966). As toxicity to humans of some tarantulas has been documented, all should be given latitude when encountered (Bettini and Brignoli 1978).

While usually shy and retiring, they may become aggressive if threatened. Stories of their jumping abilities may be exaggerated, but some arboreal species are quite capable of leaps of a meter or more over level ground (orig. obs.).

Against humans and enemies, tarantulas use their bite only in defense. They also discourage attack by flicking the hairs of the dorsum of the abdomen with a hind leg. These are urticating and may lodge in the eyes or sensitive nasal membranes of potential predators (Cooke et al. 1973).

Some Neotropical tarantulas produce a snakelike hissing sound by rubbing the surfaces of basal segments of the pedipalps against opposite surfaces of the first legs. Adults of both sexes use this form of stridulation apparently as a protective mechanism; it can be heard up to 6 meters away (Thoms 1983).

These spiders, particularly the females, are long lived. In nature, most probably mature in 5 to 10 years and may live several years thereafter.

The venom is normally used to subdue prey. Their food consists mostly of other ground-dwelling arthropods, although the largest species certainly catch and devour small vertebrates such as frogs, toads, lizards, and nestling or small birds. A famous illustration of a specimen in the act of feeding on a small bird, which appeared in Madame Merian's *Metamorphoses Insectorum Surinamensium* (1705), is probably responsible for the reputation of these spiders as ornithophages.

There are both terrestrial and arboreal tarantulas (pl. 1a). The former spend the day in burrows of their own construction or natural retreats in the soil, emerging at night to hunt or seek mates. Tree-dwelling forms hide among epiphytes or in crevices or under the loose bark of dead trees. Others form silken nests in rolled up leaves of terrestrial bromeliads, bananas or *Heliconia,* in palm spathes, or in the bristly bases of these leaves.

A bit of folklore prevailing in Mexico and Central America is the legend of the

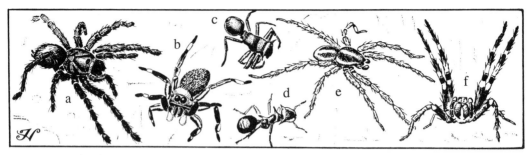

**Figure 4.2  SPIDERS.** (a) Tarantula (*Theraphosa lablondi,* Theraphosidae). (b) Jumping spider (unidentified, Salticidae). (c) Ant-mimicking jumping spider (*Aphantochilus* sp., Salticidae). (d) Ant model of ant-mimicking jumping spider (*Cephalotes* sp., Formicidae). (e) Wolf spider (*Lycosa raptoria,* Lycosidae). (f) Banana spider (*Phoneutria fera,* Ctenidae).

*matacaballo.* At one time, many people thought that hoof-and-mouth disease of cattle was caused by tarantulas. The spiders were presumed to hunt sleeping animals and take a narrow strip of hair from above the hoof for its nest building, using an acid secretion to make the hair slough off (a symptom of the disease). The site of the injury suffers infection, and the hoof may be lost. The disease is actually caused by a bacillus; tarantulas line their nests with their own silk (Gertsch 1979: 117).

The taxonomy of these spiders remains in an unsettled state. Although their higher classification has been organized (Raven 1985), only a few of the major theraphosid groups have received attention internally by modern workers (Schiapelli and Gerschman de Pikelin 1967, 1979; Gerschman de Pikelin and Schiapelli 1973). Approximately four hundred Latin American species are described, ranging through all climes from deserts to rain forest. They are much fewer and less common at the higher elevations.

### References

BETTINI, S., AND P. M. BRIGNOLI. 1978. Review of the spider families, with notes on the lesser-known poisonous forms. *In* S. Bettini, ed., Arthropod venoms. Springer, Berlin. Pp. 101–120.

BÜCHERL, W. 1971. Spiders. *In* W. Bücherl and E. E. Buckley, eds., Venomous animals and their venoms. 3. Venomous invertebrates. Academic, New York. Pp. 197–277.

COOKE, J. A. L., F. H. MILLER, R. W. GROVER, AND J. L. DUFFY. 1973. Urticaria caused by tarantula hairs. Amer. J. Trop. Med. Hyg. 22: 130–133.

ESPINOZA, N. C. 1966. Acción de veneno de *Hapalopus limensis.* Inst. Butantan Mem. 33: 799–808.

GERSCHMAN DE PIKELIN, B. S., AND R. D. SCHIAPELLI. 1966. Contribución al conocimiento de *Theraphosa lablondi* (Latreille), 1801 (Aranea: Theraphosidae). Inst. Butantan Mem. 33: 667–674.

GERSCHMAN DE PIKELIN, B. S., AND R. D. SCHIAPELLI. 1973. La subfamilia Ischnocolinae (Araneae: Theraphosidae). Mus. Argentino Cien. Nat. Bernardino Rivadavia, Insto. Nac. Invest. Cien. Nat. Entomol. Rev. 14: 43–77.

GERTSCH, W. J. 1979. American spiders. 2d ed. Van Nostrand Reinhold, New York.

HEMLEY, G. 1986. Spotlight on the red-kneed tarantula trade. Traffic (U.S.A.) 6(4): 16–17.

RAVEN, R. J. 1985. The spider infraorder Mygalomorphae (Araneae): Cladistics and systematics. Amer. Mus. Nat. Hist. Bull. 182: 1–180.

SCHIAPELLI, R. D., AND B. S. GERSCHMAN DE PIKELIN. 1967. Estudio sistemático comparativo de los géneros "*Theraphosa*" Walck., 1805, "*Lasiodora*" C. L. Koch, 1851 y "*Sericopelma*" Ausserer, 1975 (Aranae, Theraphosidae). Atas Simp. Biota Amazonica 5: 481–494.

SCHIAPELLI, R. D., AND B. S. GERSCHMAN DE PIKELIN. 1979. Las arañas de la subfamilia Theraphosinae (Araneae, Theraphosidae). Mus. Argentino Cien. Nat. Bernardino Rivadivia, Insto. Nac. Invest. Cien. Nat. Entomol. Rev. 5: 287–300.

THOMS, E. 1983. Sound production by a South

American theraphosid. Paper delivered at the annual meeting of the Entomological Society of America, Detroit.

## Jumping Spiders

Salticidae. *Spanish:* Papamoscas (Peru).

In spiders of this family, the anterior, median eyes are paired and enormously enlarged and face forward on the steeply elevated anterior portion of the cephalothorax. They afford the animal excellent binocular vision, enabling it to judge distance very well. Two pairs of additional smaller eyes are located behind the principal eyes.

Good eyesight is correlated with oversized forelegs, and short, strong hind legs, all modifications for prey capture by ambush (Forster 1982). Prey is stalked by characteristic jerking, orienting visual movements of the whole body, to within 7 to 15 centimeters distance; then a thread is attached to the substrate and the final distance jumped. The spider may then drop off the substrate on the thread line to isolate the quarry. Extensive webbing is spun only for refuge and protection of the eggs. A wide variety of insects and other arachnids are taken; some salticids attack and subdue victims considerably larger than themselves (Robinson and Valerio 1977). They even pounce on orb spiders situated in the center of their webs.

The males often have extra long chelicerae and are brightly colored, frequently in polychrome. The patterns of many are complex and resplendent in detail, including iridescent blue and green spots and scale patches. The entire dorsum of the abdomen is often solid, vivid red or orange. Salticids are well known for their eye-catching courtship displays, during which the often elaborately decorated males vigorously posture and dance in front of observant females (fig. 4.2b) (Crane 1949; see Richman 1981 for a bibliography).

Many salticids are ant mimics (Galiano 1967, Reiskind 1977). In addition to having an antlike body, they move their slender forelegs in front of the head like the probing antennae of their models and often have enlarged pedipalps that resemble ant mandibles. In Peru, ants of the genus *Cephalotes* (fig. 4.2d) are naturally well protected by their heavy armor and spined bodies. They are closely mimicked by spiders of the genus *Aphantochilus* (fig. 4.2c) which have a narrowed, formicoid waist, black color, and even a spine on the anterior part of the cephalothorax like the ant. The anterior part of the body has a form that looks much the same as the ant's head (Preston-Mafham and Preston-Mafham 1984). There is one apparent case of an ant mimic in symbiosis with the leaf-nesting ant *Tapinoma melanocephalum*, the spider using the ant's nest for support and perhaps protecting the latter from invading predatory insects (Shepard and Gibson 1972).

Salticidae is a large family with many small genera. The majority of the Neotropical species are probably still unknown.

## References

Crane, J. 1949. Comparative biology of salticid spiders at Rancho Grande, Venezuela. Pt. IV. An analysis of display. Zoologica 34: 159–215.

Forster, L. 1982. Vision and prey-catching strategies in jumping spiders. Amer. Sci. 70: 165–175.

Galiano, M. E. 1967. Salticidae (Araneae) formiciformes. VIII. Nuevas descripciones. Physis 27: 27–39.

Preston-Mafham, R., and K. Preston-Mafham. 1984. Spiders of the world. Facts on File, New York.

Reiskind, J. 1977. Ant-mimicry in Panamanian clubionid and salticid spiders (Araneae: Clubionidae, Salticidae). Biotropica 9: 1–8.

Richman, D. B. 1981. A bibliography of courtship and agonistic display in salticid spiders. Peckhamia 2: 16–23.

Robinson, M. H., and C. E. Valerio. 1977. Attacks on large or heavily defended prey by tropical salticid spiders. Psyche 84: 1–10.

Shepard, M., and F. Gibson. 1972. Spider ant symbiosis: *Continusa* spp. (Araneida: Salticidae) and *Tapinoma melanocephalum* (Hymenoptera: Formicidae). Can. Entomol. 104: 1951–1954.

## Wolf Spiders

Lycosidae. *Spanish:* Arañas lobos (General), paccha arañas (Peru). *Portuguese:* Aranhas lôbos.

Wolf spiders (Stratton 1985) are varied in size (BL 10–20 mm), hairy, usually dark brown, and fast moving. They attract attention and are much feared, although most are harmless. Large species in the genus *Lycosa,* however, may bite humans with serious consequences. The venom is cytotoxic and produces great local pain as well as lingering necrotic effects. The spiders are recognized by their moderate hairiness and the three pairs of long, heavy, black spines arming the anterior tibia. The back of the light brown cephalothorax is often marked with contrasting broad, longitudinal bars or lines.

Lycosids are otherwise of much importance ecologically as controllers of insect populations, especially on the ground, which is their usual haunt. They are vagrant hunters with good vision used in prey capture. Very few build webs; they sit and wait for other spiders and insects that they ambush, usually nocturnally. They can be located at night by the bright reflectance of their eyes in a flashlight beam. A widespread species is *Lycosa raptoria* (fig. 4.2e).

Many have pronounced aggressive and fighting behaviors as well as highly developed maternal instincts. Females carry egg sacs and spiderlings for some time. Spiny, knobbed hairs on the back of the abdomen apparently provide the stimulus and means of attachment for this form of brood care (Rovner et al. 1973).

## References

ROVNER, J. S., G. A. HIGASHI, AND R. F. FOELIX. 1973. Maternal behavior in wolf spiders: The role of abdominal hairs. Science 182: 1153–1155.

STRATTON, G. E. 1985. Behavioral studies of wolf spiders: A review of recent research. Rev. Arachnol. 6: 57–70.

## Banana Spiders

Ctenidae, Cteninae, *Phoneutria. Portuguese:* Aranhas armadeiras (Brazil). Wandering spiders.

Spiders in the genus *Phoneutria* are fairly large (10–12 cm leg span, adult females BL 35–50 mm) and powerfully built. The eight eyes are in three rows (2-4-2), the last two the largest and the two laterals of the second row the smallest. They have a short coat of grayish to pale brown pelage, with larger black spines and chelicerae that are conspicuously clothed with long red hairs. The inner surfaces of the three apical segments of the palpi are heavily fringed with hairs.

These are essentially nocturnal, vagrant spiders, wandering on the ground from evening to dawn in search of prey. They do not construct webs to entrap food. They are also very aggressive and pugnacious, rearing on the hind legs with the two pairs of forelegs raised threateningly and fangs bared (fig. 4.2f). This exposes their undersides, which are strongly darkened on either side of a contrasting pale joint membrane between the basal joint of the first two pairs of legs.

Banana spiders retreat into dark places during the day and frequently enter dwellings where they may hide themselves in clothes or shoes. They are often found among bunches of bananas, which has earned them their vernacular name.

The venom of banana spiders is considered highly toxic, and a few cases of human deaths from their bite are well documented (e.g., Trejos et al. 1971), although this has occurred only in persons who are weak or small children. Bites occur frequently among workers who handle clusters of bananas. Some characteristic symptoms are local intense pain, tachycardia, salivation, disturbed vision, sweating, priapism, and prostration (Schenberg and Pereira Lima 1978).

Two well-known species are *P. nigriventer,* from southern Brazil, and *P. fera* (fig. 4.2f), which inhabits Amazonas.

Equally large but less dangerous spiders in the genus *Ctenus* are easily confused with banana spiders but differ in the lack of dense hair brush on the inner palpal surfaces (Bücherl et al. 1964).

### References

BÜCHERL, W., S. LUCAS, AND V. DESSIMONI. 1964. Aranhas da familia Ctenidae, Subfamilia Cteninae. I. Redescrição dos gêneros *Ctenus* Walckenaer 1805 e *Phoneutria* Perty 1833. Inst. Butantan Mem. 31: 95–102.

SCHENBERG, S., AND F. A. PEREIRA LIMA. 1978. Venoms of Ctenidae. *In* S. Bettini, ed., Arthropod venoms. Springer, Berlin. Pp. 217–245.

TREJOS, A., R. TREJOS, AND R. ZELEDÓN. 1971. Aracnidismo por *Phoneutria* en Costa Rica (Araneae: Ctenidae). Rev. Biol. Trop. 19: 241–249.

### Typical Orb Web Spiders
Araneidae (= Argiopidae).

This is the most characteristic of the several families that spin orb-shaped webs (like a wagonwheel with a hub, radiating struts and concentric circular ties). Most of the spiders in the family Araneidae are fairly large, often with grossly enlarged and frequently brightly colored abdomens. The lateral eyes are at a distance from the medial, the latter forming a square. They have poor vision and locate prey caught in their webs by feeling the tension and vibration of the threads (Craig 1989). They wrap captives in sheets of silk drawn from the spinnerets by the hind legs. Several members of the family are favorite subjects of behavioral (Robinson and Olazarri 1971, Robinson and Robinson 1980) and ecological study, especially species in the genera *Araneus* and *Argiope* (Robinson and Robinson 1978), although there are rela-

tively few Neotropical species in these primarily Holarctic genera. A widespread, common Latin American araneid is *Eustala anistera* (fig. 4.3a).

*Araneus* is the larger and more widespread genus. The female's abdomen is often immense and nearly spherical and many times is marked with complex spotted or variegate patterns. Many species of the genus are nocturnal and thus are seldom seen and appreciated. There are fewer kinds of *Argiope*, but they are more conspicuous because of large size, bright colors, and diurnal habits (Levi 1968). The females sit head downward with legs oriented like an "X"; they remain in the center of their webs during the day, often flanked above and below by zigzagging sheet silk structures (stabilimenta) in the web (at the tips of the legs in the silver orb weaver). Several species are spread over many parts of the globe, such as the golden orb weaver (*A. aurantia*), which occurs only in the New World, from Mexico to Guatemala, with an ovoid black abdomen marked by irregular sublateral orange-yellow streaks; the banded orb weaver (*A. trifasciata*), whose similarly shaped abdomen is white with narrow black rings; and the silver orb weaver (*A. argentata*) (fig. 4.3c), with a compressed, marginally lobed abdomen that is red, black, and silver spotted posteriorly and solid silver basally like that of the adjoining thorax. The embryology of the last species has been studied by Tse and Tse (1980).

Natural selection has produced many variations on the orb web theme for prey capture. Bolas spiders (*Mastophora*) (fig. 4.3d), for example, diverge radically from the typical orb weavers in their hunting method, as they do not rely passively on a web to catch prey. Instead, at night, they spin a hanging line with a sticky round globule at the tip. When prey approaches, the spider swings this "bola" and catches it on the globule (Eberhard 1980). The podadora (*Mastophora gasteracanthoides*) or

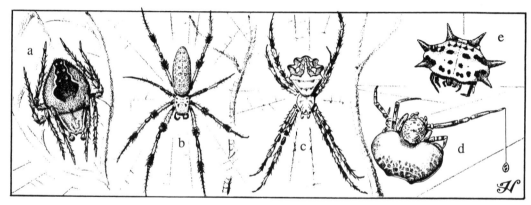

**Figure 4.3   ORB WEB SPIDERS (ARGIOPIDAE).** (a) Common orb weaver (*Eustala anistera*), female. (b) Golden silk spider (*Nephila clavipes*), female. (c) Silver orb weaver (*Argiope argentata*), female. (d) Bolas spider (*Mastophora dizzydeani*), female. (e) Spiny orb weaver (*Gasteracantha cancriformis*).

"true bolas spider," a moderate-sized (BL 11 mm), dark brown, slow-moving species with large horns on the abdomen, is known to workers in vineyards in South America who believe it bites with serious effects. It is probable, however, that only its fearsome appearance incriminates it, and some other agent is actually responsible for the lesions suffered by some individuals who encounter it. The bolas structure has been found to contain a volatile substance similar to the sex pheromone of a favored prey, owlet moths of the genus *Spodoptera*. The male moths are attracted to the vicinity of the spider whose chances of securing a meal are thus greatly enhanced (Eberhard 1977).

## References

CRAIG, C. L. 1989. Alternative foraging modes of orb web weaving spiders. Biotropica 21: 257–264.

EBERHARD, W. G. 1977. Aggressive chemical mimicry by a bolas spider. Science 198: 1173–1175.

EBERHARD, W. G. 1980. The natural history and behavior of the bolas spider *Mastophora dizzydeani* sp. n. (Araneidae). Psyche 87: 143–169.

LEVI, H. W. 1968. The spider genera *Gea* and *Argiope* in America (Araneae: Araneidae). Mus. Comp. Zool. (Harvard Univ.) Bull. 136: 319–352.

ROBINSON, M. H., AND J. OLAZARRI. 1971. Units of behavior and complex sequences in the predatory behavior of *Argiope argentata* (Fabricius): (Araneae: Araneidae). Smithsonian Contrib. Zool. 65: 1–36.

ROBINSON, M. H., AND B. C. ROBINSON. 1978. Thermoregulation in orb-web spiders: New descriptions of thermoregulatory postures and experiments on the effects of posture and coloration. Zool. J. Linnean Soc. 64: 87–102.

ROBINSON, M. H., AND B. C. ROBINSON. 1980. Comparative studies of the courtship and mating behavior of tropical araneid spiders. Pacific Ins. Monogr. 36: 1–218.

TSE, M. DO C. P., AND H. D. TSE. 1980. Notas sobre o desenvolvimento da aranha Labidognatha *Argiope argentata* (Fabricius, 1775) (Araneae, Araneidae). Rev. Brasil. Biol. 40: 249–255.

### Golden Silk Spider

Araneidae, Nephilinae, *Nephila clavipes*.
*Spanish:* Araña de oro (Costa Rica).
Golden-web spider, giant orb weaver.

Female *Nephila clavipes* are very large orb web spiders. It is not unusual to find specimens with a body length of 3 to 4 centimeters and a leg span of 5 centimeters resting head downward in the middle of their gigantic orb-shaped webs (fig. 4.3b). The web may be almost a meter square. Young individuals make a complete web; adults build only the bottom portion. The males are much smaller (BL 1 cm) and remain secreted in foliage at the edge of the web.

The web strands are exceedingly heavy and strong and are shiny gold in color. A perpendicular ladderlike structure (stabilimentum) of dense, multistranded zigzags is often constructed by a young female above and below her resting place at the hub. The function of this device is not known (Robinson and Robinson 1973). The webs are commonly occupied by other spiders (e.g., *Conopistha, Argyrodes*) that are not detected or tolerated by the web makers. They are considered kleptoparasites, stealing prey caught in the host's web. Their presence may be responsible for frequent relocation of the host to new web sites (Rypstra 1981). Robinson and Robinson (1977) observed milichiid flies of the genus *Phyllomyza* living on the cephalothorax of this *Nephila* in Panama. The flies helped themselves to the juices of the spider's prey after it was wrapped and partially digested.

This is the only Latin American species of its genus, one comprised of several widely distributed species in the Old World tropical lowlands (Lubin 1983, Robinson and Robinson 1980: 34f.). The body is elongate, the abdomen long ovoid and steeply inclined toward the front. The latter is colored olive brown with a double series of lighter, cream-colored spots along the dorsum. The legs are very long and have characteristic thick tufts of black hairs below the joints of the basal segments. The third pair of legs are shorter than the others and lack tufts. Males are similar but much smaller, with a body length of only about 4 to 8 millimeters, and their legs are untufted.

The species prefers sparse to moderate forest vegetation in which it builds its large webs in gaps and corridors where its prey is likely to pass (Moore 1977). If prey is sufficiently abundant, different females may build webs contiguous to each other in large aggregations. Prey usually consists of flies, bees, wasps, and small moths and butterflies (Robinson and Mirick 1971).

Females locate their round egg sacs on leaves at the ends of twigs at least 1 meter aboveground (Christenson and Wenzel 1980). Seasonal variation in egg production has been studied by Christenson, Wenzel, and Legum (1979). Spiderlings are communal and feed only on immobile prey, small insects and their own kind (Hill and Christenson 1981).

Recent experiments show that the species discriminates unpalatable butterflies and releases them unharmed from its web. Release is not accidental but results from a behavioral sequence specifically programmed to that end (Vasconcellos-Neto and Lewinsohn 1984).

## References

CHRISTENSON, T. E., AND P. A. WENZEL. 1980. Egg-laying of the golden silk spider, *Nephila clavipes* L. (Araneae, Araneidae): Functional analysis of the egg sac. J. Anim. Behav. 28: 1110–1118.

CHRISTENSON, T. E., P. A. WENZEL, AND P. LEGUM. 1979. Seasonal variation in egg hatching and certain egg parameters of the golden silk spider *Nephila clavipes* (Araneidae). Psyche 86: 137–147.

HILL, E. M., AND T. E. CHRISTENSON. 1981. Effects of prey characteristics and web structure on feeding and predatory responses of *Nephila clavipes* spiderlings. Behav. Ecol. Sociobiol. 8: 1–5.

LUBIN, Y. D. 1983. *Nephila clavipes* (Arāna de oro, Golden Orb-spider). *In* D. H. Janzen, ed., Costa Rican Natural History. Univ. Chicago Press, Chicago. Pp. 745–747.

MOORE, C. W. 1977. The life cycle, habitat and variation in selected web parameters in the spider, *Nephila clavipes* Koch (Araneidae). Amer. Midl. Nat. 98: 95–108.

ROBINSON, M. H., AND H. MIRICK. 1971. The predatory behavior of the golden-web spider *Nephila clavipes* (Araneae: Araneidae). Psyche 78: 123–139.

ROBINSON, M. H., AND B. C. ROBINSON. 1973. The stabilimenta of *Nephila clavipes* and the origins of stabilimentum-building in araneids. Psyche 80: 277–288.

ROBINSON, M. H., AND B. C. ROBINSON. 1977.

Associations between flies and spiders; bibio-commensalism and dipsoparasitism. Psyche 84: 150–157.

ROBINSON, M. H., AND B. C. ROBINSON. 1980. Thermoregulation in orb-web spiders: New descriptions of thermoregulatory postures and experiments on the effects of posture and coloration. Zool. J. Linnaean Soc. 64: 87–102.

RYPSTRA, A. L. 1981. The effect of klepto-parasitism on prey consumption and web relocation in a Peruvian population of the spider *Nephila clavipes*. Oikos 37: 179–182.

VASCONCELLOS-NETO, J., AND T. M. LEWINSOHN. 1984. Discrimination and release of unpalatable butterflies by *Nephila clavipes*, a Neotropical orb-weaving spider. Ecol. Entomol. 9: 337–344.

### *Spiny Orb Weavers*
Araneidae, Gasteracanthinae.

The females of this category of orb weavers (Robinson and Robinson 1980) are hard bodied, and the abdomen bears conspicuous, sharp-tipped spines, either radiating from its periphery (*Gasteracantha*) or diverging from the posterior (*Micrathena*), the latter giving the spider an arrowhead shape. Their abdomens are also colored lavishly in red, yellow, or white. Small dark depressions or pits dot the dorsum as well. The much smaller males have a cylindrical abdomen, lacking conspicuous spines (Chickering 1961). The abdomen of males in the genus *Gasteracantha* is squarish and much wider than long; in *Micrathena*, it is longer than wide. The latter is represented by about forty varied species throughout Latin America.

The usual *Gasteracantha* in the New World is the common cosmopolitan *G. cancriformis* (fig. 4.3e; Muma 1971), the genus being primarily Oriental. The species is distributed from Mexico to northern Argentina and is recognized by the presence of six spines on the abdomen. A second, poorly known species, *G. tetracantha*, has only four abdominal spines and is restricted to Puerto Rico, the Virgin Islands, and the Bahamas (Levi 1978).

The webs of *Gasteracantha* are conspicuous orbs found between the branches of shrubs and even on buildings. They are made in the morning and are usually inclined at an angle. The outer threads are decorated with flocculent tufts of silk.

### References
CHICKERING, A. J. 1961. The genus *Micrathena* (Araneae, Argiopidae) in Central America. Mus. Comp. Zool. (Harvard Univ.) Bull. 125: 392–470.

LEVI, H. W. 1978. The American orb-weaver genera *Colphepeira, Micrathena* and *Gasteracantha* north of Mexico (Araneae, Araneidae). Mus. Comp. Zool. (Harvard Univ.) Bull. 148: 417–442.

MUMA, M. 1971. Biological and behavioral notes on *Gasteracantha caucriformis* (Arachnida, Araneidae). Fla. Entomol. 54: 345–351.

ROBINSON, M. H., AND B. ROBINSON. 1980. Comparative studies of the courtship and mating behavior of tropical araneid spiders. Pacific Ins. Monogr. 36: 1–218.

### Giant Crab Spiders
Heteropodidae (= Sparassidae) and Selenopidae.

Members of these two families are commonly noticed on walls inside and outside of buildings where their large size (leg span 7–12 cm, BL 2–3 cm) and swift, sideways movements attract attention. Members of both families are known for their ability to hide in narrow cracks and crevices by day, emerging at night to catch insects. They may be especially common around exterior house lights to which moths and beetles are attracted (Muma 1953).

Typically, these spiders are very flat, with a compressed carapace and abdomen, the former almost circular or slightly wider than long. They hold their spiny legs out flat to the sides like a crab. Heteropodids have four of the total of eight eyes, selenopids six, arranged in a row along the anterior margin of the carapace.

The huntsman spider (sometimes also called the banana spider, but see *Phoneutria* above), *Heteropoda venatoria*, is the best-known species (fig. 4.4a). It is cosmopoli-

**Figure 4.4  SPIDERS AND HARVESTMEN.** (a) Huntsman spider (*Heteropoda venatoria*, Heteropodidae). (b) Black widow (*Latrodectus mactans*, Theridiidae), female. (c) South American violin spider (*Loxosceles laeta*, Loxoscelidae). (d) Harvestman (*Prionostemma* sp., Gagrellidae). (e) Harvestman (*Gonyleptus janthinus*, Gonyleptidae).

tan in distribution, probably introduced from Asia. It is very common in human habitations throughout the warm lowlands of Latin America. Here it is valued as a predator of cockroaches, and it is known at times also to kill and eat scorpions and even small bats. Some details of its life history have been elucidated (Ross et al. 1982). It is capable of producing a faintly audible buzz or hum by means of leg oscillations while coupled to the substratum by tarsal adhesive hairs; the sound apparently plays a role in courtship (Rovner 1980).

### References

MUMA, M. H. 1953. A study of the spider family Selenopidae in North America, Central America, and the West Indies. Amer. Mus. Nov. 1619: 1–55.

Ross, J., D. B. RICHMAN, F. MANSOUR, A. TRAMBARULO, AND W. H. WHITCOMB. 1982. The life cycle of *Heteropoda venatoria* (Linnaeus) (Araneae: Heteropodidae). Psyche 89: 297–305.

ROVNER, J. S. 1980. Vibration in *Heteropoda venatoria* (Sparassidae): A third method of sound production in spiders. J. Arachnol. 8: 193–200.

### Widow Spiders

Theridiidae, *Latrodectus*. *Spanish:* Viudas negras (General); arañas naranjas (Venezuela); cul rouge, 24-horas (West Indies); lucachas (Peru); guinos, pallus (Chile); huyuros micos (Bolivia); rastrojeras, arañas del lino (Argentina); arañas capulinas, po-ko-moo (Mexico); arañas bravas (southern South America). *Nahuatl:* Tzintlatlauhqueh, sing. tzintlatlauhqui (var. chintatlahuc).

Female "black widow" spiders (*Latrodectus mactans*) are medium-sized (BL 8–15 mm) and jet black, with a large, naked, globose abdomen having a characteristic reddish hourglass marking on the underside (fig. 4.4b). In other species of the genus, the background color may vary from white to reddish-brown, with beautiful red and yellow lines or spots adorning the dorsum. The males are four to five times smaller but with legs almost as long as the females'. The eight eyes are in two rows.

All are widely feared for their venomous qualities. Indeed, they bear a highly potent neurotoxic venom (Bettini and Maroli 1978). Symptoms of the bite begin with a sharp local pain that gradually moves from the wound area to other parts of the body, concentrating finally in the abdomen or legs. Other effects are nausea, dizziness, fainting, and shock, occasionally with a fatal outcome. In spite of the potential seriousness of envenomization by these spiders, they are reluctant to bite or to inject much venom so should not be considered really dangerous.

Because of their medical importance, they have been investigated more than other spiders, particularly in Argentina, where the several species are now fairly well studied embryologically (González 1981, 1984) and ecologically (Schnack et al. 1983, Estévez et al. 1984). They are controlled naturally by predaceous mud and spider wasps (Sphecidae, Pompilidae) and parasitoids among the chalcidoid wasps (*Desantisca*, Eurytomidae) and chloropid flies (*Pseudogaurax*) (Pérez Rivera 1980, Sabrosky 1966).

Black widows are shy and largely nocturnal. During the day, they rest in their finely threaded, amorphous webs, which they construct in protected, cool, dry, dark retreats. These are often soil fissures or spaces among debris, in wood piles, refuse piles, or under houses (Anderson 1972).

The taxonomy of these spiders is difficult and still unsettled owing to their great variability and overlap in structural features. Levi (1959) once recognized only three basic species, *L. mactans, L. geometricus,* and *L. curacaviensis,* but he concedes that this is an oversimplification and that there are probably several Latin American species (Levi 1983). Various other species have been described (Carcavallo 1959) and reclassifications proposed (see review of Bettini and Maroli 1978: 149f.), but the genus *Latrodectus* is in need of a complete revision using modern analytic techniques.

## References

ANDERSON, M. P. 1972. Notes on the brown widow spider, *Latrodectus geometricus* (Araneae: Theridiidae) in Brazil. Great Lakes Entomol. 5: 115–118.

BETTINI, S., AND M. MAROLI. 1978. Venoms of Theridiidae, genus *Latrodectus*. *In* S. Bettini, ed., Arthropod venoms. Springer, Berlin. Pp. 149–212.

CARCAVALLO, R. U. 1959. Una nueva *Latrodectus* y consideraciones sobre las especies del género en la República Argentina (Arach. Theridiidae). Neotropica 5: 85–94.

ESTÉVEZ, A. L., A. GONZÁLEZ, AND J. A. SCHNACK. 1984. Estadísticos vitales en especies Argentinas del género *Latrodectus* Walckenaer (Araneae, Theridiidae). II. *Latrodectus antheratus* (Babcock), *Latrodectus corallinus* Abalos y *Latrodectus diaguita* Carcavallo. Physis, Sec. C, 42(102): 29–37.

GONZÁLEZ, A. 1981. Desarrollo postembrionario de *Latrodectus mirabilis, Latrodectus corallinus* y *Latrodectus antheratus* (Araneae, Theridiidae). Physis, Sec. C, 39(97): 83–91.

GONZÁLEZ, A. 1984. Desarrollo postembrionario y evolución de los órganos mecanorreceptores de *Latrodectus diaguita* Carcavallo, y estudio de la tricobotriotaxia de *Latrodectus quartus* Abalos (Araneae, Theridiidae). Physis, Sec. C, 42(102): 1–5.

LEVI, H. W. 1959. The spider genus *Latrodectus* (Araneae, Theridiidae). Amer. Micro. Soc. Trans. 78: 7–43.

LEVI, H. W. 1983. On the value of genitalic structures and coloration in separating species of widow spiders (*Latrodectus* sp.) (Arachnida: Araneae: Theridiidae). Naturwiss. Ver. Hamburg Verh. (n.f.) 26: 195–200.

PÉREZ RIVERA, R. A. 1980. Distribución geográfica, potencial reproductivo y enemigos naturales de la viuda negra en Puerto Rico. Carib. J. Sci. 15: 79–82.

SABROSKY, C. W. 1966. Three new Brazilian species of *Pseudogaurax* with a synopsis of the genus in the Western Hemisphere (Diptera: Chloropidae). Dept. Zool., Sec. Agric., São Paulo, Pap. Avul. 19: 117–127.

SCHNACK, J. A., A. GONZÁLEZ, AND A. L. ESTÉVEZ. 1983. Estadísticos vitales en especies Argentinas del género *Latrodectus* Walckenaer (Araneae, Theridiidae). I. *Latrodectus mirabilis* Holmberg. Neotropica 29(82): 141–152.

## Violin Spiders

Loxoscelidae, *Loxosceles. Spanish:* Arañas de las rincones (Chile).

*Loxosceles* are shy, sedentary spiders that occupy a wide variety of dark, secretive habitats in natural and domestic situations, usually rock crevices or hollows under rocks, under debris and loose bark, or at cave entrances. They are common in corners and niches in adobe brick houses and other domestic structures. Their irregular webs are large, with thick, very sticky threads. They usually remain on their webs, which they continue to enlarge as long as they live.

Violin spiders carry a venom capable of severely injuring humans (Schenone and Suárez 1978). The venom's tissue-destroying capability has been well established. Clinical signs from bites range from mild necrosis to systemic reactions but rarely death. Although the venomousness of only a few species is recorded, it appears that all species of the genus are toxic. Shy and retiring, they never come forth to bite intentionally. Cases of envenomization, "loxoscelism" (Bücherl 1961), are normally only caused by specimens that have accidentally crawled into beds or onto clothing and that have bitten in defense when compressed.

These spiders are recognized by their small to medium size (BL 5–20 mm), light to medium-brown color, and long, thin legs. The back of the carapace often carries a dark outline in the shape of a violin. They are unique in having the six, equal-sized eyes forming a transverse row, in three diads. The legs and body are thickly clothed with abundant, fine, basally feathered hairs lying between long, erect, toothed hairs.

Although spiders of the genus *Loxosceles* are known from Africa, they are most diverse throughout the Americas. Here seventy-four species are found (58 in Mexico and Central America, 6 in the West Indies, 30 in South America). *Loxosceles laeta* (fig. 4.4c) is a large species. It has gained notoriety because of its tendency to live in urban settings, because it has been introduced into new areas of the world by commerce, and because of its reputation for being especially toxic. As a result, its biology has been studied in some detail (e.g., Galiano 1967, Schenone et al. 1970, Lowrie 1980). Complete bibliographies on violin spider biology and taxonomy are available (Gertsch 1967, Gertsch and Ennik 1983).

## References

Bücherl, W. 1961. Aranhas do gênero *Loxosceles* e loxoscelismo na América. Cien. Cult. 13: 213–224.

Galiano, M. E. 1967. Ciclo biológico y desarollo de *Loxosceles laeta* (Nicolet, 1849) (Araneae, Scytodidae). Acta Zool. Lilloana 23: 431–464.

Gertsch, W. J. 1967. The spider genus *Loxosceles* in South America (Araneae, Scytodidae). Amer. Mus. Nat. Hist. Bull. 136: 117–174.

Gertsch, W. J., and F. Ennik. 1983. The spider genus *Loxosceles* in North America, Central America, and the West Indies (Araneae, Loxoscelidae). Amer. Mus. Nat. Hist. Bull. 175: 264–360.

Lowrie, D. C. 1980. Starvation longevity in *Loxosceles laeta* (Nicolet) (Araneae). Entomol. News 91: 130–132.

Schenone, H., A. Rojas, and H. Reyes. 1970. Prevalence of *Loxosceles laeta* in houses in central Chile. Amer. J. Trop. Med. Hyg. 19: 564–567.

Schenone, H., and G. Suárez. 1978. Venoms of Scytodidae, genus *Loxosceles*. *In* S. Bettini, ed., Arthropod venoms. Springer, Berlin. Pp. 247–275.

## HARVESTMEN

Opiliones (= Phalangida). *Spanish:* Macacos (Mexico).

There is an inordinate diversity of Opiliones in the Neotropical Region (Kaestner 1937), where the majority of the world's approximately 5,000 species are found (Roewer 1923).

There are two dominant groups, the Cyphopalpitores and Laniatores (Martens 1986). The former, typified by the Gagrellidae, mostly exhibit a small (BL 4–6 mm), simple, oval body suspended in the center of the immensely long, slender (almost filamentous) legs, the second pair of which are usually the longest (fig. 4.4d). The pedipalps are slender with weak claws apically, and the coxae of the legs are separated ventrally by a breast plate. The Laniatores, best known in the Neotropics by the family Gonyleptidae, have relatively shorter legs of varied length and stoutness, the hind pair often longest and heaviest, with very large coxal segments and elaborate spines or other excrescences (fig. 4.4e). The pedipalps are stout, with grasping claws at the tip. The hard body is also often spined, and the coxae of the legs touch along the midline ventrally. Both groups typically possess eight legs and a prosoma divided dorsally into three parts by two transverse grooves. The abdomen is distinctly segmented and continuous with the prosoma. The three-segmented chelicerae have long blades, and the pedipalps are leglike but always much shorter than the legs.

Most species in this order live in humid retreats, beneath rocks, in tree bark crevices, and in niches on the forest floor. Others roam freely on the ground or on tree trunks or other vegetation. All prefer shade and moist conditions. Their food consists of other small invertebrates (mites, springtails, even snails). Some may take only plant detritus.

Harvestmen protect themselves in a variety of ways (Cokendolpher 1987). They are secretive, often nocturnal, cryptically colored and formed, and practice appendotomy (voluntary dropping of legs). When disturbed, they can eject defensive quinones and phenols (Roach et al. 1980) from paired repugnatorial glands on the anterior edge of the prosoma. *Vonones sayi* from Panama initially discharges a clear fluid containing quinones from its glands and the mouth and then dips the tips of its forelegs into the mixture and brushes them against the offending agent (Eisner et al. 1971).

Generally, the sexes meet fortuitously, eggs are laid on moist substrata, and no parental care is exhibited. But the males of at least one species, *Zygopachylus albomarginis* construct and guard a nest of bark detritus on fallen trees which females visit to copulate and lay their eggs (Rodríguez and Guerrero 1976).

The order is little studied in Latin America. Soares and Soares (1948, 1949, 1955) provide a catalog of most of the genera. Ringuelet's (1959) extensive review of the Argentinian fauna is of general utility. Many taxa are easily recognized by their beautiful coloration and strange body forms.

## References

COKENDOLPHER, J. C. 1987. Observations on the defensive behaviors of a Neotropical Gonyleptidae (Arachnida, Opiliones). Rev. Arachnol. 7: 59–63.

EISNER, T., A. F. KLUGE, J. E. CARREL, AND J. MEINWALD. 1971. Defense of phalangid: Liquid repellent administered by leg dabbing. Science 173: 650–652.

KAESTNER, A. 1937. Ordnung der Arachnida: Opiliones Sunderval = Weberknechte. *In* W. Kühkenthal and T. Krumbach, ed., Handbuch der Zoologie. W. Gruyter & Co., Berlin. Band 3, Hälfte 2, Lief. 9, (2) Teil. Pp. 385–496.

MARTENS, J. 1986. Die Grossgliederung der Opiliones und die Evolution der Ordnung (Arachnida). Actas 10th Cong. Int. Arach. (Jaca, Spain) 1:289–310.

RINGUELET, R. A. 1959. Los arácnidos Argentinos del orden Opiliones. Mus. Arg. Cien. Nat. "Bernardino Rivadavia," Inst. Nac. Cien. Nat. Zool. Rev. 5: 127–439, pl. 1–20.

ROACH, B., T. EISNER, AND J. MEINWALD. 1980. Defensive substances of opilionids. J. Chem. Ecol. 6: 511–516.

RODRÍGUEZ T., C. A., AND S. GUERRERO B., S. 1976. La historia natural y el comportamiento de *Zygopachylus albomarginis* (Chamberlain) (Arachnida, Opiliones: Gonyleptidae). Biotropica 8: 242–247.

ROEWER, C. F. 1923. Die Weberknechte der Erde, systematische Bearbeitung der bisher bekannten opiliones. G. Fischer, Jena.

SOARES, B. A. M., AND H. E. M. SOARES. 1948. Monografía dos gêneros de Opiliões Neotrópicos. Arq. Zool. São Paulo 5: 553–635.

SOARES, B. A. M., AND H. E. M. SOARES. 1949. Familia Gonyleptidae, continuação. Arq. Zool. São Paulo 7: 151–239.

SOARES, B. A. M., AND H. E. M. SOARES. 1955. Monografía dos gêneros de Opiliões Neotrópicos. Arq. Zool. São Paulo 8: 225–302.

## MITES AND TICKS

Acari (= Acarina, Acarida).

The mites are the most diverse and species-rich group of arachnids and also the most difficult to characterize (Krantz 1978, Hughes 1959, Flechtmann 1975). Most are very small, some minute (BL 1 to 2 mm). The body is fused into one piece, with no separation between the prosoma and unsegmented opisthosoma (abdomen), and usually has an ovoid shape, somewhat flattened in most ticks, elongate or quadrate in some mites. The mouthparts extend from a partial, headlike structure anteriorly, formed from fusion of basal pedipalpal and mouthpart segments. The first active stage of development is a "larva" that has six legs.

Once considered a natural group, the mites and ticks are now thought to have multiple origins, and at least three (possibly unrelated) major groups are presently recognized by specialists: Opilioacariformes (Notostigmata), Parasitiformes, and Acariformes. The first is a primitive assemblage, composed of comparatively large (1 mm or more), elongate, longlegged, leathery mites, somewhat resembling harvestmen (Hoffmann and Vázquez 1986). The second are medium-sized to large, well sclerotized, and with a tracheal system, opening through paired ventrolateral spiracles (stigmata). Many of these are parasitic, including the atypical ticks. The Acariformes are mostly small and have a poorly formed tracheal system, and the body is often divided into two regions by a transverse furrow. Within each category, major subgroupings are recognized according to the presence or absence and position of the stigmata: hidden (Cryptostigmata), anterior (Prostigmata), between the second and fourth coxae (Mesostigmata), near the posterior coxae (Metastigmata), dorsally (Notostigmata), or absent (Astigmata).

All these groups are well represented in Latin America (Schuster 1969). The fauna is immense, and though many thousands of species are now known, these are surely only a fraction of those that must exist. Anything approaching even a general review is well beyond the scope of this book.

Only a very few of the better-known types of special importance ecologically or economically in the American tropics can be treated here.

All biotic types are represented among mites. There are free-living and parasitic forms on animals and plants. Many parasitic forms are vectors of diseases (Oldfield 1970). Free-living forms dwell in the soil and on vegetation where they feed on sap and tissue fluids, sometimes causing considerable damage to crops (Jeppson et al. 1975). Some cause galls (Eriophyidae), and a few infest flour and other stored grain products.

There are many parasites of vertebrates, including humans. They may be of considerable medical and veterinary importance (Baker et al. 1956, Flechtmann 1973). Most are external biters (*Dermanyssus*), but a few burrow into the skin (scabies mite) or lodge deep in body cavities (nasal chiggers) or skin pores (hair follicle mite). They are often highly allergenic. *Pneumonyssus* (Halarachnidae) and members of other families infest the lungs and respiratory tracts of snakes and birds. Many are specific ectoparasites of bats (e.g., Chirodiscidae, Chirorhynchobiidae, Spelaeorhynchidae, Spinturnicidae) and other characteristic Neotropical mammals, such as *Archemyobia latipilis* in opossums (Fain et al. 1981). Pollen- and nectar-feeding species of *Rhinoseius* and *Proctolaelaps* (Ascidae) are phoretic on hummingbirds, living in the nares and using them for transportation from flower to flower (Colwell 1979). The ticks are highly modified bloodsuckers on all vertebrate orders.

Other mites live in bird and mammal nests, scavenging on the host's food or organic detritus therein. Such is *Hypoaspis dasypus*, known only from armadillo burrows (Menzies and Strandtmann 1952).

Many are associates of other arthropods (Bischoff de Alzuet 1978, Mauri and Bischoff de Alzuet 1972) in various ways, as parasites, predators, and consumers of exudates from the host. Others practice phoresy, using the host for transport only. Some of the better-known examples of the latter are *Macrocheles,* which are often very numerous on the bodies of dung beetles (Evans and Hyatt 1963). Mites of diverse groups lodge in the tympanic cavities of moths (Treat 1975). Several kinds of mites are also associated with stingless bees (Flechtmann and de Camargo 1974). The varroa mite (*Varroa jacobsonii*), a notorious apicultural pest in the Old World and now established in all major beekeeping areas in South America, is associated with the honeybee. *Arrehenurus* is a widespread aquatic genus, ectoparasitic on mosquito larvae and pupae.

Entire families, for example, Analgidae, Cheyletidae, Dermoglyphidae, are adapted to life on the feathers of birds (Gaud and Atyeo 1979). *Ophiomegistus* (Antennophoridae) are found on snakes and lizards. *Iguanacarus* (Trombiculidae) lives in the nasal fossae of the marine iguana of the Galápagos Islands (Vercammen-Grandjean 1965).

Some major groups are aquatic, such as the primarily marine superfamilies Halacaroidea and freshwater Hydrachnoidea (Cook 1980). The latter can be very abundant in tropical ponds and lakes where they form a part of the plankton community (Gliwicz and Biesiadka 1975).

## References

BAKER, E. W., T. M. EVANS, D. J. GOULD, W. B. HULL, AND H. L. KEEGAN. 1956. A manual of parasitic mites of medical or economic importance. Nat. Pest Contr. Assoc. Tech. Publ.

BISCHOFF DE ALZUET, A. 1978. Ácaros asociados a artrópodos de interés sanitario. Neotropica 24: 145–149.

COLWELL, R. K. 1979. The geographic ecology of hummingbird flower mites in relation to their host plants and carriers. Rec. Adv. Acarol. 2: 461–468.

COOK, D. R. 1980. Studies on Neotropical water mites. Amer. Entomol. Inst. Mem. 31: 1–645.

EVANS, G. O., AND K. H. HYATT. 1963. Mites of the genus *Macrocheles* Latr. (Mesostigmata) associated with coprid beetles in the collection of the British Museum (Natural History). Brit. Mus. Nat. Hist. Bull. 9: 327–401.

FAIN, A., E. MÉNDEZ, AND F. S. LUKOSCHUS. 1981. *Archemyobia* (*Nearchemyobia*) *latipilis* sp. n. (Acari: Prostigma: Myobiiidae) parasitic on marsupials in Panama and Brazil. Rev. Biol. Trop. 29: 77–81.

FLECHTMANN, C. H. W. 1973. Ácaros de importância médico-veterinária. Liv. Nobel, São Paulo.

FLECHTMANN, C. H. W. 1975. Elementos de Acarologia. Liv. Nobel, São Paulo.

FLECHTMANN, C. H. W., AND C. A. DE CAMARGO. 1974. Acari associated with stingless bees (Meliponidae, Hymenoptera) from Brazil. 4th Int. Cong. Acarol. Proc. Pp. 315–319.

GAUD, J., AND W. T. ATYEO. 1979. Co-évolution des acariens sarcoptiformes plumicoles et de leurs hôtes. Acarologia 21: 291–306.

GLIWICZ, Z. M., AND E. BIESIADKA. 1975. Pelagic water mites (Hydracarina) and their effect on the plankton community in a Neotropical man-made lake. Arch. Hydrobiol. 76: 65–88.

HOFFMANN, A., AND M. VÁZQUEZ. 1986. Los primitivos ácaros Opilioacáridos en México. Fol. Entomol. Méx. 67: 53–60.

HUGHES, T. E. 1959. Mites or the acari. Athlone, Univ. London, London.

JEPPSON, L. R., H. H. KEIFER, AND E. W. BAKER. 1975. Mites injurious to economic plants. Univ. California, Berkeley.

KRANTZ, G. W. 1978. A manual of acarology. 2d ed. Ore. St. Univ. Bookstores, Corvallis.

MAURI, R., AND A. BISCHOFF DE ALZUET. 1972. Ácaros asociados a artrópodos. Soc. Argentina Entomol. Rev. 34: 151–159.

MENZIES, G. C., AND R. W. STRANDTMANN. 1952. A new species of mite taken from nest of armadillo. Entomol. Soc. Wash. Proc. 54: 265–269.

OLDFIELD, G. N. 1970. Mite transmission of plant viruses. Ann. Rev. Entomol. 15: 343–380.

SCHUSTER, R. 1969. Die terrestrische Milbenfauna Südamerikas in zoogeographischer Sicht. *In* E. J. Fittkau, J. Illies, H. Klinge, G. H. Schwabe, and H. Sioli, eds., Biogeography and ecology in South America. 2: 741–763. Junk, The Hague.

TREAT, A. E. 1975. Mites of moths and butterflies. Cornell Univ., Ithaca.

VERCAMMEN-GRANDJEAN, P. H. 1965. *Iguanacarus,* a new subgenus of chigger mite from nasal fossae of the marine iguana in the Galápagos Islands, with a revision of the genus *Vatacarus* Southcott (Acarina, Trombiculidae). Acarologia 7 (suppl.): 266–274.

## Eriophyids

Eriophyidae. *Spanish:* Eriófidos (General), ácaros de agallas. Gall mites.

These mites are marked by structural simplification: they are wormlike and so small as to be barely visible to the unaided eye (BL 0.2 mm); only the anterior two pairs of legs are developed; a tracheal system is absent, and respiration is by cutaneous diffusion. The cuticle bears conspicuous fine rings over the whole body.

All species of eriophyids are obligate plant feeders and constitute the main family of mites injurious to cultivated plants. Physical damage is caused by mechanical feeding and inoculation of harmful viruses. Host responses to feeding are leaf curling and adventitious tissue formation, including twigs, blisters, and galls.

A serious and widespread pest species is the citrus bud mite (*Eriophyes* [= *Aceria*] *sheldoni*) (fig. 4.5a). It is known in all citrus-growing areas of Central and South America. The species infests the bud, developing blossoms, and area beneath the button of citrus fruits as well as leaf axil buds. The growing tissue may be completely destroyed or altered to form stunted or misshapen structures, especially in the fruit, which may assume grotesque shapes, becoming unfit for market.

Another very injurious species is the coconut mite (*Eriophyes guerreronis*) (Moore and Alexander 1987). Populations develop under the bracts, where their feeding causes scarring on the nuts, leading to reduced copra yields.

### Reference

MOORE, D., AND L. ALEXANDER. 1987. Aspects of migration and colonization of the coconut palm by the coconut mite, *Eriophyes guerreronis* (Keifer) (Acari: Eriophyidae). Bull. Entomol. Res. 77: 641–650.

## House Dust Mites

Pyroglyphidae, *Dermatophagoides.*

House dust, consisting of human and pet skin particles, food fragments, hair, cotton, paper, wool and synthetic fibers, soil, and like material, provides food and shelter for a community of exceedingly minute organisms dominated by arthropods, including many kinds of mites, and fungi (Van Bronswijk 1979, 1981). Among the latter are most commonly the house dust mites of the genus *Dermatophagoides*, which is represented in Latin America by both the widespread species *D. farinae* (fig. 4.5b) and the localized *D. pteronyssinus. D. neotropicalis* is a species sometimes also considered a part of this complex (Fain and Van Bronswijk 1973). As many as 3,000 such mites may inhabit a gram of dust (Arlian et al. 1979). They are more common in the dust on mattresses and bedroom floors than elsewhere in the house.

Dust has long been known to cause allergies and asthma in sensitive persons.

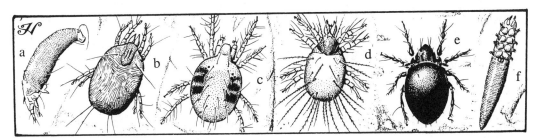

**Figure 4.5 MITES.** (a) Citrus bud mite (*Eriophyes sheldoni,* Eriophyidae). (b) American house dust mite (*Dermatophagoides farinae,* Pyroglyphidae). (c) Spider mite (*Tetranychus telarius,* Tetranychidae). (d) Mold mite (*Tyrophagus putrescentiae,* Acaridae). (e) Soil mite (*Oribatula minuta,* Oribatulidae). (f) Follicle mite (*Demodex folliculorum,* Demodicidae).

Numerous studies have reported a correlation between symptoms and the presence of house dust mites, although a direct relationship is not yet clearly established (Lecks 1973). What little has been done on these mites and their effect on health in Latin America has centered in Colombia (Mulla and Sánchez Medina 1980, Charlet et al. 1979) and Chile (Casanueva and Artigas 1985).

## References

ARLIAN, L. G., I. L. BERNSTEIN, C. L. JOHNSTON, AND J. S. GALLAGHER. 1979. Ecology of house dust mites and dust allergy. Rec. Adv. Acarol. 2: 185–195.

CASANUEVA, M. E., AND J. N. ARTIGAS. 1985. Distribución geográfica y estacional de los ácaros del polvo de habitación en Chile. Gayana. Zool. 49: 3–75.

CHARLET, L. D., M. S. MULLA, M. SANCHEZ-MEDINA, AND M. A. REYES. 1979. Species composition and population trends of mites in various climatic zones of Colombia. J. Asthma Res. 16: 131–148.

FAIN, A., AND J. E. M. H. VAN BRONSWIJK. 1973. On a new species of *Dermatophagoides* (*D. neotropicalis*) from house dust, producing both normal and heteromorphic males (Sarcoptiformes: Pyroglyphidae). Acarologia 15: 181–187.

LECKS, H. I. 1973. The mite and house dust allergy: A review of current knowledge and its clinical significance. Clin. Pediat. 12: 514–517.

MULLA, M. S., AND M. SÁNCHEZ-MEDINA, ed. 1980. Domestic acari of Colombia, bionomics, ecology and distribution of allergenic mites, their role in allergic disease. COLCIENCIAS, Bogotá.

VAN BRONSWIJK, J. E. M. H. 1979. House dust as an ecosystem. Rev. Adv. Acarol. 2: 167–172.

VAN BRONSWIJK, J. E. M. H. 1981. House dust biology for allergists, acarologists, and mycologists. NIB Publ., Zeist, The Netherlands.

## Spider Mites

Acarida, Tetranychidae. *Spanish:* Arañitas, arañas rojas (General). *Portuguese:* Ácaros de teia, ácaros rajados, ácaros vermelhos (Brazil).

These mites often live amid masses of fine silk webbing on the undersides of leaves which they spin from glands located in the prosoma, a habit earning them their vernacular name (Baker 1979, Helle and Sabelis 1985). The feeding of large numbers of these mites on commercial plants often causes severe damage. Leaves develop pale blotches and may eventually totally dry up. Plants lose their vigor and die (Huffaker et al. 1969). Consequently, many species are recognized pests in Latin America (Pritchard and Baker 1955). Second to the eriophyids, they are probably the most agriculturally important plant-feeding group of mites. A complex of species, the two-spotted spider mite group (*Tetranychus bimaculatus, T. telarius* [fig. 4.5c], and *T. cinnabarinus*) is especially troublesome to cotton, beans, and vegetable crops in general, as well as to flowers, especially in greenhouses (Vereau et al. 1978).

Spider mites are green, yellow, orange, or red and small (BL always less than 1 mm), often with dark blotches on either side of the dorsum.

## References

BAKER, E. W. 1979. Spider mites revisited—a review. Rec. Adv. Acarol. 2: 387–394.

HELLE, W., AND M. W. SABELIS, eds. 1985. Spider mites: Their biology, natural enemies and control. Vol. 1, Pt. A, 405; Pt. B, 458. Elsevier, Amsterdam.

HUFFAKER, C. B., J. A. McMURTRY, AND M. VAN DE VRIE. 1969. The ecology of tetranychid mites and their natural control. Ann. Rev. Entomol. 14: 125–174.

PRITCHARD, A. E., AND E. W. BAKER. 1955. A revision of the spider mite family Tetranychidae. Pac. Coast Entomol. Soc. Mem. 2: 1–472.

VEREAU, W. V., M. CUEVA, AND D. OJEDA. 1978. Biología de la "arañita roja del algodonero" *Tetranychus cinnabarinus* (Boisduval) (Acarina, Tetranychidae). Rev. Peruana Entomol. 21: 50–54.

## Stored Product Mites

Acariformes, various families, including Acaridae and Carpoglyphidae.

Quite a number of mite species are associated with stored food and fiber products, feeding directly on the substance or secondarily on fungi growing on it. A few are parasitoids of insects living on the substratum.

They are all very small mites (BL mostly 1 mm or less), pale and with long body setae. They may develop enormous populations and be very destructive both in industrial (warehouses, granaries, holds of ships) and household (larders, cupboards, medicine chests) environments. They may literally replace the substratum with the mass of their collective bodies. Some are known to bite humans working around stored products, causing minor rashes ("baker's itch").

The principal offending species in Latin America belonging to this category are *Tyrophagus putrescentiae* (Acaridae) (fig. 4.5d), *Carpoglyphus lactis* (Carpoglyphidae), and *Pyemotes ventricosus* (Pyemotidae).

## Reference

FLECHTMANN, C. H. W. 1983. Ácaros de importância agrícola. Liv. Nobel, São Paulo.

## Soil Mites

Oribatulatidae (= Cryptostigmata). Moss mites, beetle mites.

Soil mites (Balogh 1988) form a very large and complex group, comprised of many families and about 450 species as presently known in South America, far short of the probable actual number. Balogh (1972) lists about 270 Neotropical genera, including cosmopolitan and pan-tropical taxa. These mites are very adaptable to environmental stresses and represent one of the few types of terrestrial arthropods surviving on the harsh fringes of Antarctica and the most southerly Atlantic islands (Wallwork 1965). By their very numbers, they play an important role in the decomposition of organic substances in the soil (Williams 1941; see special habitats, chap. 2).

Soil mites, such as the common *Oribatula minuta* (fig. 4.5e), are generally very well sclerotized, dark-pigmented, hard-bodied mites, most with a body length of less than 1 millimeter. Some have enlarged mouthparts, and they often have hinged, wing-like structures extending from the sides of the body under which they can tuck their legs to form a compact ball for protection. The leg segments are sometimes swollen, giving the legs a nodular appearance.

## References

BALOGH, J. 1972. The oribatid genera of the world. Akadémiai Kiadó, Budapest.

BALOGH, J. 1988. Oribatid mites of the Neotropical Region. I. The soil mites of the world. Vol. 2.

WALLWORK, J. A. 1965. The Cryptostigmata (Acari) of Antarctica with special reference to the Antarctic Peninsula and South Shetland Islands. Pacific Ins. 7: 453–468.

WILLIAMS, E. C. 1941. An ecological study of the floor fauna of the Panamanian rain forest. Chicago Acad. Sci. Bull. 6: 63–124.

## Follicle Mites

Demodicidae, *Demodex*. *Portuguese:* Cravos (Brazil).

These mites (Desch and Nutting 1971) are all ectoparasites of mammals and remarkably modified for residence in pits in the skin; rarely, they secondarily invade the lymphatic and circulatory system.

They seem to be rigidly host specific (Nutting 1974), certain species being known from several Latin American animals, including the domestic horse (*Demodex equi*), dog (*D. canis*), sheep (*D. ovis*), pig (*D. phylloides*), goat (*D. caprae*), cat (*D. cati*), and cow (*D. bovis*). More species are being discovered in wild mammals as well, for example, guinea pigs, mice, monkeys (Lebel and Nutting 1973), and bats (Desch and Nutting 1972). The entire life cycle is spent on the host, all stages being found in skin pustules and hair follicles on the body, especially about the face (Quintero 1978). Transfer to a new host normally requires close contact, possibly in infancy or in the nest.

Humans also harbor a particular species,

*D. folliculorum* (fig. 4.5f). These are minute (BL 0.1–0.4 mm), with an elongate, almost wormlike body. The opisthosoma is transversely striated. The legs are short and stubby without setae and are located close together anteriorly (Desch and Nutting 1977). The mite inhabits the hair follicles and sebaceous glands, particularly around the nose and eyelids but sometimes elsewhere, such as the scalp. Its presence is usually innocuous but has been known to induce acnelike conditions. It is surprisingly common; probably the majority of the human population anywhere unknowingly serves as host to this species.

In animals, severe symptoms can develop when the mites interfere with secretion and when they invade the skin and bloodstream (Nutting 1975).

### References

Desch, D. E., and W. B. Nutting. 1971. Demodicids (Trombidiformes: Demodicidae) of medical and veterinary importance. 3d Int. Cong. Acarol. Proc. Pp. 499–505.

Desch, D. E., and W. B. Nutting. 1972. Parasitic mites of Surinam. VII: *Demodex longissimus* n. sp. from *Carollia perspicillata* and *D. molossi* n. sp. from *Molossus molossus* (Demodicidae: Trombidiformes); Meibomian complex inhabitants of Neotropical bats (Chiroptera). Acarología 14: 35–53.

Desch, C. E., and W. B. Nutting. 1977. Morphology and functional anatomy of *Demodex folliculorum* (Simon) of man. Acarología 19: 422–462.

Lebel, R. R., and W. B. Nutting. 1973. Demodectic mites of subhuman primates. I: *Demodex saimiri* sp. n. (Acari: Demodicidae) from the squirrel monkey, *Saimiri sciureus*. J. Parasit. 59: 719–922.

Nutting, W. B. 1974. Synhospitaly and speciation in the Decodicidae [sic] (Trombidiformes). 4th Int. Cong. Acarol. Proc. Pp. 267–272.

Nutting, W. B. 1976. Pathogenesis associated with hair follicle mites (Acari: Demodicidae). Acarologia 17: 493–506.

Quintero, M. T. 1978. Frecuencia de ácaros en especies de animales domésticos. Vet. Mex. 9: 111–114.

### Scabies Mite

Sarcoptidae, *Sarcoptes scabiei. Spanish:* Ácaro de la sarna. *Portuguese:* Acariano da sarna (Brazil).

There are several strains or subspecies of this mite (Mellanby 1943, 1985), adapted to different mammal hosts (horses, dogs, pigs, sheep, cattle) including humans, in which it can cause a general skin affliction called scabies or mange (Arlian 1989, Gordon and Unsworth 1947). The species spends its entire life cycle on the host. Females are the infective stage. They lay their eggs in tunnels made by burrowing through the subcutaneous layers of the skin. The adults live a month or more.

The burrowing and feeding (on blood) of the mites of all stages cause extreme itching. The sinuous tunnels are near the skin's surface and can be seen as delicate gray lines, on humans usually between the

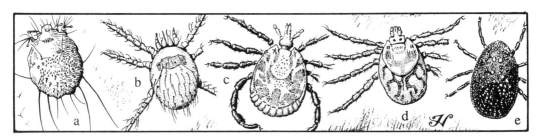

**Figure 4.6  MITES AND TICKS.** (a) Scabies mite (*Sarcoptes scabiei,* Sarcoptidae), female. (b) Common, or "sweet potato" chigger (*Eutrombicula batatas,* Trombiculidae). (c) cayenne tick (*Amblyomma cajennense,* Ixodidae), male. (d) Tropical horse tick (*Dermacentor nitens,* Ixodidae), male. (e) Fowl tick (*Argas miniatus,* Argasidae).

fingers and toes, behind the knee, and on the genitalia. Pimples or vesicles may form on the affected skin which when scratched, become infected and cause sores and scabs. Animals may develop large areas of leathery encrustments and lose most of their body hair (mange, *sarna pira*).

These are very small mites (females 0.4 mm in diameter, males 0.2 mm). They are rotund with very short, stubby legs. The third pair in the male and the third and fourth pairs in the female (fig. 4.6a) are tipped with a very long bristle; the other legs are tipped with small, stalked suckers. The integument is coarsely striated and bears spinelike projections of two sizes on the posterior half of the dorsum.

The species is cosmopolitan in distribution. Infestations spread through close contact between infected hosts. Sarcoptic mange in livestock causes reduction in meat, milk, and wool production and even the death of severely afflicted animals.

## References

ARLIAN, L. G. 1989. Biology, host relations, and epidemiology of *Sarcoptes scabiei*. Ann. Rev. Entomol. 34: 139–161.

GORDON, R. M., AND K. UNSWORTH. 1947. A review of scabies since 1939. Carib. Med. J. 9: 56–71.

MELLANBY, K. 1943. Scabies. Oxford Univ. Press, London.

MELLANBY, K. 1985. Biology of the parasite [scabies mite]. *In* M. Orkin and H. I. Maibach, eds., Cutaneous infestations and insect bites. Marcel Dekker, New York. Pp. 9–18.

## Chiggers

Trombiculidae, *Eutrombicula. Spanish:* Coloradillos, bichos colorados (General); patatas (Surinam); callo-callo, isangos (Peru); celembas (Ecuador). *Nahuatl:* Tlalzahuatl (Mexico). *Portuguese:* Bichos colorados. *Tupi-Guaraní:* Micuims (Brazil). *French:* Bêtes rouges (Cayenne). Harvest mites, red bugs, red mites.

Worldwide, various genera of the family Trombiculidae are the infamous biting mites known as chiggers (Sasa 1961, Wharton and Fuller 1952). Among the eighty-seven genera in America (Brennan and Goff 1977, 1978), the main offender is *Eutrombicula,* whose larvae attach to humans, causing severe, itching rashes (trombidiosis) and other allergenic reactions. *Parascoschoengastia* (*Euschoengastia*) *nunezi* has been responsible for dermatosis in Mexico (Andrade 1947).

In the normal life cycle, the larvae feed on mammals, birds, or reptiles (Wrenn and Loomis 1984). After engorging, they drop off permanently to continue development as free-living, predaceous terrestrial mites (nymphs and adults). The adults are fairly large (BL 2–3 mm), frequently bright red or orange, and covered with a velvety pelage of feathery hairs. They are usually found just beneath the surface of loose soil, in cracks, burrows, leaf litter, humus, or decaying wood where they feed on the eggs and early stages of other minute arthropods, such as springtails.

Structurally, the six-legged larval chiggers are recognizable under high magnification by a pair of unique knobbed or plumose hairs arising from a rectangular plate on the anterior-dorsal surface of the body (idiosoma; Goff et al. 1982, Wharton et al. 1951). The mouthparts consist of a pair of heavy piercing chelicerae and a pair of palpi, which together give the appearance of a head. Contrary to popular belief, the chigger does not burrow into the skin or imbed even this false head. Only the tips of the medial mouthpart elements are inserted. The salivary secretions contain powerful histolytic enzymes that break down deep epithelial cells. The tissue hardens around the chelicerae, forming a feeding canal (stylostome), through which the host's dissolved cells and lymph are siphoned (Allred 1954). After the chigger departs, it is the persistence of this foreign object that produces the long-lasting itching and irritation so characteristic of its bite.

There are nearly eighty species of

*Eutrombicula* in Latin America (Loomis and Wrenn 1984, Hoffmann 1970). Only a few are regular human pests, and these normally attack lizards, snakes, ground-dwelling birds, or mammals. Species of other genera with regular mammal hosts seem rarely to cause an allergic reaction when they do attach to man. None transmit pathogenic organisms. However, *Leptotrombidium*, known vectors of rickettsioses elsewhere, and *Pseudoschoengastia* can cause dermatosis and are potentially troublesome.

The most common chiggers attacking humans throughout America belong to the *Eutrombicula alfreddugesi* group, comprised of several poorly distinguished species (Williams 1946). Their larvae are abundant in transition areas between forest and grassland, along swamp margins, and from sea level to nearly 3000 meters elevation. They normally parasitize almost any terrestrial vertebrate and readily transfer to man.

Another widespread species of major medical importance throughout Latin America is the so-called sweet potato chigger, *Eutrombicula batatas* (fig. 4.6b) (Michener 1946). This is an animal of sunlit places, the unfed larvae emerging onto the surface of the soil sometimes in large numbers, especially in grassy areas. It may become very abundant around homes and villages where domestic animals, particularly chickens (Canfalonieri and de Carvalho 1973), are numerous. The normal hosts are principally birds and small terrestrial mammals.

## References

ALLRED, D. M. 1954. Observations on the stylostome (feeding tube) of some Utah chiggers. Utah Acad. Sci. Arts & Letters Proc. 31: 61–63.

ANDRADE, R. M. 1947. Trombididiasis por *Neoschöngastia nuñezi* Hoffmann, 1944. Gaceta Med. (Mexico) 77: 219–240.

BRENNAN, J. M., AND M. L. GOFF. 1977. Keys to the genera of chiggers of the Western Hemisphere (Acarina: Trombiculidae). J. Parasitol. 63: 554–566.

BRENNAN, J. M., AND M. L. GOFF, 1978. Three new monotypic genera of chiggers (Acari: Trombiculidae) from South America. J. Med. Entomol. 14: 541–544.

CANFALONIERI, U. E. C., AND L. P. DE CARVALHO. 1973. Ocorrência de *Trombicula* (*Eutrombicula*) *batatas* (L.) em *Gallus gallus domesticus* L. no estado do Rio de Janeiro (Acarina, Trombiculidae). Rev. Brasil. Biol. 33: 7–10.

GOFF, M. L., R. B. LOOMIS, W. C. WELBOURN, AND W. J. WRENN. 1982. A glossary of chigger terminology (Acari: Trombiculidae). J. Med. Entomol. 19: 221–238.

HOFFMANN, A. M. 1970. Estudio monográfico de los trombicúlidos de México (Acarina: Trombiculidae). Primera parte. Esc. Nac. Cien. Biol. México An. 18: 191–263.

LOOMIS, R. B., AND W. J. WRENN. 1984. Systematics of the pest chigger genus *Eutrombicula* (Acari: Trombiculidae). Acarology 6(1): 152–159.

MICHENER, C. D. 1946. Observations on the habits and life history of a chigger mite, *Eutrombicula batatas* (Acarina: Trombiculidae). Entomol. Soc. Amer. Ann. 39: 101–118.

SASA, M. 1961. Biology of chiggers. Ann. Rev. Entomol. 6: 221–244.

WHARTON, G. W., AND H. S. FULLER. 1952. A manual of the chiggers. Entomol. Soc. Wash. Mem. 4: 1–185.

WHARTON, G. W., D. W. JENKINS, J. M. BRENNAN, H. S. FULLER, G. M. KOHLS, AND C. B. PHILLIP. 1951. The terminology and classification of trombiculid mites (Acarina: Trombiculidae). J. Parasitol. 37: 13–31.

WILLIAMS, R. W. 1946. A contribution to our knowledge of the bionomics of the common North American chigger, *Eutrombicula alfreddugesi* (Oudemans), with a description of a rapid collecting method. Amer. J. Trop. Med. 26: 243–250.

WRENN, W. J., AND R. B. LOOMIS. 1984. Host selectivity in the genus *Eutrombicula* (Acari: Trombiculidae). Acarology 6(1): 160–165.

## Ticks

Ixodoidea. *Spanish:* Garrapatas (General); conchudas, plateadas, pinolillos (Mexico). *Portuguese:* Carrapatos, carrapatos pólvoras, carrapatos fogo, carrapatinhos (Brazil-larvae). *Nahuatl:* Mazaatemimeh, sing. mazaatémitl (Mexico).

Ticks comprise an isolated and very specialized group of mites distinguished by well-developed, multifaceted spiracles situated lateral to the third coxae or behind the fourth coxae (Obenchain and Galun 1982, Oliver 1989). The integument is leathery to hard, and they are all considerably larger than most mites (BL 2–5 mm). Their mouthparts are unique: the digits of the chelicerae have lateral teeth, and between the pedipalps is a median holdfast organ with recurved teeth (called the hypostome). The first tarsus has a dorsal heat-sensitive organ in a large pit (Haller's organ) to assist in finding vertebrate hosts on which they feed. Ticks anchor onto the host's skin by the hypostome and do not drop off until they finish feeding. If forcibly removed, the hypostome (erroneously thought to be the head by most people) may remain in the wound and fester.

After molting, ticks wait on the tips of twigs or leaves, forelegs outstretched and ready to snag any animal brushing past. On the potential host's approach, the legs wave frantically in response to various stimuli: primarily to carbon dioxide exhaled during respiration but also to vibration, body heat, or the host's shadow passing over them.

Two main families of ticks are recognized, the "hard ticks" (Ixodidae) and "soft ticks" (Argasidae). Hard ticks are flattened, and all stages bear a thick, tough plate (scutum) dorsally. This plate covers only about half the anterior end of the body in larvae, nymphs, and females; the entire body in males. It may or may not be ornamented with silver. Soft ticks are baglike, without plates in postlarval stages. The mouthparts are hidden by the overhanging body (camerostoma). They are primarily nest inhabitants and parasites of birds and small mammals (mainly bats) in semitropical and tropical areas. They are very resistant to water loss and are characteristic of desert faunas.

Ticks are ectoparasitic in all stages, feeding primarily on the blood of mammals, birds, reptiles, and amphibians. Usually,

individuals attach to different host species in their various stages (larva, nymph, adult) and feed only once on each. Some, however, remain on a single host throughout their lifetime. The body, especially that of the female, is elastic and capable of enormous distension when the stomach fills with blood or eggs. Specimens may reach the size of a grape after 5 to 6 days of feeding. Adult males of some species apparently do not feed at all.

There are many more ixodid than argasid species (114 and 58, respectively) in the Neotropics (Keirans pers. comm.). The genera are *Ixodes, Dermacentor* (including *Anocenter;* Yunker et al. 1986), *Haemaphysalis, Boophilus, Amblyomma, Rhipicephalus,* and *Aponomma.*

Several hard tick species are of special importance. The southern cattle tick (*Boophilus microplus*) is prevalent over most of Latin America. It may be very abundant in some areas (Rawlins 1979) and very troublesome to cattle, transmitting babesiosis. The tropical horse tick (*Dermacentor nitens,* fig. 4.6d) is distributed from Mexico to Argentina and on the Greater Antilles. It infests the ears of its hosts (horses mainly but also cattle, deer, and goats) where it undergoes its complete development.

*Amblyomma* is the largest genus, with fifty Neotropical species (Jones et al. 1972). Many exhibit beautiful colors and ornamentation. They are often large, flat, and almost circular in outline. Hosts are varied, usually mammals, but also reptiles, including turtles (Ernst and Ernst 1977), and often birds, in the larval and nymphal stages.

The cayenne tick (*Amblyomma cajennense,* fig. 4.6c), known as *mostacilla* or *carrapato estrela,* is a general nuisance in all parts of the Neotropics where it menaces both man and livestock (Hoffmann 1962). The larvae may swarm in thousands in grass and low herbage. Very little has been published on its natural biology (Drummond and Whetstone 1975), although methods for artificial culture have been worked out (Travassos and Vallejo-Freire 1944).

The tropical bont tick (*A. variegatum*) has been introduced from Africa to the Caribbean, where it is associated with streptothricosis, a bacterial skin disease of livestock (Garris 1987), and heartwater (caused by the rickettsia *Cowdria ruminantium*).

*Ixodes pararicinus* (once confused with the European castor bean tick, *I. ricinus*) is known widely as a parasite of cattle in the southern half of South America (Keirans et al. 1985).

Soft tick genera in Latin America are *Otobius, Antricola, Argas, Nothoaspis,* and *Ornithodoros* (Jones and Clifford 1972). Notable regional taxa are as follows. *Argas* are small and found on odd hosts. The twenty Latin American species are mostly associated with birds (owls, fowl, etc.) (Hoogstraal et al. 1979). *A. persicus* is the cosmopolitan fowl tick ("adobe tick," "tampan") and one of the most important poultry parasites. *A. miniatus* (fig. 4.6e), however, may more often be the offender in Latin America generally, *A. moreli* in Peru (Keirans et al. 1979). *A. transversus* attaches to the neck and throat skin of giant Galápagos tortoises (Hoogstraal et al. 1973). All are restricted to dry niches in desert to savanna life zones.

*Ornithodorus talaje* and *O. rudis* are common Neotropical species. They feed on wild rodents and most domestic animals and man and are vectors of the relapsing fever spirochete *Borrelia* in Guatemala, Panama, and Colombia. *O. darwini* and *O. galapagensis* have been described from the Galápagos iguanas (Kohls et al. 1969, Keirans et al. 1980). *Antricola* are found on bats or in their guano.

Members of both tick families are purveyors of numerous serious diseases of man and domestic animals (Arthur 1961, Hoogstraal 1981). Particularly serious to human health in Latin America are spotted fevers caused by rickettsial organisms (Hoogstraal 1967) and relapsing fevers

brought on by *Borellia*. Rocky Mountain spotted fever is endemic in many parts of Latin America (Mexican spotted fever, fiebre manchada, Tobia fever, São Paulo fever) where it may be transmitted by almost any tick that lives on the mammalian hosts for the pathogens. Extensive use of one major vector, *Amblyomma cajennense*, has been applied to attempts to develop a vaccine for this virulent disease (Travassos and Vallejo-Freire 1944). Tickborne viral diseases seem not to be a regional problem, but in general, these are not well studied in Latin America.

Among animals, domestic and wild, ticks transmit Texas cattle fever or red water fever (tristeza) caused by the protozoan *Babesia bigemina*, in various parts of the American tropics. The vector is usually *Boophilus microplus*. The disease destroys the red blood cells and has a very high mortality rate. Some natural control of these vectors by fire ants in Mexico has been discovered (Butler et al. 1979).

Ticks also cause local trauma and inflammation at the site of attachment and sometimes elicit a paralysis by the injection of some neurologically active substance in their saliva. Fortunately, tick paralysis (Murnaghan and O'Rourke 1978) is usually temporary, disappearing when the tick is removed, but recovery may be delayed and can cause death in young or sensitive people, especially when the bite focus is on the neck at the base of the skull.

The nonexistent tick "*Ixodes maloni*" was recorded from a Guianan tepui by Sir Arthur Conan Doyle in his fiction classic, *The Lost World* (Hoogstraal 1972).

## References

ARTHUR, D. R. 1961. Ticks and disease. Row, Peterson, Evanston, Ill.

BUTLER, J. F., M. L. CAMINO, AND T. M. PÉREZ. 1979. *Boophilus microplus* and the fire ant *Solenopsis geminata*. Rec. Adv. Acarol. 1: 469–472.

DRUMMOND, R. O., AND T. M. WHETSTONE. 1975. Oviposition of the cayenne tick, *Amblyomma cajennense* (F.), in the laboratory. Entomol. Soc. Amer. Ann. 68: 214–216.

ERNST, C. H., AND E. M. ERNST. 1977. Ectoparasites associated with Neotropical turtles of the genus *Callopsis* (Testudines, Emydidae, Batagurinae). Biotropica 9: 139–142.

GARRIS, G. I. 1987. *Amblyomma variegatum* (Acari: Ixodidae): Population dynamics and hosts used during an eradication program in Puerto Rico. J. Med. Entomol. 24: 82–86.

HOFFMANN, A. 1962. Monografía de los Ixodoidea de México. I. Rev. Soc. Mex. Hist. Nat. 23: 191–307.

HOOGSTRAAL, H. 1967. Ticks in relation to human diseases caused by *Rickettsia* species. Ann. Rev. Entomol. 12: 377–420.

HOOGSTRAAL, H. 1972. *Ixodes maloni* Doyle, 1912 (nomen nudum) (Ixodoidea: Ixodidae) parasitizing humans in Brazil. Entomol. Soc. Amer. Bull. 18: 141.

HOOGSTRAAL, H. 1981. Changing patterns of tickborne diseases in modern society. Ann. Rev. Entomol. 26: 75–99.

HOOGSTRAAL, H., C. M. CLIFFORD, AND J. E. KEIRANS. 1973. *Argas* (*Microargas*) *transversus* (Ixodoidea: Argasidae) of Galápagos giant tortoises: Description of the female and nymph. Entomol. Soc. Amer. Ann. 66: 727–732.

HOOGSTRAAL, H., C. M. CLIFFORD, J. E. KEIRANS, AND H. Y. WASSEF. 1979. Recent developments in biomedical knowledge of *Argas* ticks (Ixodoidea: Argasidae). Rec. Adv. Acarol. 2: 269–278.

JONES, E. K., AND C. M. CLIFFORD. 1972. The systematics of the subfamily Ornithodorinae (Acarina: Argasidae). V.A. revised key to larval Argasidae of the Western Hemisphere and description of seven new species of *Ornithodoros*. Entomol. Soc. Amer. Ann. 65: 730–740.

JONES, E. K., C. M. CLIFFORD, J. E. KEIRANS, AND G. M. KOHLS. 1972. The ticks of Venezuela (Acarina: Ixodoidea) with a key to the species of *Amblyomma* in the Western Hemisphere. Brigham Young Univ. Biol. Ser. 17(4): 1–40.

KEIRANS, J. E., C. M. CLIFFORD, A. A. GUGLIELMONE, AND A. J. MANGOLD. 1985. *Ixodes* (*Ixodes*) *pararicinus*, n. sp. (Acari: Ixodoidea: Ixodidae), a South American cattle tick long confused with *Ixodes ricinus*. J. Med. Entomol. 22: 401–407.

KEIRANS, J. E., C. M. CLIFFORD, AND H. HOOGSTRAAL. 1980. Identity of the nymphs and adults of the Galápagos iguanid lizard parasites, *Ornithodoros* (*Alectorobius*) *darwini* and *O*. (*A*.) *galapagensis* (Ixodoidea: Argasidae). J. Med. Entomol. 17: 427–438.

KEIRANS, J. E., H. HOOGSTRAAL, AND C. M. CLIFFORD. 1979. Observations on the subgenus *Argas* (Ixodoidea: Argasidae: *Argas*). 16. *Argas* (*A*.) *moreli*, new species, and keys to Neotropical species of the subgenus. J. Med. Entomol. 15: 246–252.

KOHLS, G. M., C. M. CLIFFORD, AND H. HOOGSTRAAL. 1969. Two new species of *Ornithodoros* from the Galápagos Islands (Acarina: Argasidae). J. Med. Entomol. 6: 75–78.

MURNAGHAN, M. F., AND F. J. O'ROURKE. 1978. Tick paralysis, *In* S. Bettini, ed., Arthropod venoms. Springer, Berlin. Pp. 419–464.

OBENCHAIN, F. D., AND R. GALUN, eds. 1982. The physiology of ticks. Pergamon, New York.

OLIVER, JR., J. H. 1989. Biology and systematics of ticks (Acari: Ixodida). Ann. Rev. Ecol. Syst. 20: 397–430.

RAWLINS, S. C. 1979. Seasonal variation in the population density of larvae of *Boophilus microplus* (Canestrini) (Acari: Ixodoidea) in Jamaican pastures. Bull. Entomol. Res. 69: 87–91.

TRAVASSOS, J., AND A. VALLEJO-FREIRE. 1944. Criação artificial de *Amblyomma cajennense* para o preparo de vacina contra a febre maculosa. Inst. Butantan Mem. 18: 1–91.

YUNKER, C. E., J. E. KEIRANS, C. M. CLIFFORD, AND E. R. EASTON. 1986. *Dermacentor* ticks (Acari: Ixodoidea: Ixodidae) of the New World: A scanning electron microscope atlas. Entomol. Soc. Wash. Proc. 88: 609–627.

# WHIP SCORPIONS

Uropygi (= Pedipalpida, in part). *Spanish:* Vinagrosos, vinegrones (General), vinagrillos (Mexico). Vinegarroons.

A whiplike flagellum extending from the tip of the abdomen gives this group its common name. Other distinctive characteristics include a longer than wide prosoma, spiderlike chelicerae, and massive pedipalps equipped with short but strong spines (Weygoldt 1972). The first legs are feelerlike and slightly longer than the others, which are relatively short and held close to the body. A pair of defensive glands that open near the anus secrete formic and acetic acid, among other com-

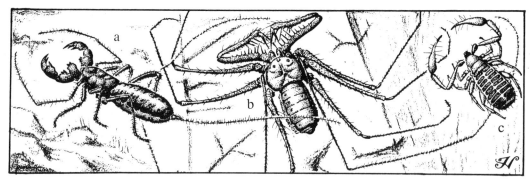

**Figure 4.7  WHIP SCORPIONS AND PSEUDOSCORPION.** (a) Vinegarroon (*Mastigoproctus giganteus,* Elyphonidae). (b) Tailless whip scorpion (*Heterophrynus longicornus,* Phrynidae). (c) Pseudoscorpion (*Chelifer cancroides,* Cheliferidae).

pounds (e.g., caprylic acid, a wetting agent). These compounds may be forcibly expelled from the anus as a means of chemical protection (Eisner et al. 1961).

These animals are nocturnal predators that live on the ground, for example, under litter and stones and in stumps (Kaestner 1968: 117–119). *Mastigoproctus giganteus* (fig. 4.7a) actively excavates tunnels in the ground where it remains during the daytime. Those most often encountered are the *vinagrillos* (so-called because of their vinegar odor when defending themselves with their chemical sprays) of the genus *Mastigoproctus.* Specimens visit the vicinity of artificial lights at night, particularly in desert areas, to feast on insects attracted to the illumination. They are large (BL to 7 cm), dark brown to black, and heavily sclerotized. In spite of their fearsome scorpionlike appearance, they do not bite or sting.

Whip scorpions prefer humid conditions and are mostly found in tropical climes. Those that inhabit arid regions appear only after rains.

There are representatives of two families in Latin America (Rowland and Cooke 1973). Most are of the family Elyphonidae, the largest genus being *Mastigoproctus* with twelve species. The one species of *Thelyphronellus* is restricted to Guyana and northeast-

ern Brazil, as is *Arnauromastigon*, the single genus and species of the other family, Hypoctonidae.

## References

Eisner, T., J. Meinwald, A. Monro, and R. Ghent. 1961. Defense mechanisms of arthropods. I. The composition and function of the spray of the whip scorpion, *Mastigoproctus giganteus* (Lucas) (Arachnida, Pedipalpida). J. Ins. Physiol. 6: 272–298.

Kaestner, A. 1968. Invertebrate zoology, arthropod relatives, Chelicerata, Myriapoda. Vol. 2. Wiley Interscience, New York.

Rowland, J. M., and J. A. L. Cooke. 1973. Systematics of the arachnid order Uropygida (= Thelyphonida). J. Arachnol. 1: 55–71.

Weygoldt, P. 1972. Geisselskorpione und Geisselspinnen (Uropygi und Amblypygi). Zeit. Kölner Zoo. 15(3): 95–107.

## TAILLESS WHIP SCORPIONS

Amblypygi (= Pedipalpida, in part). *Spanish:* Frinos (Panama). Whip spiders.

These arachnids (Weygoldt 1972) somewhat resemble the whip scorpions but differ in lacking a terminal appendage or tail and having spinous pedipalps. They are greatly flattened and have tremendously elongated forelegs; the pedipalps are not as massive as the former group but possess a formidable array of long, very sharp-tipped spines. The walking legs are long and slender and splayed out to the side. They lack spray glands.

Amblypygids (Kaestner 1968) live in tropical to subtropical areas (Beck 1968), always under fairly humid conditions, under stones, under bark and in hollow logs, in litter, and so on. They shun light and are well-known cave inhabitants. They occasionally run on the trunks of large trees or even the walls of homes at night, where they search in darkness with their sensitive, antenniform (whiplike) forelegs and snare with their spiny pedipalps. Little more is known of their habits in nature; some laboratory studies have been completed (Weygoldt 1977).

There are two suborders, Pulvillata and Apulvillata (Quintero 1986), each containing several families and many genera, found from the southern United States to the northern half of South America. Among the former, there are nine genera with varied habits. *Paracharon* and *Trichari-nus* are small, blind, and live in the nests of ants and termites (Quintero 1986). Other dominant genera are *Chirinus* in South America and *Charinides* in the Caribbean (Quintero 1983). The New World species of Apulvillata are placed in five genera: *Phrynus*, the most widespread, *Paraphrynus* (Mullinex 1975), *Heterophrynus* (fig. 4.7b), *Acanthophrynus*, and *Trichodamon* (Mello-Leitão 1935). The last is the only New World member of its otherwise Old World family, Damonidae, and is found in Brazil (Quintero 1976).

*Phrynus* are sometimes much feared by locals for their large size (BL to 2.8 cm, leg span of 20 cm), ugliness, and wrongly suspected venomousness. They are truly harmless creatures. They defend themselves only by pinching and carry no sting or poisonous bite. In this genus, the pedipalpal spines form a kind of basket for catching prey when interposed.

## References

BECK, L. 1968. Aus den Regenwaldern am Amazonas II. Natur Museum 98(2): 71–80.

KAESTNER, A. 1968. Invertebrate zoology, arthropod relatives, Chelicerata, Myriapoda. Vol. 2. Wiley Interscience, New York.

MELLO-LEITÃO, C. 1935. Sobre o gênero *Trichodamon* M.-L. Inst. Butantan Mem. 10: 297–302.

MULLINEX, B. L. 1975. Revision of *Paraphrynus* Moreno (Amblypygida: Phrynidae) for North America and the Antilles. Calif. Acad. Sci. Occ. Pap. 116: 1–80.

QUINTERO, JR., D. 1976. *Trichodamon* and Damonidae, new family status (Amblypygi: Arachnida). Brit. Arachnol. Soc. Bull. 3: 222–227.

QUINTERO, JR., D. 1980. Systematics and evolution of *Acanthophrynus* Kraepelin (Amblypygi, Phrynidae). 8th Int. Cong. Arachnol. (Vienna) Proc. Pp. 341–347.

QUINTERO, JR., D. 1983. Revision of the amblypygid spiders of Cuba and their relationships with the Caribbean and continental American amblypygid fauna. Stud. Fauna Curaçao Carib. Is. 196: 1–54.

QUINTERO, JR., D. 1986. Revisión de la clasificación de Amblypygidos pulvinados: Creación de subórdenes, una nueva familia y un nuevo género con tres nuevas especies (Arachnida: Amblypygi). 9th Int. Cong. Arachnol. (Panama City) Proc. Pp. 203–212.

WEYGOLDT, P. 1972. Geisselskorpione und Geisselspinnen (Uropygi und Amblypygi). Zeit. Kölner Zoo. 15(3): 95–107.

WEYGOLDT, P. 1977. Coexistence of two species of whip spiders (Genus *Heterophrynus*) in the Neotropical rain forest (Arachnida, Amblypygi). Oecologia 27: 363–370.

# FALSE SCORPIONS

Pseudoscorpionida. *Spanish:* Quernitos (General). Book scorpions.

False scorpions all possess scorpionlike pedipalps and general shape but differ from those arachnids most conspicuously by their much smaller size (BL of most 2–4 mm) and the lack of a segmented tail. Also missing are the scorpion's characteristic ventral pectines. The fingers of the pedipalps usually contain venom glands opening at the sharp tips. They also have silk glands, located in the prosoma, with ducts opening at the tip of the chelicera's movable article. Silk is used to make cocoonlike nests about twice the animal's size in which to molt, pass the winter, or oviposit.

There are three suborders, all with members in the Neotropics. The region's fauna is poorly known; Beier (1932) listed about 180 species, to which authorities have since added approximately 400 species; some 100 to 200 more may remain undiscovered (Mahnert pers. comm.). The Amazonian fauna is especially diverse (Mahnert 1979, Mahnert et al. 1986).

Phoresy is practiced by many species in this order, usually only by females, and is apparently important to them as a means of dispersal (Muchmore 1971). The best-known hosts are long-horned beetles, especially the harlequin beetle, under whose elytra several species travel.

The best known of these is *Cordylochernes scorpioides*, which constructs a "safety harness" of silk on the beetle's abdomen to which they cling when the host flies (Zeh and Zeh 1991). A few pseudoscorpions have become associated with civilization. *Cheiridium museorum, Chelifer cancroides* (fig. 4.7c), and *Withius piger* (in Chile) have spread recently from the Old World to South America in grain shipments in which they have secreted themselves to feed on stored product pests.

Pseudoscorpions are typically cryptic, living in soil and litter, under bark, beneath rocks, and in similar retreats. Many are cavernicolous (Muchmore 1972). There are several maritime species, living under stones and shore debris such as driftwood and stranded seaweed and in rock crevices, where they search for prey (Lee 1979).

Courtship and mating are complex behaviors much like those of scorpions. The process begins with a form of dance in which the male stops close to the female, shakes his abdomen, and waves his pincers. He then drops a thread from his ventral genital opening. This hardens into a pedestal (spermatophore) on which he leaves seminal fluid. The female straddles the pillar and draws the fluid into her sexual opening, during which process the male may advance and shake her by the legs to ensure that the fluid becomes detached within her body.

## References

BEIER, M. 1932. Pseudoscorpionida. Das Tierreich 57: 1–258, 58: 1–294.

LEE, V. F. 1979. The maritime pseudoscorpions of Baja California, México (Arachnida: Pseudoscorpionida). Calif. Acad. Sci. Occ. Pap. 131: 1–38.

MAHNERT, V. 1979. Pseudoskorpione (Arachnida) aus dem Amazonas-Gebiet (Brasilien). Rev. Suisse Zool. 86: 719–810.

MAHNERT, V., J. ADIS, AND P. F. BÜHRNHEIM. 1986. Key to the families of Amazonian Pseudoscorpiones (Arachnida). Amazoniana 10: 21–40.

MUCHMORE, W. B. 1971. Phoresy by North and Central American pseudoscorpions. Rochester Acad. Sci. Proc. 12: 79–97.

MUCHMORE, W. B. 1972. New diplosphyronid pseudoscorpions, mainly cavernicolous, from Mexico. Amer. Micros. Soc. Trans. 91: 261–276.

ZEH, D. W., AND J. A. ZEH. 1991. Novel use of silk by the harlequin beetle-riding pseudoscorpion, *Cordylochernes scorpioides* (Pseudoscorpionida, Chernetidae). J. Arach. 19: 153–154.

## SCORPIONS

Scorpionida. *Spanish:* Escorpiones (General), alacranes. *Portuguese:* Escorpiões, lacraus (Brazil). *Nahuatl:* Colomeh, sing. cótotl (Mexico).

Scorpions (Polis 1990, Williams 1987) are characterized by their eight legs, a pair of large pinchers or pedipalps (like lobster claws) at the front, and a much smaller but heavy pair of jaws (chelicerae) around the mouth. On the underside, a pair of comblike "pectines" are conspicuous structures. The main body is divided into a large front trunk segment, comprising the head, which is fused with the thorax, and basal abdominal segments. The latter is followed by a long six-segmented tail, the last article of which (the telson) is bulbous and sharp tipped, constituting the sting. The latter contains paired poison glands with ducts opening in a slit a short distance from the tip.

Scorpion sexual behavior is peculiar. The male first faces the female and grasps her pedipalps or chelicerae with his own like appendages. The partners advance and retreat in a kind of dance ("promenade a deux") for a time until a suitable place is found for mating. The male expels a sperm packet to the ground which is picked up by the female's genitalia after more "dancing." Then the female frees herself from the male and goes to a burrow or other shelter; the male goes away. It is not true that the female kills the male. The latter may even live to ultimately mate with several other females.

Females give birth to active larval scorpions. These migrate to their mother's back where they remain until their first molt, when they leave and begin independent lives. In most species, after one to three years and several more molts, they attain adulthood. A rare few of these arachnids lead a semisocial existence (Polis and Lourenço 1986).

Scorpions feed entirely on other arthropods (including other scorpions), mainly spiders and insects that they encounter and overcome during their nightly ramblings. Certain desert species may go without water for several months but will drink it readily if available; yet scorpions from humid forest environments may die in a few days if forced to go without water.

Scorpion venom produces neurotoxic circulatory and muscular effects in humans. Several species are of major medical importance (Keegan 1980); those in Latin America belong to the genera *Tityus* and *Centruroides*, both in the family Buthidae. *Tityus* (Diniz 1978, Bücherl 1978) are found in varied habitats. Many are humid forest inhabitants, living under the bark of dead trees and among ground litter. They are an entirely Neotropical genus of wide distribution continentally and are present on several of the Antilles. The genus is also the region's largest, with over 130 species (Lourenço 1978). The deadliest species is *T. serrulatus* (fig. 4.8a), which causes the demise of many people, especially very young children, in southeastern Brazil. Its venom has been characterized chemically and pharmacologically (Possani et al. 1977).

In the drier parts of Mexico and Central America, species of "bark scorpions" (*Centruroides*) are the most dangerous (Bücherl 1971, Stahnke 1978); the genus is also found in the West Indies and in parts of South America (Sissom and Lourenço 1987). Among the over fifty species (Santiago-Blay pers. comm., Stanhke and Calos 1977) is the well-known Durango scor-

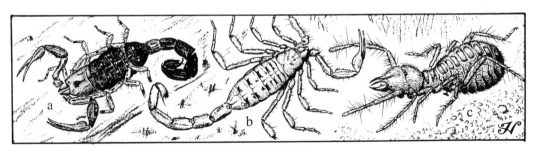

**Figure 4.8  SCORPIONS AND SUN SPIDER.** (a) Forest scorpion (*Tityus serrulatus*, Buthidae). (b) Durango scorpion (*Centruroides suffusus*, Buthidae). (c) Sun spider (*Eremobates* sp., Erematobatidae).

pion, *C. suffusus* (fig. 4.8b) (Atúnez 1950), which was responsible for 1,608 deaths between 1890 and 1926 in the city of Durango, which had a population of about 40,000 in this period. Actually, the death rate from scorpions is higher in other areas in Mexico, over 110 per hundred thousand in some years in southern states such as Colima (Mazzotti and Bravo-Becherelle 1963).

*C. limpidus* is an even more potent species and is considered the most dangerous scorpion in Mexico, responsible for more than 50 percent of the 100,000 stings per year in the country (according to Possani et al. [1980], who have also described the qualities of its venom).

From a public health standpoint, scorpions are the most important venomous animals of Mexico. During the period 1940–1949 and 1957–1958, a total of 20,352 persons were killed by scorpion stings (Mazzotti and Bravo-Becherelle 1963).

Small arboreal scorpions are also known in South America. Others, so-called field scorpions, prefer the soil of damp places and are found under stones, especially along rivers. Desert or semidesert types, like *Centruroides,* are often burrowers in places protected from the heat. Many live in old buildings, under houses, and in garages and often enter occupied premises where they come into contact with humans. True cave-dwelling scorpions are comparatively rare (Francke 1981, Lourenço & Francke 1985, Mitchell 1968). Most scorpions are continental, although a few exist on oceanic or offshore continental islands, such as Coco, the Antilles, and in the Gulf of California.

Overall, in Latin America, there are some 48 genera and more than 400 species allocated to 7 families (Lourenço pers. comm.). The largest genera are *Tityus,* dominant in South America (Lourenço 1978), and *Centruroides,* which occurs mostly in Central America and Mexico (Stahnke and Calos 1977). No general taxonomic treatment exists for the entire area, although attempts have been made to review the South American fauna and general distributional patterns (Armas 1982, Lourenço 1986, Mello-Leitão 1945). A key to the genera of Buthidae is available (Vachon 1977).

## References

ATÚNEZ, F. 1950. Los alacranes en el folklore de Durango. Priv. publ., Aguascalientes, Mexico.

BÜCHERL, W. 1971. Classification, biology and venom extraction of scorpions. *In* W. Bücherl and E. E. Buckley, eds., Venomous animals and their venoms. III. Venomous invertebrates. Academic, New York. Pp. 317–347.

BÜCHERL, W. 1978. Venoms of Tityinae. A. Systematics, distribution, biology, venomous apparatus, etc., of Tityinae; venom collection, toxicity, human accidents and treatment of stings. *In* S. Bettini, ed., Arthropod venoms. Springer, Berlin. Pp. 371–379.

DE ARMAS, L. F. 1982. Algunos aspectos zoogeográficos de la escorpionfauna antillana. Poeyana 238: 1–17.

DE MELLO-LEITÃO, C. 1945. Escorpiões Sul-Americanos. Mus. Nac. (Rio de Janeiro) Arq. 40: 7–468.

DINIZ, C. R. 1978. Venoms of Tityinae. B. Chemical and pharmacologic aspects of Tityine venoms. *In* S. Bettini, ed., Arthropod venoms. Springer, Berlin. Pp. 379–394.

FRANCKE, O. F. 1981. A new genus of troglobitic scorpion from Mexico (Chactidae, Megacorminae). Amer. Mus. Nat. Hist. Bull. 170: 23–28.

KEEGAN, H. L. 1980. Scorpions of medical importance. Univ. Mississippi, Jackson.

LOURENÇO, W. R. 1978. Sur les difficultés rencontrées dans la révision du genre *Tityus* (Scorpiones, Buthidae). Zool. Soc. London Symp. 42: 502.

LOURENÇO, W. R. 1986. Les modèles de distribution géographique de quelques groupes de scorpions néotropicaux. Soc. Biogeogr. Comp. Rend. 62: 61–83.

LOURENÇO, W. R., AND O. F. FRANCKE. 1985. Révision des connaissances sur les scorpions cavernicoles (troglobies) (Arachnida, Scorpions). Mem. Biospél. 12: 3–7.

MAZZOTTI, L., AND M. A. BRAVO-BECHERELLE. 1963. Scorpionism in the Mexican republic. *In* H. L. Keegan and W. V. Macfarlane, eds., Venomous and poisonous animals and noxious plants of the Pacific Region. Macmillan, New York. Pp. 119–131.

MITCHELL, W. R. 1968. *Typhlochactas*, a new genus of eyeless cave scorpions from Mexico (Scorpionidae, Chactidae). Ann. Speleol. 23: 753–777.

POLIS, G. A. 1990. The biology of scorpions. Stanford Univ. Press, Stanford.

POLIS, G. A., AND W. R. LOURENÇO. 1986. Sociality among scorpions. 10th Cong. Int. Arachnol. (Jaca, Spain) Actas 1: 111–115.

POSSANI, L. D., A. C. ALAGÓN, P. L. FLETCHER, JR., AND B. W. ERICKSON. 1977. Purification and properties of mammalian toxins from the venom of the Brazilian scorpion *Tityus serrulatus* Lutz and Mello. Arch. Biochem. Biophys. 180: 394–403.

POSSANI, L. D., P. L. FLETCHER, JR., A. B. C. ALAGÓN, A. C. ALAGÓN, AND J. Z. JULIÁ. 1980. Purification and characterization of a mammalian toxin from venom of the Mexican scorpion, *Centruroides limpidus tecomanus* Hoffmann. Toxicon 18: 175–183.

SISSOM, W. D., AND W. R. LOURENÇO. 1987. The genus *Centruroides* in South America (Scorpiones, Buthidae). J. Arachnol. 15: 11–28.

STAHNKE, H. L. 1978. The genus *Centruroides* (Buthidae) and its venom. *In* B. Bettini, ed., Arthropod venoms. Springer, Berlin. Pp. 277–307.

STAHNKE, H. L., AND M. CALOS. 1977. A key to the species of the genus *Centruroides* Marx (Scorpionida: Buthidae). Entomol. News 88: 111–120.

VACHON, M. 1977. Contribution à l'étude des scorpions Buthidae du nouveau monde. I. Complément à la connaissance de *Microtityus rickyi* Kj.-W. 1956 de l'île de la Trinité. II. Description d'une nouvelle espèce et d'un nouveau genre Mexicains: *Darchenia bernadettae*. III. Clé de determination des genres de Buthidae du nouveau monde. Acta Biol. Venezuelica 9: 283–302.

WILLIAMS, S. C. 1980. Scorpions of Baja California, Mexico and adjacent islands. Calif. Acad. Sci. Occ. Pap. 135: 1–12.

WILLIAMS, S. C. 1987. Scorpion bionomics. Ann. Rev. Entomol. 32: 275–295.

# SUN SPIDERS

Solpugida (= Solifuga). Wind scorpions.

Members of this order are distinguished from other arachnid groups by their huge, forward-projecting, scissorlike chelicerae. The prosoma behind the chelicerae is constricted, forming a headlike anterior portion. The pedipalps are similar to the walking legs but are elevated when in use and provided with adhesive organs at the tip. Typically, their soft bodies are covered with short, silky pelage. Most are medium-sized (BL 2–4 cm).

Sun spiders prefer warm, arid habitats and are most abundant in the Chilean, Peruvian, and Mexican deserts. Some occur at high elevations in the Andes. At night, they run rapidly over the substratum and occasionally are seen near houses, where they come to catch insects attracted to lights. They spend the daylight hours and winter months in ground burrows of their own making; some burrow in pithy or rotten wood. Although large specimens are aggressive and formidable in appearance, they are without venom-delivering capabilities and are innocuous to humans. They feed on the other ground-dwelling arthropods that they encounter on their nightly wanderings.

Latin American species are almost all found in the two families Ammotrechidae (widespread; 61 species in several genera) and Eremobatidae (far northern Mexico and Baja California; 21 species in two genera) (Muma 1970, 1976). *Eremobates* (fig. 4.8c) is a major genus. Two species, *Amacata penai* (Muma 1971) and *Syndaesia mastix* (Maury 1980), from the Atacama Desert and western Argentina, respectively, represent the family Daesiidae (= Amacataidae).

## References

MAURY, E. A. 1980. Presencia de la family Daesiidae en América del Sur con la descripción de un nuevo género (Solifugae). J. Arachnol. 8: 59–67.

MUMA, M. H. 1970. A synoptic review of North American, Central American, and West Indian Solpugida (Arthropoda: Arachnida). Arths. Fla. Neighborhood Land Areas 5: 1–62.

MUMA, M. H. 1971. The solpugids (Arachnida, Solpugida) of Chile, with descriptions of a new family, new genera, and new species. Amer. Mus. Nov. 2476: 1–23.

MUMA, M. H. 1976. A review of solpugid families with an annotated list of Western Hemisphere solpugids. West. New Mex. Univ., Off. Res. Pub. 2(1): 1–33.

## MYRIAPODS

Myriapods (Camatini 1980) are all terrestrial but take to moist to humid environments. They are elongate (often wormlike), with many similar body segments from which arise one pair of appendages each (these true segments have fused in the millipedes). The mouthparts are mandibulate and the legs several jointed, without basal lobes or subdivisions. There is reason to believe that myriapods and insects share a common multilegged ancestor.

### Reference

CAMATINI, M., ed. 1980. Myriapod biology. Academic, New York.

## MILLIPEDES

Diplopoda. *Spanish:* Milpies. *Portuguese:* Gongolos, piolhos de cobra (Brazil). *Quechua:* Pachac chaqui.

Like centipedes, which they superficially resemble, millipedes gain their name from an abundance of legs. Two pairs arise from most apparent segments, a condition created by embryonic fusion of alternate body somites, each carrying one pair of legs, a pair of tracheal openings, and a ventral nerve cord ganglion, to form double somites (diplosomites). The resulting secondary segment (diplosegment), therefore, has two ganglia and four legs and spiracles, all of which are located in the metazonite, representing the posterior of the fused somites. The prozonite, representing the anterior of the fused somites, lacks appendages and can thus be telescoped into the preceding metazonite, resulting in a more compact body form. The four anteriormost segments have one leg or no legs, a secondary loss associated with the ability to curl the head backward into a protective posture.

The head of millipedes bears one pair of short antennae, one pair of internal mandibles, used to masticate decaying wood and other vegetation, and a flattened, platelike structure, the gnathochilarium, which forms the anterior floor of the mouth. The latter consists of several small sclerites whose shape and arrangement differ among orders and are, therefore, taxonomically useful.

The subclass Diplopoda is divided into three superorders, each represented in the Neotropical region, but the vast majority belong to the superorder Helminthomorpha, containing calcified forms that are generally elongate with cylindrical or flattened bodies. The reproductive tracts open on segment 3, but in males, one or both pairs of legs on segment 7 and occasionally the anterior pair on segment 8 are modified into copulatory structures called gonopods. The gonopods transfer sperm packets to the female, and their configurations are of primary significance at the generic and specific levels. Pairs of millipedes are often encountered enraptured, stretched out or coiled together venter to venter in a many-legged copulatory embrace. After copulation, males of some large "flat-backed" species ride on the backs of females for several days (Heisler 1983).

Nine of the eleven helminthomorph orders occur in the Neotropics, but the Polydesmida, Spirostreptida, and Spirobolida are dominant. The Polydesmida are generally characterized by lateral expansions of the dorsum called "paranota," which impart a flattened appearance to the animals, hence the name "flat-backed" millipedes. Seventeen polydesmid families occur in the Neotropics (Hoffman 1979),

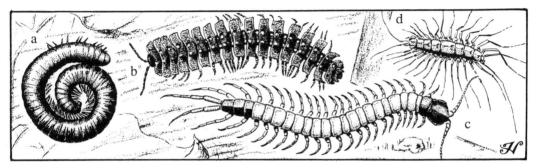

**Figure 4.9 MILLIPEDES AND CENTIPEDES.** (a) Spirobolid millipede (*Orthoporus* sp., Spirostreptidae). (b) Polydesmid millipede (*Barydesmus* sp., Platyrhacidae). (c) Giant centipede (*Scolopendra gigantea*, Scolopendridae). (d) House centipede (*Scutigera coleoptrata*, Scutigeridae).

many of which are comprised of minute, cryptic forms. The dominant families with large-bodied forms are Platyrhacidae and Chelodesmidae.

Common platyrhacid genera are *Nyssodesmus* (Costa Rica) and *Barydesmus* (Costa Rica to Peru and western Brazil; fig. 4.9b), whose adults range from 70 to 100 millimeters in length. Only slightly smaller are *Psammodesmus* (Panama to Peru) and the diverse assemblage of forms usually assigned to *Amplinus, Pycnotropis,* and *Polylepiscus* (Mexico to Peru). In the Chelodesmidae, *Chondrodesmus*, with around forty species, is abundant from Mexico to Brazil. The Spirostreptida and Spirobolida contain phenotypically similar cylindrical forms of variable length and diameter; they are distinguished by details of the head, exoskeleton, and gonopods. Nearly thirty Neotropical spirostreptid genera, in three families, are currently recognized (Hoffman 1979, Krabbe 1982), but the most widespread is *Orthoporus* (fig. 4.9a), with around forty species that range from the southwestern United States throughout most of South America to Brazil. *Vilcastreptus hoguei* is a large, dark-bodied form with pinkish legs, seen by thousands of tourists each year in the Incan ruins of Machu Picchu (Hoffman 1988).

Of the five Neotropical spirobolid families, the Rhinocricidae is dominant. These millipedes vary greatly in size and are readily recognized as belonging to the family, but a satisfactory generic arrangement has not been attained.

The Latin American millipede fauna is diverse but poorly known. Hoffman (1969) reported some 470 species from Brazil, and Loomis (1968) reported around 750 from Mexico and Central America. These are the only counts available. The total fauna doubtlessly consists of several thousand species, probably less than 20 percent of which have been discovered, much less named and described.

Although deserticolous forms exist, these are creatures mostly of dank, humid habitats. They live in rotten logs, among leaf litter, in the soil, under stones and loose tree bark, in caves, and are a major component of the forest ground fauna; a few are symbiotic with army ants (Loomis 1959).

Millipedes are harmless creatures, protected primarily by their hard exoskeleton. However, most produce droplets of caustic or noxious secretions from a series of lateral pores on most segments (Eisner et al. 1978). The liquid usually oozes out, but some large millipedes, such as the Peruvian platyrhacid *Barydesmus* (orig. obs.), the Costa Rican *Nyssodesmus python* (Heisler 1983), and *Rhinocricus lethifer,* a Jamaican rhinocricid (Loomis 1936), can forcibly squirt the fluid 1 to 4 decimeters or more.

The active chemicals in these secretions are diverse. Most orders have single-chambered glands that produce benzoquinones, cresols, or aldehydes that can blister or tan human skin (Burtt 1947). Polydesmids give off hydrogen cyanide from the outer part of two-chambered glands, from nontoxic mandelonitrile precursors in the inner compartment (Woodring and Blum 1963). Diplopods also adopt a defensive posture by coiling with the vulnerable head tucked in the center. Two Neotropical polydesmoid families can roll up into balls or spheres, an adaptive feature that has evolved independently in other polydesmoid families and another subclass in other parts of the world.

Members of the class harbor a diverse microflora in the gut which functions to release nutrients and reduce toxins from ingested plant matter (Sakwa 1974).

Discussions of the anatomy and general biology of millipedes are available in several general textbooks on arthropods (Kaestner 1968: 389–429). Refer to Loomis (1968) and Hoffman (1979) for bibliographies and taxonomy. Important regional taxonomic works have been written by Loomis (1936, 1964, 1968) and Jeekel (1963).

### References

Burtt, E. 1947. Exudate from millipedes with particular reference to its injurious effects. Trop. Dis. Bull. 44: 7–12.

Eisner, T., D. Alsop, K. Hicks, and J. Meinwald. 1978. Defensive secretions of millipeds. In S. Bettini, ed., Arthropod venoms. Springer, Berlin. Pp. 41–72.

Heisler, I. L. 1983. Nyssodesmus python (milpies, large forest-floor millipede). In D. H. Janzen, ed., Costa Rican natural history. Univ. of Chicago Press, Chicago. Pp. 747–749.

Hoffman, R. L. 1969. The origin and affinities of the southern Appalachian diplopod fauna. In P. C. Holt, ed., The distributional history of the biota of the southern Appalachians. Virginia Polytech. Inst., Blacksburg. Pp. 221–246.

Hoffman, R. L. 1979. Classification of the Diplopoda. Mus. Hist. Nat., Geneva.

Hoffman, R. L. 1988. A new genus and species of spirostreptoid millipedes from the eastern Peruvian Andes. Myriapodologica 2: 29–36.

Jeekel, C. A. W. 1963. Diplopoda of Guiana (1–5). Nat. Stud. Suriname Nederlandse Antillen 4(27): 1–157.

Kaestner, A. 1968. Invertebrate zoology. Vol. 2. Wiley Interscience, New York.

Krabbe, E. 1982. Systematik der Spirostreptidae (Diplopoda: Spirostreptomorpha). Natur. Ver. Hamburg Abh. (n.f.) 24: 1–476.

Loomis, H. F. 1936. The millipedes of Hispaniola, with descriptions of a new family, new genera, and new species. Mus. Comp. Zool. (Harvard Univ.) Bull. 80: 1–191.

Loomis, H. F. 1959. New myrmecophilous millipeds from Barro Colorado Island, Canal Zone and Mexico. Kans. Entomol. Soc. J. 32: 1–7.

Loomis, H. F. 1964. The millipedes of Panama (Diplopoda). Fieldiana: Zoology 47: 1–136.

Loomis, H. F. 1968. A checklist of the millipedes of Mexico and Central America. U.S. Natl. Mus. Bull. 266: 1–137.

Sakwa, W. N. 1974. A consideration of the chemical basis of food preference in millipedes. Zool. Soc. London Symp. 32: 329–346.

Woodring, J. P., and M. S. Blum. 1963. Anatomy and physiology of repugnatorial glands of Pachydesmus. Entomol. Soc. Amer. Ann. 56: 448–453.

## CENTIPEDES

Chilopoda. *Spanish:* Ciempiés, escolopendras (General); alacranes (Puerto Rico). *Portuguese:* Centopéias, lacraias (Brazil). *Nahuatl:* Petlazolcoameh, sing. petlazolcoatl (Mexico). Scolopenders.

Centipedes (Kaestner 1968: 356–388, Lewis 1981) are slender, very elongate, and many segmented arthropods, resembling millipedes only in a very general way. They are distinguished by bodies with only one pair of legs per segment. The latter are clearly distinct, not fused into pairs. The number of segments varies from 15 to 181; no species has an even number of pairs of legs and never, therefore, has one hundred legs, as the group's common name suggests. Centipedes possess a single pair of long, very flexible antennae and mandibulate, forward-projecting mouthparts. The anteriormost legs are modified into four-jointed, sharply pointed fangs (forcipules) connected to internal venom glands.

These ubiquitous arthropods exhibit a large size range. The majority are relatively small (BL 1–5 cm). Some very large chilopods belong to the Scolopendromorpha (Bücherl 1974). *Scolopendra gigantea* (fig. 4.9c) of the West Indies is gigantic, attaining a body length of 27 centimeters (Bücherl 1971). Because of their great size, such centipedes are greatly feared. They sometimes bite humans, causing pain and often a local inflammation but usually nothing more serious, notwithstanding horror tales in the literature (Minelli 1978). Scolopenders (*Scolopendra* spp.) are most often the offending species. It is probably a myth that they leave a wound with each leg if they crawl on bare skin, although the legs are tipped with sharp claws that may cause a prickling sensation. There is no evidence that the legs contain venom glands, although some centipedes produce noxious chemicals from other parts of the body. The chemical composition of centipede venom is still virtually unknown. Bücherl (1946, 1971), experimenting with five of the largest and most common Brazilian species on laboratory animals, concluded that the venom from actual bites, while capable of giving intense pain, was too weak to seriously harm humans. In fact, it most probably can be handled with impunity. Some even have been gathered for use as human food. Von Humboldt and Bonpland (1852: 1:157) saw Chaymas Indian children drag 45-centimeter-long centipedes out of ground burrows and devour them directly.

Centipedes are most at home in warm, humid retreats where they hide by day under stones or logs on the ground, beneath loose bark, in rotting wood, and in caves and similar niches. The house centipede (*Scutigera coleoptrata;* fig. 4.9d) is common in dwellings. Centipedes emerge at night to prey on other small surface-dwelling invertebrates, especially earthworms and insects. Some *Scolopendra* take small vertebrates (mice, lizards, toads) in captivity, and this may be their chief food in nature also, which may explain the generally large size of the members of this genus. They are quick and easily subdue their prey with a venomous bite from the anterior fangs.

The taxonomy of centipedes is still unsettled. Classifications of higher groups vary greatly, and many species remain undescribed. Four orders are usually recognized: the many-legged, wormlike, soil-dwelling Geophilomorpha; the short-bodied Lithobiomorpha, with 15 pairs of legs; the Scutigeromorpha, which has 15 pairs of very long, spiderlike legs; and the large, flattened Scolopendromorpha, with 21 or 23 pairs of legs. The last is the most familiar group, with many members belonging to the genus *Scolopendra*. The ten Neotropical species range over the northern half of South America and the Caribbean islands of Jamaica and Trinidad. Centipedes are commonly called *alacranes* in Puerto Rico, a name that should be reserved for Scorpionida (Santiago-Blay 1985).

## References

BÜCHERL, W. 1946. Ação do veneno dos escolopendromorfos do Brazil sôbre alguns animais de laboratório. Inst. Butantan Mem. 19: 181–197.

BÜCHERL, W. 1971. Venomous chilopods or centipedes. *In* W. Bücherl and E. E. Buckley, eds., Venomous animals and their venoms. 3: Venemous invertebrates. Academic, New York. Pp. 169–191.

BÜCHERL, W. 1974. Die Scolopendromorpha der neotropischen region. Zool. Soc. London Symp. 32: 99–133.

KAESTNER, A. 1968. Invertebrate zoology. Vol. 2. Wiley Interscience, New York.

LEWIS, J. G. E. 1981. The biology of centipedes. Cambridge Univ. Press, Cambridge.

MINELLI, A. 1978. Secretions of centipedes. *In*

S. Bettini, ed., Arthropod venoms. Springer, Berlin. Pp. 73–85.

SANTIAGO-BLAY, J. 1985. Aclaraciones en torno a los significantes zoológicos de la voz "alacran" en Puerto Rico. Ciencia 12(2): 43–45.

VON HUMBOLDT, A., AND A. BONPLAND. 1852 [1814–1825]. Personal narrative of travels to the equinoctial regions of America, during the year 1799–1804. 3 vols. Henry Bohn, London. Translated by Thomasina Ross.

## HEXAPODS

Possession of three pairs of legs, hence the name Hexapoda, is a constant feature of the Insecta but also of three additional primitive orders, among them the Collembola. Because of peculiarities in the anatomy of this assemblage they are separated from the true insects as the subclass Parainsecta. Larval mites and ticks also have six legs but are members of the very distinct subphylum Chelicerata. (See Evolution and Classification, chap. 1.)

## SPRINGTAILS

Collembola. *Spanish* and *Portuguese:* Colémbolos (General).

These minute to small (BL 0.3–10 mm, average 2 mm), soft-bodied insect relatives take their common name in English from a forked appendage located on the underside of the abdomen which is used as a springlike device to propel the animal upward. This form of locomotion is used when the insect is disturbed and can send it many times the length of its body into the air or to the side. Collembola also possess a tubular structure (collophore) on the midventral part of the first abdominal segment. This is tipped with a bilobed, eversible sac thought to have a function in osmoregulation. Six normal walking legs are present. The head bears four-jointed antennae and poorly formed eyes.

Springtails are often in parallel with insects because of their six legs and general form, but they differ in several fundamental ways, including the presence of muscles in all antennal segments (basal only in true insects), water-repellent substances in the cuticle (not found in insects), and complete cleavage in the embryo (polar in insects). The elongate chewing or sucking mouthparts are also uniquely withdrawn into a deep pouch in the head.

Although seldom seen and rarely appreciated, these are extremely numerous inhabitants of many different ecological regimes (Christiansen 1964). They prefer damp microhabitats where they are protected from wetting by a hydrophobic cuticle (Ghiradella and Radigan 1974) and are highly characteristic soil and litter animals. The collembolan fauna and population densities of the midlatitude tropical habitats, however, are much less diverse than that of temperate regions, probably because of the comparatively poor litter layer.

They are found often among decomposing plant matter, on which they feed directly, or on associated fungi and algae. A few occur along the seashore, even within the intertidal zone, and are really marine organisms. Others live on the surface of fresh water and even on snow. A few are symbiotic in ant and termite nests. Some eke out a bare existence on Antarctic shores (Rapoport 1971). A considerable cave fauna also exists (Palacios 1983). There are also economically important forms that attack mushrooms, onions, tomatoes, sugarcane, and alfalfa.

Studies on the taxonomy and biology of the Latin American collembolan fauna are only beginning. It is estimated that the approximately 800 known species (Mari Mutt and Bellinger 1989) may represent only 25 percent of the total number actually living in the region. The Latin American fauna has representatives of all the world's twelve common families (Palacios 1990), plus a rare new family (Coenaletidae) that has re-

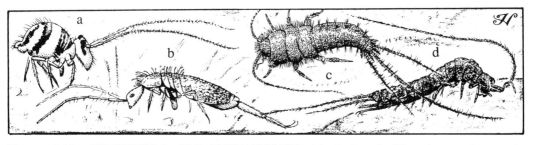

**Figure 4.10 SPRINGTAILS AND THYSANURANS.** (a) Springtail (*Temeritas surinamensis,* Sminthuridae). (b) Springtail (*Ctenocyrtinus prodigus,* Entomobryidae). (c) Silverfish (*Ctenolepisma longicaudata,* Lepismatidae). (d) Rock hopper (*Neomachillelus scandens,* Meinertellidae).

cently been found on the island of Guadeloupe (Bellinger 1985). Representative species of the large Latin American fauna are *Temeritas surinamensis* (Sminthuridae; fig. 4.10a) and *Ctenocyrtinus prodigus* (Entomobryidae; fig. 4.10b).

### References

BELLINGER, P. F. 1985. A new family of Collembola (Arthropoda, Tracheata). Carib. J. Sci. 21: 117–123.

CHRISTIANSEN, K. 1964. Bionomics of Collembola. Ann. Rev. Entomol. 9: 147–178.

GHIRADELLA, H., AND W. RADIGAN. 1974. Collembolan cuticle: Wax layer and antiwetting properties. J. Ins. Physiol. 20: 310–306.

MARI MUTT, J. A., AND P. F. BELLINGER. 1989. Catalog of Neotropical Collembola. Flora and Fauna, Gainesville.

PALACIOS, J. G. 1983. Collemboles cavernicoles du Mexique. Pedobiologia 25: 349–355.

PALACIOS, J. G. 1990. Diagnosis y clave para determinar las familias de los Collembola de la región Neotropical. Man. Guias Est. Microartr. (UNAM, Mexico) 1: 1–15.

RAPOPORT, E. H. 1971. The geographical distribution of Neotropical and Antarctic Collembola. Pacific Ins. Monogr. 25: 99–118.

## THYSANURANS

Thysanura, sensu lat.

Thysanurans are considered to comprise the most primitive order of true insects because of their complete winglessness, weak sclerotization, and presence of vestigial, jointed appendages on the underside of the abdomen. Immatures grow gradually into adults without appreciable change in form (i.e., no metamorphosis). They also bear three long, many-jointed tails extending from the tip of the abdomen, and the equal thoracic segments have rather large lateral lobelike expansions. They are generally small (most BL 10–15 mm). The body is covered with scales but may be bare except for numerous fine bristles.

There are two suborders, the silverfish (Zygentoma) and the bristletails (Microcoryphia) (Remington 1954). These categories have been variously recognized and named by taxonomists, and some confusion remains on the best way to classify the group.

Members of both groups are widely distributed and mostly secretive, occurring in all sorts of general habitats such as in rotten wood, in rock crevices, and in humus and ground litter. Some have highly specialized habits, living in bird's nests, associating with social insects, or living in caves (Wygodzinsky 1967). Many have been transported throughout the world by human traffic. Free-living thysanurans, especially those requiring a warm environment, often become adapted to domiciles.

### References

REMINGTON, C. L. 1954. The suprageneric classification of the order Thysanura (Insecta). Entomol. Soc. Amer. Ann. 47: 277–286.

WYGODZINSKY, P. 1967. On the geographical distribution of the South American Microcoryphia and Thysanura (Insecta). *In* D. Deboutteville and E. Rapoport, eds., Biologie de L'Amérique Australe. 3: 505–524. Ed. Cent. Nat. Recher. Sci., Paris.

## Silverfish

Zygentoma ( = Lepismatoidea; formerly Thysanura, in part). *Spanish:* Pececitos de plata, pescaditos plateados (General). *Portuguese:* Traças dos livros (Brazil).

Silverfish (so called because of their shiny, slick appearance) are familiar cosmopolitan household pests. However, the classic domestic species, *Lepisma saccharina,* has been found in Latin America only sporadically in the cooler highlands of Brazil, Bolivia, and Argentina. They apparently do not tolerate humid, tropical conditions. *Lepisma wasmanni* is a soil-dwelling resident of the lomas of coastal Peru but seems to have been introduced from the Mediterranean region.

The long-tailed house silverfish (*Ctenolepisma longicaudata,* fig. 4.10c) is the species most commonly encountered indoors in Latin America (Wygodzinsky 1967). It has a thick body covering of slick, slate-colored scales. It usually occurs in damp situations, but its precise range in the region is unknown. *Stylifera gigantea* is also common but seldom occurs indoors. There are many other species in various genera, but knowledge of their distribution and habits is rudimentary (Wygodzinsky 1967).

Members of this suborder are characterized by their cylindrical or somewhat flattened shape, thorax that is not arched, mouthparts that are directed forward, and small eyes that are set widely apart. The lateral lobes of the thorax are only slightly expanded. The group as a whole requires warmer environments than the bristletails, which prefer cooler, usually mountainous climates.

Three of the world's four families are found in Latin America (Paclt 1963, 1967; Wygodzinsky 1972). The Lepismatidae is the largest and most diverse and contains several domestic types as well as many native species. Several of the latter are associated symbiotically with social insects, as are the Nicoletiidae. Members of the latter family are subterranean, often inhabiting caves (Mexico and Cuba) and living in ant and termite nests. Cave forms lack eyes and integumentary pigment. The Maindroniidae, with its single bizarre genus, *Maindronia,* has only been encountered under drying seaweed along arid Peruvian and Chilean coasts. These silverfish have a very elongate, pigmented, unscaled body.

## References

PACLT, J. 1963. Thysanura Fam. Nicoletiidae. Genera Insectorum 216: 1–58.

PACLT, J. 1967. Thysanura Fam. Lepidotrichidae, Maindroniidae, Lepismatidae. Genera Insectorum 218: 1–86.

WYGODZINSKY, P. 1959. Thysanura and Machilida of the Lesser Antilles and northern South America. Stud. Fauna Curaçao Carib. Is. 36: 28–49.

WYGODZINSKY, P. 1967. On the geographical distribution of the South American Microcoryphia and Thysanura (Insecta). *In* D. Deboutteville and E. Rapoport, eds., Biologie de L'Amérique Australe. 3: 505–524. Ed. Cent. Nat. Recher. Sci., Paris.

WYGODZINSKY, P. 1972. A review of the silverfish (Lepismatidae, Thysanura) of the United States and the Caribbean area. Amer. Mus. Nov. 2481: 1–26.

## Bristletails

Microcoryphia (= Archaeognatha, Machiloidea, Machilida). Rock jumpers.

While silverfish tend to be cylindrical or slightly flattened, bristletails are compressed laterally. The head is rotated downward so that the mouthparts project ventrally, and the thorax is strongly arched. The eyes are well developed and contiguous, with many facets. The lobular sides of the thorax are large and appressed to the sides. Also, in contrast to silverfish, bristletails are much less seen, completely wild insects, living secretly in all sorts of dry or moist hideaways, under stones and bark, in leaf litter, and among rocks at the seashore. They are terrestrial or littoral, nocturnal or crepuscular, and are highly active, running swiftly and, if threatened, jumping violently to escape capture.

The largest and most widespread Neotropical genus is *Neomachillelus* (fig. 4.10d), which inhabits spaces under tree bark and the soil surface. It is found over the entire area except for Patagonia (Wygodzinsky 1959). The southern temperate regions are inhabited by *Machiloides* and *Machilinus*. The latter tends toward arid zones. The genus *Meinertellus* is represented by several species in northeastern South America.

## Reference

WYGODZINSKY, P. 1959. Thysanura and Machilida of the Lesser Antilles and northern South America. Stud. Fauna Curaçao Carib. Isl. 36: 28–49.

# 5 ORTHOPTEROIDS AND OTHER ORDERS

Most of this chapter treats those orders closely associated taxonomically with the Orthoptera. For convenience, several additional orders, although not related, are included also.

## ORTHOPTEROIDS

Orthopterodea.

This is an assemblage of primitive insects that have close affinities with the order Orthoptera. The classification of the major groups of orthopteroids is very unsettled. Kevan (1977) provides an exhaustive and authoritative scheme, although tentative. The term "orthopteroids" traditionally refers to several orders with similar generalized body form, moderate to large size, ovoid head with mandibulate mouthparts, usually narrow, long, somewhat thickened fore wings, enlarged prothorax (often prolonged backward), and leathery (fore) or membranous (hind) wings (when present) with a complex, reticulate vein pattern. Internally, the presence of very numerous Malpighian tubules is also characteristic.

With the major exception of mantids and some species and groups (e.g., Decticinae) that occasionally take prey or feed on insect carcasses, orthopteroids are basically all vegetarians. Nymphs of some groups often visit flowers; although they feed on petals and anthers, they may benefit the plant by effecting pollination (Schuster 1974).

Orthopteroids employ a wide variety of protective devices, often highly specialized and used in all combinations and multiples. Relatively few are involved in mimicry, but a great many are cryptically colored, some very realistically, to mimic leaves, sticks, stones, and other inert objects. This is often combined with startling displays, involving eyespots or aposematic color fields, coupled with repugnant secretions and ominous sounds. Most have spiny legs and powerful jaws, both of which are employed in active defense. Nocturnal or very secretive habits afford safety to others, and most can either jump or fly to escape persistent attack (Robinson 1969).

Sound production is particularly well developed in the orthopteroids, primarily by males for the attraction of females. Both sexes are well equipped with hearing organs (tympana) located in various parts of the body. Males may sing alone or exhibit "chorusing," that is, calling by two or more conspecific males, involving interactions with each other. Chorusing takes many forms, depending on the precise timing of the multiple sounds (Greenfield and Shaw 1983). In some primitive phaneropterine katydids, the female has a series of minute knoblike stridulatory organs on the wing base which are used to answer calling males.

## References

GREENFIELD, M. D., AND K. C. SHAW. 1983. Adaptive significance of chorusing with special reference to the Orthoptera. *In* D. T.

Gwynne and G. K. Morris, eds., Orthopteran mating systems. Westview, Boulder, Colo. Pp. 1–27.

KEVAN, D. K. McE. 1977. Suprafamilial classification of "orthopteroid" and related insects; a draft scheme for discussion and consideration. Lyman Entomol. Mus. Res. Lab. (McGill Univ.), Mem. 4, Spec. Publ. 12: Appendix, 1–26.

ROBINSON, M. H. 1969. The defensive behaviours of some orthopteroid insects from Panama. Royal Entomol. Soc. London Trans. 121: 281–303.

SCHUSTER, J. C. 1974. Saltatorial Orthoptera as common visitors to tropical flowers. Biotropica 6: 138–140.

## Long-Horned Orthopteroids

Grylloptera (= Ensifera)

These are mainly the katydids and crickets with long antennae of well over thirty segments. The auditory organs are located on the fore tibiae and the stridulatory organs on the basal portions of the fore wings.

# KATYDIDS

Tettigoniidae. *Spanish:* Esperanzas (General), pulpones (Costa Rica), langostas verdes (Argentina), saltamontes nocturnos (Panama), grillos voladores (Peru). *Portuguese:* Esperanças.

Katydids are also called long-horned grasshoppers and indeed resemble those orthopterans except for their very long, many-segmented, whiplike antennae. They may be fully winged and capable of flight, or they may have very short wings, the hind wing completely lost and the fore wings stubby. When wings are present, an area at the base of the thickened fore wing is often modified in the males into a stridulatory or sound-producing organ. One wing has a roughened ridge (file), which is rubbed against an opposing sharp edge (scraper) on the other. The action sets both wings into rapid vibration and produces the familiar chirping, buzzing, lisping, clicking, or snapping sounds of these noisy creatures. Small tympana are located in slots on either side of the base of the front tibiae and in pockets on the sides of the thorax near the hind edge of the thoracic shield.

The female ovipositor may be strongly compressed for incising leaves and wood or valvelike for inserting eggs in the ground. Compressed forms are short and strongly curved or long and swordlike.

This is a large family in Latin America, with some 1,350 known species; this number probably will increase by 30 percent or more ultimately (Nickle pers. comm.). The subfamilies may be recognized by use of Rentz's key (1979), which recognizes broader categories than Kevan's (1977) classification. Several new genera and species among the shield-backed katydids from Chile and Argentina have been added to the fauna by Rentz and Gurney (1985). They show affinities with others of the subfamily from Australia and western North America (ibid., 70).

The common English name of these insects derives from the song of the North American species, *Pterophylla camellifolia*, which sounds like the plaintive phrase, "kate-she-did," "kate-did-she-did," or "katy-did." A folktale exists concerning the accountability of a fictitious lady in the death of a lover who spurned her. Local names in Spanish and Portuguese mean "hope," in reference to the green color of so many species in the family, the symbolic color of this emotion.

Katydids are an extremely important link in vertebrate food chains, as they are utilized by many birds, bats, monkeys, lizards, and snakes. Consequently, they have evolved a rich array of antipredator defenses (Belwood 1990). Forest katydids in Panama are major prey for foliage-gleaning bats. To some degree, the bats have adapted to reduced acoustic production by these katydids, who use an alternative form of communication ("substrate transmitted tremulation"), rather than the usual singing, to attract mates (Belwood and Morris 1987).

Like other orthopteroids, katydids possess a rich repertoire of defense tactics. A device not exhibited by others but used by the katydid *Ancistrocercus* in Costa Rica and *Eremopedes colonialis* in Mexico (Rentz 1972: 54) is association with wasps (Downhoper and Wilson 1973).

## References

BELWOOD, J. J. 1990. Anti-predator defences and ecology of Neotropical forest katydids, especially the Pseudophyllinae. *In* W. J. Bailey and D. C. F. Rentz, The Tettigoniidae: Biology, systematics and evolution. Springer, Berlin. Pp. 8–26.

BELWOOD, J. J., AND G. K. MORRIS. 1987. Bat predation and its influence on calling behavior in Neotropical katydids. Science 238: 64–67.

DOWNHOPER, J. F., AND D. E. WILSON. 1973. Wasps as a defense mechanism of katydids. Amer. Midl. Natur. 89: 451–455.

KEVAN, D. K. McE. 1977. Suprafamilial classification of "orthopteroid" and related insects; a draft scheme for discussion and consideration. Lyman Entomol. Mus. Res. Lab. (McGill Univ.), Mem. 4, Spec. Publ. 12: Appendix, 1–26.

RENTZ, D. C. F. 1972. Taxonomic and faunistic comments on decticine katydids with the description of several new species (Orthoptera: Tettigoniidae: Decticinae). Acad. Nat. Sci. Philadelphia Proc. 124: 41–77.

RENTZ, D. C. F. 1979. Comments on the classification of the orthopteran family Tettigoniidae, with a key to subfamilies and description of two new subfamilies. Austr. J. Zool 27: 991–1013.

RENTZ, D. C. F., AND A. B. GURNEY. 1985. The shield-backed katydids of South America (Orthoptera: Tettigoniidae, Tettigoniinae) and a new tribe of Conocephalinae with genera in Chile and Australia. Entomol. Scandinavica 16: 69–119.

## Broad-winged Katydids

Tettigoniidae, Pseudophyllinae.

Not all pseudophyllines have broad wings as their common name implies, nor are they leaflike as indicated by their scientific name. Perhaps a majority of the approximately 600 described Neotropical species are small to medium, spindle-shaped insects that sing little and do not look like leaves. The adults of members of some tribes, especially the following, do have very broad fore wings and closely resemble leaves.

### *Leaf Katydids*

Tettigoniidae, Pseudophyllinae, primarily the tribe Pterochrozini. *Spanish:* Esperanzas hojas (General). *Portuguese:* Bichos folhas (Brazil).

Many katydids in this group of just under one hundred species appear incredibly like the leaves among which they live. They are green to brown in general color with an outline shape, even with drip tip, and markings imitating in every detail the color and structure of leaves, complete to midribs, side veins, and even transparent

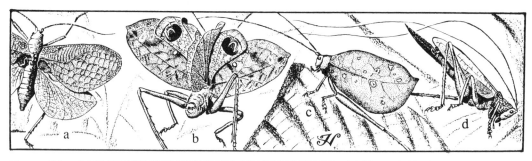

**Figure 5.1   BROAD-WINGED KATYDIDS (TETTIGONIIDAE).** (a) Tananá (*Thliboscelus hypericifolius*). (b) Eye-winged Katydid (*Pterochroza ocellata*). (c) Leaf katydid (*Cycloptera speculata*). (d) Long-winged leaf katydid (*Cocconotus* sp.).

and discolored areas resembling insect feeding holes, leaf miner tunnels, chewed edges, and mold spots (pl. 3g)!

This resemblance is further enhanced with behavior, for those species that dwell in living trees often walk with a slow undulating gait that makes them look like a leaf gently swaying in the breeze. Inhabitants of litter and dead vegetation tend to sit very still, angled on their sides and fitting perfectly with the brown, dried leaves on the forest floor. The latter tend also to be very large, approaching the sizes of the parts of plants they resemble. *Celidophylla albimacula* (BWL 8 cm) is a rare, gigantic species from Central America (Hogue 1979) with yellow-green wings which looks like a large withered leaf.

The principal genera are *Pterochroza, Mimetica, Tanusia, Cycloptera,* and *Typophyllum* (Vignon 1930). A Brazilian species made famous by Henry Walter Bates in the related tribe Pterophyllini, among which are also many leaf-shape forms, is the tananá (*Thliboscelus hypericifolius = Chlorocoelus tanana;* fig. 5.1a).

*The males produce a very loud and not unmusical noise by rubbing together the overlapping edges of their wing-cases. The notes are certainly the loudest and most extraordinary that I ever heard produced by an orthopterous insect. The natives call it the Tananá, in allusion to its music, which is a sharp, resonant stridulation resembling the syllables ta-na-ná, ta-na-ná, succeeding each other with little intermission. It seems to be rare in the neighborhood. When the natives capture one, they keep it in a wicker-work cage for the sake of hearing it sing. (Bates 1892: 129)*

*Pterochroza ocellata* (pl. 1b) is an Amazonian species with greenish or brownish-red, brown-black-splotched fore wings that camouflage it well on dead leaves. But if this first line of defense fails, it fans out its hind wings to display large threatening eyespots at their outer corners (fig. 5.1b). The leaf pattern is especially well developed in genera like *Cycloptera* (fig. 5.1c). When folded, the lower portion of the wings are often rolled under and hide the insect's abdomen. In repose, the tips of the fore wings project beyond or are equal to the length of the hind wings.

The nymphs are also leaf- or twiglike, often with very compressed bodies and flattened leg segments and erect crests on the back of the prothorax. They rest on the upper surfaces of leaves with their long legs spread spiderlike and bodies pressed close to the surface.

Because they are arboreal and nocturnal and so difficult to see even during the light of day, these katydids are seldom collected, and little is known of their biology. Those rare adults that come to artificial lights constitute the majority of specimens in museum collections.

Members of the related tribe Cocconotini, primarily the genus *Cocconotus* (fig. 5.1d), are also large (BWL 8–9 cm) and have very long antennae, two to three times the body length. They habitually rest in rolled leaves of bananas, heliconias, gingers, cannas, and the like, head outward, only with the tips of the antennae exposed, testing the outside world while the rest of the insect remains hidden. They are mostly dark gray-brown and elongate, their wings rolled around the body.

Some members of the subfamily are wingless. A giant species found in the canopy of the Peruvian rain forest is *Panoploscelus* (tribe Eucocconotini). It is well protected by its great size and strength and sharply spined legs.

## References

BATES, H. W. 1892. The naturalist on the River Amazons. John Murray, London.
BEIER, M. 1960. Orthoptera, Tettigoniidae (Pseudophyllinae II). Das Tierreich 74: 1–396.
BEIER, M. 1962. Orthoptera, Tettigoniidae (Pseudophyllinae I). Das Tierreich 73: 1–468.
HOGUE, C. L. 1979. A third specimen of *Celidophylla albimacula* (Orthoptera: Tettigoniidae)

and remark on the emergence of Diptera from insect carrion. Entomol. News 90: 151.

Vignon, M. P. 1930. Recherches sur les sauterelles-feuilles de l'Amérique tropicale. Première partie. Révision du groupe des "Pterochrozae" (Phasgonuridae, Pseudophyllinae). Mus. Natl. Hist. Natur. (Paris), 6 Ser., Arch. 5: 57–214.

### Narrow-winged Katydids

Tettigoniidae, Phaneropterinae.

Many of these katydids are leaf mimics but not quite so remarkably as those belonging to the broad-winged group (fig. 5.2d). The fore wings are generally longer and narrower and more plainly marked, green to brown. In a few species, the hind wings are brightly colored, such as *Vellea*, which has a broad scarlet zone basally. When the wings are folded, the tips of the hind pair project beyond the apices of the fore pair. Females have a globose head and short, upturned ovipositor. The outer part of the fore tibia in cross section is square and with a flat or slightly concave, dorsal surface. The thoracic auditory pockets are large and exposed.

Most narrow-winged katydids are medium-sized to small (BWL 25–50 mm), but *Steirodon* (formerly *Peucestes*) is gigantic (BWL 10–12 cm) and possesses a coarsely serrate corona around the periphery of the prothoracic shield (fig. 5.2a) (Nickle 1985). Nymphs of the genus are strongly compressed and have a conspicuous dark ocellate spot laterally, on the wing pad.

The group is phytophagous and arboreal. Some habitually rest on lichen-covered tree trunks, blending in perfectly on account of their own mottled white, green, and black body markings. The resemblance is increased by lobular excrescences on the legs and body exactly like the foliose form of the plant. An excellent example is *Dysonia fuscifrons* (fig. 5.2b) from Mexican cloud forests where the perpetually humid atmosphere supports a lush growth of epiphytes (Dampf 1939). Members of the tribe Pleminiini (e.g., *Championica*) are large, with mottled green and brown markings that camouflage them on moss- and lichen-covered tree bark where they habitually rest.

In Brazil, *Scaphura* and *Aganacris* (fig. 5.2c) mimic tarantula hawk wasps (*Pepsis*) and are not only colored the same, with deep blue body and rust or dark wings, but have a waspish attitude as well, walking jerkily and rapidly and waving their orange-tipped antennae. At the approach of danger, they also flutter their wings and turn up the abdomen in mockery of an attempt to sting.

The group is unique in possessing the only katydids whose females answer the calling songs of the males with a signal, though weak, of their own (Nickle and Carlysle 1975). The interval between the male and female sounds is important in determining the correct association and drawing the male to the female. The chro-

**Figure 5.2  NARROW-WINGED KATYDIDS (TETTIGONIIDAE).** (a) Giant narrow-winged leaf katydid (*Steirodon* sp.). (b) Lichen-mimicking katydid (*Dysonia fuscifrons*). (c) Tarantula hawk-mimicking katydid (*Aganacris* sp.). (d) Common leaf-mimicking katydid (undetermined).

mosomes of a number of Neotropical species of this subfamily have been described by Ferreira (1977).

The subfamily contains approximately 600 named species in Latin America.

### References

DAMPF, A. 1939. Un caso de fitomimetismo en un ortóptero Mexicano. Esc. Nac. Cien. Biol. Anal. 1: 525–533.

FERREIRA, A. 1977. Cytology of Neotropical Phaneropteridae (Orthoptera-Tettigoniidae). Genetica 47: 81–86.

NICKLE, D. A. 1985. A new steirodont katydid from Colombia (Orthoptera: Tettigoniidae). Entomol. News 96: 11–15.

NICKLE, D. A., AND T. C. CARLYSLE. 1975. Morphology and function of female sound producing structures in ensiferan Orthoptera with special emphasis on the Phaneropterinae. Int. J. Ins. Morph. Embryol. 4: 159–168.

## Cone-Headed and Meadow Katydids
Tettigoniidae, Conocephalinae, Copiphorini and Conocephalini.

Many of these closely related groups of 140 to 150 species of katydids have powerfully developed jaws adapted for seed eating or, in some cases, predation. Their name, however, derives from the strongly projecting anterior portion of the head; this may be only a simple convexity in meadow katydids (Conocephalini) or an elongate cone with a tubercle at its base (this often notched below) in the cone-headed katydids (Copiphorini).

The meadow katydids, such as *Conocephalus* (fig. 5.3a), are relatively small (BWL seldom over 25 mm). They frequent marshes or verdant areas, usually living close to the ground and in aggregations. Their stridulatory activity is mainly diurnal or crepuscular. Cone-headed katydids are larger (BWL often 5–6 cm) and occupy diverse habitats, but many are associated with grasses whose seeds they eat. Their two-part songs (ticks, giving way to buzzes) are heard from dusk to dawn and are often of the chorusing type. Males stridulate for several hours each evening from exposed perches, apparently in sexual competition. Some nonsinging males also remain close to the singer, taking advantage of his call to steal confused females that come near (Greenfield 1983).

One South American cone-head, *Copiphora* (fig. 5.3b), is fairly large (BWL 4–6 cm) and has bright blue, red, and yellow colors on the abdomen. These are brought into view when the insect is disturbed, usually by tipping the head down against the substratum and elevating the rear part of the body upward; the wings also may be opened or raised to better expose these colors, warning would-be predators of presumably noxious body fluids.

*Neoconocephalus* (fig. 5.3d) is another common Neotropical cone-head, whose large size (BWL 5–6 cm) and characteristic

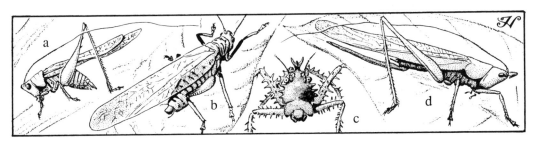

**Figure 5.3  KATYDIDS (TETTIGONIIDAE).** (a) Meadow katydid (*Conocephalus* sp.). (b) Multicolored katydid (*Copiphora* sp.). (c) Spike-headed katydid (*Panacanthus* sp.), nymph. (d) Cone-headed katydid (*Neoconocephalus* sp.).

loud, penetrating calls have made them attractive subjects for collectors and field studies of behavior (Greenfield 1990). Many of the species are barely distinguishable morphologically but can be separated readily by the distinctive sound patterns of the males (Walker and Greenfield 1983).

The enormous head of the nymph of the spike-headed katydid (*Panacanthus;* fig. 5.3c) has bulbous eyes and is grotesquely adorned with a peripheral crown of large teeth and a multispiked horn on the forehead. A South American genus, *Coniungoptera,* has been recently added to the cone-head group (Rentz and Gurney 1985). It is associated with *Nothophagus* in the southern beech forests and is most closely related to an Australian genus, another example of an amphinotic distribution.

## References

GREENFIELD, M. C. 1983. Unsynchronized chorusing in the cone-headed katydid *Neoconocephalus affinis* (Beauvois). J. Anim. Behav. 31: 102–112.

GREENFIELD, M. C. 1990. Evolution of acoustic communication in the genus *Neoconocephalus:* Discontinuous songs, synchrony, and interspecific interactions. *In* W. J. Bailey and D. C. F. Rentz, The Tettigoniidae: Biology, systematics and evolution. Springer, Berlin. Pp. 71–97.

RENTZ, D. C., AND A. B. GURNEY. 1985. The shield-backed katydids of South America (Orthoptera: Tettigoniidae, Tettigoniinae) and a new tribe of Conocephalinae with genera in Chile and Australia. Entomol. Scandinavica 16: 69–119.

WALKER, T. J., AND M. D. GREENFIELD. 1983. Songs and systematics of Caribbean *Neoconocephalus* (Orthoptera: Tettigoniidae). Amer. Entomol. Soc. Trans. 109: 357–389.

# CRICKETS

Gryllidae. *Spanish:* Grillos (General).
*Portuguese:* Grilos (Brazil).

Although many orthopteroids produce highly audible and complex sounds, it is the crickets that are best known for their musical talents. Males can make a variety of notes, most characteristically, a series of short, pulsed chirps or continuous soft trilling, more melodious and less rasping than the calls of katydids. The sound is produced by the vibration of membranous areas of the fore wings. These are set into motion when the scraper near the base of one wing is drawn across the opposing file of the other.

This acoustic behavior plays a major role in pair formation. Generally, the male gives a species-specific calling song to which females respond by approaching him. Males of ground-dwelling types call with the wings raised at about a 40-degree angle to the long body axis; those inhabiting vegetation hold their fore wings up at a right angle to the body axis. These positions are related to sound-producing efficiency in their habitats (Forrest 1982).

The Central American long-legged *Amphiacusta maya,* found during the day in hollow trees and under overhanging banks, mates in groups consisting of both sexes and subadult nymphs. Male courtship chirping appears to be more of a warning to other males than a signal to females, to reduce fighting that reduces successful mating frequency (Boake 1984). The habits of related species, such as *A. annulipes* (fig. 5.4a), are mostly known. There is evidence that crickets also respond with evasive behavior to the ultrasonic emanations of bats, one of their chief predators (Doherty and Hoy 1985).

Crickets may be flattened or cylindrical in cross section. Most are small to medium-sized (BL 5–25 mm) but often have very much enlarged, muscled hind femora, making them good jumpers. The number of tarsal segments in the legs is reduced to three, not five like most other orthopteroids. The female's ovipositor is tubular and the cerci frequently large and conspicuous.

There are three major cricket types: (1) "bush crickets," which are small to medium-sized, brownish, and arboreal (several sub-

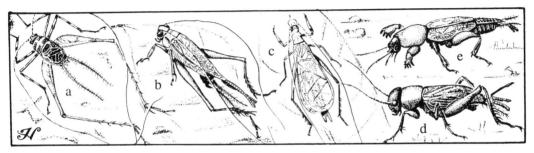

**Figure 5.4  CRICKETS.** (a) Long-legged cricket (*Amphiacusta annulipes,* Gryllidae). (b) Bush cricket (*Eneoptera surinamensis,* Gryllidae). (c) Tree cricket (*Oecanthus* sp., Gryllidae), male. (d) Ground cricket (*Gryllus* sp., Gryllidae). (e) Mole cricket (*Scapteriscus* sp., Gryllotalpidae).

families; a widespread species is *Eneoptera surinamensis;* fig. 5.4b); (2) "tree crickets" (Oecanthinae, especially *Oecanthus,* fig. 5.4c, and *Neoxabea*), which are also found in vegetation but are delicate, pale green, and translucent, with slender bodies and legs and almost horizontal heads, males having broad, oval wings and thoracic glands that produce a mating pheromone (Walker 1967); (3) "ground and field crickets" (Gryllinae and other subfamilies), which are small to large and robust and generally live on the ground.

Among the last group are some species of native *Gryllus,* many going under the aggregate name *G. assimilis* (fig. 5.4d) (Aguilar and Sáenz 1970). The related European house cricket (*Acheta domesticus*) has been introduced into Latin American houses from Europe but is not generally common. Already widespread and increasing its range rapidly is the Indian house cricket (*Grylloides supplicans*).

Probably because of their familiarity and the pleasant sounds emitted by the domestic species, numerous superstitions and folk beliefs have built up around crickets. They are almost universally considered a sign of good luck, although sometimes a harbinger of death. In Caraguatatuba, Brazil, a black cricket in a room is a signal of sickness, a gray one a sign of money, and green, of hope. In Rio Grande do Sul, Brazil, killing a cricket is thought to bring rain (Lenko and Papavero 1979).

In contrast to these mostly positive ways of viewing crickets, the leanings of these insects toward population explosions and their taste for crops stands them in poor stead with the farmer. Nocturnal and attracted by light and food, plagues of millions may invade fields and nearby urban areas, consuming leaves and covering everything, even piling up in drifts many centimeters deep in protected places.

The habits of native crickets are not well known among the poorly studied Latin American fauna. Crickets are generally herbivorous on grasses, herbs, and other plants but take insect prey opportunistically. Young tree crickets may be entirely predaceous on scale insects, aphids, and other small insects and thus may be of considerable benefit as natural control of these pests.

The family is large, although only some 500 species are now listed for Latin America (Chopard 1967–68); many species are awaiting description. As a group, they are widely distributed and are found in almost all life zones.

## References

AGUILAR, P. G., AND D. SÁENZ. 1970. Algunas variaciones morfológicas en el grillo común de la costa central. Rev. Peruana Entomol. 13: 76–86.

BOAKE, C. R. B. 1984. Natural history and acoustic behavior of a gregarious cricket. Behaviour 89: 241–250.

CHOPARD, L. 1967–68. Gryllides. *In* M. Beier,

ed., Orthopterorum catalogus. W. Junk, Gravenhage.

DOHERTY, J., AND R. HOY. 1985. The acoustic behavior of crickets: Some views of genetic coupling, song recognition, and predator detection. Quart. Rev. Biol. 60: 457–472.

FORREST, T. G. 1982. Acoustic communication and baffling behaviors of crickets. Fla. Entomol. 65: 33–44.

LENKO, K., AND N. PAPAVERO. 1979. Insetos no folclore. Sec. Cult. Cien. Tech, São Paulo.

WALKER T. J. 1967. Revision of the Oecanthinae (Gryllidae: Orthoptera) of America south of the United States. Entomol. Soc. Amer. Ann. 60: 784–796.

## Mole Crickets

Orthoptera, Gryllotalpidae, *Scapteriscus* and *Neocurtilla*. *Spanish:* Grillotopos (General), changas (Puerto Rico), perritos de monte, playacuros (Peru). *Portuguese:* Grilos toupeiros, frades, macacos, cachorrinhos, daguas, paquinhas (Brazil). Ground puppies.

Mole crickets (fig. 5.4e), as their name implies, live in subterranean galleries. They are aptly specialized for digging by the modified forelegs, which are powerful excavating tools. The segments are massive and powerfully muscled and bear heavy spines on the under margin to form cutting chisels and scrapers. Other identifying features of these insects are their cricket-like form, fairly large size (BL 3–4 cm), medium brown color, and abbreviated fore wings beyond the tips of which the twisted hind wings project for some distance. Also, the entire surface of the body is covered with a short, sparse, yellowish velvet.

Female mole crickets deposit their eggs in a cluster in the enlarged chamber at the end of a side gallery. The nymphs remain underground, feeding directly on herbaceous plant roots or on stems of plants that they pull down into their underground passages. The adults feed in the same way but may also leave the burrow at night to disperse and find mates. They are frequently attracted to artificial lights. Their food most often consists of seedlings or small crops growing in the friable soil they prefer for their diggings. Some species take considerable animal matter as food as well and should be considered omnivores (Castner and Fowler 1984a).

Although superficially similar, many distinct species in various genera (especially *Scapteriscus*) occur in Latin America with varied life-styles (Fowler and de Vasconcelos 1989). The best-known mole cricket, because of its taste for commercial plantings, is the *changa* of Puerto Rico (Barrett 1902, Thomas 1928). Formerly thought to be one species, *Scapteriscus "vicinus,"* (actually *didactylus* but misidentified by all early authors), it may be comprised of two or more as indicated by the existence of populations with divergent mating songs; one was named *S. imitatus* (Nickle and Castner 1984). These and *S. abbreviatus* have been introduced to Puerto Rico from South America (Castner and Fowler 1984b) and have become pests of turf and agriculture. *Scapteriscus oxydactylus* is the largest South American mole cricket and is a pest of rice and other crops cultivated along the banks of the Amazon River when the waters are receding. Some natural control is wrought by the cricket's parasitoid enemies in the sphecid wasp genus *Larra* (Castner 1984, Castner and Fowler 1984b).

Male mole crickets use songs to call mates (Forrest 1983). The sound is a broken trill at a pulse rate of about 57 to 68 per second and of low frequency. It is emitted by the insect underground, either from closed burrows or through special funnel-shaped openings to augment the sound ("acoustic horns") (Bennet-Clark 1970). Females occasionally also make sounds of short duration and for unknown purposes (Ulagaraj 1976). Stridulation is accomplished by friction between scraper and file elements on the fore wings, much like true crickets (Bennet-Clark 1970).

## References

BARRETT, O. W. 1902. The changa, or mole cricket (*Scapteriscus didactylus* Latr). Puerto Rico Agric. Exper. Sta. Bull. 2: 1–19.

BENNET-CLARK, H. C. 1970. The mechanism and efficiency of sound production in mole crickets. J. Exper. Biol. 52: 619–652.

CASTNER, J. L. 1984. Suitability of *Scapteriscus* spp. mole crickets [Ort.: Gryllotalpidae] as hosts of *Larra bicolor* [Hym.: Sphecidae]. Entomophaga 29: 323–329.

CASTNER, J. L., AND H. G. FOWLER. 1984a. Gut content analysis of Puerto Rican mole crickets (Orthoptera: Gryllotalpidae: *Scapteriscus*). Fla. Entomol. 67: 479–481.

CASTNER, J. L., AND H. G. FOWLER. 1984b. Distribution of mole crickets (Orthoptera: Gryllotalpidae: *Scapteriscus*) and the mole cricket parasitoid *Larra bicolor* (Hymenoptera: Sphecidae) in Puerto Rico. Fla. Entomol. 67: 481–484.

FORREST, T. G. 1983. Phonotaxis and calling in Puerto Rican mole crickets. (Orthoptera: Gryllotalpidae). Entomol. Soc. Amer. Ann. 76: 797–799.

FOWLER, H. G., AND H. L. DE VASCONCELOS. 1989. Preliminary data on life cycles of some mole crickets (Orthoptera, Gryllotalpidae) of the Amazon Basin. Rev. Brasil. Entomol. 33: 134–141.

NICKLE, D. A., AND J. L. CASTNER. 1984. Introduced species of mole crickets in the United States, Puerto Rico, and the Virgin Islands (Orthoptera: Gryllotalpidae). Entomol. Soc. Amer. Ann. 77: 450–465.

THOMAS, W. A. 1928. The Puerto Rican mole cricket. U.S. Dept. Agr., Farm. Bull. 1561: 1–8.

ULAGARAJ, S. M. 1976. Sound production in mole crickets (Orthoptera: Gryllotalpidae: *Scapteriscus*). Entomol. Soc. Amer. Ann. 69: 299–306.

## Short-Horned Orthopteroids

Orthoptera (= Caelifera)

Grasshoppers make up the majority of this group, which is characterized by short antennae with less than thirty segments. Auditory organs, when present, are located on the tergum of the first abdominal segment. Stridulatory structures are on the distal portions of the wings.

# GRASSHOPPERS

Acrididae. *Spanish:* Saltamontes, saltarines, saltones (General); tucuras (Argentina). *Nahuatl:* Chapultins, sing. chapolin (Mexico). *Quechua:* Chlli cutu. *Portuguese:* Gafanhotos.

Grasshoppers are so familiar that they hardly need description. The characteristics of enlarged hind femur (containing the jumping muscles) combined with short, stout antennae always distinguish them from similar orthopteroids. The dorsal part of the prothorax forms a broad collar, folding over the segment on the sides, and is often crested middorsally. Both sexes also possess an auditory organ, visible as a circular membrane, on either side of the basal abdominal segment. Most have fully developed wings, the fore wing being elongate and parallel sided, the hind wing broad, fanlike, and reticulate veined. Many types have reduced or virtually no wings and are flightless, particularly some high Andean (*Melanoplus;* Roberts 1973) and rain forest (Rowell 1978) types.

The female uses her short ovipositor to excavate holes in the soil into which she deposits up to one hundred elongate eggs in a mass cemented together with a viscid secretion. Both nymphs and adults feed on vegetation and when abundant, are devastating crop pests. Certain species, especially in the genus *Schistocerca*, form enormous migrating swarms that can destroy fields over wide areas. These are the locusts, most famous in the Old World but represented by a species in Latin America no less imposing or devastating.

Although grasshoppers have been used as sustenance by peoples in Africa, the Middle East, and other parts of the world, both in history and today, they seem to be of relatively minor importance as a food in tropical America. Exceptions in the past were the natives of several West Indian islands who, as Martyr (Bodenheimer

1951: 301) records, ate grasshoppers and stored them for trade. Padre Florian Pauche, who lived with the Mocovíes Indians of Argentina in the eighteenth century, described their practice of catching, cooking, and eating the locusts (*Schistocerca*) that periodically plagued their land (Liebermann 1948).

Grasshoppers were important in the diet of the Aztecs; one species was called *acáhapali*, arrow, in Nahuatl, because of its shape and because it made a distinctive buzz in flight (Curran 1937). According to Kevan (1977), the word *chapulli* apparently referred to many edible species, particularly *Sphenarium*, and formed the basis for many place-names in central Mexico during the period of the Aztec civilization. A hill in Mexico City, famous to the Aztecs and where Maximilian's castle is located, was named *chapultepec*, combining the word for grasshopper with hill (*tepeque*). Grasshoppers also appear as design motifs in the art of early Middle American cultures (Böning 1971).

Grasshoppers live in all habitats and generally have highly specialized life forms to adapt to prevailing environmental conditions. They seem to prefer open habitats and are most numerous and diverse in grasslands. There are indications that understory species decrease as one enters tropical forests (Brodey 1975), but distinct canopy (Roberts 1973, Descamps 1976) and understory (Rowell 1978) assemblages are evident. There are even successional stages of these categories following clearing of forest (Amedegnato and Descamps 1980).

Many grasshoppers can make sounds by frictional contact between various parts of the body, usually one of the hind leg segments against the fore wings. In the winged lubbers, rapid closing of the hind wings brings serrate veinlets on this wing in contact with scraper veins on the fore wing, producing a snapping or rattling sound. A cracking is often heard from band wings in flight. It is thought to be caused by the partial folding and rapid expansion of the hind wing (Uvarov 1966: 176f.).

In addition to the locusts (*Schistocerca*), which mass in flight, the nymphs of some lubbers also are migratory on foot. Young lubbers also form dense groups on the tops of plants. They are more brightly colored than the adults and are certainly distasteful, a fact advertised more effectively en masse than individually. Lowland and lower montane forest grasshoppers have been found to have narrow feeding preferences (Rowell et al. 1984). Probably, many such feeding specialists are sequestering toxic chemicals from their hosts (Rowell 1978).

Grasshoppers are essentially land insects, but a few have semiaquatic habits in South America. The best-studied examples are *Marellia remipes*, *Paulinia acuminata* (fig. 5.5b), and *Cornops aquaticum* (Pauliniidae), which live on broad, floating leaves of water lilies and other aquatic plants. They frequently fall into the water after a hasty jump. They show clear morphological and behaviorial adaptations to life in water including paddle-shaped hind legs in some species (Bentos and Lorier 1991). Females often insert their eggs into the stems of water plants (*Cornops*) or place them on the undersides of floating leaves (*Marellia* and *Paulinia*) (Carbonell 1959, 1964).

The so-called band-wing grasshoppers sport brightly colored red, blue, purple, green, yellow, or orange hind wings. To this group belong the common, widespread, arid-land species, *Trimerotropis pallidipennis* (fig. 5.5a), recognized by its translucent yellow hind wings. The fore wings are usually cryptically colored to match the gravel or sandy soil on which the species habitually rests. Some grass-loving types (*Achurum;* fig. 5.5d) are very slender and elongate, better to hide among the blades on which they rest.

With their hind legs held out at right

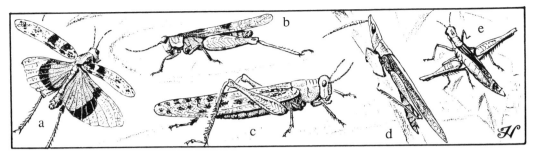

**Figure 5.5    GRASSHOPPERS.** (a) Band-wing grasshopper (*Trimerotropis pallidipennis,* Acrididae). (b) Aquatic grasshopper (*Paulinia acuminata,* Pauliniidae). (c) American locust (*Schistocerca picei-frons,* Acrididae). (d) Grass-mimicking grasshopper (*Achurum sumichrasti,* Acrididae). (e) Eumastacid grasshopper (*Eumastax* sp., Eumastacidae).

angles to the body, *Eumastax* (Eumasta-cidae; fig. 5.5e) bask on leaves in the warm pools of sunlight that flood into the forest undergrowth below holes in the canopy (pl. 1c). The sunbeams excite the irides-cence in the greens, blues, and yellows on the bodies of these forms, making them glow like jewels.

Uvarov (1977: 421f.) presents a faunistic summary of Latin American grasshoppers; Amedegnato (1974) reviews the genera. The geographic and evolutionary history of the regional forms has been traced by various authors (Carbonell 1977, Amede-gnato and Descamps 1979).

## References

AMEDEGNATO, C. 1974. Les genres d'Acridiens néotropicaux, leur classification par familles, sous-familles et tribus. Acrida 3: 193–203.

AMEDEGNATO, C., AND M. DESCAMPS. 1979. History and phylogeny of the Neotropical acridid fauna, Metaleptea. Soc. Panamer. Acridiología 2(1): 1–10.

AMEDEGNATO, C, AND M. DESCAMPS. 1980. Evolution des populations d'Orthopteres d'Amazonie du nord-ouest dans les cultures traditionnelles et les formations secondaires d'origine anthropique. Acrida 9: 1–33.

BENTOS, A., AND E. LORIER. 1991. Acridomorfos acuáticos (Orthoptera, Acridoidea). I. Adapta-ciones morfológicas. Rev. Brasil. Entomol. 35: 631–653.

BODENHEIMER, F. S. 1951. Insects as human food. Junk, The Hague.

BÖNING, K. 1971. Mesoamerikanische Heu-schreckendarstellungen. Anz. Schädlings-kunde Pflanzenschutz 44: 185–189.

BRODEY, K. 1975. A study of grasshopper species composition in primary and secon-dary growth in Costa Rica. Entomol. News 86: 207–211.

CARBONELL, C. S. 1959. The external anatomy of the South American semiaquatic grasshop-per *Marellia remipes* Uvarov (Acridoidea, Pau-liniidae). Smithsonian Misc. Coll. 137: 61–97.

CARBONELL, C. S. 1964. Habitat, etología y ontongenia de *Paulinia acuminata* (Dg.), (Ac-ridoidea, Pauliniidae) en el Uruguay. Soc. Uruguayo Entomol. Rev. 6: 39–48.

CARBONELL, C. S. 1977. Origin, evolution, and distribution of the Neotropical acridomorph fauna (Orthoptera): A preliminary hypothe-sis. Soc. Entomol. Argentina Rev. 36: 153–175.

CURRAN, C. H. 1937. Insect lore of the Aztecs. Nat. Hist. 39: 196–203.

DESCAMPS, M. 1976. La faune dendrophile néotropicale. I. Revue des Proctolabinae (Orth. Acrididae). Acrida 5: 63–167.

KEVAN, D. K. McE. 1977. The American Pyrgo-morphidae (Orthoptera). Soc. Entomol. Ar-gentina Rev. 36: 3–28.

LIEBERMANN, J. 1948. Curiosidades históricas acerca de la langosta. Alm. Min. Agric. Ganad. Buenos Aires 23: 434–438.

ROBERTS, H. R. 1973. Arboreal Orthoptera in the rain forests of Costa Rica collected with insecticide: A report on the grasshoppers (Acrididae), including new species. Acad. Nat. Sci. Philadelphia Proc. 125: 49–66.

ROWELL, C. H. F. 1978. Food plant specificity in Neotropical rain-forest acridids. Entomol. Exper. Appl. 24: 651–662.

ROWELL, C. H. F., M. ROWELL-RAHIER, H. E. BAKER, G. COOPER-DRIVER, AND L. D. GÓMEZ. 1984. The palatability of ferns and the ecology of two tropical forest grasshoppers. Biotropica 15: 207–216.

UVAROV, B. 1966, 1977. Grasshoppers and locusts: A handbook of general acridology. 2 vols. Cambridge Univ. Press, Cambridge.

## Locusts

Acrididae, Cyrtacanthacridinae, *Schistocerca. Spanish:* Langostas, langostas voladoras (General). *Portuguese:* Gafanhotos de praga.

Although not all easily distinguished, twenty to thirty kinds of *Schistocerca* inhabit tropical America. They are all large (BL 4–6 cm), slender grasshoppers with expansive wings that extend well beyond the tip of the abdomen when folded. Females are much larger than the males. The fore wings of both sexes are varied in coloration but are generally dull, yellowish with irregular brownish spots; the hind wings are more or less pelucid.

These forms are variable in behavior as well as in coloration. Certain of them produce large swarms and migrate periodically, like the infamous Old World migratory locust (*S. gregaria*) to which they are closely related. Some of this variation may be caused by developmental influences, especially crowding, leading to so-called phases. There is considerable confusion in the literature regarding the exact species status, interrelationships, and genetic significance of all these forms and phases. In one taxonomic study based on external morphology, the migratory types have been relegated to a single subspecies (*S. paranensis*) within a widespread species, *S. americana,* wherever they occur in Central or South America (Dirsh 1974), while others classify the locusts differently (Harvey 1981). Hybridization experiments, however, show the picture to be still more complex: at least *S. piceifrons* (fig. 5.5c), *S. americana,* and *S. cancellata* should be considered separate species (Harvey 1979, Jago et al. 1982), the first being the true swarming pest (Harvey 1983).

Whatever its correct name, this locust is well known in Latin America for its ravages to crops and rangeland from prehistory to modern times. Descriptions of the invasions rival the stories of the Old World species recounted in the Bible and other historical writings. While crossing the dry pampas of Argentina in March 1835, Charles Darwin (1962 [1845]: 330–331) wrote,

> Shortly before arriving at the village and river of Luxan, we observed, to the south, a ragged cloud of a dark reddish brown color. At first we thought it was caused by some great fire on the neighboring plains, but we soon found that it was a swarm of locusts. . . . The sound of their wings was as the sound of chariots of many horses running to battle. . . . The sky, seen through the advanced guard, appeared like a mezzotinto engraving; but the main body was impervious to sight. . . . When they alighted, they were more numerous than the leaves in the field, and the surface became reddish instead of green.

These events are known in many parts of the region, most commonly and regularly in northern Argentina and southern Brazil but also to a limited extent in coastal Chile and the Andean mountain valleys of Peru, Ecuador, and Colombia as well as in scattered localities in Central America and Mexico.

The annual patterns of breeding are strongly influenced by the weather and its seasonal variations. Breeding is confined to the period of summer rains. Outbreaks occur in the driest parts of the species' range, triggered by the occasional abundant rainfalls that foster the insect's high reproductive capacity (Hunter and Cosenzo 1990, Waloff and Pedgley 1986).

In most southern areas, there are two annual generations, beginning with mating and oviposition in November and December. The nymphs hatch and are active and growing from December to April. They become adults through April and May, when the migrations to alternative breeding grounds take place. Here a second mating and egg laying ensues, and the young of this generation produce a whole new brood of adults by September to November which migrates back to the areas of original habitat (Daguerre 1953).

Other *Schistocerca* species and subspecies overlap the range of the true locusts in diverse parts of Latin America, all of a totally sedentary nature. These include endemic species on many isolated islands in the West Indies, Bermuda, and the Galápagos and Revillagigedo archipelagos.

### References

DAGUERRE, J. B. 1953. Vida de la langosta voladora. Reun. Com. Interamer. Perm. Antiacrid (Puerto Alegre), 1952: 55–79.

DARWIN, C. 1962 [1845]. The voyage of the Beagle. Edited by L. Engel. Doubleday-Amer. Mus. Nat. Hist, New York.

DIRSH, V. M. 1974. Genus *Schistocerca*. Junk (Ser. Entomol. 10), The Hague.

HARVEY, A. W. 1979. Hybridization studies in the *Schistocerca americana* complex. I. The specific status of the Central American Locust. Biol. J. Linnean Soc. 12: 349–355.

HARVEY, A. W. 1981. A reclassification of the *Schistocerca americana* complex. Acrida 10: 61–77.

HARVEY, A. W. 1983. *Schistocerca piceifrons* (Walker), the swarming locust of tropical America: A review. Bull. Entomol. Res. 73: 171–184.

HUNTER, D. M., AND E. L. COSENZO. 1990. The origin of plagues and recent outbreaks of the South American locust, *Schistocerca cancellata* (Orthoptera: Acrididae) in Argentina. Bull. Entomol. Res. 80: 295–300.

JAGO, N. D., A. ANTONIOU, AND J. P. GRUNSHAW. 1982. Further laboratory evidence for the separate species status of the South American locust (*Schistocerca cancellata* Serville) and the Central American locust (*Schistocerca piceifrons piceifrons*) (Acrididae, Cyrtacanthacridinae). J. Nat. Hist. 16: 763–768.

WALOFF, Z., AND D. E. PEDGLEY. 1986. Comparative biogeography and biology of the South American locust, *Schistocerca cancellata* (Serville), and the South African desert locust, *S. gregaria flaviventris* (Burmeister) (Orthoptera: Acrididae): A review. Bull. Entomol. Res. 76: 1–20.

### Lubber Grasshoppers

Romaleidae.

This grasshopper family is found only in the New World and consists currently of forty-eight genera and at least double that number of species, distributed over all the Neotropical Region but not on the Antilles except Cuba (Rehn and Grant 1959). The group is of very diverse morphology but is generally made up of quite large, heavy-bodied forms, including the biggest of the world's grasshoppers, *Tropidacris* (fig. 5.6a), with body lengths of 8 to 9 centimeters. They are fully winged to apterous, and many have partially developed wings never used for flight. The integument is often granulate or with pronounced, tuberculate points. The thoracic disk usually has a

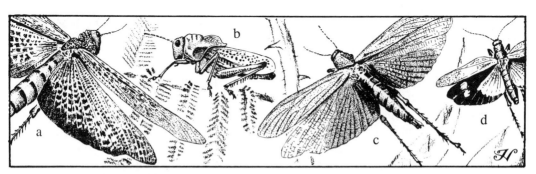

**Figure 5.6   LUBBER GRASSHOPPERS (ROMALEIDAE).** (a) Giant grasshopper (*Tropidacris cristata*). (b) Lubber grasshopper (*Taenopoda varipennis*). (c) Rainbow-winged grasshopper (*Titanacris gloriosa*). (d) Independence grasshopper (*Chromacris speciosa*).

definite crest but may be wide or simple, like that of other acridids.

The vernacular name of these grasshoppers refers to the terrestrial habits of the majority of species, which is dictated by their flightlessness ("landlubbers"). Unable to fly, they rely on other modes of self-protection, often releasing repugnant secretions from glands or the mouth and emitting threatening sounds. Release of noxious liquids or foams from the mesothoracic spiracle can be accompanied by a hissing and buzzing of the wings in an imposing threat display.

These are abilities of the common genus *Taenopoda*, for example, which is represented by several species, including the well-known *T. eques*, in the drier parts of northern Mexico, down through Central America to Panama, where *T. varipennis* (fig. 5.6b) takes over. They are usually fond of mimosas and acacias, their primary food plants. *Taenopoda* are moderately large (BL 6–7 cm) and have only slightly diminutive wings. The fore wings are basically green, with fine, yellow reticulation or dark blotches. The nymphs are black with red gilding and are gregarious. They apparently rely on unpalatability for protection, being so conspicuous in color and vulnerable on exposed, herbaceous vegetation.

Another northern species is *Brachystola magna*, which prefers grasslands on the Mexican plateau and is confined to a life on the ground because of its almost complete winglessness and inability to climb. The fore wing is only a small disk, the hind wing not much more than a vestigial fan.

Lubbers of the genus *Chromacris* are usually a glossy green with yellow markings and striking red or yellow wings. The hind wings are various shades of red, orange, or yellow with contrasting black bands. They occur in the humid portions of the American tropics from Mexico to Argentina. All show a preference for solanaceous and composite plants and have gregarious juveniles like other romaleids. *C. speciosa* (fig. 5.6d) is the most wide-ranging and variable species. It is green and yellow, the colors of the Brazilian flag, and has been dubbed the "independence grasshopper" (*gafanhoto da independência*) in the republic (Ohaus 1990: 233, as *C. miles;* Roberts and Carbonell 1982).

## References

OHAUS, F. 1990. Bericht über eine entomologische Reise nach Centralbrasilien. Stettiner Entomol. Zeit. 61: 164–273.

REHN, J. A. G., AND H. J. GRANT, JR. 1959. A review of the Romaleinae (Orthoptera; Acrididae) found in America north of Mexico. Acad. Nat. Sci. Philadelphia Proc. 111: 109–271.

ROBERTS, H. R., AND C. S. CARBONELL. 1982. A revision of the grasshopper genera *Chromacris* and *Xestotrachelus* (Orthoptera, Romaleidae, Romaleinae). Calif. Acad. Sci., Ser. 4, Proc. 43: 43–58.

### Giant Grasshoppers

Romaleidae, *Tropidacris, Titanacris.*
Birdwing grasshoppers, rainbow grasshoppers.

*Tropidacris* and *Titanacris* are unusually large, fully winged grasshoppers. The first genus contains only three species, which are 10 to 13 centimeters from head to wing tips and have a wingspan of 25 centimeters or more. A widespread but not common species is *Tr. cristata*. The second genus, in which there are seven species (e.g., *T. velazquezii, T. gloriosa;* Descamps and Carbonell 1985), is only slightly smaller, 10 to 11 centimeters in length and with a wingspan of up to 23 centimeters.

Aside from their great size, these grasshoppers are spectacular in flight or when crepitating from perches, when they display their brilliantly colored hind wings. In *Tropidacris*, these wings are generally red with a solid black marginal zone giving way to a finely checkered gray or bluish pattern toward the wing base. The fore wings are gray-green, splotched with gray, and the posterior part of the pronotum is flat and yellow and green speckled. The hind wing of *Titanacris gloriosa* (fig. 5.6c) is a veritable rainbow, bright blue basally, grading to crimson anteriorly, and then green over the apical portion. The fore wings are uniformly leaf green, as is the pronotum; the latter has a serrated crest running down the entire middle to distinguish it from *Tropidacris*, whose crest runs over the anterior half of the pronotum only.

In spite of their conspicuousness, not much has been recorded regarding the life habits of these enormous orthopterans. They are occasionally found on shrubby vegetation but more normally inhabit the crowns of trees and are particularly active on hot days, stridulating and readily taking flight. *Tr. cristata* has been encountered on plants of the genus *Quassia* (Simaroubaceae) which contain repellent chemicals that the insect may sequester (Rowell 1983). The immatures are gregarious. Young (pers. comm.) reports the species common on *Erythrina* (Papilionaceae) in Costa Rica.

## References

DESCAMPS, M., AND C. S. CARBONELL. 1985. Revision of the Neotropical arboreal genus *Titanacris* (Orthoptera, Acridoidea, Romaleidae). Soc. Entomol. France Ann. (n.s.) 21: 259–285.

ROWELL, H. F. 1983. *Tropidacris cristata* (saltamonte o Chapulín gigante, giant red-winged grasshopper). *In* D. H. Janzen, ed., Costa Rican natural history. Univ. Chicago Press, Chicago. Pp. 772–773.

## Jumping Sticks

Proscopiidae. *Spanish:* Palitos vivientes de antenas cortas (General). *Portuguese:* Chicos magros, gafanhotos de marmeleiro, gafanhotos de jurema, María moles (Brazil).

Jumping sticks are found only on the South American continent, as far north as Panama, and on the Caribbean island of Bonaire. Their biology is incompletely unknown. Specimens are usually seen resting on vegetation, their cryptic stick form and colors giving them a measure of camouflage, especially among grasses, which are a dominant habitat. Some information regarding internal anatomy and cytology has been provided by Ferreira (1978).

There are 153 named species in seventeen genera, but many more probably exist. Some known species may be found to be composed of multiple species when details of their anatomy, such as in the genitalia, are studied (Descamps 1973). *Apioscelis* (fig. 5.7a) is a typical species.

These orthopterans (Carbonell 1977, Mello-Leitão 1939) are sticklike in shape, fairly large (adult BL 5–10 cm), and easily mistaken for walkingsticks (Phasmatodea) (Liana 1972). The slightly dilated hind femur, elongate thorax, and vertical, cone-shaped grasshopper-type head distinguish them, however. Also, the antennae are short and have few segments like the grasshoppers, which are their closer relatives but from which they differ by having a vertical head and an elongate prothorax that lacks a dorsal shield. The portion of the head anterior to the eyes may be extremely prolonged in some genera. Wings are completely missing in almost all and at most are minute vestiges, and they lack tympanic and stridulatory organs. They are mostly dull colored, solid olive to brown, although some have black, red, or yellow markings.

## References

CARBONELL, C. S. 1977. Superfam. Proscopoidea, Fam. Proscopiidae. *In* M. Beier, ed., Orthopterum Catalogus. Junk, The Hague. Pp. 1–29.

DESCAMPS, M. 1973. Notes préliminaires sur les genitalia de Proscopoidea (Orthoptera Acridomorpha). Acrida 2: 77–95.

FERREIRA, A. 1978. Contribuição ao estudo da evolução dos Proscopiidae (Orthoptera-Proscopoidea). Studia Entomol. 20: 221–233.

LIANA, A. 1972. Etudes sur les Proscopiidae (Orthoptera). Polska Akad. Nauk, Inst. Zool., Ann. Zool. 29: 381–459.

MELLO-LEITÃO, C. 1939. Estudio monográfico de los Proscópidos. Mus. La Plata Rev., Nov. Ser., Zool. 8(1): 279–449.

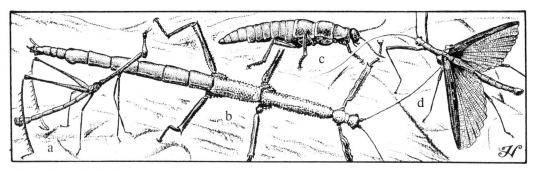

**Figure 5.7  JUMPING STICK AND WALKINGSTICKS.** (a) Jumping stick (*Apioscelis* sp., Proscopiidae). (b) Giant walkingstick (*Phibalosoma phyllinum*, Phibalosomatidae). (c) Chinchemoyo (*Paradoxomorpha crassa*, Anisomorphidae). (d) Winged walkingstick (*Pseudophasma* ?, Pseudophasmatidae).

## WALKING STICKS

Phasmatodea (= Phasmatoptera, Phasmida, Phasmodea). *Spanish:* Bichos palitos, zacatones (General); chinchemoyos, palotes (Chile); campamochas (Mexico); palitos vivientes (Peru). *Portuguese:* Bichos pau. Phasmids.

A greatly attenuated, sticklike body and legs are the hallmarks of these common but seldom seen insects. Their resemblance to twigs and leaves is so perfect that they are usually discovered in their natural haunts only by accident. Human eyes can play time and again over a specimen in its resting place and never see it. Some in central South America are among the largest insects as measured by length. *Bactridium grande* (female BL 26 cm) is the biggest; *Phibalosoma phyllinum* (fig. 5.7b) and *Otocrania aurita* both are over 20 centimeters long.

There are both winged (fig. 5.7d) and wingless stick insects, a tendency toward the latter more in females than in males. When present, the hind wings only are well formed, and then they are large, fanlike, and efficient flight organs. The fore wings are much reduced, often only scalelike cups or elongate leathery plates.

In general, the group (Bedford 1978) shows a number of interesting biological as well as structural features, but little specific information is available on the Neotropical species (Willig et al. 1986, Zapata and Torres 1970). All feed on plants and are apparently fairly host specific. Parthenogenesis occurs in a number of European forms and presumably also occurs in regional mantids. An unexplained characteristic of some female walkingsticks is a bright red color at the base of the fore femur. Individuals can regenerate lost limbs, and those of some species can change color to match their background.

Defensive behaviors, coupled with their cryptic forms and colors, are well developed and parallel to some extent those of mantids. They include (1) rocking motions in which the body is swayed from side to side by flexing of the legs at the upper joints, an action thought to enhance resemblance to plant parts moving in the breeze; (2) active escape by dropping or flight; (3) flashing of wings to startle or display aposematic colors; (4) catalepsy, or freezing or feigning death; (5) sound production by wing rattling; (6) fighting with the legs; and (7) release of repugnant or caustic chemicals. *Paradoxomorpha* (fig. 5.7c) can fire an aerosol spray that can blister human skin. The substance has been analyzed as containing ethyl ether or orthoformic acid and is produced by a pair of immense glands in the thorax that exit through an opening on either side behind the head (Moreno 1940). This is a rather robust stick insect, fairly common in parts of southern Argentina and Chile where it is known by several vernaculars in Quechua (*chinchemoyo,* "stinking chest") and Mapuche (*tabolongo, chirindango, fitquilén*) (Schneider 1934).

A single species may employ a variety of these tactics. *Pterinoxylus spinulosus* from Panama enhances its resting posture resemblance to sticks by closely folding the mid- and hind legs, which then look like small shoots from the main stick (Robinson 1968).

Phasmid eggs are curiously shaped, looking like seeds with a hard shell and either smooth and shiny and unicolorous or heavily sculptured and patterned. All have an operculum. They are placed carefully on the host plant or dispersed by flicking movements of the abdomen or scattered indiscriminately.

Stick insects are found in all habitats but are most abundant and diverse in humid forests. Two species, *Bostra scabrinota* and *Libethra minuscula,* are typical of the lomas or seasonal herblands of the coastal Peruvian desert. Developmental stages exhibit conspicuous changes of color corresponding to the flourishing and waning of this vegetation that they mimic. They pass the long dry season as eggs on the soil surface (Aguilar 1970).

Four ceratopogonid gnats in the subgenus *Microhelea* of *Forcipomyia* attach to stick insects and suck their blood. With the exception of one Indonesian example, these phasmid parasites, called "stick ticks," are known only from tropical America (Wirth 1971).

The order has been newly reclassified into six families (Bradley and Galil 1977) and its world distribution reviewed (Gunther 1953). A high percentage, perhaps 30 percent of more than 2,500 world species, live in Latin America, including the West Indies (Moxey 1972).

## References

AGUILAR, P. G. 1970. Los "palitos vivientes de Lima." I. Phasmatidae de las lomas. Rev. Peruana Entomol. 13: 1–8.

BEDFORD, G. O. 1978. Biology and ecology of the Phasmatodea. Ann. Rev. Entomol. 23: 125–149.

BRADLEY, J. C., AND B. S. GALIL. 1977. The taxonomic arrangement of the Phasmatodea with keys to the subfamilies and tribes. Entomol. Soc. Wash. Proc. 79: 176–208.

GUNTHER, K. 1953. Uber die taxonomische Gliederung und die geographische Verbreitung der Insektenordnung de Phasmatodea. Zeit. Entomol. 3: 541–563.

MORENO, A. 1940. Glándulas odoríferas en *Paradoxomorpha.* Mus. La Plata Not. (Zool. 45) 5: 319–323.

MOXEY, C. F. 1972. The stick-insects (Phasmatodea) of the West Indies: Their systematics and biology. Ph.D. diss., Harvard Univ. [Not seen.]

ROBINSON, M. H. 1968. The defensive behavior of *Pterinoxylus spinulosus* Redtenbacher, a winged stick insect from Panama. Psyche 75: 195–207.

SCHNEIDER, C. O. 1934. Las emanaciones del chinchemoyo *Paradoxomorpha crassa* (Blanch, Kirby). Rev. Chilena Hist. Nat. 38: 44–46.

WILLIG, M. R., R. W. GARRISON, AND A. J. BAUMAN. 1986. Population dynamics and natural history of a Neotropical walkingstick, *Lamponium portoricensis* Rehn (Phasmatodea: Phasmatidae). Texas J. Sci. 38: 121–137.

WIRTH, W. W. 1971. A review of the "stick-ticks," Neotropical biting midges of the *Forcipomyia* subgenus *Microhelea* parasitic on walkingsticks (Diptera: Ceratopogonidae). Entomol. News 82: 229–245.

ZAPATA, S., AND E. TORRES. 1970. Biología y morfología de *Bacteria granulicollis* (Blanchard) (Phasmida). Univ. Chile, Cen. Est. Entomol. Publ. 10: 23–42.

# COCKROACHES

Blattodea (= Blattaria). *Spanish:* Cucarachas. *Portuguese:* Baratas. *Quechua:* Utiuti.

Cockroaches are regarded with disgust by nearly everyone. Although widely believed to carry disease, their importance as mechanical vectors is probably overrated

(Roth and Willis 1957). Some evidence to the contrary does exist (Gazivoda and Fish 1985). It is a shame that a few unsavory types have given a bad name to an otherwise diverse and biologically fascinating group of insects, one begging investigation in Latin America where a large and distinctive fauna exists which is practically unknown.

Cockroaches are among the most primitive and ancient of winged insects. Their origin dates back 250 million years to the Carboniferous period. Fossil records show that they were very abundant at that time and very little different in structure from their modern descendants.

Nocturnal cockroaches are mostly oval, much flattened and dull-colored, with pliable wings, all adaptations for life in dark, confined spaces. Others are teardrop shaped, convex, with brightly hued, hard, beetlelike wings; these features relating them to an exposed, diurnal existence on vegetation, tree bark, and tree trunks.

All cockroaches have a well-developed, thin layer of grease or wax on the outside of the cuticle which protects them from desiccation and gives them their slick feel. The head is horizontal but bent backward so that the mouthparts project toward the rear, being situated almost between the bases of the front legs. The head is often entirely concealed from above by a large, widely expanded, and flattened disklike shield (the pronotum), out from under which project the very long, whiplike antennae. Wings are usually present and large with a highly complex, reticulate venation. They are often abbreviated in the female and in many species may be absent altogether from both sexes. The fore wing is narrow and elongate and more rigid in texture than the hind wing, which is broad, fanlike, and membranous. The legs are well spined. A pair of elongate, segmented cerci project from the apex of the abdomen. These appendages are richly endowed with external sense organs adapted for perception of vibrations, sound, and air movements.

Female cockroaches give birth in different ways. They normally encapsulate their eggs in groups of few to many within a hard, darkly pigmented case (ootheca), which is deposited in the proper environment. Some retain the case in the birth canal where it often remains partly extruded until the young hatch. Others keep it entirely within the abdomen, and the nymphs hatch inside the mother, spending up to several days in a kind of uterus before being born alive.

Cockroaches actively secrete a variety of exocrine substances, such as sex lures, aggregation stimulants, and defensive compounds of all kinds (Roth and Alsop 1978). The glands that produce these chemicals are located either in the head (mandibular glands) or on the abdomen, usually on the back of the more posterior segments.

A great deal is known about the anatomy, physiology, and natural history of the few common domestic and semidomestic forms (Cornwell 1968, Guthrie and Tindall 1968), but information is scarce concerning these aspects in the majority of wild species. One study (Schal and Bell 1986) indicates a vertical segregation of species in the layered vegetation of forests. Many genera are represented in bromeliad "terraria" (Rocha and Lopes 1976); *Epilampra* (fig. 5.8a) has been observed to swim well (Crowell 1946), even having tubular spiracles at the rear of the abdomen, possibly functioning as elementary snorkels. Several species mimic other insects. For example, lycid beetles of the genus *Calopteron* are mimicked by the cockroaches *Holocampsa* and *Paratropes*, lampyrid beetles *Cratomorphus* and *Aspisoma* by *Achroblatta luteola* (figs. 5.8b, 5.8c), and the giant fungus beetle *Erotylus* by *Plectoptera*. Most of these are diurnal and commonly seen running on forest understory vegetation; the firefly mimics are nocturnally active like their models, the similarity be-

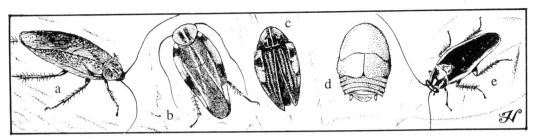

**Figure 5.8  COCKROACHES.** (a) Aquatic cockroach (*Epilampra* sp., Epilampridae). (b) Firefly-mimicking cockroach (*Achroblatta luteola,* Atticolidae). (c) Firefly model of firefly-mimicking cockroach (*Aspisoma* sp., Lampyridae). (d) Myrmecophilous cockroach (*Myrmecoblatta* sp., Atticolidae). (e) Leaf cockroach (*Pseudomops* sp., Blatellidae).

ing effective, however, during the daytime when the insects are at rest and visible to predators. The lycid mimic *Paratropes* may be responsible for pollination of plants in the canopy of Central American rain forests (Perry 1978).

The fat body of cockroaches is packed with intracellular bacteria (bacteroids) that contribute to the synthesis of amino acids used in metabolism. The alimentary canal of some wood-feeding types also contain symbiotic protozoa, which, like those of certain termites, assist in the digestion of cellulose (Roth and Willis 1960).

Ecological niches occupied by tropical American cockroaches are extremely varied. Many of these insects are associated casually with vegetation, usually seen sitting on the upper surfaces of leaves (*Plectoptera*), or are disposed to feed on the fruit, leaves, bark, roots, and other parts of living plants of particular species; others consume the wood of rotting logs. *Pseudomops* (fig. 5.8e) are common, small, brightly colored forms (head orange and wings and pronotum shining dark purple, the latter margined with yellow) that perch on the leaves of low plants in forests and clearings, hopping from place to place in the daytime. *Blaberus* depends on the excrement or dead bodies of other creatures and lives in animal burrows, hollow trees, or caves. A number of species live symbiotically with ants (Bolivar 1905), such as the minute *Attiphila,* found in the fungus gar-

dens of leaf cutter ants (*Atta*), and *Myrmecoblatta* (fig. 5.8d), a genus associated with carpenter ants (Deyrup and Fisk 1984). Cockroaches have been found in all climes, from hot deserts and cold mountaintops to warm, humid lowland rain forests. Everywhere they are important reducers of leaf litter and wood. In the inundation forests of Amazonia, *Epilampra* feeds on dead leaves and insect carcasses. In any one area, as much as 5.6 percent of the yearly leaf fall and other organic detritus may be turned over by members of this genus (Irmler and Furch 1979). There is some tendency for adult cockroaches in lowland forests to disperse their activity over a long time, thereby possibly reducing interspecific competition in nonseasonal locations; their occurrence does not follow this pattern where well-defined wet and dry seasons are present (Wolda and Fisk 1981).

A number of cockroaches are capable of sound production. They accomplish this by rubbing the abdomen against the wings (*Blaberus*), by stridulation of roughened areas on the thorax (*Panchlora*), or by other devices (Roth and Hartman 1967).

Of the 3,500 known species in the world, probably over 30 percent are found in Latin America (Princis 1962–1971). No general guide to their classification is available, although there are some good local treatments with broad applicability (Fisk and Wolda 1979, Bruijning 1959).

## References

BOLIVAR, I. 1905. Les blattes myrmécophiles. Schweizerischen Entomol. Ges. Mitt. 11: 134–141.

BRUIJNING, C. F. A. 1959. The Blattidae of Surinam. Stud. Fauna Suriname Guianas 2(4): 1–103.

CORNWELL, P. B. 1968. The cockroach. Vol. 1. Hutchinson, London.

CROWELL, H. H. 1946. Notes on an amphibious cockroach from the Republic of Panama. Entomol. News 57: 171–172.

DEYRUP, M., AND F. FISK. 1984. A myrmecophilous cockroach new to the United States (Blattaria: Polyphagidae). Entomol. News 95: 183–185.

FISK, F. W., AND H. WOLDA. 1979. Keys to the cockroaches of central Panama. Stud. Neotrop. Fauna Environ. 14: 177–201.

GAZIVODA, P., AND D. FISH. 1985. Scanning electron microscope demonstration of bacteria on tarsi of *Blatella germanica*. New York Entomol. Soc. J. 93: 1064–1067.

GUTHRIE, D. M., AND A. R. TINDALL. 1968. The biology of the cockroach. Arnold, London.

IRMLER, U., AND K. FURCH. 1979. Production, energy, and nutrient turnover of the cockroach *Epilampra irmleri* Roch e Silva and Aguilar in a Central-Amazonia inundation forest. Amazoniana 6: 497–520.

PERRY, D. R. 1978. *Paratropes bilunata* (Orthoptera; Blattidae): An outcrossing pollinator in a Neotropical wet forest canopy. Entomol. Soc. Wash. Proc. 80: 656–657.

PRINCIS, K. 1962–1971. Blatariae. *In* M. Beier, ed., Orthopterorum Catalogus. Junk, The Hague. Pts. 3, 4, 6, 7, 8, 11, 13, 14.

ROCHA E SILVA ALBUQUERQUE, I., AND S. M. RODRIGUES LOPES. 1976. Blattaria de bromélia (Dictyoptera). Rev. Brasil. Biol. 36: 873–901.

ROTH, L. M., AND D. W. ALSOP. 1978. Toxins of Blattaria. *In* S. Bettini, ed., Arthropod venoms. Springer, Berlin. Pp. 465–487.

ROTH, L. M., AND H. B. HARTMAN. 1967. Sound production and its evolutionary significance in the Blattaria. Entomol. Soc. Amer. Ann. 60: 740–752.

ROTH, L. M., AND E. R. WILLIS. 1957. The medical and veterinary importance of cockroaches. Smithsonian Misc. Coll. 134(10): 1–147.

ROTH, L. M., AND E. R. WILLIS. 1960. The biotic associations of cockroaches. Smithsonian Misc. Coll. 141: 1–470.

SCHAL, C., AND W. J. BELL. 1986. Vertical community structure and resource utilization in Neotropical forest cockroaches. Ecol. Entomol. 11: 411–423.

WOLDA, H., AND F. W. FISK. 1981. Seasonality of tropical insects. II. Blattaria in Panama. J. Anim. Ecol. 50: 827–838.

## Domestic Cockroaches

Less than thirty-five species of cockroaches live in intimate closeness with humans. They frequent homes, restaurants, hotels, and all our dwellings, where they scavenge food leavings and whatever edible organic matter they can find (Cornwell 1968). All are Old World in origin, probably carried to Latin America in the last four centuries aboard ships. Some less adaptable species have spotty distributions, reflecting to some degree their points of arrival, while others have spread virtually everywhere.

Probably the most widespread and frequently seen is the American cockroach, or *cucarachón* (*Periplaneta americana;* fig. 5.9a), which infests warm dwellings throughout the world (Bell and Adiyodi 1981). It prefers a heated, moist environment and is common outdoors in tropical portions of America as well as indoors. It is often found in sewers. A medium-sized cockroach (BWL 28–44 mm), its body is shining reddish-brown, with a paler yellow area around the edge of the head shield. The fully developed wings extend well beyond the abdomen in the male but only just overlap the abdomen in the female. Females produce up to ninety free oothecae with about twelve eggs each.

A related and similarly ubiquitous species is the Australian cockroach (*Periplaneta australasiae;* fig. 5.9b), which is a general pest in cooler areas wherever moist artificial environments prevail. About the same size (BWL 30–35 mm) and facies as *P. americana,* it differs primarily in having well-defined, bright yellow "shoulders" (elongate marks at the outer bases of the fore wings). Its biology is also similar to that of the American cockroach, but it seems not to be a denizen of sewers. The

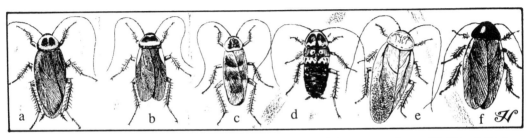

**Figure 5.9  COCKROACHES.** (a) American cockroach (*Periplaneta americana*, Blattidae). (b) Australian cockroach (*Periplaneta australasiae*, Blattidae). (c) Brown-banded cockroach (*Supella longipalpa*, Blatellidae). (d) Harlequin cockroach (*Neostylopyga rhombifolia*, Blattidae). (e) Madeira cockroach (*Leucophaea maderae*, Oxyhaloidae). (f) Surinam cockroach (*Pycnoscelus surinamensis*, Pycnoscelididae).

original home of both of these *Periplaneta* species was probably tropical Africa.

The brown-banded cockroach (*Supella longipalpa*, sometimes called *S. supellectilium*; fig. 5.9c) is small (BWL 10–14 mm), with complete wings in the male and shortened wings in the female. It is buff in general color but with two suffuse, transverse black or brown bands. It is highly domiciliary, taking up residence in furniture and forming colonies in drawers, behind pictures on the wall, on bookshelves, and in like places. Here the adults and nymphs hide by day, emerging at night to feed on any available fodder, including glue, sizing on books, wallpaper, food scraps, and even plain paper.

The harlequin cockroach (*Neostylopyga rhombifolia*; fig. 5.9d) is medium-sized (BL 20–25 mm) and completely without wings in both sexes, although the fore wings are represented by small stubs below the outer corners of the head shield. It has a striking color pattern of deep yellow marbling on a shining, brownish-black background. Its biology is largely unstudied, although it is common both indoors and out.

The Madeira cockroach (*Leucophaea maderae*; fig. 5.9e), known as *barata cascuda* in Brazil, has become widely established around the Carribean, where it often is a serious pest in homes and warehouses, and in many parts of South and Central America. In tropical environments, it lives out-

doors and is common in sugarcane and banana plantations. It is a fairly large cockroach (BWL 4–5 cm), with ample wings in both sexes. It is pale brown to tawny olive in general color, the fore wings marked with a dark spatter pattern over most of their surface; the basal and anterior marginal areas are contrastingly plain, except for a dark linear arc running obliquely across the wing base. Adults are slow moving but capable of defending themselves well with an offensive odor. They also stridulate. Females bear twenty-five to thirty live young at a time.

Although first found in South America, the Surinam cockroach (*Pycnoscelus surinamensis*; fig. 5.9f), known as *barata de pau podre* in Brazil, is most likely of Oriental origin. Because it is parthenogenetic, even unmated females can start thriving populations wherever they happen to be carried, and the species has become established widely in Latin America. Away from civilization, it occurs under stones and loose litter and can burrow superficially into the soil. It is medium-sized (BWL 18–24 mm), shining brown, with dark shoulder streaks and a contrasting black head shield. Wings are complete in both sexes. It often has a pale band along the anterior margin of the head shield, the posterior margin of which is strongly sinuate.

Another domiciliary cockroach from Africa is the lobster cockroach (also called the

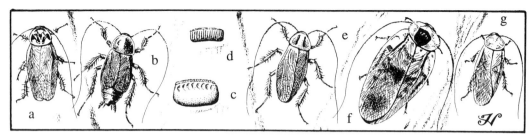

**Figure 5.10 COCKROACHES.** (a) Cinereous cockroach (*Nauphaeta cinerea,* Oxyhaloidae). (b) Oriental cockroach (*Blatta orientalis,* Blattidae). (c) Oriental cockroach egg case. (d) German cockroach egg case. (e) German Cockroach (*Blatella germanica,* Blatellidae). (f) Death's-head cockroach (*Blaberus giganteus,* Blaberidae). (g) Cuban cockroach (*Panchlora nivea,* Panchloridae).

cinereous cockroach, *Nauphaeta cinerea;* fig. 5.10a), which derives its name from a lobster-shaped design on the head shield. It is large (BWL 25–29 mm), with well-developed, ash-colored wings that are just short of covering the abdomen in both sexes. Males stridulate when courting females. The females form oothecae with twenty-six to forty eggs that hatch as she extrudes the capsule from her brood sac.

The original home of the Oriental cockroach (*Blatta orientalis;* fig. 5.10b) is North Africa. Its body is shining black, except for the reddish-brown short wings of the male; these organs are mere stubs in the female. It is a medium-sized species (BWL 20–24 mm) that produces up to eighteen large oothecae (5 by 10 mm) (fig. 5.10c). The latter contain rather oversized eggs that develop free of the parent. The species is of local occurrence in dwellings but is often found outdoors where the climate is moderate.

Another very widespread vagrant cockroach is the German cockroach (*Blattella germanica;* fig. 5.10e). It is small (BWL 10–15 mm) and pale buff in color, with distinct, parallel, wide bands on the head shield. Wings are nearly completely formed, only the tip of the abdomen being left exposed in both sexes. The species prefers a warm, moist ambience and is therefore common in kitchens, larders, and restaurants, rarely in bedrooms. It has the capacity to build up enormous populations when food is adequate. The oothecae are dropped free by the female and contain on the average thirty-seven eggs (fig. 5.10d). The species is considered an indoor pest, but in warm environments, heavy infestations may overflow outdoors.

**References**

BELL, W. J., AND K. G. ADIYODI, eds. 1981. The American cockroach. Chapman Hall, London.
CORNWELL, P. B. 1968. The cockroach. Vol. 1. Hutchinson, London.

**Giant Cockroaches**
Blaberidae, Blaberinae, Blaberini, *Blaberus. Spanish:* Cascudas (General); cucarachas mandingas, cucarachas mama, cargatables (Peru). Cockroaches of the Divine Face, death's-head cockroaches.

Apart from being very big (the largest species, *B. colosseus,* reaches a BWL to 8 cm), the giant cockroaches are recognized by their fully developed, shiny, light brown fore wings, these usually marked with a large squarish spot in the middle and a diffuse spot sometimes on the outer third. There can be also an elongate thin dark bar in the shoulder area of the wing, paralleling the front margin. In addition, they have a sharply defined, square to trapezoidal black patch resting against the hind margin in the center of the oval head shield in which some see the image of a human face, or death's-head.

This genus contains twelve very similar but separate species, readily distinguished only by the structure of the genitalia (Roth 1969). All normally inhabit caves, rock crevices, hollow trees, and cavities under loose tree bark. *Blaberus parabolicus* (*tranga bakkas*) lives under houses in Surinam and becomes a pest indoors when attracted to light at night (Bruijning 1959). Adults and the yellow-spotted, trilobitelike nymphs are often found together and exhibit subsocial behavior (Gautier 1978, Schal 1983). Males assume a dominant rank in dense groups, their status distinguishable by their erect posture and aggressiveness. An aggregation pheromone, containing undecane and tetradecane and other compounds of unknown function (Brossut et al. 1973), is secreted from mandibular glands in all stages and by both sexes. Females also produce a volatile sex pheromone that acts as a primary releaser of complex and lengthy male courtship activities. The repertoire includes intense intermale fighting over calling females (Wendelken 1977).

Nymphs are adept at burrowing in the soil or rotting wood to escape predation. Adults lead more exposed lives and are likelier to avoid harm by flying, releasing offensive odors, or kicking with their sharp-spined legs (Crawford and Cloudsley-Thompson 1971). They are often attracted to artificial light during their nocturnal wanderings.

The food of giant cockroaches consists of organic detritus that accumulates in their dank niches, including bat guano, rotting wood and fruits, seeds and other decomposing vegetation, and dead insects or animals at times.

The two common species, *Blaberus craniifer* and *B. giganteus* (fig. 5.10f) (Schal 1983), are easy to maintain in culture and are used in many laboratories for physio-

logical research (Lefeuvre 1960). Adults can live for as long as twenty months (Piquett and Fales 1953). Females internally incubate their oothecae; the latter contain about thirty to forty eggs.

Other very large Neotropical cockroaches are the four species of South American *Megaloblatta* (Blatellidae, Nyctiborinae). They often are over 8 centimeters in length. A Colombian specimen, whose overall length (head to wing tips) was measured at 10 centimeters (Gurney 1959), is the largest cockroach recorded.

## References

Brossut, R., P. Dubois, and J. Rigaud. 1973. Le grégarisme chez *Blaberus craniifer:* Isolement et identification de la phéromone. J. Ins. Physiol. 20: 529–543.

Bruijning, C. F. A. 1959. The Blattidae of Surinam. Stud. Fauna Suriname Guianas 2(4): 1–103.

Crawford, C. S., and J. L. Cloudsley-Thompson. 1971. Concealment behavior of nymphs of *Blaberus giganteus* L. (Dictyoptera: Blattaria) in relation to their ecology. Rev. Biol. Trop. 18: 53–61.

Gautier, J. Y. 1978. Le comportement social de *Blaberus colosseus* en milieu naturel; plasticité du système social. Ins. Soc. 25: 289–301.

Gurney, A. B. 1959. The largest cockroach (Orthoptera, Blattoidea). Entomol. Soc. Wash. Proc. 61: 133–134.

Lefeuvre, J. C. 1960. A propos de *Blabera craniifer* Burmeister 1838 (Insecte dictyoptère). Soc. Scien. Bretagne Bull. 35: 145–161.

Piquett, P. G., and J. H. Fales. 1953. Life history of *Blaberus giganteus*. J. Econ. Entomol. 46: 1089–1090.

Roth, L. M. 1969. The male genitalia of Blattaria. I. *Blaberus* spp. (Blaberidae: Blaberinae). Psyche 76: 217–250.

Schal, C. 1983. *Blaberus giganteus* (cucaracha, giant cockroach, giant drummer, cockroach of the Divine Face) and *Xestoblatta hamata* (cucaracha). *In* D. H. Janzen, ed., Costa Rican natural history. Univ. Chicago Press, Chicago. Pp. 693–696.

Wendelken, P. W. 1977. The evolution of courtship phenomena in *Blaberus* and related genera with reference to sexual selection. Diss. Abstr. B37(8): 3816.

## Green Cockroaches

Panchloridae, Panchlorini, *Panchlora.*

The best known of the forty or so species in this genus (Gurney and Roth 1972) are pale, translucent green, or some shade of gray or cream, the fore wings also with some dark mottling. They are medium-sized (BWL 25–30 mm) and elongate-ellipsoidal in outline shape. Both sexes have complete wings and a smooth, un-marked head shield. Occasional pink individuals turn up, the color an apparent symptom of viral infection (Roth and Willis 1960). Curiously, the virus, *Serratia marcescens,* is also a human pathogen causing septicemia with a high mortality (Appel pers. comm.).

These conspicuous cockroaches are common throughout Latin America. Some species, such as *Panchlora nivea* (formerly *cubensis*) (fig. 5.10g), are transported widely by commerce, especially in shipments of tropical fruits. In nature, they live in the rotting trunks of palms where they feed on the decomposing brown fibers of the trunk, through which they make tunnels.

Females make a thin-walled, pale ootheca that may be extruded during formation and then retracted into a broad sac where the eggs develop. When the embryos mature, the ootheca is reextruded. The nymphs free themselves from their developing membranes as the ootheca is forced out and drop to the substratum (Roth and Willis 1958).

### References

GURNEY, A. B., AND L. M. ROTH. 1972. A generic review of the cockroaches of the subfamily Panchlorinae (Dictyoptera, Blattaria, Blaberidae). Entomol. Soc. Amer. Ann. 65: 521–532.

ROTH, L. M., AND E. R. WILLIS. 1958. The biology of *Panchlora nivea,* with observations on the eggs of other Blattaria. Amer. Entomol. Soc. Trans. 83: 195–208.

ROTH, L. M., AND E. R. WILLIS. 1960. The biotic associations of cockroaches. Smithsonian Misc. Coll. 141: 1–470.

## MANTIDS

Mantodea (= Mantoidea, Manteodea). *Spanish:* Mantidos, adivinadores (General); tata dios, mamboretás, comepiojos (Argentina); Santa Teresas (Peru). *Portuguese:* Louvas a Deus (Brazil). Preying mantids, praying mantids. Mantises.

Although many other orthopteroids take insects as food, they are basically vegetarian. The mantids, however, are all rapacious carnivores. The principal feature identifying them is their means of catching prey—the highly specialized, raptorial forelegs. The femur is large, carrying powerful and quick-acting muscles for closing the spined, grasping tibia against it, to form a vise from which escapes are few. Mantids wait in ambush for passing insect prey, often with their foreparts erect and foreleg segments closed and held in a praying position, for which they are called "praying mantids."

This and other humanoid behavior, such as the way they follow motion with their mobile, triangular heads and the large, "pupiled" eyes, has generated much folklore. Mantid means "soothsayer" or "diviner," and they are believed by some people to possess occult powers or to be worthy of being regarded as sacred. In Amazonia, the sex of an unborn child can be learned from a mantid placed nearby the expectant mother (*pōe mesa*). If it just moves its forelegs, the infant will be female; if the insect jumps onto someone, a male can be expected (Lenko and Papavero 1979: 11–12)

Mantids are mostly fairly large (BWL 3–15 cm), with an elongate prothorax and otherwise slender body and walking legs. Adults usually bear full wings, although the females' wings are often abbreviated. They are cryptically colored in leafy greens and browns. The dead leaf mantid (*Acanthops falcataria;* fig. 5.11a) is a common

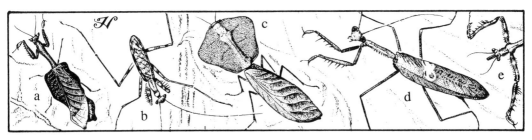

**Figure 5.11　MANTIDS (MANTIDAE).** (a) Dead leaf mantid (*Acanthops falcataria*). (b) Bark mantid (*Liturgusa* sp.). (c) Leaf mantid (*Choeradodis rhombicollis*). (d) Common mantid (*Stagmotoptera* sp.). (e) Horned mantid (*Vates* sp.).

example of the latter. It has broad, brown wings with the exact crinkly texture and twisted shape of a dead leaf. Others are marked with lichenose patterns and frequent tree trunks in moist forests, well camouflaged against the encrusted bark (e.g., *Liturgusa;* fig. 5.11b).

The leaf mantid (*Choeradodis rhombicollis;* fig. 5.11c) is unique in form (BWL 65–70 mm), the prothoracic dorsum being greatly expanded laterally to form a flat, rhomboid plate covering the head completely. The wings are wide and the whole depressed. With its green color and strongly veined wings, it looks incredibly like a living leaf.

Female mantids flood their eggs at the time they are laid with a whitish, frothy secretion from glands off the oviduct. They attach the mass to tree trunks, limbs, rocks, and other rigid substrata, often in very exposed situations. The secretion dries to form a hard, protective encasement for the eggs. Each species' case has its own characteristic shape.

Methods of defense employed by mantids have been well studied (Crane 1952, Robinson 1969). There are four general strategies: (1) resemblance to inanimate objects, including cryptic structural and color mimicry of leaves, sticks, bark, and so on, combined with stillness or swaying; (2) active flight, including dodging, jumping and dropping, threat and flying; (3) startling displays, consisting primarily of facing the enemy with wings raised and forelegs splayed apart; and (4) active attack by

striking with the forelegs. Some mantids have an imperfect eyespot in the middle of the hind wing. When the insect is molested, the wings are elevated and these spots threateningly displayed. Many also expose a dark marking on the inner surface of the fore femur when the legs are spread in a threat posture. Few adult mantids are involved in true mimicry complexes. In Belize, *Mantoida maya* nymphs, in their earliest stages, are very small and have the shape and attitude of *Camponotus* ants (Jackson and Drummond 1974).

The Neotropical Region is rich in its variety of mantids. The latest reviews (Beier 1933–1935, Giglio-Tos 1927) record approximately 300 species in some 74 genera. Many species certainly await discovery, and the group needs a comprehensive monograph. Some local reviews (e.g., Beebe et al. 1952) are helpful for the more common types. Dominant genera are *Stagmomantis* and *Stagmotoptera* (fig. 5.11d), which are large and usually green; *Liturgusa*, which are very common, small (BWL 2–3 cm), flattened forms that run actively on tree trunks; slender, sticklike *Angela* and *Vates* (fig. 5.11e), which have a sharp spine on the forehead and leaflike flanges on the mid- and hind leg segments.

### References

Beebe, W., J. Crane, and S. Hughes-Schrader. 1952. An annotated list of the mantids (Orthoptera, Mantoidea) of Trinidad, B.W.I. Zoologica 37: 245–258, Pl. I–VIII.

BEIER, M. 1933–1935. Mantodea. General Insectorum 196: 1–37; 197: 1–10; 198: 1–9; 200: 1–32; 201: 1–10; 203: 1–146.

CRANE, J. 1952. A comparative study of innate defensive behavior in Trinidad mantids (Orthoptera, Mantoidea). Zoologica 37: 259–293.

GIGLIO-TOS, E. 1927. Orthoptera, Mantidae. Das Tierreich 50: 1–707.

JACKSON, J. F., AND B. A. DRUMMOND III. 1974. A Batesian ant-mimicry complex from the Mountain Pine Ridge of British Honduras, with an example of transformational mimicry. Amer. Midl. Nat. 91: 248–251.

LENKO, K., AND N. PAPAVERO. 1979. Insetos no folclore. Conselho Estad. Artes Cien. Hum., São Paulo.

ROBINSON, M. H. 1969. The defensive behaviour of some orthopteroid insects from Panama. Royal Entomol. Soc. London Trans. 121: 281–303.

## EARWIGS

Dermaptera. *Spanish:* Tijeretas.
*Portuguese:* Tesouras.

Earwigs are most easily recognized by the pincers or forcepslike cerci borne on the end of the abdomen. Their tips are strongly incurved, and they often have internal teeth. The cerci, normally larger in males than females, are short to almost as long as the body. They are used to capture and manipulate prey as well as for defense.

Dermaptera are otherwise monotonously similar, somber-colored insects, most small to medium-sized (BL 1–2.5 cm), with chewing mouthparts, short threadlike or beadlike antennae, and unspecialized, similar legs. The fore wings are thickened and leathery and very short, meeting in a straight line down the back, almost like the elytra of some beetles. The hind, or flight wings are membranous and roughly circular and fold fanlike along many radially arranged creases. Appreciable anatomical variation in their body structure occurs primarily in the length and shape of the cerci, for example, in *Metresura ruficeps,* they are considerably longer than the body (fig. 5.12c). Some species have reddish (*Carcinophora americana;* fig. 5.12a) or yellow (*Doru lineare;* fig. 5.12b) areas on the fore wings.

Scientifically, this is one of the most neglected orders of insects, even though they are commonplace. Earwigs are, on the whole, not popular owing to their appearance and their unwelcome presence in gardens and homes. While they do cause some economic damage to stored food, roots, and shoots of tender young vegetables, some are predaceous and help to control populations of other more serious pests.

Urban earwigs are all adventives, thought to have made their way from Europe to Latin America with the earliest colonists,

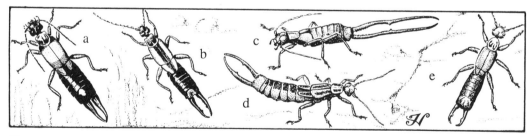

**Figure 5.12  EARWIGS.** (a) Giant earwig (*Carcinophora americana,* Anisolabiidae). (b) Lined earwig (*Doru lineare,* Forficulidae). (c) Earwig (*Metresura ruficeps,* Anisolabiidae). (d) Shore earwig (*Labidura riparia,* Labiduridae). (e) Flat earwig (*Sparatta pelvimetra,* Sparattidae).

possibly in the ballast of ships or on nursery stock. One of the most widespread is the ring-legged earwig (*Euborellia annulipes*), so called because of the faint dark band around the femora and tarsi of each leg. Also characteristic are the dark antennae with the third to fifth subapical segments contrastingly pale. It is medium-sized (BL 1.5 cm) and wingless. The forceps of both sexes are very short and stout.

The shore (or striped) earwig (*Labidura riparia;* fig. 5.12d) is another nearly ubiquitous species. It is much larger than the other adventive species (BL 2.5 cm) and much lighter colored, generally tan and with dark, longitudinal stripes on the wings and dorsum of the abdomen, these markings contrasting sharply with the lighter adjacent areas. The forceps of males are almost as long as the abdomen, slender, and smoothly incurved. The species prefers damp habitats near water, including the sea. It is a general predator and scavenger and thus is to be considered a beneficial insect. It is also a good flier and often attracted to lights (Gross and Spink 1971).

The maritime earwig (*Anisolabis maritima*) also is a cosmopolitan species but is not so widespread as the foregoing as it occurs only near the seashore. It is fairly large (BL 2 cm), wingless, and with a shiny black to dark brown body and pale yellowish legs. The forceps are short, stout, and strongly incurved in the male.

It is strange that the European earwig (*Forficula auricularia*), so common almost everwhere else, has not succeeded in invading Latin America. It is known there from only a few, far nothern localities.

In the Neotropics, a much more varied and less-known native fauna lives in all climes and situations (Brindle 1968). The 300 or so remaining species, in fifty-eight genera, mostly belong to the family Labiidae (Reichardt 1968–1971; Steinmann 1973, 1975). The biggest genus is *Marava*.

The Pygidicranidae is a family of primitive earwigs, containing several South American species, whose closest relatives are in southern Africa (Brindle 1984).

The native earwigs dwell in all sorts of damp, secluded habitats, under stones, in rotten wood, in abandoned termite nests, in cracks in rocks, and the like. *Sparatta* (fig. 5.12e) is particularly flattened as an adaptation for living under bark. Earwigs are generally more conspicuous in damp forests, but there are desert and mountain dwellers as well. In contrast to their dull-colored semidomestic counterparts, many are marked with bright patterns, often spotted red or yellow.

Typically, earwigs are active at night, when they forage for food. Although most are omnivorous, a number appear to be at least partly carnivorous, feeding primarily on other insects. They use their forceps to seize and hold the victim and curve the abdomen forward to access the mandibles.

## References

BRINDLE, A. 1968. The Dermaptera of Surinam and other Guyanas. Stud. Fauna Suriname Guyanas 36: 1–60.

BRINDLE, A. 1984. The Esphalmeninae (Dermaptera: Pygidicranidae): A group of Andean and southern African earwigs. Syst. Entomol. 9: 281–292.

GROSS, H. R., JR., AND W. T. SPINK. 1971. Flight habits of the striped earwig *Labidura riparia*. Entomol. Soc. Amer. Ann. 64: 746–748.

REICHARDT, H. 1968–1971. Catalogue of New World Dermaptera (Insecta). Dept. Zool. Sec. Agric. (São Paulo), Pap. Avul. Zool. 21: 183–193 (Pt. I. Introduction and Pygidicranoidea); 22: 35–46 (Pt. II. Labioidea, Carcinophoridae); 23: 83–109 (Pt. III. Labioidea, Labiidae); 24: 161–184 (Pt. IV: Forficuloidea); 24: 221–257 (Pt. V: Additions, corrections, bibliography and index).

STEINMANN, H. 1973. A zoogeographical checklist of world Dermaptera. Fol. Entomol. Hungarica 26: 145–154.

STEINMANN, H. 1975. Suprageneric classification of Dermaptera. Acad. Sci. Hungaricae Acta Zool. 21: 195–220.

# TERMITES

Isoptera. *Spanish:* Comejenes, hormigas blancas, palomillas de San Juan, polillas de la madera (General). *Portuguese:* cupins (sing. cupim), formigas brânças, formigas de asa, tucurus, aleluias do cupim (alates) (Brazil).

Termites are the analogues in tropical soils of earthworms in temperate regions. All Neotropical termites, except Kalotermitidae, live in the soil or maintain a close connection between nest and soil (pl. 1f). Their physical burrowing to construct nests and digestion of plant structural material (cellulose) add significantly to soil fertility and earn termites a place in nature as very beneficial insects, notwithstanding the few species that damage man's structures (Snyder 1924). Their role as soil animals has been studied more in other regions (Lee and Wood 1971) but is certainly similar in Latin America.

Recently, it has been realized that termites also have the potential to alter the environment in other ways, principally by releasing large amounts of methane, carbon dioxide, and hydrogen gases into the atmosphere as by-products of cellulose digestion. The greatest emissions come from natural tropical wet savanna and areas of human disturbance, such as cleared, burned, and cultivated lands, where abundant wood resources are available (Zimmerman et al. 1982).

Termites live in nests (termitaria) of their own construction. The form and location of these structures vary among the different kinds of termites. They may be wholly subterranean with mounds covering or linked to subsoil chambers by tunnels or arboreal masses off the ground but with runways communicating with the soil surface. Mounds may be large, rising 3 to 4 meters aboveground and forming conspicuous edifices in the landscape, particularly noticeable in open, flat country (Lacher et al. 1986). Such are the nests of *Cornitermes cumulans,* common in pastures, cultivated land, and savannas in southern Brazil (Redford 1984). Arboreal nests may also be large, obvious, ovoid structures, but these are lodged on branches of trees and shrubs.

The materials used for construction depend on the termites' feeding habits and availability. They usually consist of clay soil, excreta, and plant fragments mixed with saliva.

The nest generally has an inner labyrinth of chambers, including special central rooms for the royal pair, where eggs are laid and the brood is raised (and in some cases, where food is stored and fungus cultivated). This is surrounded by a protective wall, itself sometimes perforated with galleries that lead to the exterior. There may also be long tunnels running to the surface, along the ground, or on the trunks and main limbs of trees. There is no clear correlation between nest architecture and termite systematics (Noirot 1977).

It is difficult to generalize about the biology of the fauna because so very little is known about only a few species (Matthews 1977, Araújo 1970). *Amitermes* constructs nests with a very tall portion aboveground, some with umbrellalike lateral projections that function as rain-shedding devices. *Syntermes* (fig. 5.13a) is restricted to South America and is conspicuous because of the large size of individuals of most species (BL up to 17 mm or more) and the enormous volcano-shaped nests of some. Workers cut fragments of leaves and grass stalks and transport them to undergound galleries where they become stores for later consumption. *Neocapritermes* soldiers (fig. 5.13b) carry large, asymmetrical mandibles that may be snapped crosswise explosively, emitting an audible click and driving the sharp angulate tips into the skin of any animal holding it.

Nests of many species provide an inviting abode for an assemblage of other higher animals, including nesting birds.

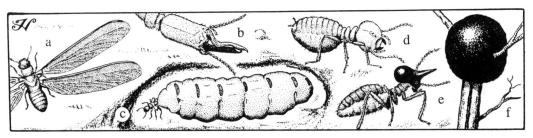

**Figure 5.13  TERMITES.** (a) Common tropical termite (*Syntermes dirus,* Termitidae), reproductive male. (b) Crooked jaw termite (*Neocapritermes braziliensis,* Termitidae), head of worker. (c) Common tropical termite (*Syntermes* sp., Termitidae), physogastric queen from macropterous female. (d) Nasute termite (*Nasutitermes costalis,* Termitidae) worker. (e) Nasute termite soldier. (f) Nasute termite nest.

Many reptiles, such as legless lizards, feed on termites and occupy termitaria almost continuously. In Amazonia, in the forest away from normal sunlit riverbanks, females of the crocodilian *Paleosuchus trigonatus* find ground nests a convenient incubator for their eggs (Magnusson et al. 1985). According to locals, the Amazonian tortoise (*Geochelone*) also employs nests in this manner (orig. obs.). Many mammals specialize on termites as food, particularly the so-called anteaters. They eat more termites than ants, tearing open the nests with their powerful front claws.

There are some true commensal termitophiles also among the insects and arthropods. These are other termites, silverfish, scale insects (*Termitococcus,* Margarodidae), and even tiger beetles that live in intimate association with termite colonies. Many beetles, particularly some rove beetles, are highly modified for life among termites (Kistner 1969). Abondoned nests of one kind of termite can be taken over by termites of a different species.

As is true of termites in other parts of the world, there are those habitually found in man-made structures that can cause considerable damage when they feed on important wooden members. Some of these, such as *Cryptotermes brevis* (Wolcott 1957) and *Coptotermes havilandi,* are introduced from other places in the world, but some indigenous species are major pests as

well, for example, *Incisitermes snyderi* and *Coptotermes niger.*

Termites are social insects and exhibit complex group behavior similar to that of ants and social bees and wasps. They are fundamentally different, however, in having gradual metamorphosis and thus not having to provide for helpless larvae and pupae. Individuals comprising a colony are really members of a large family of sibling progeny started by a single paired winged reproductive male (fig. 5.13a) and female. The female queen is accompanied throughout her life by the attendant male and may live many years, growing in size to tremendous proportions (fig. 5.13c). The offspring consist of morphologically and functionally different castes that may be produced from newly hatched immatures, depending on the requirements of the colony. Control of this development depends on pheromones given off by members of the reproductive castes.

Sterile castes are the workers and soldiers, wingless individuals in which the growth of the reproductive organs is suppressed. The former are the most numerous and generalized in form. They are responsible for all foraging activity and care for the eggs, larvae, and queen.

Soldiers have similar bodies but highly modified mouthparts and an enlarged, strongly sclerotized head. There are two well-defined types, those with large prominent mandibles and the nasutes with a snoutlike prolongation on the front of the head and vestigial mandibles. Both actively defend the nest against attackers, by biting and pinching or by the emission of viscid, noxious secretions (Deligne et al. 1981; Prestwich 1983a, 1983b). The head of the soldiers is so modified that they cannot feed themselves.

Depending on the species, the colony's primary food is dead wood and other vegetative parts of plants (usually dry), roots, humus, dung, and fungi, although the last are not cultured in Neotropical forms as in some in Africa. To a large extent, termites are dependent on unique, symbiotic flagellate protozoa and bacteria for digestion of cellulose and other complex polysaccharides in their diet. Members of the family Termitidae lack the protozoa, although bacteria are present which assume the same digestive function.

In times past in Brazil, large, hard termitaria have been hollowed out and used for ovens (Southy, in Cowan 1865: 134). The same nests were pulverized and used as a kind of cement to make "concrete" floors for the early settlers of this land.

There is just one review of Neotropical termites (Araújo 1970). In the Neotropical region, there are sixty-two genera, containing 408 species (Araújo 1970, 1972, 1977). Most belong to the family Termitidae. General information on these insects is available in Krishna and Weesner (1969–70). Very useful bibliographies of termite literature to 1978 have been compiled by Snyder (1956, 1961, 1968) and Ernst and Araújo (1986).

## References

Araújo, R. L. 1970. Termites of the Neotropical Region. In K. Krishna and F. M. Weesner, Biology of termites. 2: 527–576. Academic, New York.

Araújo, R. L. 1972. Súmula faunistica dos Isoptera americanos. Cien. Cult. 24(3): 253–256.

Araújo, R. L. 1977. Catálogo dos Isoptera do novo mundo. Acad. Brasileira Ciên., Rio de Janeiro.

Cowan, F. 1865. Curious facts in the history of insects. Lippincott, Philadelphia.

Deligne, J., A. Quennedey, and M. S. Blum. 1981. The enemies and defense mechanisms of termites. In H. R. Hermann, Social insects. 2: 1–76. Academic, New York.

Ernst, E., and R. L. Araújo. 1986. A bibliography of termite literature 1966–1978. Wiley & Sons, Somerset, N.J.

Kistner, D. H. 1969. The biology of termitophiles. In K. Krishna and F. M. Weesner, Biology of termites. 1: 525–557. Academic, New York.

Krishna, K., and F. M. Weesner. 1969–70. Biology of termites. 2 vols. Academic, New York.

Lacher, Jr., T. E., I. Egler, C. J. R. Alho, and M. A. Mares. 1986. Termite community composition and mound characteristics in two grassland formations in central Brazil. Biotropica 18: 356–359.

Lee, K. E., and T. G. Wood. 1971. Termites and soils. Academic, London.

Magnusson, W. E., A. P. Lima, and R. M. Sampaio. 1985. Sources of heat for nests of Paleosuchus trigonatus and a review of crocodilian nest temperatures. J. Herpet. 19: 199–207.

Matthews, A. G. A. 1977. Studies on termites from the Mato Grosso State, Brazil. Acad. Brasileira Cien., Rio de Janeiro.

Noirot, C. 1977. Nest construction and phylogeny in termites. 8th Int. Union Study Soc. Ins. (Netherlands 1977) Proc. Pp. 177–180.

Prestwich, G. D. 1983a. Chemical systematics of termite exocrine secretions. Ann. Rev. Ecol. Syst. 14: 287–311.

Prestwich, G. D. 1983b. The chemical defenses of Termites. Sci. Amer. 249: 78–87.

Redford, K. H. 1984. The termitaria of Cornitermes cumulans (Isoptera, Termitidae) and their role in determining a potential keystone species. Biotropica 16: 112–119.

Snyder, T. E. 1924. Damage by termites in the Canal Zone and Panama and how to prevent it. U.S. Dept. Agric. Dept. Bull. 1232: 1–25.

Snyder, T. E. 1956. Annotated, subject-heading bibliography of termites 1350 B.C. to A.D. 1954. Smithsonian Misc. Coll. 130: 1–305.

Snyder, T. E. 1961. Supplement to the annotated, subject-heading bibliography of termites 1955 B.C. to A.D. 1960. Smithsonian Misc. Coll. 143: 1–137.

Snyder, T. E. 1968. Second supplement to the annotated subject-heading bibliography of termites 1961–1965. Smithsonian Misc. Coll. 152(3): 1–188.

Wolcott, G. N. 1957. Inherent natural resistance of woods to the attack of the West Indian dry-wood termite, *Cryptotermes brevis* Walker. J. Agric. Univ. Puerto Rico 41: 259–311.

Zimmerman, P. R., J. P. Greenberg, S. D. Wandiga, and P. J. Crutzen. 1982. Termites: A potentially large source of atmospheric methane, carbon dioxide, and molecular hydrogen. Science 218: 563–565.

## Nasute Termites

Termitidae, Nasutitermitinae, *Nasutitermes.*
*Spanish:* Comejénes común (General).

The large (1–2 m diameter), round or ovoid, dark brown nests of these ubiquitous tropical termites punctuate the moist lowland landscape throughout Latin America (fig. 5.13f, pl. 4a). They are constructed of a paste of chewed wood and fecal cement put into place by myriad workers. The paste hardens into a carton substance that is soft and papery on the outside and increasingly harder toward the inside. The outer envelope is continuous, forming a protective shell for the labyrinthine interior. Nests undergo continual expansion during their existence (Jones 1979).

Runways, covered with this same carton substance, lead from the main nest to remote foraging sites. These tunnels end abruptly on the surfaces of dead wood which the workers penetrate to feed. The presence of these runways is what distinguishes termite nests from the other arboreal paper or carton ant and wasp nests with which they are often confused. Not all species build such elevated structures; some, such as *Nasutitermes fulviceps*, live on the soil surface and are partly subterranean (Talice et al. 1969).

The tunnels may also lead to fence posts, telephone poles, and the foundations of houses, whose substance is devoured with equal gusto. Consequently, these termites are considered pests and often require control. The evidence for a beneficial role for *Nasutitermes* as nitrogen fixers is inconclusive (Bradley et al. 1983).

The soldiers are best known and are usually seen when they appear at the surface of nests that are being damaged. They are small (BL 3–4 mm), with brownish, pigmented bodies and dark brown heads bearing a conspicuous, elongate beak on the front (fig. 5.13e). They also possess an effective chemical defense. When a break occurs in the nest surface, they immediately swarm to the site and remain there until the intruder leaves or until they are ravished. As a means of repelling attackers, they ooze or squirt an irritating, thick, entangling substance (nasute glue) from their long pointed snouts. This sticky, smelly secretion is produced by large glands in the head and contains volatile terpenoids (Prestwich 1982). It is used mostly against other insect invaders of the colony but is effective also in warding off the attacks of termitophagous vertebrates, including anteaters (*Tamandua, Myrmecophaga*). These chemicals are topical irritants that affect the skin and mucous membranes of the nostrils and mouth (Lubin and Montgomery 1981). Incursions by animals and the formation of nest holes by trogons, parakeets, and other birds often do considerable damage to nests and may force abandonment by their occupants. Ants of the genus *Azteca* are sometimes found living within *Nasutitermes* nests and may exclude the owners from their rightful home.

The workers (fig. 5.13d) are about the same size as the soldiers but have a round head, are pale, and generally remain deep within the nest, caring for young and the queen, even during times of outward threat. Although workers have no special weapons, they can bite effectively and sometimes join with the soldiers in aggressive encounters, especially against other termites (Thorne 1982).

Queens may attain great size (BL 3–6 cm) by the enlargement of the abdomen with fat and eggs. Normally, there is one per colony, but *Nasutitermes corniger* is known to be facultatively polygynous (multiple queens) (Thorne 1982).

Members vary greatly in number, usually from five thousand to six thousand, depending on age, species, and health of the colony. But much more numerous populations are possible. Some *Nasutitermes corniger* nests may contain 800,000 to a million individuals (Thorne and Noirot 1982, Laffitte and Aber de Szterman 1976).

Reproduction is seasonal. Swarming of the winged male and female reproductive stages usually occurs after the first showers of the incipient rainy season and may continue for many months into the wet part of the year.

Many of the sixty-seven Neotropical species of *Nasutitermes* place their nests in trees, or at least off the ground in shrubby vegetation. The walls of the rare terrestrial nests are shown to contain nutrients in excess of the surrounding soil, thus concentrating substrate richness for plant growth and contributing to patchy vegetation patterns in the Amazon Basin (Salick et al. 1983, Goodland 1965). General information on the genus is sparse (Lubin 1983, Araújo 1970).

## References

Araújo, R. L. 1970. Termites of the Neotropical Region. *In* K. Krishna and F. M. Weesner, Biology of termites. 2: 527–576. Academic, New York.

Bradley, R. S., L. A. de Oliveira, and A. Gomez Bandeira. 1983. Nitrogen fixation in *Nasutitermes* in central Amazonia. *In* P. Jaisson, Social insects in the tropics. 1st Int. Symp. (Cocoyoc 1980), Int. Union Stud. Soc. Ins. and Soc. Mexicana Entomol. Proc. 2: 235–244.

Goodland, R. J. A. 1965. On termitaria in a savanna ecosystem. Can. J. Zool. 46: 641–650.

Jones, R. J. 1979. Expansion of the nest of *Nasutitermes costalis*. Ins. Soc. 26: 322–342.

Laffitte, S. and A. Aber de Szterman. 1976. Comportamiento interespecifico en *Nasutitermes fulviceps* (Silvestri, 1901) con otras termites. Rev. Biol. Uruguay 4: 59–65.

Lubin, Y. D. 1983. *Nasutitermes* (comején, hormiga blanca, nasute termite). *In* D. H. Janzen, ed., Costa Rican natural history. Univ. Chicago Press, Chicago. Pp. 743–744.

Lubin, Y. D., and G. G. Montgomery. 1981. Defenses of *Nasutitermes* termites (Isoptera, Termitidae) against Tamandua anteaters (Edentata, Myrmecophagidae). Biotropica 13: 66–76.

McMahan, E. A. 1970. Radiation and the termites at El Verde. *In* H. T. Odum, ed., A tropical rain forest: A study of irradiation and ecology at El Verde, Puerto Rico. U.S. AEC, Washington, D.C. Pp. E-105–122.

Prestwich, G. D. 1982. From tetracycles to marcocycles: Chemical diversity in the defense secretions of nasute termites. Tetrahedron 38: 1911–1919.

Salick, J., R. Herrera, and C. F. Jordan. 1983. Termitaria: Nutrient patchiness in nutrient-deficient rain forests. Biotropica 15: 1–7.

Talice, R. V., S. L. Mosera, R. Caprio, and A. M. S. de Sprechmann. 1969. Estructura de los termiteros. Univ. Uruguay, Dept. Biol. Gen. Exper. Publ. 2: 1–20.

Thorne, B. L. 1982. Termite-termite interactions: Workers as an agonistic caste. Psyche 89: 133–150.

Thorne, B. L. 1982. Polygyny in termites: Multiple primary queens in colonies of *Nasutitermes corniger* (Motschuls) (Isoptera: Termitidae). Ins. Soc. 29: 102–117.

Thorne, B. L., and C. Noirot. 1982. Ergatoid reproduction in *Nasutitermes corniger* (Motschulsky): Isoptera, Termitidae. J. Ins. Morph. Embryol. 11: 213–226.

# WEB SPINNERS

Embiidina (= Embioptera, Embiodea). Embiids.

While the majority of embiids (Ross 1970) are secretive and unknown except to the specialist, some species are very conspicuous because of the extensive webs they construct on tree trunks and limbs. At times, almost the entire boles of large trees may be covered with these filmy mats that show little organized structure save branching galleries. It is within these passages that the web spinners live, and they may be seen through the walls as they move back and forth.

**Figure 5.14   INSECTS OF VARIOUS ORDERS.** (a) Web spinner (*Clothoda urichi*, Clothodidae). (b) Barklouse (*Poecilopsocus iridescens*, Psocidae). (c) Booklouse (*Liposcelis bostrychophila*, Lipuscelidae). (d) Black hunter (*Leptothrips mali*, Phlaeothripidae). (e) Greenhouse thrip (*Heliothrips haemorrhoidalis*, Thripidae).

The silk forming these labyrinths issues from glands in the basal segment of the forelegs. This segment is inflated in nymphs and adults and clearly distinguishes these insects from all others. Other identifying features are the usually complex, asymmetrical genitalia of the male, and, in alate species (many lack wings altogether or have only small alar buds), wings that have pigmented bands along the veins alternating with clear stripes. Web spinners are mandibulate with filiform antennae with numerous segments.

A web spinner's body is elongate and very supple. Flexible wings and short legs allow it to move with great ease, even backward as easily as forward, through its galleries. These insects are well protected by their ability to retreat deeply within their silken tent, which not only forms a physical barrier to entry by such primary enemies as ants but hides them from the eyes of larger predators.

Embiids are gregarious. Typical colonies consist of a single female living in the midst of its brood (Edgerly 1988). Although they share in common a complex of galleries, they should be considered subsocial, for they lack castes, division of labor, or other characteristics of the true insect societies. At least one Trinidadian web spinner, *Clothoda urichi* (fig. 5.14a), is facultatively communal (Edgerly 1987).

The food of embiids consists of bark, dead leaves, moss, lichens, and other organic matter of plant origin that they find in their immediate habitat.

Although Latin America is a major center of evolution of the order, the embiid fauna has been studied only to a limited degree (Ross 1943, 1944, 1984). Several hundred may actually exist, but only about 150 species in five families are presently described. The family Clothodidae is confined to South America (including Trinidad and Panama). *Chelicera* (Anisembiidae) is the dominant genus, with many species in semiarid environments.

Web spinners are found in widely varied habitats, from humid forests to deserts. One species is known from the Galápagos Islands, another from the fog-dampened lomas of coastal Peru (Ross 1966). Around human habitations, the most common is *Oligotoma saundersii*, a "weed species," spread by man from India. Its males are attracted to lights.

## References

EDGERLY, J. S. 1987. Colony composition and some costs and benefits of facultatively communal behavior in a Trinidadian webspinner *Clothoda urichi* (Embiidina: Clothodidae). Entomol. Soc. Amer. Ann. 80: 29–34.

EDGERLY, J. S. 1988. Maternal behavior of a webspinner (Order Embiidina): Mother-nymph associations. Ecol. Entomol. 13: 263–272.

Ross, E. S. 1943. Métodos de recolección, crianza y estudio de los Embiópteros (Ins. Embioptera). Rev. Entomol. 14: 441–446.

Ross, E. S. 1944. A revision of the Embioptera

of the New World. U.S. Natl. Mus. Proc. 94: 401–504.

of the New World. U.S. Natl. Mus. Proc. 94: 401–504.

Ross, E. S. 1966. A new species of Embioptera from the Galápagos Islands. Calif. Acad. Sci. Proc. (Ser. 4) 34: 499–504.

Ross, E. S. 1970. Biosystematics of the Embioptera. Ann. Rev. Entomol. 15: 157–171.

Ross, E. S. 1984. A classification of the Embiidina of Mexico with descriptions of new taxa. Calif. Acad. Sci. Occ. Pap. 140: 1–50.

# HEMIPTEROIDS

## PSOCIDS

Psocoptera (= Corrodentia). Barklice.

Psocids (New 1987) are free-living insects that feed on microflora and organic debris on surfaces of vegetation or on other surfaces. The range of food includes fungi (hyphae and spores), yeasts, lichens, or fragments of animal or vegetable matter. Most are arboreal (Thornton 1985) and are found on the bark or leaves of trees, but many also occur in ground litter. There tends to be a higher proportion of leaf frequenters in the canopy than near the ground in Neotropical forests that have been sampled for psocids (Broadhead and Evans 1979, Broadhead and Wolda 1985, Wolda and Broadhead 1985). Some dwell in the nests of mammals or birds, but none are parasitic like their close relatives, the biting and sucking lice. Members of the family Trogiidae make sounds by drumming the abdomen against the substrate. Psocids are frequently gregarious as nymphs or adults or both and may even group together under a communal web.

Several cosmopolitan types are commonly found indoors under humid conditions and are considered household pests. These are sometimes called "booklice" (fig. 5.14c) because of their habit of feeding on paper, sizing, and glue in book bindings (Broadhead 1946). They are minute insects (BL 1 mm or less) but are usually noticeable against a white paper background. All are similar wingless (or near wingless) forms, brownish in color, and with slender legs. There are several species whose proper names are confused in the literature. The widely used name *Liposcelis divinatoria* has been declared invalid by psocid taxonomists; *L. bastrychophila* is the species to which it formerly referred (Lienhard 1990). These insects are apparently widespread in Latin America, but the real extent of their occurrence has not been documented because of a lack of collecting and the uncertainty of their identification.

Psocids are related to lice (Lyal 1985) and are louselike in general appearance, but adults usually have wings that are held rooflike over the abdomen when at rest. The wings have few veins, and the fore pair are much larger than the hind pair. Polymorphism is common in some families, the usual alternate form involving the reduction or loss of wings. A unique development is the bulging clypeal region on the front of the round head, which is unusually movable at the neck for an insect. The prothorax is reduced. The legs are slender and simple with a reduced number of tarsal segments.

Although some common pest species are minute, most wild psocids are small (BL 1–2 mm) and drably colored, gray or brown, frail insects. In humid tropical lowland forests, there are some much larger and quite colorful forms. *Poecilopsocus iridescens*, Psocidae (fig. 5.14b), of Amazonian Peru is approximately 12 millimeters long, with dark blue, white, and red wing markings and long antennae. Possibly, they are mimics of mirid or reduviid bugs (Mockford pers. comm.).

This is a much larger order in Latin America than indicated by published lists (e.g., Smithers 1967). At present, there are at least 780 species described in 96 genera and several hundred more that are certain to be found (Mockford pers. comm.). Some speciose, typical regional genera are *Thrysophorus*, *Ceratipsocus*, and *Graphocaecilius*.

A great many species certainly are undiscovered in the tropical portions of the region. As with other insect groups, there is a progressive increase in psocid diversity from temperate to tropical forests (Broadhead 1983). Some show amphinotic Gondwanaland distributions with close relatives in Australia (e.g., *Drymopsocus*, Elipsocidae), but the southern temperate groups are mostly endemic and distinct from the rest of America to the north (New 1987).

## References

BROADHEAD, E. 1946. The book louse and other library pests. Brit. Book News 68: 77–81.

BROADHEAD, E. 1983. The assessment of faunal diversity and guild size in tropical forests with particular reference to the Psocoptera. *In* S. L. Sutton, T. C. Whitmore, and A. C. Chadwick, eds. Tropical rain forest: Ecology and management. Blackwell, Oxford. Pp. 107–119.

BROADHEAD, E., AND H. A. EVANS. 1979. The diversity and ecology of Psocoptera in tropical forests. 4th Int. Symp. Trop. Ecol. [Panama] Acta 1: 185–196.

BROADHEAD, E., AND H. WOLDA. 1985. The diversity of Psocoptera in two tropical forests in Panama. J. Anim. Ecol. 54: 739–754.

LIENHARD, C. 1990. Revision of the western Palaearctic species of *Liposcelis* Motschulsky (Psocoptera: Liposcelidae). Zool. Lb. Sys. 117: 117–174.

LYAL, C. H. C. 1985. Phylogeny and classification of the Psocodea, with particular reference to the lice (Psocodea: Phthiraptera). Syst. Entomol. 10: 145–165.

NEW, T. R. 1987. Biology of the Psocoptera. Oriental Ins. 21: 1–109.

SMITHERS, C. N. 1967. A catalogue of the Psocoptera of the world. Austr. Zool. 14: 1–145.

THORNTON, I. W. B. 1985. The geographical and ecological distribution of arboreal Psocoptera. Ann. Rev. Entomol. 30: 175–196.

WOLDA, H., AND E. BROADHEAD. 1985. Seasonality of Psocoptera in two tropical forests in Panama. J. Anim. Ecol. 54: 519–530.

## THRIPS

Thysanoptera.

These are all very small insects (BL of most 1–2 mm), although "giant" forms are found in the tropical forests of Latin America. The largest is the Peruvian *Dasythrips regalis*, which reaches a body length of 12 millimeters. The order is only recently becoming well known generally (Ananthakrishnan 1984, Lewis 1973).

Thrips are characterized structurally mostly by their unique wings; both pairs are very slender and elongate, without well-defined or extensive venation and with very long hair fringes. Many species are wingless, however, and other features, such as the asymmetric mouthparts located on a conical beak on the underside of the head, must be called on to define them. Only the mandible of the left side is developed and is used to punch holes in the epidermis of plants to release the sap, which is then sucked up. They also have a protrusible, saclike pad at the apex of each leg.

Most thrips live on plants, from which they take their liquid nourishment. The banana flower thrips (*Frankliniella parvula*; Harrison 1963) and others are often common on flowers, where their feeding may result in injured fruit; some hide in curled leaves (called *queima* in Brazil) or galls, which are caused by their feeding. The vegetarians may be very numerous and cause extensive economic damage to commercially valuable plants directly or by introducing pathogenic microorganisms. A large number of species are associated with the coconut palm (Sakimura 1986).

A very different group lives on dead twigs and among leaf litter and soil where they are predaceous on other minute insects and mites or feed on the fungi (hyphae and spores) associated with the early stages of decay (Mound 1977). Some of the predatory types are considered beneficial when they attack pests. An example is the black hunter (*Leptothrips mali*; fig. 5.14d), which takes all sorts of injurious insects, including aphids, scale insects, mites, and other thrips. Some species feed on termites. Thrips also are pollinators, for example, the banana flower thrips, which frequents the flowers of cacao in Trinidad (Billes 1941).

Development in some members of the order exhibits a parallel with that of the holometabolous insects, the last nymphal instar being quiescent and resembling a pupa.

The higher classification of the order has recently been clarified (Mound et al. 1980). Most of the eight families have Latin American representatives. However, primarily only the species with pest status are known. The majority belong to the two ubiquitous families Thripidae and Phlaeothripidae. These include some cosmotropical species such as the greenhouse thrips (*Heliothrips haemorrhoidalis;* fig. 5.14e), citrus thrips (*Scirtothrips*), tobacco and cotton thrips (*Frankliniella* and *Thrips*), gladiolus thrips (*Taeniothrips simplex*), and banana thrips (*Chaetanaphothrips*).

Native species are very poorly known in the region. A bizarre Mexican and Jamaican thrip is *Arachisothrips*, in which the leading edge of the fore wing is ballooned into a hollow, peanut-shaped outgrowth with a reticulate surface (Stannard 1952). It lives in rain forest ground cover, but the adaptiveness of this strange structural feature is unknown. One Brazilian species, with its nearest relative in Singapore, comprises the aberrant family Uzelothripidae. *Franklinothrips vespiformis* and relatives mimic various genera of ants in Mexico (Johansen 1983).

## References

ANATHAKRISHNAN, T. N. 1984. Bioecology of thrips. Indira, Oak Park, Mich.

BILLES, D. J. 1941. Pollination of *Theobroma cacao* L. in Trinidad, B.W.I. Trop. Agric. (Trinidad) 18: 151–156.

HARRISON, J. O. 1963. Notes on the biology of the banana flower thrips *Frankliniella parvula*, in the Dominican Republic. Entomol. Soc. Amer. Ann. 56: 664–666.

JOHANSEN, R. M. 1983. Nuevos estudios acerca del mimetismo en el género *Franklinothrips* Back (Insecta: Thysanoptera), en Mexico. Inst. Biol. Univ. Nac. Aut. México Ann. (Ser. Zool. 1) 53: 133–156.

LEWIS, T. 1973. Thrips, their biology, ecology and economic importance. Academic, London.

MOUND, L. A. 1977. Species diversity and the systematics of some New World leaf litter Thysanoptera (Phlaeothripinae; Glytothripini). Syst. Entomol. 2: 225–244.

MOUND, L. A., B. S. HEMING, AND J. M. PALMER. 1980. Phylogenetic relationships between the families of recent Thysanoptera (Insecta). Zool. J. Linnean Soc. 69: 111–141.

SAKIMURA, K. 1986. Thrips in and around the coconut plantations in Jamaica, with a few taxonomic notes (Thysanoptera). Fla. Entomol. 69: 348–363.

STANNARD, L. J. 1952. Peanut-winged thrips. Entomol. Soc. Amer. Ann. 45: 327–330.

# NERVE-WINGED INSECTS

Neuroptera.

The most characteristic feature of adult neuropterans is their well-developed wing venation, with highly complex vein-branching patterns, end twigging of the main veins, diverse polygonal cells in the middle areas, and frequent stair-step pathways of many vein branches. A few small groups are atypical, however, in having much simplified venation or wings very much reduced in size (brachypterous). Four equal size and shape wings are the rule; they are held roofwise over the body when not in use for flight. The wing membrane is usually clear but may sometimes be brown pigmented or rarely may have bright color spots and fields; it is completely whitish opaque in the dusty wings (Coniopterygidae). The mouthparts are mandibulate, and the head possesses long, many-segmented, filiform antennae and well-formed compound eyes. A few families have well-developed ovipositors.

Most neuropteran adults are rather drably colored in cryptic greens, browns, and grays. Several unrelated types, however, form an aposematic mimetic complex in southern South America, all having a similar pair of bright red stripes on the prothorax near the openings of scent glands that produce a noxious, skunklike (contains skatol) odor. Some also have conspicuous patterns on the wings. Among these are ant lions (Myrmeleontidae), such as *Dimares, Glenurus,* and *Maracandula,* some chrysopids, and mantispids. The mantispid genus *Climaciella* mimics large social wasps, whereas the genus *Anchieta* resembles stingless bees. The largest mantispid in the world, *Drepanicus gayi* from Chile, looks like a green katydid. Ant lions are cryptically patterned to match their daytime resting sites (Stange 1970).

The general form of the larvae varies greatly, but many resemble ground beetle larvae. Some have peculiarly modified elongate, curved jaws, formed like tongs to grasp insect prey. Each has an internal canal through which the juices of the food are siphoned. Many larvae are hairy or spiny.

The larvae of some Chrysopidae bear hooked bristles on their backs to which they fix minute bits and pieces of debris to give them a kind of camouflage (a habit that parallels the decorator crabs in coral reefs and the North American *Chrysopa slossonae,* which attaches bits of the wax secretions of its woolly aphid prey to its body to mask it from recognition by the ants that protect this aphid (Eisner et al.

1978). Such "trash-carrying lacewings" belong to the genera *Leucochrysa* and *Ceraeochrysa.* Larval habitats vary considerably and include vegetation, sandy soil, bark crevices, and cavities under objects on the ground. A few ant lion larvae (*Glenurus*) are also trash carriers.

Hardly anything has been published on the biology of the immatures of the Neotropical fauna except for a few beneficial types such as the Chrysopidae (Nuñez 1989a, 1989b) and Hemerobiidae, the larvae of which are voracious predators of homopterous pests (aphids and scale insects, primarily) and thus of considerable value as biocontrol agents (New 1975). Species of *Chrysoperla* (fig. 5.15b) are even reared in insectaries to be broadcast on crops for this purpose. Mantispids (fig. 5.15a) are spider egg predators (Birabén 1960) or bee parasites (Parker and Stange 1965).

Penny (1977) lists approximately 950 Latin American species of Neuroptera. These are distributed among eleven families, the largest and most common of which are the ant lions (Myrmeleontidae), dusty wings (Coniopterygidae), and green lacewings (Chrysopidae). There are three neuropteran subfamilies found only in the Neotropical Region: the Platymantispinae, of uncertain affinity but usually placed in the Mantispidae, are strange subterranean predators (Parker and Stange 1965); the

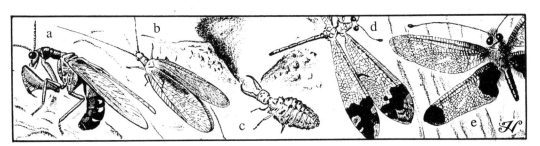

**Figure 5.15    NEUROPTEROUS INSECTS.** (a) Mantispid (*Climaciella* sp., Mantispidae). (b) Green lacewing (*Chrysoperla* sp., Chrysopidae). (c) Ant lion (*Myrmeleon* sp., Myrmeleontidae), larva in pit. (d) Ant lion (*Glenurus peculiaris,* Myrmeleontidae). (e) Owlfly (*Corduleceris maclachlani,* Ascalaphidae).

Brucheiserinae, two species of dusty wings found under rocks and possibly flightless; and finally the Albardinae, containing one unique, highly modified owlfly species (Penny 1985).

## References

BIRABÉN, M. 1960. Mantispa parásita en el cocón de *Metepeira*. Neotropica 6: 61–64.

EISNER, T., K. HICKS, M. EISNER, AND D. S. ROBINSON. 1978. "Wolf-in-sheep's-clothing" strategy of a predaceous insect larva. Science 199: 790–794.

NEW, T. R. 1975. The biology of Chrysopidae and Hemerobiidae (Neuroptera), with reference to their usage as biocontrol agents: A review. Royal Entomol Soc. London Trans. 127: 115–140.

NUÑEZ, E. 1989a. Chrysopidae (Neuroptera) del Perú y sus especies más comunes. Rev. Peruana Entomol. 31: 69–75.

NUÑEZ, E. 1989b. Cíclo biológico y crianza de *Chrysoperla externa* y *Ceraeochrysa cincta* (Neuroptera, Chrysopidae). Rev. Peruana Entomol. 31: 76–82.

PARKER, F. D., AND L. A. STANGE. 1965. Systematic and biological notes on the tribe Platymantispini and the description of a new species of *Plega* from Mexico. Can. Entomol. 97: 604–612.

PENNY, N. D. 1977. Lista de Megaloptera, Neuroptera e Raphidioptera do México, América Central, ilhas Caraíbas e América do sul. Acta Amazonica 7 (4) suppl.: 1–61.

PENNY, N. D. 1985 [1983]. Neuroptera of the Amazon Basin. Pt. 9. Albardiinae. Acta Amazonica 13(3–4): 697–699.

STANGE, L. A. 1970. Revision of the ant-lion tribe Brachynemurini of North America. Univ. Calif. Publ. Entomol. 55: 1–192.

## Ant Lions

Neuroptera, Myrmeleontidae.

These are probably the best-known neuropterans, not because of the adults but because of the work of the larvae, the familiar, funnel-shaped ant lion pits commonly seen in fine sandy soil. These are constructed in places protected from rain: under staircases and by the edges of elevated buildings, beneath overhanging rocks or logs, and near the bases of trees. They may be very numerous and occur in large, concentrated groups at times (McClure 1976). Only larvae of the genus *Myrmeleon* make these death traps, into which ants or other small, terrestrial insects fall (Wilson 1974). (Pit making is found in a few other groups [*Brachynemurus* in Argentina and Mexico], but tubes extend downward from the pits.) The escape of the hapless insect is prevented by the constantly failing sloped sides of the pit. The ant lion at the bottom also flicks sand onto the struggling prey to dislodge it and cause it to tumble down to the neck of the funnel. The larva waits there, buried, only its ice-tong jaws projecting above the surface. These formidable structures close on the prey, the tips piercing its body, and take its blood through internal canals. These larvae only move backward, a trait shared by the giant *Vella* larvae that often prey on them (Stange pers. comm.). The meandering subsurface burrows of the latter common Neotropical genus are often evident in loose sandy areas.

Other ant lion larvae merely burrow in sand or loose soil, searching for other subterranean insects to be captured directly and eaten. Many can dig rapidly in reverse by repeatedly arching the abdomen forward under the body. Some ant lions that live in dry tree holes (e.g., some *Glenurus;* Miller and Stange 1983) or on tree bark or bare rock surfaces (*Navasoleon;* see Miller and Stange 1985) have diminished or lost their ability to dig backward.

Ant lion larvae (fig. 5.15c) are small (BL 5-10 mm), ovoid creatures with a boxlike head and typical neuropteran sickle-shaped jaws. They usually have enlarged bristles at the rear end, employed in rapid digging, and have their eyes on stalks. The abdomen is bulbous and the anterior portion steep, so that the head and thorax emerge very low.

Adult myrmeleontids are mostly fairly

large (BL 3–4 cm), with a very long, slender abdomen, sometimes twice as long as the wings in the male. The wing venation is very complex: there is a very long cell under the stigma spot in both wings; the conspicuous vein forking in the middle of the wings near the base looks identical in the fore wing and hind wing but actually involves totally different crossveins (Stange 1970); the subcostal area of the wing is the only part lacking cross veins. The antennae are short, thickened, and expanded at the tip into a club. Males of many species have claspers at the tip of the abdomen and a peculiar brush at the base of the hind wing. Most are dully marked with brown and black speckling on the body and transparent wings, but a few have conspicuously marked wings, such as *Morocordula apicalis* from southern Mexico and species of *Glenurus* (fig. 5.15d) and Dimarini (Stange 1989).

There are 35 genera in Latin American containing 224 species (Penny 1977).

## References

McClure, M. S. 1976. Spatial distribution of pit-making ant lion larvae (Neuroptera: Myrmeleontidae): Density effects. Biotropica 8: 179–183.

Miller, R. B., and L. A. Stange. 1983. The antlions of Florida: *Glenurus gratus* (Say) (Neuroptera: Myrmeleontidae). Fla. Dept. Agric. Entomol. Circ. 251: 1–2.

Miller, R. B., and L. A. Stange. 1985. Description of the antlion larva *Navasoleon boliviana* Banks with biological notes (Neuroptera; Myrmeleontidae). Neuroptera Intl. 3: 119–126.

Penny, N. D. 1977. Lista de Megaloptera, Neuroptera e Raphidioptera do México, América Central, ilhas Caraíbas e América do sul. Acta Amazonica 7(4) suppl.: 1–61.

Stange, L. A. 1970. Revision of the ant-lion tribe Brachynemurini of North America. Univ. Calif. Publ. Entomol. 55: 1–192.

Stange, L. A. 1989. Review of the New World Dimarini with the description of a new genus from Peru (Neuroptera: Myrmeleontidae). Fla. Entomol. 72: 450–461.

Wilson, D. S. 1974. Prey capture and competition in the ant lion. Biotropica 6: 187–193.

## Owlflies

Neuroptera, Ascalaphidae.

Looking somewhat like dragonflies but actually related to ant lions, these are medium-sized to large (BL 2–4 cm, BWL to 10 cm) neuropterans with reticulate wing venation. The wings are often infused with brown or black, although they are usually crystalline except for the small stigmal spot near the anterior edge of the tip. They have large, many-faceted eyes, and the antennae are characteristically long and filamentous, with a large, flat, terminal knob, somewhat like those of butterflies. The anterior part of the head, thorax, and legs is frequently hairy. Males of some species have a pronounced elongate dorsal process on the second abdominal segment.

Owlfly larvae are nonburrowers, living openly on leaves, on tree trunks, or on the ground. They are similar to those of ant lions but lack the enlarged digging claw on the hind leg. Almost all species have hairy, fingerlike processes on the body, which are rare in ant lions (Henry 1976). They are slow-moving and usually wait in ambush for prey. The pupa is protected by a weak cocoon woven entirely of silk by the mature larva. Another peculiarity of the group is the laying of deformed eggs (repagula) around the fertile ones (New 1971).

Biological information on the family in the Neotropics is meager. Adults of the genus *Corduleceris* (fig. 5.15e), unique in having strongly dimorphic sexes, have been observed to aggregate at the tips of tree branches overhanging streams (Covell 1989, Hogue and Penny 1989). The purpose of this behavior is unknown, although the clusters may be composed of sleeping individuals. Single adults often rest on twigs, head downward, with wings and the antennae held closely parallel to the surface, the abdomen sharply erect. They are usually excellent fliers, although they sometimes elect to catch other insects while foraging from leaf surfaces rather than on the wing. There are both nocturnal and diurnal species. The former come to artificial light occasionally.

The family is fairly diverse in Latin America. Penny (1977) lists 94 species in 15 genera, the largest and most common of which are *Ululodes* and *Ameropterus*. There are two large groups separated by the form of the eye, those with split eyes, that is, with the eye partially divided by a transverse groove, and those with entire eyes. A third subfamily, represented in the New World only by *Albardia furcata* in Brazil, is peculiar in the very short antennae and woolly abdomen of the adult (Penny 1983).

## References

COVELL, JR., C. V. 1989. Aggregation behavior in a Neotropical owlfly, *Cordulecerus maclachlani* (Neuroptera: Ascalaphidae). Entomol. News 100: 155–156.

HENRY, C. S. 1976. Some aspects of the external morphology of larval owlflies (Neuroptera: Ascalaphidae), with particular reference to *Ululodes* and *Ascaloptynx*. Psyche 83: 1–31.

HOGUE, C. L., AND N. D. PENNY. 1989. Aggregations of Amazonian owlflies (Neuroptera: Ascalaphidae: *Cordulecerus*). Acta Amazonica 18: 359–361

NEW, T. R. 1971. Ovariolar dimorphism and repagula formation in some South American Ascalaphidae. J. Entomol. 46: 73–77.

PENNY, N. D. 1977. Lista de Megaloptera, Neuroptera e Raphidioptera do México, América Central, ilhas Caraíbas e América do sul. Acta Amazonica 7(4) suppl.: 1–61.

PENNY, N. D. 1983. Neuroptera of the Amazon Basin. Acta Amazonica 13: 697–699.

# 6 AQUATIC ORDERS

Aquatic insect life in Latin America is diverse and abundant in both running and standing inland waters and, to a limited extent, in the seas surrounding the continental and island areas (see Ecosystems, chap. 2). All the major categories of water-dwelling insects are represented and widely distributed except in the nutrient-poor black water and clear water systems. Relative to those carrying white waters, the major black water drainages of the Guiana (e.g., Rio Negro) and Brazilian (e.g., Rio Tapajos) shields of South America are well known as faunistically depauperate ("hungry rivers") with regard to insects as well as vertebrates (Junk and Furch 1985: 15). Amazonia is still largely unexplored with respect to aquatic insects, as is most of the Andes. Many taxa found in streams in the southern portion of South America are very ancient and show affinities to the faunas of the Australian region and southern Africa (Gondwanaland and amphinotic distributions) (Illies 1969).

Families of insects in mostly terrestrial orders that have adapted to life in fresh water are discussed elsewhere (see water bugs, chap. 8; water beetles, chap. 9; water midges, punkies, mosquitoes, etc., chap. 11). Those orders totally adapted to life in fresh water are the subject of this chapter.

Two of these orders, the Ephemeroptera and Odonata, form the Paleoptera, considered to be the stem group and the most ancient of the winged insects, from which all higher insects evolved. Evidence for this comes partly from the lack of a flexing mechanism in the wing articulation and the reliance of the immatures on gills for respiration, not atmospheric air, the latter being a secondary adaptation in aquatic insects. Possibly the most primitive of the Neoptera are the Plecoptera, as indicated by their generalized body structure. The Megaloptera and Trichoptera are higher Neoptera but are thought to occupy positions basal to the phylogeny of the neuropteroid and panorpoid orders for the same reasons (see Evolution and Classification, chap. 1).

The larvae and nymphs (sometimes called naiads) of these orders inhabit waters of all descriptions, including saline and thermal waters, but none are marine as are some species of the aquatic families discussed elsewhere. They are active but remain within the boundaries of their vastly varied microhabitats. Dobsonfly and caddis fly pupae are submerged, often in protective cases or cocoons, but may be terrestrial and even active, having functional muscles for movement of legs and mouthparts. The adults remain close by the habitats of the immatures and are mostly predaceous or do not feed.

Current and comprehensive introductions to the biology and zoogeography of Latin American aquatic insects, including extensive literature citations, are to be found in an important three-part collaborative treatment edited by Hurlbert and others (1977, 1981, 1982); for important treatments of local faunas, see also Vanzolini (1964) for Brazil only and Roldán (1988) for Colombia.

## References

HURLBERT, S. H. ed. 1977. Biota acuática de sudamérica austral. San Diego State Univ., San Diego.

HURLBERT, S. H., G. RODRÍGUEZ, AND N. DIAS DOS SANTOS, eds. 1981. Aquatic biota of tropical South America. Pt 1. Arthropoda. San Diego State Univ., San Diego.

HURLBERT, S. H., AND A. VILLALOBOS FIGUEROA, eds. 1982. Aquatic biota of Mexico, Central America and the West Indies. San Diego State Univ., San Diego.

ILLIES, J. 1969. Biogeography and ecology of Neotropical freshwater insects, especially those from running waters. In E. J. Fittkau, J. Illies, H. Klinge, G. H. Schwabe, and H. Sioli, eds., Biogeography and ecology in South America. 2: 685–708. Junk, The Hague.

JUNK, W. J., AND K. FURCH. 1985. The physical and chemical properties of Amazonian waters and the relationships with the biota. In G. T. Prance and T. E. Lovejoy, Amazonia. Pergamon, Oxford. Pp. 3–17.

ROLDÁN, G. 1988. Guía para el estudio de los macroinvertebrados acuáticos del Departamento de Antioquia. Fondo Fen Colombia/COLCIENCIAS, Univ. Antioquia, Bogotá.

VANZOLINI, P. E., ed. 1964. História natural de organismos aquáticos do Brasil: Bibliografia comentada. Fund. Amparo Pesq. Est. São Paulo, São Paulo.

# MAYFLIES

Ephemeroptera (= Ephemerida).
*Portuguese:* Siriruias (Brazil).

This is considered the most primitive of all the living orders of winged insects. Only mayflies undergo a molt after acquiring functional wings. The wings are incapable of being folded rearward and often possess a very complex, netlike venation, both ancestral characters indicative of early origin. The hind wing is always much smaller than the triangular fore wing and in many cases is lost altogether.

Adult mayflies are further recognized by their strongly developed eyes, particularly in the males. These are often divided into upper and lower portions of larger and smaller ommatidia. The antennae are mere stylets, but the two or three terminal sensory filaments (cerci plus median caudal filament) are very long, extending from the tip of the abdomen. Mouthparts are vestigial and the legs weak or reduced, or even vestigial in *Campsurus* (fig. 6.1a). Adults of most species are dull colored. Those in the widespread genus *Thraulodes* (fig. 6.1c) may have striking patterns; male *Tricorythodes* are black with milky wings. The wings of the males in many species have partially maculate wings.

The biology of mayflies has been extensively studied by entomologists and limnologists (Flannagan and Marshall 1980). The body form of the aquatic nymphs differs much among families but usually little from that of its own adult. Eyes, antennae, and mouthparts are well formed, as are the cerci and median caudal filament. Conspicuous also are four to seven pairs of articulated, lateral, platelike gills (often double plates) on most of the abdominal segments. Nymphs also display varied and sometimes bizarre shapes as specializations to their different submerged aquatic habitats. Slender, cylindrical types are strong, "minnowlike" swimmers, occupying still water in ponds and stream pools and sometimes mountain torrents. There are flattened forms that lodge between and under rocks, and some have splayed legs that cling to exposed rock surfaces in fast currents (e.g., *Thraulodes*, fig. 6.1d). Others are robust, with heavy, shovel-shaped, spiny legs and head used for burrowing in bottom muds or sand (e.g., *Campsurus*, fig. 6.1b).

The nymphs of most species are herbivores or scavengers, taking detritus and aquatic microorganisms, especially diatoms. A minority, such as *Chiloporter* and *Chaquihua* in Chile and Argentina, are predatory on other small aquatic invertebrates (Edmunds pers. comm.).

On maturing, the nymphs transform

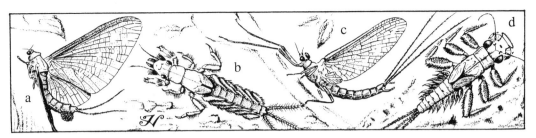

**Figure 6.1   MAYFLIES.** (a) Legless mayfly (*Campsurus albicans,* Polymitarcyidae). (b) Legless mayfly (*Campsurus* sp.) nymph. (c) Tropical mayfly (*Thraulodes* sp., Leptophlebiidae). (d) Tropical mayfly (*Thraulodes* sp.) nymph.

into the alate, flying, but sexually immature subimago (the "dun"). The transformation usually takes place at the water's surface but may also occur below the water or after the nymph has crawled out of the water onto some object. This stage soon metamorphoses into a reproductive adult (imago). Subimagos have infuscate wings and the integument covered with microspines to distinguish them from the glassy-winged, glossy-surfaced imagos (Edmunds 1988).

Imagos are ephemeral, their active lives lasting only a few hours or, at most, days. They do not feed and spend most of their short existence on the wing, mating and egg laying. They generally remain near their breeding grounds where they are seen flying or resting. Many species swarm, some (as in the genera *Tortopus* and *Campsurus*) in such great numbers as to constitute a plague, especially when drawn to street lights or house lights in urbanized areas. Thousands of individuals, mostly females, pile up in the streets or press indoors to make themselves a nuisance. Within these swarms, the sexes copulate in flight, and females fall into the water as they release their eggs.

Mayflies in all stages form a large part of the diet of freshwater fish and small riparian birds. They are also eaten by many types of carnivorous aquatic insects. The importance of some prolific species in food chains has caused them to be referred to as "insect cattle" (e.g., *Callibaetis*).

Mayfly classification and evolutionary history are complex (Edmunds 1972). There are 13 families, perhaps containing, when all are discovered, about 500 or more species in all of Latin America, extrapolating from Hubbard's (1982) list of 300 currently named species. Their poor abilities to disperse make them useful as biogeographic indicators. Although some common genera in South America may extend to North America, members of several groups in southern South America show closer relationships with species from other austral areas (Australia, New Zealand, and southern Africa) than with species to the north. Examples include the genera *Metamonius, Chiloporter, Chaquihua,* and *Siphlonella* of the Siphlonuridae and *Nousia* of the Leptophlebiidae (Pescador and Peters 1982, 1985). For general reviews of the order in Latin America, see Hubbard and Peters (1977, 1981) and Edmunds (1982). An important taxonomic paper is the review of the Neotropical Leptophlebiidae by Savage (1987).

## References

EDMUNDS, JR., G. F. 1972. Biogeography and evolution of Ephemeroptera. Ann. Rev. Entomol. 17: 21–42.

EDMUNDS, JR., G. F. 1982. Ephemeroptera. *In* S. H. Hurlbert and A. Villalobos Figueroa, eds., Aquatic biota of Mexico, Central America and the West Indies. San Diego State Univ., San Diego. Pp. 242–248.

EDMUNDS, JR., G. F. 1988. The mayfly subimago. Ann. Rev. Entomol. 33: 509–529.

FLANNAGAN, J. F., AND K. E. MARSHALL, eds. 1980. Advances in Ephemeroptera biology. Plenum, New York.

HUBBARD, M. D. 1982. Catálogo abreviado de Ephemeroptera da América do sul. Pap. Avul. Zool. (São Paulo) 34(24): 257–282.

HUBBARD, M. D., AND W. L. PETERS. 1977. Ephemeroptera. In S. H. Hurlbert, ed., Biota acuática de sudamérica austral. San Diego State Univ., San Diego. Pp. 165–169.

HUBBARD, M. D., AND W. L. PETERS. 1981. Ephemeroptera. In S. H. Hurlbert, G. Rodriguez, and N. Dias dos Santos, eds., Aquatic biota of tropical South America. Pt 1. Arthropoda. San Diego State Univ., San Diego. Pp. 55–63.

PESCADOR, M. L., AND W. L. PETERS. 1982. Four new genera of Leptophlebiidae (Ephemeroptera: Atalophlebiinae) from southern South America. Aquatic Ins. 4: 1–19.

PESCADOR, M. L., AND W. L. PETERS. 1985. Biosystematics of the genus Nousia from southern South America (Ephemeroptera: Leptophlebiidae: Aetalophlebiinae). Kans. Entomol. Soc. J. 58: 91–123.

SAVAGE, H. M. 1987. Biogeographic classification of the Neotropical Leptophlebiidae (Ephemeroptera) based upon geological centers of ancestral origin and ecology. Stud. Neotrop. Fauna Environ. 22: 199–222.

## DRAGONFLIES AND DAMSELFLIES

Odonata.

These are familiar insects, always found near water, although the powerful and untiring wings of dragonflies may take them on long journeys. They are easily recognized by their elongate bodies, four similar, many-veined wings (with a dark spot on the leading edge near the tip, the pterostigma), and bulbous eyes with an enormous number of minute ommatidia. Between the eyes arise the tiny, bristlelike antennae. The thoracic segments are angled obliquely so that their dorsal surfaces form an incline.

Dragonflies are distinguished from damselflies principally by their more robust and usually larger bodies and their habit of extending the wings out to the sides when at rest. Some damselflies are also large, but they are always slender. Most damselflies fold their wings together back over the abdomen when not in use. The wings are also abruptly narrowed and slender at the base in contrast to the broadly based dragonfly wings. The nymphs of dragonflies are also more heavily built than those of the damselflies. The former have a broad, tapering abdomen, tipped with short spinose processes (fig. 6.2b); their gills are located internally in folds of the rectum. Damselfly nymphs have elongate, slender abdomens, bearing three, conspicuous, finlike terminal gills (fig. 6.3b, e).

The body and wings of odonates are very often highly decorated with bright or gaudy colors. The tints of the body are transient and quickly disappear from dead specimens; but those of the wings persist as spot and band patterns or broad fields, most often over the basal half or third of the wing. In some, the entire wing may be colored, often in glossy red, orange, or

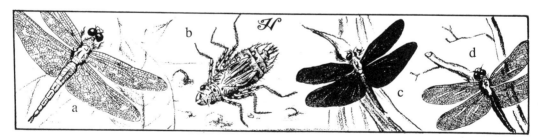

**Figure 6.2 LIBELLULID DRAGONFLIES (LIBELLULIDAE).** (a) Globetrotter (*Pantala flavescens*). (b) Globetrotter nymph. (c) Black wing (*Diastatops dimidiata*). (d) Amber wing (*Perithemis indensa*).

blue. The coloration of the male frequently differs from that of the female, and it is sometimes difficult to know they are of the same species until they are seen copulating.

During the pairing process, the sexes are peculiarly joined in a tandem configuration unique to this order of insects. The male's copulatory structures (genital fossa) are situated on the undersides of the second and third abdominal segments at a considerable distance from the true sexual aperture at the tip of the abdomen. Prior to mating, he transfers spermatozoa from the gonopore to the penis in the fossa. During mating, the male grasps the female behind the head (dragonflies) or thorax (damselflies) with his strong, tonglike cerci while she bends her abdomen forward to the fossa and receives the sperm from the penis. Couples fastened together in this manner are commonly seen resting on vegetation by the water's edge, or even in flight.

Adult biology, such as flight patterns and competition for prey and hunting space, in Latin America has not been well studied but is generally the same as that observed elsewhere. Some special adaptations for surviving dry periods have been noted (Morton 1977).

Immature odonates are all aquatic, the nymphs being found in all sorts of running and still water environments: ponds, shallow stream and lake margins, stream pools, tree holes, and tank plants. The nymphs are insectivorous. They capture prey with a mantislike grab of the enlarged and elongated, extensile lower lip (labium). The nymphs show structural adaptations in response to the diverse ecological niches they occupy, but these are less extreme than those of mayflies. Burrowers tend to be short and broad, bottom sprawlers flattened and long-legged (fig. 6.2b), and swimmers slender and streamlined. Probably only about 15 percent of the immatures of the Latin American species are known.

The literature on Latin American odonates is reviewed by Dias dos Santos (1981) and Paulson (1977, 1982). A species list is provided by Davies and Tobin (1984–85).

## References

DAVIES, D. A. L., AND P. TOBIN. 1984–85. The dragonflies of the world: A systematic list of the extant species of Odonata. 2 vols. Int. Soc. Odonatologia, Utrect.

DIAS DOS SANTOS, N. 1981. Odonata. *In* S. H. Hurlbert, G. Rodríguez, and N. Dias dos Santos, eds., Aquatic biota of tropical South America. Pt. 1. Arthropoda. San Diego State Univ., San Diego. Pp. 64–85.

MORTON, E. S. 1977. Ecology and behavior of some Panamanian Odonata. Entomol. Soc. Wash. Proc. 79: 273.

PAULSON, D. R. 1977. Odonata. *In* S. H. Hurlbert, ed., Biota acuática de sudamérica austral. San Diego State Univ., San Diego. Pp. 170–184.

PAULSON, D. R. 1982. Odonata. *In* S. H. Hurlbert and A. Villalobos Figueroa, eds., Aquatic biota of Mexico, Central America and the West Indies. San Diego State Univ., San Diego. Pp. 249–277.

# DRAGONFLIES

Anisoptera. *Spanish:* Zurcidores, caballitos del diablo, caballitos de San Pedro, matacaballos, libélulas, aguacilas (General); chispiaguas (Colombia); matapiojos (Chile). *Portuguese:* Cavalos de cão, lavandeiras, lavabundas, pitos, cavalinhos de judeu (Brazil).

Dragonflies (Corbet 1962, 1980) are highly opportunistic predators that hunt singly or sometimes in aggregations, taking other aquatic insects flying over still water ponds, marshes, and the margins of lakes and streams (Young 1980). Their aerial acrobatics are spectacular; they can hover, dart forward instantly, and even fly backward with consummate ease and speed, feats rivaled by few other aerial creatures. In many species, most of their daylight hours of activity are spent on the wing, and so highly modified is the body for flying that they come to rest only occasionally to perch and to spend the night in sleep. The legs are useless for walking, forming instead a basketlike aerial sieve to "strain" insects out of the air.

Dragonflies usually oviposit while they are flying, dipping the tip of the abdomen below the water's surface as the eggs are extruded. The submerged eggs then drift to the bottom. Some have more specialized habits, including dropping the eggs in the water from some height, or directly placing them onto leaves, rocks, or mud, or inserting them into plant tissue (Paulson 1969).

Dias dos Santos (1981) records 705 species of dragonflies in 99 genera from the Neotropics, but many undescribed species are certain to exist. Most belong to the family Libellulidae. A few taxa in the southern parts of South America, such as *Phenes raptor* (Petaluridae), *Gomphomacromia chilensis* (Corduliidae), and the family Petaliidae, have their closest relatives in other southern continents. Dragonflies are mainly tropical, however, and the group tends to be poorly represented in the higher latitudes of Chile and Argentina. The strong flight capabilities of the adults, some of which are migratory, often carry them great distances. A few species are commonly found far out at sea, and some have colonized isolated oceanic islands. The cosmopolitan globetrotter (*Pantala flavescens*) (fig. 6.2a), which is capable of flying for hours at a time, has been collected on Coco Island and the mid-Atlantic Ilha Trindade.

Some Latin American dragonflies are distinctive and widespread. The following examples all belong to the family Libellulidae.

The ruby tail (*Libellula herculea*) is a medium-sized species (WS 8 cm) that stands out conspicuously against the vegetation on which it is often seen resting because of its brilliant, almost glowing, scarlet abdomen. The rest of the body is black. The sides of the thorax are covered with a gray pruinosity. The ubiquitous ferruginous skimmer (*Orthemis ferruginea*) is a dusky red-bodied, moderately large (WS 88 mm) species.

The "amber wings" (*Perithemis*) (fig. 6.2d) are small (WS 5 cm) and have partially or completely amber-tinted wings. There may also be brown cross bands or spots on the wings. Similar in size are the "black wings" (*Diastatops;* fig. 6.2c) that have solid black wings and a bright to dull red abdomen.

Another conspicuous group of small dragonflies are the "butterfly dragonflies" (*Zenithoptera*). These resemble the black wings in size (WS 4–5 cm) and obscure wings, but there are clear streaks intruding from the anterior wing margin just beyond the halfway point which run transversely into the dark field. Also, it is the habit of the butterfly dragonflies to slowly open and close the wings while perched, in the manner of a sunning butterfly.

## References

CORBET, P. S. 1962. A biology of dragonflies. Quadrangle, Chicago.

CORBET, P. S. 1980. Biology of Odonata. Ann. Rev. Entomol. 25: 189–217.

DIAS DOS SANTOS, N. 1981. Odonata. *In* S. H. Hurlbert, G. Rodríguez, and N. Dios dos Santos, eds., Aquatic biota of tropical South America. Pt. 1. Arthropoda. San Diego State Univ., San Diego. Pp. 64–85.

PAULSON, D. R. 1969. Oviposition in the tropical dragonfly genus *Micrathyria* (Odonata, Libellulidae). Tombo 12: 12–16.

YOUNG, A. M. 1980. Observations on feeding aggregations of *Orthemis ferruginea* (Fabricius) in Costa Rica (Anisoptera: Libellulidae). Odonatologica 9: 325–328.

## DAMSELFLIES

Zygoptera. *Spanish:* Doncellas (General), chupajeringas (Peru).

Damselflies are much smaller and weaker fliers than dragonflies. They always stay

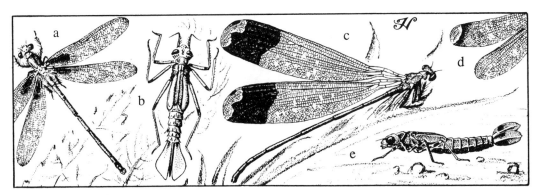

**Figure 6.3    DAMSELFLIES.** (a) Ruby wing (*Hetaerina* sp., Calopterygidae), male. (b) Ruby wing (*Hetaerina americana*) nymph. (c) Giant damselfly (*Magaloprepus coerulatus,* Pseudostigmatidae). (d) Giant damselfly (*Mecistogaster* sp.), wing tips of male. (e) Common damselfly (*Argia vivida,* Coenagrionidae) nymph.

very close to the habitats of their nymphs, flitting from perch to perch in search of small insects, mosquitoes, and other flies on which to feed. Their bodies are often multicolored, but the wings, with a few exceptions, are entirely transparent.

There are 786 species of damselflies in 94 genera recorded for the Neotropics by Dias dos Santos (1981). The actual number will certainly go higher after the faunas of Amazonia and other remote tropical areas are more fully explored. These species belong to a dozen families, the largest and most common by far being the Coenagrionidae, dominant genera of which are *Argia* (fig. 6.3e), *Acanthagrion,* and *Telchasis. Hetaerina* (Calopterygidae) is also common and easily recognized by brilliant, deep red wing bases ("ruby spots") in the male (fig. 6.3a, b) (Eberhard 1986, Garrison 1990, Williamson 1923). Four families are exclusive to Latin America (Pseudostigmatidae, Polythoridae, Perilestidae, and Heliocharitidae). There are no relictual taxa with amphinotic or Gondwanaland distributions, the Patagonian species having more northern relatives.

### References

Dias dos Santos, N. 1981. Odonata. *In* S. H. Hurlbert, G. Rodríguez, and N. Dias dos Santos, eds., Aquatic biota of tropical South America. Pt. 1. Arthropoda. San Diego State Univ., San Diego. Pp. 64–85.

Eberhard, W. G. 1986. Behavioral ecology of the tropical damselfly *Hetaerina macropus* Selys (Zygoptera: Calopterygidae). Odonatologica 15: 51–60.

Garrison, R. W. 1990. A synopsis of the genus *Hetaerina* with descriptions of four new species (Odonata: Calopterygidae). Amer. Entomol. Soc. Trans. 116: 175–259.

Williamson, E. B. 1923. Notes on the habitats of some tropical species of *Hetaerina* (Odonata). Mus. Zool. Univ. Michigan Occ. Pap. 130: 1–45.

### Giant Damselflies

Pseudostigmatidae, *Megaloprepus* and *Mecistogaster. Spanish:* Helicópteros, chinchilejos (Peru).

Members of the small family Pseudostigmatidae inhabit mature forests from northern Mexico to southern Brazil. Members of the genera *Megaloprepus* and *Mecistogaster* are very large and showy insects. The sight of one of these gossamer creatures slowly flying through the trees, its wings tracing a blurred whirl above its body, is one of the loveliest of the rain forest. Some natives believe them to be human spirits that have recently become disembodied and insist that they be left unmolested (Klots and Klots 1959). These are enormous damselflies with somber body colors but gaudy patterns on the wing tips. *Megaloprepus* contains only one species, *M. coerulatus* (fig. 6.3c), which has a greater wingspan (60–75 mm) than most species of *Mecistogaster* (usually 60–65 mm, but up to 170 mm) but a shorter abdomen (76–96 mm compared to up to 135 mm). The wings of *Megaloprepus* are marked subapically with a broad, dark purple band and are suffused basally with white (in the males), while in *Mecistogaster*, orange, yellow, or red color fields are present near the wing tips (fig. 6.3d). Wing coloration sometimes varies within a species due to age or seasonal differences. *Mecistogaster* often has no light-colored spot, but all the members of the family are characterized, without exception, by a multicelled pterostigma. Size has been found to vary greatly in the adults owing to differences in the quality of the nymphal habitats (Fincke 1984).

The nymphs are among the few Zygoptera that employ bromeliad tanks and tree holes for development. Young (1981) has observed *Megaloprepus* ovipositing in an accumulation of water in a depression in the trunk of a fallen tree. Female *Mecistogaster* are known to throw their eggs singly onto the water surface in such container habitats. The very long, slender abdomen appears to be an adaptation to facilitate this procedure (Machado and Martínez 1982).

For food, the adults specialize on kleptoparasitic spiders that live in the webs of orb weavers (Araneidae such as *Nephila* and *Gasteracantha*) (Young 1980). In flight, they approach the web and flutter before it briefly prior to darting in to pluck out the prey, which they carry to a nearby perch to devour.

Males of the sexually dimorphic *Megaloprepus* hold mating territories around water-filled tree holes for up to two months, defending them from conspecific males and permitting only females with whom they have mated to enter and oviposit in the holes (Fincke 1984).

## References

Fincke, O. M. 1984. Giant damselflies in a tropical forest: Reproductive biology of *Megaloprepus coerulatus* with notes on *Mecistogaster* (Zygoptera: Pseudostigmatidae). Adv. Odonatol. 2: 13–27.

Klots, A. B., and E. B. Klots. 1959. Living insects of the world. Doubleday, Garden City.

Machado, A. B. M., and A. Martínez. 1982. Oviposition by egg-throwing in a zygopteran, *Mecistogaster jocaste* Hagen, 1869 (Pseudostigmatidae). Odonatologica 11: 15–22.

Young, A. M. 1980. Feeding and oviposition in the giant tropical damselfly *Megaloprepus coerulatus* (Drury) in Costa Rica. Biotropica 12: 237–239.

Young, A. M. 1981. Notes on the oviposition microhabitat of the giant tropical damselfly *Megaloprepus coerulatus* (Drury) (Zygoptera: Pseudostigmatidae). Tombo 23: 17–21.

## STONEFLIES

Plecoptera.

This is a small order of very ancient and primitive insects whose nymphs are typical inhabitants of streams. The adults remain near the nymphal habitat and are frequently found resting on rocks or boulders in midstream (hence their common name) or on nearby vegetation and tree trunks.

Adult stoneflies are soft bodied, small to medium-sized (most with length to tips of folded wings 1–2 cm, but some to 4–5 cm), and rather elongate in overall form. Except for a few with stubby or no wings, all possess two pairs of complete wings with a fairly complex venation: the fore wing is long, with parallel sides; the hind wing has a large, fanlike area posteriorly. Stoneflies are drably colored in browns and grays, except for some South American Eustheniidae that have wings splashed with bright reds and yellows. The antennae and cerci (when present) are both filiform, the former much longer than the latter. Mouthparts are mandibulate but with weak elements. Between the two suborders, the Arctoperlaria and Antarctoperlaria, there is a basic difference in food habits corre-

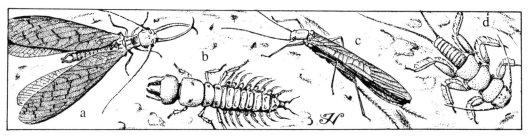

**Figure 6.4   DOBSONFLIES AND STONEFLIES.** (a) Dobsonfly (*Corydalus cornutus,* Corydalidae). (b) Dobsonfly (*Corydalus cornutus*) larva. (c) Stonefly (*Anacroneuria* sp., Perlidae). (d) Stonefly (*Anacroneuria* sp.) nymph.

lated with two mouthpart types. In the Arctoperlaria, the lobes of the labium are long and flexible and well structured for carnivory; in the Antarctoperlaria, these lobes are reduced, relatively inflexible, and used for chewing plant tissues. (The suborders Filipalpia and Setipalpia, used widely in earlier literature, are no longer recognized as phylogenetically logical subdivisions of the order.)

The immatures resemble the adults, except that they lack wings (fig. 6.4d). They may have conspicuous external filamentous gill tufts ventrally on various parts of the body, vestiges of which remain in adults and are of taxonomic utility. Nymphs of the South American *Pelurgoperla* have long, sticky dorsal hairs in which debris becomes entangled, rendering them inconspicuous amid the bottom trash of their habitat.

Except for a few limited studies (Illies 1964), the biology of Latin American stoneflies is not well investigated (Hynes 1976). Most species breed only in cool, running water or cold mountain lakes. Adults are sluggish and often difficult to see when they are at rest on their similarly colored substrata. They fly readily but slowly and feebly.

The Latin American stonefly fauna is fairly extensive and diverse (Benedetto 1974). There are more than 170 species in 46 genera distributed among 7 families (Illies 1966, Zwick 1973). Because of their considerable geologic age and significant fossil record, the order offers a wealth of facts for phylogenetic and zoogeographic analysis on a worldwide basis (Illies 1965). The more primitive Antarctoperlaria are mainly austral and exhibit an amphinotic distribution, being found in southern South America as well as in Australia and New Zealand (Eustheniidae, Gripopterygidae, Austroperlidae). Diamphipnoidae is restricted entirely to the southern Andes. The arctoperlarian *Neonemura illiesi* (Notonemouridae) is also southern, but its nearest relatives are distributed much like the preceding family and in South Africa (Gondwanaland distribution). These patterns indicate the origin and early diversification of the Plecoptera in Gondwanaland. A few boreal members of this suborder have invaded Latin America, except the West Indies, by way of a southward dispersion from North America. These are certain Perlidae, including the dominant genus *Anacroneuria* (74 species) (fig. 6.4c), which has moved over the entire continental area, and the genus *Amphinemura* (Nemouridae), which extends only to central Mexico. General information and literature on the order in Latin America can be found in Baumann (1982), Illies (1977), and Froehlich (1981).

## References

Baumann, R. W. 1982. Plecoptera. *In* S. H. Hurlbert and A. Villalobos Figueroa, eds., Aquatic biota of Mexico, Central America and the West Indies. San Diego State Univ., San Diego. Pp. 278–279.

BENEDETTO, L. 1974. Clave para la determinación de los Plecópteros sudamericanos. Stud.
Neotrop. Fauna 9: 141–170.

FROEHLICH, C. G. 1981. Plecoptera. *In* S. H.
Hurlbert, G. Rodríguez, and N. Dias dos
Santos, eds., Aquatic biota of tropical South
America. San Diego State Univ., San Diego.
Pp. 86–88.

HYNES, H. B. N. 1976. Biology of the Plecoptera. Ann. Rev. Entomol. 21: 135–153.

ILLIES, J. 1964. The invertebrate fauna of the
Huallaga, a Peruvian tributary of the Amazon
River, from the sources down to Tingo Maria.
Int. Ver. Limnol. Verh. 15: 1077–1083.

ILLIES, J. 1965. Phylogeny and zoogeography
of the Plecoptera. Ann. Rev. Entomol. 10:
117–140.

ILLIES, J. 1966. Katalog der rezenten Plecoptera. Das Tierreich 82: i–xxx, 1–632.

ILLIES, J. 1977. Plecoptera. *In* S. H. Hurlbert,
ed., Biota acuática de sudamérica austral. San
Diego State Univ., San Diego. Pp. 185–186.

ZWICK, P. 1973. Insecta: Plecoptera. Phylo-
genetisches System und Katalog. Das Tierreich 94: i–xxxii, 1–465.

## DOBSONFLIES

Megaloptera. *Spanish:* Perros del agua
(larvae, Mexico).

This is a small order, related to the
Neuroptera (Penny 1977, 1983). There are
two families, the large Corydalidae (BL 8–
12 cm, wingspan to 16 cm) and the much
smaller Sialidae (BL to 25 mm). The Corydalidae ("dobsonflies") are widespread and
dominant in mainland Latin America (only
a single Antillean species on Dominica),
with 7 genera and 47 species. *Protochauliodes* and *Archichaulides* are found in southern Chile and have their congeners in
Australia. The North American *Corydalus
cornutus* (fig. 6.4a) is known as far south as
Panama, from where a second, closely
related species, *C. armatus*, continues into
South America. The Sialidae ("alderflies")
are represented by just 7 localized species
in a single genus and are not discussed
further here.

Like neuropterans, dobsonflies have
two pairs of membranous wings with complete venation, although the multiplicity of
spurious veinlets and cross veins found in
the Neuroptera is lacking. Except for minor differences in the branching of veins
and being slightly narrower, the fore wings
are similar to the hind wings. The body is
cylindrical, elongate, and somewhat soft
and flexible. Legs are short and similar on
each thoracic segment. In both the adults
and larvae, the mouthparts are mandibulate and with elements, especially the mandibles, strongly developed. Males of some
species of *Corydalus* have extremely long,
tusklike jaws (Glorioso 1981). All dobsonflies, except the lemon yellow *Chloronia*
(Penny and Flint 1982), are brown to dull
gray or black. Males of *Platyneuromus* (=
*Doeringia*) have a flattened lateral expansion on the head behind the compound
eyes.

The larvae (hellgrammites) (fig. 6.4b)
are similar in body structure to the adults
but are slightly flattened and have conspicuous fingerlike gills laterally on the
abdomen. These gills may be bare or
fringed, and there may be a tuft of accessory gill filaments at the base of most of the
primary appendages. The last abdominal
segment bears a pair of large prolegs, each
with a lateral filament and large claws at
the tips.

Dobsonflies are ubiquitous aquatic insects, generally associated with clean, cold,
mountain streams. The adults remain near
such streams and are rarely seen except
when they come to lights at night or when
flying, as they do occasionally on cool,
overcast days or at dusk. Females attach
their eggs in large, single-layered clusters
to objects near or overhanging the water.
These white, flattened masses are sometimes conspicuous on dark-colored boulders or tree trunks. The larvae are active
predators, catching and eating a variety of
other aquatic insects that they find under
and around debris and rocks. They leave
the water to pupate. Pupae have free,

muscled legs and are able to walk and use their mandibles to bite in defense.

In the Peruvian Amazon region, dobsonfly larvae, devoured raw or cooked, are collected by the natives for food (Murphy pers. comm.). Called *perros de agua*, they are considered venomous and much feared in parts of Mexico.

The literature of the Latin American members of this order is reviewed by Flint (1977) and Penny (1981, 1982).

## References

FLINT, JR., O. S. 1977. Neuroptera. *In* S. H. Hurlbert, ed., Biota acuática de sudamérica austral. San Diego State Univ., San Diego. Pp. 187–188.

GLORIOSO, M. J. 1981. Systematics of the dobsonfly subfamily Corydalinae (Megaloptera: Corydalidae). Syst. Entomol. 6: 253–290.

PENNY, N. D. 1977. Lista de Megaloptera, Neuroptera e Raphidioptera do México, América Central, ilhas Caraíbas e América do sul. Acta Amazonica 7(4) suppl.: 1–61.

PENNY, N. D. 1981. Neuroptera. *In* S. H. Hurlbert, G. Rodríguez, and N. Dias dos Santos, eds., Aquatic biota of tropical South America. Pt. 1. Arthropoda. San Diego State Univ., San Diego. Pp. 89–91.

PENNY, N. D. 1982. Neuroptera. *In* S. H. Hurlbert and A. Villalobos Figueroa, eds., Aquatic biota of Mexico, Central America and the West Indies. San Diego State Univ., San Diego. Pp. 280–282.

PENNY, N. D. 1983. Neuroptera of the Amazon Basin. Pt. 7. Corydalidae. Acta Amazonica 12: 825–837.

PENNY, N. D., AND O. S. FLINT. 1982. A revision of the genus *Chloronia* (Neuroptera: Corydalidae). Smithsonian Contrib. Zool. 348: 1–27.

# CADDISFLIES

Trichoptera.

Caddisflies populate all types of freshwater habitats in Latin America, both in running waters (springs, streams, waterfalls, seeps, rivers) and standing waters (lakes, ponds, marshes, pools). Some *Phylloicus* develop in the leaf axils of bromeliads. A few *Atanatolica* move out to the moist margins of seeps and are virtually terrestrial (Holzenthal 1988).

The immatures are somewhat similar to lepidopterous larvae but with the abdominal prolegs restricted to the terminal segment and bearing anal claws. All types produce silk and are free-living, make fixed retreats, or construct portable cases for themselves. The retreat makers use silk for fabricating shelters and food-trapping nets. The case makers use silk as a matrix to bind together the material composing their cases, such as leaf fragments, twigs, and sand grains. Some cases are constructed entirely of silk. Although some case types are very characteristic, usually of genera, others, such as the common tapering, cylindrical, sand grain case, are made by a number of species in a variety of genera and families. Larvae of the leptocerid genus *Triplectides* will pick up and use empty cases of other caddisflies or make their own by tunneling a small stick. Most cases are tubular, but others may have the form of a snail shell (*Helicopsyche*) or may be somewhat flattened, four sided, or purselike (Hydroptilidae). Dense masses of the long, tusk-shaped cases of *Atanatolica* and *Grumichella* (Leptoceridae) and *Grumicha* (Sericostomatidae) are often seen clinging to rock faces in waterfalls. These cases are made entirely of silk or have fine sand incorporated in their walls (Holzenthal 1988, Müller 1880).

The part of the larva protected by the case is generally weakly sclerotized in case makers. The free-living types are well pigmented and thick skinned throughout.

Among still water species, the larval food consists primarily of algae, fungi, and decaying organic matter and occasionally living plant tissue. Some free-living larvae, especially those in fast water, are predaceous on other aquatic invertebrates.

Adult caddisflies are mothlike, with body and wings clothed in short, easily detached, hairlike scales. At rest, they hold their fore wings rooflike over the body at a steep angle. The antennae are usually very long and filamentous. This stage lives near the larval habitats, and the diurnal species are often seen resting on rocks and vegetation by the water's edge. Many of the nocturnal species are attracted to artificial light.

The order is well studied generally, but very little ecological or natural history information is available specifically on Neotropical representatives (Flint 1977, 1981; Bueno-Soria and Santiago Fragoso 1982). McElravy and others (1981, 1982) compared the diversity of species at a nonseasonal site in Panama to that of Nearctic streams and found it not significantly higher but with relatively less variation over long time periods.

The order's taxonomy in the region is likewise very incomplete. According to Flint (pers. comm.), there are probably between 3,000 and 4,000 species, although only about 1,200 to 1,500 are now described (Fischer 1960–1973). Flint (1983) also provides a key to the South American families that is of general use for all of Latin America. Through the lowlands, the most common species are found in the families Leptoceridae (e.g., the black-and-white speckled *Nectopsyche;* fig. 6.5d, e) and Hydropsychidae (e.g., pale green *Leptonema albovirens;* fig. 6.5a, b). In Patagonian habitats, there are many Limnephilidae (especially Dicosmoecinae). The dominant

free-living, actively predatory, Holarctic genus, *Rhyacophila,* is absent and replaced by the Hydrobiosidae (e.g., *Atopsyche callosa;* fig. 6.5c). The family Anamolopsychidae is restricted to the far southern regions, as are the Helicophidae, Kokiriidae, Philorheithridae, Stenopsychidae, Tasimiidae, and a number of genera in other families (e.g., Leptoceridae: *Hudsonema, Notalina;* Holzenthal 1986a, 1986b). All show a distinct relationship to taxa in Australia and New Zealand. Lepidostomatidae are present in montane areas of Mexico and Central America but absent from South America.

In parts of Brazil, the larval cases (called *curubixá* or *grumixá*) are admired by the Indians as artworks of nature and used to adorn clothing and the body. Some watercourses are even given these names because of the common occurrence of caddis flies in them (Ihering 1968: 328).

### References

BUENO-SORIA, J., AND S. SANTIAGO FRAGOSO. 1982. Trichoptera. *In* S. H. Hurlbert and A. Villalobos Figueroa, eds., Aquatic biota of Mexico, Central America and the West Indies. San Diego State Univ., San Diego. Pp. 398–400.

FLINT, JR., O. S. 1977. Trichoptera. *In* S. H. Hurlbert, ed., Biota acuática de sudamérica austral. San Diego State Univ., San Diego. Pp. 249–253.

FLINT, JR., O. S. 1981. Trichoptera. *In* S. H. Hurlbert, G. Rodríguez, and N. Dias dos

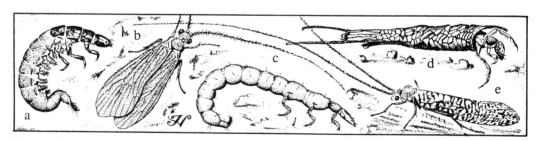

**Figure 6.5 CADDISFLIES.** (a) Hydropsychid caddisfly (*Leptonema* sp., Hydropsychidae) larva. (b) Hydropsychid caddisfly (*Leptonema albovirens*). (c) Hydrobiosid caddisfly (*Atopsyche callosa,* Hydrobiosidae) larva. (d) Leptocerid caddisfly (*Nectopsyche* sp., Leptoceridae), larva in case. (e) Leptocerid caddisfly (*Nectopsyche punctata*).

Santos, eds., Aquatic biota of tropical South America. Pt. 1. Arthropoda. San Diego State Univ., San Diego. Pp. 221–226.

FLINT, JR., O. S. 1983. Studies of Neotropical caddisflies. XXXIII. New species from austral South America (Trichoptera). Smithsonian Contrib. Zool. 377: 1–100.

FISCHER, F. C. J. 1960–1973. Trichopterorum catalogus. 15 vols. Nederlandsche Entomol. Ver., Amsterdam.

HOLZENTHAL, R. W. 1986a. Studies in Neotropical Leptoceridae (Trichoptera). VI. Immature states of *Hudsonema flaminii* (Navás) and the evolution and historical biogeography of Hudsonemini (Triplectidinae). Entomol. Soc. Wash. Proc. 88: 268–279.

HOLZENTHAL, R. W. 1986b. The Neotropical species of *Notalina*, a southern group of long-horned caddisflies (Trichoptera: Leptoceridae). Syst. Entomol. 11: 61–73.

HOLZENTHAL, R. W. 1988. Studies in Neotropical Leptoceridae (Trichoptera). VIII. The genera *Atanatolica* Mosely and *Grumichella* Müller (Triplectidinae: Grumichellini. Amer. Entomol. Soc. Trans. 114: 71–128.

IHERING, R. VON. 1968. DICIONÁRIO DOS ANIMAIS DO BRASIL. ED. UNIV. BRASILIA, SÃO PAULO.

McELRAVY, E. P., V. H. RESH, H. WOLDA, AND O. S. FLINT, JR. 1981. Diversity of adult Trichoptera in a "non-seasonal" tropical environment. 3d Int. Symp. Trichoptera, Ser. Entomol., Proc. 20: 149–156.

McELRAVY, E. P., H. WOLDA, AND V. H. RESH. 1982. Seasonality and annual variability of caddisfly adults (Trichoptera) in a "non-seasonal" tropical environment. Arch. Hydrobiol. 94: 302–317.

MÜLLER, F. 1880. Sobre as casas construidas pelas larvas de insetos Trichoptera da Provincia de Santa Catharina. Mus. Nac. Rio de Janeiro Arch. 3: 99–134, 210–214, pls. 8–11.

# 7 ECTOPARASITIC ORDERS

External parasitism is a pervasive adaptation determining some major evolutionary lines; all members of three entire orders of insects are obligatory ectoparasites of vertebrates (Askew 1971, Marshall 1981). They live on the host superficially, sometimes burrowing into its skin or penetrating exterior cavities but completely tied to it (Nelson et al. 1975). Such is the way of life of the Mallophaga (chewing lice), Anoplura (sucking lice), and Siphonaptera (fleas). This chapter treats these orders independently, separate from those families and lesser aggregations of ectoparasitic insects among other orders which are discussed in their places next to their free-living relatives (see scabies mite, chiggers, etc., chap. 4; bedbugs, bat bedbugs, etc., chap. 8; parasitic rove beetles, chap. 9; bat tick flies, louse flies, chap. 11). (For phylogenetic reasons, the louse orders are combined by some authors into one, the Phthiraptera, or even joined with the Psocoptera; they are treated here as separate, the most familiar arrangement.)

All these ectoparasites exhibit similar adaptive structures and ecological strategies to life on the skin and among the pelage or plumage of vertebrate animals. Most obvious is a flattening of the body, either laterally (compression), as in the case of fleas, or dorsoventrally (depression), as in lice, bedbugs, bat bedbugs, and others. This condition gives the insect a minimal profile and streamlining to hide and move over the body unhindered by hairs or feathers. Correlated with this general body remodeling, in many cases, is a reduction or complete loss of wings and eyes. Linear sets (combs) of flat structures that function like the barbs of an arrowhead, allowing rapid forward progress between obstructions but inhibiting motion backward as might be caused by preening by the host (Marshall 1980), are de novo developments in these groups. Most also have holdfast devices of some kind, including heavy grasping clawlike tarsi, sticky secretions, or even suction cups. The mouthparts may bear recurved hooks or spines to anchor them into the host during feeding. The food of ectoparasites consists of the blood, lymph, skin, hair, and other skin products of the host.

Because of the close attachment of most ectoparasites to their host, their lack of wings, and scant mobility in most cases, dispersion is accomplished primarily by direct body contact between hosts. Phoresy is also occasionally practiced by several groups, especially by Mallophaga. They affix themselves by their mandibles to the wing veins or hairs of other winged insects (most often biting flies and louse flies) that are visiting their hosts and thus become transported to a new host.

Ectoparasites evidently all have evolved from free-living ancestors, as indicated by their body structure and host associations. The sclerites of the adult flea thorax, for example, still bear the muscles and sclerotic braces of a winged insect, although in a vestigial and nonfunctional state. The larvae of fleas resemble those of primitive nonparasitic Diptera found in the nests of vertebrates. It is easy to imagine how a dependence on a warm-blooded host could have developed from living in such closeness, followed by structural adaptations.

The lowland ectoparasite fauna of Middle and South America is remarkably uniform in taxonomic composition. Much greater endemism is experienced by highland groups, except those using cosmopolitan domestic animals as hosts. Locally domesticated hosts, however, may harbor endemic species (Escalante 1981). Species of both categories are strongly isolated ecologically from their North American counterparts (Wenzel 1972) and show independent evolution. The systematics of ectoparasites has been used to provide clues to the relationships of host taxa. The study of patterns of infestation with lice tends to confirm modern views on the affinities of South American land mammals (Vanzolini and Guimarães 1955).

Important regional studies of ectoparasite faunas have been conducted for Panama (Wenzel and Tipton 1966) and Venezuela (various authors 1972, 1975–1976).

## References

ASKEW, R. R. 1971. Parasitic insects. American Elsevier, New York.

ESCALANTE, J. A. 1981. Ectoparásitos de animales domésticos en el Cusco. Rev. Peruana Entomol. 24: 123–125.

MARSHALL, A. G. 1980. The function of combs in ectoparasitic insects. *In* R. Traub and H. Starke, Fleas. Balkema, Rotterdam. Pp. 79–87.

MARSHALL, A. G. 1981. The ecology of ectoparasitic insects. Academic, London.

NELSON, W. A., J. E. KEIRANS, J. F. BELL, AND C. M. CLIFFORD. 1975. Host-ectoparasite relationships. J. Med. Entomol. 12: 143–166.

VANZOLINI, P., AND L. GUIMARÃES. 1955. Lice and the history of South American land mammals. Rev. Brasil. Entomol. 3: 13–45.

VARIOUS AUTHORS. 1972, 1975–76. Ectoparasites of Venezuela. A series of articles on ectoparasite groups published by Brigham Young Univ., Sci. Bull., Biol. Ser. Vols. 17 (1972) and 20 (1975–76).

WENZEL, R. L. 1972. Some observations on the zoogeography of Middle and South American ectoparasites. J. Med. Entomol. 9: 589.

WENZEL, R. L., AND V. J. TIPTON, eds. 1966. Ectoparasites of Panama. Field Mus. Nat. Hist., Chicago.

## CHEWING LICE

Mallophaga. Feather lice, bird lice.

As its name implies, this order is characterized by having mandibulate mouthparts. Chewing lice consume feathers, hairs, and other cutaneous material, including blood, if accessible (dried from wounds, etc.), and sebaceous secretions. These are more heavily sclerotized than the sucking lice, with well-defined abdominal sclerites and comparatively rigid body. In the majority of species, the head is relatively large, wider than the prothorax, and freely movable. Both of the two suborders are found in Latin America: the Ischnocera have filiform antennae, exposed on the sides of the head, and vertically biting mandibles but lack maxillary palpi; the Amblycera (Clay 1970) have short antennae concealed in pockets on the undersides of the head, mandibles that work laterally, and maxillary palpi.

Both bird- and mammal-infesting species are found among the two groups. They are often very host-specific, especially the Ischnocera, some even being confined to a particular area of the body of birds. A few hosts support a diversity of lice. Tinamous, for example, carry no less than twelve mallophagan genera (Carriker 1953–1962). None live on bats, marine mammals, or lagomorphs. The world's seven widespread families of Mallophaga, plus two endemic families, Abrocomophagidae on the rat chinchilla (Emerson and Price 1976) and Trochiliphagidae on hummingbirds, are represented in Latin America. Mammal-associated Amblycera are almost entirely confined to marsupials

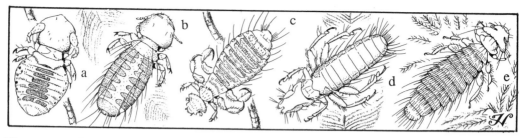

**Figure 7.1  CHEWING LICE.** (a) Cat louse (*Felicola felis*, Trichodectidae). (b) Bird louse (*Paragoniocotes mirabilis*, Philopteridae). (c) Oval guinea pig louse (*Gyropus ovalis*, Gyropidae). (d) Giant bird louse (*Laemobothrion opisthocomi*, Laemobothriidae). (e) Fowl louse (*Menacanthus stramineus*, Menoponidae).

and rodents in the Neotropics; those of the Ischnocera infest placental mammals.

Among the Ischnocera, the Philopteridae is the largest family, with diverse species on birds, including the characteristic Neotropical parrots and macaws (*Paragoniocotes*, fig. 7.1b) and others. Mammals are the hosts of Trichodectidae, including *Felicola* on felines (fig. 7.1a) and *Geomydoecus* on fossorial rodents (Werneck 1945). Sloths, monkeys, kinkajous, coatis, and other uniquely Neotropical mammals have their own chewing lice as well (*Lymeon*, *Cebidicola*, *Trichodectes*, and *Neotrichodectes*, respectively). To date, Mallophaga have not been found on the tapir, capybara, anteater, or armadillo.

From the Amblycera, the Gyropidae are well developed in Latin America; there are several small genera on wild pigs and rodents. *Gyropus ovalis* (fig. 7.1c) and *Gliricola porcelli* are the most common of several guinea pig lice. Bird lice of the family Menoponidae are diverse in species and habits. *Piagetiella bursaepelecani* lives in the throat pouch of pelicans. Laemobothriidae are typical of water birds and birds of prey. A primitive species, *Laemobothrion opisthocomi* (fig. 7.1d), parasitizes the likewise primitive hoatzin. The Ricinidae parasitize songbirds. The Trimenoponidae live on marsupials and rodents.

A few introduced, cosmopolitan species are pests of domestic animals. These include mainly the fowl lice, *Menacanthus*

*stramineus* (fig. 7.1e) and *Menopon gallinae*, (Ancona 1935*b*); the ox, goat, and donkey lice, *Bovicola;* and the pigeon louse, *Columbicola columbae* (Ancona 1935*a*). The latter affect their hosts only when very numerous, when they cause aggravation from their persistent presence as well as skin irritation from feeding.

Aside from host associations, little is known of the biology of chewing lice in Latin America. Many passerine bird hosts have been observed indulging furiously in the habit of "anting" with members of the ant subfamily Formicinae. They squat near an ant nest and passively allow the ants to crawl onto their plumage or place them there with their beaks. Agitation from preening movements causes the ants to release formic acid vapors that apparently act as a repellent to any chewing lice present.

Worldwide, the number of mallophagan species described exceeds 5,000, which is probably less than 10 percent of the species that await discovery. In Latin America, there are several thousand species to be named. Important literature in the study of these lice is provided by Emerson (1967), Werneck (1948), and von Kéler (1960).

## References

ANCONA, L. 1935*a*. Contribución al conocimiento de los piojos de los animales de México. I. *Columbicola columbae*. Inst. Biol. Univ. Nac. Auc. México Anal. 5: 342–351.

ANCONA, L. 1935*b*. Contribución al conocimiento de los piojos de los animales de México. II. *Menopon gallinae* Linn. Inst. Biol. Univ. Nac. Aut. México Anal. 6: 53–62.

CARRIKER, JR., M. A. 1953–62. Studies in Neotropical Mallophaga. XII. Lice of the tinamous. Pts. 1–2. Rev. Bras. Biol. 13: 209–224, 324–346; pts. 3–4, Bol. Entomol. Venezolana 11: 3–30, 97–131; pts. 5–7, Rev. Brasil. Biol. 21: 205–216, 325–338, 373–384; 22: 433–448 (1962).

CLAY, T. 1970. The Amblycera (Phthiraptera: Insecta). Brit. Mus. Nat. Hist. (Entomol.) Bull. 25: 73–98.

EMERSON, K. C., ed. 1967. Carriker on Mallophaga. Posthumous papers, catalog of forms described as new, and bibliography. U.S. Natl. Mus. Bull. 248: 1–150.

EMERSON, K. C., AND R. D. PRICE. 1976. Abrocomophagidae (Mallophaga: Amblycera), a new family from Chile. Fla. Entomol. 59: 425–428.

VON KÉLER, S. 1960. Bibliographie der Mallophagen. Zool. Mus. Berlin Mitt. 36: 146–403.

WERNECK, F. L. 1945. Os tricodectideos dos roedores (Mallophaga). Inst. Oswaldo Cruz Mem. 42: 85–150.

WERNECK, F. L. 1948. Os malófagos de mamíferos. 2 vols. Ed. Inst. Oswaldo Cruz, Rio de Janeiro.

## SUCKING LICE

Anoplura. Mammal lice.

Sucking lice resemble biting lice but are immediately distinguished by modification of the mouthparts into stylets for sucking blood from their mammal hosts. None are found on birds. In general, they also are less well sclerotized than other lice, the abdomen having an elastic cuticle, with small, dorsal abdominal sclerites, allowing for considerable distension during feeding.

Hosts include most major groups of placental mammals, excepting, most notably, the bats, anteaters, and aquatic groups, although some highly specialized forms infest marine pinnipeds. The latter are adapted for resisting cold and submersion in water by a dense covering of hydrophobic, scalelike bristles and spiracular valves for closing off the tracheal system when the host dives. Host specificity is less rigid than among the Mallophaga.

Latin American species number approximately 43, classified in only a few of the 15 world families (Kim and Ludwig 1978). Two are major pests of the human body, the human louse and crab louse (see below). Others are widespread forms that infest domestic animals, such as *Haematopinus*, especially the hog louse (*Haematopinus suis*, fig. 7.2a); five species of *Solenopotes* which live on cattle (another on deer); and *Linognathus peddalis* (Linognathidae), which parasitizes sheep in South America (Kim and Weisser 1974). *Microthoracicus mazzai* (fig. 7.2c), *M. praelongiceps*, and *M. minor* are found on the llama; the last species is also a parasite of the alpaca, while the second also occurs on the vicuña (Ferris 1951).

**Figure 7.2  SUCKING LICE.** (a) Hog louse (*Haematopinus suis*, Haematopinidae). (b) Wild pig louse (*Pecaroecus javalii*, Haematoponidae). (c) Llama louse (*Microthoracicus mazzai*, Linognathidae). (d) Human louse (*Pediculus humanus*, Pediculidae). (e) Crab louse (*Phthirus pubis*, Pediculidae) and egg (nit).

The remainder are found on wild mammals of many sorts. *Pecaroecus* (Haematopinidae, fig. 7.2b) lives on peccaries (Babcock and Ewing 1938). The genera *Hoplopleura* and *Polyplax* (Hoplopleuridae) are widespread on rodents.

### References

BABCOCK, O. G., AND H. E. EWING. 1938. A new genus and species of Anoplura from the peccary. Entomol. Soc. Wash. Proc. 40: 197–210.

FERRIS, G. F. 1951. The sucking lice. Pacific Coast Entomol. Soc. Mem. 1: 1–320.

KIM, K. C., AND H. W. LUDWIG. 1978. The family classification of the Anoplura. Syst. Entomol. 3: 249–252.

KIM, K. C., AND C. F. WEISSER. 1974. Taxonomy of *Solenopotes* Enderlein, 1904, with redescription of *Linognathus panamensis* Ewing (Linognathidae: Anoplura). Parasitology 69: 107–135.

## PRIMATE LICE

### Human Louse

Pediculidae, *Pediculus humanus*. *Spanish:* Piojo, liendres (eggs). *Nahuatl:* Atémitl. *Quechua:* Usa. *Portuguese:* Piolho, piolho ladro, lêndeas (eggs). *Tupi-Guaraní:* Muquirana. Nits (eggs).

The human louse (Buxton 1947) has been an intimate companion of humans as long as we have existed, and they may have lived on our subhuman progenitors. Ewing (1926) recognized distinct subspecies of human louse (not now recognized, see below), associated with the Negroid, Caucasoid, Asian, and American Indian host races. The last were described from specimens taken from the scalps of 4,000-year-old, pre-Columbian mummies found in Peru (see also Weiss 1932). These originally distinct types have formed hybrids with the contemporary melding of humanity and are no longer recognizable, but their existence indicates that the species is to be considered indigenous to all areas of human occupation.

The human louse is well established in Latin American history (Hoeppli 1969). A louse plays a part in the Popol-Vuh, the very famous book of Mayan mythology. The story tells of Ixmucané, a goddess in the shape of an old woman who sends an important message by a louse to her grandsons. Chronicles of the Mexican conquest referred to bags of lice in the treasure houses of Moctezuma, but these may have been stocks of dried cochineal bugs. Oviedo, speaking of the household of the Aztec emperor, mentions priests who were wearing long hair full of lice that they were catching and eating while murmuring prayers. Actual lice were collected by the Inca rulers of Peru as a tax that sufficed from the poor and destitute (de la Vega 1979). A curious belief persisted among early travelers to the New World with regard to lice. It was thought that lice disappeared about the longitude of the Azores on the voyage west and reappeared on the return to Europe (Hogue 1981).

Two human louse subspecies are distinguished, based on the tendency of populations to segregate either on the head ("head lice," *Pediculus humanus capitis*) or body ("body lice," *P. h. corporis*). They are scarsely distinguishable structurally, they hybridize readily, and many even cross into opposite habitats and show only some behavioral differences so that they are obviously variants of one species.

The species (fig. 7.2d) is small (BL 2–3 mm) and gray to brown; blooded individuals usually show a large, dark globule in the abdomen representing the coagulated last meal. The flattened body is elongate, the abdomen elliptical in outline. The antennae are clearly visible projecting from the anterolateral margins of the spherical head. The single sharp pointed claws at the tips of the stout legs are also conspicuous.

Infestations of the human louse (pediculosis) are common throughout the world and remain an accepted way of life among peasants and indigents as well as peoples whose cultures do not emphasize bodily cleanliness. People normally feel extreme discomfort when lice are biting, and the skin becomes sensitive to continued feeding, developing red papules and rashes and possibly becoming hardened and deeply pigmented ("vagabond's disease"). It is still a common sight among Indians and in the Andes to see mothers grooming the hair of their children and removing lice and nits and usually eating them. Many other ways of killing or discouraging lice are practiced, using plant extracts (Lenko and Papavero 1979: 120f.). They are also used in making remedies.

The human louse is a vector of epidemic typhus (*tabardillo* in Mexico; classical typhus), a very serious disease found throughout the world caused by *Rickettsia prowazekii*. In Latin America, it is most prevalent in colder climates where the body louse, the primary carrier, is present. It is a poverty-associated disease and also a common result of social disorder during wars or following catastrophe. It currently has receded but remains a potential threat to human health.

This species of *Pediculus* is restricted to *Homo sapiens*. Members of the subgenus *Parapediculus* live on New World spider monkeys (*Ateles*). Human lice are distinguished by having flat lateral plates on the abdomen instead of lobed as in the monkey lice and have much heavier setae and a more weakly sclerotized cuticle.

Human lice are transmitted by close body contact, although they cling to clothing and may leave the body and range over bedding and thus gain access to new individuals. Infestations are usually heaviest when persons are crowded into sleeping quarters and where bathing and cleaning facilities are inadequate. In tropical climates, body lice tend to disappear, probably because of their sensitivity to high skin temperatures, but head lice remain.

The life cycle is simple. After hatching, the nymphs begin sucking blood at once and feed frequently throughout life, both night and day. They reach maturity after three molts in 16 to 18 days. Females mate and begin to attach their eggs to hairs and clothing a day or two after reaching adulthood. If unfed, these lice very soon die.

## References

BUXTON, P. A. 1947. The louse: An account of the lice which infest man, their medical importance and control. 2d ed. Arnold, London.

DE LA VEGA, G. 1979. The Incas: The royal commentaries of the Inca Garcilaso de la Vega. Translated by M. Jolas. Lib. A B C, Lima.

EWING, H. E. 1926. A revision of the American lice of the genus *Pediculus*, together with a consideration of the significance of their geographical and host distribution. U.S. Natl. Mus. Proc. 68: 1–30.

HOEPPLI, R. 1969. Parasitic diseases in Africa and the Western Hemisphere, early documentation and transmission by the slave trade. Acta Trop. suppl. 10: 1–240.

HOGUE, C. L. 1981. Commentaries in cultural entomology. 2. The myth of the louse line. Entomol. News 92: 53–55.

LENKO, K., AND N. PAPAVERO. 1979. Insetos no folclore. Conselho Est. Artes Ciên. Hum., São Paulo.

WEISS, P. 1932. Restos humanos de Cerro Colorado. Rev. Mus. Nac. (Lima) 1(2): 90–102.

## Crab Louse

Pediculidae, *Phthirus pubis*. *Spanish:* Ladilla (General). *Portuguese:* Chato (Brazil). Pubic louse.

This bloodsucking louse (Payot 1920) inhabits the human body along with the human louse but seldom strays from the pubic region. It is distinct in form, crablike, with a short broad body and very stout laterally projecting legs (fig. 7.2e). It is smaller than the human louse (BL 1.5–2 mm) but about the same color.

Although it infests the genital area primarily, the crab louse occasionally migrates to the armpits or other hairy parts of the body—the eyelids, beard, eyebrows. Individuals tend to be stationary, remaining attached for days at one point with the mouthparts inserted in the skin. Continued defecation by the louse during this time results in the accumulation of excretory material around the site.

The life cycle (Nutall 1918) requires approximately a month. The female deposits eggs on the coarser hairs of the host. Nymphs and adults feed almost continuously, causing skin rashes and intense itching. Transmission is by physical contact, most often sexual intercourse, but the louse may spread on contaminated towels, blankets, and clothing.

### References

NUTALL, G. H. F. 1918. The biology of *Phthirus pubis*. Parasitology 10: 383–405.

PAYOT, F. 1920. Contribución à l'étude du *Phthirus pubis* (Linné, Leach). Soc. Vaud. Sci Nat Bull. 53: 127–161.

## FLEAS

Siphonaptera. *Spanish* and *Portuguese:* Pulgas. *Nahuatl:* Tecpintin, sing. tecpin (Mexico). *Quechua:* Pique kuna.

Fleas are all too familiar insects. Several species infest pets and domestic animals and bite humans. As bloodsuckers, they transmit diseases, among the more serious, murine typhus (*Rickettsia typhi*) and plague (*Yersinia pestis*) (see Medical Entomology, chap. 3).

These are small (BL 1–6 mm), dark brown insects with sclerous bodies that are wingless and strongly compressed laterally. The head projects forward as a rounded shield and bears short, three-segmented, clubbed antennae and sometimes dark ocular spots. The head, thorax, and anterior abdominal segments may also have rows of stiff, flat, scalelike bristles (ctenidia) that aid in mobility through hairs or feathers. The legs and body have prominent setae. The mouthparts are elongated and supremely adapted for bloodsucking with three saberlike stylets and associated palpi and channeled labium. The legs, especially

the posterior pair, are much larger and powerfully constructed and give the fleas incredible leaping capabilities (Rothschild et al. 1973).

Fleas are usually confined to animals that customarily burrow and have a den, ground nest, or habitual resting place. Insectivores and rodents are common hosts, but fleas are ectoparasites of owls, swallows, domestic fowl, rodents, bats, and many other mammals (except aquatic ones). Host specificity varies, some flea species being catholic in their tastes, others highly restricted to one host species (da Costa Lima and Hathaway 1946).

Unlike the other ectoparasite orders discussed in this chapter, fleas have complete metamorphosis. The larval stage lives in the soil and organic debris that collects in the host's nest, feeding on cast-off blood, feces, and scurf from the animals living there. After attaining full growth, the larva (fig. 7.3b) spins a cocoon in which it pupates (fig. 7.3e). The pupa is inactive and passes a developmental period of a few days to several months while metamorphosing into the adult flea. Adults remain on the hosts most of the time but must go elsewhere, especially when hosts leave the nest or perish. Eggs are placed randomly, but the timing of egg production is dependent on the reproductive cycle of the host. Copulation and oviposition occur most often prior to nesting, egg laying, or parturition of the vertebrates on which the fleas' life depends.

Blood is the only adult food. When passing into the stomach, the blood courses through a check valve (proventriculus) lined with hundreds of prominent spines. After the flea has engorged, the valve prevents a back flow of fluid into the foregut. This structure may become clogged with plague bacilli that develop after a blood meal on an infected rat or human. The organisms are regurgitated forward during subsequent feeding attempts and enter the bloodstream of a

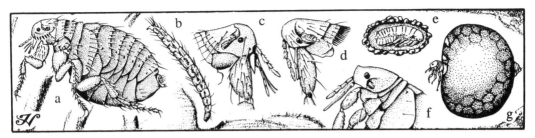

**Figure 7.3 FLEAS.** (a) Cat flea (*Ctenocephalides felis,* Pulicidae). (b) Cat flea larva. (c) Oriental rat flea (*Xenopsylla cheopis,* Pulicidae). (d) Rodent flea (*Dasypsyllus lasius,* Dolichopsyllidae). (e) Cat flea, pupa in cocoon. (f) Chigoe (*Tunga penetrans,* Tungidae), male. (g) Chigoe, gravid female.

new individual, where the disease can develop.

In Latin America, the flea fauna is diverse and well developed, with a total number of species probably in excess of five hundred, but flea taxonomy in this part of the world is incomplete (Johnson 1957, Traub 1950). The flea fauna of South America is notable for its high percentage of endemic higher taxa. Some, like the Stephanocircidae (helmeted fleas) and Pygiopsyllidae, found elsewhere only in the Australasian region, suggest that the original hosts of fleas were marsupials (Wenzel 1972). Other primarily South American flea families are Malacopsyllidae and Rhopalopsyllidae. The former live on edentates and carnivores; they have enlarged tarsal claws and a leathery integument; the latter are parasites of rodents and a variety of other mammals.

As with other ectoparasites, fleas may be grouped according to those associated with household and husbanded animals, sometimes secondarily attacking humans and probably mostly introduced from the Old World, and indigenous species on wild mammals and wild birds.

Humans suffer mostly from the species found on animals kept in and around homes, because of their shared abode. The dog flea (*Ctenocephalides canis*) and cat flea (*C. felis,* fig. 7.3a) are primary in this respect. The Oriental rat flea (*Xenopsylla cheopis,* fig. 7.3c), another cosmopolitan species, ranges freely from its hosts, often

biting people; it is the premier vector of plague. Other domestic rodent fleas are the mouse flea (*Leptopsylla segnis*) and the northern rat flea (*Nosopsyllus fasciatus*). The human flea (*Pulex irritans*) has a wide range of hosts among animals whose lives are encouraged by human activity and is another excellent transmitter of plague, including two unusual types of the disease found in Ecuador, vesicular (virola pestosa) and tonsillar (angina pestosa) forms.

The sticktight flea (*Echidnophaga gallinacea*) attaches to unfeathered skin on the head of domestic fowl, causing considerable agitation. It is a relative of the burrowing flea (see below) but rarely infests humans.

Bat fleas form the family Ischnopsyllidae and mainly associate with insectivorous chiropteran species. Their droppings provide a better nutrient medium on roost floors than those of the fruit-eating or carnivorous bats. *Ceratophyllus* (Ceratophyllidae) and *Dasypsyllus* (Dolichopsyllidae, fig. 7.3d) are common genera mainly associated with wild birds.

## References

DA COSTA LIMA, A. AND C. R. HATHAWAY. 1946. Pulgas: Bibliographia, catálogo e animais por elas sugados. Inst. Ozwaldo Cruz Monogr. 4: 1–522.

JOHNSON, P. T. 1957. A classification of the Siphonaptera of South America. Entomol. Soc. Wash. Mem. 5: 1–299.

ROTHSCHILD, M., Y. SCHLEIN, K. PARKER, C. NEVILLE, AND S. STERNBERG. 1973. The flying leap of the flea. Sci. Amer. 229(5): 92–100.

TRAUB, R. 1950. Siphonaptera from Central America and Mexico. Fieldiana Zool. Mem. 1: 1–127.

WENZEL, R. L. 1972. Some observations on the zoogeography of Middle and South American ectoparasites. J. Med. Entomol. 9: 589.

## BURROWING FLEA

Tungidae, *Tunga penetrans. Spanish:* Nigua, chique (General). *Portuguese:* Bicho de pé, bicho de porco, jatecuba (Brazil). *Indian:* Sika, tunga (Brazil). *French:* Chique. Chigoe, jigger, sand flea.

Unlike other fleas, females of this species work their way into the skin of their hosts and encyst, enlarging to several times their original size, up to the size of a small pea (3–5 mm diameter) (fig. 7.3g). The male, which is mobile and active throughout its life, copulates with the female after she has completely penetrated the skin. Females may enter skin anywhere on the body but most often burrow under the toenails or into the soles of the feet of humans where the pressure of their growth causes great discomfort. Nodular, ulcerating swellings result. Persons going barefoot in infested areas (usually where pigs, the usual alternate hosts, are common) are likeliest to pick up the fleas in this way. They are removed only precariously with the tip of a sterile needle; if the insect's body is ruptured, infection and immune reactions with serious consequences may follow, including gangrene and loss of the appendage. Many of the early settlers of Brazil lost their feet to this insect in a dreadful manner. To be infested is referred to as *cambado* in that country.

The original home of this flea appears to be America (Hoeppli 1969: 169f.). This is based on the evidence of much earlier accounts of the species from tropical America (dating to 1526 by Oviedo) than from Africa, where it is not definitely recorded until the mid-1700s. Also, anthropomorphic clay vessels from pre-Incan Peru show feet unmistakably infested by the flea. Although there are earlier reports of this flea in Africa, a known introduction occurred in 1872 by a British ship traveling from Rio de Janeiro to Angola, whence it spread rapidly along the West African coast and subsequently disseminated to the interior. The early explorers of the Amazon Basin found it a terrible nuisance: "It is a great happiness that its legs have not the elasticity with those of [other] fleas; for could this insect leap, every animal body would be filled with them; and, consequently, both the brute and human species be soon extirpated by the multitude of these insects" (de Ulloa 1758). The pest became so well known to the Brazilians that a body of folklore and poetry has evolved around them (Lenko and Papavero 1979).

Indians have effective methods for the extraction of encapsulated fleas. The most ancient technique utilized fish spines, wood splinters, and plant spines to pry them out physically. Later, pointed knives and needles were employed. Certain women and children in the interior of Brazil became adept at removing the fleas and received considerable status for this specialized profession. Various other remedies relying on the application of poultices of all kinds are ineffectual (ibid.).

The burrowing flea is a small (BL 1.0 mm), reddish-brown flea with a somewhat stubby appearance from the contracted thoracic segments (fig. 7.3f). The large head and thorax completely lack combs. The outline of the head is angular, and the mouthparts are overly large and stiff, with a four-segmented palpus. The stiff elongate mandibles are used to slash open the skin to permit the entry of the body when the female inserts herself into a new locus.

Both males and females suck blood, but only the females burrow into skin. Besides pigs, other domestic and wild animals serve as hosts (dogs, rodents, and burrowing owls). All are fossorial and probably pick up the fleas from the ground when digging in it. The larvae are typically flealike in form; they may hatch within the sinus formed by the mother but usually drop to the ground to develop.

The species is tropical, ranging only a few degrees north and south of the equator, on the continent and in the West Indies. It may be found in abundance in dry, sandy places, in animal enclosures, and on ranches and farms, especially where pigs are raised. Other known species of *Tunga* are of no medical significance.

## References

DE ULLOA, A. 1758. Relación histórica del viaje a la América Meridional. 4 vols. Madrid.

HOEPPLI, R. 1969. Parasitic diseases in Africa and the Western Hemisphere, early documentation and transmission by the slave trade. Acta Trop. suppl. 10: 1–240.

LENKO, K., AND N. PAPAVERO. 1979. O bicho-do-pé—nasce fêmea, morre macho. *In* K. Lenko and N. Papavero, Insetos no folclore. Cons. Est. Artes Ciên. Hum., São Paulo. Pp. 479–498.

# 8 BUGS

Hemiptera. *Spanish:* Chinches. *Portuguese:* Percevejos.

In English, the term "bug" has a precise meaning for insects of this order, in addition to its general application to all kinds of small insectoid creatures. The origin of the word is obscure. Theories suggest that it comes either from the Old English or Welsh *bwg*, meaning a goblin or ghost, or from the Arabic *buk,* a longstanding and widespread name for the infamous bedbug, *Cimex lectularius* (Usinger et al. 1966: 5).

True bugs (Weber 1968) are all recognizable by their mouthparts, which are modified for sucking fluids, such as plant sap, nectar, and insect or vertebrate blood. There are two pairs of sclerotized, flexible stylets (modified mandibles and maxillae) lying in a groove in a one- to four-segmented labium. Together, these structures form a proboscis, always arising on the front of the head but flexible and when not in use, projecting toward the posterior between the forelegs.

There are two suborders, separated most evidently by wing structure. The first is the Heteroptera ("uneven wings"), with fore wing divided midway into two areas of radically different textures: a basal thick and rigid part and an apical papery and flexible portion. This type of wing is referred to as a hemielytron and is characteristic of the heteropterans. The other suborder is the Homoptera ("uniform wings"), which have homogeneous, diaphanous, or parchmentlike fore wings. Because it is a widespread habit of the Homoptera to produce waxy secretions from integumentary glands, I use the convenient and appropriate common name "wax bugs" here. Wings are partially or totally absent in many members of both suborders, especially among animal and plant ectoparasites or forms with secluded life-styles.

## References

USINGER, R. L., J. CARAYON, N. T. DAVIS, N. UESHIMA, AND H. E. MCKEAN. 1966. Monograph of the Cimicidae. Vol. 7. Entomol. Soc. Amer. (Thomas Say Foundation), College Park, Md.

WEBER, H. 1968. Biologie der Hemipteren. Asher, Amsterdam.

## HETEROPTERANS

Hemiptera, Heteroptera.

In this suborder of bugs, the fore wing is composed of two distinct areas as described above. The thickened basal portion may be subdivided further into elongate triangles, an anterior, basal corium, and a small, posterior clavus. The soft, papery membrane makes up a third or more of the apical part of the wing. Veins are apparent in the membrane but rarely in the other areas.

Less apparent but highly characteristic of the suborder are repugnatorial or stink glands that are present in most families. In the adult, these open ventrally on the thorax, often through spoutlike processes, near the base of the middle pair of legs. Scent glands also are present in the nymphs but open on the back of the abdomen. Although they may attract others of the same species, having aggregating or alarm-calling functions, the products of these glands are usually foul smelling and repugnant. Their chemical composition varies greatly and is complex. In coreids and relatives (Aldrich and Yonke 1975), they contain varied aliphatic acids, aldehydes, alcohols, and esters. These chemicals readily volatilize into the environment and seem to have a repellent effect on other insects and animals. They may also be rubbed or squirted on an assailant to burn the skin or other vulnerable tissues. In Trinidad and the Guianas, the "pepper flies" (*Amnestus*) may accidentally enter the eye, where they liberate a strong, caustic secretion that causes an agonizing, persistent burning comparable to that of cayenne pepper (Myers 1934).

Assassin bugs possess formidable bites in which they inject toxins in their saliva that can cause humans excruciating pain (Ryckman 1979, Ryckman and Bentley 1979). Two major heteropteran groups are medically significant as disease vectors or noxious bloodsuckers: the kissing bugs and the bedbugs.

Most heteropterans are phytophagous, sucking sap and tissue liquids from all parts of plants. Some kinds spend much of their lives exposed on tree trunks where their cryptic colors and rough integuments afford them camouflage against similarly appearing bark. Wetting and darkening of the substratum by rain would destroy this protection, but the cuticle of at least two species of bark bugs (Aradidae) absorb water and undergo corresponding color change, thus preserving this protective device (Silberglied and Aiello 1980).

A recurring phenomenon in heteropteran families is mimicry of other insects, especially of ants (fig. 8.1a, b). Adults or nymphal stages may be involved. There are numerous examples. A species of *Barberiella* (Miridae) is a member of a Batesian complex of several different insects and a spider that mimics *Camponotus planatus*, an aggressive ant in Honduras (Jackson and Drummond 1974). The Brazilian bug *Thaumastaneis montandoni* (Pyrrhocoridae) (Hussey 1927), many *Arhaphe* (Largidae), and a *Hyalymenus* species (Alydidae) (da Costa Lima 1940: 90) have ant-shaped bodies, but the specific models are yet unknown.

Some heteropterans are also myrmecophiles, termitophiles, or arachnophiles. In South America, nymphs of *Arachnocoris albomaculatus* (Nabidae) live in company

**Figure 8.1 HETEROPTERANS.** (a) Ant mimicking bug (*Hyalymenus* sp., Alydidae). (b) Ant model for ant mimicking bug (*Camponotus* sp., Formicidae). (c) Giant big-legged bug (*Pachylis pharaonis*, Coreidae). (d) Leaf-legged bug (*Diactor bilineatus*, Coreidae). (e) Rice stinkbug (*Oebalus poecilus*, Pentatomidae). (f) Conchuela (*Chlorochroa ligata*, Pentatomidae).

with certain spiders and generally have the appearance of their host, with long slender legs and slightly bulbous abdomen (Myers 1925).

A bizarre nymphal mirid (possibly *Paracarnus*) is found in Cuba (China 1931). It bears a four-part, peduncled prothoracic horn, giving it a resemblance to certain treehoppers (genus *Cyphonia*?), with which it may have some association, possibly as a predator.

Although primarily treating the Ecuadorian fauna, Froeschner (1981) provides a useful compendium of the suborder for Latin America.

### References

ALDRICH, J. R., AND T. R. YONKE. 1975. Natural products of abdominal and metathoracic scent glands of coreoid bugs. Entomol. Soc. Amer. Ann. 68: 955–960.

CHINA, W. E. 1931. A remarkable mirid larva from Cuba, apparently belonging to a new species of the genus *Paracarnus*, Dist. (Hemiptera, Miridae). Ann. Mag. Nat. Hist. (ser. 10) 8: 283–288.

DA COSTA LIMA, A. 1940. Hemípteros, insectos de Brasil. Vol. 4. Escuela Nac. Agron., Ser. Didac., Rio de Janeiro.

FROESCHNER, R. C. 1981. Heteroptera or true bugs of Ecuador: A partial catalog. Smithsonian Contrib. Zool. 322: 1–147.

HUSSEY, R. F. 1927. On some American Pyrrhocoridae, Brooklyn Entomol. Soc. Bull. 22: 227–235.

JACKSON, J. F., AND B. A. DRUMMOND III. 1974. A Batesian ant-mimicry complex from the mountain pine ridge of British Honduras, with an example of transformational mimicry. Amer. Midl. Nat. 91: 248–251.

MYERS, J. G. 1925. Biological notes on *Arachnocoris albomaculatus* Scott (Hemiptera; Nabidae). N.Y. Entomol. Soc. J. 33: 136–146, pl. 1.

MYERS, J. G. 1934. Field observations on some Guiana insects of medical and veterinary interest. Trop. Agric. 11: 279–283.

RYCKMAN, R. E. 1979. Host reactions to bug bites (Hemiptera, Homoptera), a literature review and annotated bibliography. Calif. Vect. Views 26(1–2): 1–24.

RYCKMAN, R. E., AND D. G. BENTLEY. 1979. Host reactions to bug bites (Hemiptera, Homoptera): A literature review and annotated bibliography. Calif. Vect. Views 26(3–4): 25–30.

SILBERGLIED, R., AND A. AIELLO. 1980. Camouflage by integumentary wetting in bark bugs. Science 207: 773–775.

## BIG-LEGGED BUGS

Coreidae. Squash bugs.

Big-legged bugs are common and typical Heteroptera, generally similar to stink bugs and seed bugs but distinguished by the numerous complex veins in the membranous portion of the fore wing which the others lack. Although many have all slender legs, the hind legs are often enlarged and at times very stout, especially the femur, which may be greatly swollen; the tibia also is sometimes wide and flattened, as in the so-called leaf-legged bugs. Because of this common hypertrophy of the legs and their diverse food, I have coined the common name used here which seems more widely applicable and descriptive than "squash bugs," better restricted to a few genera such as *Anasa*.

Most are small to medium-sized bugs (BL 10–15 mm), but some are large and robust, such as *Pachylis* (fig. 8.1c), which may be over 30 millimeters long. They usually are dull brown or greenish, but quite a few display red or yellow lines on the fore wings, and the dorsum of the abdomen is often colored brilliantly in red or yellow. These parts of the body are displayed when the bug is molested in what are believed to be warning signals of the foul tastes and smells many of them possess. The nymphs also form aggregations, which enhance the noxious effect. Nymphs of *Thasus acutangulus*, for example, are vivid orange, yellow, and black and group together on twigs of *Pithecellobium* trees in Central America. When alarmed, they pulsate violently, spray jets of anal fluid, and exude a noxious secretion over the back of the abdomen. *Paryphes blandus* nymphs react similarly, giving off a pungent odor.

A few coreids are predaceous, but most

feed on plants, many, such as the injurious genus *Anasa,* on vines of Cucurbitaceae (Brailovsky 1985). A common pest of the passion vine (*Passiflora*) is *Diactor bilineatus* (fig. 8.1d), a leaf-legged bug (*percevejo do maracujá*). It is colorfully marked with yellow spotted black leg expansions and a green dorsum with an orange chevron; the nymphs are bizarre, with black legs and wing buds, white collar, and orange abdomen with large black buttons.

An estimate of the number of Neotropical coreid species cannot be made at present; there are at least one hundred genera.

## References

ALDRICH, J. R., AND M. S. BLUM. 1978. Aposematic aggregation of a bug (Hemiptera: Coreidae): The defensive display and formation of aggregations. Biotropica 10: 58–61.

BRAILOVSKY, H. 1985. Revisión del género *Anasa* Amyot-Serville (Hemiptera-Heteroptera-Coreidae-Coreinae-Coreini). Inst. Biol. Univ. Nac. Aut. Mexico Monogr. 2: 1–266.

YOUNG, A. M. 1980. Notes on the interaction of the Neotropical bug *Paryphes blandus* Horvath (Hemiptera: Coreidae) with the vine *Anguria warscewiczii* Hook F. (Curcurbitaceae). Brenesia 17: 27–42.

# STINKBUGS

Pentatomidae. *Spanish:* Conchuelas (General). *Portuguese:* Persevejos do mato, Marias fetidas (Brazil).

As their name implies, stinkbugs have well-developed and very effective repugnatorial glands and are known as much by their smell as by their physical attributes. Among the latter are a broadly oval or shield-shaped body with a large, triangular central plate (scutellum) in the middle of the back. The lateral corners of the prothorax and posterolateral corners of the abdominal segments are often prolonged or sharp pointed. The legs are always simple and similar in size and length. Most are plain colored, green or brown, but some, especially in the immature stages, are painted in bright hues of red, blue, and yellow (pl. 1d).

Both the nymphs and adults suck sap and may be very abundant on their host plants (Young 1984). The latter are found among diverse taxa. Because of their habits of feeding on cultivated plants, many species are pests. Such is the rice stinkbug (*Oebalus poecilus;* fig. 8.1e), the bane of rice in many areas (*pulga d'anta, chupador, tamanjuá*). Other injurious species are the conchuela (*Chlorochroa* = *Petidia ligata;* fig. 8.1f) and several members of the genus *Euschistus* and *Mormidea*. Some, however, are considered beneficial because they attack other insects which may be enemies of cultigens.

Nymphs, especially the younger instars, are often gregarious. Subsocial behavior is exhibited by the Colombian *Antiteuchus tripterus* (Discocephalinae). Its females guard both their eggs and first instar nymphs against predators and parasitoid wasps of the family Scelionidae (Eberhard 1975).

Species of Pentatomidae (*jumiles*) are sold in the Mexican villages of Cuautla and Taxco for human consumption. Marketed live in paper cones in handful lots, they have a piquant taste and are believed to alleviate liver, kidney, and stomach ailments when ingested (orig. obs., Ancona 1932, 1933).

There are more than a thousand species of stinkbugs in some ninety genera in South America alone.

## References

ANCONA, L. 1932. Los jumiles de Taxco (Gro.), *Atizies taxcoensis,* spec. nov. Inst. Biol. Univ. Mexico Anal. 4: 149–162.

ANCONA, L. 1933. Los jumiles de Cuautla, *Euschistus zopilotensis* Distant. Inst. Biol. Univ. Mexico Anal. 4: 103–108.

EBERHARD, W. G. 1975. The ecology and behavior of a subsocial pentatomid bug and two scelionid wasps: Strategy and counter strategy in a host and its parasites. Smithsonian Contrib. Zool. 205: 1–39.

Young, A. M. 1984. Phenological patterns in reproduction of *Senna fructicosa* (Mill.) Irwin and Barneby (Caesalpinaceae) and a pod associate, *Pellaea sticta* (Dallas) (Heteroptera: Pentatomidae), in Costa Rican tropical rain forest. Kans. Entomol. Soc. J. 57: 413–422.

## SEED BUGS

Lygaeidae.

This family is composed of elongate, small to medium-sized bugs (BL 3–15 mm) characterized by a wing membrane with only four or five veins. The antennae and beak are four segmented. Most are drab, but many display reds, yellows, and other bright colors that are doubtlessly aposematic in function (Sillén-Tullberg et al. 1982). Polymorphism in wing size occurs; usually the hind pair is atrophied or lost, and the membranous tip of the fore wing may be reduced.

Members of this family feed largely on the mature seeds of various plants, a large number from figs (*Ficus*) (Slater 1972). The orange- and black-banded milkweed bugs, *Oncopeltus*, form a conspicuous group (Ojeda 1973). These infest milkweed plants, especially the widespread weed *Asclepias curassavica*, on the seed pods of which they may be found in clusters (Root and Chaplin 1976). These and *Lygaeus* (and probably others) feed on many asclepiadaceous plants, which contain cardiac glycosides. The bugs probably sequester these toxic alkaloids for their own protection. *Oncopeltus fasciatus* (fig. 8.2a) is widely cultured as a laboratory animal for research in physiology and other areas (Feir 1974).

Several Central American lygaeids exhibit very strong resemblances to beetles (coleoptery), with shell-like, coreaceous hemielytra, which meet evenly along the midline (Slater 1985). The functional significance of the resemblance is not known.

A few species are crop pests. The infamous chinch bug (*Blissus leucopterus;* fig. 8.2b) causes considerable damage to rice in Peru. Others are beneficial. The big-eyed bug (*Geocoris punctipes*) is an efficient predator of red spider mites on cotton plantations.

There are about 450 species of this family in Latin America (Slater 1964).

### References

Feir, D. 1974. *Oncopeltus fasciatus:* A research animal. Ann. Rev. Entomol. 19: 81–96.

Ojeda, D. 1973. Contribución al estudio del género *Oncopeltus* Stål (Hemiptera: Lygaeidae). Rev. Peruana Entomol. 16: 88–94.

Root, R. B., and S. J. Chaplin. 1976. The lifestyles of tropical milkweed bugs, *Oncopeltus* (Hemiptera: Lygaeidae), utilizing the same hosts. Ecology 57: 132–140.

Sillén-Tullberg, B., C. Wiklund, and T. Järvi. 1982. Aposematic coloration in adults and larvae of *Lygaeus equestris* and its bearing on Müllerian mimicry: An experimental study on predation on living bugs by the great tit *Parus major*. Oikos 39: 131–136.

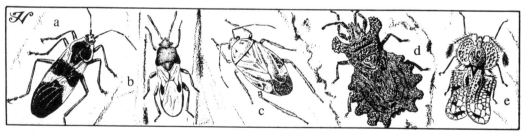

**Figure 8.2 HETEROPTERANS.** (a) Large milkweed bug (*Oncopeltus fasciatus*, Lygaeidae). (b) Chinch bug (*Blissus leucopterus*, Lygaeidae). (c) Tarnished plant bug (*Lygus lineolaris*, Miridae). (d) Lunate flat bug (*Dysodius lunatus*, Aradidae). (e) Cotton lace bug (*Corythucha gossypii*, Tingidae).

SLATER, J. A. 1964. A catalogue of the Lygaeidae of the world. 2 vols. Univ. Conn. Storrs.

SLATER, J. A. 1972. Lygaeid bugs (Hemiptera: Lygaeidae) as seed predators of figs. Biotropica 4: 145–151.

SLATER, J. A. 1985. A remarkable new coleopteroid lygaeid from Colombia (Hemiptera: Heteroptera). Int. J. Entomol. 27: 229–234.

# PLANT BUGS

Miridae. Leaf bugs.

The best-known plant bugs are the many species that cause direct harm from their feeding. Damage often takes the form of rough, hard, gall-like lesions (stigmonosis) on the surface of valued leaves (especially on tobacco), fruit (commonly cacao), or tender stems. Chief among the offending genera are *Lygus*, *Engytatus*, and *Monalonion*. Although members of the family are widely recognized for their depredations among field crops, many are beneficial predators of aphids and other injurious insects. A widespread crop pest is the tarnished plant bug (*Lygus lineolaris*, fig. 8.2c).

For the most part, these are small (BL less than 10 mm in most), elongate bugs with the hemielytra characteristically "broken" or angled sharply downward at the base of the membrane. A small notch (cunneus) at this point, only one or two cells in the veins of the hemielytral membrane, and lack of ocelli are also family characteristics. Their bodies are slightly flattened, soft, and frequently have brightly colored, elongate stripes or harlequin patterns. Many species, especially in the subfamily Mirinae, are ant mimics (see Heteropterans, above).

This is the largest family of heteropterans, with several thousand species in the Neotropics (Carvalho 1957–1960).

## References

CARVALHO, J. C. M. 1957–1960. A catalogue of Miridae of the world (Hemiptera). Mus. Nac. Rio de Janeiro Arch., Pt. 1, 44: 1–158; Pt. 2, 45:1–216; Pt. 3, 47: 1–161; Pt. 4, 48: 1–384; Pt. 5, 51–194.

# LUNATE FLAT BUG

Aradidae, Mezirinae, *Dysodius lunatus*.

The extraordinarily depressed shape of this bug (fig. 8.2d) and its family relatives constitute an adaptation primarily for life on or under the bark of dead trees. These bugs also possess tremendously long (maxillary) stylets for sucking the juices of fungi that grow well in their habitat. These structures are coiled up within the head when not in use.

The integument is fluted, scored, tuberculate, and very coarsely roughened, which, with its somber colors, gives the insect such a resemblance to the substrata of lichens, eroded wood, and such mottled surfaces that it is not easily seen. Specimens are usually found exposed on the bark of trees but are detected with difficulty because of their cryptic form and color (Silberglied and Aiello 1980).

This species is the best-known representative of 472 species of Aradidae in the Neotropics (Kormilev and Froeschner 1987, Usinger and Matsuda 1959). It is very large by family standards (BL to 15 mm) and has scallops on the margins of the abdomen and the sides of the prothorax, the latter greatly produced anteriorly as broad, platelike lobes, rounded on the outer margins and extending forward well beyond the level of the eyes. Adults have complete wings.

## References

KORMILEV, N. A., AND R. C. FROESCHNER. 1987. Flat bugs of the world: A synonymic list (Heteroptera: Aradidae). Entomography, Sacramento.

SILBERGLIED, R., AND A. AIELLO. 1980. Camouflage by integumentary wetting in bark bugs. Science 207: 773–775.

USINGER, R. L., AND R. MATSUDA. 1959. Classification of the Aradidae (Hemiptera-Heteroptera). Brit. Mus. Nat. Hist., London.

## LACE BUGS

Tingidae.

In this family (Drake and Davis 1960), the entire dorsal surface, including the wings, has taken on an alveolar or reticulate appearance. Like tiny panes of glass, transparent membranes enclose the spaces between a complex lacework of cells. These may form inflated sacs or broad, winged expansions on the sides and rear of the prothorax, the latter extending forward over the head. The surface of some may be profusely spined in addition. Because they are generally small (BL usually less than 5 mm), one may appreciate this structure only when the bug is beneath the microscope.

Because of its economic importance, the family has received considerable taxonomic attention (Drake and Ruhoff 1960). All of the approximately 615 Latin American species (Drake and Ruhoff 1965) live on the foliage of trees and shrubs. Ordinarily, they congregate on the undersurfaces of leaves, and their sap-sucking, especially by members of the large genus *Corythucha,* occasionally causes harm to crops such as cotton. *C. gossypii* (fig. 8.2e) is a widely distributed injurious species with many hosts among the cultivated plants (Leonard and Mills 1931).

Blind, beetlelike members of the subfamily Vianaidinae are atypically smooth, only the surface being lightly punctate. They live symbiotically with ants underground (Drake and Davis 1960).

### References

DRAKE C. J., AND N. T. DAVIS. 1960. The morphology, phylogeny, and higher classification of the family Tingidae, including the description of a new genus and species of the subfamily Vianaidinae (Hemiptera: Heteroptera). Entomol. Amer. 39: 1–100.

DRAKE, C. J., AND R. A. RUHOFF. 1960. Lace-bug genera of the world (Hemiptera: Tingidae). U.S. Natl. Mus. Bull. 112: 1–105.

DRAKE, C. J., AND R. A. RUHOFF. 1965. Lace bugs of the world: A catalog (Hemiptera: Tingidae). U.S. Natl. Mus. Bull. 213: 1–634.

LEONARD, M. D., AND A. S. MILLS. 1931. Observations on the bean lace bug in Puerto Rico. Dept. Agric. Puerto Rico J. 15: 309–323.

## ASSASSIN BUGS

Reduviidae.

These are mostly medium to large bugs of varied form but always with a narrowly attached, elongate head that bears a flexible, segmented beak and fairly long antennae. At rest, the beak fits neatly into a pronounced furrow on the underside of the head which has microscopically visible cross striations on its inner surface. In many species, the forelegs are modified for grasping small insect prey. The prothorax is trapezoidal and often bears sharp spines at its outer corners. The abdomen is usually broad and expands markedly beyond the margins of the folded wings. Typical examples are found in the widespread genera *Apiomerus* (fig. 8.3a) and *Arilus.* The widespread cogwheel bug (*chinche crestada, Arilus carinatus*) is large (BL 20 mm) and easily recognized by a prominent, vertical, semicircular crest on the back of the thorax which is margined with coarse teeth (fig. 8.3b).

Members of the subfamily Emesinae, called thread-legged bugs, are sticklike (fig. 8.4c) and often mistaken for phasmids (Wygodzinsky 1966). Some of these are known to steal prey caught in spiders' webs.

Many assassin bugs are brightly colored, with red or yellow legs and spots that appear on the back of the abdomen when the wings are raised. The brightly colored, spiny nymphs are a common sight on low vegetation in forests. Their brilliant markings (pl. 3d) obviously warn enemies of their venomous and painful bite (aposematic). Many such species participate in Müllerian mimicry complexes with other insects. Remarkable examples are certain *Spiniger* that closely resemble tarantula

**Figure 8.3    ASSASSIN BUGS (REDUVIIDAE).** (a) Assassin bug (*Apiomerus lanipes*). (b) Cogwheel bug (*Arilus carinatus*). (c) Tarantula hawk model for tarantula hawk-mimicking assassin bug (*Pepsis* sp., Pompilidae). (d) Tarantula hawk-mimicking assassin bug (*Spiniger* sp.). (e) Trash-gathering assassin bug (*Salyavata variegata*).

hawks (spider wasps of the genus Pepsis) (fig. 8.3c, d) as well as *Hiranetix* and *Graptocleptes,* which are colored and behave like band-winged, ichneumonid wasps. *Notocyrtus vesiculosus* mimics the stingless bee, *Trigona fulviventris,* both species having a black head, thorax, and legs, and orange-yellow abdomen (Jackson 1973). *Apiomerus pictipes* (Johnson 1983) has been observed in Mexico feeding on *Trigona* bees, which it resembles and which it seems to be attracted by some agent, possibly a chemical lure (Weaver et al. 1975).

A species with another deceptive hunting strategy is *Salyavata variegata* (fig. 8.3e), whose nymph lives in nasute termite nests in Costa Rica. For camouflage, it scratches off bits of nest and plasters them to hairlike projections on its back. It also "fishes" for termites, using the empty carcasses of previous prey as a lure. Workers are attracted to their own dead as a source of protein; when a termite takes the "bait," the bug drops it and grabs the fresh prey (McMahan, 1982, 1983).

Assassin bugs, as their name suggests, are fiercely predaceous. All feed on blood from other insects. (Those of the subfamily Triatominae suck vertebrate blood; see Kissing Bugs, below.) They attack mercilessly, using their raptorial forelegs to grasp and hold their victims while they stab with their mouth stylets. Chemicals contained in the saliva immobilize their captives and can cause considerable pain when

used as a defensive "sting" against vertebrates and humans. Because of their insectivorous habits, assassin bugs are important controllers of crop pest and natural insect populations.

Just over a thousand species of Reduviidae are recorded from Latin America (Wygodzinsky 1949).

### References

JACKSON, J. F. 1973. Mimicry of *Trigona* bees by a reduviid (Hemiptera) from British Honduras. Fla. Entomol. 56: 200–202.

JOHNSON, L. K. 1983. *Apiomerus pictipes* (reduvio, chinche asesina, assassin bug). *In* D. H. Janzen, ed., Costa Rican natural history. Univ. Chicago Press, Chicago. Pp. 684–687.

MCMAHAN, E. A. 1982. Bait-and-capture strategy of a termite-eating assassin bug. Ins. Soc. 29: 346–351.

MCMAHAN, E. A. 1983. Bugs angle for termites. Nat. Hist. 92(5): 40–47.

WEAVER, E. C., E. T. CLARKE, AND N. WEAVER. 1975. Attractiveness of an assassin bug to stingless bees. Kans. Entomol. Soc. J. 48: 17–18.

WYGODZINSKY, P. 1949. Elenco sistemático de los Reduviiformes Americanos. Univ. Nac. Tucumán Publ. 473, Monogr. 1: 1–102.

WYGODZINSKY, P. 1966. A monograph of the Emesinae (Reduviidae, Hemiptera). Amer. Mus. Nat. Hist. Bull. 133: 1–614.

### Kissing Bugs

Reduviidae, Triatominae, *Panstrongylus, Triatoma,* and relatives. *Spanish:* Chirimachas (Peru), vinchucas (Argentina, Chile), pitos (Venezuela), chinches voladoras (Mexico), chinches yurú pucú (Paraguay). *Portuguese:* Bichos de parede, chupanças, pinchadores, barbeiros, fincões (Brazil). Conenose bugs.

These Reduviidae are of major medical importance in the New World tropics because of the role of several species as vectors of Chagas' disease (Zeledón and Rabinovich 1981). This is a very debilitating and frequently fatal syndrome caused by the protozoan *Trypanosoma (Schizotrypanum) cruzi.* The pathogens are introduced by the bug when it feeds, not directly through the bite but in liquid fecal material that it habitually releases following engorgement. The microorganisms in these droppings enter through the perforation made by the bite or through the mucous membranes. The site of inoculation may be anywhere on the skin, but infection commonly occurs at the outer corner of the eye. The parasites multiply rapidly and localize eventually in vital internal organs; after years of chronic disease, the patient often succumbs. Chagas' disease is restricted to the New World ("American trypanosomiasis"), and although the pathogen occurs in its bug and intermediate wild mammal hosts over a much wider area, it is prevalent principally in dry areas of marginal agriculture throughout southern Mexico, around the periphery of the Amazon Basin (but apparently not within, even though vectors are present; Barbosa

de Almeida 1971), and across the middle of South America. It is estimated that 13 to 14 million people in South America suffer from the disease at this time.

In 1909, Carlos Chagas first discovered the flagellate in the triatomine bug, *Panstrongylus megistus* (fig. 8.4a). Since that time, several other species in this subfamily of bloodsucking bugs have been incriminated as transmitters. The most effective vector seems to be the widely distributed, often domestic, *Triatoma infestans* (Rabinovich 1972). Other important general vectors are *T. dimidiata* in Central America, Ecuador, and Peru and *Rhodnius pallescens* in Panama. Several species of *Panstrongylus* (Lent and Jurberg 1975) and *Triatoma* and *Rhodnius prolixus* (fig. 8.4b) are localized vectors in different parts of Latin America (Lent and Wygodzinsky 1979: 135f.). Many more triatomines have been found naturally infected with the parasite.

Kissing bugs are mostly medium to fairly large reduviids (BL 10–45 mm). The head is cylindrical, about three times as long as wide, and has a constricted, cone-shaped anterior portion from which the three-segmented beak arises. A membranous connection between the second and third rostral segments, permitting flexure of the third segment during bloodsucking, is a unique condition among bugs. The abdomen is concave dorsally and widely expanded laterally. In many species, these expansions are conspicu-

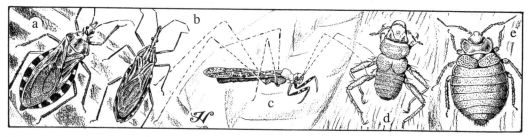

**Figure 8.4  HETEROPTERANS.** (a) Kissing bug (*Panstrongylus megistus,* Reduviidae). (b) Kissing bug (*Rhodnius prolixus,* Reduviidae). (c) Thread-legged bug (*Empicoris rubromaculatus,* Reduviidae). (d) Bat bug (*Hesperoctenes* sp., Polyctenidae). (e) Bedbug (*Cimex lectularius,* Cimicidae).

ously banded alternately with light and dark colors. The general body color is black or brown, with well-defined harlequin or variegated patterns of light yellow, orange, or red.

The main biological feature of kissing bugs is their obligate, vertebrate blood-feeding habits. All species need this blood to complete their life cycle. Their primary habitat is the nest of their hosts and nearby confined refuges (under bark, rock crevices, crowns of palms, etc.) where they hide during the day, emerging at night to feed on vertebrates visiting these niches. Many species are domestic, occurring in poultry pens and stock corrals and habitations, especially mud or adobe huts.

Adults rarely fly but occasionally are attracted to lights at night. Most have a long life cycle, 300 days on the average from egg to adult; some two years. Others complete development in less than a year and may have two to three generations per annum. There are five nymphal stages. When disturbed, many species release a pungent odor from metathoracic glands (Schofield and Upton 1978). The active chemical is butyric acid, which is repugnant to humans and other animals, including ants. Other aspects of the behavior of Triatominae are reviewed by Schofield (1979).

Because of their voracious appetites for human blood, kissing bugs were well known to the indigenous Americans and have attracted attention historically since early colonial days. Fray Reginaldo de Lizárraga was the first to describe the bugs in 1590 (Abalos and Wygodzinsky 1951).

The subfamily contains five tribes with 13 genera and 111 species in the Western Hemisphere, whose vector importance, classification, and structure have been thoroughly reviewed (Lent and Wygodzinsky 1979, Usinger et al. 1966). Bibliographies of Chagas' disease and lists of species of the kissing bugs of Latin America are also available (Ryckman 1984, 1986; Ryckman and Blankenship 1984; Ryckman and Zackrison 1987).

*Rhodnius prolixus* colonies are now widely used as experimental animals in many laboratories as a result of the pioneering work of insect physiologist V. B. Wigglesworth.

## References

ABALOS, J. W., AND P. WYGODZINSKY. 1951. La vinchuca: Folklore y antecedentes históricos. Cien. Invest. 7: 472–475.

BARBOSA DE ALMEIDA, F. 1971. Triatomíneos da Amazonia. Acta Amazonica 1: 89–93.

LENT, H., AND J. JURBERG. 1975. O gênero *Panstrongylus* Berg, 1879, com um estudo sôbre a genitalia externa das especies (Hemiptera, Reduviidae, Triatominae). Rev. Brasil. Biol. 35: 379–438.

LENT, H., AND P. WYGODZINSKY. 1979. Revision of the Triatominae (Hemiptera, Reduviidae), and their significance as vectors of Chagas' disease. Amer. Mus. Nat. Hist. Bull. 163: 123–520.

RABINOVICH, J. E. 1972. Vital statistics of Triatominae (Hemiptera: Reduviidae) under laboratory conditions. I. *Triatoma infestans* Klug. J. Med. Entomol. 9:351–370.

RYCKMAN, R. E. 1984. The Triatominae of North and Central America and the West Indies: A checklist with synonymy (Hemiptera: Reduviidae, Triatominae). Soc. Vector Ecol. Bull. 9: 71–83.

RYCKMAN, R. E. 1986. The Triatominae of South America: A checklist with synonymy (Hemiptera: Reduviidae: Triatominae). Soc. Vector Ecol. Bull. 11: 199–208.

RYCKMAN, R. E., AND C. M. BLANKENSHIP. 1984. The Triatominae and Triatominae-borne trypanosomes of North and Central America and the West Indies: A bibliography with index. Soc. Vector Ecol. Bull. 9: 112–430.

RYCKMAN, R. E., AND J. L. ZACKRISON. 1987. A bibliography to Chagas' disease, the Triatominae and Triatominae-borne trypanosomes of South America (Hemiptera: Reduviidae: Triatominae). Soc. Vector Ecol. Bull. 12: 1–464.

SCHOFIELD, C. J. 1979. The behavior of Triatominae (Hemiptera: Reduviidae): A Review. Bull. Entomol. Res. 69: 363–379.

SCHOFIELD, C. J., AND C. P. UPTON. 1978. Brindley's scent-glands and the metasternal scent-glands of *Panstrongylus megistus* (Hemiptera, Reduviidae, Triatominae). Rev. Brasil. Biol. 38: 665–678.

USINGER, R. L., P. WYGODZINSKY, AND R. E. RYCKMAN. 1966. The biosystematics of Triatominae. Ann. Rev. Entomol. 11: 309–330.

ZELEDÓN, R., AND J. E. RABINOVICH. 1981. Chagas' disease: An ecological appraisal with special emphasis on its insect vectors. Ann. Rev. Entomol. 26: 101–133.

## Parasitic Bugs

Bugs in two heteropteran families have lost their wings and become totally adapted for a parasitic existence on the bodies, or in the nests, of birds and mammals. Among these, the Polyctenidae, or bat bugs, are the more extremely modified structurally and biologically, for life on bats. They are small (BL 3 to 5 mm), lack eyes, are flattened, and have comblike rows of flattened spines on the body (fig. 8.4d). The females are viviparous (Ryckman and Casdin 1977). A single genus, *Hesperoctenes*, is present in the New World (Ueshima 1972).

The cimicids, or bedbugs, are much better known because of the two species that associate closely with mankind. They are not well adapted to cling to fur or leathers but are temporary visitors to the body of the host, normally bats and various birds. Most common and best known are the introduced domestic bedbugs (*Cimex lectularius* and *C. hemipterus*) on bats, chickens, and humans throughout America (see below). Also of economic importance are the parrot bedbug (*Psitticimex*), on parrots in northern Argentina, the Mexican chicken bug (*Haematosiphon inodorus*), on chickens in Mexico (Lee 1955), and the Brazilian chicken bug (*Ornithocoris toledoi*), which, prior to its control with organic insecticides, was a pest on chickens in southeastern South America.

Twelve genera of cimicids are found in Latin America. The absence of native species in Central America, where only the domestic pests are represented, is a curious distribution pattern among New World bedbugs (Ryckman et al. 1981, Usinger et al. 1966).

Cimicids are larger than polyctenids (BL 4–5 mm), more or less oval, and slightly flattened. The abdomen is soft and capable of enormous enlargement with blood at the time of engorgement. They are flightless, but padlike, vestigial fore wings persist in most species.

A unique development in this family is the presence of a secondary copulatory receptacle (spermalege) in the female, situated on one side, at the base of the abdomen. The male genitalia are asymmetrical and possess a formidable, swordlike penis. Males perforate the integument with this intromittent organ in one or more places, usually through an external fold in an intersegmental membrane over the receptacle. The sperm is then delivered to an underlying pocket that communicates with the genital ducts. Copulation with the female's normal terminal genitalia never occurs.

### References

LEE, R. D. 1955. The biology of the Mexican chicken bug, *Haematosiphon inodorus* (Duges) (Hemiptera: Cimicidae). Pan-Pacific Entomol. 31: 47–61.

RYCKMAN, R. E., D. G. BENTLEY, AND E. F. ARCHBOLD. 1981. The Cimicidae of the Americas and oceanic islands: A checklist and bibliography. Soc. Vector Ecol. Bull. 6: 93–142.

RYCKMAN, R. E., AND M. A. CASDIN. 1977. The Polyctenidae of the world, a checklist with bibliography. Calif. Vector Views 24: 25–31.

UESHIMA, N. 1972. New World Polyctenidae (Hemiptera), with special reference to Venezuelan species. Brigham Young Univ. Sci. Bull. (Biol. Ser.) 17: 13–21.

USINGER, R. L., J. CARAYON, N. T. DAVIS, N. UESHIMA, AND H. E. McKEAN. 1966. Monograph of Cimicidae. Vol. 7. Entomol. Soc. Amer. (Thomas Say Foundation), College Park, Md.

## BEDBUGS

Cimicidae, Cimicinae, *Cimex lectularius* and *C. hemipterus. Spanish:* Chinches de cama (General). *Portuguese:* Percevejos das camas (Brazil). *Nahuatl:* Texcantin, sing. texcan (Mexico). Wall lice, red coats.

Human bedbugs are truly cosmopolitan insects, having followed humans over most of the world (Anon. 1973). The two anthrophilic species have been well known since antiquity in the Old World, where their original hosts were probably bats. They first took to man in southern Europe or the Middle East in the case of the common bedbug (*C. lectularius*) or the Orient in the case of the tropical bedbug (*C. hemipterus*) and surely spread to the New World with the first explorers and immigrants. Early published mentions of the species are always suspect owning to the chroniclers' use of *chinche* and Aztec *texcan* not only for common bedbugs but for other arthropods (Hoeppli 1969: 183).

These bugs are ravenous nocturnal blood seekers cohabiting with man, hiding in cracks, mattresses, and so on, and behind furniture during the day. They emerge at night to join us in our beds and to fill their guts with our life fluids. In unkempt premises, hundreds may be found in hideaways, and their obnoxious, sickeningly sweet "nest odor" can be overwhelming. The chemicals responsible for the odor may have alternative attractant and repellent functions toward other bedbugs or animals (Levinson and Barllan 1971).

Bedbug bites cause varied reactions in humans. In some, there is marked swelling and irritation; in others, no effect. Although many experiments have been conducted to incriminate these two species as natural disease vectors, the results have virtually always been negative. They can transmit several pathogens under laboratory conditions, however.

Bedbugs are small, ovate, and much flattened except when engorged, and then they are almost spherical. Adults are reddish-brown and covered with minute bristles. Their legs are slender and fairly long. The lateral expansions of the prothorax of the common bedbug (fig. 8.4e) are more extensive and flattened than those of the tropical bedbug, and its color is lighter. The former also tends to be more urban and extends into higher latitudes than the latter, which is a rural species and restricted to the middle latitudes in the Americas.

Information on all aspects of the natural history of these species is summarized in the masterful monograph by Usinger et al. (1966).

## References

ANONYMOUS. 1973. The bed-bug. 8th ed. Brit. Mus. Nat. Hist. Econ. Ser. 5: 1–17.

HOEPPLI, R. 1969. Parasitic diseases in Africa and the Western Hemisphere, early documentation and transmission by the slave trade. Acta Trop. suppl. 10: 1–240.

LEVINSON, H. Z., AND A. R. BARLLAN. 1971. Assembling the alerting scents produced by the bedbug *Cimex lectularius* L. Experientia 27: 102–103.

USINGER, R. L., J. CARAYON, N. T. DAVIS, N. UESHIMA, AND H. E. MCKEAN. 1966. Monograph of the Cimicidae. Vol. 7. Entomol. Soc. Amer. (Thomas Say Foundation), College Park, Md.

## Water Bugs

As has occurred secondarily in the evolution of Coleoptera and other orders, several families of Heteroptera have adopted aquatic habits and are modified for swimming, diving, respiring, and feeding while submerged. Two basic life-styles have evolved: the "semiaquatic bugs" (Andersen 1982) live on the surface film and are generally less adapted to the medium than the truly "aquatic bugs" that spend the majority of their time beneath the surface.

These are insects of great importance in the ecology of all watery habitats, from small trickles, highland streams, and ponds to the margins of the seas.

Although specifically treating the Guyana region, the review by Nieser (1975) is widely applicable. The general Latin American literature is compiled in the bibliographies of Bachman (1977), Nieser (1981), and Polhemus (1982).

### References

ANDERSEN, N. M. 1982. The semiaquatic bugs (Hemiptera, Gerromorpha), phylogeny, adaptations, biogeography and classification. Entomonograph 3: 1–455.

BACHMANN, A. O. 1977. Heteroptera. *In* S. H. Hurlbert, ed., Biota acuática de sudamérica austral. San Diego State Univ., San Diego. Pp. 189–212.

NIESER, N. 1975. The water bugs (Heteroptera: Nepomorpha) of the Guyana Region. Stud. Fauna Suriname Guyanas 59: 1–310.

NIESER, N. 1981. Hemiptera. *In* S. H. Hurlbert, G. Rodríguez, and N. Dias dos Santos, eds., Aquatic biota of tropical South America. Pt. 1. Arthropoda. San Diego State Univ., San Diego. Pp. 100–128.

POLHEMUS, J. T. 1982. Hemiptera. *In* S. H. Hurlbert and A. Villalobos Figueroa, eds., Aquatic biota of Mexico, Central America and the West Indies. San Diego State Univ., San Diego. Pp. 288–292.

## GIANT WATER BUGS

Belostomatidae, Lethocerinae, *Lethocerus*.
  *Spanish:* Cucarachas del agua (General).
  *Portuguese:* Baratas d'agua (Brazil).
  Electric light bugs, toe biters.

Belonging to a cosmopolitan family, this genus has speciated prolifically in Latin America. There are nearly twenty distinct species, mostly in the genus *Lethocerus* (Menke 1963). The two species *L. maximus* (fig. 8.5a) and *L. grandis* are among the largest of insects with a body length up to 11.5 centimeters and weight of 15 to 25 grams.

All are much flattened, ellipsoid in outline, and shiny dark brown. The head is rigidly fixed to the thorax and does not rotate; its frontal portion projects strongly forward between the eyes. Straplike, retractile respiratory appendages are borne at the tip of the abdomen. The latter penetrate the surface film when the bug ascends, rearward, to take in air. Most of the store is carried under the wings in a cavity created by the depressed abdomen.

These are rapacious predators, catching all sorts of other aquatic invertebrates and even small vertebrates like tadpoles and fish with their raptorial forelegs, then killing them with a vicious stab of the mouth stylets. The rostrum acts like a hypodermic syringe, injecting saliva that both immobilizes and digests the organisms on which they feed (Picado 1937). The bugs wait patiently in ambush among plants or debris in the water, relying on remaining motionless and on their cryptic coloration to escape detection by their

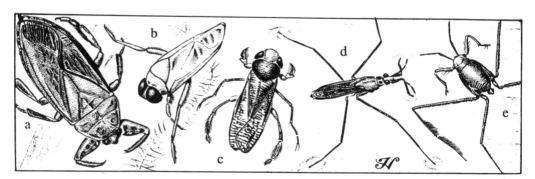

**Figure 8.5  WATER BUGS.** (a) Giant water bug (*Lethocerus maximus*, Belostomatidae). (b) Back swimmer (*Buenoa pallens*, Notonectidae). (c) Salt marsh water boatman (*Trichocorixa reticulata*, Corixidae). (d) Common water strider (*Gerris remigis*, Gerridae). (e) Sea strider (*Halobates micans*, Gerridae).

prey. Female giant water bugs lay their eggs on emergent aquatic vegetation, not on the back of males as is characteristic of other genera in the family.

Specimens may be common at times near their well-vegetated, marshy pond and lakeshore habitats. On warm evenings during their dispersal season, they sometimes accumulate under electric lights and attract a great deal of attention (Lanzer 1975). The biology of *L. maximus* has been investigated in some detail in Trinidad (Cullen 1969) and it has been maintained in the laboratory for studies on the physiology of its flight muscles (Barros 1973).

## References

BARROS S., M. C. 1973. Mantenção da barata d'agua gigante (gen. *Lethocerus*) no laboratório. Univ. São Paulo, Inst. Biol. Mar., Bol. Zool. Biol. Mar. (Nov. Ser.) 30: 613–623.

CULLEN, M. J. 1969. The biology of giant water bugs (Hemiptera: Belostomatidae) in Trinidad. Royal Entomol. Soc. London Proc. A 44: 123–136.

LANZER, M. E. B. 1975. Nota prévia sôbre o comportamento de *Belostoma* Latreille, 1807 e *Lethocerus* Mayr, 1853 em aquário e no meio ambiente. Iheringia (Ser. Div.) 4: 47–50.

MENKE, A. S. 1963. A review of the genus *Lethocerus* in North and Central America, including the West Indies (Hemiptera: Belostomatidae). Entomol. Soc. Amer. Ann. 56: 261–267.

PICADO T., C., 1937. Estudo experimental sôbre o veneno de *Lethocerus del-pontei* (DeCarlo) (Hemiptera-Belostomidae). Inst. Butantan (São Paulo) Mem. 10: 303–310, figs. 1–3.

# BACK SWIMMERS

Notonectidae.

Back swimmers are so named because of their habit of swimming upside down. They are also recognized by their long, oarlike hind legs that pull in unison when the insect swims. The body is slender, boat shaped, and often colored with white, red, or greenish areas, although it is usually all dull brown. The fore tarsus is visibly two segmented, slender, and bare. Like their relatives, the giant water bugs, they have large eyes and a sharp beak with potent powers; their bite is very painful and often is the salvation of a specimen in the collector's clutches. Also like their relatives, they are predaceous, but the size of their prey is smaller, consisting of fly and beetle larvae, fish fry, and crustacea.

There are between 300 and 400 species, mostly in the dominant genera *Notonecta, Martarega,* and *Buenoa.* These are distributed almost everywhere where suitable sluggish water habitats are available but show considerable selectivity to specific niches determined by water conditions and prey availability (Gittleman 1975).

*Notonecta* are the largest (BL of most 12–15 mm) and must rise frequently to take on oxygen from the atmosphere; males are silent. *Buenoa* are smaller (maximum BL 11 mm) and possess hemoglobin containing cells that store oxygen and assist them in remaining submerged for longer periods than other notonectids; males stridulate by means of a ridged protuberance on the inside of the fore tibia. *B. pallens* (fig. 8.5b) is a widespread species. *Martarega* are the smallest (maximum BL 10 mm) and are most commonly found at water's edge in slow-moving rivers and do not stridulate.

Since earliest times, Mexicans around Lake Texcoco and Chalco near Mexico City have collected *Notonecta unifasciata* (*axayácatl*) and associated water boatmen, which they consume in various forms (Ancona 1933). The eggs (*ahuautli, axayácatl, etc.*) *are cooked into a kind of bread (hautlé);* adults (*ahuatle, bledo delagua*) may be ground and mixed with saltpeter to make a salty hash (Bodenheimer 1951).

Refer to Bachmann (1977), Nieser (1981), and Polhemus (1982) for bibliographies on the regional fauna.

## References

ANCONA, L. 1933. El ahuatle de Texcoco. Inst. Biol. Univ. Nac. Mexico Ann. 4: 51–69.

BACHMANN, A. O. 1977. Notonectidae. *In* S. H. Hurlbert, ed., Biota acuática de sudamérica austral. San Diego State University, San Diego. Pp. 193–195.

BODENHEIMER, F. S. 1951. Insects as human food. Junk, The Hague.

GITTLEMAN, S. H. 1975. The ecology of some Costa Rican backswimmers (Hemiptera: Notonectidae). Entomol. Soc. Amer. Ann. 68: 511–518.

NIESER, N. 1981. Notonectidae. *In* S. H. Hurlbert, G. Rodríguez, and N. Dias dos Santos, eds., Aquatic biota of tropical South America. Pt. 1. Arthropoda. San Diego State University, San Diego. Pp. 115–117.

POLHEMUS, J. T. 1982. Notonectidae. *In* S. H. Hurlbert and A. Villalobos Figueroa, eds., Aquatic biota of Mexico, Central America and the West Indies. San Diego State University, San Diego. Pp. 306–308.

# WATER BOATMEN

Corixidae.

Water boatmen are superficially similar to back swimmers but always swim right side up and differ in many anatomical details: the body is cylindrical or slightly flattened, the mouthparts are short, cone shaped, and unsegmented, and the male genitalia are asymmetrical. The fore wings are often marked with fine, transverse zigzagging dark lines. The tarsus of the foreleg is composed of a single broad, hairy segment. They vary considerably in size (BL 3–10 mm).

These water bugs choose habitats like the back swimmers, although they tend to prefer more stagnant or torpid waters where they spend most of their time at the bottom. Some, such as *Trichocorixa reticulata* (fig. 8.5c), tolerate saline conditions and may be extremely abundant in inland salt pools or estuaries on the continent and in the Galápagos Islands.

Corixid food and feeding habits are varied. A common method of feeding is by sieving edible particles from bottom debris. They also may ingest other living or dead aquatic invertebrates if small enough or in a decomposed state.

Males chirp by rasping areas of small pegs on the base of the fore femur against the sharp edge of the mouth beak.

The family is diverse in the Neotropics, with approximately 117 species in 14 genera.

As with notonectids (see above), some corixids (*Corisella*) are found in Mexican markets. Tons of these dried insects are shipped abroad as bird or fish fodder, and they are gathered also for human use, both as eggs or adults (*mosco, moschitos*). Eggs are laid in enormous numbers on reeds placed in the water by the gatherers. Like those of *Notonecta*, they are made into a fish-flavored cake, also called *huatlé* or *ahuahutl* (Bodenheimer 1951: 295f.).

Refer to Bachmann (1977), Nieser (1981), and Polhemus (1982) for bibliographies on the Latin American fauna.

## References

BACHMANN, A. O. 1977. Corixidae. *In* S. H. Hurlbert, ed., Biota acuática de sudamérica austral. San Diego State Univ., San Diego. Pp. 191–193.

BODENHEIMER, F. S. 1951. Insects as human food. Junk, The Hague.

NIESER, N. 1981. Corixidae. *In* S. H. Hurlbert, G. Rodríguez, and N. Dios dos Santos, eds., Aquatic biota of tropical South America. Pt. 1. Arthropoda. San Diego State Univ., San Diego. Pp. 117–119.

POLHEMUS, J. T. 1982. Corixidae. *In* S. H. Hurlbert and A. Villalobos Figueroa, eds., Aquatic biota of Mexico, Central America and the West Indies. San Diego State Univ., San Diego. Pp. 308–310.

# WATER STRIDERS

Gerridae. *Spanish:* Zapateros (Argentina). Pond skaters.

These are oval to elongate bugs with four-segmented antennae and large, globular eyes. Their other distinctive features are slender spiderlike legs, with tarsal claws inserted before the tip of the last segment. The body is covered with a velvety pile of hairs that are hygrophobic and make the insect resistant to wetting should it submerge. Their colors range from black to brown with occasional silvery markings. Wings may be absent, abbreviated, or fully developed.

Water striders (Andersen and Polhemus 1976) all are "semiaquatic bugs," spending their entire lives moving jerkily to and fro on the surface film. They feed on the blood of insects and invertebrates that fall onto the surface film or reside on it. A common species is *Gerris remigis* (fig. 8.5d).

One group of striders has colonized the open ocean. The best known of these, the "sea skaters," belong to the genera *Halobates* (Herring 1961, Cheng 1985) and *Rheumatobates* (Cheng and Lewin 1971), which generally resemble freshwater striders except for a grossly reduced abdomen and complete absence of wings. Just a few species live on the waters of the ocean off both coasts of America and among associated oceanic islands. *Halobates micans* (fig. 8.5e) and *H. robustus* are common inhabitants of small bays in the Galápagos Islands. Little is known of their biology. Coastal species lay their eggs on rocks; pelagic species may oviposit on floating objects (wood, feathers) and are even known to attach eggs to birds that have been resting on the waves. Their food consists of pelagic, surface-dwelling animals like jellyfish.

Refer to Bachmann (1977), Nieser (1981), and Polhemus (1982) for bibliographies on the regional fauna.

### References

ANDERSEN, N. M., AND J. T. POLHEMUS. 1976. Water-striders (Hemiptera: Gerridae, Veliidae, etc.). *In* L. Cheng, ed., Marine insects. North-Holland, Amsterdam. Pp. 187–224.

BACHMANN, A. O. 1977. Gerridae. *In* S. H. Hurlbert, ed., Biota acuática de sudamérica austral. San Diego State Univ., San Diego. Pp. 204–206.

CHENG, L. 1985. Biology of *Halobates* (Heter-

optera: Gerridae). Ann. Rev. Entomol. 30: 11–135.

CHENG, L., AND R. A. LEWIN. 1971. An interesting marine insect, *Rheumatobates aestuarius* (Heteroptera: Gerridae), from Baja California, Mexico. Pacific Ins. 13: 333–341.

HERRING, J. L. 1961. The genus *Halobates* (Hemiptera: Gerridae). Pacific Ins. 3: 223–305.

NIESER, N. 1981. Gerridae. *In* S. H. Hurlbert, G. Rodríguez, and N. Dias dos Santos, eds., Aquatic biota of tropical South America. Pt. 1. Arthropoda. San Diego State Univ., San Diego. Pp. 125–128.

POLHEMUS, J. T. 1982. Gerridae. *In* S. H. Hurlbert and A. Villalobos Figueroa, eds., Aquatic biota of Mexico, Central America and the West Indies. San Diego State Univ. San Diego. Pp. 319–323.

## HOMOPTERANS

Hemiptera, Homoptera. Wax bugs.

This suborder is a large one with many families, several of which are strictly Neotropical. Opinions differ as to its internal classification because of the diversity of basic body forms and extreme biological and anatomical adaptations. Some unifying characteristics are sucking mouthparts similar to those of heteropterans but with the elements usually consolidated into a short beak set so far back on the head as to appear to arise from behind the forelegs and piercing stylets often very long and coiled in the body when not in use. Wings, when present, are homogenous in texture, usually membranous but often thickened and fairly rigid and with few veins (although some have a reticulate pattern). There are two pairs of wings except for male scale insects, which have only the fore pair developed.

I use the name "wax bugs" to refer to the suborder because of the almost universal presence of wax-producing glands in its constituent families. These glands are variously developed on different parts of the body and always open exteriorly on the cuticle. Their secretions are extruded to the outside to take many forms for diverse functions, some well defined, such as the protective scale covering of armored scale insects, and others with unclear significance, such as the massive plumes trailing from the abdomen of some large fulgorids. Silk- and lac-producing glands are also present.

All are sap feeders and often very specific in their preferences for the sap of certain plants (Johnson and Foster 1986). They remove this fluid, sometimes in such quantity, because of enormous populations that may develop, that they kill or seriously injure their hosts. For this reason, and also because many are very efficient carriers of plant pathogens, wax bugs are plant pests of prime importance. Injurious species are found in almost all families but especially among the aphids, scale insects, and leafhoppers. Unlike their heteropteran relatives, no wax bugs have adapted to feeding on vertebrate blood or have become aquatic.

A mutualistic relationship exists between some hymenopterans and many wax bugs, especially aphids and scale insects (Letourneau and Choe 1987). The latter secrete "honeydew," a carbohydrate-rich overflow from the alimentary canal or integumentary glands which is greedily consumed and even specifically solicited, especially by ants. The ants attend the bugs, protecting them from predators, dispersing them, and even building shelters for them (Way 1953).

## References

JOHNSON, L. K., AND R. B. FOSTER. 1986. Associations of large Homoptera (Fulgoridae and Cicadidae) and trees in a tropical forest. Kans. Entomol. Soc. J. 59: 415–422.

LETOURNEAU, D. K., AND J. C. CHOE. 1987. Homopteran attendance by wasps and ants: The stochastic nature of interactions. Psyche 94: 81–91.

WAY, J. T. 1953. Mutualism between ants and honeydew producing Homoptera. Ann. Rev. Entomol. 8: 307–344.

# CICADAS

Cicadidae. *Spanish:* Chicharras (General); coyuyos, cigarras (Argentina).
*Portuguese:* Cigarras.

Among homopterans, cicadas are famous for their sound-producing abilities. (A few others have vocal organs but none so well developed or capable as those of Cicadidae, and nothing is known of their function in Neotropical forms.)

On the back of the first abdominal segment there is a pair of exposed (or protected by a fold of the body wall), taut membranes (tymbals) that may be made to rapidly vibrate by well-developed oblique muscles. The tymbals vibrate extremely fast, producing a loud, sometimes strident, steady or pulsating buzz or siren scream that varies in pitch, pulse, intensity, duration, or other acoustic quality according to species. One type (tentatively determined as the very widespread *Quesada gigas;* fig. 8.6a) in park trees in Caracas emits a fairly deafening, throbbing wail. Some species are silent or produce only interrupted clicks.

Both sexes have auditory tympani anteriorly on the underside of the abdomen, concealed beneath large protective plates (not to be confused with the tymbals). Males often congregate on trees in forests and join their voices in intense synchronized chorusing that acts as a call to assemble females for mating. This aggregative behavior seems to be exhibited in secondary habitats and on trees that are the nymphal hosts (Young 1980). The latter are varied hardwoods and palms (Young 1973). Peaks of singing intensity occur in many species at dawn and dusk (Young 1981b).

On trees, females insert masses of eggs in twigs or fronds with an ovipositor adapted for slitting bark. In many species, oviposition is limited to dead trees and fronds in the forest understory, while others put their eggs into dead grasses near

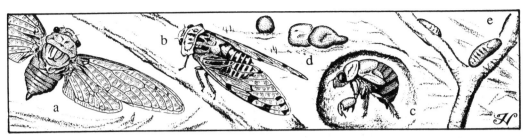

**Figure 8.6** **WAX BUGS.** (a) Giant cicada (*Quesada gigas,* Cicadidae). (b) Spotted cicada (*Zammara smaragdina,* Cicadidae). (c) Cicada nymph (Cicadidae). (d) Ground pearls (*Margarodes formicarum,* Margarodidae), females. (e) Axin (*Llaveia axin,* Margarodidae).

the ground. The nymphs drop to the soil and burrow deeply. They commonly spend two (but some, several to many) years underground, feeding on sap from the host's roots. On maturity, they crawl out of the ground and up onto tree trunks and limbs to transform. Fixed to such surfaces, their cast skins, split widely down the back, are a common sight. The nymphs of at least one Amazonian cicada (*Fidicina chlorogena;* Ginzberger 1934) constructs tall (20–30 cm), hollow, mud chimneys in which to pass their final days before becoming adults.

The imagos are presumed to be short-lived, perhaps existing from a few weeks to a few months. They imbibe fluids from the xylem tissue of their hosts with a sturdy, jointed proboscis. Most are active in the forest canopy, although some species frequent the trunks of trees close to the ground. Few additional details of tropical cicadan biology are available, and these are mostly from species in Costa Rica (Young 1981*a*). Others (Bartholomew and Barnhart 1984) have studied flight metabolism in the Central American *F. manifera.* It can fly at a body temperature of 22° C., but takeoff must be preceded by warming up by body movements. During active flight, the body heats up to 33° C. They do not jump but are easily excited and frightened into flight.

Adults are easily distinguished as a group by their robust form large size (BL of most 20–50 mm), and membranous wings, the anterior pair of which are almost twice as long as the posterior and extending well beyond the end of the body when folded. A transversely ridged swelling on the face and the singing organs of the male also are unique features of the family. The lateral margins of the thorax have winglike flanges in some, for example, *Zammara* (fig. 8.6b). The bodies of most cicadas are marked with green or white against a black background. The wing veins are usually dark, the color usually staining only the adjoining membrane, although the usually transparent membranous areas may be splotched with brown patterns as well.

Nymphs (fig. 8.6c) are stout-bodied, brownish, molelike creatures with outsized, powerful fore tibiae and otherwise heavy legs adapted for digging. The abdomen is stubby and curved downward.

Cicadas seldom cause economic damage to plants with their feeding and oviposition. They have been recorded molesting fruit trees and coffee in Brazil (da Fonseca 1945).

A complete bibliography on the family to 1956 is available (Metcalf 1962). The species are cataloged by Metcalf (1963*a*, 1963*b*, 1963*c*) and Duffels and van der Laan (1985). The number of described Latin American species is roughly estimated at over 800, in some 80 or so genera; the former figure may represent as few as 60 percent of those that are to be found (Moore pers. comm.).

## References

BARTHOLOMEW, G. A., AND M. C. BARNHART. 1984. Tracheal gases, respiratory gas exchange, body temperature and flight in some tropical cicadas. J. Exper. Biol. 111: 131–144.

DA FONSECA, J. P. 1945. As cigarras do cafeeiro e seu combate. Bol. Agr. 46: 297–304.

DUFFELS, J. P., AND P. A. VAN DER LAAN. 1985. Catalogue of the Cicadoidea (Homoptera, Auchenorhyncha) 1956–1980. Junk (Series Entomologica), The Hague.

GINZBERGER, A. 1934. Die Bauten der Larve der Singzikade *Fidicina chlorogena* Wlk. Anz. Akad. Wiss. Wien 71: 55.

METCALF, Z. P. 1962. A bibliography of the Cicadoidea (Homoptera: Auchenorhyncha). Entomol. Dept., N.C. Agric. Exper. Sta. Pap. 1373: 1–229.

METCALF, Z. P. 1963a. Cicadoidea, Cicadidae, Tibiceninae. In Z. P. Metcalf, General catalogue of the Homoptera. N.C. State Coll., Raleigh. Fasc. 8, pt. 1, sec. I.

METCALF, Z. P. 1963b. Cicadoidea, Cicadidae, Gaeaninae and Cicadinae. In Z. P. Metcalf, General catalogue of the Homoptera. N.C. State Coll., Raleigh. Fasc. 8, pt. 1, sec. 2.

METCALF, Z. P. 1963c. Cicadoidea, Tibicenidae. In Z. P. Metcalf, General catalogue of the Homoptera. N.C. State Coll., Raleigh. Fasc. 8, pt. 2.

YOUNG, A. M. 1973. Cicada populations on palms in tropical rain forest. Principes Palm Soc. J. 17: 3–9.

YOUNG, A. M. 1980. Observations on the aggregation of adult cicadas (Homoptera; Cicadidae) in tropical forests. Can. J. Zool. 58: 711–722.

YOUNG, A. M. 1981a. Notes on the population ecology of cicadas (Homoptera: Cicadidae) in the Cuesta Angel forest ravine of northeastern Costa Rica. Psyche 88: 175–195.

YOUNG, A. M. 1981b. Temporal selection for communicatory optimization: The dawn-dusk chorus as an adaptation in tropical cicadas. Amer. Nat. 117: 826–829.

## SCALE INSECTS

Coccoidea. *Spanish:* Cochinillas, coccídeos (General). *Portuguese:* Cochonilhas, coccídeos (Brazil).

Wax glands are especially well developed in these usually minute to small (BL 1–2 mm) homopterans and chiefly function to secrete a protective shell or covering for the insect. This takes a scalelike form in many groups. All stages and both sexes have legs with segmented tarsi tipped with a single claw.

Females are quite unlike the males in form and life-style. The latter are active, midgelike, free-living forms with one pair of wings and no feeding beak. Most, but not all, females live their entire adult lives as amorphous bags of tissue without functional legs or wings, remaining attached to their food plants by their incredibly long, hairlike feeding stylets, usually under a protective scale or interred in a hard encrustment. They lay their eggs under or within their casements. The eggs hatch into tiny mobile crawling nymphs (crawlers) that molt and either settle down to a sessile life as females or remain active and develop into flying males.

There are several different subgroups of scale insects, each with its own structural and secretory characteristics. Female "giant scale insects" (*cochinillas de cola, Margarodidae*) have well-formed legs and antennae and cover their bodies in white, waxy, overlapping plates that build up to form large (to 2 cm), hardened spheres. The insects are subterranean, feeding on roots around which they form grapelike clusters. These cystlike structures ("ground pearls") at one time were collected and made into bead necklaces in parts of the Caribbean, especially those of *Margarodes formicarum* (fig. 8.6d), a species first found associated with ants in the Lesser Antilles (Guilding 1830).

*Llaveia axin* (fig. 8.6e) is another margarodid type from acacias and the hog-plum tree (*Spondias purpurea*) in Yucatán which were used for centuries by Mesoamerican Indians. The bright orange females, each almost 3 centimeters long, were crushed and kneaded together into a wax ball that formed the base for cosmetics and medicinals and an ingredient of hard waxy finishes applied to early Mayan and Aztec pottery and gourds. Names used for this insect by natives of the area in the past include *ni-in, nije, axin, oji,* and *tuch-cuy* (Edwards 1970, Jenkins 1970).

Giant scale insects are also notoriously hardy and long lived. The published record for vitality in an insect is for specimens of *Margarodes vitium* that remained alive without feeding for seventeen years (Ferris 1919). The species is a pest on grapes in dry areas of South America. Apparently, this ability to lie dormant for long periods is an adaptation to dryness, the adults emerging normally after the first rains.

Other very large coccoids are the leathery "tortoise scales" (Coccidae), which have a convex shape and are covered with a thin layer of wax but no scale (BL 1–20 mm). An example is *Neolecanium sallei* (Wheeler 1913), found on coral trees (*Erythrina*) in Central America.

Included in this group are the "fluted scales," so called because of the ridged form of the wax accumulations that build up beneath the body. The most well known representative is the cottony cushion scale (*Icerya purchasi;* fig. 8.7a–c), a major enemy of citrus throughout America.

Female "armored scales" (*escamas*—Diaspididae) typically live under a flattened, disklike scale of wax and cast nymphal skins. This is the largest and probably most injurious group, populations often exploding on their hosts; many are notorious pests of orchard and shade trees and other cultivars. Particularly bad are the California red scale (*Aonidiella aurantii;* fig. 8.7d) and San Jose scale (*Quadraspidiotus perniciosus*) (Marín 1987).

Mealybugs (Pseudococcidae and Eriococcidae) are aberrant scale insects. They are motile and without armor, being covered with a thick, crusty or flaky, white layer of wax instead. They also have long, taillike filaments projecting from the posterior (fig. 8.7f). A serious and widespread pest of pineapple is the pineapple mealybug (*Dysmicoccus brevipes*). Another group is the "cochineal bugs," discussed below.

Many scale insects produce copious quantities of honeydew, a source of polysaccharide for multitudes of insects near the base of food chains in ecosystems (Salas and Jirón 1977).

## References

EDWARDS, J. G. 1970. Giant margarodid scales from Yucatán. Pan-Pacific Entomol. 46: 68.

FERRIS, G. F. 1919. A remarkable case of longevity in insects (Hem., Hom). Entomol. News 30: 27–28.

GUILDING, L. 1830. An account of *Margarodes,* a new genus of insects found in the neighborhood of ants nests. Linnaean Soc. London Trans. 9: 912–914.

JENKINS, K. D. 1970. The fat-yielding coccid, *Llaveia,* a monophlebine of the Margarodidae. Pan-Pacific Entomol. 46: 79–81.

MARÍN, L. R. 1987. Biología y morfología de la "escama de San José" *Quadraspidiatus perni-*

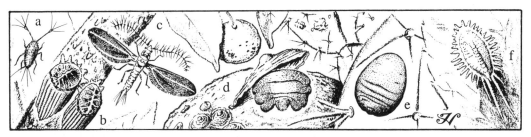

**Figure 8.7  WAX BUGS.** (a) Cottony cushion scale (*Icerya purchasi,* Coccidae), first stage nymph ("crawler"). (b) Cottony cushion scale, females. (c) Cottony cushion scale, adult male. (d) California red scale (*Aonidiella aurantii,* Diaspididae), female and scale. (e) Cochineal bug (*Dactylopius coccus,* Dactylopidae), female. (f) Longtailed mealybug (*Pseudococcus longispinus,* Pseudococcidae).

ciosus (Comst.). Rev. Peruana Entomol. 29: 81–87.

SALAS, S., AND L. F. JIRÓN. 1977. Simbiosis entre cochinillas de cola (Homoptera: Margarodidae) y otros insectos. II. Estructura del sistema y significado ecológico. Brenesia 10/11: 57–64.

WHEELER, W. M. 1913. A giant coccid from Guatemala. Psyche 20: 31–33.

## Cochineal Bugs

Dactylopidae, *Dactylopius*. *Spanish:* Cochinillas. *Portuguese:* Cochonilhas. *Nahuatl:* Nocheztli = "prickly-pear fruit blood."

The story of cochineal is one of the most fascinating, involved, and historically important of any relating to human use of an insect product (Donkin 1977). The insect was found by the Spaniards on their first arrival in Mexico and soon exploited as a source of an intense vermilion dye that became widely known. In fact, during the early colonial period, it became the most widely traded and, next to gold and silver, the most valuable product of the Spanish Indies. The pigment, called cochinealin or carminic acid (a tetracyclic carboxylic acid or anthraquinone), is concentrated in the insect's fat body and is repugnant to ants and probably to parasites that would pretend to infect this insect (Goetz and Meinwald 1980). The substance was thought to have medicinal value in addition to its use as a textile dye and cosmetic tint.

The name "cochineal" refers to the dye and "cochineal insect" or "cochineal bug" to the insect from which it is extracted. The latter consists of two major types, a larger, "domesticated" species (*D. coccus*), which shows evidence of a long association and partial dependence on human caretaking, and eight smaller "wild" and self-sufficient species, chiefly *D. tomentosus*. These are exposed scale insects that live openly and without protective scale coverings on the pads of *Opuntia* cacti, from which they suck sap.

Since earliest times in Mexico, long before the Conquest, the bugs have been collected, dried (*grana*), and processed in various ways into a dyestuff. Typically, the grana was ground into a fine powder that is soaked in hot water, to which natural acids such as citrus juice may be added (Ross 1986). The importance of the bugs is evidenced by the story in Torquemada of bags of "lice" (most certainly of dried cochineal bugs or *talegas*) stored in Moctezuma's treasure house and representing tribute from his subjects (Cowan 1865: 316f.).

Culture and harvesting was best developed in Central Mexico among the Aztecs and remained almost entirely in the hands of the Indian population (*nopaleros*) until very late in history. A monopoly on supply of the product was maintained by the Spanish for 250 years, but this was eventually broken by entrepreneurs anxious to profit by an industry closer to markets in Europe, India, around the Mediterranean, and even in Australia. The host was easily introduced to these areas but was followed by successful establishment of the insect only in a few places, notably, in the Canary Islands.

Evidence is strong that a similar, though less extensive, industry (from wild *Dactylopius*) thrived among the ancient Peruvians among whom the Quechua word *macno* was used for the color and its source. The historical relationships, if any, between the Mexican and South American uses of cochineal is not known.

Although its importance suffered a worldwide decline since the advent of modern synthetic dyes, cochineal has continued to be in demand in many areas of Latin America as a natural food coloring and pharmaceutical. It is even enjoying a general revival in response to the popular demand for natural substances for human consumption (Marín and Cisneros 1985, Moran 1981).

These insects, of course, are still common everywhere that *Opuntia* grows and are most readily recognized by the thick, stringy wax masses in which they live. These white shelters are conspicuous against the green background of the cactus pads that they may damage by their feeding. In fact, *D. opuntiae* has been used to control prickly pear cactus in California, India, South Africa, and elsewhere (probably the first organism used deliberately and successfully in biological control) (Moran 1981).

The insects themselves are similar to mealybugs but are smooth, with only minute surface features; they are deep, dark red and have a shiny, waxy surface. Those seen on the cacti are the nymphs (BL 2–3 mm) and adults of females (BL 4 mm), which lack wings and antennae and are bound to a small area of the plant because of their practically useless legs (fig. 8.7e). The motile, winged males are rarely observed. The minute active crawlers develop long waxy filaments, climb high on the plant, and expose themselves to the wind, which disperses them.

### References

Cowan, F. 1865. Curious facts in the history of insects. Lippincott, Philadelphia.

Donkin, R. A. 1977. Spanish red: An ethnogeographical study of cochineal and the *Opuntia* cactus. Amer. Phil. Soc. Trans. 67: 1–84.

Goetz, M., and J. Meinwald. 1980. Red cochineal dye (carminic acid): Its role in nature. Science 208: 1039–1042.

Marín L., R. Cisneros, and F. Cisneros. 1985 [1983]. Factores que deben considerarse en la producción de la "cochinilla de carmín" *Dactylopius coccus* (Costa) en ambientes mejorados. Rev. Peruana Entomol. 26: 81–83.

Moran, V. C. 1981. Belated kudos for cochineal insects. Antenna 5: 54–58.

Ross, G. N. 1986. The bug in the rug. Nat. Hist. 95: 66–73.

## PLANTHOPPERS

Fulgoroidea.

This is a large and highly diverse group in the tropics throughout the world (Nault and Rodríguez 1985). The Neotropical species are not well known but already number some 498 genera and 2,794 species (O'Brien pers. comm.), and new forms are being found continually. They are mostly small (BL 10 mm or less) and difficult to characterize anatomically because of their varied body types. Fairly constant features, however, are antennae that arise on the sides of the head beneath the eyes and a small plate (tegula) covering the base of the wing like an epaulet. Also, the inner corners of the coxae of all three legs meet along the midline ventrally. Many have convex, inflated or elongated (sometimes greatly so) frontal areas on the head.

The many families are very diverse in body form (O'Brien and Wilson 1985); among the more conspicuous are the wedge-shaped Acanaloniidae and Flatidae. In these, the wings are held vertically at rest and are shaped like half disks, flat along the back, convex along the ventral edge. Many are colorful, greenish, roseate, or whitish and form cryptic, leaflike or flowerlike clusters along plant stems. The Derbidae resemble small, white butterflies, resting with wings outstretched; the fore wing is triangular and much larger than the hind as in the Lepidoptera.

Members of the Fulgoridae are often large (Hogue et al. 1989, Johnson and Foster 1986), and some produce elaborate trailing plumes of white wax from the abdomen. The dragon-headed bugs (*Fulgora*) belong to this group (see below). In such species as *Cerogenes auricoma* (WS 7 cm, BWL 3.5 cm; blackish fore wing with red base, hind wing white with black apex; fig. 8.9c), *Pterodictya reticularis* (WS 5.5 cm, BWL 3 cm; greenish, semitransparent, reticulate wings; fig. 8.9b), and *Phenax variegata* (WS 9 cm, BWL 4 cm; mottled, black and white wings; fig. 8.8a), these excrescences are dense and featherlike and reach several centimeters in length behind the body. The tail of *C. auricoma* ("flying

**Figure 8.8  PLANTHOPPERS (FULGORIDAE).** (a) Variegate giant planthopper (*Phenax variegata*). (b) Red-dotted planthopper (*Lystra strigata*). (c) Dragon-headed bug (*Fulgora laternaria*).

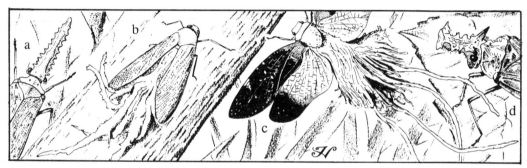

**Figure 8.9  PLANTHOPPERS (FULGORIDAE).** (a) Saw-nosed planthopper (*Cathedra serrata*). (b) Reticulate planthopper (*Pterodictya reticularis*). (c) Flying mouse (*Cerogenes auricoma*). (d) Wart-headed planthopper (*Phrictus diadema*).

mouse") (Arnaud 1970) may be over 10 centimeters long. It is sometimes observed resting on the trunks and branches of oaks (*Quercus conspersa*) in southern Mexico and readily takes flight when disturbed. *Phrictus* (Caldwell 1945) are also fairly large (WS 5.5 cm, BWL 4 cm), snouted bugs like *Fulgora*, but the irregular proboscis is wartlike and trident apically (fig. 8.9d); they have bright red hind wings basally and erect, black horns over the eyes. These do not produce extravagant wax excrescences like some of their relatives. Another moderate-sized (WS 4.5 cm, BWL 2 cm) but very common planthopper is the snoutless *Lystra strigata*, which has a black fore wing with white specks and a red-streaked

hind margin (fig. 8.8b); it produces modest wax trailers that curve upward.

## References

Arnaud, Jr., P. H. 1970. "Flying mouse" identified as *Cerogenes auricoma* (Burmeister) (Homoptera: Fulgoridae). Pan-Pacific Entomol. 46: 68.

Caldwell, J. S. 1945. Neotropical lanternflies of the genus *Phrictus* in the United States National Museum, with descriptions of four new species. U.S. Natl. Mus. Proc. 96: 177–184.

Hogue, C. L., T. Taylor, A. Young, and M. E. Platt. 1989. Egg masses and first instar nymphs of some giant Neotropical planthoppers (Homoptera: Fulgoridae). Rev. Biol. Trop. 37: 221–226.

Johnson, L. K., and R. B. Foster. 1986. Associations of large Homoptera (Fulgoridae and

Cicadidae) and trees in a tropical forest. Kans. Entomol. Soc. J. 59: 415–422.

NAULT, L. R., AND J. G. RODRÍGUEZ, eds. 1985. The leafhoppers and planthoppers. Wiley, Chichester.

O'BRIEN, L. B., AND S. W. WILSON. 1985. Planthopper systematics and external morphology. *In* L. R. Nault and J. G. Rodríguez, eds. The leafhoppers and planthoppers. Wiley, New York. Pp. 61–102.

## Dragon-Headed Bugs

Fulgoridae, *Fulgora. Spanish:* Chicharras machacuy, c. machaca, c. machacú (Peru); portas-linternas (Argentina); mariposas caiman. *Portuguese:* Cigarras víboras, c. cobras, cobras do ar, c. de asa (Brazil). *Tupi-Guaraní:* Jaquirana-bóias (jequitirana-bóias, gitirana-bóias, tiram-bóias, jakyranam-bóias, etc.) (Brazil). Alligator-headed bugs, peanut-headed bugs, lantern bugs.

In place of its many existing names, all of which I believe are inappropriate, I am introducing here a new common name for this insect based on the shape and mimetic pattern of the large head protuberance that I believe actually simulates the up-turned head of a medium-sized, arboreal reptile (Hogue 1985). The row of squarish spots along either side of the structure is very similar to the scales bordering the margins of the mouth of such reptiles, and the basal prominence on top, with its central orbicular spot, is similar to the bulbous eye of *Plica, Enyaliodes, Anolis,* and others. The bug's large size (BWL 7 cm, WS 13 cm), elongate form, and mottled green and white wings all assist in this deception. The lizards falsely recognize bugs as their own kind and therefore avoid eating them. A theory that the resemblance is with caimans (Poulton 1924) is untenable since the bugs do not go near water and remain mostly high on tree trunks rather than on the ground.

This is an insect with a lengthy and fascinating history. There are less than a dozen similar species, distributed over most of the Neotropical Region in lowland dry to wet forests (Metcalf 1947, Brailovsky and Beutelspacher 1978). The best known, *F. laternaria,* was named for its alleged ability to luminesce. This precept was probably concocted by Nehemiah Grew from reading Mouffet's account of headlight beetles (*Pyrophorus*) in *Insectorum theatrum* (1634). The myth has been perpetuated in innumerable publications, including an often-copied account by the famous early naturalist, Maria Sybilla Merian, in her *Metamorphoses insectorum surianamensium* pubished in 1705 (China 1924).

Other fanciful traditions surround this homopteran (Lenko and Papavero 1979). In most of South America, it is considered deadly venemous. Its resemblance to a reptile may be the source of this idea. Locals shun it even though in truth it is completely harmless. Even the formidable centimeter-long beak it tucks between its forelegs when not in use is never used except to suck sap from the host trees. Also widespread is the story that anyone bitten by this creature must have sexual relations within twenty-four hours or suffer a horrible death. The latter seems to be more of a ruse invented recently by local machos and used to their personal advantage than a valid folktale (Hogue and Lamas 1990).

The folklore surrounding these bugs has overshadowed our real knowledge of their biology (Janzen and Hogue 1983). In spite of the frequency about which they are written and discussed, little is recorded regarding their true habits. Dragon-headed bugs are now known to rest on the trunks of several different kinds of trees that are resinous or have bitter sap. Some hosts are *Simarouba amara* and *Simaba* (Simaroubaceae), *Hymenaea* (Fabaceae), and *Zanthoxylum* (Rutaceae) (Hogue 1985), from which the bugs may sequester toxic or noxious substances. This may constitute a secondary line of defense behind their reptilian mask. When disturbed, they also protect themselves by flashing large eyespots near the outer corners of the hind wings much like eyed saturniid moths.

Adults (fig. 8.8c, pl. 1e) remain inactive or lethargic during the day but fly at night, occasionally being attracted to artificial light. Unlike other large tropical fulgorids, they do not develop wax plumes that trail from the abdomen. Rather, they have a thin, white, powdery bloom covering most of the whole body which sometimes accumulates in greater quantities at the rear of the abdomen.

The immatures are seen rarely. According to Hingston (1932: 288–290), the females immerse their eggs in a frothy substance that hardens around them, forming a structure similar to a mantid egg case. The nymph resembles the adult in the possession of the inflated head structure but is wingless and much smaller (Hagmann 1928).

The related *Cathedra serrata* (fig. 8.9a) is sometimes confused with *Fulgora* (Squire 1972). It is easily distinguished by its smaller size (WS 6.5 cm, BWL 5.5 cm) and slender, straight head protuberance with conspicuous coarse teeth projecting from the sides in a manner much like the beak of a sawfish. It has the dark hind wing colorfully marked with a large orange orbicular spot apically.

## References

BRAILOVSKY, H., AND C. R. BEUTELSPACHER. 1978. Una nueva especie de *Fulgora* (Linneo (Homoptera: Fulgoridae) de Mexico. Inst. Biol. Univ. Nac. Auton. México Ser. Zool. An. 49: 175–182.

CHINA, W. E. 1924. On the luminosity of *Laternaria phosphorea* L. Entomol. Soc. London Trans. 1924: xlix–lii.

HAGMANN, G. 1928. A larva de *Laternaria phosphorea* L. Mus. Nac. Rio de Janeiro Bol. 4(3): 1–6.

HINGSTON, R. W. G. 1932. A naturalist in the Guiana forest. Longmans Green, New York.

HOGUE, C. L. 1985. Observations on the plant hosts and possible mimicry models of "lantern bugs": (*Fulgora* spp.) (Homoptera: Fulgoridae). Rev. Biol. Trop. 32: 145–150.

HOGUE, C. L., AND G. LAMAS. 1990. The love bug. Americas 42: 24–26.

JANZEN, D. J., AND C. L. HOGUE. 1983. *Fulgora laternaria* (Machaca, peanut-head bug, lantern fly). *In* D. H. Janzen, ed., Costa Rican natural history. Univ. Chicago Press, Chicago. Pp. 726–727.

LENKO, K., AND N. PAPAVERO. 1979. Insetos no folclore. Conselho Estad. Art. Cien. Hum. Sec. Cult. Cien. Tech., São Paulo.

METCALF, Z. P. 1947. Fulgoroidea, Fulgoridae. *In* Z. P. Metcalf, ed., General catalogue of the Hemiptera. Smith College, Northampton, Mass., Fasc. IV, pt. 9.

POULTON, E. B. 1924. The terrifying appearance of *Laternaria* (Fulgoridae) founded on the most prominent feature of the alligator. Entomol. Soc. London Trans. 1924: xliii–xlix, Pl. A.

SQUIRE, F. A. 1972. Entomological problems in Bolivia. Pest Art. News Serv. 18: 239–268.

## TREEHOPPERS

Membracidae. *Spanish:* Saltones torritos, saltarines (General).

Most treehoppers are small (BL 2–12 mm). If they were much bigger, they would be the hobgoblins of the insect world because they exhibit a tremendous variety of bizarre body forms, all the result of modifications of the shape of the dorsal shield of the prothorax. The function of this structural elaboration has been much debated. Evidence indicates that a variety of purposes are served (Boulard 1973), including cryptic mimicry (especially the thorn-shaped types), sexual display, aposematic coloration, and self-amputation (apophysectomy; the structure is loosely attached and easily detaches when jostled) (Mann 1912). The presence of numerous sensory hairs on its surface suggests a role in perception of odor, air currents, or sound (Wood and Morris 1974), but this is not likely a definitive purpose since such sense organs are widely placed and not specialized in structure or distribution.

These homopterans are otherwise basically similar to leafhoppers, except their wings are partly concealed by the enlarged prothorax and they have stouter legs. Also there are only two ocelli, and the bristlelike antennae arise anterior to the eyes.

The nymphs are commonly very spiny, especially along the middle of the back. They often form aggregations after hatching in company with their maternal parent. In many species, the latter practices brood care, hovering over her eggs and living among the nymphs until they mature (Wood 1979, Hinton 1977). Aggregations are sometimes composed of mixed species.

The majority of treehoppers are actually solitary, but presocial behavior has developed in some subfamilies, possibly out of mutualism with ants in lowland, wet tropical forests (Wood 1984). Nymphs secrete copious honeydew, which attracts ants and other insects seeking this nutritious substance. Treehoppers thus provide sustenance for aggressive ant species that guard them from insectivores (Wood 1984). In Brazil, *Aetalion reticulatum* (a species in the related family Aetalionidae) is served by two protectors alternately: stingless bees deter enemies during the day and are replaced by *Camponotus* ants at night (Castro 1975).

The family is well represented in Latin America. There are about 190 genera containing approximately 1,250 species

(Deitz 1975; Metcalf and Wade 1963, 1965; Funkhouser 1927). A correlation seems to exist between a low number of species per genus in lowland regions and lessened host specialization (Wood and Olmstead 1984).

Because of their strange shapes, variety, and colors, treehoppers have attracted the attention of many entomologists and artists who have portrayed them elegantly (Buckton 1903, Fowler 1894–1909). They exhibit an infinite variety of structure, some of which tends to sort into a few generalized types, depending on the basic form of the pronotum. Some of these are the single thorn shape (*Umbonia;* fig. 8.10a); double thorn shape (*Hemikyptha*); compressed wedge types (*Membracis;* fig. 8.10f); carriers of clusters of four spherical swellings on stalks (*Cyphonia, Bocydium* [fig. 8.10c], *Eucyphonia*); bulbous form (*Combophora;* fig. 8.10d); ball-with-handle form, or very long with a bulbous swelling posteriorly (*Heteronotus;* fig. 8.10b); crested, or branching spongiform excrescences (*Spongophorus;* fig. 8.10e); and fusiform (*Polyglypta*) types. Treehoppers are not notable wax producers, although the egg masses of some species are sometimes covered in waxy secretions.

Membracids are seldom injurious. *Metcalfiella monogramma* (*periquito del aguacate*) sometimes damages avocados in Mexico (Camacho 1944). Biological information

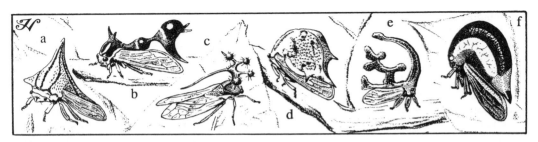

**Figure 8.10   TREEHOPPERS (MEMBRACIDAE).** (a) Thorn mimic (*Umbonia spinosa*). (b) Spine mimic (*Heteronotus flavomaculatus*). (c) Small treehopper (*Bocydium* sp.). (d) Inflated treehopper (*Combophora* sp.). (e) Fungiform treehopper (*Spongophorus* sp.). (f) Crested treehopper (*Membracis* sp.).

on the family generally is scant. The life history strategies of some Neotropical species have been described (Wood 1984).

## References

BOULARD, M. 1973. Le pronotum des Membracides: Camouflage sélectionné ou orthogénése hypertélique? Mus. Nat. Hist. Natur. (Paris) Bull. (ser. 3) 83: 145–165.

BUCKTON, G. B. 1903. A monograph of the Membracidae. Lovell Reeve, London.

CAMACHO, A. D. 1944. El periquito del aguacate. Fitófilo (México) 3(4): 3–54.

CASTRO P. R. C. 1975. Mutualismo entre *Trigona spinipes* (Fabricius, 1793) e *Aethalion reticulatum* (L. 1767) em *Cajanus indicus* Spreng. na presença de *Camponotus* supp. Cien. Cult. 27(5): 537–539.

DEITZ, L. L. 1975. Classification of the higher categories of the New World treehoppers (Homoptera: Membracidae). N.C. Agric. Exper. Sta. Tech. Bull. 225: 1–177.

FOWLER, W. W. 1894–1909. Rhyncota: Hemiptera-Homoptera. Biología Centr. Amer. Vol. 2, pt. 1.

FUNKHOUSER, W. D. 1927. Membracidae, general catalogue of the Hemiptera. Smith College, Northampton, Mass., Fasc. 1.

HINTON, H. E. 1977. Subsocial behavior and biology of some Mexican membracid bugs. Ecol. Entomol. 2: 61–79.

MANN, W. M. 1912. A protective adaptation on a Brazilian membracid. Psyche 19: 145–147.

METCALF, Z. P., AND V. WADE. 1963. A bibliography of the Membracoidea and fossil Homoptera (Homoptera: Auchenorhyncha) [sic]. N.C. State Univ., Raleigh.

METCALF, Z. P., AND V. WADE. 1965. General catalogue of the Homoptera: A supplement of Fascicle I: Membracidae of the General catalogue of Hemiptera. Membracoidea. Sec. 1–11. N.C. State Univ., Raleigh.

WOOD, T. K. 1979. Sociality in the Membracidae (Homoptera). Entomol. Soc. Amer. Misc. Publ. 11: 15–22.

WOOD, T. K. 1984. Life history patterns of tropical membracids (Homoptera: Membracidae). Sociobiology 8: 299–344.

WOOD, T. K, AND G. K. MORRIS. 1974. Studies on the function of the membracid pronotum (Homoptera). I. Occurrence and distribution of articulated hairs. Can. Entomol. 106: 143–148.

WOOD, T. K., AND K. L. OLMSTEAD. 1984. Latitudinal effects on treehopper species richness (Homoptera: Membracidae). Ecol. Entomol. 9: 109–115.

## FROGHOPPERS

Cercopidae. Spittlebugs.

The squat, froglike appearance and frothy secretions of the nymphs give these homopterans their common names. One to several individuals may live in masses of this whitish, bubbly substance ("cuckoo spit") produced from the anus and from a mucilagenous excretion from hypodermal glands on the seventh and eighth abdominal segments. The foam provides protection from enemies and desiccation (fig. 8.11c).

Adult cercopids, which do not produce spittle, are very similar to leafhoppers but tend to be broader and flatter; many are larger (BL 10–15 mm) and have only a few stout spines on the hind tibia instead of one or two rows of small spines. Members of the genus *Tomapsis* are large and conspicuous, with black body and wings, the latter with transverse red, orange, or yellow bands (fig. 8.11a).

Both stages feed on plant sap, and several species damage useful plants. The best known is the sugarcane froghopper (*Aeneolamia varia saccharina;* fig. 8.11b).

## LEAFHOPPERS

Cicadellidae. *Spanish:* Chicharritas, cigarritas, loritos (General). *Portuguese:* Cigarrinhas.

All members of this varied family (Hamilton 1983, Nielson 1985) are small (BL less than 10 mm). They are slender in form and usually have thickened fore wings that display all the colors in dull uniform to highly variegate and bright patterns (e.g., the yellow-spotted *Amblyscartidia albofasciata;* fig. 8.11d, and multicolored *Baleja flavoguttata;* fig. 8.11e). A distinguishing anatomical characteristic is a double row of spines along the hind tibia. They are active jumpers, and adults fly readily. Both

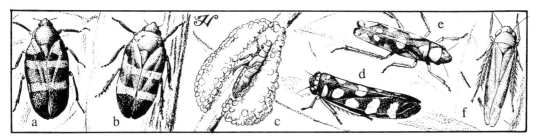

**Figure 8.11 FROGHOPPERS (CERCOPIDAE) AND LEAFHOPPERS (CICADELLIDAE).** (a) Froghopper (*Tomapsis inca*). (b) Sugarcane froghopper (*Aeneolamia varia saccharina*). (c) Froghopper nymph in froth nest. (d) Leafhopper (*Amblyscartidia albofasciata*). (e) Leafhopper (*Baleja flavoguttata*). (f) Bean leafhopper (*Empoasca kraemeri*).

young and adults have the curious habit of running sideways.

All are plant feeders, piercing plant tissues and withdrawing sap. In so doing, they often cause damage by drying up the host. Like aphids, they also play an important role as vectors of plant diseases, particularly those caused by viruses (Maramorosch and Harris 1979), most often on species in the grass family. Such are members of the genus *Dalbulus*, which are the principal vectors of pathogens belonging to the corn stunt disease complex (Gámez and León 1985). Another very injurious genus in Latin America is *Empoasca,* with species (especially the very widespread *lorito verde, E. kraemeri;* fig. 8.11f) that damage beans directly by their feeding (van Schoonhoven et al. 1985, Wilde et al. 1976). In earlier literature, *E. kraemeri* was confused with the North American *E. fabae,* from which it and a number of other Neotropical species have been segregated.

This is a large and diverse assemblage in Latin America with approximately 7,500 species in 25 subfamilies.

### References

Gámez, R., and P. León. 1985. Ecology and evolution of a Neotropical leafhopper-virus-maize association. *In* L. R. Nault and J. G. Rodríguez, eds., The leafhoppers and planthoppers. Wiley, Chichester. Pp. 331–350.

Hamilton, K. G. A. 1983. Classification, morphology and phylogeny of the family Cicadellidae (Rhyncota: Homoptera). *In* W. J. Knight, N. C. Pant, T. S. Robertson, and M. R. Wilson, eds. First international workshop on leafhoppers and planthoppers of economic importance. Commonwealth Inst. Entomol., London. Pp. 15–37.

Maramorosch, K., and K. F. Harris. 1979. Leafhopper vectors and plant disease agents. Academic, New York.

Nielson, M. W. 1985. Leafhopper systematics. *In* L. R. Nault and J. G. Rodríguez, eds. The leafhoppers and planthoppers. Wiley, Chichester. Pp. 11–39.

van Schoonhoven, A., G. J. Hallman, and S. R. Temple. 1985. Breeding for resistance to *Empoasca kraemeri* Ross and Moore in *Phaseolus vulgaris. In* L. R. Nault and J. G. Rodríguez, eds., The leafhoppers and planthoppers. Wiley, Chichester. Pp. 405–422.

Wilde, G., A. van Schoonhoven, and L. Gómez. 1976. The biology of *Empoasca kraemeri* on *Phaseolus vulgaris*. Entomol. Soc. Amer. Ann. 69: 442–444.

## APHIDS

Aphididae. *Spanish:* Afidos, pulgones de plantas (General). *Portguese:* Afideos, pulgões de plantas (Brazil).

Aphids normally live on plants whose sap they siphon with their elongate proboscis. A few are fed by ants with which they are symbiotically associated, and many can live in the soil for a long time without feeding. Feeding may be negligible by a few individuals or so extensive in massive populations that the host's growth is severely

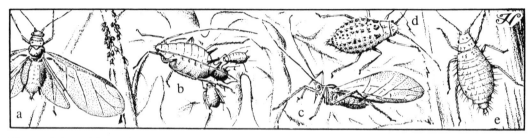

**Figure 8.12   APHIDS (APHIDIDAE).** (a) Cotton aphid (*Aphis gossypii*), winged sexual female. (b) Cotton aphid, wingless viviparous female with young. (c) Pea aphid (*Acyrthosiphon pisum*). (d) Spotted alfalfa aphid (*Therioaphis maculata*), wingless female. (e) Yellow sugarcane aphid (*Sipha flava*), wingless viviparous female.

perturbed. Some aphids can cause the formation of galls.

Unlike most insects, aphids lack Malpighian tubules and excrete nitrogenous wastes in the form of ammonia instead of uric acid. The large volume of water in their diet and internal symbiotic microorganisms help flush out this substance through the gut and detoxify it.

In their feeding, aphids may wreak havoc on crops. They reduce plant vitality by extracting significant quantities of sap but are even more serious threats as vectors of pathogenic viruses. Most of the aphid pests of the region are European in origin and occur on introduced crops; some important examples are the cotton aphid (*Aphis gossypii*, figs. 8.12a, b), pea aphid (*Acyrthosiphon pisum*, fig. 8.12c), spotted alfalfa aphid (*Therioaphis maculata*, fig. 8.12d), and yellow sugarcane aphid (*Sipha flava*, fig. 8.12e).

About 97 percent of the known plant viruses are transmitted by insects and mites, 60 percent of these by aphids (Harris and Maramorosch 1977). The virus-inoculating abilities of the native aphid fauna are unknown. In Latin America, several different *Aphis* and a few other introduced species transmit potato leaf roll virus (*Corium solani*) and mosaic viruses of sugarcane (*Marmor sacchari*), beans (*M. phaseoli*), and crucifers (*M. cruciferarum*). Another virus disease of prime importance

is "tristeza," a plague on citrus in Brazil caused by *C. viatorum,* which is carried by several native aphid species.

Aphid body forms and reproductive state vary according to the phase or type of life cycle. The latter is complex and consists basically of alternating generations of wingless, highly fecund, parthenogenetic, live-bearing females during favorable seasons which convert to sexual, winged phases at least once a year at those times appropriate for aerial mating, egg laying, and dispersal.

In tropical climes, mating usually occurs during the wet, warm season; in dry climes, usually during nonrainy, warm periods. Deviations from the basic pattern are various, and many cycles are very complex. Alternation of hosts may accompany the change in reproductive types. Most have seven or more generations a year, the duration of each depending on environmental conditions.

The wax glands of aphids are localized on a pair of unique, posteriorly projecting, tubular structures called cornicles. They also produce honeydew droplets from the anus, often under stimulation by ants, which avidly seek this substance as food. Symbiotically, ants actively protect and facilitate the presence of aphids by attacking predators and disseminating young and even building shelters. A basis for this practice has been suggested (Kloft 1959):

in the posterior facies presented by the aphid, the cornicles simulate antennae and the anus, the mouth of another ant. The nursing ant is fooled to respond to the aphid as if it were another ant proffering food.

Structurally, aphids are a fairly homogeneous assemblage, small (BL usually less than 4 mm), soft bodied, and ovoid in shape. In color, most are unicolorous green, yellow, reddish, or black. Wings, which are present only in the sexually active, dispersing forms, are diaphanous and with few veins.

Because of their economic importance, aphids have attracted a great deal of study. As a result, good, modern treatments of their biology (Dixon 1973, 1977), taxonomy (Eastop and Ris Lambers 1976), and

methods of study (van Emden 1972) and a bibliography (Smith 1972) are available.

## References

DIXON, A. F. G. 1973. Biology of aphids. Inst. Biol. Stud. Biol. 44: 1–58.

DIXON, A. F. G. 1977. Aphid ecology. Ann. Rev. Ecol. Syst. 8: 329–353.

EASTOP, V. F., AND D. H. RIS LAMBERS. 1976. Survey of the world's aphids. Junk, The Hague.

EMDEN, H. F. VAN, ED. 1972. Aphid technology. Academic, New York.

HARRIS, K. F., AND K. MARAMOROSCH. 1977. Aphids as virus vectors. Academic, New York.

KLOFT, W. 1959. Versuch einer Analyse der trophobiotischen Beziehungen von Ameisen zu Aphiden. Biolog. Zentralblatt 78: 863–870.

SMITH, C. F. 1972. Bibliography of the Aphididae of the world. N.C. Agric. Exper. Sta. Tech. Bull. 216: 1–717.

# 9 BEETLES

Coleoptera. *Spanish:* Escarabajos, mayates (Central America, Mexico); ron ron, chocorones (Nicaragua). *Nahuatl:* Mayameh, sing. mayatl (Mexico). *Portuguese:* Besouros, cascudos, pisões (Brazil).

These commonplace insects form the largest order of organisms, about 70 percent of the insects and 40 percent of all animals, with well over 300,000 currently named species (Blackwelder 1944–1957). The number of Neotropical species is unknown; some estimates are astronomical, but the rich variety still hidden in the vast American rain forests, even by conservative standards, will certainly amaze us when it is fully appreciated. In an average hectare of Panamanian lowland forest alone, it is estimated that there are 11,410 host-specific species of beetles on seventy kinds of trees (Erwin 1982).

Their tremendously varied body forms and sizes adapt them to all climes and niches, high to low, moist to dry, terrestrial to aquatic (Crowson 1981, Evans 1975, Reitter 1961). Most are free-living, but a few are parasitic (see parasitic rove beetles) or commensals (Barrera 1969) on rodents. The smallest known beetle lives in fungus fruiting bodies and is the Mexican feather-winged beetle (Ptiliidae), *Nanosella fungi*, with a body length of only 0.4 millimeters; the largest is the forest monster *Dynastes hercules* (Scarabaeidae) of Central America, with males up to 16 centimeters long (including horn). Some very large beetles

can regulate their body temperatures metabolically like mammals and birds (Bartholomew and Casey 1977).

In spite of their diversity, almost all adult beetles are alike in possessing a hard exoskeleton, well-developed mandibulate mouthparts, and fore wings modified into horny or leathery shells (called elytra) without a trace of venation; when closed, their inner margins meet tightly to form a straight suture down the back. The elytra are useless for flight but protect the abdomen and hind wings when in repose.

Many large beetles have outsized horns or mandibles with one surface densely covered with short, stiff, fine hairs. The function of this arrangement is unproven but may form clamps with frictional surfaces to hold and accommodate the irregular and slippery surface of the females, which they carry following or during courtship rituals.

Several beetle types also possess large serrate mandibles (Cerambycidae, *Macrodontia*) or opposing, toothed horns (Scarabaeidae, *Dynastes*, *Golofa*) whose sawlike appearance has inspired a widespread belief that they are used to sever twigs. Naive people visualize the beetle clasping the twig with its "saws" and flying around and around it, gradually cutting it in two. The name *aserradores* or *serradores* (also *torneadores*) has thus been applied to them, although nothing of the sort actually takes place; some long-horned beetles do actually girdle and sever twigs but through a chewing action of the mandibles, not by sawing in the manner just described. These names are nevertheless appropriate for the larvae of long-horned beetles, which are true "sawyers," boring through wood while feeding and developing.

The larvae are greatly varied in shape and habits. The thoracic legs and head are generally well developed, the latter with mandibulate mouth structures similar to those of the mature insect.

A considerable number of beetles in the larval and adult stages are economic pests. They attack crops and produce of all categories, and a few are even medically significant as intermediate hosts for human parasites or bearers of vesicating substances. The many predaceous types are beneficial regulators of populations of other insects.

A good number of species are large, strangely shaped, and brilliantly colored and thus attract attention. They have been long appreciated, even revered by people in ancient times, and today command high prices from tourists and collectors. The larvae of many, especially the larger Scarabaeidae, Curculionidae, and Cerambycidae, are avidly sought for food (*papás, suri* in Peru). Indian groups use parts of many in the making of jewelry (see fig. 1.13) (Klausnitzer 1981).

### References

BARRERA, A. 1969. Notes on the behaviour of *Loberopsyllus traubi*, a cucujoid beetle associated with the volcano mouse, *Neotomodon alstoni* in Mexico. Entomol. Soc. Wash. Proc. 71: 481–486.

BARTHOLOMEW, G. A., AND T. M. CASEY. 1977. Endothermy during terrestrial activity in large beetles. Science 195: 882–883.

BLACKWELDER, R. E. 1944–1957. Checklist of the Coleopterous insects of Mexico, Central America, the West Indies, and South America. U.S. Natl. Mus. Bull. 1944: Pt. 1, 185: 1–188; 1944: Pt. 2, 185: 189–341; 1945: Pt. 3, 185: 343–550; 1946: Pt. 4, 185: 551–763; 1947: 185: 765–925; 1957: 185: i–vii, 927–1492.

CROWSON, R. A. 1981. The biology of the Coleoptera. Academic, London.

ERWIN, T. L. 1982. Tropical forests: Their richness in Coleoptera and arthropod species. Coleop. Bull. 36: 74–75.

EVANS, G. 1975. The life of beetles. Hafner, New York.

KLAUSNITZER, B. 1981. Beetles. Simon & Schuster, New York.

REITTER, E. 1961. Beetles. Putnam's Sons, New York.

## GROUND BEETLES

Carabidae, Carabinae. *Spanish:*
Escarabajos terrestres. *Portuguese:*
Besouros terestres.

This is one of the largest and most common beetle families, respresented in the Neotropics by at least 4,400 described species (Erwin et al. 1979). Carabids generally shun light and live in protected niches on the ground, beneath stones, logs, or other objects, and scurry to protection if exposed. However, some are gaudily colored, diurnal, and run about exposed on logs. Such are *Lebia*, whose larvae prey on the immatures of similar-appearing leaf beetles, and *Eurycoleus*, which enter into mimicry complexes (probably Müllerian) with fungus beetles (*Priotelus*) and darkling beetles (*Poecilesthus*) (Erwin and Erwin 1976).

A great many carabids are arboreal, some even restricted to epiphytic bromeliads high above the ground (*Agra;* fig. 9.1a). Cavernicolous species, well known elsewhere, are not common in the American tropics. The grass-feeding *Notiobia peruviana* (fig. 9.1b) is one of the most widespread, extending through the high elevation grasslands of the entire length of the Andean cordillera (Noonan 1981). The genus *Selenophorus* (fig. 9.1d) is also very common. In lowland tropical areas, the majority of species are restricted to very limited microhabitats (Erwin 1979*b*).

Both larvae (fig. 9.1c) and adults are predaceous, feeding on other insects, and may be beneficial natural control agents

**Figure 9.1   GROUND AND TIGER BEETLES (CARABIDAE).** (a) Canopy carabid (*Agra* sp.). (b) Grass carabid (*Notiobia peruviana*). (c) Ground beetle, larva. (d) Ground beetle (*Selenophorus* sp.). (e) Caterpillar hunter (*Calosoma alternans*). (f) Tiger beetle (*Cicindela* sp.), larva. (g) Tiger beetle (*Cicindela carthagena*). (h) Tiger beetle (*Megacephala* sp.). (i) Tiger beetle (*Odontocheila* sp.). (j) Tiger beetle (*Pseudoxychila bipustulata*).

(Allen 1977). The genus *Calosoma* (fig. 9.1e) is composed of large, conspicuous beetles that prey on lepidopteran larvae ("caterpillar hunters," or *tesoureiros* in Brazil).

Adults are recognized by the narrow, elongate head that is always narrower than the first thoracic segment; the antenna are attached between the eyes and base of the mandibles. Most are shiny black and medium-sized (BL 5–20 mm), but *Enceladus gigas* reaches 7 centimeters.

Because they are ubiquitous, these beetles provide good material for ecological and zoogeographic study and have received considerable attention from biologists in Latin America (Erwin 1979a, 1979b; Reichardt 1979). Although most have completely developed wings and fly well, flightless forms are particularly abundant on mountains and West Indian islands (Darlington 1970). Their general biology (mostly temperate forms) is covered by Thiele (1977).

The classification of Neotropical carabids is complex (Reichardt 1977). There are certainly hundreds of undescribed species.

## References

ALLEN, R. T. 1977. *Calosoma (Castrida) alternans granulatum* Perty: A predator of cotton leaf worms in Bolivia (Coleoptera: Carabidae: Carabini). Coleop. Bull. 31: 73–76.

DARLINGTON, P. M. 1970. Carabidae on tropical islands, especially the West Indies. Biotropica 2: 7–15.

ERWIN, T. L. 1979a. The American connection, past and present, as a model blending dispersal and vicariance in the study of biogeography. *In* T. L. Erwin, G. E. Ball, D. R. Whitehead, and A. L. Halpern, eds., Carabid bettles, their evolution, natural history, and classification. Proc. 1st Int. Cong. Carabidology. Junk, The Hague. Pp. 355–367.

ERWIN, T. L. 1979b. Thoughts on the evolutionary history of ground beetles: Hypotheses generated from comparative faunal analyses of lowland forest sites in temperate and tropical regions. *In* T. L. Erwin, G. E. Ball, D. R. Whitehead, and A. L. Halpern, eds., Carabid beetles, their evolution, natural history, and classification. Proc. 1st Int. Cong. Carabidology. Junk, The Hague. Pp. 539–587.

ERWIN, T. L., G. E. BALL, D. R. WHITEHEAD, AND A. L. HALPERN, eds. 1979. Carabid beetles, their evolution, natural history, and classification. Proc. 1st Int. Cong. Carabidology. Junk, The Hague.

ERWIN. T. L., AND L. J. M. ERWIN. 1976. Relationships of predaceous beetles to tropical forest wood decay. Pt. II. The natural history of Neotropical *Eurucoleus macularis* Chevrolat (Carabidae: Lebiini) and its implications in the evolution of ectoparasitoidism. Biotropica 8: 215–224.

NOONAN, G. R. 1981. South American species of the subgenus *Anisotarsus* (genus *Notiobia* Perty: Carabidae: Coleoptera). Pt. I. Taxonomy and natural history. Milwaukee Publ. Mus. Contrib. Biol. Geol. 44: 1–84.

REICHARDT, H. 1977. A synopsis of the genera of Neotropical Carabidae (Insecta: Coleoptera). Quaest. Entomol. 13: 346–493.

REICHARDT, H. 1979. The South American carabid fauna: Endemic tribes and tribes with African relationships. *In* T. L. Erwin, G. E. Ball, D. R. Whitehead, and A. L.

Halpern, eds., Carabid beetles, their evolution, natural history, and classification. Proc. 1st Int. Cong. Carabidology. Junk, The Hague. Pp. 319–325.

Thiele, H. U. 1977. Carabid beetles in their environments. Springer, Berlin.

# TIGER BEETLES

Carabidae, Cicindelinae. *Portuguese:*
Tigres velozes (Brazil).

Tiger beetles (Peña 1969, Pearson 1988) are aggressive and efficient predators both as adults and as larvae, hence their common name. They are small to middle-sized beetles (BL 6–40 mm), with a distinctively large head (wider than the prothorax, which is itself much narrower than the rest of the body), conspicuous, bulbous eyes, and scythelike, grossly toothed mandibles. They often have strong patterns and in many instances display very beautiful iridescent green, red, or blue color. Active and agile on long slender legs, they are quick to flee when approached.

Several of the 500 or so Neotropical species are fairly conspicuous, especially the large *Pseudoxychila* (fig. 9.1j), which are dull, velvety blue-green or blue, with a single, round white or reddish spot in the center of each elytron. They are flightless and diurnal, running on the ground, usually near streams, in a manner similar to that of velvet ant females (Mutillidae), which they apparently mimic. Their biology has been partly elucidated (Palmer 1983).

*Megacephala* (fig. 9.1h, pl. 1g) called *caballitos de siete colores* in Peru, are also common tiger beetles but are nocturnal and multimetallic colored. *Odontocheila* (fig. 9.1i) are mostly small (BL 8–10 mm) and dull brown. They are usually seen resting on the upper leaf surfaces of forest understory vegetation during the day (Pearson 1983) between hunting forays to the ground (Pearson and Anderson 1985).

Males make a mysterious waving display with a foreleg (Palmer 1981), possibly a courtship ritual. *Cicindela* (fig. 9.1g) are usually marked with contrasting elytral patterns and are common on the open banks of watercourses and lakes or along forest paths and clearings. *Ctenostoma* are very active and live in the canopy of lowland forests; unlike other cicindelines whose larvae are terrestrial, their larvae develop in rotting logs (Zikán 1929).

Most larval tiger beetles are somewhat wormlike and position themselves head upward in vertical burrows in mud (fig. 9.1f), awaiting the approach of other insects and small ground-dwelling invertebrates that they capture by reaching out and grasping them with their mandibles. They have an excessively large head, which, with the prothoracic shield, is held at a right angle to the body to form an effective plug for the burrow mouth. With a hooklike spine on the humped, fifth abdominal segment, the larvae anchor themselves in their tubes and avoid being pulled out by large prey. They leave their burrows at times and creep to new sites, using their legs to drag the long abdomen behind. Adults have also been observed nesting in similar burrows in large numbers (Wille and Michener 1962).

Although commonly treated as a distinct family, tiger beetles are considered a subfamily of Carabidae by modern coleopterists.

## References

Palmer, N. 1981. Notes on the biology and behavior of *Odontochila mexicana*. Cicindela 13(3/4): 29–36.

Palmer, M. K. 1983. *Pseudoxychila tarsalis* (Abejón tigre, tiger beetle). *In* D. H. Janzen, ed., Costa Rican natural history. Univ. Chicago Press, Chicago. Pp. 765–766.

Pearson, D. L. 1983. Patterns of limiting similarity in tropical forest tiger beetles (Coleoptera: Cicindelidae). Biotropica 12: 195–204.

Pearson, D. L. 1988. Biology of tiger beetles (Coleoptera: Cicindelidae). Ann. Rev. Entomol. 33: 123–147.

PEARSON, D. L., AND J. J. ANDERSON. 1985. Perching heights and nocturnal communal roosts of some tiger beetles (Coleoptera: Cicindelidae) in southeastern Peru. Biotropica 17: 126–129.

PEÑA, L. E. 1969. Notes on the Cicindelidae of Chile. Cicindela 1(2): 1–7.

WILLE, A., AND C. D. MICHENER. 1962. Inactividad estacional de *Megacephala sobrina* Dejean (Coleoptera: Cicindelidae). Rev. Biol. Trop. 10: 161–165.

ZIKÁN, J. F. 1929. Zur Biologie der Cicindeliden Brasiliens. Zool. Anz. 82: 269–414.

## DARKLING BEETLES

Tenebrionidae. *Spanish:* Cucarachas martina (Peru), pinacates (Mexico).

This is an extremely varied group, but most are dark-colored, dull black or brown beetles, characterized by four-segmented hind tarsi, the other tarsi being five segmented. The antennae are usually threadlike or slightly to strongly clubbed, and they have notched eyes. The edge of the prothorax is often ridged or sharp, and the elytra are smooth to ridged, with lines of pits or crudely roughened. They are common in all habitats, in rotton wood, under logs, in fungi, or on the ground, from moist wooded areas to deserts. Both adults and larvae are omnivores and scavengers, feeding on decaying vegetation, fungi, and other organic matter of all kinds.

There are about 3,000 species in the Neotropical Region. Some are unusual, such as the attic beetles (*Zophobas;* fig. 9.2a), whose larvae feed in the bat guano in caves and attics of houses. In this species, cannibalism is a way of life for their larvae, which must retreat far from the colony for safety during pupation (Tschinkel 1981). Among the larger Neotropical beetle groups, this family tends most to have many members adapted to dry, even desert habitats. In these, a large, subelytral cavity is thought to provide a pocket of air, insulating them against the heat of the sun. Many are flightless, long lived, and nocturnal, hiding in the soil during the day.

This family is so well developed in Chile that it has served as the basis for classifying the country's entomological faunal regions, according to one author (Peña 1966).

### References

PEÑA, L. E. 1966. A preliminary attempt to divide Chile into entomofaunal regions based on the Tenebrionidae (Coleoptera). Postilla 97: 1–17.

TSCHINKEL, W. R. 1981. Larval dispersal and cannibalism in a natural population of

**Figure 9.2  DARKLING (TENEBRIONIDAE) AND FUNGUS (EROTYLIDAE) BEETLES.** (a) Attic beetle (*Zophobas* sp.). (b) Horned darkling beetle (*Tauroceras* sp.). (c) Giant darkling beetle (*Mylaris* sp.). (d) Darkling beetle (*Strongylium* sp.). (e) Ma'kech (*Zopherus chilensis*). (f) Desert darkling beetle (*Gyriosomus* sp.). (g) Desert darkling beetle (*Nyctelia* sp.). (h) Desert darkling beetle (*Proacis bicarinatus*). (i) Desert darkling beetle (*Scotobius gayi*). (j) Fungus beetle mimicking darkling (*Cuphotes* sp.). (k) Fungus beetle model for darkling mimic (*Cypherotylus* dromedarius, Erotylidae). (l) Fungus beetle (*Erotylus* sp., Erotylidae).

*Zophobas atratus* (Coleoptera: Tenebrionidae). J. Anim. Behav. 29: 990–996.

# GIANT DARKLINGS

Tenebrionidae, Tenebrioninae, Coelometopini, *Mylaris* (formerly *Nyctobates*).

*Mylaris* (fig. 9.2c) are among the giants of the darkling beetle family, some spectacular species being nearly 5 centimeters long. They are elongate with a semispherical prothorax that is decidedly narrower than the elytra. The latter are parallel sided anteriorly, tapering posteriorly in a gradual arc, and are strongly ridged or punctate longitudinally. The legs are long and slender, and the antennae are beadlike.

Members of this genus dwell in humid lowland forests, seeking out drier microhabitats, such as clearings, dry logs, and the like, where the adults are usually seen perambulating during the daytime. Near Iquitos, Peru, I have observed the latter feeding on bracket fungi.

*Tauroceras*, with diverging horns on the head of the males (fig. 9.2b), are related to *Mylaris* and are also very large tenebrionids, ranging over most of Latin America. Larvae, probably belonging to this genus, were described by Spilman (1963). They are also very large, straight, cylindrical, and heavily sclerotized, especially the ninth abdominal segment. His specimens from Jamaica were found in rotting wood. Ohaus (1900: 227) records larvae and pupae of both genera from rotten wood in Brazil.

## References

OHAUS, F. 1900. Bericht über eine entomologische Reise nach Central Brasilien. Stettinger Entomol. Zeit. 61: 164–273.

SPILMAN, T. J. 1963. On larvae, probably *Tauroceras*, from the Neotropics (Coleoptera: Tenebrionidae). Coleop. Bull. 17: 58–64.

# DESERT DARKLINGS

Tenebrionidae, Tenebrioninae, Nycteliini, Praocini, Scotobiini, etc.

The arid coastal areas of Chile and Peru, including the great Atacama, and the dry steppes of southern South America in general are home to an assemblage of endemic darkling beetles specifically adapted to hot, dry places, much like members of their tribal relatives in the deserts of southern Africa and the southwestern United States.

They are small to moderate-sized (BL 15–25 mm) tenebrionids, mostly dull black and hard shelled. They tend to have compact, oval to elongate-oval body forms, though in endless variety of detail; the strange flattened, spider-legged or sickle-jawed types found in other parts of the world are lacking. The immature stages are almost totally unknown in the majority of species, and there is little recorded on adult biology (Peña 1963).

At times, some species are found in great abundance. Their appearance coincides with wet years. At other times, they are scarce or absent altogether. They are active beetles, scurrying over the sand or sparsely vegetated ground in search of mates or food, which consists of ephemeral grasses, herbaceous plants, or almost any organic debris (ibid.).

There are numerous species, most in dominant large genera such as *Scotobius* (Scotobiini), *Proacis* (Proacini), and *Nyctelia* and *Gyriosomus* (Nycteliini).

*Scotobius* (fig. 9.2i) are fairly elongate, with longitudinal, smooth or coarsely cornicled ridges, the space between often transversly corrugate, and the edges of the prothorax may be flanged (Kulzer 1955). Members of the genus *Proacis* (fig. 9.2h) are wide bodied and with both smooth to ridged elytra (Kulzer 1958). Many *Nyctelia* (fig. 9.2g) are obovate, with deep, transverse corrugations on the elytra (Kulzer 1963). *Gyriosomus* (fig. 9.2f) typically sport oblique, white, streaked elytral patterns of dense short setae (Kulzer 1959).

## References

KULZER, H. 1955. Monographie der Scotobiini: Zehnter Beitrag zur Kenntnis der Tenebrioniden. Mus. G. Frey Entomol. Arb. 6: 383–485, Pls. XIX–XXIV.

KULZER, H. 1958. Monographie der Südamerikanischen Tribus Praocini (Col.). 16. Beitrag zur Kenntnis der Tenebrioniden. Mus. G. Frey Entomol. Arb. 9: 1–105.

KULZER, H. 1959. Neue Tenebrioniden aus Südamerika (Col.). 18. Beitrag zur Kenntnis der Tenebrioniden. Mus. G. Frey Entomol. Arb. 10: 523–567, Pls. XI–XII.

KULZER, H. 1963. Revision der Südamerikanischen Gattung Nyctelia Latr. (Col. Tenebr.). 24. Beitrag zur Kenntnis der Tenebrioniden. Mus. G. Frey Entomol. Arb. 14: 1–71, Pls. I–VI.

PEÑA, L. E. 1963. Las Nyctelia (Coleoptera: Tenebrionidae). Mus. G. Frey Entomol. Arb. 14: 72–75.

## STRONGYLINE DARKLINGS

Tenebrionidae, Tenebrioninae, Strongylini.

Some of the world's most spectacular darkling beetles belong to this tribe, which attains its greatest diversity in the wet, tropical parts of South America. They are mostly medium to large (BL 1–3 cm) and often brilliantly hued in metallic blue, green, or shiny black base colors. Many species are spotted with bright red, yellow, orange, or blue. A number of dull black and white species are strongly convex and broad (*Cuphotes* fig. 9.2j), resembling and entering into mimetic associations with giant fungus beetles (*Cypherotylus*).

*Strongylium* (fig. 9.2d) is a common genus with approximately 320 species in the New World tropics whose bright colors and unusual patterns often attract attention. Some species show strong sexual dimorphism, the males having lobster claw-shaped fore tibae that must have a forceful

but unknown function in copulation (Triplehorn 1985).

## Reference

TRIPLEHORN, C. A. 1985. A remarkable example of sexual dimorphism in *Strongylium* (Coleoptera: Tenebrionidae). Coleop. Bull. 39: 25–27.

## MA'KECH

Tenebrionidae, Zopherinae, Zopherini, *Zopherus chilensis*.

Like all members of the genus *Zopherus*, the ma'kech (fig. 9.2e) is a fairly large (BL 34–40 mm) beetle with an extremely thick, hard integument. Despite its hardness, the elytron can be scored easily with a sharp instrument, having the physical properties of an extremely dense wax. The very rough or granulate back of the beetle is largely white with dark rectangular or irregular black blotches showing through. The head is retracted into the thorax.

Little is recorded regarding its life history. Adults are found on the bark of dead trees in whose wood the larvae probably mine (Burke 1976). Other species in the same genus have been reared from specimens found in rotten wood and the larvae described (Doyen and Lawrence 1979). It is believed that adults seldom, if ever, eat and can live a year or more without food, but this is unproven. Adults feed freely in the laboratory on cereal food provided them (Doyen pers. comm.).

The species' distribution runs from southern Mexico to Venezuela and Colombia (Triplehorn 1972). Local people collect them in Yucatán and make them into a curious, living pendant by gluing baubles to their backs and fixing them to a short pin and chain. The novelty of a tethered jewel beetle on the lapel never fails to attract attention. Such specimens are sold locally as good luck charms and traditional reminders of an ancient Yucatecan legend.

A young Mayan prince was saved from capture by his lover's guards by being turned into this beetle by the Moon Goddess. In the language of the time, *ma'kech* meant "thou art a man," as uttered by the maiden in appreciation of his courageousness in overcoming obstacles in their path to love; it also meant "does not eat," in reference to the prince's (and beetle's) ability to endure prolonged fasts (Wright 1956). Although many specimens make their way abroad in tourists' luggage, it is doubtful that the species could be established elsewhere and become a pest.

### References

Burke, H. R. 1976. The beetle, *Zopherus nodulosus haldemani:* Symbol of the Southwestern Entomological Society. Southwest. Nat. 1: 105–106.

Doyen, J. T., and J. F. Lawrence. 1979. Relationships and higher classification of some Tenebrionidae and Zopheridae (Coleoptera). Syst. Entomol. 4: 333–337.

Triplehorn, C. A. 1972. A review of the genus *Zopherus* of the world (Coleoptera: Tenebrionidae). Smithsonian Contrib. Zool. 108: 1–24.

Wright, N. P. 1956. A living luck charm. Nat. Hist. 65: 410–411, 445–446.

## GIANT FUNGUS BEETLES

Erotylidae, Erotylinae.

Species in the genus *Erotylus* (fig. 9.2l) are the most frequently observed members from this diverse family. They are moderate-sized (BL 2–3 cm) beetles with an elliptical body and often strongly convex (even conical) clytra. The latter are brightly colored with red or yellow spots or zigzagging colored lines running transversely across a black field.

The family has been neglected by biologists, and little is known regarding its natural history. The beetles are diurnal, traversing the ground or walking about on logs and stones. Adults and larvae live in fungi on the bark of jungle trees. Female

*Pselaphicus giganteus* from Trinidad guard newly hatched larvae and guide them to their first fungus meal. They also synchronize egg laying with fungal development and select logs with fungi beginning to grow (O'Toole and Preston-Mafham 1985). The larvae of some *Erotylus* are known to cluster when pupating under fallen trees, the pupae hanging like butterfly chrysalids from the shed larval skins (orig. obs.).

Some species (*Cypherotylus dromedarius*) are mimicked by tenebrionid beetles (*Cuphotes immaculipes*) (figs. 9.2j, k), and other genera (*Priotelus*) by ground beetles (*Eurycoleus*) (Erwin and Erwin 1976).

Erotylids are protected by an apparently foul-tasting liquid exuded from the knee joints and anus and by their strange flopping actions followed by dropping to the ground and feigning death.

### References

Erwin, T. L., and L. J. M. Erwin. 1976. Relationships of predaceous beetles to tropical forest wood decay. Pt. II. The natural history of Neotropical *Eurycoleus macularis* Chevrolat (Carabidae: Lebiini) and its implication in the evolution of ectoparasitoidism. Biotropica 8: 215–224.

O'Toole, C., and K. Preston-Mafham. 1985. Insects in camera: A photographic essay on behaviour. Oxford Univ. Press, Oxford.

## WATER BEETLES

Beetles of a number of families are totally aquatic in all stages. Both larvae and adults live their lives submerged, in a variety of running and standing water habitats, although the adults leave the water to fly to new localities if their homes dry up. Adults take oxygen from the atmosphere, storing it in air bubbles under the elytra or among the body's surface hairs; larvae likewise tap surface air by means of snorkellike posterior breathing tubes. When molested, whirligigs and predaceous diving beetles discharge a milky secretion from abdominal glands which has a fruity odor and apparently serves as a protective measure against other aquatic predators, such as fish. Hydrophilids apparently lack this ability.

Larvae are varied in form and preferred habitats (Bertrand 1972). Virtually all leave the water to form their pupal cells.

### Reference

BERTRAND, H. P. I. 1972. Larves et nymphes des coléoptères aquatiques du globe. Paillart, Paris.

## PREDACEOUS DIVING BEETLES

Dytiscidae. *Spanish:* Escarabajos acuáticos depredores (General).

This and the following are the most common families of beetles leading submerged lives in all aquatic habitats, for example, ponds, streams, vegetated lake margins, marshes, even hot springs.

Dytiscids (Moroni and Bachmann 1977; Spangler 1981, 1982) are broad but streamlined, boat-shaped beetles. Like water scavenger beetles, they have flattened hind legs fringed with hairs, but the threadlike antennae are much longer than the palpi. When swimming, the hind legs move together, oarlike; hydrophilids move them alternately. They are usually brown or black, but some have yellow or whitish spots or other markings. There are over 550 Neotropical species. Most are small to medium (BL 1.5–15 mm), but there are large (BL to 35 mm) species in the genera *Cybister* (fig. 9.3a) and *Megadytes* (fig. 9.3c). These are black to dark olive green and often have a wide buff to yellow marginal border. The elytra may be smooth or deeply striated.

All active stages are highly predaceous; the larvae (fig. 9.3b), known as water tigers, have pronounced, channeled, sickle-like jaws with which they grasp prey and through which they suck its body juices. They lack abdominal gills but have a pair or three long apical appendages. Adults of many species produce milky secretions containing defensive steroids from glands on the prothorax (Miller and Mumma 1974).

These beetles do not have the ventral film of air held by hydrophilids, being almost hairless underneath. Because the major spiracles are on the abdomen, dytiscids come to the surface for air with the tail up.

### References

MILLER, J. R., AND R. O. MUMMA. 1974. Seasonal quantification of the defensive steroid titer of *Agabus seriatus* (Coleoptera: Dytiscidae). Entomol. Soc. Amer. Ann. 678: 850–852.

MORONI, J., AND A. C. BACHMANN. 1977. Dytiscidae. *In* S. H. Hurlbert, ed., Biota acuática de Sudamérica austral. San Diego State Univ., San Diego. Pp. 217–225.

**Figure 9.3  WATER BEETLES.** (a) Predaceous diving beetle (*Cybister* sp., Dytiscidae). (b) Predaceous diving beetle, larva. (c) Predaceous diving beetle (*Megadytes giganteus*, Dytiscidae). (d) Water scavenger beetle (*Tropisternus lateralis,* Hydrophilidae). (e) Water scavenger beetle (*Berosus* sp., Hydrophilidae). (f) Giant water scavenger beetle (*Hydrophilus insularis,* Hydrophilidae). (g) Whirligig beetle (*Gyretes* sp., Gyrinidae). (h) Giant whirligig beetle (*Gyrinus* sp., Gyrinidae).

SPANGLER, P. J. 1981. Dytiscidae. *In* S. H. Hurlbert, G. Rodríguez, and N. Dias dos Santos, eds., Aquatic biota of tropical South America. Pt. 1. Arthropoda. San Diego State Univ., San Diego. Pp. 136–148.

SPANGLER, P. J. 1982. Dytiscidae. *In* S. H. Hurlbert and A. Villalobos Figueroa, eds., Aquatic biota of Mexico, Central America and the West Indies. San Diego State Univ., San Diego. Pp. 335–343.

# WATER SCAVENGER BEETLES

Hydrophilidae.

Adults of this family are omnivorous scavengers of organic detritus in ponds, lakes, marshes, and streams. Their food is principally of plant origin, but they occasionally eat dead animals. They are distinguished from their most similar aquatic relatives, the predaceous water beetles (Dytiscidae), by their prominent maxillary palps that are longer than the antennae. The antennae are short and clubbed and the hind legs flattened, with a fringe of hairs. Most are dark brown or black, rarely patterned. The body is strongly convex, and the thorax of some possesses a mid-ventral keel, which projects posteriorly as a sharp, free spine. They are poorer swimmers than dytiscids and move their legs alternately, as in walking. They also appear silvery beneath from a thin layer of air held there by a thick pile. Air is captured at the surface, head up, with the help of the antennae and tapped with thoracic spiracles.

The larval feeding habits, in contrast to those of the adult, are strictly predaceous. They are voracious pursuers of other aquatic invertebrates, including small vertebrates, such as fish and tadpoles. Varied in form, they mostly lack abdominal appendages and have short or toothed mandibles.

The most conspicuous of the 360 Neotropical species belong to the genera *Hydrophilus* (fig. 9.3f), *Tropisternus* (fig. 9.3d), and *Berosus* (fig. 9.3e). Species in *Hydrophilus* are much larger than the others (BL 40–50 mm) and are occasionally noticed around electric lights near bodies of water. The biology and literature pertaining to the family in Latin America are reviewed by Bachmann (1977) and Spangler (1981, 1982).

## References

BACHMANN, A. O. 1977. Hydraenidae y Hydrophilidae. *In* S. H. Hurlbert, ed., Biota acuática de sudamérica austral. San Diego State Univ., San Diego. Pp. 231–237.

SPANGLER, P. J. 1981. Hydrophilidae. *In* S. H. Hurlbert, G. Rodríguez, and N. Dias dos Santos, eds., Aquatic biota of tropical South America. Pt. 1. Arthropoda. San Diego State Univ., San Diego. Pp. 163–175.

SPANGLER, P. J. 1982. Hydrophilidae. *In* S. H. Hurlbert and A. Villalobos Figueroa, eds., Aquatic biota of Mexico, Central America and the West Indies. San Diego State Univ., San Diego. Pp. 355–363.

# WHIRLIGIG BEETLES

Gyrinidae. *Spanish:* Mamatetas (Costa Rica, *Dineutus*).

Whirligigs skate and gyrate in groups on the water in stream eddies, in ponds, and at the edges of lakes. They are supremely adapted for life on the surface film with their flat, oval bodies, divided eyes, which enable them to see above and below simultaneously, and long forelegs, with which they snatch prey that has fallen into their paths. Their hind legs are not fringed with hairs but are flattened like oars and function as efficient propellers. All are solid black, some with metallic blue-green or coppery reflections, and varied in size (BL 2–15 mm). They are diurnally active. If frightened, they dive quickly and remain submerged for some time.

The larvae roam the bottom searching out and preying on a variety of other small aquatic invertebrates. They are long and slender and bear conspicuous featherlike gills on the sides and tip of the abdomen.

These common beetles are found almost everywhere in Latin America to about 42 degrees south latitude. The genera *Gyretes* (fig. 9.3g) and *Gyrinus* (fig. 9.3h) contain the majority of the approximately 340 species.

When threatened or handled, whirligigs many emit a fruity-smelling, liquid discharge. The secretion emanates from glands near the anus and contain norsesquiterpenes, chemicals shown to be toxic or to act as deterrents to fish and the common newt (Newhart and Mumma 1978).

For general information and a bibliography, see Bachmann (1977) and Spangler (1981, 1982).

## References

BACHMANN, A. D. 1977. Gyrinidae. *In* S. H. Hurlbert, ed., Biota acuática de Sudamérica austral. San Diego State Univ., San Diego. Pp. 227–231.

NEWHART, A. T., AND R. O. MUMMA. 1978. High-pressure liquid chromatographic techniques for the separation and quantification of norsesquiterpenes from gyrinids. J. Chem. Ecol. 4: 503–510.

SPANGLER, P. J. 1981. Gyrinidae. *In* S. H. Hurlbert, G. Rodríguez, and N. Dias dos Santos, eds., Aquatic biota of tropical South America. Pt. 1. Arthropoda. San Diego State Univ., San Diego. Pp. 155–163.

SPANGLER, P. J. 1982. Gyrinidae. *In* S. H. Hurlbert and A. Villalobos Figueroa, eds., Aquatic biota of Mexico, Central America and the West Indies. San Diego State Univ., San Diego. Pp. 349–355.

## ROVE BEETLES

Staphylinidae.

Although an immense group with more than 5,000 regional species, rove beetles in general are conservative in body form, elongate, and usually easily recognized by their very short truncated elytra barely reaching to the base of the abdomen. Most have well-developed hind wings and fly well. The head is large, frequently almost the size of the prothorax. They are active beetles and often elevate the tip of the abdomen threateningly when disturbed, although they have no sting.

Their life-styles as adults and larvae vary tremendously, but most are predaceous and inhabit organic matter. Many are tolerant of salt and dwell at the ocean shore near the tide line (*Bledius;* fig. 9.4a). Some bizarre species live symbiotically with ants and termites; these exhibit odd modifications of form to resemble their hosts. Some 210 species in 79 genera are known to occupy army ant colonies alone, many (*Dioploeciton, Cryptomimus,* etc.; Seevers 1965) that outwardly look amazingly similar to the worker ants: elongate legs; bulbous abdomen, like

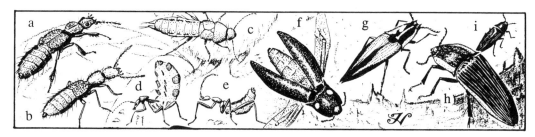

**Figure 9.4   ROVE (STAPHYLINIDAE) AND CLICK (ELATERIDAE) BEETLES.** (a) Rove beetle (*Bledius* sp., Staphylinidae). (b) Whiplash beetle (*Paederus irritans*, Staphylinidae). (c) Parasitic rove beetle (*Amblyopinus* sp., Staphylinidae). (d) Termitophilic rove beetle (*Termitogaster* sp., Staphylinidae). (e) Myrmecophilic rove beetle (*Ecitophya* sp., Staphylinidae). (f) Headlight beetle (*Pyrophorus nyctophanus*, Elateridae). (g) Click beetle (*Semiotus* sp., Elateridae). (h) Giant click beetle, *Chalcolepidius bonplanni*, Elateridae). (i) Wireworm beetle (*Conoderus* sp., Elateridae).

the gaster; and compressed anterior abdominal segments, like the nodes. A typical genus in the Neotropics associated with the familiar *Eciton* army ants is *Ecitophya* (fig. 9.4e). They also have specialized glands for secreting pheromonelike substances that pacify their benefactors, whom they groom and feed. They feed on booty and the larval brood of the host ants (Akre and Rettenmeyer 1966). Although many have typical rove beetle shapes, some may resemble silverfish or be trilobitelike in body form (*Termitonannus*), and a common feature of termitophilous forms is oddly shaped body outgrowths (*Spirachtha*); many hold the abdomen erect or bent forward over the thorax, as in *Termitogaster* (fig. 9.4d), a widespread genus that lives in the nests of nasute termites (Seevers 1957).

Blackwelder (1943) has published a major taxonomic work on the family in the West Indies.

### References

Akre, R. D., and C. W. Rettenmeyer. 1966. Behavior of Staphylinidae associated with army ants (Formicidae: Ecitonini). Kans. Entomol. Soc. J. 39: 745–782.

Blackwelder, R. E. 1943. Monograph of the West Indian beetles of the family Staphylinidae. U.S. Natl. Mus. Bull. 182: 1–658.

Seevers, C. H. 1957. A monograph on the termitophilous Staphylinidae (Coleoptera). Fieldiana Zool. 40: 1–334.

Seevers, C. H. 1965. The systematics, evolution and zoogeography of staphylinid beetles associated with army ants (Coleoptera, Staphylinidae). Fieldiana Zool. 47: 137–351.

## WHIPLASH BEETLES

Staphylinidae, Paederinae, Paederini, *Paederus. Spanish:* Zorritos (Peru), bichos de fuego (Argentina); corallilos (Guatemala); jallapas (Bolivia); picajuis (Venezuela). *Portuguese:* Potós, potós pimentas, trepas moleque (Brazil).

When these beetles walk on human skin, they sometimes deposit a caustic liquid from glands on the last abdominal segment which produces pustulent dermatitis (Frank and Kanamitsu 1987), although more often the effects are caused by hemolymph from accidentally crushed beetles (Guillén 1989). If the insect enters the eye, the effects are extremely severe and may result in loss of sight. Often, the injury is accompanied by fever and other systemic disorders that have been documented from numerous cases (Ojeda 1967). Agricultural workers in Paraguay have suffered from the abundance of these beetles in their environment (Dallas 1939).

Adults can be seen running on shrubs in warm, humid climes. They are slow moving, but when agitated, they elevate the abdomen in a threatening posture. Their food usually consists of small dead arthropods, although they sometimes attack and overpower other live but lethargic insects.

These are small (BL 12–14 mm), elongate, slender beetles, slightly flattened and of a deep, metallic black color (fig. 9.4b). Typical of their family, the elytra are short and truncate, covering only about a third of the abdomen. They have a globular prothorax and similarly shaped head of about the same size. Immatures of the Latin American species are apparently unknown.

Just over eighty species inhabit the Latin American region.

### References

Dallas, E. D. 1939. Coleópteros que origem dermatitis en al Republica de Argentina. 7th Int. Cong. Entomol. Proc. 2: 678–682.

Frank, J. H., and K. Kanamitsu. 1987. *Paederus*, sensu lato (Coleoptera: Staphylinidae): Natural history and medical importance. J. Med. Entomol. 24: 155–191.

Guillén, Z. 1989. Lesiones cutáneas producidas por *Paederus irritans* (Coleoptera, Staphylinidae) sobre animales de laboratorio. Rev. Peruana Entomol. 31: 31–35.

Ojeda, D. 1967. Estudo sobre un caso de dermatitis producida por *Paederus irritans* Chapin (Col.: Staphylinidae). Rev. Peruana Entomol. 10: 28–31.

## PARASITIC ROVE BEETLES

Staphylinidae, Staphylininae,
  Amblyopinini.

Because ectoparasitism is a rare phenomenon among beetles, these few rove beetle species that live on rodents and marsupials are extraordinarily interesting (Machado-Allison and Barrera 1972, Seevers 1955). They exhibit a high degree of host specificity, and species have been found to be associated with a large variety of mammals. Numerous species in five genera occur mostly in mountainous areas from Central America throughout the Andes to southern Chile and Argentina and across the continent in southeastern Brazil (Ashe and Timm 1988). Many more species are thought to exist. (It is odd that outside Latin America, the tribe is represented only by one genus in Australia and Tasmania.)

All are fairly typical rove beetles in their dark brown color and general elongate form, showing no obvious structural adaptations for ectoparasitism (fig. 9.4c). Some have a flat, shovel-shaped head and heavy setae on the underside of the abdomen which may have some function as yet unknown, possibly one associated with parasitism. They are distinguished from relatives by a pair of conspicuous lobes projecting from the tip of the abdomen, each bearing large setae at the apex.

Adults attach firmly to mice and opossums, usually on the hindquarters or near the anus. They have been seen to bury their mandibles in the skin, although they do not feed on the host but are predators on the host's ectoparasites. Observations of the immatures are lacking.

Rodent host genera are mainly myomorphs and hystricormorphs, including many peculiar South American types such as guinea pigs, chinchillas, pacas, and capybaras, although mice and rats in such genera as *Cyemomys* and *Oryzomys* are more usual. Marsupial hosts are all opossums in the genus *Didelphis*.

### References

ASHE, J. H., AND R. M. TIMM. 1988. *Chilambyopinus piceus,* a new genus and species of Amblyopinine (Coleoptera: Staphylinidae) from southern Chile, with a discussion of Amblyopinine generic relationships. Kans. Entomol. Soc. J. 61: 46–57.

MACHADO-ALLISON, C. E., AND A. BARRERA. 1972. Venezuelan Amblyopinini (Insecta: Coleoptera: Staphylinidae). Brigham Young Univ. Sci. Bull. 17: 1–14.

SEEVERS, C. H. 1955. A revision of the tribe Amblyopinini: Staphylinid beetles parasitic on mammals. Fieldiana Zool. 37: 211–264.

## CLICK BEETLES

Elateridae. *Spanish:* Apretadores (Chile), tuco tuco (Argentina), mayates saltadores (Central America).
  *Portuguese:* Tem tem, tec tec, salta martims (Brazil).

Ventrally, adults of these common beetles have a spine on the prothorax which fits into a groove on the mesothorax. A flexible union between the segments allows the former to be forcibly snapped into the latter, causing the beetle to jerk suddenly and, with a clicking sound, catapult up to several centimeters into the air. The action is the same whether the beetle is resting on its back or standing on its feet.

Click beetles are also recognized by their elongate, parallel-sided shape. The posterior corners of the prothorax project as sharp points. Most are small (BL 8–10 mm), but many Neotropical species, especially those in the genus *Chalcolepidius* (fig. 9.4h), are giants, 4 to 5 centimeters long. These are usually dull blue or green, with longitudinal light streaks on the elytra and bordering the prothorax and a slightly tapered blunt posterior end. *Semiotus* (fig. 9.4g) are similar but a little smaller (BL 3 cm), shiny yellow, and with wing covers pointed at the rear.

The larvae are wormlike, smooth, and with a hard exoskeleton. They burrow in the soil seeking roots and tubers, which they penetrate. These "wireworms" (particularly in the genera *Conoderus* [fig. 9.4i] and *Aeolus*) can do considerable economic damage to crop plants. Others work rotten wood and are beneficial as reducers.

More than 1,800 species inhabit the Neotropics and are widely distributed in almost every habitat. The luminescent headlight beetles are the most widely appreciated.

## HEADLIGHT BEETLES

Elateridae, Pyrophorini, *Pyrophorus*.
   *Spanish:* Cocujos (General), carbunclos (Costa Rica), cucubanos (Puerto Rico).
   *Portuguese:* Pirilampos (Brazil). Flying candles, fire beetles, peeny wallys.

Headlight beetles are large click beetles, notable in their unique ability among click beetles to produce light. An intense glow emanates from two round luminescent organs on the prothorax (the "headlights") and a broad area on the underside of the first abdominal segment. In flight, both sexes produce a brilliant blue-green streak of light that dazzles the onlooker.

The original genus has been broken up so that the hundred or so species originally placed in *Pyrophorus* are now distributed among 17 genera; only 26 species remain in the genus *Pyrophorus* (Costa 1976). They all occur in forests from southern Mexico to southeastern Brazil (40° S) and the West Indies.

The brightness and constancy of their light is legendary (Perkins 1869). Peter Martyr's *History of the West Indies* (1516) notes a number of uses to which these beetles were put in the sixteenth century: "the Islanders . . . go with their good will by night with 2 Cucuji tyed to the great toes of their feete: for the travailer goeth better by direction of the light of the Cucuji, then if he brought so many candles with him." Humboldt noted that a dozen of these beetles placed in a perforated gourd sufficed as a reading lamp (Allen and Wooton 1963). In 1535, Oviedo noted, "The Indians in fun stained their hands and faces with a paste made from those 'cocuyos' to scare others not familiar with the ruse." It is said that Sir Thomas Cavendish on his celebrated voyage to the West Indies in 1634, desisted on his first landing, seeing the light of *Pyrophorus* on shore and thinking they were Spanish soldiers. Guenther (1931: 228) adds further anecdotes: "Wasmann facetiously suggested that it might one day be possible to prepare pills of 'cucujin,' which scholars might swallow in order to be able to work by the light of their own bodies. I have been told that burglars have sometimes rubbed their faces with the green luminous substance, in order to frighten the inmates of the house they were entering."

Martyr started the widespread myth that these beetles catch mosquitoes and were useful in keeping houses free of these pests. In reality, the adults are phytophagous, feeding on rotting fruit and plant exudates.

Adults are attracted to artificial light, a flashlight, or a glowing cigarette. In the early days of Hispaniola, the natives collected them by going out at night with a burning coal (Martyr 1516).

Headlight beetles are fairly large (BL 2–4 cm), typically elaterid in form, that is, elongate, the prothorax with toothed rear angles (fig. 9.4f). All are uniformly dark brown, except for the two prothoracic spots, and have serrate antennae.

Mature larvae and pupae also luminesce. The former display bright round spots laterally and transverse zones on the dorsal and ventral plates of all segments except the prothorax, which glows on the margins of its dorsal plate. The larvae (Casari-Chen 1986) are predaceaous soil dwellers, feeding on scarab and other beetle larvae (Costa 1970).

## References

ALLEN, I. M., AND A. WOOTTON. 1963. Man's use of fire-flies for light. Entomol. Mon. Mag. 99: 27–30.

CASARI-CHEN, S. A. 1986. Larvas de Coleoptera da região Neotropical. XV. Revisão de Pyrophorini (Elateridae, Pyrophorinae). Rev. Brasil. Entomol. 30: 323–357.

COSTA, C. 1970. Genus *Pyrophorus*. 3. Life history, larva and pupa of *Pyrophorus punctatissimus*, Blanchard (Col., Elateridae). Mus. Zool., Univ. São Paulo, Pap. Avul. Zool. 23: 69–76.

COSTA, C. 1976. Speciation and geographical patterns in *Pyrophorus* Billberg, 1820 (Coleoptera, Elateridae, Pyrophorini). Mus. Zool., Univ. São Paulo, Pap. Avul. Zool. 29: 141–154.

GUENTHER, K. 1931. A naturalist in Brazil. Allen & Unwin, London.

MARTYR, P. 1516. [1912]. Decades (Alcala de Henares). Trans. by F. A. MacNutt as De Orbe novo: The eight decades of Peter Martyr d'Anghera. 2: 310–313. Putnam's Sons, New York.

PERKINS, G. A. 1869. The cucuyo; or, West Indian fire beetle. Amer. Nat. 2: 422–433.

# FIREFLIES

Lampyridae. *Spanish:* Luciérnagas, ciegos (General). *Portuguese:* Vagalumes, cagalumes, cagafogos, mamoás (Brazil). *Quechua:* Ayañahui (adult), pichin kuro (larva) (Peru).

For romanticists, fantasists, poets, and mythmakers, fireflies have always provided inspiration (Lenko and Papavero 1979). These unique insects are invariably afforded a beneficient position for their service in punctuating the dreaded darkness with happy light, like the stars to which they are allegorically attached:

> The Pleiads, rising thro' the mellow shade,
> Glitter like a swarm of fireflies tangled in a silver braid.
>
> Tennyson, "Locksley Hall"

The Aztecs called their sonnets fireflies, tiny lights in the great darkness, little truths within the ignorance surrounding them (Nicholson 1959). In some cultures, however, they are held in suspicion or considered bad omens, for example, among the Quechua-speaking people of the Andes, who refer to them as "ghost eyes" (*añañahui*).

In reality, fireflies are soft-bodied, elongate beetles with well-developed luminous organs in the terminal abdominal segments (fifth segment in female, fifth and sixth in male). The white or pale green light given off is produced by a chemical change of a special type of luciferin ("firefly luciferin," a carboxylic acid) in the cells of these organs (Buck 1948). The tissues of the light organs are penetrated by numerous fine tracheae, giving them ready access to the air and allowing the insect to regulate light emission by controlling the oxygen supply (Ghiradella 1983). In the presence of air and an enzyme called luciferase, the luciferin is oxidized, and a cold light is produced (Herring 1978).

Latin American species have been little studied, but much can be extrapolated from extensive knowledge available from work in North America. The basic function of sequenced light flashed from males and females is known in some fireflies to be a reciprocal courtship signal, bringing together the sexes of the same species (Lloyd 1983). The widespread genus *Photinus* (fig. 9.5a) has been studied in the most detail (Lloyd 1965, Soucek and Carlson 1975). The males fly in the evening, emitting short bursts of light for precise durations and intervals. Females, perched on vegetation, perceive this display and respond with their own lights. On seeing this answer, the male turns toward the female. The pattern is repeated until the male lands near the female, finally contacting and mating with her. So irresistible is the power of these light dances that beetles may be attracted

**Figure 9.5  BEETLES.** (a) Firefly (*Photinus* sp., Lampyridae), male. (b) Railroad worm (*Phrixothrix* sp., Phengodidae), larva. (c) Predatory firefly (*Photuris* sp., Lampyridae). (d) Firefly (*Lucidota* sp., Lampyridae). (e) Passalus beetle (*Passalus* sp., Passalidae), larva. (f) Passalus beetle, adult. (g) Chilean stag beetle (*Chiasognathus granti,* Lucanidae), female. (h) Chilean stag beetle, male.

by anyone imitating the paradigm with a flashlight.

Other kinds of the hundreds of firefly species practice similar systems. Some female *Photuris* (fig. 9.5c) are carnivorous, catching other species in flight (Lloyd and Wong 1983) or even going so far as to mimic the signals of *Photinus* females to falsely attract their males, which they capture and summarily devour (Lloyd 1965, 1984). *Photuris* males may even mimic these hapless *Photinus* males to increase their chances of finding a mate (Lloyd 1981). Additional functions for the light displays (Lloyd 1969), such as prey attraction (Lloyd 1984), and advertisement of unpalatability and illumination (Lloyd 1968), have been proposed.

Other characteristics of the adults are a broadly expanded disk-shaped prothoracic shield that totally hides the head, curved mandibles perforated by a tubular canal, and large eyes. They also possess a pungent odor and acrid-tasting body fluids, which may afford them some protection from vertebrate predators, and at least one genus, *Cratomorphus*, apparently is a Batesian mimicry model (see cockroaches, chap. 6). For *Photinus*, compounds called lucibufagins (related to the toxic steroids of certain toads) have been identified (Eisner et al. 1978). Lampyrids are host to a number of parasites and miscellaneous predators (Lloyd 1973).

The larvae are elongate, tapered, and strongly segmented, most of the segments with a flat, platelike, laterally expanded dorsum. They are predaceous like the adults and likewise luminescent. Sexually mature females of some species retain the larval body form even after maturing, still producing a bright light (larviform glowworms) to attract males.

The world fauna is composed of about 2,000 species, of which a possible 800 to 1,000 (many presently undescribed) occur in the Neotropics. Jamaica alone is home to at least 50 species (Lloyd 1969). They are principally inhabitants of moist woodlands. Dominant genera are *Photuris, Photinus,* and *Lucidota* (fig. 9.5d).

The well-known "railroad worms" of the moist forests of Paraguay, Argentina, and southeastern Brazil, so called because of their double row of yellow lights, a pair on each segment of the wingless body with a red head light, are the adult larviform females of *Phrixothrix* (fig. 9.5b), members of the family Phengodidae, which is related to the lampyrids (Tiemann 1970). Behavioral evidence suggests that the light organs are neurally controlled like those of fireflies (Halverson et al. 1973). They are activated when the animal is disturbed and when it is engaged in active prey capture.

### References

BUCK, J. B. 1948. The anatomy and physiology of the light organ in fireflies. New York Acad. Sci. Ann. 49: 397–482.

EISNER, T., D. F. WIEMER, L. W. HAYNES, AND J. MEINWALD. 1978. Lucibufagins: Defensive steroids from the fireflies *Photinus ignitus* and *P. marginellus* (Coleoptera: Lampyridae). Natl. Acad. Sci. Proc. 75: 905–908.

GHIRADELLA, H. 1983. Permeable sites in the firefly lantern tracheal system: Use of osmium tetroxide vapor as a tracer. J. Morph. 177: 145–156.

HALVERSON, R. C., J. F. CASE, J. BUCK, AND D. TIEMANN. 1973. Control of luminescence in phengodid beetles. J. Ins. Physiol. 19: 1327–1339.

HERRING, P. J., ed. 1978. Bioluminescence in action. Academic, New York.

LENKO, K., AND N. PAPAVERO. 1979. Vagalumes—Poeira das estrelas. *In* K. Lenko and N. Papavero, Insetos no folclore. Cons. Est. Artes Cien. Hum., São Paulo. Pp. 379–406.

LLOYD, J. E. 1965. Aggressive mimicry in *Photuris*: Firefly femmes fatales. Science 149 (3684): 653–654.

LLOYD, J. E. 1968. Illumination, another function of firefly flashes? Entomol. News. 79: 265–268.

LLOYD, J. E. 1969. Signals and systematics of Jamaican fireflies: Notes on behavior and on undescribed species (Coleoptera: Lampyridae). Entomol. News 80: 169–176.

LLOYD, J. E. 1973. Firefly parasites and predators. Coleop. Bull. 27: 91–106.

LLOYD, J. E. 1981. Mimicry in the sexual signals of fireflies. Sci. Amer. 245(1): 139–145.

LLOYD, J. E. 1983. Light in the summer darkness. *In* 1984 Yearbook Sci. Future, Encyclopaedia Brit., Chicago. Pp. 188–201.

LLOYD, J. E. 1984. Occurrence of aggressive mimicry in fireflies. Fla. Entomol. 67: 368–376.

LLOYD, J. E., AND S. R. WONG. 1983. Nocturnal aerial predation of fireflies by light-seeking fireflies. Science 22: 634–635.

NICHOLSON, I. 1959. Firefly in the night: A study of ancient Mexican poetry and symbolism. Faber and Faber, London.

SOUCEK, B., AND A. D. CARLSON. 1975. Flash pattern recognition in fireflies. J. Theor. Biol. 55: 339–352.

TIEMANN, D. L. 1970. Nature's toy train, the railroad worm. Natl. Geogr. 138: 56–67.

## PASSALUS BEETLES

Passalidae, Passalinae. *Spanish:* Pin-pin (Guatemala). Bessbugs, betsy bugs, patent-leather beetles.

In wooded areas, almost any damp, rotting log or stump may harbor a passalus beetle family or two. These are highly developed subsocial insects with adults and maturing larvae living together in groups (Schuster and Schuster 1985). Both stages produce a repertoire of sounds that apparently serve a number of purposes in their complex life histories. Prevalent signals express disturbance, aggression, and courtship (Schuster and Schuster 1971, Schuster 1983). Some fourteen different acoustic signals have been recognized, associated with eleven behavioral actions, the most known for any arthropod species (Schuster 1983).

The adults stridulate by rubbing rough places on the undersides of the elytra with the hind wings or spiny areas on the hind legs against the abdomen (Reyes-Castillo and Jarman 1983). The sounds are strong and perceptible to the human ear as high-pitched squeaks. Faint sounds are also emitted by the larvae, but with the hind legs, which are modified into pawlike stumps used to scrape over a file on the coxa of the midleg (Reyes-Castillo and Jarman 1980). These sounds are similar to those of their parents, yet their function remains unknown.

Adults feed on decaying wood and their own feces that has decomposed through microbial action. The larvae also eat the latter material and wood chewed into small pieces by the adults (Mason and Odum 1969, Gray 1946).

The beetles are easily recognized by their elongate, large prothorax, ridged elytra, shiny black color, and curved antennae tipped with a club composed of three to five flat plates (fig. 9.5f). Many also have a short horn on the top of the head and somewhat depressed elytra. They range in size from 1.2 to 10 centimeters in length. Some have hind wings reduced to thin straps used only for stridulation and useless for flight. Many species can fly, but this is seldom witnessed. Larvae (Costa and Ruy V. da Fonseca 1986) are pale, with a well-developed, pigmented head and slightly enlarged posterior (fig. 9.5e). They are unique in the reduction of the hind legs to form a part of the stridulatory device.

The family is well developed throughout the Neotropics, except the southern Chilean temperate forests. It is divided into numerous genera, all in the New World subfamily Passalinae, containing 278 species (Schuster and Reyes-Castillo 1981, Reyes-Castillo 1979). The larvae are keyed to genera by Schuster and Reyes-Castillo (1981).

Most occur in moist forests, at high as well as low elevations, although some abound in dry woodlands and even in special niches such as the nest of desert ants and oil bird caves (Schuster 1978). *Ptichopus* are associated with leaf cutter ants (*Atta*), commonly living in detritus piles or in the chambers formed by these ants (Schuster 1984). The most common microhabitat is a rotting log in intermediate stages of decay. Hardwoods are the usual hosts, but some inhabit the broken-down wood of conifers and palms.

Like other beetles, passalids are infested with ectoparasitic and phoretic mites but in greater variety than any other family of insect hosts. Trichomycetes fungi are found in the gut of both larvae and adults and may be involved in cellulose digestion (Lichtwardt 1986: 231–232, 306).

## References

Costa, C., and C. Ruy V. da Fonseca. 1986. Larvae of Neotropical Coleoptera. XIII. Passalidae, Passalinae. Rev. Brasil. Entomol. 30: 57–78.

Gray, I. E. 1946. Observations on the life history of the horned passalus. Amer. Midl. Nat. 35: 728–746.

Lichtwardt, R. W. 1986. The Trichomycetes (Fungal associates of arthropods). Springer, Berlin.

Mason, W. H., and E. P. Odum. 1969. The effect of coprophagy on retention and bioelimination of radionucleides by detritus-feeding animals. *In* D. J. Nelson and F. C. Evans, eds., 2d Natl. Symp. Radioecol. (Ann Arbor) Proc. Pp. 721–724.

Reyes-Castillo, P. 1979. Coleoptera, Passalidae: Morfología y división en grandes grupos: Géneros Americanos. Fol. Entomol. Mexicana 20–22: 1–240.

Reyes-Castillo, P., and M. Jarman. 1980. Some notes on larval stridulation in Neotropical Passalidae (Coleoptera: Lamellicornia). Coleop. Bull. 34: 263–270.

Reyes-Castillo, P., and M. Jarman. 1983. Disturbance sounds of adult passalid beetles (Coleoptera: Passalidae) structural and functional aspects. Entomol. Soc. Amer. Ann. 76: 6–22.

Schuster, J. C. 1978. Biogeographical and ecological limits of New World Passalidae (Coleoptera). Coleop. Bull. 32: 21–28.

Schuster, J. C. 1983. Acoustical signals of passalid beetles: Complex repertoires. Fla. Entomol. 66: 486–496.

Schuster, J. C. 1984. Passalid beetle (Coleoptera: Passalidae) inhabitants of leaf-cutter ant (Hymenoptera: Formicidae) detritus. Fla. Entomol. 67: 175–176.

Schuster, J. C., and P. Reyes-Castillo. 1981. New World genera of Passalidae (Coleoptera): A revision of larvae. Escuela Nac. Cien. Biol. Mexico An. 25: 79–116.

Schuster, J. C., and L. B. Schuster. 1971. Un esbozo de señales auditivas y comportamiento de Passalidae (Coleoptera) del Nuevo Mundo. Rev. Peruana Entomol. 14: 249–252.

Schuster, J. C., and L. B. Schuster. 1985. Social behavior in passalid beetles (Coleoptera: Passalidae): Cooperative brood care. Fla. Entomol. 68: 266–272.

## CHILEAN STAG BEETLE

Lucanidae, Chiasognathinae,
   *Chiasognathus granti. Spanish:* Ciervo volante, cantábria, cacho de cabra (Chile). *Mapuche:* Llico-llico (Chile).

This beetle achieved notoriety as a result of its inclusion in Darwin's discussion of sexual selection in his *The Descent of Man.* He knew the species from his visit to southern Chile, the male of which he described as "bold and pugnacious," and "when threatened he faces round, opens his great jaws, and at the same time stridulates loudly. But the mandibles were not strong enough to pinch my finger so as to cause actual pain" (Darwin 1871). It has now been determined that the enormously long, toothed mandibles of the males (fig. 9.5h), with hooked tips and an angular bend downward at midlength, are used not for injuring enemies but as forceps for plucking rival males from their perches in trees and hurtling them to the ground below (Eberhard 1980). The beetle's extra long forelegs assist in this process. According to Joseph (1928), some fights end in death by decapitation for the loser, his thorax and head being separated by the viselike jaws of the victor.

The male is otherwise a typical, medium-large beetle (BL 8 cm, excluding the mandibles, which are about as long as the body), with an olive prothorax and reddish-brown elytra, both regions being highly polished and with a pearly sheen. The females are smaller than the males and lack the enlarged mandibles (fig. 9.5g).

The species is fairly common in southern Chile and Argentina, appearing January to April. Most facts on its biology are recorded by Joseph (1928). Both sexes congregate around oozing wounds on the trunks of *Nothofagus* and *Weinmannia* trees, from which they feed. Very large numbers may appear in odd years. Males fly with facility at dusk and into the night.

Females deposit their eggs in the soil. The larvae (Cekalovic and Castro 1983) are subterranean and feed externally on the roots of shrubs and other plants. They form spacious cells for pupation, 30 centimeters or less below the soil surface.

## References

CEKALOVIC, T., AND M. CASTRO. 1983. *Chiasognathus granti* Stephens, 1831 (Coleoptera Lucanidae), descripción de la larva y nuevas localidades para la especie. Soc. Biol. Concepción Bol. 54: 71–76.
DARWIN, C. 1871. The descent of man, and selection in relation to sex. 2 vols. J. Murray, London.
EBERHARD, W. G. 1980. Horned beetles. Sci. Amer. 242(3): 166–182.
JOSEPH, C. 1928. El *Chiasognathus grandtii* Steph. Rev. Univ. (Univ. Católica Chile, Santiago) 13(5–6): 529–535.

## SCARABS

Scarabaeidae. *Spanish:* Escarabajos (General). *Portuguese:* Escaravelhos, cascudos (Brazil, adults). Pãos de galinha, joães torresmo, bichos bolas, bichos gordos (Brazil, larvae).

The most famous members of this large family (Morón 1984) are the sacred scarabs (*Scarabaeus* and other genera) of ancient Egypt. The American tropics are rich in species, many of which are well known for their great size and curious habits. Most belong to five subfamilies: the dung scarabs (Scarabaeinae = Coprinae, 1,100 species); horned scarabs (Dynastinae, 620 species); June beetles or chafers (Melolonthinae, 1,500 species); flower scarabs (Cetoniinae, 200 species); and the shiny scarabs (Rutelinae, 1,200 species).

All scarabs are identifiable by the club of the straight antennae which is composed of three to nine flat plates that may be spread apart. Most are small to medium-sized beetles, but some are among the largest insects known.

Scarab larvae are all pale grubs ("white grubs") with a well-developed head, jaws, and thoracic legs and a C-shaped body. Most are plant feeders, usually on the roots, but others eat dung, rotting organic matter, carrion, fungi, and so on. Those larvae of large species that develop in rotting palm logs are avidly sought for food by natives of many areas; some (including the larvae of palm weevils) are even sold in the marketplace in Iquitos, Peru (*papás*). Adults generally are leaf, fruit, nectar, and flower eaters but also feed on decaying organic material (dung, carrion, etc.).

**Reference**

MORÓN, R. 1984. Escarabajos, 200 milliones de
años de evolución. Insto. Ecol., Mus. Hist.
Nat., México.

## DUNG SCARABS

Scarabaeidae, Scarabaeinae (= Coprinae).
*Nahuatl:* Mayameh, sing. máyatl.

Feeding on feces is a fundamental feature
of the biology of this scarab subfamily
which determines behavior, distribution,
morphology, and development. Adults are
attracted to fresh animal excrement by its
odor and feed directly on it or remove
portions on which to lay their eggs and
provide for larval nutrition (Halffter and
Edmonds 1982). The fact that this food
source abounds principally in grasslands
has largely determined the prevalence of
these beetles there, in association with
large grazing mammals and their predators, although it would be incorrect to
assume that they are not abundant in kind
and number in forest habitats as well
(Howden and Nealis 1975). Here, especially smaller types in several genera (e.g.,
*Canthidium, Eurysternus,* fig. 9.6a) are commonly observed perching on leaves in low
vegetation (Howden and Nealis 1978). Several functions for this behavior have been
suggested, including resource-partitioning
strategy, assessment of predator density
nearby, mimicry display, and thermoregulation (Young 1984), but none has been
conclusively demonstrated.

Various members of this group digress
from coprophagy and utilize carrion (necro-
phagy) or decomposing vegetable matter
(saprophagy) as food. A very specialized
example of the latter is the consumption of
debris that accumulates in the nests of leaf
cutter ants (*Atta*) by such genera as *Liatongus*
and *Onthophagus.* An extraordinary feeding
specialization is that of a Brazilian *Canthon*
that attacks ants of the same genus (Navajas
1950). A few even inhabit the hair of sloths
(*Uroxys* and *Trichillum;* Ratcliffe 1980) and
monkeys (*Glaphyrocanthon*).

Adults are small to large, well-armored
beetles that are very compactly built, many
almost spherical in shape. The antennal
club has only three segments. Most possess
a shovel-shaped head and fossorial forelegs, useful also in cutting, molding, and
burying dung. Many are colored brightly
in metallic green, blue, or coppery hues,
some stunningly so.

The eyeless, heavy-jawed larvae are C-
shaped, as is typical in the scarabs, but they
also have a characteristic "hump," or dorsal enlargement of the middle abdominal
segments. The projecting hump acts as an
anchor, aiding the rotational movement of
the larva inside the cavity it creates when
feeding within its food (Halffter and Matthews 1966). Detailed studies of the biol-

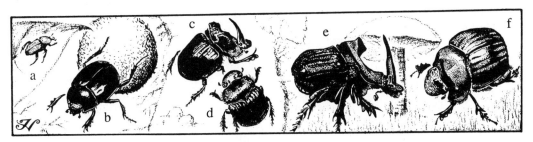

**Figure 9.6    DUNG SCARABS (SCARABAEIDAE).** (a) Perching dung beetle (*Eurysternus deplanatus*). (b) Dung roller (*Canthon smaragdulum*). (c) Dung digger (*Phanaeus demon*), male. (d) Dung
digger, female. (e) Giant dung digger (*Coprophanaeus lancifer*). (f) Black dung beetle (*Dichotomius
carolinus*).

ogy and taxonomy of the New World dung scarab larvae have been published (Edmonds and Halffter 1972, Halffter and Edmonds 1982).

Dung beetles are well represented in the New World. Dominant are Coprini, especially the genera *Dichotomius* and *Phanaeus*, and Scarabaeini, particularly *Canthon*. Several of the other approximately seventy-five genera are also speciose.

This interesting group has attracted a great deal of attention by collectors and specialists who are rapidly exposing its biology (Halffter and Matthews 1966; Howden and Young 1981; Peck and Forsyth 1982; Peck and Howden 1984; Wolda and Estribí 1985).

## References

EDMONDS, W. D., AND G. HALFFTER. 1972. A taxonomic and biological study of the immature stages of some New World Scarabaeinae (Coleoptera: Scarabaeidae). Esc. Nac. Cien. Biol., Mexico, An. 19: 85–122.

HALFFTER, G., AND W. D. EDMONDS. 1982. The nesting behavior of dung beetles (Scarabaeinae), an ecological and evolutive approach. Insto. Ecol. (Mus. Hist. Nat., Mexico City) 10: 1–184.

HALFFTER, G., AND E. G. MATTHEWS. 1966. The natural history of dung beetles of the subfamily Scarabaeinae (Coleoptera, Scarabaeidae). Fol. Entomol. Mexicana 12–14: 1–312.

HOWDEN, H. F., AND V. G. NEALIS. 1975. Effects of clearing in a tropical rain forest on the composition of the coprophagous scarab beetle fauna (Coleoptera). Biotropica 7: 77–83.

HOWDEN, H. F., AND V. G. NEALIS. 1978. Observations on height of perching in some tropical dung beetles (Scarabaeidae). Biotropica 10: 43–46.

HOWDEN, H. F., AND O. P. YOUNG. 1981. Panamanian Scarabaeinae: Taxonomy, distribution, and habits (Coleoptera, Scarabaeidae). Amer. Entomol. Inst. Contrib. 18(1): 1–204.

NAVAJAS, E. 1950. Manifestações de predatismo em Scarabaeidae do Brasil e alguns datos bionômicos de *Canthos virens* (Mannh.) (Col. Scarabaeidae). Cien. Cult. 2: 284–285.

PECK, S. B., AND A. FORSYTH. 1982. Composition, structure, and competitive behavior in a guild of Ecuadorian rain forest dung beetles (Coleoptera: Scarabaeidae). Can. J. Zool. 60: 1624–1634.

PECK, S. B., AND H. F. HOWDEN. 1984. Response of a dung beetle guild to different sizes of dung bait in a Panamanian rain forest. Biotropica 16: 235–238.

RATCLIFFE, B. C. 1980. New species of Coprini (Coleoptera; Scarabaeidae: Scarabaeinae) taken from the pelage of three toed sloths (*Bradypus tridactylus* L.) (Edentata: Bradypodiae) in central Amazonia with a brief commentary on scarab-sloth relationships. Coleop. Bull. 34: 337–350.

WOLDA, H., AND M. ESTRIBÍ. 1985. Seasonal distribution of the large sloth beetle *Uroxys gorgon* Arrow (Scarabaeidae; Scarabaeinae) in light traps in Panama. *In* G. G. Montgomery, ed., The evolution and ecology of armadillos, sloths and vermilinguas. Smithsonian Inst., Washington, D.C. Pp. 319–322.

YOUNG, O. P. 1984. Perching of Neotropical dung beetles on leaf surfaces: An example of behavioral thermoregulation? Biotropica 16: 324–327.

## Dung Rollers

Scarabaeidae, Scarabaeinae, Scarabaeini, *Canthon. Spanish:* Ruedacacas (General), escarabajos peloteros (Argentina). *Portuguese:* Rola-bostas, pilulares (Brazil). Tumble bugs, ball-rolling dung beetles, telocoprids.

These are the well-known dung-ball rollers (fig. 9.6b). The beetle does not dig a burrow prior to forming the ball but pushes the soil away from under the ball after it is made and buries it simultaneously with the digging of the burrow.

The balls may be eaten directly by the adults (food balls), or a single egg may be laid in it and consumed by the larva (brood balls). The ball is always made at the site of the food source before it is rolled. Ball-rolling techniques are highly developed. The usual rolling position for a single beetle is head downward, the forelegs held on the ground and the other legs on the ball. Two beetles may cooperate in rolling, usually one assuming a pushing and the other a pulling position.

## Dung Diggers

Scarabaeidae, Scarabaeinae, Coprini, *Phanaeus. Spanish:* Mierderos (Peru). *Quechua:* Ismatanga (Peru). Burrowing dung beetles, paracoprids.

In this entirely New World genus and its relatives, females first excavate burrows beneath or near the food source. They then extricate a morsel of dung from the main mass and transport it to the underground chamber, sometimes with the aid of a male, carefully model it into a sphere or pear-shaped unit, and provide it with an outer shell of soil and a single egg. *Phanaeus* do not roll finished balls of dung on the surface of the ground.

These are mostly brightly colored beetles with metallic green, blue, or reddish-purple coppery integuments. Males (fig. 9.6c) have a single, pointed, and erect short to very long head horn and usually also heavy ridges or conical protuberances on a broad prothoracic shield. The latter is steep and somewhat excavated in most. Females lack the horn and ridges (fig. 9.6d). The elytra are strongly grooved. The majority are medium-sized (BL 1.5–2 cm), although the genus *Coprophanaeus* (subgenus *Megaphanaeus*) contains some large, bulky species (BL to 5 cm; fig. 9.6e).

See the references under Dung Beetles for natural history of the genus. Reproductive biology is discussed in detail by Halffter and López (1977), and a detailed anatomical and phylogenetic study of the group was made by Edmonds (1972).

### References

EDMONDS, W. D. 1972. Comparative skeletal morphology, systematics and evolution of the phanaeine dung beetles (Coleoptera: Scarabaeidae). Univ. Kans. Sci. Bull. 49: 731–874.

HALFFTER, G., AND Y. LÓPEZ. 1977. Development of the ovary and mating behavior in *Phanaeus.* Entomol. Soc. Amer. Ann. 70: 203–213.

## Black Dung Beetle

Scarabaeidae, Scarabaeinae, Coprini, *Dichotomius carolinus. Spanish:* Rueda caca (Costa Rica).

Adults of this very common dung beetle (Howden 1983) are not ball rollers. They normally divide up cattle and horse droppings into irregular fragments and take them directly into a burrow. The dung may be eaten in this form by adults or packed into a mass at the end of a burrow on which an egg is laid.

These are fairly large beetles (BL 22–30 mm), dull black, and very convex in shape (fig. 9.6f). The elytra bear deep grooves that, toward the posterior, are frequently packed with dry soil from their digging activities. They are nocturnal and often attracted to electric light in lowland areas of Mexico and Central America.

### Reference

HOWDEN, H. F. 1983. *Dichotomius carolinus colonicus* (rueda caca, dung beetle). *In* D. H. Janzen, ed., Costa Rican natural history. Univ. Chicago Press, Chicago. Pp. 713–714.

## HORNED SCARABS

Scarabaeidae, Dynastinae. *Spanish:* Cucarrones (Colombia).

Horned scarabs may weigh up to 40 grams and be as bulky as a human fist. Including the anteriorly projecting horns, many attain body lengths of 12 to 13 centimeters. In these, the development of the horns varies much through disproportionate (allometric) growth, although they tend to sort into two extreme size categories (Eberhard 1987). Smaller individuals normally have very poorly formed horns; the larger individuals have tremendously elongated projections; a few fall in between. These are classed as minors, mediums, and majors, like the classes of ant workers based on overall body size and relative size of the jaws (Eberhard 1982).

Such appendages are known to be an outcome of sexual selection, for females usually lack them almost completely. Horns for grasping and prying, largeness, and great strength favor the males, who employ these attributes in combat for territory and mating rights, although minor males ("satellite males," see centris bees) may use their quickness and relative inconspicuousness to sneak by larger rivals preoccupied with fighting and successfully mate (Eberhard 1980, 1982).

The taxonomy of this major group has been reviewed by Endrödi (1966, 1984).

## References

EBERHARD, W. G. 1980. Horned beetles. Sci. Amer. 242(3): 166–182.

EBERHARD, W. G. 1982. Beetle horn dimorphism: Making the best of a bad lot. Amer. Nat. 119: 420–426.

EBERHARD, W. G. 1987. Use of horns in fights by the dimorphic males of *Aegopsis nigricollis* (Coleoptera, Scarabaeidae, Dynastinae). Kans. Entomol. Soc. J. 60: 504–509.

ENDRÖDI, S. 1966. Monographie der Dynastinae (Coleoptera, Lamellicornia). I. Teil. Staat. Mus. Tierkunde Dresden Entomol. Abhand. 33: 1–457.

ENDRÖDI, S. 1984. The Dynastinae of the world. Junk (Ser. Entomol. 28), The Hague.

## Hercules Beetles

Scarabaeidae, Dynastinae, Dynastini,
*Dynastes. Spanish:* Tijeras (General).
*French:* Scieurs de long (Guadeloupe).

Males of the large beetles in this genus are easily recognized by their very long, down-curved medial prothoracic horn whose length, plus that of the body, makes them some of the largest beetles in the world. Specimens as long as 17 centimeters are known. Without the horn, the body length is considerably less, to 8 centimeters, but still great enough to rival the rhinoceros beetles for the record. Their live weight may approach 40 grams. Females lack horns of any size (fig. 9.7b), but their bodies are nearly as massive as those of the males (BL to 8 cm).

The strength of the larger species is also prodigious. Males are capable of exerting tremendous force in the closure of the prothoracic horn against the horn of the head. In one experiment, a live beetle was observed to lift a 2-kilogram weight with its head horns, exerting a force of 140 Newtons (or $1.4 \times 10^7$ dynes) (Jarman and Hinton 1974). A male I once kept in a bird cage demonstrated its power by escaping at will, simply by bending the heavy metal wires apart with the force of its body.

The elytra of the best-known species, *D. hercules* (fig. 9.7a, b), are usually olive-yellow, although this color may change to black and back again within a few minutes. It is not clear how much control the beetle (usually males) has over this change; it seems to be a function more of ambient humidity. A spongy yellow layer beneath the outer cuticle layer when saturated allows the black underlayers to show; when dry, the layer reflects yellow (Hinton and Jarman 1973).

The horns of the males are definitely used as grappling devices in combat over females, who may be carried off by the victor. (See horned beetles, above). William Beebe (1947) describes such beetles eloquently.

Males also stridulate when stimulated, during combat or artificially, making a soft "zizzing squeak" or "huff." The abdomen is moved so that a roughened area on either side of the enlarged last segment is rubbed against the inner apices of the elytra (Jarman and Hinton 1974).

These are essentially forest dwellers. Adults are attracted to sap oozing from wounded trees (especially palms) and by sweet fruits. They are nocturnal and arrive at electric lights around which their loud and powerful flight can be appreciated.

**Figure 9.7    HORNED SCARAB BEETLES (SCARABAEIDAE).** (a) Hercules beetle (*Dynastes hercules*), male. (b) Hercules beetle, female. (c) Rhinoceros beetle (*Megasoma* sp.), larva. (d) Elephant beetle (*Megasoma elephas*), male.

Females oviposit in crevices in the bark of hosts, which are *Licania ternatensis* (Rosaceae), *Amanoa caribaea* (Euphorbiaceae), *Inga* (Fabaceae), and other trees (Verrill 1907, Gruner and Chalumeau 1977).

Legend has it in Guadeloupe that *D. hercules* can cut a tree limb by grasping it with the horns and flying around and around until it is severed or sap is caused to flow, on which it then feeds (Gruner and Chalumeau 1977). (See long-horned beetles, below.) In some regions, the mythical aphrodisiacal powers of ground rhinoceros horn are attributed to the horns of these beetles as well (orig. obs.).

There are four species of *Dynastes* in the Neotropics (*hercules, neptunus, satanas* and *hyllus*) (Dechambre 1980). The prothorax of the first three is black; that of *hyllus* is gray green. Only *D. hercules* has normal terminal tarsal segments in the legs. In *D. neptunus* and *D. satanas* the last tarsal segments are dilated, but in the former the accessory prothoracic horns originate well below the base of the main horn; in *satanas*, these horns arise on the same level as the major horn.

*Dynastes hercules* and *D. neptunus* are widespread throughout Middle America, including the northern half of South America; *D. satanas* is Bolivian only; and the relatively small, slightly horned *D. hyllus* is from Central Mexico (Dechambre 1980, Morón 1987).

*D. hercules* on the island of Guadeloupe harbors a mite parasite (Costa 1976).

## References

BEEBE, W. 1947. Notes on the Hercules beetle, *Dynastes hercules* (Linn.), at Rancho Grando, Venezuela, with special reference to combat behavior. Zoologica 32: 109–116.

COSTA, M. 1976. *Dynastaspis hercules* sp. n., a new gamasine mite associated with the Hercules beetle in Guadeloupe. Acarologia 18: 187–193.

DECHAMBRE, R.-P. 1980. Le genre *Dynastes* (Coleoptera Scarabaeoidea Dynastidae). Soc. Sci. Nat. (France) Bull. 28: 5–10.

GRUNER, L., AND F. CHALUMEAU. 1977. Biologie et élevage de *Dynastes h. hercules* en Guadeloupe (Coleoptera, Dynastinae). Soc. Entomol. France Ann. (n.s.)13: 613–624.

HINTON, H. E., AND G. M. JARMAN. 1973. Physiological colour change in the elytra of the Hercules beetle, *Dynastes hercules*. J. Ins. Physiol. 19: 533–549.

JARMAN, G. M., AND H. E. HINTON. 1974. Some defense mechanisms of the Hercules beetle, *Dynastes hercules*. J. Entomol. (Ser. A) 49: 71–80.

MORÓN, M. A. 1987. Los estados inmaduros de *Dynastes hyllus* Chevrolat (Coleoptera: Melolonthidae: Dynastinae): Con observaciones sobre su biología y el crecimiento alométrico del imago. Fol. Entomol. Mexicana 72: 33–74.

VERRILL, A. H. 1907. Description of a new species or sub-species of Hercules beetles from Dominica Island, B.W.I., with notes on the habits and larvae of the common species and other beetles. Amer. J. Sci. (Ser. 4) 24: 305–308.

## Rhinoceros Beetles

Scarabaeidae, Dynastinae, Dynastini,
  *Megasoma. Spanish:* Papasos (Peru),
  cornizuelos (Costa Rica), congarochos
  (Venezuela), bobutes (Andes).

From the standpoint of bulk, males of certain species of these beetles are the biggest insects in the world. Large living specimens of *M. actaeon* may weigh 30 grams. From head to the apex of the abdomen, some *M. elephas* individuals measure up to 8 centimeters; including the head horn, they attain lengths of 13 centimeters or more and can weigh 35 grams. Curiously, these beetles behave much like small mammals in their ability to metabolically increase their body temperature when the air cools (Morgan and Bartholomew 1982).

Their common name refers to the long, slender, upcurved, rhinoceroslike head horn of the males. The apical bifurcation of the horn distinguishes them from other very large regional horned scarabs in which this armament always has a simple apex or is down curved. The prothoracic shield may also bear a central horn and triangular lateral horns. Females lack horns altogether.

There are seven species in the genus, found in undisturbed lowland forest throughout the tropical and subtropical regions of Latin America (Hardy 1972). Males of the three best known are easily separated: the elephant beetle, *M. elephas* (fig. 9.7d), has a velvety brown, textured integument; *M. actaeon* is dull black with forward-pointing thoracic horns; *M. mars* is shiny black with divergent thoracic horns.

The larvae of these monstrous beetles also are insect behemoths themselves. They are typical curved scarab grubs but when mature may reach lengths to 13 centimeters with the body extended (fig. 9.7c). Some attack small living palms, but most feed in the pulp of dead palm trunks

(*Cocos, Mauritia,* etc.) and probably take three to four years to mature. Because *M. elephas occidentalis* is considered a pest of young *Licistona chinensis* (Reitter 1961: 42), it has been studied in some detail (Morón 1977).

Adults may be discovered on palm inflorescences and fruit; they also come to artificial lights during their nocturnal flights. In northern Brazil and other parts of Amazonia, various tribes make amulets or fetishes from parts of *Megasoma.* The horns symbolize sexual and physical power and are believed to increase potency and protect one from disease. Like those of the rhinoceros and hercules beetles, the head horns are ground and taken by some unenlightened people in the hope of improving their sexual powers. The horns are useful, in fact, only to the beetles as fighting instruments. Males engage in combat for rights to a female or feeding site (Beebe 1944) in a manner similar to Hercules beetles.

### References

BEEBE, W. 1944. The function of secondary sexual characters in two species of Dynastidae (Coleoptera). Zoologica 29: 53–58, Pls. I–V.

HARDY, A. R. 1972. A brief revision of the North and Central American species of *Megasoma* (Coleoptera: Scarabaeidae). Can. Entomol. 104: 765–777.

MORGAN, K. R., AND G. A. BARTHOLOMEW. 1982. Homeothermic response to reduced ambient temperature in a scarab beetle. Science 216: 1409–1410.

MORÓN, M. A. 1977. Description of the third-stage larva of *Megasoma elephas occidentalis* Bolívar y Pieltain et al. (Scarabaeidae: Dynastinae). Coleop. Bull. 31: 339–345.

REITTER, E. 1961. Beetles. Putnam's Sons, New York.

## Golofas

Scarabaeidae, Dynastinae, Oryctini,
  *Golofa. Spanish:* Torneadores (General),
  toritos (Peru), aserradores (Venezuela).

This is a fairly large genus (Dechambre 1979, Voirin 1979) of moderate to large (BL

**Figure 9.8 SCARAB BEETLES (SCARABAEIDAE).** (a) Caliper beetle (*Golofa porteri*), male. (b) Caliper beetle, female. (c) Ox beetle (*Strategus aloeus*), male. (d) Flower scarab (*Gymnetis holocericea circumdata*). (e) Green fruit beetle (*Cotinus mutabilis*).

without horn 2–8 cm), brown to black, horned scarabs. Dechambre (1979) recognizes twenty species, ranging throughout the moist, forested regions of the Neotropics. Males of all species have a single, median, elongate prothoracic horn, which opposes an equally well-developed head horn. The former is very erect but down curved at the tip, which may be variously widened or spade shaped in the different species. The function of the prothoracic horn has not been seen but may act with the head horn as a clamp for transporting females.

The caliper beetle (*G. porteri*) is the best-known golofa. Large males (fig. 9.8a) have exceedingly long, slender horns and remarkably elongate forelegs with large tarsi sporting thick growths of golden hair on their undersurfaces. These elaborations (lacked by the female, fig. 9.8b) undoubtedly function as direct weapons in fights between males. Accompanied by stridulation and displays of the hairy areas, they may also be used to intimidate opponents (Eberhard 1977).

Adults of the torito de la caña (*G. aegeon*) feed on young sugarcane plants and at times are pests in Peru (Wille 1952). Larvae of *G. eacus* attack the roots of corn plants in the same country (Ochoa 1980).

In the Sierra Nevada de Santa Marta of Colombia, these beetles have been seen battling for possession of single shoots of bamboolike grasses (*chusquea*) on which they feed (Howden and Campbell 1974).

Battles have been described by Eberhard (1980):

> When two males confront each other, each holds on to the support with its middle and hind legs, wraps its long front legs around the other male's body and then tilts its prothorax and lowers its head so that the head horn is inserted under the other male's body. To begin an attack one of the intertwined beetles rakes its front legs sharply across its opponent's middle and hind legs. This action apparently serves to tear the opponent's legs from the support, and an instant later the attacker jerks its head up to throw the opponent off the support.

Because of the scythelike, toothed prothoracic horns, *G. porteri*, like male Hercules beetles, is alleged to be a sawyer of tree limbs.

### References

DECHAMBRE, R.-P. 1979. Le genre *Golofa* (Col. Dynastidae). Soc. Sci. Nat. (France) Bull. 23: 1–11.

EBERHARD, W. G. 1977. Fighting behavior of male *Golofa porteri* beetles (Scarabaeidae: Dynastinae). Psyche 84: 292–298.

EBERHARD, W. G. 1980. Horned beetles. Sci. Amer. 242(3): 166–182.

HOWDEN, H. F., AND J. M. CAMPBELL. 1974. Observations on some Scarabaeoidea in the Colombian Sierra Nevada de Santa Marta. Coleop. Bull. 28: 109–114.

OCHOA, O. 1980. Ciclo biológico de *Golofa eacus* Burmeister (Coleoptera: Scarabaeidae), nueva plaga del maíz. Rev. Peruana Entomol. 23: 141–142.

Voirin, K. 1979. Détermination des espèces du genre *Golofa* Hope (Coleoptera Melolonthidae Dynastinae). Soc. Sci. Nat. (France) Bull. 23: 6–8.

Wille, J. E. 1952. Entomología agrícola del Perú. 2d ed. Min. Agric., Lima.

## Ox Beetles

Scarabaeidae, Dynastinae, Oryctini, *Strategus.*

Ox beetles are large (BL 20–80 mm), robust, highly polished, reddish-brown to black beetles. The males generally have two moderately long, forward-pointing prothoracic horns and an equally long or longer, erect head horn, much like *Megasoma* males, but the head horn is not bifurcate at the tip and is much less developed. The size of these horns, however, is a variable characteristic of individuals in the genus, as with other scarabs. It is evident that the males defend small individual reproductive and feeding territories, for which they fight, attacking with their pronotal horns, continuously up to two or three hours before one or the other desists (Morón 1976). Females are similar to males, except the anterior projections are only slightly developed and differ from *Megasoma* females in the convex rather than concave prothorax.

The genus contains thirty-one species with distributions covering all of Latin America, including the West Indies (Ratcliffe 1976). The most widespread species is *S. aloeus* (fig. 9.8c), adults of which commonly come to artificial light during their nocturnal flights.

The larvae apparently normally feed on the decaying wood or pith of *Agave* and various trees, including palms. Adults feed on the juices of the larval hosts, occasionally burrowing into tree or cane trunks at ground level. Several are of economic importance because of the attacks of their larvae on the roots of sugarcane, mangoes, date palms, wax palms, oil palms, cacao, and pineapple. The coconut rhinoceros beetle (*S. oblongus = quadrifoveatus*) is a major pest of coconut. The principal damage results from the adults feeding on the germinal tissues of young trees. The sugarcane rhinoceros beetle (*S. talpa = barbigerus*) once was considered to be a similar enemy of sugarcane in Puerto Rico, but larvae actually have been found to feed only on rotting wood.

The larvae of this genus are occasionally eaten by aboriginals in Guyana and elsewhere (Bodkin 1919).

### References

Bodkin, G. E. 1919. Notes on the Coleoptera of British Guiana. Entomol. Mon. Mag. 55: 210–219.

Morón, M. A. 1976. Notas sobre la conducta combativa de *Strategus julianus* Burmeister (Coleoptera, Melolonthidae, Dynastinae). Inst. Biol., Univ. Nac. Autón. México, Ser. Zool., An. 47: 135–142.

Ratcliffe, B. C. 1976. A revision of the genus *Strategus* (Coleoptera: Scarabaeidae). Univ. Nebr. State Mus. Bull. 10: 93–204.

## FLOWER SCARABS

Scarabaeidae, Cetoniinae, Gymnetini.

The most obvious members of this group are *Gymnetis,* slightly flattened beetles with a pentagonal shape (viewed from above) and colorful patterns of dark streaks or splotches on a dusky brown or greenish background (fig. 9.8d). The scutellum is also minute and sunken behind a triangular prolongation of the prothorax.

The similarly shaped but solid-colored, dark green to brown fruit beetles comprise the related genus *Cotinis* (18 species; fig. 9.8e). These feed on ripe fruits and are sometimes considered economic pests in orchards. The larvae are subterranean, feeding on the roots of grasses and buried humic matter.

The active buzzing flight of these beetles attracts attention, and they are often seen visiting flowers on warm, sunny days. They are probably significant pollinators.

Both genera are of general Latin American distribution and inhabit a wide variety of environments.

## PRECIOUS METAL SCARABS

Scarabaeidae, Rutelinae, Rutelini, *Plusiotis* and *Pelidnota.*

These two, closely related genera contain some of the world's most beautiful insects because of their polished metal colors. A true "gold bug" is *Plusiotis batesi* (fig. 9.9a), which is solid, white gold over the entire body. Other species are pure silver (*Plusiotis chrysargyrea*), burnished copper (*Pelidnota virescens*), and shining steel (*Pelidnota sumptuosa;* fig. 9.9b). In other species, the surfaces are not shining, but more like porcelain. They are no less brilliant and strikingly colorful and may have metallic reflections or streaks of silver contrasting with a green background (Hardy 1975).

There are about 200 species of these living jewels widely distributed over tropical America, from the United States frontier to Peru. Most are montane, living especially in cloud forests toward the south and in dry forests in the north. They all seem to be nocturnal and are usually only seen when attracted to artificial lights.

Virtually nothing is known of the life histories of the members of the group. The larvae of some have been found feeding on the roots of oaks and pines in Mexico and Central America. Hosts in South America remain undiscovered. The adults are said to feed on oak foliage where these trees grow (Morón 1981). There is one unconfirmed record of *Plusiotis chrysargyrea* from the roble tree (*Tecoma pentaphylla* = *Tabebuia pentaphylla*) in Costa Rica (Boucard 1878).

### References

BOUCARD, A. 1878. Notes on some Coleoptera of the genus *Plusiotis,* with descriptions of three new species from Mexico and Central America. Zool. Soc. London Proc. 1878: 293–296, pl. xvi, figs. 1–5.

HARDY, A. R. 1975. A revision of the genus *Pelidnota* of America north of Panama (Coleoptera: Scarabaeidae; Rutelinae). Univ. Calif. Publ. Entomol. 78: 1–43.

MORÓN, M. A. 1981. Descripción de dos especies nuevas de *Plusiotis* Burmeister, 1844 y discusión de algunos aspectos zoogeográficos del grupo de especies "costata" (Coleoptera, Melolonthidae, Rutelinae). Fol. Entomol. Mexicana 49: 49–69.

### Green-Gold Beetle

Scarabaeidae, Rutelinae, *Chrysophora chrysochlora.*

The green-gold beetle is a fairly common scarab. The male (BL 3.8 cm) (fig. 9.9c) is decidedly larger than the female (BL 2.9 cm) (fig. 9.9d). The male also is unique in possessing tremendously elongate, heavy, curved hind legs that bear outsized spurs on the insides of the tips of the tibia and enlarged tarsal claws. The function of

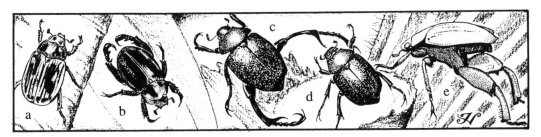

**Figure 9.9  SCARAB BEETLES (SCARABAEIDAE).** (a) Gold beetle (*Plusiotus batesi*). (b) Precious metal beetle (*Pelidnota sumptuosa*). (c) Green-gold beetle (*Chrysophora chrysochlora*), male. (d) Green-gold beetle, female. (e) Big-legged scarab (*Chrysina* sp.).

these specializations is unknown, but they are presumably used to confine and manipulate the female during courtship and mating. Hypertrophication of the male hind legs in this manner also occurs in other ruteline genera, such as *Chrysina* (fig. 9.9e), *Heterosternus, Paraheterosternus, Macropoidelimus,* and *Macropoides* (Morón 1983).

Both sexes have brilliant green elytra with gold or crimson reflections and a coarse granulate texture; the prothorax is similar but smoother. The legs are smooth, shiny purple to bronze.

Adults are found on leaves and flowers of arborescent vegetation in the tropical lowlands of Peru and Ecuador. The larva has not been described.

The natives of the upper Río Napo make ear ornaments and necklaces from the elytra as they do from pieces of other shiny, colorful beetles (Reitter 1961).

### References

Morón, M. A. 1983. A revision of the subtribe Heterosternina (Coleoptera, Melolonthidae, Rutelinae). Fol. Entomol. Mexicana 55: 31–101.

Reitter, E. 1961. Beetles. Putnam's Sons, New York.

## OTHER SCARABS

Males of several additional horned scarabs (Dynastinae) are renowned for their large size and elaborately developed horns. The black pan beetle (*Enema pan*), aside from its facetious name, is noteworthy for its enormity (BL 5 cm), tanklike shape, and forward-curving thoracic horn into whose apical fork a great head horn fits (fig. 9.10a). The great horned scarab (*Megaceras jasoni = chorinaeus*) is similar, but the base of the prothoracic projection is massive, with wide set apical horns (fig. 9.10b).

June or May beetles (*jobotos, fogotos, gallinas ciegas*), Melolonthinae, Melolonthini, comprise an enormous assemblage (about 340 species, mostly in the genus *Phyllophaga;* fig. 9.10c) of nondescript, brown, medium-sized (BL 8–12 mm), ovoid scarabs, familiar everywhere around electric lights on warm nights and on the leaves of many kinds of plants. Their larvae ("white grubs") attack the roots of their hosts, often graminaceous plants such as sugarcane, corn, and sorghum (King 1985).

Cockchafers (*Macrodactylus;* fig. 9.10d) are also melolonthines and well-known depredators of plants. Many of the approximately ninety species feed as adults on crops and ornamentals (especially grapes, roses, coffee, citrus). They are recognized by their long, spiny legs, bearing oversized tarsal claws. Their larvae, like those of other scarabs, are subterranean feeders on the roots of the adult hosts.

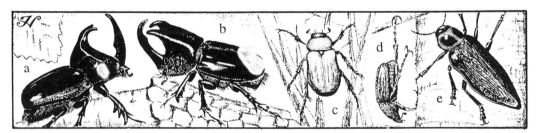

**Figure 9.10   SCARAB (SCARABAEIDAE) AND WOOD BORING (BUPRESTIDAE) BEETLES.** (a) Pan beetle (*Enema pan*), male. (b) Great horned scarab (*Megacerus jasoni*), male. (c) June beetle (*Phyllophaga portiricencis*). (d) Cockchafer (*Macrodactylus* sp.). (e) Giant metallic ceiba borer (*Euchroma gigantea*, Buprestidae).

**Plate 1. LATIN AMERICAN INSECTS.**

a.

b.

a. Tarantula (undetermined arboreal species)

b. Broad-winged leaf katydid (*Pterochroza ocellata*) in threat posture (photograph by James L. Castner)

c.

d.

c. Eumastacid grasshopper (*Eumastax* sp.)

d. Stinkbug (*Edessa* sp., Pentatomidae)

**Plate 1. (continued)**

e.

f.

e. Dragon-headed bug (*Fulgora laternaria*)

f. Termites swarming on the ground, probably *Syntermes* sp.

g.

h.

g. Tiger beetle (*Megacephala* sp.)

h. Giant metallic ceiba wood borer (*Euchroma gigantea*)

**Plate 2. LATIN AMERICAN INSECTS.**

a.

b.

a. Harlequin beetle (*Acrocinus longimanus*) (photograph by George Dodge)

b. Tortoise beetle (*Cyclosoma mirabilis*)

c.

d.

c. Window-winged saturnian (*Rothschildia erycina*)

d. Larva of eyed saturnian, *Automeris* (photograph by George Dodge)

**Plate 2. (continued)**

e.

f.

e. Larva of undetermined tiger moth (Arctiidae) (photograph by James N. Hogue)

 f. Achilles morpho (*Morpho achillaena*)

g.

h.

g. Larva of owl butterfly (*Caligo* sp.)

h. Bivouac of colony of the army ant, *Eciton hamatum*

**Plate 3. ECOLOGY OF LATIN AMERICAN INSECTS.**

a.

b.

a. "Tamshi." *Dinoponera gigantea* killed by fungal infection. Fruiting bodies of fungus fully developed.

b. Kelep ant (*Ectatomma tuberculatum*) tending extrafloral nectary of *Inga* sp.

c.

d.

c. Formicarium of *Cordia* sp. housing a colony of aztec ants (*Azteca* sp.)

d. Aposematically colored nymph of unidentified assassin bug

**Plate 3. (continued)**

e.                                    f.

e. Viper worm (*Hemeroplanes ornatus*) in threatening posture (photograph by George Dodge)

f. Müllerian mimicry cluster of moths and butterflies with "tiger" pattern. Left to right—top row: *Chetone angulosa* (Arctiidae, Pericopinae), *Dismorphia amphiona* (Pieridae), *Castnia* sp. (Castniidae); middle row: *Lycorea halia* (Nymphalidae, Danainae), *Papilio zagreus* (Papilionidae), *Consul fabius* (Nymphalidae, Nymphalinae); *Melinaea ethra* (Nymphalidae, Ithomiinae), *Eresia phillyra* (Nymphalidae, Nymphalinae), *Heliconius ismenius* (Nymphalidae, Heliconiinae)

g.                                    h.

g. Broad-winged katydid leaf mimic (undetermined species)

h. *Cyclocephala* (Scarabaeidae) pollinator in flower of *Victoria amazonica*

**Plate 4. ECOLOGY OF LATIN AMERICAN INSECTS.**

a.

b.

a. Arboreal nest of nasute termite (*Nasutitermes* sp.)

b. Social sphecid nest (*Microstigmus comes*)

c.

d.

c. Paper wasp nest (*Polybia* sp.) (photograph by George Dodge)

d. Paper wasp nest (*Polybia scutellaris*)

**Plate 4. (continued)**

e.    f.

e. Nest of bell wasp (*Chartergus chartarius*)

f. Parasol wasp nest (*Apoica pallens*)

g.

h.

g. Aztec ant nests (*Azteca trigona*) associated with nests of the yellow-rumped cacique

h. Ant garden

## Reference

KING, A. B. S. 1985. Factors affecting infestation by larvae of *Phyllophaga* spp. (Coleoptera: Scarabaeidae) in Costa Rica. Bull. Entomol. Res. 75: 417–427.

## METALLIC WOOD BORERS

Buprestidae. Flat-headed borers (larvae).

The larvae of this family, along with those of long-horned beetles, are common wood borers. They are distinguished from the former entirely cylindrical forms by their unique, flattened, disk-shaped thorax, which is decidedly wider than the narrow abdomen.

Adults are often highly metallic colored, coppery or bronzy, green, blue, or reddish. They are robust, hard shelled and extremely varied in size (BL 3–70 mm), with an elongate, streamlined body and strongly tapering elytra posteriorly. They feed on foliage and bark; some are twig girdlers or leaf miners, and many frequent blossoms to feed on pollen and petals. They are sun loving and are habitually active on hot days during the midday hours when they are often seen walking on freshly downed logs and tree limbs.

Approximately 3,200 species are known from the Neotropics, where they occur in all major habitats from wet forest to desert, alpine barrens to the seacoast.

## GIANT METALLIC CEIBA BORER

Buprestidae, *Euchroma gigantea*. *Spanish:* Catzo (Ecuador). *Portuguese:* Mãe do sol, ôlho do sol (Brazil).

This species, common from Mexico to Argentina, is the largest in its family (BL 6–7 cm) and like its relatives, brilliantly colored in glowing metallic hues (Hespenheide 1983) (fig. 9.10e, pl. 1h). The rugose elytra are generally shining green but pro-fused throughout with red; the back of the prothorax bears two large black spots side by side. Freshly emerged specimens are covered with a yellowish, waxy powder.

The larva is elongate (BL to 10 cm), with typical flattened thorax. It bores in the relatively soft wood of dead trees in the family Bombacaceae, most often the giant ceiba (*Ceiba pentandra*) but also balsa (*Ochroma*), *Bombacopsis,* and *Pseudobombax.* Adults are usually collected while walking on or flying about the trunks of these trees on warm days (Bondar 1926).

Adults have been used as food by the Tzeltal-Mayan Indians of Mexico (Hubbell 1979). And their attractive and durable elytra are made into ornaments by many tribes in diverse areas.

### References

BONDAR, G. 1926. A biologia de *Euchroma gigantea* L. Correio-agricola (Bahia) 4: 192–193.

HESPENHEIDE, H. A. 1983. *Euchroma gigantea* (Euchroma, giant metallic ceiba borer). *In* D. H. Janzen, ed., Costa Rican natural history. Univ. Chicago Press, Chicago. P. 719.

HUBBELL, P. 1979. Adult beetles as food. Coleop. Bull. 33: 91.

## LADYBIRD BEETLES

Coccinellidae. *Spanish:* Marias, mariquitas (General); chinitas (Chile); catitas, loritos, vaquitas (Argentina). *Portuguese:* Joaninhas. Lady beetles, ladybugs.

Ladybirds (Hodek 1973) are familiar small beetles (BL 5–10 mm). The body is hemispherical to elongate and the head quite small and covered by the prothorax. The head bears heavy mandibles and slightly clubbed antennae. The elytra are often brightly colored, most often deep red but may be black, yellow, pink, or yellow and usually with contrasting spots, stripes, or irregular patterns of black or red. Some rounded species are difficult to distinguish superficially from similar leaf beetles; they

**Figure 9.11 BEETLES.** (a) Ladybird beetle (*Cycloneda sanguinea*, Coccinellidae). (b) Southern squash beetle (*Epilachna tredecimnotata*). (c) Lacewing beetle (*Calopteron brasiliense*, Lycidae). (d) Longhorn beetle mimic of lacewing beetle (*Thelgetra* sp., Cerambycidae). (e) Wasp moth mimic of lacewing beetle (*Correbia* sp., Arctiidae). (f) Heteropteran mimic of lacewing beetle (*Oncopeltus* sp., Lygaeidae). (g) Cacao borer (*Xyleborus ferrugineus*, Scolytidae). (h) Coffee borer (*Hypothenemus hampei*, Scolytidae).

have three major segments in the tarsi rather than four as in that family. Larvae are elongate, with long legs and a warty integument and often are pilose, sometimes with long branched hairs.

The adults and larvae are common on vegetation where they prey on other insects, such as aphids and mealybugs, and therefore are considered beneficial to agriculture (Szumkowski 1955). Some species are of specific importance as biological control agents against scales (Bartlett 1939), such as *Rodolia cardinalis* against the cottony cushion scale (*Icerya purchasi*). A few species in the genus *Epilachna* (Gordon 1976) are herbivorous and are actually pests on crops such as beans (Mexican bean beetle, *E. varivestis*), squash and other cucurbits (southern squash Beetle, *E. tredecimnotata*, fig. 9.11b), or melons (melon beetle, *E. paenulata*).

There are well over a thousand coccinellid species in Latin America, but they are generally poorly known there. The most widespread species, *Cycloneda sanguinea* (fig. 9.11a), has solid red elytra.

### References

BARTLETT, K. A. 1939. The collection in Trinidad and southern Brazil of coccinellids predatory on scales. 6th Pacific Sci. Cong. Proc. 4: 339–343.

GORDON, R. D. 1976. A revision of the Epilachninae of the Western Hemisphere (Coleoptera: Coccinellidae). U.S. Dept. Agric. Tech. Bull. 1493: 1–409.

HODEK, I. 1973. Biology of Coccinellidae. Academia, Czechoslovak Acad. Sci., Prague.

SZUMKOWSKI, W. 1955. Observaciones sobre la biología de algunos Coccinellidae (Coleoptera). Bol. Entomol. Venezolana 11: 77–96.

## NET-WINGED BEETLES

Lycidae.

Net-winged beetles possibly represent one of the most ancient living models for mimetic insects. Their noxious body contents (expelled by voluntary bleeding), expansive colorful wing covers, and gregarious and sluggish habits make them ideal for this role. They are resembled in shape, color pattern, and behavior by members of more orders of insects than any other. A widespread lycid pattern is anterior, medial, and posterior dark cross bands on a yellow-orange background, such as in *Calopteron* (fig. 9.11c), which is resembled by *Thelgetra* (fig. 9.11d) and *Lycoplasma* (Cerambycidae), *Correbia lycoides* (Arctiidae, Ctenuchinae) (fig. 9.11e), *Oncopeltus fasciatus* (Lygaeidae) (fig. 9.11f), and others.

Most studies on Neotropical mimicry complexes based on the lycid model deal only with partial components (Darlington 1938, Parsons 1940); complete series of Batesian and Müllerian components may be similar to those cited from northern Mexico (Emmel 1965, Linsley et al. 1961). The latter is called the *Lycus fernandezi* complex: in this series, the lycid is large (BL 19–18 mm), mainly orange-yellow, with the tips of the elytra, antennae, and legs black. It gathers on flowers for feeding and mating. Strongly resembling it in the same geographic areas are another lycid (*Lycus arizonensis*), a long-horned beetle (*Elytroleptus apicalis*), a smoky moth (Zygaenidae, *Seryda constans*) and lithosiid moths (*Ptychoglene coccinea* and *P. phrada*). In other complexes elsewhere, these insect types may be joined by other beetles, such as click beetles (Elateridae), false blister beetles (Oedemeridae), and soldier beetles (Cantharidae), as well as tiger and wasp moths (Arctiidae), seed bugs (Lygaeidae), sawflies and ichneumonids (Hymenoptera), and robber flies (Diptera), all being of a similar appearance focused on a central lycid type.

The chemical basis for lycid unpalatability is not known, but the beetles contain yellow or pink body fluids similar to those in other chemically protected insects. The fluid may be exuded by the insects when mishandled; a slight amount of pressure on the body causes membranes on the ventral surface to rupture, releasing the fluid.

Adults of this family are typified by semitransparent, finely reticulate, soft elytra. They have a flattened form, and the wings are often widened (especially posteriorly). The prothorax is flat and shieldlike and the antennae often serrate.

Little is known of their biology aside from aspects of mimicry. Both adults and larvae of some species are known to be predatory, the former living under bark of dead trees. This is a diverse group with just over 700 species in Latin America.

## References

DARLINGTON, JR., P. J. 1938. Experiments on mimicry in Cuba, with suggestions for future study. Entomol. Soc. London Trans. 87: 681–695.

EMMEL, T. C. 1965. A new mimetic assemblage of lycid and cerambycid beetles in central Chiapas, Mexico. Southwest. Nat. 10: 14–16.

LINSLEY, E. G., T. EISNER, AND A. B. KLOTS. 1961. Mimetic assemblages of sibling species of lycid beetles. Evolution 15: 15–29.

PARSONS, C. T. 1940. Observations in Cuba on insect mimicry and warning coloration. Psyche 47: 1–7.

# BARK AND AMBROSIA BEETLES

Scolytidae and Platypodidae. *Spanish:* Taladrillos (Argentina).

Bark beetles (Scolytidae) and ambrosia beetles (Platypodidae) are all small (BL 1–5 mm) and uniformly black, dark brown, or light brown in color. They are cylindrical in shape and often have strongly punctate or apically sculptured wing covers. The short, elbowed antennae, tipped with a knob composed always of only the three terminal segments, are also characteristic. The two families are closely related and are distinguished only by minute structural differences; the head of platypodids is normally wider than the thorax and more exposed dorsally than in scolytids. Tarsal segment 1 in platypodids is as long as segments 2–5 combined, while in scolytids, segments 1–3 are about equal in length.

The larvae all are small, white, and legless. The body is arched slightly and the anterior end somewhat larger in diameter than the posterior. The head is lightly sclerotized.

All are borers in woody plants, both as adults and larvae. Biologically, however, they may be divided into two types, not corresponding directly with the taxonomic families: the so-called bark beetles feed directly on wood, both as larvae and adults (most scolytids); the ambrosia beetles cultivate special types of fungi in their tunnels and feed on these (many scolytids and all platypodids). The latter predominate in the lowland tropics of the New World, where bark beetles are relatively unimportant as pests. The bark beetles are more conspicuous in the higher latitudes, particularly in the coniferous forests of Mexico and Central America and the southern beech forests of Chile and Argentina.

Many species of Scolytidae are serious pests of trees, both hardwoods and conifers. The damage in tropical areas is caused by the staining of sapwood by fungal growths (more so by the ambrosia beetles than by bark beetles), in addition to physical destruction of the wood by the tunneling activities of both biological types. Injured or unthrifty trees and cut timber are most susceptible to attack, but some species infest living trees, mining the cambium and often girdling and killing major branches or the entire tree. Damage is often first noticed, particularly in conifers, when copious amounts of sap appear at the site of entry. Healthy trees are often able to ward off infestation by flooding out the colonizing beetles. The tops of affected trees often turn brown and wilt.

The burrows of Scolytidae are restricted largely to the phloem and are two-dimensional. They form varied geometrical patterns on the inside of the bark and on the surface of the sapwood, from simple straight or meandering lines to biramous, stellate, multiramous, or branching designs. Each pattern is usually characteristic of a species or genus and thus of value in identification. Platypodids burrow deep within the sapwood.

A widespread, very injurious ambrosia beetle is *Platypus parallelus*, which riddles logs left after cutting and prior to removal to the mill (Wood pers. comm.). In addition to attacking forest trees, scolytids are common pests of woody crops such as cacao (*Xyleborus;* fig. 9.11g) and coffee (*Hypothenemus hampei, broca do café, barrenador del café;* fig. 9.11h) (Quezada and Urbina 1987).

The study of these beetles in Latin America is in an embryonic state (Atkinson and Equihua 1986, Haack et al. 1989, Martinez and Atkinson 1986). There are no comprehensive publications, although portions of the fauna have been described in monographs (Scolytidae in Wood 1982, Platypodidae in Schedl 1972). The fragmentary literature refers to somewhat more than 1,500 known scolytid and about 250 platypodid species. This may be no more than 50 percent of the actual total faunas of each family.

## References

ATKINSON, T. H., AND E. EQUIHUA. 1986. Biology of the Scolytidae and Platypodidae (Coleoptera) in a tropical deciduous forest at Chamela, Jalisco, Mexico. Fla. Entomol. 69: 303–310.

HAACK, R. A., R. F. BILLINGS, AND A. M. RICHTER. 1989. Life history parameters of bark beetles (Coleoptera: Scolytidae) attacking West Indian Pine in the Dominican Republic. Fla. Entomol. 72: 591–603.

MARTÍNEZ, A. E., AND T. H. ATKINSON. 1986. Annotated checklist of bark and ambrosia beetles (Coleoptera: Scolytidae and Platypodidae) associated with a tropical deciduous forest at Chamela, Jalisco, Mexico. Fla. Entomol. 69: 619–635.

QUEZADA, J. R., AND N. E. URBINA. 1987. La broca del fruto del cafeto, *Hypothenemus hampei*, y su control. *In* J. Pinochet, Plagas y enfermedades de carácter epidémico en cultivos frutales de la región Centroamericana. CATIE, Panama. Pp. 48–59.

SCHEDL, K. E. 1972. Monographie der Familie Platypodidae Coleoptera. Junk, The Hague.

WOOD, S. L. 1982. The bark and ambrosia beetles of North and Central America (Coleoptera: Scolytidae), a taxonomic monograph. Great Basin Nat. Mem. 6: 1–1359.

## LONG-HORNED BEETLES

Cerambycidae. *Spanish:* Toritos, asserradores, serradores, corta palos (General, adults); chichis (Cuba); taladradores, taladros, trozadores (Central America, larvae). *Portuguese:* Serras pão (Brazil).

Most of the approximately 5,200 known Neotropical longhorns (Linsley 1959*a*,

Zajciw 1976) are mundane, small beetles (BL less than 10 mm), but many are among the largest and most spectacular of insects. The greatly elongated antennae give them their common name. The antennae of the males are at least half the length of the body; in many, two to three or more times the length. Female antennae are somewhat shorter. The family is also recognized by an elongate shape, eye notches into which the antennae are inserted, and apparently four-segmented tarsi, the terminal segment of which is usually large, flattened, and heart shaped (there are actually five segments, but the fourth is tiny and concealed in the notch of the fifth).

The mandibles are frequently very heavy in the larger species, even grotesquely enlarged, and always powerful for gnawing wood. These jaws often bear teeth, which has given rise to the widespread myth that they saw tree branches in two by hooking the jaws around them and flying around and around (*serradores*). A few cerambycids are true girdlers (e.g., *Oncideres;* fig. 9.12a), but the females slowly gnaw only the outer wood of the piece, seldom completely severing it, and oviposit in the distal portion. The action kills the branch beyond the wound, and it sometimes breaks and falls to the ground, making it suitable for the larvae to feed. The grinding sound of this activity can sometimes be heard several meters away.

The larvae are elongate, cylindrical, although slightly enlarged in the thoracic region, and conspicuously segmented or lobed (Duffy 1960). The thoracic legs are minute or absent, but the head and jaws are strongly developed and directed forward for boring through solid or decayed wood. The wood is ingested and forms the larva's food. Larvae can destroy logs and stored lumber, and some kill healthy trees and are therefore considered forestry pests. Under natural conditions, most are probably more beneficial as reducers of dead wood, returning the components of cellulose to the soil. Also, they are among those wild foods gathered by natives from rotten logs.

These beetles exhibit many intricate protective devices (Silberglied and Aiello 1976). For example, the long segments of the antennae are at times armed with very sharp barbs. Many species are cryptically marked like lichens or tree bark, and some present a reversed illusion by the arrangements of lines near the rear of the elytra and the long trailing antennae (tergiversation). Some also emit distasteful or foul-smelling secretions (Zikán and Zikán 1946). Certain species produce squeaking noises (*toca violas, visitas,* Brazil), which probably serve to frighten or distract predators.

Batesian mimicry among adults is also highly developed and widespread among groups, the most common models being

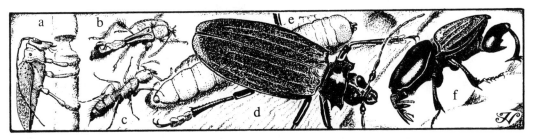

**Figure 9.12  LONG-HORNED BEETLES (CERAMBYCIDAE).** (a) Serrador (*Oncideres sara*). (b) Ant-mimicking longhorn (*Acyphoderes sexualis*). (c) Ant model for ant-mimicking longhorn (*Pachycondyla* sp., Formicidae). (d) Titanic longhorn (*Titanus giganteus*). (e) Titanic longhorn, larva. (f) Mole beetle (*Hypocephalus armatus*).

stinging ants, bees and wasps, or other unpalatable beetles, especially the net-winged beetles (Lycidae), leather-winged beetles (Cantharidae), and leaf beetles (Chrysomelidae) (Linsley 1959b). There are even cases of double mimicry, such as by *Acyphoderes sexualis* to *Paraponera clavata* or *Pachycondyla* sp. (Formicidae) when walking (fig. 9.12b, c) but to wasps when in flight (Silberglied and Aiello 1976).

Allometric growth in cerambycids, as in horned scarabs, is especially pronounced in the subfamily Prioninae, to which most of the following species belong. Major males, consequently, may grow outlandishly long forelegs or large jaws.

### References

DUFFY, E. A. J. 1960. A monograph of the immature stages of Neotropical timber beetles (Cerambycidae). Brit. Mus. Nat. Hist., London.

LINSLEY, E. G. 1959a. Ecology of the Cerambycidae. Ann. Rev. Entomol. 4: 99–138.

LINSLEY, E. G. 1959b. Mimetic form and coloration in the Cerambycidae (Coleoptera). Entomol. Soc. Amer. Ann. 52: 125–131.

SILBERGLIED, R. E., AND A. AIELLO. 1976. Defensive adaptations of some Neotropical longhorned beetles (Coleoptera, Cerambycidae): Antennal spines, tergiversation, and double mimicry. Psyche 83: 256–262.

ZAJCIW, D. 1976. Introdução ao estudo dos longicórneos do Brasil (Coleoptera, Cerambycidae). Soc. Entomol. Brasil An. 4:116–121.

ZIKÁN, J. F., AND W. A. ZIKÁN. 1946. A insetofauna do Itatiaia e da Mantiqueira, Coleoptera-Cerambycidae. Min. Agr. (Rio de Janeiro) Bol. 33: 1–50.

## TITANIC LONGHORN

Cerambycidae, Prioninae, Prionini,
*Titanus giganteus.*

This gargantuan species, which lives in the northern Amazonian rain forest, is the largest longhorn and one of the world's largest beetles (fig. 9.12d). The biggest specimens may measure fully 20 centimeters in body length, with 10-centimeter-long antennae and powerful, massive jaws that can snap wooden pencils in half. They are solid dark brown to black and with the elytra lightly ridged longitudinally.

These great insects were considered exceedingly rare until recent years, when their habitat and area of occurrence became known. Now specimens are collected regularly by natives to sell to dealers who supply souvenir hunters and amateur entomologists. In 1914, a male fetched 2,000 gold marks, the equivalent then of US$476 (Reitter 1961), and today prime specimens command no less a price.

An exciting hunt for these beetle behemoths in Amazonian Brazil is related by Zahl (1959). A presumed mature larva (fig. 9.12e) was found and photographed. It is like an enormous sausage, more than 20 centimeters long and 3 centimeters in diameter. It was found feeding in a decomposing log and presumably required several years to mature because of the fantastic size it had reached (Zahl pers. comm.). Its host trees and other details of life history remain totally unknown.

Linnaeus named and described the species in 1771, not from specimens but from a figure by D'Aubenton in a set of illustrations for a bird encyclopedia published in 1765 (Reitter 1961:32).

### References

REITTER, E. 1961. Beetles. Putnam's Sons, New York.

ZAHL, P. 1959. Giant insects of the Amazon. Natl. Geogr. 115(4): 632–669.

## GIANT JAWED SAWYERS

Cerambycidae, Prioninae, Ancistrotini,
*Macrodontia. French:* Mouches scieur de long, mouches cafe (Guadeloupe).
Sawyer beetles.

Six species make up this genus of unique, large longhorns, but only three are at all well known. *Macrodontia cervicornis* (fig.

**Figure 9.13  LONG-HORNED BEETLES (CERAMBYCIDAE).** (a) Giant jawed sawyer (*Macrodontia cervicornis*). (b) Big jawed sawyer (*Macrodontia dejeani*). (c) Bearded imperious sawyer (*Callipogon barbatum*), male. (d) Giant imperious sawyer (*Callipogon armillatum*), male.

9.13a) is the largest (BL to 15 cm) and is more widespread than *M. flavipennis* or *dejeani* (fig. 9.13b) (BL of both reaching only 9 cm). It has a brown and black patterned prothorax, head, legs, and jaws and very irregular, dark longitudinal elytral markings. The latter two species, of local occurrence in the Caribbean and South America, are solid purplish-black anteriorly and with dark legs and elytra striped with neat lines. All have similar enormous, incurved mandibles with uniform internal teeth, one of which is enlarged just beyond half the length; *Macrodontia* tend to be somewhat flattened and wide also, in contrast to most other more cylindrical cerambycids.

Because of their serrate jaws, they are among those beetles erroneously thought to sever twigs (*serradores*). They allegedly clasp the branches of the coffee plant, for instance, with the mandibles and then gyrate around until the limb is cut through. And they are said to produce a characteristic buzzing sound when doing so (Duffy 1960: 53). Only fragmentary information is available regarding the biology of these giants (Bondar 1926). The adults are nocturnal and sometimes come to artificial lights where they are often captured for sale to collectors, by whom they are in great demand.

The larvae, which are likewise very large (up to 21 cm in *M. cervicornis*), create extensive galleries more than a meter long and 10 centimeters wide in the hearts of dead and dying softwood trees, such as the coconut palm (*Cocos*), *Attalea*, and *Ceiba pentandra*. The Ungurahui palm (*Jessenia weberbauri*) is a host in Peru (Paprzycki 1942). Hosts of *M. dejeani* alone are *Acrocomia* and *Acacia decurrens* (Duffy 1960: 51–54).

Morphologically, the larvae are unique among the family in having the thoracic and abdominal segments densely velured and the integument brown instead of white as is usual. In Brazil, they are avidly sought by the natives for food (Netolitzky 1920).

Another large species in the same tribe is *Ancistrotus cummingi*, known in Chile as *madre de la culebra* (mother of snakes), probably from the elongate form of its larva.

### References

BONDAR, G. 1926. A biologia e a larva do bezouro *Macrodontia cervicornis*. Chacaras e Quintaes (São Paulo) 34: 33–35.

DUFFY, E. A. J. 1960. A monograph of the immature stages of the Neotropical timber beetles (Cerambycidae). Brit. Mus. Nat. Hist., London.

NETOLITZKY, F. 1920. Käfer als Nahrungs- und Heilmittel. Koleop. Rund. 8: 21–26.

PAPRZYCKI, P. 1942. Datos para la captura y crianza del mas grande de los cerambícidos "*Macrodontia cervicornis*" en la selva peruana. Mus. Hist. Nat. Javier Prado (Lima) Bol. 6: 349–351.

## IMPERIOUS SAWYERS

Cerambycidae, Prioninae, Callipogonini, *Callipogon*.

The six species comprising this genus are widely distributed, wholly Neotropical cerambycids, impressive for their rich amber or chestnut-colored elytra and great size (BL of most 10–11 cm). One species, *Callipogon armillatum* (fig. 9.13d) is very large, almost as long (BL to 15 cm) as the titanic longhorn.

They are all fairly elongate, often have strong spines on the margins of the prothorax, and the forelegs are slightly longer than the other legs. Also, they have small but heavy incurved or upturned jaws. In major males of several species, the mandibles are grossly enlarged. Those of *C. barbatum* (fig. 9.13c) and *C. senex* are massive, forked structures bearing a dense vestiture of reddish-brown hairs on their inner surfaces. The elytra in these are variable, solid brown or with conspicuous, broad, longitudinal streaks of white pubescence.

The larvae bore into the heartwood of pines in Central America and a wide variety of hardwoods in all areas. They are considered serious timber pests.

## METALLIC LONGHORNS

Cerambycidae, Prioninae, Prionini, *Psalidognathus* and *Callichroma*.

Longhorns in the genus *Psalidognathus* (fig. 9.14a) are large (BL 5–7 cm) and beautifully colored in metallic tones of blue, green, and purple. The jaws of the males are long and the elytra somewhat thin and flexible. The females are without flight wings, some even with shortened elytra as well. They never fly, remaining on the ground or at the foot of the host trees, which are true cedars (*Cedrus*) (Duffy 1960: 69). The males of some fly at dusk or

at night and are readily attracted to lights. Other males (e.g., *P. modestus*) have been seen during the day flying in the Costa Rican cloud forest (Giesbert pers. comm.).

The closely related genera *Callichroma*, *Plinthocoelium* (fig. 9.14e), and *Schwarzerion* are smaller (BL 2–3 cm) and much more streamlined, compact longhorns but also brilliantly colored in metallic greens and blues and also characterized by a strongly flattened hind femur and tibia. Both sexes are fully winged and fly readily. They are unusual in the possession of scent glands that give off a pungent odor, described as pleasing and vanillalike (Robertson pers. comm.). According to Welling (pers. comm., via Chemsak), the natives on the Yucatán Peninsula eat these beetles as an aphrodisiac.

The larva of one common species in Trinidad, *Callichroma velutinum*, attacks dead logs of the balata tree (*Manilkara bidentata*). It feeds under the bark when young but at maturity, burrows deeply through the sapwood and heartwood where it matures and pupates. Because the wood of this tree is used for railroad ties, the beetle is considered a pest (Duffy 1960: 162–167).

### Reference

DUFFY, E. A. J. 1960. A monograph of the immature stages of the Neotropical timber beetles (Cerambycidae). Brit. Mus. Nat. Hist., London.

## HARLEQUIN BEETLE

Cerambycidae, Lamiinae, Acrocinini, *Acrocinus longimanus*. *Spanish:* Escarabajo arlequín. *Portuguese:* Arlequim da mata (Brazil). *French:* Mouche bagasse. Jak tree borer (Trinidad).

The gaudy, harlequin pattern of black, olive, and coral crescents and bars and the outlandishly long forelegs of the males im-

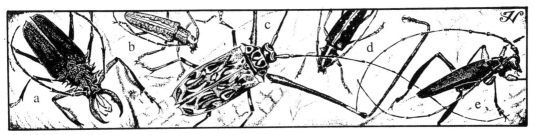

**Figure 9.14   LONG-HORNED BEETLES (CERAMBYCIDAE).** (a) Metallic longhorn (*Psalidognathus friendi*). (b) Three-lined fig tree borer (*Neoptychodes trilineatus*). (c) Harlequin beetle (*Acrocinus longimanus*), male. (d) Longhorn beetle (*Taeniotes scalaris*). (e) Metallic longhorn (*Plinthocoelium* sp.).

mediately identify this large longhorn (BL 4.5–7.5 cm) (fig. 9.14c, pl. 2a). This pattern, although conspicuous on a neutral background, affords the beetle considerable camouflage on the lichen-covered or mottled tree trunks that are its favored resting places. It is also a very strong, spiny creature and difficult to hold. Some early authors have said that the large lateral spines of the prothorax can rotate as if on ball bearings, but this is erroneous (Bates 1861).

Individuals are strongly attracted to exudates or the sap seeping from wounded trees, especially the bagasse tree (*Bagassa guianensis,* Moraceae), a fact well known to collectors in the past who deliberately scored the trees to attract large numbers of the beetles (Wood 1883: 244). The species is widespread throughout the forested portions of the Neotropics. (A specimen once turned up in Portugal in 1977 where it was believed to have strayed without human assistance [Lemos Pereira 1978]).

Females often select trees infested with bracket fungi for oviposition (Chemsak 1983). Some say that the very long forelegs of the adults are used to brachiate among the branches of vegetation, much like a spider monkey, but this is doubtful (Duffy 1960: 231), their normal function probably being concerned with mating. The beetle is active both day and night, but apparently it flies after dark and is occasionally observed around electric lights.

The larvae bore under the bark of dead trunks of many hardwoods, such as *Ficus,*

which seems to be preferred, *Chorisia, Enterolobium,* and jak fruit (*Artocarpus*) (Duffy 1960: 228). Live trees are seldom attacked, so the species should be considered a reducer of dead wood rather than a forest pest. Little else is recorded regarding its natural history (Tippmann 1951).

Pseudoscorpions of the family Chernetidae (*Cordylochernes* and *Lustrochernes*) are almost invariably found riding under the wing covers of the harlequin beetle. They utilize the beetle host not only for transportation but to prey on phoretic or parasitic mites also living on it. *Cordylochernes scorpioides* assumes a special posture and makes "beckoning" movements with its pedipalps when encountering the beetle and quickly tries to grasp it and hide under the elytra (Beck 1968).

One author (Reitter 1961: 33) thinks the hieroglyphlike pattern on the elytra might be copied on the shields of South American Indians.

## References

BATES, H. W. 1861. Contributions to an insect fauna of the Amazon Valley. Coleoptera: Longicornes. Ann. Mag. Nat. Hist. (3)8: 40–52.

BECK, L. 1968. Aus den Regenwäldern am Amazonas. I. Natur. Mus. 98(1): 24–32.

CHEMSAK, J. A. 1983. *Acrocinus longimanus* (Arlequín, harlequin beetle). *In* D. H. Janzen, ed., Costa Rican natural history. Univ. Chicago Press, Chicago. Pp. 678–679.

DUFFY, E. A. J. 1960. A monograph of the immature stages of Neotropical timber beetles (Cerambycidae). Brit. Mus. Nat. Hist., London.

LEMOS PEREIRA, A. B. 1978. Sobre o apareci-mento em Portugal de um Coleóptero longicórnio exótico, do género *Acrocinus*. Inst. Zool. "Dr. Augusto Nobre." Publ. 138: 11–16.
REITTER, E. 1961. Beetles. Putnam's Sons, New York.
TIPPMANN, F. F. 1951. Eine Harlekinade am Rio Tulumayo. Entomol. Zeit. 61: 137–142, 147–152, 155–157.
WOOD, J. G. 1883. Insects abroad. Routledge Sons, London.

## THREE-LINED FIG TREE BORER

Cerambycidae, Lamiinae, Monochamini, *Neoptychodes trilineatus.*

Adults of this medium-sized longhorn (BL 22–29 mm females, 14–21 mm males) (fig. 9.14b) are commonly seen resting on the leaves and feeding on the bark of *Ficus* and other tropical trees in the genera *Alnus, Morus, Chlorophora,* and of mango, which are the larval hosts as well (Horton 1917). They are clearly recognized by their rather long antennae (held out laterally from the body and about 2.5 times the length of same) and gray and orange-splotched elytra, usually with three irregular, whitish, linear markings that run the length of the body. There are also two lateral fine lines running from the antennal bases and one irregular broader median line extending from the base of the prothorax along the length of the wing covers.

The many members of the closely related genus *Taeniotes* (fig. 9.14d) are similar in appearance and have comparable habits. They are distinguished by the presence of sharp spines on the sides of the prothorax, which *Neoptychodes* lacks. Both genera are sometimes orchard pests; the larvae infest the branches and boles of host trees. Damage from these is sometimes identifiable by the frass ejection holes found at regular intervals along the galleries.

### Reference

HORTON, J. R. 1917. Three-lined fig-tree borer. J. Agric. Res. 11: 371–382, pl. 35–37.

## MOLE BEETLE

Cerambycidae, Anoploderminae, Hypocephalini, *Hypocephalus armatus. Portuguese:* Vaqueiro, carocha, Iá-Iá de cintura (Brazil).

The range of this curious beetle is extremely local, apparently confined within the radius of about 50 kilometers of the village of Condeuba in Bahia, Brazil. It is a fossorial insect, but its burrowing purposes still elude entomologists, and its early stages are poorly known. The larvae are probably root feeders, judging from the subterranean habits of the adults. Nearly all specimens that have been found have been males, crawling over open ground in December.

Both sexes are without flight wings but otherwise are similarly shaped, fusiform and with a strongly constricted "waist" between the ovoid, smooth, shiny black prothorax, and rhomboid, rough-surfaced, dark brown wing covers, the latter tapering to a point posteriorly (fig. 9.12f). The hind legs are much larger and heavier than the others; the head and its appendages are reduced. The beetle is fairly large (BL 4–5 cm).

Soon after its discovery, the species became a prize for beetle collectors. They were caught by locals in the vicinity of Condeuba and sold abroad. At one time, specimens were also caught by native women, who tied them with a ribbon to their babies' cradles for decoration and to amuse the infants (Gounelle 1905).

### Reference

GOUNELLE, E. 1905. Contribution a l'étude des moeurs d'*Hypocephalus armatus* (Col.). Soc. Entomol. France Ann. 74: 105–108.

# LEAF BEETLES

Chrysomelidae. *Spanish*: San Juanes, pololos (Chile, metallic green types).

This is an immense family with over 12,000 species in the Neotropics. Adults range in size from very small (BL 1.5 mm) to moderate (BL 20–22 mm) and are of varied body form. Although some species resemble long-horned beetles (all tarsi with apparently four segments), most are oval or much flattened, and they never have antennae longer than half their body length. They are often brightly colored, spotted or striped, and many are brilliant metallic green, blue, or gold. Many are good jumpers; some tiny species may have greatly developed hind legs for hopping and are called "flea beetles" (Alticinae, *pulgas saltonas*) (Scherer 1983).

All are plant feeders in both the larval and adult phases and sometimes are serious agricultural pests, for example, the large Neotropical genus *Diabrotica* (asparagus beetles, *catarinitas, vaquitas, vaquinhos*) (fig. 9.15a). The larvae are stem borers; the adults attack principally the new leaves and may riddle them with holes or completely strip away the tissue between the veins (skeletonizing). Such leaves are a common sight among otherwise healthy lush vegetation in the tropical lowlands (Carroll 1978). Others are leaf miners.

Larvae are frequently gregarious. Under stress, some leaf beetles exude a yellow fluid presumed to render them noxious or poisonous to predators.

Most of the nineteen subfamilies are represented in Latin America (Seeno and Wilcox 1982). The tortoise beetles (Cassidinae) and the mining leaf beetles (Hispinae), discussed below, contain the most common species. Some species in the genus *Diabrotica* (Galerucinae; Smith and Lawrence 1967) are also conspicuous and are agricultural pests. Although several species of the well-known aquatic subfamily Donaciinae live on floating and emergent water plants in Mexico, Central America, and Cuba, none are known to occur in South America.

## References

Carroll, C. T. 1978. Beetles, parasitoids and tropical morning glories: A study in host discrimination. Ecol. Entomol. 3: 79–85.

Scherer, G. 1983. A diagnostic key for the Neotropical alticine genera (Coleoptera: Chrysomelidae: Alticinae). Entomol. Arbeit. Mus. Frey 31/32: 1–89.

Seeno, T. N., and J. A. Wilcox. 1982. Leaf beetle genera (Coleoptera: Chrysomelidae). Entomography 1: 1–221.

Smith, R. F., and J. F. Lawrence. 1967. Clarification of the status of the type specimens of Diabrocticites (Coleoptera, Chrysomelidae, Galerucinae). Univ. Calif. Publ. Entomol. 45: 1–168.

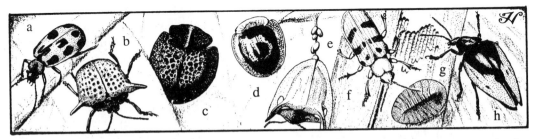

**Figure 9.15  LEAF BEETLES (CHRYSOMELIDAE).** (a) Spotted cucumber beetle (*Diabrotica undecempunctata*). (b) Horned tortoise beetle (*Omocerus eximius*). (c) Tortoise beetle (*Stolas cyanea*). (d) Target tortoise beetle (*Charidotis circumducta*). (e) Tortoise beetle (*Acromis spinifex*), male (from female eggs on filament). (f) Rolled-leaf beetle (*Chelobasis bicolor*). (g) Rolled-leaf beetle, larva. (h) Giant leaf beetle (*Pseudocalaspidea cassidea*).

# TORTOISE BEETLES

Chrysomelidae, Cassidinae.

In these beetles, the outline of the body is oval to nearly circular, the edges flared out and flattened, enabling the beetles to appress themselves tightly to a smooth leaf surface; when disturbed, head and appendages tuck out of sight like those of a tortoise (pl. 2b). The larvae are also flat but very spiny and bear a long forked tail to which cast skins, excrement, and debris become attached, giving the insect an assumed measure of camouflage.

Adults are frequently semitransparent, partially hued with radiant golden, silvery, or greenish colors. A common type are the target tortoise beetles such as *Charidotis circumducta* (fig. 9.15d) and *Coptocycla arcuata*, which sport a series of concentric circular dark markings on the back.

There are many large tortoise beetles in lowland wet forests. The genera *Stolas* (fig. 9.15c) and *Polychalca* contain many species almost 2 centimeters long, with strongly pitted or roughened, deep green, blue, or red splotched elytra; *Omocerus* (fig. 9.15b) and *Tauroma* are similar but also have prominent horns extending laterally from the bases of the elytra. *Acromis* (formerly known as *Selenis*) species are more bizarre, the males with the anterolateral corners of the elytra expanded with pointed, winglike anterior projections that are used in combat with other beetles (fig. 9.15e) (Windsor 1987).

Some tortoise beetles exhibit brood care. *Acromis* females attach their eggs to a filament made of secretions (fig. 9.15e). The thread with eggs hangs from a leaf, and the female sits at the top until the larvae hatch (Fiebrig 1910: 166), and they are then assiduously guarded until they are mature (Windsor 1987). Under the elytral expansion of *Omaspides pallidipennis*, the young larvae hide until they are better able to fend for themselves (Ohaus 1900: 230).

## References

FIEBRIG, K. 1910. Cassiden und Cryptocephaliden Paraguays. Ihre Entwicklungsstadien und Schutzvorrichtungen. Zool. Jahrb. 12: 161–264, pls. 4–9.

OHAUS, F. 1900. Bericht über eine entomologische Reise nach Central Brasilien. Stettiner Entomol. Zeit. 61: 164–273.

WINDSOR, D. M. 1987. Natural history of a subsocial tortoise beetle, *Acromis sparsa* Boheman (Chrysomelidae, Cassidinae in Panama). Psyche 94: 127–150.

# MINING LEAF BEETLES

Chrysomelidae, Hispinae. *Spanish:* Abejónes del platanillo (Costa Rica).

Larvae of this subfamily are either leaf miners or surface feeders. They are very flat, smooth, and oval, somewhat resembling cockroach nymphs or water pennies (the larvae of the aquatic beetle family Psephenidae). The integument is also strongly sclerotized. The adults have hard, elongate bodies, a large head visible from above, and coarsely sculptured, sometimes almost reticulate, elytra. In some, the outer posterior corners of the elytra are protracted into points.

One group, the so-called rolled-leaf hispine beetles (Harvey 1988; Strong 1977, 1983), are almost exclusively feeders on members of the ginger order (Zingiberales, *Heliconia*, *Zingiber*, and relatives). Species like *Chelobasis bicolor* (fig. 9.15f) spend their entire life cycle in the young, scroll-like, rolled leaves of heliconias (Seifert 1982), ginger, and marantas and occur virtually nowhere else. The larvae (fig. 9.15g) stay out of sight between the leaf layers near the bottom; adults feed on tissue near the top of the leaf. Adults and larvae feed at night by scraping the leaf surface, crawling forward with each scoop of the mandibles, and defecating, thus leaving a linear trail often littered with fecal pellets. They do not puncture the leaf. Because of the low nitrogen content in the *Heliconia* leaves, larval development of these beetles is prolonged and may require over two hundred days. As many as eight species may intermingle at a single site, leading to narrow host specificity (Strong 1982). *Xenarescus monocerus* larvae, of Venezuela, feed first on young *Heliconia* inflorescences, apparently to avoid competition with other species, moving later to rolled leaves to complete their development (Seifert and Seifert 1979).

*Pseudocalaspidea cassidea* (fig. 9.15h) is a very large hispine (BL 25 cm), whose teardrop-shaped body is margined broadly in red, leaving a large black triangular area centrally. Its larva is unknown, although adults are common on understory vegetation in humid forests.

## References

HARVEY, R. 1988. The ecology of arthropod communities associated with *Heliconia* leaf-curls in Tambopata Wildlife Preserve, southeastern Peru. The Entomologist 107: 11–19.

SEIFERT, R. P. 1982. Neotropical *Heliconia* insect communities. Quart. Rev. Biol. 57: 1–28.

SEIFERT, R. P., AND F. H. SEIFERT. 1979. Utilization of *Heliconia* (Musaceae) by the beetle *Xenarescus monocerus* (Oliver) (Chrysomelidae: Hispinae) in a Venezuelan forest. Biotropica 11: 51–59.

STRONG, JR., D. R. 1977. Rolled-leaf hispine beetles (Chrysomelidae) and their Zingiberales host plants in Middle America. Biotropica 9: 156–169.

STRONG, JR., D. R. 1982. Potential interspecific competition and host specificity: Hispine beetles on *Heliconia*. Ecol. Entomol. 7: 217–220.

STRONG, JR., D. R. 1983. *Chelobasis bicolor* (abejón del platanillo, rolled-leaf hispine). *In* D. H. Janzen, ed., Costa Rican natural history. Univ. Chicago Press, Chicago. Pp. 708–711.

## SEED BEETLES

Bruchidae.

These are small (BL 1–10 mm) beetles with a distinctive boxlike or egg shape, slightly broader posteriorly, the head concealed, and with a short, broad, almost weevillike snout. The very short antennae are clubbed or sawtoothed and the elytra abbreviated, exposing the tip of the abdomen.

The larvae feed on seeds, and family members are the principal inhibitors of sexual reproduction in many plants, thus acting as strong selective agents in seed evolution. Although rather host specific, they may not necessarily contribute to plant species richness through their selective efforts (Janzen 1980). Females oviposit on the developing flowers or pods or directly onto newly exposed seeds in dehiscing fruit, and the larvae bore through into the seed. A single larva develops in a single seed; some species may glue several seeds together as a pupal chamber or leave the seed and pupate in a cocoon. Pupation occurs in the seed or seed cavity in the fruit. Legumes are common hosts (Johnson 1985), but many other plants are used as well, such as *Cordia*, *Sesbania*, palms, *Pithecellobium* (Janzen 1983), *Acacia*, and *Enterolobium*. The adult leaves the seed through a typically round exit hole. This stage is not known to feed on seeds but probably subsists on nectar and pollen. Bruchids sometimes are called seed "weevils" (*gorgojos, gorgulhos*), but it seems best to reserve this term for true weevil (Curculionidae) predators of seeds.

Many of the five hundred Neotropical species, especially *Acanthoscelides* (Johnson 1983, 1990) (fig. 9.16a) and *Callosobruchus*, are of economic importance as they severely reduce seed productivity in peas, beans, alfalfa, and so on.

According to Bondar (1928), larvae of a bruchid (*Pachymerius nucleorum*) are frequently found infesting the nuts of various palms and are coveted as food by residents of Bahia, Brazil. The insects, called *bichos de coco*, are eaten and relished along with the palm fruits. In figurative speech in many parts of the country, a sly or cunning man is known as a "bicho de coco."

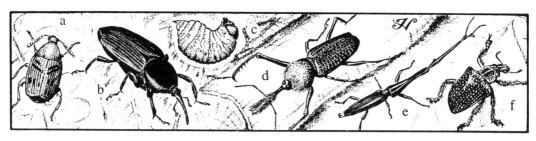

**Figure 9.16 BEETLES.** (a) Seed beetle (*Acanthoscelides* sp., Bruchidae). (b) Palm weevil (*Rhynchophorus palmarum,* Curculionidae). (c) Palm weevil, larva. (d) Bearded weevil (*Rhinostomus barbirostris,* Curculionidae). (e) Brentid weevil (*Brentus anchorago,* Brentidae). (f) Jeweled weevil (*Entimus imperialis,* Curculionidae).

## References

BONDAR, G. 1928. O bicho do côco. Extr. Cor. Agric. 6(1): 1–18. [Not seen.]

JANZEN, D. H. 1980. Specificity of seed-attacking beetles in a Costa Rican deciduous forest. J. Ecol. 68: 929–952.

JANZEN, D. H. 1983. *Merobruchus columbinus* (gorgojo de cenizero, rain-tree bruchid). *In* D. H. Janzen, ed., Costa Rican natural history. Univ. Chicago Press, Chicago. Pp. 738–739.

JOHNSON, C. D. 1983. Ecosystematics of *Acanthoscelides* (Coleoptera: Bruchidae) of southern Mexico and Central America. Entomol. Soc. Amer. Misc. Publ. 56: 1–24, figs. 1–596.

JOHNSON, C. D. 1985. Potential useful tropical legumes and their relationships with bruchid beetles. 1: 206–210. *In* K. C. Misra, ed., Ecology and resource management in tropics. Int. Soc. Trop. Ecol., Varanasi, India.

JOHNSON, C. D. 1990. Systematics of the seed beetle genus *Acanthoscelides* (Bruchidae) of northern South America. Amer. Entomol. Soc. Trans. 116: 297–618.

## WEEVILS

Curculionidae. *Spanish:* Gorgojos (General), picudos (Central America). *Portuguese:* Gorgolhos.

Weevils are immediately recognized by their snout, an elongation of the anterior portion of the head, which carries the elbowed antennae at the sides and mouthparts at the tip. The mouthparts include powerful mandibles used for drilling holes in seeds and nuts for feeding and oviposition.

The family is the largest of any in the animal or plant kingdom, with more than 50,000 species worldwide; there are at least 12,000 species in Latin America, and certainly many hundreds are still unknown to science (O'Brien and Wibmer 1981, 1982, 1984a, 1984b). A considerable number of species are injurious to agriculture, among them the famous boll weevil (*Anthonomus grandis*). This weevil and close relatives are found throughout Middle America and parts of South America (Burke et al. 1986).

## References

BURKE, H. R., W. E. CLARK, J. R. CATE, AND P. A. FRYXELL. 1986. Origin and dispersal of the boll weevil. Entomol. Soc. Amer. Bull. 32: 228–238.

O'BRIEN, C. W., AND G. J. WIBMER. 1981. An annotated bibliography of keys to Latin American weevils, Curculionidae sensu lato (Coleoptera: Curculionidae). Southwest. Entomol. suppl. 2: 1–58.

O'BRIEN, C. W., AND G. J. WIBMER. 1982. Annotated checklist of the weevils (Curculionidae sensu lato) of North America, Central America, and the West Indies (Coleoptera: Curculionidae). Amer. Entomol. Inst. Mem. 34: 1–382.

O'BRIEN, C. W., AND G. J. WIBMER. 1984a. An annotated bibliography of keys to Latin American weevils, Curculionidae sensu lato (Coleoptera: Curculionidae) (Supplement I). Southwest. Entomol. 9: 279–285.

O'BRIEN, C. W., AND G. J. WIBMER. 1984b. Annotated checklist of the weevils (Curculionidae sensu lato) of North America, Central America, and the West Indies—Supplement 1. Southwest. Entomol. 9: 286–307.

# PALM WEEVIL

Curculionidae, Curculioninae,
*Rhynchophorus palmarum*. *Spanish*:
Mayate prieto (Mexico). *Portuguese*:
Aramandaia (Brazil).

One of the world's largest weevils (BL 3–4.9 cm) (fig. 9.16b), the palm weevil frequently draws attention when seen crawling on the trunks of its palm hosts. It is completely shiny black and has a snout about as long as the body is wide and strongly grooved wing covers that are slightly shorter than the abdomen. It is a powerful insect and difficult to dislodge from objects it grasps.

The species is well known throughout Latin America as a depredator of coconut palms, by direct feeding and as a vector of the injurious red-ring nematode (*Rhadinapelenchus cocophilus*) in many parts of its range (Griffith 1987). Females place their eggs, one at a time, in an incision made with the beak at the base of the leaf rachis and in stems. The larvae (fig. 9.16c) bore in via the opening and eventually become large (BL 45–60 mm), fat, white, curved grubs with a short fusiform tail and large mahogany brown head. The back plate of the first thoracic segment is small. They develop while mining in the trunk usually of sick or dead trees but also in young, healthy trees, often killing them. The species is considered the most destructive pest of coconut palms in the West Indies and Central America (Wilson 1963). It is also linked with red-ring viral disease and nematode pests. Pupation occurs on the exterior of stems and leaves in a cocoon made of interlacing fibers cut from the interior of the stem; the larva draws these tightly around itself before pupating. A fair amount is known about its biology because of its economic importance (González and Camino 1974, Hagley 1965). Stages overlap in development time, and it is possible to find all in one tree or any at all times of the year. The total life cycle varies greatly, requiring from 30 to 100 days.

This weevil uses many other palms as hosts, such as palmetto palms (*Sabal*), *Acrocomia, Attalea, Mauritia,* and oil palm (*Elaeis guineensis*), on which it is considered a pest. It also feeds on other plants with fibrous stems such as papaya, large grasses (*Gynerium*), and sugarcane. It usually develops within young plants, from which it has traditionally been harvested as a food item by natives. These food larvae, called *suri* in the Peruvian Amazon and *grou-grou* (or *gru-gru*) elsewhere, may be eaten raw but are more usually fried in a pan with oil and salt (Cowan 1865: 69, DeFoliart 1990, orig. obs.).

A related and similar species, *Rhynchophorus cruentatus* is also a plant pest in Latin America (Wattanapongsiru 1966). This and the palm weevil have been reared in the laboratory (Giblin-Davis et al. 1989).

## References

Cowan, F. 1865. Curious facts in the history of insects. Lippincott, Philadelphia.

DeFoliart, G. 1990. Hypothesizing about palm weevil and palm rhinoceros beetle larvae as traditional cuisine, tropical waste recycling, and pest and disease control on coconut and other palms. Food Ins. Newsl. 3: 1, 3, 4, 6.

Giblin-Davis, R. M., K. Gerber, and R. Griffith. 1989. Laboratory rearing of *Rhynchophorus cruentatus* and *R. palmarum* (Coleoptera: Curculionnidae). Fla. Entomol. 72: 480–488.

González, A., and M. Camino. 1974. Biología y hábitos del mayate prieto de la palma de coco, *Rhynchophorus palmarum* (L.), en la Chontalpa, Tab. Fol. Entomol. Mexicana 28: 13–19.

Griffith, R. 1987. Red ring disease of coconut palm. Plant Dis. 71: 193–196.

Hagley, E. A. C. 1965. On the life history and habits of the palm weevil, *Rhychophorus palmarium*. Entomol. Soc. Amer. Ann. 58: 22–28.

Wattanapongsiru, A. 1966. A revision of the genera *Rhynchophorus* and *Dynamis* (Coleoptera: Curculionidae). Dept. Agric. Sci. Bull. (Bangkok) 1: 1–328.

Wilson, M. E. 1963. Investigations into the development of the palm weevil *Rhynchophorus palmarum* (L.). Trop. Agric. 310: 185–196.

## BEARDED WEEVIL

Curculionidae, Rhynchophorinae, Sipalini, *Rhinostomus* (= *Rhina*) *barbirostris*. *Portuguese:* Broca do tronco do coqueiro (Brazil, larva).

Uniquely, this weevil's snout possesses a dense vestiture of erect, reddish-brown hairs, somewhat resembling a bottle brush. Males use this pubescent proboscis apparently to gain sexual favors from females, whom they stroke assiduously with it.

This black species is fairly large (BL 3.5–5 cm), with a cylindrical body, ridged wing covers, and extra long forelegs, the tibiae of which are heavily spined on the underside (fig. 9.16d). It is usually encountered walking on tree trunks, which males patrol in search of females that are in the act of oviposition. Females drill holes in the bark of newly fallen palm trees, place eggs in these, and then seal them with a gluelike secretion. If multiple males find a female thus engaged, they will spar with their beaks, trying to dislodge each other. The winner of these battles mates with the female (Eberhard 1980).

The species is widespread and moderately common in moist forests throughout Central and South America. Its larval hosts are various palms, including the coconut, which it often damages severely by ovipositing in the stipes that the developing larvae later destroy. Larvae also mine the trunks of mature trees (Bondar 1922).

The larva is moderately large (BL 20–50 mm), cylindrical (not tapered like that of the palm weevil), and curved strongly just behind the thorax. Its head is yellow, and on the back of the first thoracic segment is a large yellowish plate with small incisions on its rear margin. The terminal segments are greatly restricted and retractable within the rest of the body.

### References

BONDAR, G. 1922. Broca do tronco do coqueiro. *In* G. Bondar, Insectos nocivos e molestias do coqueiro (*Cocos nucifer*) no Brasil. Imp. Off. Est. Bahia, Bahia. Pp. 18–31.
EBERHARD, W. G. 1980. Horned beetles. Sci. Amer. 242(3): 166–182.

## JEWELED WEEVILS

Curculionidae, Leptosinae, Entimini, *Entimus*.

These are fairly large (BL 12–40 mm), spectacularly colored weevils (fig. 9.16f), with the black elytra much wider than the prothorax and spangled with numerous brilliant green, gold, or blue scales in longitudinal rows (gray-white hairs replace the scales in one species), which gives them a spectral sparkle, like emeralds or sapphires. There is also a green midline stripe on the prothorax, while the remainder of the body and legs are flecked with tiny green scales. On the elytra, the scales are situated in pits or other depressions. Many species also have granules or tubercles on the dorsal surface. All are winged, the beak is short and robust, and the legs are hairy, especially in males.

Because of their lovely and bright colors, they have been used to make jewelry like other regional beetles. In his day, Cowan (1865) related,

> At Rio Janeiro, the brilliant Diamond Beetle, Entimis nobilis, *is in great request for broches for gentlemen, and ten piasters are often paid for a single specimen. In this city many owners send their slaves out to catch insects, so that now the rarest and most brilliant species are to be had at a comparatively trifling sum. For these splendid insects there is a general demand; and their wing cases are now sought for the purpose of adorning the ladies of Europe—a fashion, it is said, which threatens the entire extinction of this beautiful tribe.*

These weevils have survived and are still prized items in tourist curio displays sold in Rio de Janeiro, Bogotá, and Lima.

The five species are exclusively Neotropical and distributed throughout Central and South America, except Chile and southern Argentina (Vaurie 1951). They are typically lowland or coastal, preferring humid climates.

Most of what is known regarding their biology comes from a work by Bruch (1932) on *Entimus nobilis*. He found it breeding in the tubercular roots of *Stigmaphyllon littorale* (Malpighiaceae) along the banks of the Ríos Plata and Paraná. Adults feed on the leaves of this plant, although they are recorded also on the leaves of bombacaceous trees (*Ceiba, Chorisia, Bombax*) (Bruch 1932).

The female *E. nobilis* doubles the leaves of its host, sticking the edges together with a viscous substance, to form a cupola for its eggs. After hatching, the young larvae drop to the ground and burrow through the soil in search of tubers in which to complete their development. The mature C-shaped larvae are strongly wrinkled.

### References

BRUCH, C. 1932. Metamorfosis de *Entimus nobilis* Oliv. (Coleopt., Curculionidae). Rev. Entomol. 2: 179–185.

COWAN, F. 1865. Curious facts in the history of insects; including spiders and scorpions. Lippincott, Philadelphia.

VAURIE, P. 1951. Revision of the genus *Entimus* with notes on other genera of Entimini (Coleoptera: Curculionidae). Rev. Chilena Entomol. 1: 147–170.

## BRENTIDS

Brentidae. Giraffe beetles.

These relatives of the true weevils are typified by an elongate, very slender form. Most are black with parallel, longitudinal yellow lines on the wing covers, and the body surface is very smooth and slippery. They are generally small to medium-sized (BL 8–50 mm), but there is considerable variation in size among individuals of the same species due to the different environmental conditions experienced by the larvae. The larvae are known to bore under the bark of dead or dying hardwoods, but the detailed biologies of few species have been elucidated. *Brentus anchorago* (fig. 9.16e) is one of the most common and widespread species, occurring in lowland dry to wet forests. The adults are frequently found in great numbers (up to 400 per square meter) under the loose bark of dead *Bursera, Simarouba,* and *Pseudobombax* and logs of other trees (Johnson 1983a).

Promiscuous reproductive aggregations often occur on host trees, and intense competition and fighting ensues between males for females and between females for suitable oviposition areas. Larger males seem to have greater mating success than smaller ones. Females lay their eggs in holes bored with the jaws at the tip of their beak into the surface of dead trees (Johnson 1983b).

### References

JOHNSON, L. K. 1983a. *Brentus anchorago* (bréntido, brentid beetle). *In* D. H. Janzen, ed., Costa Rican natural history. Univ. Chicago Press, Chicago. Pp. 701–703.

JOHNSON, L. K. 1983b. Reproductive behavior of *Claeoderes bivittata* (Coleoptera; Brentidae). Psyche 90: 135–149.

# 10 MOTHS AND BUTTERFLIES

Lepidoptera.

This order is immediately recognized by the presence of minute scales on the wing membranes (the easily detached "dust" that imparts color to them) and a long, siphoning proboscis that coils up like a watch spring under the head when not employed in imbibing nectar and other liquids, such as the juices of rotting fruits (in some, the mouthparts are vestigial). Division of the Lepidoptera into butterflies ("Rhopalocera") and moths ("Heterocera") is a misleading, although useful, dichotomy. The former actually represent only one small side branch of the order that is specialized for diurnal existence. Correlated with flying in the sun are the expansive, often gaily colored wings, although the shade dwellers often have subdued or transparent wings. The larvae of most butterflies pupate in naked, suspended chrysalids, and the adults typically have knobbed antennae and lack a well-defined device for coupling the fore and hind wings. Moths, for the most part, are nocturnal and have relatively larger bodies with somber-colored wings. Their larvae pupate in cocoons or in cells in the ground, and most possess special structures for hooking the hind wing to the fore wing. Antennae are threadlike or plumed but rarely knobbed. These distinctions meet with many exceptions (many moths are brightly colored and diurnal, for example) but are generally adequate to characterize the groups, unequal and contrived as they may be.

Butterflies and moths exhibit many unusual behavioral traits as adults or larvae. The strange attraction of the latter to lights is still poorly understood (see moths, below). Adults of both groups also seek moisture and some nutrients from mud and moist sand (especially that laced with urine or soap residues), bird droppings, dung and carrion, rotting fruits and fungi, and even human perspiration. Butterflies are best known for this habit because of their visibility during the day, but moths of many kinds also "puddle" and siphon similarly unsavory liquids at night (Adler 1982, Becker 1983, Downes 1973). This activity is almost exclusively limited to males, which suggests a need for acquiring some substance (perhaps amino acids and sodium; Arms et al. 1974, Adler and Pearson 1982) of unknown importance in their sex lives.

Many adult butterflies and moths are known to require specific chemicals in their mating displays or to render them toxic or distasteful to predators (Brower 1984). The former are usually metabolized into volatile perfumes, dispersed from various secretory tissues (Barth 1960) at the base of hairlike scales located either on eversible organs on the body (coremata, chaetosemata) or directly on the wing surfaces (androconia). These substances are pyrrolizidine alkaloids that the insects obtain from droplets exuding from withered or injured plant tissues (Pliske 1975a). Such are greedily sought by many Ithomiinae, Danainae, Ctenuchinae, and Arctiinae, which also are major pollinators of these plants (Pliske 1975*b*). Species in the borage family (*Heliotropium fedegoso, Tournefortia*), the composite family (*Eupatorium, Senecio*), pea family (*Crotalaria*), and dogbane family (*Parsonsia*) are now known to provide these essential chemicals, but there may be others. A phenylacetaldehyde in the bladder flower (*Araujia sericofera*, Asclepiadaceae) also attracts moths (Cantelo and Jacobson 1979).

Although the variety of lepidopteran colors and their arrangements are almost infinite, and the majority of these seem to act as camouflage, a great many otherwise segregate into basic types that are widely imitated among the families. These primary patterns are broadly recognized as follows: "tiger," basically orange with yellow, brown, red, and black streaks (pl. 3f); "red," generally deep orange, with oblique bars of black and yellow spots toward the apex of the fore wing; "blue," nearly all black (or bluish-black), often with broad, oblique fore wing or hind wing spots of red, blue, or green; and "transparent," clear or pale membranes and dark veins. These sometimes merge or overlap, and there are secondary types of all, such as the "zebra" striped variant of the "tiger," with simple alternating, longitudinal yellow or orange bars on a black background (zebra butterfly) or the "orange bordered" variant of the transparent which has broad, black-bordered margins to the wings (ithomiines), and others specifically resembling various models such as lycid beetles and vespid wasps. The latter "wasp types" with transparent, veined, or smoky wings may represent distinct categories evolved as Batesian mimics but are possibly related to the "transparent" and "blue" primary variants. All four possibly came into being, at least in butterflies, as imitations of other distasteful or stinging insects, but some researchers have postulated their origins as cryptic images, blending with the interplay of light and vegetation and the background against which the butterfly flies and is visible to predators (Papageorgis 1975). Transparent forms fly in dappled sunlight near the ground in the understory, yellow, black, and orange-striped (tiger) forms fly a bit higher (7–13 m), blue forms fly in the upper canopy, and orange and yellow forms fly above the canopy and at the forest margin.

A few patterns mimic nocturnal models and are apparently effective in obtaining protection for their bearers when at rest during the day: *Opharus* (arctiid moth) to *Pyrophorus* (elaterid beetle); *Endobrachus revocans* and relatives (megalopygid moths) and *Cratoplastis* (= *Automolis*) *diluta* (arctiid moths) to *Achroblatta luteola* (cockroach).

These patterns are employed widely in several unrelated families in Müllerian mimicry complexes with toxic or noxious chemical qualities: among the butterflies, these are commonly found in the Heliconiinae, Ithomiinae, Danainae, Acraeinae, Papilionidae, and more rarely, in the Riodininae, Nymphalinae, and Pieridae. The moth families with such colors are the Arctiidae (Ctenuchinae and Pericopinae), Zygaenidae, Dioptidae, Castniidae, and Agaristidae. The chemicals are usually cardiac glycosides and alkaloids obtained during the larval feeding periods (Rothschild 1972) or as adults (Brown 1984). They have an unpleasant, musky smell, even to humans, and are dissolved in yellow-colored body fluids that are bitter tasting. Like the adults, lepidopteran larvae exhibit a similar variety of defensive colors and patterns, mimetic and otherwise (Haviland 1925, Nentwig 1985). Toxic substances may also be contained in body fluids and tissues or produced by specialized poison glands (Quiroz 1978).

Lepidopterous larvae are the familiar caterpillars, with an elongate, cylindrical body equipped with several (usually four intermediate and one anal) pairs of stumpy walking legs in addition to small, segmented thoracic legs used primarily as aids in feeding. The integument is naked and smooth or variously adorned with fine to coarse hairs or elaborate horns or tubercles. The head may also bear a pair of long spines in butterfly caterpillars, but these are never present on the head of moth larvae, although a prothoracic pair will often project forward over the head and seem to arise from it. Caterpillars lead an absolutely different life from the adults and are much more ecologically diverse (Janzen 1988). Those of a large number of species are of agricultural importance when abundant on cultivated plants (Margheritis and Rizzo 1965). They may be so numerous at times that their dry droppings fall like raindrops, making an audible patter on dry leaves on the ground.

Pupae are usually simple, drably colored, and hidden in most moths or variously shaped and colored to resemble leaves and other objects and suspended nude in the environment, as in the chrysalids of butterflies. Many tropical butterfly pupae are adorned with gold or silver markings that resemble reflective droplets of rainwater or dew.

The Lepidoptera form an immense order (one-tenth of all animal species) and are particularly well developed in the Neotropics (Heppner in press), where at least 50 percent of the world fauna resides (Holloway 1984). In addition to being divided into butterflies and moths, the order is often informally broken in "macrolepidoptera" (families of larger species) and "microlepidoptera" (families of small moths, although many of these may actually be larger than some smaller macrolepidoptera). There are an estimated 43,000 species of Neotropical macrolepidoptera (Watson and Goodger 1986). No complete identification work exists, although Seitz (1907–1939) covers most of some major groups of macrolepidoptera. The *Atlas of Neotropical Lepidoptera* (Heppner 1981–) is now being issued in parts and will ultimately provide illustrations and basic information on all the species. In addition to general entomology treatments (see chap. 1), works of fundamental significance are by Aurivillius and Wagner (1911–1939) and Forbes (1939, 1942). Some additional useful or popular general treatments are those of Bourquin (1945) and Raymond (1982).

Generally common and conspicuous, the Lepidoptera are the subject of much experimental research in the quest for knowledge of general biological principles (Silberglied 1977, Common 1970). Quite a number are of medical importance, both beneficially (source of drugs) and as pathological agents (urtication) (Lamas and Pérez 1987). The ubiquity and visibility of the order has also attracted the incorporation of certain species into the cultural practices of some Latin American peoples (Beutelspacher 1989).

Adult butterflies and moths often have an entourage of parasitic or commensal mites living on their bodies (Treat 1975).

## References

ADLER, P. H. 1982. Soil- and puddle-visiting habits of moths. Lepidop. Soc. J. 36: 161–173.

ADLER, P. H., AND D. L. PEARSON. 1982. Why do male butterflies visit mud puddles? Can. J. Zool. 60: 322–325.

ARMS, K., P. FENNY, AND R. C. LEDERHOUSE. 1974. Sodium: Stimulus for puddling behavior by tiger swallowtail butterflies, *Papilio glaucus*. Science 185: 372–374.

AURIVILLIUS, C., AND H. WAGNER. 1911–1939. Lepidopterorum catalogus. Pts. 1–94. Junk, Berlin. Began publishing again in 1987 as "New Series," J. Heppner, ed., by Brill, Leiden.

BARTH, R. 1960. Orgãos odoríferos dos Lepidópteros. Min. Agric., Serv. Flor., Parq. Nac. Itatiaia (Rio de Janeiro) Bol. 7: 1–157.

BECKER, V. O. 1983. ¿Por que as borboletas se juntam nos lugares úmidos? Brasil Florestal 13(53): 49–50.

BEUTELSPACHER, C. R. 1989. Las mariposas entre los antiguos Mexicanos. Fondo Cult. Econ., Mexico.

BOURQUIN, F. 1945. Mariposas Argentinas: Vida, desarrollo, costumbres y hechos curiosos de algunos lepidópteros argentinos. Pub. by author, Buenos Aires.

BROWER, L. P. 1984. Chemical defence in butterflies. *In* R. I. Vane-Wright and P. R. Ackery, eds., The biology of butterflies. Academic, London. Pp. 109–134.

BROWN, JR., K. S. 1984. Adult-obtained pyrrolizidine alkaloids defend ithomiine butterflies against a spider predator. Nature 309: 707–709.

CANTELO, W. W., AND M. JACOBSON. 1979. Phenylacetaldehyde attracts moths to bladder flower and to blacklight traps. Environ. Entomol. 8: 444–447.

COMMON, I. F. B. 1970. Lepidoptera. *In* CSIRO, ed., The insects of Australia, a textbook for students and research workers. Melbourne Univ., Carleton. Pp. 765–866.

DOWNES, J. A. 1973. Lepidoptera feeding at puddle-margins, dung, and carrion. Lepidop. Soc. J. 27: 89–99.

FORBES, W. T. M. 1939. The Lepidoptera of Barro Colorado Island, Panama. Mus. Comp. Zool. (Harvard Univ.) 85(4): 97–322, pls. 1–8.

FORBES, W. T. M. 1942. The Lepidoptera of Barro Colorado Island, Panama. No. 2. Mus. Comp. Zool. (Harvard Univ.) 90(2): 265–406, pls. 9–16.

HAVILAND, M. D. 1925. Defensive colour and pattern in four caterpillars from British Guiana. Royal Entomol. Soc. London Trans. 73: 575–578.

HEPPNER, J. B., ed. 1981–. Atlas of Neotropical Lepidoptera: An illustrated catalog of described Neotropical species. Brill, Leiden.

HEPPNER, J. B. In press. Lepidoptera family classification: A guide to the higher categories, world diversity and literature resources of the butterflies and moths. Flora and Fauna, Gainesville.

HOLLOWAY, J. D. 1984. The larger moths of the Gunung Mulu National Park: A preliminary assessment of their distribution, ecology, and potential as environmental indicators. Sarawak Mus. J. 30(51): 149–190.

JANZEN, D. H. 1988. Ecological characterization of a Costa Rican dry forest caterpillar fauna. Biotropica 20: 120–135.

LAMAS, G., AND E. PÉREZ. 1987. Lepidópteros de importancia médica. Diagnóstico 20: 121–125.

MARGHERITIS, A. E., AND H. F. E. RIZZO. 1965. Lepidópteros de interés agrícola. Ed. Sudamericana (Coll. El Mundo Agric., Ser. Plag. Enferm.), Buenos Aires.

NENTWIG, W. 1985. A tropical caterpillar that mimics faeces, leaves and a snake (Lepidoptera; Oxytenidae: *Oxytenis naemia*). J. Res. Lepidop. 24: 136–141.

PAPAGEORGIS, C. 1975. Mimicry in Neotropical butterflies. Amer. Sci. 63: 522–532.

PLISKE, T. E. 1975a. Attraction of Lepidoptera to plants containing pyrrolizidine alkaloids. Environ. Entomol. 4: 455–473.

PLISKE, T. E. 1975b. Pollination of pyrrolizidine alkaloid-containing plants by male Lepidoptera. Environ. Entomol. 4: 474–479.

QUIROZ, A. D. 1978. Venoms of Lepidoptera. *In* S. Bettini, ed., Arthropod venoms. Springer, Berlin. Pp. 555–611.

RAYMOND, T. 1982. Mariposas de Venezuela. Ed. Corpoven, Caracas.

ROTHSCHILD, M. 1972. Colour and poisons in insect protection. New Scient. (11 May): 318–320.

SEITZ, A., ed. 1907–1939. The Macrolepidoptera of the world (Kernen, Stuttgart). Vols. 5–8.

SILBERGLIED, R. E. 1977. Communication in the Lepidoptera. *In* T. A. Sebeok, ed., How animals communicate. Indiana Univ. Press, Bloomington. Pp. 361–402.

TREAT, A. E. 1975. Mites of moths and butterflies. Cornell Univ. Press, Ithaca.

WATSON, A., AND D. T. GOODGER. 1986. Catalogue of the Neotropical tiger-moths. Brit. Mus. Nat. Hist. Occ. Pap. Syst. Entomol. 1: 1–71.

# MOTHS

Lepidoptera, "Heterocera." *Spanish:* Mariposas nocturnas (frequently mariposas), palomas, palomillas, polillas (small pest species). *Portuguese:* Mariposas, bruxas (Brazil); traças (small pest species).

The vast majority of Lepidoptera are considered moths, the uneven distinction from butterflies notwithstanding. They abound in tremendous diversity throughout Latin America, possibly in part because of the higher variety of niches, especially food types, available in the tropics there compared to temperate areas (Ricklefs and O'Rourke 1975).

Except for some popular groups, such as the saturniids and sphingids (Janzen

1984) and economic species, their biology is much less known than that of butterflies. The attraction of moths to lights is a familiar phenomenon but one whose causes are still poorly understood. Various theories have been proposed either to explain how the insect arrives at the light or why it is initially attracted. It has been postulated that moths, using fixed celestial beacons such as the moon and attempting to maintain a constant angle relative to the beacon's position, inevitably fly a spiral path into the artificial illumination, which is mistaken for a distant natural point of light (von Buddenbrock 1917). They also may seek light as an indication of a place of safety, perceiving a dark area next to the light and flying to it in an attempt to escape (Hsiao 1973). It has also been suggested that the cause of the orientation of moths and other night-flying insects toward artificial light is the need to find open flight routes; the unnatural radiance from the light source is a bright spot interpreted by the insect as open space (Mazokhin-Porshnyakov 1969). Another, more novel theory holds that the light source electrically stimulates chemicals in the air nearby. These are substances important to the moth's biology and ultimately attract it when perceived with antennal receptors, the eyes having no direct role at all (Callahan 1975).

Night-flying moths encounter lower ambient temperatures than diurnal insects and have evolved internal control of body temperature to permit activity even in very cold climates. Thoracic flight temperatures, several degrees centigrade above the air, are generated by muscle vibration and maintained by insulation from a thick, integumentary scale vestiture, among other factors (Bartholomew and Heinrich 1973, Bartholomew et al. 1981).

Many moths are brightly colored and are generally unpalatable to birds, their colors and actions acting as signals of distastefulness due to poisons or noxious substances contained in their bodies (Blest 1957). Similarly, numerous species also obtain protection from predators by mimicking wasps or bees (Collins and Watson 1983).

Unlike the diurnal butterflies, which rely to a large degree on visual cues, moths appear more often to find floral food sources mainly by odor at night (Brantjes 1978). Mate encounter and recognition is also mediated by odor or pheromonal clues secreted by either sex (Greenfield 1981), more than by color or activity patterns.

Some moths have evolved sound production. They produce trains of high-pitched clicks when touched or when exposed to ultrasonic pulses such as those of echo-locating bats. Such predators usually avoid or reject clicking moths more strongly than those that remain silent. The sound acts as a kind of auditory aposematic signal by unpalatable moths, such as arctiids and ctenuchids (Dunning 1968). High-frequency sounds emitted by noctuids may "jam" bat echolocation and aid in the moth's escape during bat attacks (Fullard et al. 1979). These same moths may also hear such sounds from foraging bats and make evasive flight movements (Roeder and Treat 1961, Fenton and Fullard 1981).

Many moths are harmful, principally in the larval stage, to agriculture and even directly to humans. In many areas, the word "moth" itself is synonymous with the destructive clothes moth. Among the injurious Heterocera are many urticating caterpillars with venomous or allergenic spines and hairs (Weidner 1936, Jörg 1939, Pesce and Delgado 1971). A Venezuelan saturniid (probably *Lonomia achelous*) injects a powerful anticoagulant and can cause serious bleeding in humans (Marsh and Arocha-Piñango 1971); the toxin involved is more reminiscent of hemorrhagic snake venom than the substance usually injected by urticating lepidopterous larvae.

## References

BARTHOLOMEW, G. A., AND B. HEINRICH. 1973. A field study of flight temperatures in moths in relation to body weight and wing loading. J. Exper. Biol. 58: 123–135.

BARTHOLOMEW, G. A., D. VLECK, AND C. M. VLECK. 1981. Instantaneous measurements of oxygen consumption during pre-flight warm-up and post-flight cooling in sphingid and saturniid moths. J. Exper. Biol. 90: 17–32.

BLEST, A. D. 1957. The evolution of protective displays in the Saturnioidea and Sphingidae (Lepidoptera). Behaviour 11: 257–309.

BRANTJES, N. B. M. 1978. Sensory responses to flowers in night flying moths. Linnaean Soc. Symp. Ser. 6 (The pollination of flowers by insects): 13–19.

CALLAHAN, P. S. 1975. Tuning in to nature. Devin-Adair, Old Greenwich, Conn.

COLLINS, C. T., AND A. WATSON. 1983. Field observations on bird predation on Neotropical moths. Biotropica 15: 53–60.

DUNNING, D. C. 1968. Warning sounds of moths. Zeit. Tierpsychol. 25: 129–138.

FENTON, M. B., AND J. H. FULLARD. 1981. Moth hearing and the feeding strategies of bats. Amer. Sci. 69: 266–275.

FULLARD, J. H., M. B. FENTON, AND J. A. SIMMONS. 1979. Jamming bat echolocation: The click of arctiid moths. Can. J. Zool. 57: 647–649.

GREENFIELD, M. D. 1981. Moth sex pheronomes: An evolutionary perspective. Fla. Entomol. 64: 4–17.

HSIAO, H. S. 1973. Flight paths of night-flying moths to light. J. Ins. Physiol. 19: 1917–1976.

JANZEN, D. H. 1984. Two ways to be a tropical big moth: Santa Rosa saturniids and sphingids. Oxford Surv. Evol. Biol. 1: 85–140.

JÖRG, M. E. 1939. Dermatosis lepidopterianas (Segunda nota). Nov. Reun. Soc. Argentina Pat. Reg. (Mendoza) 3: 1617–1639.

MARSH, N. A., AND C. L. AROCHA-PIÑANGO. 1971. Observations on a saturniid moth caterpillar causing severe bleeding in man. Royal Entomol. Soc. London Proc. 36(2): 9–10.

MAZOKHIN-PORSHNYAKOV, G. A. 1969. Insect vision. Plenum, New York.

PESCE, H., AND A. DELGADO. 1971. Poisoning from adult moths and caterpillars. In W. Bücherl and E. E. Buckley, eds., Venomous animals and their venoms. 3: 119–156. Academic, New York.

RICKLEFS, R. E., AND K. O'ROURKE. 1975. Aspect diversity in moths: Temperate-tropical comparison. Evolution 29: 313–324.

ROEDER, K. D., AND A. E. TREAT. 1961. The detection and evasion of bats by moths. Amer. Sci. 49: 135–148.

VON BUDDENBROCK, W. 1917. Die Lichtkompass-bewegungen bei den Insekten, inbesondere den Schmetterlingsraupen. Heidelberg. [Not seen.]

WEIDNER, H. 1936. Beiträge zu einer Monographie der Raupen mit Gifthaaren. Zeit. Angew. Entomol. 23: 432–484.

# WILD SILK MOTHS

Saturniidae. Giant silk moths, saturnians.

These elegant moths inspire admiration and awe because of their great size (WS up to 20 cm, as in *Arsenura ponderosa*) and sumptuous color designs on broad wings. They are most often seen around electric lights, to which most are strongly attracted. Males bear highly developed, featherlike antennae that are extremely sensitive to the airborne molecules of the female sexual pheromones. With these organs, they may detect a female over a distance of several kilometers and fly upwind to find her. The female's antennae may also be feathered (but much less so than the male's) or merely pectinate or simple.

Adults of all species lack functional mouthparts and do not feed, relying on the food stores accumulated by their voracious larvae. Saturniid caterpillars are robust (some immense), obese creatures when mature. Most bear rows of swollen tubercles or spines, the latter often tipped with urticating barbs or hairs that can cause painful skin rashes on the human skin. Food plants are extremely diverse, even within species, and belong to such disparate families as Anacardiaceae, Rubiaceae, Fabaceae, and Flacourtiaceae for the Neotropical species.

Prior to pupation, saturniid caterpillars commonly spin a rigid, silken cocoon that is suspended from the branches or leaves of the host plant or placed in refuse on the ground (Raymundo 1919); many pupate without silk in an earthen cell. The shape of cocoons is characteristic of the species, as is the quality and quantity of silk incorporated into them. Those of one species (*Rothschildia aurota*) have even been produced under culture for their silk, although not very successfully. It is much coarser and therefore not as desirable as that from the domestic silk moth (*Bombyx mori*). The latter belongs to a distinct family (Bombycidae) and has been cultivated in various parts of Latin America, with varied success.

Saturniids utilize several strategies of defense, including simple cryptic pattern (*Loxolomia*), distastefulness, which is signaled by brightly colored organs (*Dirphia*), toxic abdominal hairs (*Hylesia*, female only), and elaborate threat displays involving false eyespots (*Automeris*) (Blest 1960). Many species experience high levels of parasitism in the larval and pupal stages from chalcid wasps, tachinid flies, and others, against which their protective ploys seem to have little effect.

There are more than 800 species in the region (Lemaire pers. comm.), most of which are broadly reviewed in the works of Lemaire (1978) and Michener (1952). Some local treatments of saturniid faunas (Janzen 1982, Lemaire and Venedictoff 1989) contain useful biological and zoogeographic information as well as present a cross section of the types found in the Neotropics.

### References

BLEST, A. D. 1960. A study of the biology of saturniid moths in the Canal Zone Biological Area. Smithsonian Ann. Rep. 1959: 447–464.
JANZEN, D. H. 1982. Guía para la identificación de mariposas nocturnas de la familia Saturniidae del Parque Nacional Santa Rosa, Guanacaste, Costa Rica. Brenesia 19/20: 255–299.
LEMAIRE, C. 1978. Les Attacidae Américains. (The Attacidae of America [= Saturniidae]: Attacinae.) C. Lemaire, Neuilly-sur-Seine.
LEMAIRE, C., AND N. VENEDICTOFF. 1989. Catalogue and biogeography of the Lepidoptera of Ecuador. I. Saturniidae. With a description of a new species of *Meroleuca* Packard. Allyn Mus. Entomol. Bull. 129: 1–60.
MICHENER, C. D. 1952. The Saturniidae (Lepidoptera) of the Western Hemisphere: Morphology, phylogeny and classification. Amer. Mus. Nat. Hist. Bull. 98: 335–502.
RAYMUNDO, B. 1919. Notícia sobre alguns lepidópteros serígenos do Brasil. Rev. Tribunais, Rio de Janeiro.

### Regal and Imperial Moths

Saturniidae, Ceratocampinae
    (= Citheroniinae), *Citheronia* and *Eacles*.
    *Portuguese:* Caveiras (Brazil). Horned
    devils (larvae).

One provocative common name of regal moths (*Citheronia*), "horned devils," comes from the appearance of their larvae. These giant (BL to 10 cm), varicolored (often orange or red), fearsome creatures have enormous, curved, hornlike protuberances on all the thoracic segments (fig. 10.1b). The naked skin is marked with contrasting oblique bars or other strongly contrasting patterns. The natural host plants are not entirely known, but larvae of various species have been found on wild cotton and *Pinus* and reared in captivity on Peruvian pepper (*Schinus molle*, Anacardiaceae).

These two genera are the showiest of around 160 neotropical species in the subfamily. Adult *Citheronia* (fig. 10.1a) are fairly large, stout moths (WS 8–10 cm) with rich umber bodies and hind wings; the ground colors of the fore wings are gray to black, the veins reddish-brown and the intervenous areas marked with elongate to oval white or cream spots. The abdominal segments are thinly banded by cream basally. The fore wings tend to be somewhat more elongate in this genus, in contrast to the broader, slightly falcate wings of *Eacles*, imperial moths. Adults of *Eacles* also often have more blotched wing patterns and faint ocellar spots in the middle of each wing, although they are otherwise very similar in most of the species (fig. 10.1c). Some also are strongly yellow at the base of the wing.

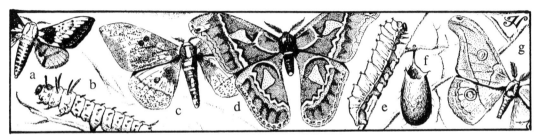

**Figure 10.1   WILD SILK MOTHS (SATURNIIDAE).** (a) Regal moth (*Citheronia laocoon*), male. (b) Regal moth, larva. (c) Imperial moth (*Eacles imperialis decoris*). (d) Window-winged saturnian (*Rothschildia orizaba*), male. (e) Window-winged saturnian, larva. (f) Window-winged saturnian, cocoon. (g) Copaxa (*Copaxa lavendera*).

The larvae of imperial moths are little different from those of *Citheronia* except for a lack (in most but not all) of the hornlike processes on the first thoracic segment and the presence of very evident, fairly long, white body hairs (in some but not all). Natural food plants for these are also mostly a mystery, but a few are known to be broad-leaved trees and shrubs, *Pinus*, and *Cochlospermum*, although the larvae thrive on many kinds of woody plants, including cultivated varieties such as guava, *Spondias*, and many others.

About 20 species of *Citheronia* and 16 species of *Eacles* occur throughout Latin America (Lemaire 1988).

**Reference**

LEMAIRE, C. 1988. The Saturniidae of America: Ceratocampinae. Ed. C. Lemaire, Neuilly-sur-Seine.

**Window-Winged Saturnians**

Saturniidae, Saturninae, Saturnini,
 *Rothschildia. Spanish:* Cuatro ventanas.
 *Nahuatl:* Itzpapálotl (Mexico).

Large, translucent, triangular or oval areas in each wing immediately identify this genus of large (WS 11–13 cm) saturnians (fig. 10.1d, pl. 2c). The wings otherwise display rippling shades of soft brown, broken by an irregular white line traversing both fore and hind wing just outside the triangles. The wings also have light margins. The abdomen is marked laterally with a chainlike series of white spots.

Approximately twenty-five species are known, occurring from Mexico to Argentina. Sixteen species dwell in Andean mountain forests (Lemaire 1978). Some confusion in their identification is caused by environmentally controlled color polymorphisms (Janzen 1984). Rust-colored or chocolate forms of a few species occur. The time of flight of these generally follow seasonal shifts in background color, against which the moths are cryptic. During hot, dry conditions, when the vegetation is relatively pale, bleached, or sparse, those with the lighter shade are seen; when cool, wet conditions prevail, and lush, deep green leaves are abundant, the darker types are present.

Mature larvae (fig. 10.1e) are large (79–85 mm), mostly green, some with light segmental rings or bands, transverse black bars, or a light-colored longitudinal ridge running along the sides; the typical saturnian tubercles are usually present but very small and bear harmless spines. They feed on a wide variety of deciduous trees and shrubs in nature, including *Spondias*, *Jatropha*, *Sapium*, *Baccharis*, *Croton*, and *Jacaranda*. They also are minor pests on cultivated plants, such as manioc, cashew, and castor bean (*Ricinus communis*), and can be reared on many common garden trees, including willow, ash, *Prunus* spp., and

*Ligustrum* (d'Almeida 1957; Urban and Lucas de Oliveira 1972). As with most large saturnians, the mortality rate in natural populations is very high, nearly 90 percent in a Salvadoran species studied by Quezada (1967).

The cocoons (fig. 10.1f) are hard and suspended by a petiole attached to one side of the upper end. Cocoons of *R. aurota* were once utilized as a source of commercial silk in Brazil (Girard 1874, Ribeiro 1948).

To the Aztecs of central Mexico, these moths were identified with fire because of the gray, wedge shape wing spots that reminded them of obsidian or flint (with which fire is started) and called *itzpapálotl* (*itzlis* = flints + *papálotl* = moth). The moth's undulating flight and wavy lines in the wing pattern were also likened to the dancing flames. Evidence for this symbolism is found in numerous bas-reliefs and designs (see fig. 1.11) found in the architecture of these people (Hoffmann 1918).

## References

D'ALMEIDA, R. F. 1957. Brevas notas sôbre o gênero *Rothschildia* Grote, 1897 (Lepidoptera, Saturniidae). Mus. Nac. Rio de Janeiro Bol. (Nov. Ser.) 171: 1–47.

GIRARD, M. M. 1874. Le ver a soie Brésilien. Soc. Acclimatation (Ser. 3) Bull. 1: 183–203.

HOFFMANN, C. C. 1918. Las mariposas entre los antiguos mexicanos. Cosmos 1. Unpaginated.

JANZEN, D. H. 1984. Weather-related color polymorphism of *Rothschildia lebeau* (Saturnidae). Entomol. Soc. Amer. Bull. 30: 16–20.

LEMAIRE, C. 1978. *Rothschildia* Grote. *In* C. Lemaire, Les Attacidae Americains (The Attacidae of America [= Saturniidae]). Ed. C. Lemaire, Neuilly-sur-Seine. Pp. 29–103.

QUEZADA, J. R. 1967. Notes on the biology of *Rothschildia ? aroma* (Lepidoptera: Saturniidae), with special reference to its control by pupal parasites in El Salvador. Entomol. Soc. Amer. Ann. 60: 595–599.

RIBEIRO, B. L. 1948. Contribuição para o conhecimento da bionomia de "*Rothschildia aurota*" (Cramer, 1775) (Lepidoptera, Saturnidae). Rev. Brasil. Biol. 8: 127–141.

URBAN, D., AND B. LUCAS DE OLIVEIRA. 1972. Contribuição ao conhecimento da biologia de *Rothschildia jacobaeae* (Lepidoptera, Saturniidae). Acta Biol. Paranaense 1: 35–49.

## Copaxas

Saturniidae, Saturniinae, Saturniini, *Copaxa. Spanish:* Canelas (Peru).

This is a diverse genus of thirty entirely Neotropical, mountain-dwelling saturnians characterized by multiple transparent, rounded to elongate spots in the center of each wing (fig. 10.1g), or a simple spot (Lemaire 1978). The fore wings are usually falcate at the tip, especially in the male, and mostly uniform, somber, warm brown, variously broken by darker, straight lines running obliquely between the spots and wavy lines external to the hind wing spot. Most are fairly large (WS 9–13 cm) and are nocturnal fliers, often attracted to artificial light.

Not many immatures are described. The larvae of most feed on plants in the family Lauraceae, are cryptically colored, and are gregarious; the exception are larvae of *Copaxa cydippe*, who feed on pine, are bold green-and-white striped, and are solitary (Wolfe 1988). The larva of *C. moinieri* in Costa Rica is a green color very similar to that of the leaves of its food plant, *Ocotea veraguensis* (Lauraceae). It is spined but not urticating. Pupation occurs in an open mesh cocoon attached to a branch or leaf of the host well above the ground (Janzen 1982: 277–278). The fully grown larvae of *C. decrescens* from Brazil are also green, with a light lateral line; the young larvae are gregarious on the food plant, which is avocado (*Persea americana*, Lauraceae) (Dias 1988).

## References

DIAS, M. M. 1988. Estágios imaturos de *Copaxa decrescens* Walker, 1855 (Lepidoptera, Saturniidae). Rev. Brasil. Entomol. 32: 263–271.

JANZEN, D. H. 1982. Guía para la identificación de mariposas nocturnas de la familia Saturniidae del Parque Nacional Santa Rosa, Guanacaste, Costa Rica. Brenesia 19/20: 255–299.

Lemaire, C. 1978. *Copaxa* Walker. *In* C. Lemaire, Les Attacidae Américains (The Attacidae of America [= Saturniidae]). Ed. C. Lemaire, Neuilly-sur-Seine. Pp. 147–205.

Wolfe, K. L. 1988. Aspectos inusuales de la biología de *Copaxa cydippe* Druce (Lepidoptera: Saturniidae). Fol. Entomol. Mexicana 75: 47–54.

## Eyed Saturnians

Saturniidae, Hemileucinae, Hemileucini, *Automeris* and relatives. *Portuguese:* Olhos de pavão (Brazil).

This is a group (Lemaire 1971–1974) of 180 large (WS 5–15 cm) and beautiful moths, almost all of which have a conspicuous eyespot in the center of each hind wing; the spot has a black center and is surrounded first by multicolored rings in various bright hues and then by one complete and two partial, wavy, concentric circles alternating with black (fig. 10.2a). In the best-known genus, *Automeris* (and also *Automerella* and *Pseudautomeris*), spot colors are bluish or pink but never red or orange, as is found in the other related genera comprising this group, such as *Hyperchiria* and *Gamelia*. Most species inhabit mountain forests (between 600–800 m elevation); relatively few live in the warm lowlands. The fore wings are almost unicolorous buff, but most have a faint circular spot about halfway along the wing near the leading edge and a pair of diagonal bars running across the middle of the wing.

The larvae (fig. 10.2b, pl. 2d) are mostly green, usually with colored or black bands running along the back and sides; some have variegated color patches of cream or pink. In one common genus (*Leucanella*), the skin is solid black and the spines contrasting bright yellow. They are all densely clothed with branched spines, especially dorsally (in place of the usual tubercles of other Saturniinae), the fine branches of which are articulated and tipped with small spinules that are highly venomous to the touch. These caterpillars are catholic in their tastes, various species choosing trees and shrubs in several unrelated families. Some recorded hosts are coral trees (*Erythrina*), figs (*Ficus*), oaks (*Quercus*), *Ceanothus*, pepper (*Schinus*), tamarind (*Tamarindus*), *Hibiscus*, and cashew (*Anacardium*) (Lemaire 1971–1974). Pupation occurs in a silken cocoon placed among leaves on the ground, among twigs, or in detritus.

At rest, these moths resemble dead leaves, their drab upper wings completely hiding the eye-mimicking pattern of the lower wings. If disturbed, they elevate the fore wings, suddenly exposing the false eyes. The behavior has been experimentally observed to cause fright in potential predators, which doubtlessly are intimidated by a supposed adversary larger than themselves (Blest 1957). Other protective devices include immobility ("playing dead") and production of noxious secretions from the abdomen, well developed in *Hyperchiria*,

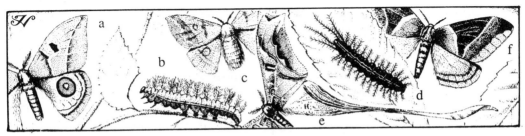

**Figure 10.2    WILD SILK MOTHS (SATURNIIDAE).** (a) Eyed saturnian (*Automeris illustris*), male. (b) Eyed saturnian (*Automeris* sp.), larva. (c) Hylesia (*Hylesia lineata*), female. (d) Hylesia, larva. (e) Swallow-tailed moth (*Copiopteryx semiramis banghaasi*). (f) Dirphia (*Dirphia avia*).

and the abundant hairs on the body of most species (Blest 1963).

The adults remain inactive during the day, taking flight at night in pursuit of mates, oviposition, and dispersion, although males of at least three species are known to seek females during the late morning hours (Marquis 1984). They are often attracted to artificial light.

## References

BLEST, A. D. 1957. The function of eyespot patterns in the Lepidoptera. Behaviour 11: 209–256.

BLEST, A. D. 1963. Longevity, palatability and natural selection in five species of New World saturniid moth. Nature 197: 1183–1186.

LEMAIRE, C. 1971–1974. Révision de genre *Automeris* Hübner et des genres voisins, biogéographie, éthologie, morphologie, taxonomie (Lep. Attacidae). Mus. Nat. Hist. Nat. Mem., Nouv. Ser. A, Zool. 68: 1–232, 79: 233–422, 92: 423–576.

MARQUIS, R. J. 1984. Natural history of a tropical daytime-flying saturniid: *Automeris phrynon* Druce (Lepidoptera: Saturniidae: Hemileucinae). Kans. Entomol. Soc. J. 57: 529–533.

## Hylesias

Saturniidae, Hemileucinae, Hemileucini, *Hylesia. Spanish:* Bichos quemadores (larvae).

Because of their small size for the family (WS 3 cm) and their usually uniform, drab, grayish coloration, members of this saturnian genus do not attract the attention their larger relatives do (fig. 10.2c). Some do have small or rudimentary eyespots and exhibit protective behaviors as adults and larvae that are worthy of considerable interest.

The caterpillars (fig. 10.2d) are covered with a thick growth of branching, venomous spines, the anteriormost of which is especially long. Individuals of some species are gregarious, and dozens are sometimes seen in tight communal masses on tree trunks during the day. This habit plus a head-flicking movement evoked in re-sponse to sharp sounds, such as might be made by an approaching predator, enhance the use of the spines. The high-pitched whine of parasitic wasps hovering over them also causes this response and prevents oviposition by these perpetual enemies (Hogue 1972). Other species live communally within a silken pouch that also serves for pupation (see below). Communal larvae apparently move in procession toward dusk to the crown of the host plant for feeding.

The number of food plant species is large and varied. They are almost all trees belonging to such families and genera as Mapighiaceaae (*Byrsonima*), Fabaceae (*Inga*), Bignoniaceae (*Tabebuia*), and Sapindacaceae (*Urvillea*).

Pupation occurs in aggregations in silk pouches (Wolfe 1988) or singly amid leaf litter on the ground. The female deposits her eggs around twigs in a close, rounded mass of defensive nettling hairs pulled from her abdomen.

When they are disturbed from their resting places, the female moths fall over on their sides and recurve the abdomen strongly, erecting a dense field of poisonous hairlike scales on their dorsum, making them distasteful to birds and insectivorous mammals. The abdominal scales of the male are nontoxic; the female incorporates them into her egg masses for their protection (Lamy and Lemaire 1983).

Adults of *Hylesia canitia* (probably misidentified; actually *H. metabus*, a notoriously bad urticator) have an especially bad reputation near the Caripito oil fields in the Orinoco delta, Venezuela. Sharp hairlike scales in the anal tufts of this species become easily detached and embedded in human skin. They are responsible for outbreaks of dermatitis ("butterfly itch," "Caripito itch") aboard tankers lying at anchor in the many channels. Lights at night attract large numbers of adults. The scales may be disseminated throughout the ship through its forced-air ventilation system (Goethe et al. 1967). When abundant around electric lights during population peaks, they are occasionally the cause of mass dermatitis in the populations of various towns (Allard and Allard 1958). In French Guiana, this affliction is known as "papillonite."

The genus is well developed only in the New World tropics, where at least one hundred species have evolved (Janzen 1984). The most complete life history is known for the common *Hylesia lineata* as a result of studies by Daniel Janzen (1984) in Costa Rica.

### References

ALLARD, H. F., AND H. A. ALLARD. 1958. Venomous moths and butterflies. Wash. Acad. Sci. J. 48: 18–21.

GOETHE, H., R. BRETT, AND H. WEIDNER. 1967. "Butterfly Itch," eine Schmetterlingsdermatose am Bord eines Tankers. Zeit. Tropenmed. Parasit. 18: 5–15.

HOGUE, C. L. 1972. Protective function of sound perception and gregariousness in *Hylesia* larvae (Saturniidae: Hemileucinae). J. Lepidop. Soc. 26: 33–34.

JANZEN, D. H. 1984. Natural history of *Hylesia lineata* (Saturniidae, Hemileucinae) in Santa Rosa National Park, Costa Rica. Kans. Entomol. Soc. J. 57: 490–514.

LAMY, M., AND C. LEMAIRE. 1983. Contribution à la systématique des *Hylesia*: Étude au microscope électronique à balayage des "fléchettes" urticantes. Soc. Entomol. France Bull. 88: 176–192.

WOLFE, K. L. 1988. *Hylesia acuta* (Saturniidae) and its aggregate larval and pupal pouch. J. Lepidop. Soc. 42: 132–137.

### Dirphias

Saturniidae, Hemileucinae, Hemileucini, "*Dirphia.*"

A group of closely related genera, informally known as "dirphias," are close relatives of the hylesias but are easily distinguished by their generally large size (WS up to 12 cm) and more elaborately patterned wings; there is usually a Y-shaped mark in the center of the fore wing of most species which is replaced by a small dot in the drably colored species (fig. 10.2f). Like hylesias, they curl their abdomen when molested, but, although hairy, they are not urticating. Rather, they are tough skinned and possess noxious body chemicals that repel attackers. This fact is signaled by bright red, orange, or yellow intersegmental bands on the abdomen which are exposed by the hyperextended dorsum. The antennae are also conspicuously colored and held erect during these warning displays.

Little is recorded regarding the immatures. Few have been reared, and this has been accomplished on substitute food plants. The natural hosts remain unknown, except for *D. avia* in Costa Rica whose hosts are *Cedrela odorata* and *Hymenaea courbaril* (Janzen 1982). The moderately large (BL 7–10 cm) larvae of dirphias are very similar to those of eyed saturnians and hylesias in their rich adornment of spines with radiating branches, all with venomous tips. They may (Janzen 1982:274) or may not be gregarious in the later stages, however. Ground colors of the body are from black or brownish to light gray, and they are often dark lined or marked with blotches. Pupation occurs singly in a flimsy, papery cocoon among dry leaves on the ground (Gardiner 1974) or in the ground in a cell (*Paradirphia*).

### References

GARDINER, B. O. C. 1974. The early stages of various species of the genus *Dirphia* (Saturniidae). J. Res. Lepidop. 13: 101–114.

JANZEN, D. H. 1982. Guía para la identificación de mariposas nocturnas de la familia Saturnidae del Parque Nacional Santa Rosa, Guanacaste, Costa Rica. Brenesia 19/20: 255–299.

### Tailed Saturnians

Saturniidae, Arsenurinae
(= Rhescyntinae).

This is a varied group, with a series of species displaying longer and longer tail-like extensions from the apex of the hind wing. The tails are small and triangular in *Arsenura* males (giving the hind wing a truncated outline posteriorly), moderately long and square-tipped in *Dysdaemonia* males, still longer and curved outward in *Paradaemonia*, and extremely long and slender, with gnarled, spatulate tips in *Copiopteryx* (fig. 10.2e), the so-called swallow-tailed moths. Others generally lack any suggestion of a tail at all (*Loxolomia* and *Rhescyntis*). All are large, some extremely so (WS to 20 cm), and with leaflike or harlequin patterns of buff, dark brown, and other subdued colors. Both fore and hind wings in *Dysdaemonia* have small, paired, round, transparent windows.

The subfamily is exclusively Neotropical and contains about fifty-seven species (Lemaire 1980) that range from Mexico to northern Argentina and southeastern Brazil. Most species are decidedly tropical, occurring below 1,500 meters mainly in Amazonia and southern Brazil.

The early stages of only a few species have been discovered. Younger larvae are adorned with a complete set of hornlike tubercles, those dorsolaterally on the third thoracic segment and the single dorsal one of the eighth abdominal segment being especially large. As they grow, the smaller tubercles disappear, but the larger develop into great protuberances. All arrive at maturity unarmed, losing their tubercles at the last molt. Recorded food plants are *Virola* (Myristicaceae) for *Rhescyntis* (Vázquez 1965); *Annona, Bombax, Chorisia,* and *Bombocopsis* for others (Janzen 1982, Martins Dias 1978). The pupa lies in an earthen cell unprotected by a cocoon.

### References

Janzen, D. H. 1982. Guía para la identificación de mariposas nocturnas de la familia Saturniidae del Parque Nacional Santa Rosa, Guanacaste, Costa Rica. Brenesia 19/20: 255–299.

Lemaire, C. 1980. Les Attacidae Américains (The Attacidae of America [= Saturniidae]), Arsenurinae. Ed. C. Lemaire, Neuilly-sur-Seine.

Martins Dias, M. 1978. Morfologia e biologia de *Dysdaemonia boreas* (Cramer, 1775) (Lepidoptera, Adelocephalidae). Rev. Brasil. Entomol. 22: 83–90.

Vázquez, L. 1965. *Rhescyntis (Rhescyntis) septentrionalis* sp. n. y algunas observaciones sobre su ecología y biología (Lepidoptera: Saturniidae-Rhescyntinae). Inst. Biol., Univ. Nac. Autón. México, Ann. 36: 203–213.

## SPHINX MOTHS

Sphingidae. *Spanish* and *Portuguese:* Esfinges. *Spanish:* Gusanos con cachos, gusanos cornudos (General, larvae). Hummingbird moths, hawkmoths (adults). Hornworms (larvae).

The most characteristic feature of this family of large (WS 5–15 cm) moths is their hovering flight, with elongate wings vibrating at high frequency (25–40 per second), in front of tubular flowers from which they siphon nectar with extra long proboscises. (Wing and proboscis length variations among species are prime indications of adult foraging ecology [Bullock and Pescador 1983].) The action exactly mimics the feeding method of hummingbirds and like that of the latter, is responsible for pollination of many deep-throated Neotropical flowers. Indeed, a symbiotic relationship exists between plant and moth, and some orchids and plants, like the tubular flowered, fragrant *Posoqueria longiflora* (Rubiaceae), may propagate sexually only with the help of long-tongued sphingids. Because of their strong flight, sphinx moths also may be important long-distance pollinators of lowland forest plants that are found in low densities. Here they most probably fly in the canopy and are maximally abundant during the first few months of the rainy season (Wolda 1980)

Larval behavior, too, is unique. Caterpillars are large (some gigantic; BL up to 15 cm) and, when disturbed, customarily draw their head under their thorax and elevate the latter in an erect posture somewhat resembling the famous Egyptian Sphinx edifice, hence the family name. Most species also possess a spine or thorn on the last abdominal segment, which, although lacking the necessary rigidity to pierce even the most vulnerable tissue, and entirely without venom, probably deters many predators with its menacing appearance. The spine is replaced with an eyespot in some *Eumorpha* (= *Pholus*) or a long whiplike tail in mature larvae of certain genera (*Isognathus*). The size and shape of the horns and the color pattern also may change drastically with development of sphingid larvae.

The family is very well developed in Latin America, with about 308 species (Heppner in press; Schreiber 1978). In spite of the strong dispersal abilities of the adults and the migration tendencies of several, probably most species tend to remain localized in fairly restricted geographic areas (Cary 1951). Several evolutionary dispersal centers have been postulated (Schreiber 1973), but these are of questionable veracity as they are based on biased data. The habits of the adults in some localities (Young 1968; Wolda 1980; Stradling and Legg 1983; Laroca and Mielke 1975) are partially known and but for a few studies (Moss 1912, 1920), the early stages and ecology of most remain a mystery. Flight energetics of forty-four Panamanian species have been studied by Bartholomew and Epting (1975).

## References

Bartholomew, G. A., and R. J. Epting. 1975. Rates of post-flight cooling in sphinx moths. *In* D. M. Gates and R. B. Schmerl, eds., Perspectives of biophysical ecology. Springer, New York. Pp. 405–415.

Bullock, S. H., and A. Pescador. 1983. Wing and proboscis dimensions in a sphingid fauna from western Mexico. Biotropica 15: 292–294.

Cary, M. M. 1951. Distribution of Sphingidae (Lepidoptera: Heterocera) in the Antillean-

Caribbean region. Amer. Entomol. Soc. Trans. 77: 63–129.

Heppner, J. B. In press. Lepidoptera family classification: A guide to the higher categories, world diversity and literature resources of the butterflies and moths. Flora and Fauna, Gainesville.

Laroca, S., and O. H. H. Mielke. 1975. Ensaios sobre ecologia de comunidade em Sphingidae na Serra do Mar, Paraná, Brasil (Lepidoptera). Rev. Brasil. Biol. 35: 1–19.

Moss, A. M. 1912. On the Sphingidae of Peru. Zool. Soc. London Trans. 20: 73–134.

Moss, A. M. 1920. Sphingidae of Para, Brazil. Nov. Zool. 17: 333–424.

Schreiber, H. 1973. Ausbreitungszentrum von Sphingiden (Lepidoptera) in der Neotropis. Amazoniana 4: 273–281.

Schreiber, H. 1978. Dispersal centres of Sphingidae (Lepidoptera) in the Neotropical Region. Junk, The Hague.

Stradling, D. J., and C. J. Legg. 1983. Observations on the Sphingidae (Lepidoptera) of Trinidad. Bull. Entomol. Res. 73: 201–232.

Wolda, H. 1980. Fluctuationes estacionales de insectos en el trópico: Sphingidae. 6th Cong. Soc. Colombiana Entomol. "Socolen" Mem. Pp. 11–58.

Young, A. M. 1968. Notes on a community ecology of adult sphinx moths in Costa Rican lowland tropical rain forest. Caribbean J. Sci. 12: 151–163.

## Fig Sphinx

Sphingidae, Macroglossinae, Dilophonotini, *Pachylia ficus.*

This is a large sphinx moth (WS 12 cm) that exhibits both brown and green color phases (fig. 10.3a). The fore wing is almost uniformly colored save a triangular, light spot at the wing tip and a small circular dark spot midway near the leading edge. The hind wing is a pale shade of the foreground color but transversely twice black banded; it always bears a conspicuous white point in the fringe on the hind margin near the base.

The species' larva also has two basic color phases, either orange-brown or pale green, both variously banded. Usually, there is a yellowish dorsolateral stripe, and the dorsum is a light shade of the same color. There are ill-defined, oblique stripes

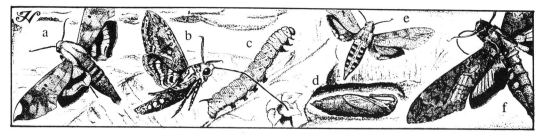

**Figure 10.3  SPHINX MOTHS (SPHINGIDAE).** (a) Fig sphinx (*Pachylia ficus*). (b) Tobacco Hornworm (*Manduca sexta*). (c) Tobacco Hornworm, larva. (d) Tobacco Hornworm, pupa. (e) Ashy Sphinx (*Erynnis ello*). (f) Giant sphinx (*Cocytius antaeus*).

on the sides. The posterior horn is stubby and slightly curved at the tip. Food plants are all species of fig (*Ficus* and others in the family Moraceae), and the moth may be very common where the host grows.

Pupation occurs in a considerable cocoon of silk under dead leaves or among grass and stones near the roots of the larval host. The pupal tongue case is fused to the body.

### Tobacco Hornworm,

Sphingidae, Sphinginae, Sphingini, *Manduca sexta*. *Spanish:* Marandová de las solanáceas (larva). *Portuguese:* Mandarová do tomate (Brazil, larva). Tomato sphinx, tomato worm (larva).

A major pest of solanaceous crops (tomato, tobacco, potato; Yamamoto and Fraenkel 1960) throughout the region, the larva of this species (fig. 10.3c) is a common sight to the home gardener and commercial farmer alike. It is large when full grown (BL 8 cm) and green, including the head, with diagonal white bars on the sides. It has a well-developed horn and fine, short, stiff hairs all over the body. It forms a pupa with free proboscis case ("jug handled") in a chamber deep in the soil (fig. 10.3d).

The adult is moderately large (WS 12 cm), with gray, bark-patterned fore wings and light gray hind wings crossed by several irregular black bars (fig. 10.3b). A row of six orange-yellow spots line each side of the abdomen.

A common, similar species with its range partly coinciding with the tomato sphinx is the pink-spotted hawkmoth (*Agrius* [= *Herse*] *cingulatus*). The adult is distinguished easily by the pink hind wings and abdominal spots.

### Reference

Yamamoto, R. T., and G. S. Fraenkel. 1960. The specificity of the tobacco hornworm, *Protoparce sexta*, to solanaceous plants. Entomol. Soc. Amer. Ann. 53: 503–507.

### Ashy Sphingids

Sphingidae, Macroglossinae, Dilophonotini, *Erynnis*. *Portuguese:* Gervão, mandarová da mandioca (Brazil, larva).

These medium-sized (WS 8–9 cm) sphingids are characterized in the adult stage by mottled, ashy gray fore wings and contrasting pink to reddish-yellow black-bordered hind wings (fig. 10.3e). They are widespread wanderers, especially *E. ello* (Beutelspacher 1967), which locally is very common around city lights at night and often turns up far from its area of origin.

Mature caterpillars are dichromatic, with green and gray-brown types. In either, the color is frequently broken by a dorsolateral lemon yellow stripe running the entire length of the body. The thorax, behind the flattened head, also may have a rectangular dark brown mark dorsally. The thorax is swollen and the terminal horn vestigial, represented only by a bare nipple. Typically, food plants are toxic wild spurges (Euphorbiaceae: *Sebastiana, Cnidoscolus*), dogbanes (Apocynaceae), and milkweeds (Asclepiadaceae: *Sarcostemma, Philibertia*), including ornamentals such as poinsettia and allamanda and crops such as papaya, guayaba, and manioc on which they are considered minor pests (Bellotti and Arias 1978). Larvae of *E. ello*, utilizing *Cnidoscolus* as a host (*mala mujer*), avoid the plant's urticating hairs and latex defenses by first grazing the former from the petiole and gnawing the petiole, effectively stopping the flow of the latex to the leaf, on which it then feeds (Dillon et al. 1983).

Pupation occurs in a stout web spun among leaves on the surface of the ground. The pupa itself is shining black, sometimes splashed with orange; its proboscis is fused to the ventral surface.

## References

BELLOTTI, A., AND B. ARIAS. 1978. Biology, ecology and biological control of the cassava hornworn (*Erynnis ello*). *In* Cassava Protection Workshop, CIAT (Cali, Colombia). Pp. 227–232.

BEUTELSPACHER, C. 1967. Estudio morfológico de *Erynnis ello* (L.), 1758 (Lepidoptera: Sphingidae). Inst. Biol., Univ. Nac. Autón. México, Ser. Zool., Ann. 1: 59–74.

DILLON, P. M., S. LOWRIE, AND D. McKEY. 1983. Disarming the "Evil Woman": Petiole constriction by a sphingid larvae circumvents mechanical defenses of its host plant, *Cnidoscolus urens* (Euphorbiaceae). Biotropica 15: 112–116.

## Giant Sphinx
Sphingidae, Sphinginae, Sphingini,
  *Cocytius antaeus*.

One of the largest sphingids in the world, this has a wingspan of 15 to 16 centimeters (fig. 10.3f). It also has one of the longest tongues of any sphinx in the world, fully 11 centimeters when extended, which is apparently adapted for tapping nectar from very deep-throated flowers. The slightly larger, related Neotropical species *Amphimoeca walkeri* actually holds the world's record for the longest tongue; it measures over 28 centimeters from base to tip (Amsel 1938).

This species' fore wings are patterned like tree bark with gray and black lines on a dark brownish background; the hind wings are translucent in the center, so that the veins are visible, orange basally and black bordered. Three yellow spots mark each side of the abdomen at its base.

Its larva is immense, reaching a body length of 14 centimeters or more when full grown, and generally uniform light greenish with a mauve line edged with white down the back. It also has faint, oblique lateral lines, the most posterior of which, extending from the base of the caudal horn, is conspicuous, cream-colored, and much larger than the others. The skin is covered with short, reddish hairs and has a well-developed, rough-surfaced pink and gray horn. It feeds on Annonaceae, including custard apple (*Annona glabra*).

The enormous (8 cm long) pupa has a short free tongue case and develops in an earthen cavity near tree roots.

## Reference

AMSEL, H. G. 1938. Die Schwärmer mit dem längsten Rüssel. Entomol. Rundsch. (Stuttgart) 55(15): 165–167.

## Frangipani Sphinx
Sphingidae, Macroglossinae,
  Dilophonotini, *Pseudosphinx tetrio*.

The gaudy, large (BL up to 15 cm, weight 15 g) caterpillar of this species (fig. 10.4b), with its clown pattern of yellow bands on a black background and bright red head, often attracts attention, the more so when it groups with others of its kind and forms long processions along tree branches and trunks. It is recognized also by orange-vermilion tail flaps and a prothoracic shield plus a long, whiplike, black tail. These colors are thought to advertise an objection-

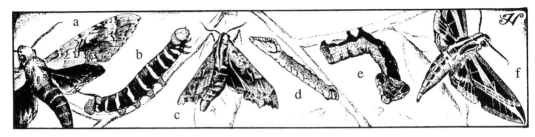

**Figure 10.4 SPHINX MOTHS (SPHINGIDAE).** (a) Frangipani sphinx (*Pseudosphinx tetrio*). (b) Frangipani sphinx, larva. (c) Viper worm (*Hemeroplanes ornatus*), adult. (d) Viper worm, larva (resting position). (e) Viper worm, larva (threat posture). (f) Harlequin sphinx (*Eumorpha fasciata*).

able taste or disposition to predators; if molested, it jerks the forepart of its body violently back and forth, bites, and rubs the substrate with the side and back of the head. An alternate theory justifies the pattern as mimicking that of a coral snake (Janzen 1980). It is well known and feeds at night on popular ornamental garden plants, including frangipani (*Plumeria rubra*) and allamanda (*Allamanda*). Its natural hosts are poorly known; "caucho de monte," a wild rubber, is cited as one.

The adult is large (WS 10–15 cm), with fore wings marked with shades of gray in a bark pattern (fig. 10.4a). The hind wings are uniformly dark gray, and the abdomen has wide dark gray bands interrupted by pale gray rings. The streamlined, elongate pupa, with appressed proboscis case, lies in the soil near the surface, covered by sparse silk threads.

The species' general biology has been reviewed by Janzen (1983) and Santiago-Blay (1985).

### References

JANZEN, D. H. 1980. Two potential coral snake mimics in a tropical deciduous forest. Biotropica 12: 77–78.

JANZEN, D. H. 1983. *Pseudosphinx tetrio* (oruga falso-coral, frangipani sphinx). *In* D. H. Janzen, ed., Costa Rican natural history. Univ. Chicago Press, Chicago. Pp. 764–765.

SANTIAGO-BLAY, J. A. 1985. Notes on *Pseudosphinx tetrio* (L.) (Sphingidae) in Puerto Rico. J. Lepidop. Soc. 39: 208–214.

## Viper Worms

Sphingidae, Macroglossinae,
 Dilophonotini, *Hemeroplanes*.

The behavior and appearance of several Neotropical sphinx larvae are incredibly similar to those of small arboreal snakes. These all belong to the genus *Hemeroplanes*, but the different species appear to mimic different snakes. The best known is *H. ornatus*, found widely through the wet forests of Amazonia and northward to Trinidad and the Caribbean borderlands (Hogue 1982). The larva (Moss 1920) is large at maturity (BL 10 cm), gray-green ventrally and pale mottled dorsally (fig. 10.4d). This is the reverse of normal countershading, an arrangement in keeping with the larva's habit of walking upside down on horizontal twigs. The relationship, however, is correct for a snake that normally travels upright on such branches.

When disturbed by a potential predator, the larva reacts violently (fig. 10.4e), arching the anterior part of its body toward the attacker (pl. 3e). At the same time, the thorax is inflated, assuming a triangular shape like a viper's head. Large, dark circular spots on either side appear as eyes, and a row of square pale spots on either side mimic perfectly the scale rows along the border of the mouth of the serpent. The effect may be enhanced by a swaying motion, like that of a snake on alert. The sham is so near perfect that even entomologists may be reluctant to pick up specimens without studying them closely first.

The models for this extreme case of Batesian mimicry are not known but are speculated to be small species of arboreal *Bothrops,* of which there are several in coincident geographic areas. A likely candidate for *H. ornatus* is *B. bilineatus,* which lives on shrubs and saplings such as the moth chooses as hosts for its larvae. Known food plant species are all found in the dogbane family (Apocynaceae) and include *Zschokkea, Echites, Amblyanthera versicolor,* and *Lacmellea aculeata.*

The adults (fig. 10.4c) are medium-sized moths (WS 7–8 cm), with jagged wing margins; their color is dark velvety brown except for conspicuous, forked silver streaks in the middle of the fore wings.

The caterpillars of a few species of *Eumorpha,* most notably, *E. labruscae,* mimic the snake form similarly, although less convincingly. A "blinking" eye (maroon spot with a black "pupil" that palpitates) replacing the anal horn in the mature stage, however, is an innovative feature of the deception (Curio 1965). These larvae are found on *Ampelopsis* (Vitaceae) in Amazonia.

### References

Curio, E. 1965. Die Schlangenmimikry einer Südamerikanischen Schwärmerraupe. Natur und Museum 95: 207–211.

Hogue, C. L. 1982. La viboruga: El extraño insecto que se parece a una víbora. Geomundo 6: 308–309.

Moss, A. M. 1920. Sphingidae of Para, Brazil. Nov. Zool. 17: 333–424.

### Harlequin Sphingids

Sphingidae, Macroglossinae,
Philampelini, *Eumorpha* (= *Pholus*).
*Portuguese:* Esfinges palhação (Brazil).

The several species in this widespread, entirely Neotropical genus are noteworthy for their large size (WS 13–14 cm). The fore wing pattern is a harlequin patchwork of light and dark browns (some green, as in *E. fasciata,* fig. 10.4f), a prominent and consistent mark being a dark, rectangular bar on the posterior margin about midway. The hind wings are sometimes pinkish, yellowish, or infused with blue. There are also identifying dark triangles on either side of the back of the thorax.

The caterpillars are stout bodied, and the horn is often replaced by a button after the last molt. The third thoracic segment is very large, and the head and first two thoracic segments are retracted into it when the larva is disturbed. Food plants are members of the grape (Vitaceae), dogbane (Apocynaceae), and evening primrose (Onagraceae) families. Pupation occurs in a subterranean cell.

## GREEN URANIAS

Uraniidae, *Urania. Spanish:* Colipatos verde (Costa Rica). Green page (Trinidad).

Two common Latin American species in this genus of day-flying moths (*Urania fulgens* and *leilus*) are well known for their migratory habits. Large-scale movements of hundreds of thousands of individuals take place synchronously every few years, usually between early August and late November. Flights of *U. fulgens* are seen moving generally in an eastward or southeastward direction through Central America from Mexico as far south as northern Colombia (Young 1970); *U. leilus* occurs in South America, and its migrations are recorded in Trinidad, the Guianas, and Venezuela (Smith 1972) where it generally moves eastward and southward. Little is known of the origins, precise routes, or function of this phase of the moth's ecology. Because the moths that comprise the migrating hordes are both male and female, the latter predominating and carrying eggs, and are freshly emerged (Odendaal and Ehrlich 1985), these flights may

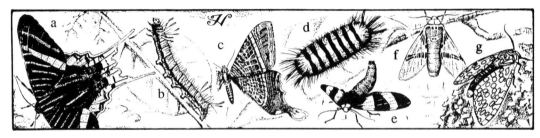

**Figure 10.5    MOTHS.** (a) Green urania (*Urania leilus,* Uraniidae). (b) Green urania, larva. (c) Eyetail (*Nothus luna,* Sematuridae). (d) Tiger moth (unidentified, Arctiidae), larva. (e) Tiger moth (*Viviennea moma*). (f) Tiger moth (*Idalus herois*). (g) Tiger moth (*Hypercompe decora*), male.

be unidirectional dispersions in response to diminished food or other resources in their usual breeding territories (Smith 1982).

These are large moths (WS 10 cm), much resembling swallowtail butterflies, with their black and iridescent green-striped pattern and long, flexible, white-fringed tails (fig. 10.5a). The green of the bars is bronzy tan in the males; bright but unreflective in the females. They differ from other moths in their filamentous (but not clubbed) antennae and wing veins forming much smaller major cells in the wings. Their mouthparts are fully functional, and adults are often seen puddling on wet sand or mud in company with their counterparts among the butterflies; at other times, they take nectar from flowers, being especially fond of the white, fluffy flowers of mimosoid legumes such as *Inga.*

*Urania* larvae (fig. 10.5b) are medium-sized (BL about 5 cm) and generally black-and-white banded but irregularly and variously so in different individuals. Generally, there is a heavier wavy band around the middle of each abdominal segment, this edged with white anteriorly and often with white blotches on the sides below; a few additional black bands may be present between the main bands. There also are long, fine, black hairs over most of the body; those of the thorax and posteriormost abdominal segments are extra long and with spatulate, curled tips. The

head is red with black spots (Guppy 1907). They feed only on vines and trees in the genus *Omphalea* (Euphorbiaceae).

Pupation occurs within a sandwich of two leaves fastened by silk from the larva. The pupa is light yellowish-brown, glossy, with irregular black dots on the abdomen and discrete, black, longitudinal lines on the wing cases.

## References

GUPPY, L. 1907. Life history of *Cydimon* (*Urania*) *leilus,* L. Entomol. Soc. London Trans. 1907: 405–410.

ODENDAAL, F. J., AND P. R. EHRLICH. 1985. A migration of *Urania fulgens* (Uraniidae) in Costa Rica. Biotropica 17: 46–49.

SMITH, N. G. 1972. Migrations of the day-flying moth *Urania* in Central and South America. Caribbean J. Sci. 12: 45–48.

SMITH, N. G. 1982. Population eruptions and periodic migration in the day-flying moth *Urania fulgens.* *In* E. G. Leigh, Jr., A. S. Rand, and D. M. Windsor, The ecology of tropical forest: Seasonal rhythms and long-term changes. Smithsonian Inst. Press, Washington, D.C. Pp. 331–334.

YOUNG, A. M. 1970. Notes on a migration of *Urania fulgens* (Lepidoptera: Uraniidae) in Costa Rica. Entomol. Soc. J. 78: 60–70.

## EYETAILS

Sematuridae, *Nothus* (= *Sematura*).

Many of the thirty-five Neotropical species of these delicately built moths are easily confused with saturnians because of their large size (WS 75 mm) and similar brown markings (fig. 10.5c). They also have tails extending from the hind wing like some Saturniidae, but these are spatulate and with a conspicuous eyespot in the rounded, apical portion unlike any in that family. Their fine, hairlike antennae, rather than the feathery elaborations, also clearly distinguish them. They are actually more closely related to the day-flying rainbow moths (Uraniidae) but are nocturnally active. They often come to rest with the fore and hind wings separated, the latter more or less rolled around the abdomen.

Virtually nothing is published regarding the early stages or biology of this genus.

## Mimetic Moths

Along with several kinds of butterflies, several related groups of slow-flying, diurnal, rubbery-bodied moths are brightly colored and participate in Müllerian mimicry complexes. External resemblances among species in the same pattern series may be so great that even specimens in hand are difficult to distinguish. Certain characteristics of the veining in the wings (often to be seen only with a magnifying lens) will separate the families. As an aid to the reader's understanding of the differences, they are summarized here, prior to the individual discussions of the families.

Smoky moths (Zygaenidae) are the easiest to separate because of the presence of a vein running lengthwise through the middle of the large, central cells in both fore and hind wing. This is a vestigial condition, revealing the family's comparatively primitive phylogenetic position.

The remaining groups segregate primarily on the basis of the branching patterns of the anteriormost set of veins in the hind wing. In the wasp moths (Arctiidae, Ctenuchinae), this vein appears to have a single fork (bipartite) due to the complete absence of the normally present first vein (subcosta). In all the remaining groups, this vein is tripartite, with the first vein being completely separate, even at its base, in the dioptid moths (Dioptidae). Additionally, in this family, the branching points of the outermost forked veins in the hind wing occur near the wing edge, beyond the outermost part of the large central cell. These veins are mostly separate basally, only a short length being fused at a point near to or before the middle of the cell bordering it behind, in the flag moths (Arctiidae, Pericopinae).

In the tiger moths (Arctiidae, Arctiinae), forester moths (Agaristidae), and giant day-flying moths (Castniidae), the fusion of these two anterior veins is complete almost for the entire distance along the length of the cell in all, and they are distinguished by other venational features, for example, the closeness of the origin of the next vein in the hind wing (following the anterior triple vein just referred to) to its neighbor behind. It is close in the tiger moths, more widely separated (equidistant from the vein preceding) in the forester moths and giant day-flying moths. The latter two families are distinguished by the large central cell of the hind wing which is open (i.e., not closed off by a cross vein). Also, all the veins tend to run parallel to each other in the latter, while many are divergent in the former. Both have clubbed antennae, much more so in the Castniidae.

In other families such as the owlet moths (Noctuidae), sphingids (Sphingidae), pyralids (Pyralidae), and measuring worm moths (Geometridae), in which nocturnalism and cryptic wing patterns are the rule, a few mimetic types occur, but these are clearly separable by gross features mentioned in the discussions of these groups. It should be stressed also that all the members of the families placed in the "mimetic moth" category above are not diurnal and highly colored. Many tiger moths, for example, are drably or cryptically marked and fly by night.

# TIGER MOTHS

Arctiidae, Arctiinae. *Spanish:* Gatas peludas (larvae). *Portuguese:* Largatas cabeludas (Brazil, larvae).

Certainly these are the most flamboyantly colored and varied single family of moths in the Neotropics. Unbelievably garish and gorgeous, polychromatic combinations of florid hues are juxtaposed in kaleidoscopic arrays on the fore wings of many species (seldom also on the hind wings, which are more usually drably monochromatic). These color arrangements have protective functions, either as camouflage or warnings of distastefulness. In the former category are mimics of lichens (*Hypercompe* = *Ecpantheria*, fig. 10.5g), dead leaves (*Bertholdia*), or bark (*Leucanopsis* = *Halysidota*); there are also tessellated (*Idalus*, fig. 10.5f; *Eucereon*) or disruptive patterns (*Viviennea*, fig. 10.5e; *Paranerita*).

Wing waving displays, cataleptic seizures, extrusions of hair tufts, reflexive frothing, and bleeding of yellow body fluid from the neck are direct defenses that they advertise by bright (aposematic) colors. Many species even emit high-frequency sounds by distorting ridges on the thoracic wall when roughly handled (Blest et al. 1963), and all such moths are summarily rejected by monkeys, bats (Dunning 1968), birds, toads, and other potential predators (Blest 1964).

Few life histories of the Neotropical species are known. The larvae are all densely clothed with nonurticating hairs that arise from small wartlike swellings on the cuticle. These hairs are usually stiff and dark but may be soft, tufted, and of other colors, often white or red and even bicolored to produce a banded pattern (fig. 10.5d, pl. 2e). The larvae feed on varied plant types, perhaps favoring Compositae, and many species are polyphagous. Some prefer plants containing pyrrolizidine alkaloids (such as *Crotalaria*, *Heliotropium*, and *Sene-*

*cio*) or other toxic compounds (*Euphorbia*, *Solanum*) in European and presumably also the Neotropical species (Rothschild et al. 1979). Pupae are hidden in a loose silken cocoon into which the larval hairs are usually incorporated.

According to a recent count (Watson and Goodger 1986), there are 1,939 described Latin American species of tiger moths classed in 174 genera, but there are many more species to be discovered. They are highly diverse in form (Watson 1971, 1973) and pattern, making them somewhat difficult to recognize as a group, but nearly all have a tymbal (sound-producing organ). The diurnal mimetic species can be placed in their proper subfamily by the tymbal and vein pattern in the hind wing (see mimetic moths, above).

A few tiger moths are apparently specific Batesian mimetics of other distasteful insects. With wings at rest, *Opharus bimaculatus* closely resembles the luminescent headlight beetles (*Pyrophorus*). Both are elongate and generally dark brown; the twin glowing spots on the beetle are simulated by large, circular cream spots in the thorax of the moth. Other species are themselves distasteful and clearly members of Müllerian mimicry complexes. An example is *Cratoplastis*, with species (especially *diluta*) that resemble the cockroach *Achroblatta luteola* and members of the firefly beetle genus *Cratomorphus* (orig. obs.).

## References

BLEST, A. D. 1964. Protective display and sound production in some New World arctiid and ctenuchid moths. Zoologica 49: 161–181.

BLEST, A. D., T. S. COLLETT, AND J. D. PYE. 1963. The generation of ultrasonic signals by a New World arctiid moth. Royal Soc. Proc. B 158: 196–207.

DUNNING, D. C. 1968. Warning sounds of moths. Zeit. Tierpsychol. 25: 129–138.

ROTHSCHILD, M., R. T. APLIN, P. A. COCKRUM, J. A. EDGAR, P. FAIRWEATHER, AND R. LEES. 1979. Pyrrolizidine alkaloids in arctiid moths (Lep.) with a discussion on host plant relationships and the role of these secondary plant substances in the Arctiidae. Biol. J. Linnean Soc. (London) 12: 305–326.

WATSON, A. 1971. An illustrated catalog of the Neotropic Arctiinae types in the United States National Museum (Lepidoptera: Arctiidae). Pt. I. Smithsonian Contrib. Zool. 50: 1–361.

WATSON, A. 1973. An illustrated catalog of the Neotropic Arctiinae types in the United States National Museum (Lepidoptera; Arctiidae). Pt. II. Smithsonian Contrib. Zool. 50: 1–160.

WATSON, A., AND D. T. GOODGER. 1986. Catalogue of the Neotropical tiger-moths. Brit. Mus. Nat. Hist. Occ. Pap. Syst. Entomol. 1: 1–71.

## Wasp Moths

Arctiidae, Ctenuchinae (= Ctenuchidae, Amatidae, Euchromiidae, Syntomidae).

This is a special group of tiger moths that often display gaudy combinations of bright, often metallic, colors on their anterior (sometimes posterior also) wings. The fore wings generally are more elongate, and they often have tufted or paddlelike expansions of scales on the hind legs which apparently imitate the pendant hind legs of real wasps in flight. They are mostly diurnal. The vein branching pattern in the hind wing also is slightly different (see mimetic moths, above). Like other tiger moths, wasp moths utilize several elaborate protective devices, such as playing dead, sound production, bad taste, emission of foul substances, and color and behavioral displays (Blest 1964). A few species possess long scaled "pigtails," structures of unknown function, trailing from the apex of the abdomen.

Many wasp moths are active during the day and mimic wasps, bees, and beetles, including some truly remarkable imitators of vespid wasps in the genus *Pseudosphex*. They have clear, plaited wings that they hold erect and are narrow waisted and aggressive, all features precisely copying vespid structure and behavior (Beebe and Kennedy 1957: 148–150). *Macrocneme* (fig. 10.6b), *Pseudopompilia,* and other species have wings with solid shades very much like those of pepsis wasps, which they resemble also in general form. Others, for example, *Correbidia assimilis* and *Correbia,* are variously black or dark blue and yellow banded or spotted in patterns copying those of net-winged beetles (Lycidae), which are protected by their yellow, oily, repugnant body fluids. Slow, fluttering flight in all of these enhances their resemblance to their models.

Adults of many species of wasp moths are attracted to *Heliotropium, Senecio,* and *Vernonia,* from which it is presumed they extract compounds important in their lives (Moss 1947), but no details are available yet concerning what these substances are or the ways in which they are used.

Common nocturnal species comprise the large genus *Eucereon* (Arctiinae), whose members are recognized by their tessel-

**Figure 10.6 MOTHS.** (a) Smoky moth (*Harrisina tergina,* Zygaenidae). (b) Wasp moth (*Macrocneme chrysitis,* Arctiidae). (c) Flag moth (*Dysschema leucophaea,* Arctiidae), larvae. (d) Dioptid moth (*Dioptis restricta,* Dioptidae). (e) Flag moth (*Daritis howardi,* Arctiidae), male. (f) Flag moth (*Daritis howardi*), female. (g) Giant day-flying moth (*Castnia licoides,* Castniidae). (h) Giant day-flying moth, larva.

lated, cryptic fore wings and red or blue marked abdomens. They can produce sound when disturbed, like some tiger moths (Blest 1964).

The larvae, very few of which are known (Beebe 1953), are similar to those of tiger moths but somewhat less hairy and with softer, even silky setae that often arise in tufts and may be extra long, especially at the extremities of the body. The portion of the integument that is exposed is often marked with colored spots or lines or both, unlike the neutral skin of the tiger moth larvae. Known food plants are chiefly monocots, including common types such as grasses and *Canna*, but some dicots are eaten as well (Moss 1947). One pest species that attacks bananas is *Antichloris viridis* (Field 1975). Pupation takes place in a loose, fragile cocoon in litter or is hidden under loose bark on tree trunks.

The subfamily contains over 2,100 species on the Neotropics, its area of maximum diversity.

## References

BEEBE, W. 1953. A contribution to the life history of the euchromid moth, *Aethria carnicauda* Butler. Zoologica 38: 155–160.

BEEBE, W., AND R. KENNEDY. 1957. Habits, palatability and mimicry in thirteen ctenuchid moth species from Trinidad, B.W.I. Zoologica 42: 147–158.

BLEST, A. D. 1964. Protective display and sound production in some New World arctiid and ctenuchid moths. Zoologica 49: 161–181.

FIELD, W. D. 1975. Ctenuchid moths of *Ceramidia* Butler, *Ceramidiodes* Hampson, and the *caca* species group of *Antichloris* Hübner. Smithsonian Contrib. Zool. 198: 1–45.

Moss, A. M. 1947. Notes on the Syntomidae of Pará, with special reference to wasp mimicry and fedegoso, *Heliotropium indicum* (Boraginaceae), as an attractant. The Entomologist 80: 30–35.

## Flag Moths
Arctiidae, Pericopinae.

Pericopines are medium-sized to large (WS 4–10 cm). All are diurnal and display mimetic wing patterns. They are named "flag moths" here because, like banners, their generally broad wings display sharply defined bands or fields of contrasting bright colors. Anatomical features that distinguish them are unusually large, hood-like flaps (tympanic hoods) arising dorsally at the base of the abdomen and extending over the rear portion of the thorax, in addition to the hind wing vein branches discussed above (see mimetic moths).

Adults are well protected by noxious body fluids; they can also emit a yellow spume from the neck, making them unpleasant even to handle. Before these lines of defense take effect, they may be mauled by a predator, but their bodies are rather rubbery and can withstand considerable trauma without much damage. Flag moth color patterns are often sexually dimorphic (as in *Daritis howardi* = *thetis*, fig. 10.6e, f) and probably mimetic, the two sexes possibly to different models. Male *Dysschema jansoni* are presumed mimics of aposematic *Parides* swallowtail butterflies, the females of ithomiines (Aiello and Brown 1988).

The immatures and biology of few flag moths are known. The larvae resemble those of tiger moths, with abundant hair, and they are brightly colored and gregarious (fig. 10.6c). They feed on various Compositae, some of which contain toxic or repugnant chemicals (*Senecio*, *Vernonia*) that are probably sequestered and responsible for the adult's unpalatability. When molested, they have been observed to thrash about and exude droplets of fluid (Young 1981). The pupae are metallic colored and placed in a loose, baggy cocoon.

There are 369 species in the Neotropical region (Watson and Goodger 1986).

## References

AIELLO, A., AND K. S. BROWN, JR. 1988. Mimicry by illusion in a sexually dimorphic, day-flying moth, *Dysschema jansoni* (Lepidoptera: Arctiidae: Pericopinae). J. Res. Lepidop. 26: 173–176.

WATSON, A., AND D. T. GOODGER. 1986. Catalogue of the Neotropical tiger-moths. Brit. Mus. Nat. Hist. Occ. Pap. Syst. Entomol. 1: 1–71.

YOUNG, A. M. 1981. Notes on the moth *Pericopis leucophaea* Walker (Lepidoptera: Pericopidae) as a defoliator of the tree *Vernonia patens* H.B.K. (Compositae) in northeastern Costa Rica. New York Entomol. Soc. J. 89: 204–213.

## SMOKY MOTHS

Zygaenidae (including Pyromorphidae). Burnets.

Most smoky moths are small (WS 25–30 mm) and wasplike with elongate, darkly infuscate wings, resembling some of the smaller wasp mimics in the Arctiidae. A few are orange, orange barred, or otherwise colored like Müllerian models among these families and net-winged beetles. They are easily distinguished, however, by the presence of conspicuous hairy, padlike organs on the head above the compound eye (chaetosemata) plus the venational differences outlined above (see mimetic moths).

The adults are diurnal and frequently seen taking nectar from flowers (fig. 10.6a) or cruising close to the ground amid low vegetation. As far as is known, they do not acquire pyrrolizidine alkaloids from plants but do take up cyanic glycosides. Collectors often find them difficult to kill in their cyanide bottles.

The larvae are small (seldom longer than 1 cm), slightly hairy, and feed gregariously, often side by side in military rows. They commonly remove only the cells between the veins, leaving just a skeleton of the leaf. Pupation occurs in a strong, elongate cocoon placed among leaves.

The family is poorly represented in the New World, with only 140 species in all of Latin America (Tarmann 1984).

### Reference

TARMANN, G. 1984. Generische revision der amerikanischen Zygaenidae mit Beschreibung neuer Gattungen und Arten (Insecta: Lepidoptera). Entomofauna, suppl. 2, 1: 1–176, 2: 1–153.

## DIOPTID MOTHS

Dioptidae.

These mimetic moths are medium-sized (WS 3–4 cm), mostly broad-winged and more delicately structured than those in the other families. They are positively distinguished, however, only by the form of the hind wing vein branching as already described (see mimetic moths, above). The diurnal forms display bright colors, as do many night-flying species (e.g., *Dioptis restricta,* fig. 10.6d), although the majority of the latter tend to be drab insects.

There are about five hundred regional species. Their classification and biology are incompletely investigated (Prout 1918). The few known larvae have an outsized, rounded head and naked skin banded with varied colors. Food plants of *Josia,* one common genus, are *Aristolochia* and *Passiflora.* Pupae are well pigmented and chrysalidlike and either suspended naked from the posterior or attached to the inside of a sparsely woven cocoon amid live leaves on the host plant or in leaf litter on the ground.

### Reference

PROUT, L. B. 1918. A provisional arrangement of the Dioptidae. Novit. Zool. 25: 395–429.

## GIANT DAY-FLYING MOTHS

Castniidae, *Castnia. Spanish:* Catarinetas (Peru). *Portuguese:* Brocas gigantes (Brazil, larvae).

These, the largest of the mimetic moths (WS 8–18 cm), are easily distinguished from similarly colored types by the hind wing vein structure (see mimetic moths, above). They also have conspicuously clubbed antennae. They are very fast, evasive day or crepuscular fliers and are difficult to catch. The body is heavy and tough and the wing scales coarse, characteristics apparently countering their violent flight, which often takes them through heavy vegetation as indicated by the commonness with which worn or damaged specimens are seen. Males of some species are said to have well-developed scent glands on the midtarsus or underside of the first two abdominal segments, the latter producing a thick, dark substance (Jordan 1922), but this has not been confirmed by repeated observations. Seitz (1913) notes that they exhibit territorial behavior and return regularly to feeding and resting sites.

The larvae of the larger species may be enormous (BL to 8–9 cm) and are typical borers (fig. 10.6h). The later instars are nearly naked, with only minute bristles on the back of the segments and with reduced legs and pale skin. They live on monocots, penetrating and mining fruits, cane stalks, banana stems, the roots and stems of bromeliads, and the pseudobulbs of orchids (Moss 1945). These habits have made some agricultural pests. *Castnia licoides* (= *licus;* fig. 10.6g) is a pest of sugarcane in Trinidad, Guyana, and other places (Skinner 1929), and *C. cyparissias* (= *daedalus*) attacks oil palm in Peru (Korytkowski and Ruiz 1979). Unlike the spherical eggs of most other Lepidoptera, those of castniids are usually elongate, like grains of rice.

Only one Neotropical genus, *Castnia* (Strand 1913), seems clearly definable, although Oiticica (1955) divided the 160 species between over twenty genera. *Microcastnia* was recently recognized by Miller (1980). The family begs biological study and has received only minimal taxonomic treatment to the present (Houlbert 1918, Miller 1972).

## References

HOULBERT, C. 1918. Révision monographique de la sous-famille des Castniinae. Etudes Lépidop. Comp. 15: 5–730, pls. 437–462.

JORDAN, K. 1922. The scent-organ of certain mimetic Castniidae. Entomol. Soc. London Trans. 1922: xci.

KORYTKOWSKI, C. A., AND E. R. RUIZ A. 1979. El barreno de los racimos de la Palma aceitera, *Castnia daedalus* (Cramer), Lepidopt.: Castniidae, en la plantación de Tocache-Peru. Rev. Peruana Entomol. 22(1): 49–62.

MILLER, J. Y. 1972. Review of the Central American *Castnia inca* Complex (Castniidae). Allyn Mus. Bull. 6: 1–13.

MILLER, J. Y. 1980. Studies in the Castniidae. III. *Microcastnia*. Allyn Mus. Bull. 60: 1–15.

MOSS, A. M. 1945. The *Castnia* of Pará, with notes on others (Lep. Castniidae). Royal Entomol. Soc. London Proc. B 14: 48–52.

OITICICA, J. 1955. Revisão dos nomes genericos Sul Americanos da subfamilia Castniinae (Lepidoptera, Castniidae). Rev. Brasil. Entomol. 3: 137–167.

SEITZ, A. 1913. Castniidae. *In* A. Seitz, ed., The Macrolepidoptera of the world. II Div.: The Macrolepidoptera of the American region, the American Bombyces and Sphinges. 6: 5–7. Kernen, Stuttgart.

SKINNER, H. M. 1929. The giant moth borer of sugarcane. (*Castnia Licus*—Drury). Trop. Agric. 7 (suppl., Jan.): 1–8.

STRAND, E. 1913. *Castnia. In* A. Seitz, ed., The Macrolepidoptera of the world. II Div.: The Macrolepidoptera of the American region, the American Bombyces and Sphinges. 6: 7–19. Kernen, Stuttgart.

## OWLET MOTHS

Noctuidae. *Spanish:* Pájaros nocturnas (General). Millers.

This is the most familiar and the largest family of moths; yet it is poorly known in Latin America, and many species await discovery. For this reason, the number of species can only be guessed at about 10,000. They form a varied assemblage, occurring throughout all the geographic regions and in most habitat types. The majority are nocturnal and are usually seen when they come to artificial lights.

Most are medium-sized (WS 2–4 cm) and cryptically marked like bark or stones in drab browns, grays, and black. The hind wings are often unlike the fore wings, usually monocolorous, often translucent, but highly colored in some ("underwings"). A relative few have brilliant colors and are diurnal (and even have eversible odoriferous glands, coremata) like wasp moths and tiger moths. The colors may have roles as warning signals or in mimicry complexes and imply possible sequestering of toxins from food plants by the larvae (e.g., in *Cydosia*, Erastrinae). One or two small circular or kidney-shaped spots are often present in the center of the fore wings. Some nocturnal species have gaudy colors, which no doubt protect them aposematically or cryptically when they are at rest during the day. When folded, the wings are ordinarily held rooflike over the body, but some repose with wings flat on the substrate to the sides or even closely appressed over the back like those of a butterfly.

The wings are relatively small in proportion to the robust body. The antennae are simple and filiform, and the proboscis is normally fully developed. The first two veins in the hind wing which are fused at the base only for a short distance before continuing separately to the outer wing margin also serve to identify these moths.

The larvae of many family members are agricultural pests. These are the so-called cutworms (*cortadores, rôscas, tierreros*) that clip young plants off near ground level, the "loopers" (*medidores*) that inch along with an arching of the body and eat the leaves of vegetable crops, the "armyworms" (*orugas militar, largartas militar*), and the "fruit worms" and "leaf worms" that consume fruits and leaves (*trepadores*) generally. Some are "borers" (*barrenadores, brocas*) in fruits and other plant tissues.

Generally, the larvae are smooth and naked (a few are hairy) and drably colored, except for contrasting longitudinal lines or stripes. Aside from the loopers, which have only three pairs, all have five pairs of abdominal walking legs. Pupation usually takes place in the soil or among ground litter, in a simple cell without silk lining or cocoon.

The closely related forester moths (Agaristidae) are mostly diurnal and colorful mimetic moths. In addition to the venational features (see mimetic moths), their antenna has a terminal club, often with a curved prolongation. Their larvae are hairy and feed mostly on plants in the families Vitaceae and Onagraceae.

## Cutworms and Armyworms

Noctuidae. *Spanish:* Gusanos de tierra, cuncunillas (Chile); tierreros (Central America).

The larvae of many owlet moths pass the day just below the ground surface or secreted among leaves or other litter. They emerge at night to feed on young plants, which they bring down by cutting them off neatly near the base. They are catholic in their tastes and destroy a wide variety of vegetable crops. Generally, they eat both wild and cultivated plants and may survive on the former near fields during fallow periods, ready to move into new plantings soon after they begin to grow. Those that have a special affinity for grasses and that may invade crops in massive marching hoards are the armyworms. When disturbed, these larvae customarily curl up, with the head tucked inside, and lie on one side, playing dead.

The more important species in Latin America from the agricultural standpoint are discussed in the next few sections. It will not be possible to identify them with surety from the descriptions because of the many similar relatives in the same or other genera, such as *Euxoa, Mamestra, Polia, Prodenia,* and *Mocis* (Valencia and Valdivia 1973).

## Reference

VALENCIA, L., AND R. VALDIVIA. 1973. Noctuideos del Valle de Ica, sus plantas hospederas y enemigos naturales. Rev. Peruana Entomol. 16: 94–101.

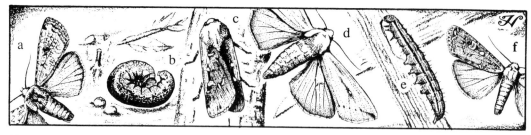

**Figure 10.7  OWLET MOTHS (NOCTUIDAE).** (a) Variegated cutworm (*Peridroma saucia*). (b) Variegated cutworm, larva. (c) Granulate cutworm (*Agrotis subterranea*). (d) Armyworm (*Pseudaletia unipuncta*). (e) Armyworm, larva. (f) Beet armyworm (*Spodoptera exigua*).

### Variegated Cutworm
Noctuidae, Agrotinae, *Peridroma saucia.*

In the adult of this large (WS 4–5.2 cm) cutworm, the fore wing is generally medium brown and marked with paired transverse lines, these abruptly darkened where they contact the anterior margin (fig. 10.7a). There are also faint, inner circular and outer kidney-shaped spots just forward of the wing's center.

The larva (fig. 10.7b) is identified by distinct pale yellow dots on the midline of the back of most segments and frequently by a W-shaped, dark mark on the back of the eighth abdominal segment. It prefers garden crops and greenhouse plants as well as the foliage, buds, and fruits of various trees, vines, and ornamentals. It may break out in phenomenal numbers. The species is a pest generally over most of Latin America.

### Agrotis Cutworms
Noctuidae, Agrotinae, *Agrotis* (= *Feltia*)
*malefida, ipsilon,* and *subterranea.*
*Spanish:* Cachazudos (Cuba, Central America), caballadas (Peru). *Portuguese:* Lagartas rôscas (Brazil).

This is a complex of three closely similar species: the black (or greasy) cutworm (*gusano cortador negro, gusano trozador, tierrero*) = *A. ipsilon*); the pale-sided cutworm (*gusano cortador costado claro*) = *A. malefida;* and the granulate cutworm (*gusano cortador cuerudo*) = *A. subterranea.*

The adults of all have fairly long fore wings with distinct round and kidney-shaped spots. The spots are connected by a black bar in *Agrotis subterranea* (fig. 10.7c); in *A. malefida,* the spots are separate but the inner one is underscored by a black square or triangle that is lacking in *A. ipsilon. A. ipsilon* also has strong zigzagging lines outside the kidney-shaped spot which are lacking or obscure in the other species. These species differ in size also: *A. subterranea* is the smallest (WS 4 cm), *A. malefida* is slightly larger (WS 4.5 cm), and *A. ipsilon* is the largest (WS 5 cm).

The larvae are also very nearly the same—pale, dirty gray cutworms with numerous fine granules on the skin. That of *A. ipsilon* tends to have darker sides and a greasy appearance. In all, the head has a reticulate pattern on the sides and is darkly barred on the front on either side of the median, triangular plate. Hosts are cotton, tobacco, corn, and many other field crops and greenhouse plants.

These species range throughout practically all agricultural areas of Latin America.

### Armyworm
Noctuidae, Hadeninae, *Pseudaletia
unipuncta. Spanish:* Gusano soldado.
*Portuguese:* Lagarta do trigo (Brazil).

The moderate-sized (WS 3.5–4.7) adult of this leaf worm (fig. 10.7d) has a pale brown fore wing that is evenly colored except for scattered, minute, black flecks and a small, light spot near the center bisecting a fine black dash; a thin dark line also runs inward obliquely from the tip of the wing.

The smooth skin of the larva (fig. 10.7e) varies from bright red through pinkish (rarely) to pale or dark gray (commonly), overlain by interrupted, fine longitudinal lines and dark flecks. The head is pale greenish-brown, finely mottled with darker brown.

This is an especially injurious pest of small grains, corn, rice, and forage grasses, but the larva eats a great variety of other plants as well. Very large populations may sometimes infest fields, where the rustling sound of their feeding and movement over the ground can be heard from some distance as they move in and devour every blade in sight.

The species is wide-ranging in the Neotropics, where it is not easily distinguished from several similar but less injurious species in the genus (Franclemont 1951). It was at times placed in the genera *Cirphis* and *Leucania* (Marcovitch 1958). A related injurious species in southern Brazil and northern Argentina is *Pseudaletia adultera*.

## References

Franclemont, J. G. 1951. The species of the *Leucania unipuncta* group, with a discussion of the generic names for the various segregates of *Leucania* in North America. Entomol. Soc. Wash. Proc. 53: 57–85.

Marcovitch, S. 1958. Biological studies on the armyworm, *Pseudaletia unipuncta* (Haworth), in Tennessee (Lepidoptera: Noctuidae). Tenn. Acad. Sci. J. 33: 263–347.

### Beet Armyworm

Noctuidae, Agrotinae, *Spodoptera exigua*.
*Spanish:* Cogollero del maíz (Peru).
*Portuguese:* Curuquerê dos capinzais (Brazil).

This is a smallish noctuid (WS 2 cm) with well-defined circular and kidney-shaped spots in the slender fore wing (fig. 10.7f). The circular spot has a dirty, pale yellow or yellow-orange center.

Its larva feasts on the leaves of rice, corn (cob kernels), sorghum, citrus, and many other plants. It is pale or olive green with a dark dorsal stripe bordered by a yellow stripe; the entire underhalf is pale yellow or cream.

The species ranges from northern Mexico to Nicaragua and is found on some Antillean islands whence it appears to be spreading southward (Todd and Poole 1980).

## Reference

Todd, E. L., and R. W. Poole. 1980. Keys and illustrations for the armyworm moths of the noctuid genus *Spodoptera* Guenée from the Western Hemisphere. Entomol. Soc. Amer. Ann. 73: 722–738.

### Lateral Lined Armyworm

Noctuidae, Agrotinae, *Spodoptera latifascia*.
*Spanish:* Gusano cortedor de líneas laterales (General).

The fore wings of the adult of this species have a harlequin pattern of white streaks, plus a conspicuous orange (male) median patch, against a variegated brown and black smeared background. The hind wing is white. It is an average-sized owlet moth (WS 4–4.8 cm).

The larva has a more robust body and a relatively smaller head than that of other cutworms. Older larvae are black to light brown with a dorsal row of black triangular spots, diminishing toward the rear; the lateral lines are generally faint or absent. It is a day feeder, devouring young plants of cotton, skeletonizing the leaves of tobacco, and consuming many other crops, such as beans, chile, maize, and vegetables.

The species (commonly cited as *Prodenia ornithogalli,* actually a different species limited to the north; Comstock 1965) occurs throughout Latin America, where it is the most serious and widespread of several species of similar "armyworms" (Todd and Poole 1980). Also a general pest is the fall armyworm (*gusano cogollero, pelon, lagarta* do cartucho, lagarto do milho—*Spodoptera frugiperda*), which feeds on a wide range of crops (Andrews 1988, Peairs and Saunders 1979).

### References

ANDREWS, K. L. 1988. Latin American research on *Spodoptera frugiperda* (Lepidoptera: Noctuidae). Fla. Entomol. 71: 630–653.

COMSTOCK, J. A. 1965. Ciclo biológico de *Prodenia ornithogalli* Guenée (Lepidoptera: Noctuidae). Inst. Biol. Univ. Nac. Autón. México, An. 36: 199–202.

PEAIRS, F. B., AND J. L. SAUNDERS. 1979. The fall armyworm, *Spodoptera frugiperda* (J. E. Smith), a review. CEIBA 23: 93–113.

TODD, E. L., AND R. W. POOLE. 1980. Keys and illustrations for the armyworm moths of the noctuid genus *Spodoptera* Guenée from the Western Hemisphere. Entomol. Soc. Amer. Ann. 73: 722–738.

### Corn Earworm

Noctuidae, Heliothidinae *Helicoverpa* (= *Heliothis*) *zea. Spanish:* Gusano del choclo, mazorquero (Peru, Chile). *Portuguese:* Lagarta da espiga. Cotton bollworm, tomato fruit worm, false tobacco cutworm.

This is the worm normally found esconsed in the silk of a corn cob amid a collection of feces and damaged kernels. Although a common pest of corn, the larva is omnivorous and damages the flowers and fruits of many other plants, both cultivated (tomato, cotton, peas, beans, tobacco, etc.) and wild (*Malva,* Malvaceae; *Desmodium,* Leguminosae; *Ludwigia,* Onagraceae; etc.).

It is a moderate-sized noctuid (WS 3.2–4.5 cm) (fig. 10.8a). The pale, creamy tan ground color of the adult fore wing is always overlain with transverse wavy lines outside of a dark spot about midway near the leading edge; the lines vary considerably in intensity from pale and obscure to dark and distinct. The hind wings are white, except for dark veins in the center and a broad dark marginal border.

Its larva (fig. 10.8b) varies from brownish-magenta to pale green, with paired dorsal dark lines and swollen black spots at the bases of the body hairs. Irregular dorsolateral and lateral pale bands are also present, delineating a wide, lateral, yellowish or cream band. The pale ground color of the head is heavily mottled with orange-brown.

The species has spread over most of Central and South America and the Greater Antilles; it is apparently absent from most of Amazonia and the extreme south of Patagonia (Hardwick 1965).

### Reference

HARDWICK, D. F. 1965. The corn earworm complex. Entomol. Soc. Can. Mem. 40: 1–247.

### Cotton Leaf Worm

Noctuidae, Catocalinae, *Alabama argillacea. Spanish:* Algodonero (General, larva). *Portuguese:* Curuquerê do algodoeiro (Brazil, larva).

The adult (fig. 10.8c) of this very serious and widespread pest has reddish or clay brown fore wings. Fine, obscure, zigzag lines cross the wing perpendicularly, and a small, dark, oval spot marks the outer third near the center; the usually round and kidney-shaped spots are lacking. The hind wings are often pinkish or olive tinted. It is a little smaller than most noctuid agricultural pests (WS 3–3.5 cm).

The first pair of abdominal legs of the larva are reduced. Its body color is generally green, with or without black lines running the length of the body, the most dorsal the broadest, the more lateral ones narrower and broken.

The larva eats the leaves of cotton and is a major pest in all Latin American countries where this crop is grown (Habib 1977). It also feeds on a number of other malvaceous plants.

### Reference

HABIB, M. E. M. 1977. Contribution to the biology of the American cotton leafworm *Alabama argillacea* (Hübner) (Lepid., Noctuidae). Zeit. Angewan. Entomol. 84: 412–418.

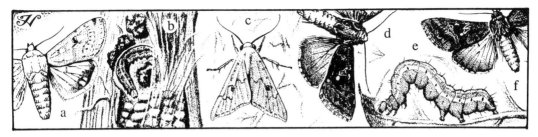

**Figure 10.8 OWLET MOTHS (NOCTUIDAE).** (a) Corn earworm (*Helicoverpa zea*). (b) Corn earworm, larva. (c) Cotton leaf worm (*Alabama argillacea*). (d) Cabbage looper (*Trichoplusia ni*). (e) Cabbage looper, larva. (f) Upsilon looper (*Rhachiplusia ou*).

## Loopers

Noctuidae, Plusiinae. *Spanish:* Agrimensores. *Portuguese:* Medidoras.

The first two abdominal legs are reduced or missing in the larvae of loopers, forcing them to inch along with a looping movement, arching the body upward and bringing the hindmost legs forward to meet the thoracic appendages, then reaching out with the forepart of the body to take the next step. In this respect, they resemble the caterpillars of the Geometridae, but they always retain vestigial walking legs on the third and fourth abdominal segments. Like cutworms, they are catholic in their tastes. An important general review of this group for North America, but including many species of Latin America, has been published (Eichlin and Cunningham 1978).

### Reference

EICHLIN, T. D., AND H. B. CUNNINGHAM. 1978. The Plusiinae (Lepidoptera: Noctuidae) of America north of Mexico, emphasizing genitalic and larval morphology. U.S. Dept. Agric., Agric. Res. Serv., Tech. Bull. 1567: 1–122.

### *Cabbage Looper*

Noctuidae, Plusiinae, *Trichoplusia ni*. *Spanish:* Gusano medidor de la col.

The fore wing of the adult cabbage looper (Shorey et al. 1962, Sutherland and Sutherland 1972) (fig. 10.8d) is basically dull brown with complex black mottling. The usual round and kidney-shaped spots are absent but are replaced by another double spot design, the outer part of which is a small silver oval, the inner part an uneven broad, U-shaped mark. It is a smaller species among leaf worms (WS 3–3.6 cm).

The larva (fig. 10.8e) is generally green, darker on the back between lateral pale yellow or cream lines. Its first two abdominal legs are very small and peglike.

This is a ravager, particularly of cruciferous (cabbage) and composite (lettuce, etc.) plants, but it feeds on many other hosts as well among crop and ornamental plants. The species has been widely referred to as *Autographa brassicae*.

### References

SHOREY, H. H., L. A. ANDRES, AND R. L. HALE, JR. 1962. The biology of *Trichoplusia ni* (Lepidoptera: Noctuidae). I. Life history and behavior. Entomol. Soc. Amer. Ann. 55: 591–597.

SUTHERLAND, D. W. S., AND A. V. SUTHERLAND. 1972. A bibliography of the cabbage looper, *Trichoplusia ni* (Hübner) 1800–1969. Entomol. Soc. Amer. Bull. 18: 27–45.

### *Upsilon Looper*

Noctuidae, Plusiinae, *Rachiplusia ou*. Gray looper.

Resembling the cabbage looper but with a gray ground color, the portion of the fore wing of this species basal to the main transverse zigzag line is dark shaded (fig. 10.8f). The inner part of the central wing spot is shaped like a narrow, uneven "U" lying on its side, its outer part a small silver oval. The wing also exhibits a slightly iridescent sheen on the outer half. The hind wing is orange tinged. It is average-sized (WS 3.1–4.1 cm).

The larva lacks any vestige of the first two abdominal legs. Its body color is more or less pale green all over, except for narrow white or yellowish longitudinal lines, the most conspicuous running just above the spiracles. This is an omnivorous feeder found throughout Latin America, where it is injurious mostly to tobacco and clover.

The species name, often mispelled as "*nu*," is not to be confused with "*oo*," formerly used for the soybean looper (*Pseudoplusia includens*), another widespread pestiferous species.

## Birdwing Moths

Noctuidae, Ophiderinae, *Thysania agrippina* and *T. zenobia*. *Portuguese:* Imperadores (Brazil).

Large specimens of this great moth (fig. 10.9a) have one of the greatest wingspans of any lepidopteran in the world (WS to 30 cm). Because of their immense size, their presence always fosters awe and disbelief. In flight, the insect is easily mistaken for a bird.

The fore wings are much longer than the hind wings; both are scalloped along their outer margins. The pale gray, almost white ground color is crossed by numerous diagonal wavy and zigzag dark gray lines. These lines blend with vertical crevices in the bark

of trees on which the moth often rests in a horizontal position. The typical noctuid kidney-shaped and round spots are present in the fore wing but overpowered by the lined pattern. Although it is much like a saturnian in its large size, it has fine, threadlike, rather than feathery, antennae.

A related but slightly smaller (WS 11 cm) species is Zenobia's birdwing moth (*Thysania zenobia*, imperador rosa, Brazil) (fig. 10.9b). Its wings are marked similarly to those of the birdwing, but males have heavy, dark bars passing through the anterior third of the fore wing parallel to the leading margin and a like bar a short distance inside the hind margin of the hind wing. The female lacks the former but possesses the latter.

Both of these moths are common forest dwellers. The latter sometimes succumbs to the wandering instinct and shows up in the southern part of the United States. Apparently nothing is known of the early stages of either species.

## Black Witch

Noctuidae, Ophiderinae, *Ascalapha odorata*. *Spanish:* Mariposa de la muerte (Mexico), pirpinto de la yeta (Argentina). *Nahuatl:* Miquipapálotl, tepanpapalotl (Mexico). *Quechua:* Taparaco (Peru). *Mayan:* X-mahan-nail (Yucatán).

This moth (fig. 10.9c) is very common throughout the New World tropics, where it

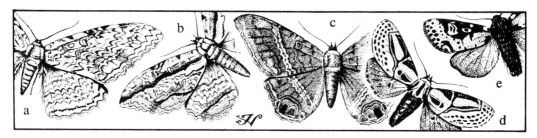

**Figure 10.9   OWLET MOTHS (NOCTUIDAE).** (a) Birdwing moth (*Thysania agrippina*), male. (b) Zenobia's birdwing moth (*Thysania zenobia*), male. (c) Black witch (*Ascalapha odorata*), female. (d) Hieroglyphic moth (*Diphthera festiva*). (e) Spanish moth (*Xanthopastis timais*).

readily comes to house and street lights at night. It also is attracted by the odor of rotting fruit. Because of its large size (WS 20–25 cm) and dark, batlike appearance, it often attracts attention and sometimes causes alarm. It is regarded by the superstitious as a harbinger of death and is known in Mexico by the Indians since Aztec times as *mariposa de la muerte*, or *miquipapálotl* (Nahuatl: *miqui* = death, black; *papálotl* = moth), for it is believed that when there is a sickness in a house and this moth enters, the sick person dies (Hoffmann 1918). The same belief prevails in Peru. On the Yucatán Peninsula, the habit of entering buildings is the basis of the Mayan name *x-mahan-nail* (*mahan* = to borrow + *nail* = house).

The adults are somewhat variable in size, depending on larval nutrition. Females are larger than the males and otherwise recognizable by a generally lighter color and a contrasting, white, transverse band crossing the wings. The upper surfaces of the wings of both sexes are otherwise dark brown with fine wavy or zigzag lines and conspicuous eyespots near the leading edge of the fore wing (smaller) and at the posterior apex of the hind wing (larger and double). The body is evenly dark brown and without scale tufts. Another characteristic color feature is a violet iridescent sheen that may be seen with oblique light; this is much more noticeable in the female because of the paler ground color.

The early stages are fairly well known (Bourquin 1947; Comstock 1936). The larva at maturity is very large (BL 6 cm) and has stout proportions. It is widest at the fourth (first abdominal) segment and tapers abruptly posteriorly. The ground color is gray or gray-brown, heavily mottled with black. There is a wide, middorsal, longitudinal band of light gray that expands on the tenth (seventh abdominal) segment into a subtriangular area. There is also a broken, undulating lateral band through the spiracles. The head is black or brownish-black dorsally.

Food plants normally consist of various leguminous plants, of which the following have been recorded: *Cassia fistula, Gymnocladus dioica, Acacia decurrens, Pithecellobium unguiscate, Inga,* and *Samanea saman* (Fabaceae). Doubtlessly, other related genera and species will be found to host the species. The larvae feed during the night and rest during the day on the bark, usually in depressions, utilizing their cryptic pattern as protection from predators. Pupation occurs in a cocoonlike accumulation of leaves on the ground or in crotches between large tree branches.

A remarkable feature of this moth is its migratory habit. It is a strong flier and turns up in scattered localities every year in the United States where it is not known to be resident (Sala 1959). Specimens, usually worn males, appear regularly in California, Kansas, and even New York and southern Canada. Most such occurrences are in the late summer and fall (August to October) and indicate a northward movement from breeding areas in Mexico or possibly farther south. No studies have been conducted to determine migration routes. Neither is it now known whether there is a southward migration in the Northern Hemisphere, reciprocal migration at the southern end of the species' distribution, or other more complex patterns of movement within the Neotropics.

The species was known as *Erebus odora* in earlier literature and also as *Otosema odorata* (Oiticica 1962).

## References

Bourquin, F. 1947. Metamorfosis de "*Erebus odoratus*" (Linné) 1758 (Lep. Het. Noctuidae). Acta Zool. Lilloana 3: 239–248.

Comstock, J. 1936. Notes on the early stages of *Erebus odora* L. (Lepidopt.). So. Calif. Acad. Sci. Bull. 35: 95–98.

Hoffmann, C. C. 1918. Las mariposas entre los antiguos Mexicanos. Cosmos 1.

Oiticica, T. 1962. Nome atual da espécie *P.* [*Halaena*] *Bombyx odorata* Linnaeus, 1758. Mus. Nac. Rio de Janeiro Arq. 52: 137–144.

SALA, F. P. 1959. Possible migration tendencies of *Erebus odora* and other similar species. J. Lepidop. Soc. 13: 65–66.

## Hieroglyphic Moth

Noctuidae, Ophiderinae, *Diphthera festiva*.

Because of its gaudy color pattern, the hieroglyphic moth (fig. 10.9d) often calls attention to itself when resting on walls near lights to which it has been attracted. The anterior half of the fore wing is basically yellow-orange with contrasting metallic blue lines marking off triangular areas; the posterior half is pale yellow, with black spots on the outer portion. It belongs to the family Noctuidae, among whose members it is of average size (WS 3.7–4.8 cm). The striped, slate blue larvae feed on various hardwood trees and sweet potato vines. They pupate in a weak cocoon made of coarse silk and plant fragments, placed in the branches of the food plant (Benjamin 1922).

The species was known formerly as *Noropsis hieroglyphica*.

### Reference

BENJAMIN, F. H. 1922. Early stages of *Noropsis hieroglyphica* Cram. (Lepidoptera, Noctuidae). Entomol. News 33: 277–278.

## Spanish Moth

Noctuidae, Hadeninae, *Xanthopastis timais*.

The Spanish moth (fig. 10.9e) is a very widely distributed, injurious species. The voracious larvae devastate the leaves of ornamental flowers of the family Amaryllidaceae (*Amaryllis, Narcissus*, etc.), often killing the plant (Biezanko and de Souza Guerra 1971). The caterpillars (BL 40–50 mm) are basically black but with numerous small, oval, milky white spots and with scattered tubercles (Bourquin 1935).

The adults (WS 4–4.5 mm) are easily recognized by the hairy black body and pinkish fore wing, the latter with a black triangular field anteriorly, in the center of which is a yellow spot with black markings. The hind wing is black.

### References

BIEZANKO, C. M., AND M. DE SOUZA GUERRA. 1971. Contribuição ao estudo de *Xanthopastis timais* Stoll, 1782, uma praga importante de Amarilidáceas (Lepidoptera, Heterocera, Noctuidae). Arq. Mus. Nac. Rio de Janeiro 54: 267–272.

BOURQUIN, F. 1935. Metamorfosis de *Xanthopastis timais* Cr. (Lep. Noct.). Rev. Soc. Entomol. Argentina 7: 195–201.

## PROMINENTS

Notodontidae

The prominents are diverse and varied in the Neotropics, where approximately 1,500 species occur (Heppner in press). They are mostly medium to large (WS 4–8 cm) and stout and often have angularly lobed or elongate fore wings. Most are drab gray or brown, but some are cryptically or disruptively marked (e.g., *Cliara croesus;* fig. 10.10a); the hind wing is patterned like the fore wing. Their larvae (fig. 10.10b) are peculiar in their habit of arching backward strongly when disturbed, elevating both the anterior and posterior extremities. The posterior end may mimic a reptile head or have a pair of long, whiplike tails. The body may be lobed or bear odd-shaped excrescences dorsally.

### Reference

HEPPNER, J. B. In press. Lepidoptera family classification: A guide to the higher categories, world diversity and literature resources of the butterflies and moths. Flora and Fauna, Gainesville.

## LAPPET MOTHS

Lasiocampidae.

Lappet moths form another group of about 650 species (Heppner in press) of stout, medium-sized (WS usually 4–6 cm),

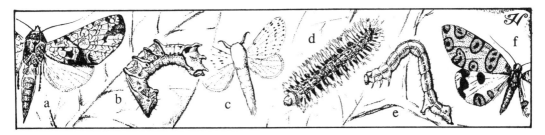

**Figure 10.10.   MOTHS.** (a) Prominent moth (*Cliara croesus,* Notodontidae). (b) Prominent moth (unidentified), larva. (c) Lappet moth (*Euglyphis cribraria,* Lasiocampidae). (d) Lappet moth (unidentified), larva. (e) Measuring worm (unidentified, Geometridae), larva. (f) Polka dot moth (*Pantherodes pardularia,* Geometridae).

drab or pale moths with bodies densely clothed with compact hairlike scales. A white species is *Euglyphis cribraria* (fig. 10.10c). Many species rest with the fore wings folded rooflike but with the hind wings splayed out flat beyond the edges of the fore wings. Females of many have reduced or vestigial wings and do not leave the location of their emergence from the pupa. The abdomen tends to be somewhat larger than in related groups. Their caterpillars usually possess dense growths of fine hairs, generally distributed over the body or in series along the sides which contact the substrate, eliminating the shadow of the body and aiding in camouflage. The hairs are venomous and can cause a rash on human skin.

Some such larvae are brightly colored, advertising their toxic nature, and have the habit of grouping together, either exposed or in large silken tents. Silk swaths gathered from the large hammock-net cocoons of *Gloveria psidii* (= *Sagana sapotoza*) and pasted together to form a kind of hard cloth, or paper, were an important trade item in Mexico at the time of Moctezuma II (= "*Bombyx madroña*"). (See domestic silk moth.)

### Reference

HEPPNER, J. B. In press. Lepidoptera family classification: A guide to the higher categories, world diversity and literature resources of the butterflies and moths. Flora and Fauna, Gainesville.

## MEASURING WORM MOTHS

Geometridae.

Measuring worms ("inchworms," "*medidores*") are the caterpillars of the family Geometridae, a large family (4,452 species; Heppner in press) in the Neotropics of mostly small to medium (WS 1–3 cm), drably or cryptically marked moths. Their wings are generally large in proportion to the body and display a great range of outline shape and color; perhaps most are brown or gray, but there are many green species. A few adults are mimetic, such as *Atyria dicroides*, which resembles net-winged beetles and their other lepidopteran mimics, or common and conspicuous, such as the polka dot moth (*Pantherodes pardalaria;* fig. 10.10f) with large blue-gray spots, rimmed and centered with black, on a yellow background.

Larvae typically lack all but the posteriormost two pairs of abdominal legs and advance by alternately releasing hold by these and the thoracic legs and looping the body (fig. 10.10e). Many look like sticks or stems with their wrinkled skins and gnarled or spined bodies. When disturbed, their resemblance to twigs is enhanced by their habit of freezing rigidly in an extended posture.

### Reference

HEPPNER, J. B. In press. Lepidoptera family classification: A guide to the higher categories, world diversity and literature resources of the butterflies and moths. Flora and Fauna, Gainesville.

## FLANNEL MOTHS

Megalopygidae. *Spanish:* Gusanos pollos (Colombia), plumillas (Puerto Rico, larvae). *Portuguese:* Lagartas de fogo, lagartas cabeludas, ursos, tatoranas (Brazil, larvae). *Tupi-Guaraní:* Tatorana, tatá-ranás, sassuranas (larvae). *Jívaro:* Bayuca (Peru, Ecuador, larvae). *Quechua:* Cuy (Peru, larvae). Puss caterpillars.

These are stout, small to medium-sized (WS usually 25–35 mm) moths, with long, hairlike pelage of loose, soft scales on the body (Hopp 1935). Their mouthparts are vestigial, and they do not feed.

The larger genera, *Podalia* and *Megalopyge* (fig. 10.11a), are drab brown, but the others, like *Trosia*, are roseate, yellow, or white with pink, red, or yellow markings.

The larvae are much better known than the adults because of the fiery sting they produce in contact with the skin. They are of two types, one covered densely and completely with long, red, orange, or white flowing hair that almost invites petting (fig. 10.11c). But hidden within this soft pelage are rigid, highly toxic nettling bristles. There are six pairs of abdominal legs (on segments 2–7) and an anal pair; legs 2 and 7 have no foot-hooks (crochets). In the sec-

ond type, only the lateral hairs are long; the dorsal, long hairs are sparse and arise from two parallel rows of wartlike tubercles (fig. 10.11b). These are responsible for many cases of caterpillar-caused skin rashes in Latin America. The operculate cocoons are hard and parchmentlike, made of a secretion of the caterpillar mixed with body hairs. They may be grouped in dense, single-layered masses on tree trunks and large branches. The larvae may aggregate for pupation, forming a communal cocoon of silk and body hairs.

The adults of some species may have colors and patterns mimicking other insects (*Endobrachys revocans* and relatives are very similar to the tiger moth, *Cratoplastis diluta*, and cockroach, *Achroblatta luteola*).

### Reference

Hopp, W. 1935. Megalopygidae. *In* A. Seitz, ed., The Macrolepidoptera of the world: The American Bombyces and Sphinges. 6: 1071–1101. Kernen, Stuttgart.

## SHAG MOTHS

Limacodidae (= Eucleidae, Cochlididae). *Spanish:* Cornegachos (Peru), gusanos ratón (Central America, larvae). *Portuguese:* Lagartas aranha (Brazil, larvae). *Jívaro:* Bayucas (Peru, larvae).

The adults of this family (Dyar 1935) are drab (sometimes green or with silver

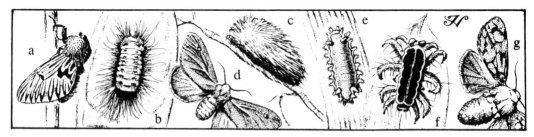

**Figure 10.11    MOTHS.** (a) Flannel moth (*Megalopyge lanata,* Megalopygidae). (b) Flannel moth, larva. (c) Flannel moth (*Megalopyge* sp.), larva. (d) Shag moth (*Acharia nesea,* Limacodidae). (e) Shag moth (*Acharia* sp.), larva. (f) Monkey slug (*Phobetron hipparchia,* Limacodidae), larva. (g) Monkey slug, adult.

marks), nondescript, and medium-sized (WS usually 15–30 mm) nocturnal moths (fig. 10.11d, g), for the most part, with robust bodies and relatively small wings. Their mouthparts are reduced, and they do not feed.

The early stages, in contrast, are highly distinctive. Larvae are sluglike, without apparent segmentation or well-defined walking legs, the usual lepidopteran foot-hooks (crochets) even lacking. They always have short bristles or hairs. The head is concealed by retraction into the thorax. Among the many, two common body types often seen are the green to yellow "saddle backs" (*Acharia* = *Sibine,* many species), with short lobes in a peripheral series and longer lobes at either extreme dorsally (fig. 10.11e), and the brown, hairy, "spider" type (*Phobetron,* "monkey slug," *P. hipparchia*), with long and short pairs of lateral fingerlike lobes (fig. 10.11f) (Young 1986). Both bear intensely toxic, urticating spines on their body lobes. They also receive protection from their spiderlike appearance, although they look like a dead leaf to some. These larvae are major sources of caterpillar dermatitis in Latin America. There are also many species with smooth larvae that lack urticating hairs.

Larval food is varied and includes many common plant varieties, such as bananas and palms on which they may be major pests; *Stenoma cecropia* is a defoliator of oil palm (Genty 1978). The silken cocoon is placed on the host, often in groups, and is a well-made, tough, ovoid structure, usually light brown or white. It may have contrasting dark spots, ostensibly mimicking the emergence holes of parasites, thus discouraging the feeding attempts of birds.

A family related to this, with interesting sluglike larvae, is the Dalceridae (Miller in press, Orfila 1961). These are small to medium-sized moths (WS usually 1–3 cm), similar to limacodids, with white, yellow, or orange ground colors, bipectinate antennae tapering gradually to the tip, and no proboscis. The larvae (*lagartas gelatinosas,* in Brazil) are covered with translucent, sticky, gelatinous, conical tubercles. They are usually rare but can be pests of tree and shrub crops, especially *Dalcerina tijucana* (often incorrectly cited as *Zadalcera fumata*) on citrus.

### References

DYAR, H. 1935. Limacodidae. *In* A. Seitz, ed., The Macrolepidoptera of the world: The American Bombyces and Sphinges. 6:1104–1139. Kernen, Stuttgart.

GENTY, P. 1978. Morphologie et biologie d'un Lépidoptère défoliateur du palmier à huile en América latine, *Stenoma cecropia* Meyrick. Oleagineux 33: 421–427.

MILLER, S. E. In press. Revision of the Neotropical moth family Dalceridae (Lepidoptera). Mus. Comp. Zool. Bull.

ORFILA, R. N. 1961. Las Dalceridae (Lep. Zygaenoidea) Argentinas. Rev. Invest. Agric. (Castelar, Argentina) 15: 249–264.

YOUNG, A. M. 1986. Notes on a Costa Rican "monkey slug" (Limacodidae). J. Lepidop. Soc. 40: 69–71.

## DOMESTIC SILK MOTH

Bombycidae, *Bombyx mori. Spanish:*
   Gusano de seda (larva). *Portuguese:*
   Bicho da sêda (Brazil, larva).

Little needs to be said regarding this species as a part of the Neotropical fauna. It is a totally domesticated animal, cultured since ancient times in the Orient for the production of silk, and quite incapable of surviving without intense human protection and care.

The species has been introduced into various parts of Latin America repeatedly by capitalists with visions of establishing profitable industries. For a variety of reasons, most often the high cost of maintaining cultures and processing the fiber in this labor-intensive endeavor, no scheme has long succeeded or survives today.

A history of sericulture in Latin America has not yet been written in full, although events in Mexico are well known (Borah 1943). The first seed was brought apparently by the Spaniards to Hispaniola in the opening years of the sixteenth century. Cortés introduced the moth to Mexico in 1523 where the silk industry was encouraged and where, aided by cheap Indian labor, it thrived. Political and economic competition from the Old World brought on a decline even before the end of the century from which it never recovered, in spite of revival by the Bourbon rule in the early 1800s. Introductions to other countries followed separately and at disparate times (see Lamas and Lamas 1980, for Peru, and Adames 1945, for Brazil).

It is a curious fact that Cortés was presented a silken cloth by Moctezuma which was made from a so-called native silk (*seda silvestre, seda de la mixteca*), called *temictli* (or *icheatzin, xochiaietlan*). Silk textile manufacture was already practiced by the Aztecs in central Mexico, the material being obtained from two indigenous lepidopterous species (Cowan 1865, Hoffman 1910). One was a lasiocampid moth (*Gloveria psidii = Sagana sapotoza*) with large hairy caterpillars that spun an enormous baglike, silken nest among the limbs of the host, guayaba or "encino" (*Psidium guajava*), from which they wander to feed. Pupation takes place in a small tight cocoon in the middle of the nest.

The other silk producer was the madrone butterfly (*mariposa del madroño*), *Eucheira socialis*, a pierid. The larvae are likewise gregarious and construct a compact whitish tissue like a bag of silk among the branches of the madrone (*Arbutus*) tree. They seek refuge in this sac during the day, leaving at night to eat leaves. The chrysalids are attached inside the bag.

In both cases, the Aztec artisans cut up the large sacs, piecing together the resulting swatches into larger pieces of "fabric." The fibers composing them were not unwound and woven into textiles as with the cocoons of *Bombyx mori* or other wild saturniid types, such as *Rothschildia* (see window-winged moths, above).

## References

ADAMES, G. E. 1945. Silks from South America. *In* C. M. Wilson, New crops for the New World. Macmillan, New York. Pp. 193–210.

BORAH, W. 1943. Silk raising in colonial Mexico. Ibero-Americana 20: 1–169.

COWAN, F. 1865. Curious facts in the history of insects. Lippincott, Philadelphia.

HOFFMANN, C. C. 1910. Humboldt's Nachrichten über die in Mexiko einheimischen Seiden spinnen Raupen, unter specieller Bearbeitung des von ihm erwähnten Madroño-Falters, *Eucheira socialis* Westw. *In* E. Wittlich, H. Beyer, and R. C. Damm y Palacio, Wissenschaftliche Festschrift zu Enthüllung des Deiten Seiner Majestät Kaiser Wilhelm II, dem Mexikanischen Volke zum Iubiläum seinen Unabhängigkeit gestift Humboldlt-Denkmals. Müller Hnos., México. Pp. 149–173.

LAMAS, G., AND J. M. LAMAS. 1980. Introducción a la historia de la entomología en el Perú. III. Albores de la entomología económica. Rev. Peruana Entomol. 23: 32–37.

## BAGWORM MOTHS

Psychidae. *Spanish:* Gusanos de cesto. *Portuguese:* Bichos de cêsto.

Bagworms are so named from the sac in which the larva and female moth spend their entire lives (fig. 10.12b). This is a tough case of silk into which are often incorporated bits of plant material, twigs, leaves, and so on, which camouflage it as well as give it physical strength. Only the head and thoracic legs project from the bag as the larva moves about while hanging from the undersides of its host parts. It is quick to withdraw if threatened. The bag is enlarged by the growing larva, protecting it from harm and serving as well for a cocoon for the pupa. Larval development, especially of the larger species, is prolonged and requires numerous supernumerary molts. Bag shape and arrangement

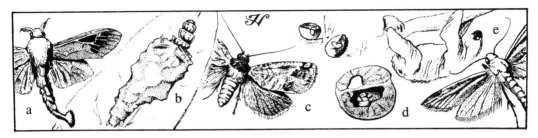

**Figure 10.12    MOTHS.** (a) Bagworm moth (*Oiketicus kirbyi*, Psychidae). (b) Bagworm, larva in bag. (c) Mexican jumping bean moth (*Cydia deshaisiana*, Tortricidae). (d) Mexican jumping bean moth, larva in seed. (e) Sloth moth (*Cryptoses choloepi*, Pyralidae).

of foreign objects are often diagnostic of species.

The adult male that escapes from the bag is fully winged and disperses in search of the flightless females that remain encapsulated. The appendages of the females of most species are vestigial, and the body is a largely amorphous membranous structure. The wing scales of the males of many are loosely attached and lost after emergence, leaving a portion of the wing membrane transparent. Mating takes place either inside or outside the bag. The former is possible because of the long telescoping abdomen of the male, which is inserted deeply inside the bag to come in contact with the female's genitalia. The eggs are normally deposited within the bag also.

There are approximately seventy Neotropical species of Psychidae (Davis 1964). The family is not well known, and new species are certain to appear as study continues. The largest genus is *Oiketicus*, with the widespread species *O. kirbyi* (fig. 10.12a) ranging throughout most of Central and South America and the West Indies. *O. platensis* is a pest of ornamentals in Argentina (de Briano et al. 1985).

Larvae are polyphagous, with wide host ranges, although they show a definite reluctance to change food abruptly during development. Some are injurious because of their attacks on ornamental plants.

In the past in Mexico, a sack containing the bag of a species of *Oiketicus* was hung on the front of a child's cotton garment for its magical effects. It was called *quahquahuini*, or "woodsman worm," and was thought to ensure a good supply of firewood (Giordano and Beutelspacher 1989).

### References

DE BRIANO, A. E., I. S. DE CROUZEL, E. SAINI, AND V. LASAIGUES. 1985 [1983]. Observaciones sobre el ciclo biológico del "bicho de cesto" (*Oiketicus platensis* Berg, 1883—Lepidoptera: Psychidae). Rev. Peruana Entomol. 26: 51–58.

DAVIS, D. R. 1964. Bagworm moths of the Western Hemisphere. U.S. Nat. Mus. Bull. 244: 1–233.

GIORDANO, C. R., AND C. R. BEUTELSPACHER. 1989. El quahquahuini o gusano leñador, y sus implicaciones magicas en un textil tradicional de la Sierra Norte de Puebla, México. Anthropologicas 3: 21–27.

## MEXICAN JUMPING BEAN MOTH

Tortricidae, *Cydia deshaisiana. Spanish:*
   Brincador (Mexico, bean with larva).
   Jumping beans, Devil's beans (seeds with larva).

In the región of the Río Mayo in southern Sonora and Chihuahua, Mexico, a species of tropical shrub called yerba de flecha (*Sebastiana pavoniana*, Euphorbiaceae) produces angular beanlike pods, some of which become infested with the larvae of this moth (Berg 1891) (fig. 10.12d). Females place the eggs on the pod early in its development, in the spring. The pods, each containing a single larva, later dry and fall to the ground. The larva lines the pod's interior with silk, and as the sun heats it, it becomes agitated and grasps the wall with its legs and snaps its body. This causes the pod to jerk. The higher the temperature, the more vicorously it jerks. The action apparently allows the larva to find a suitable crack or crevice out of the heat, which might kill it. The larva weighs about the same as the shell of the bean, so it can generate sufficient momentum to move it.

These "jumping beans" (Hutchins 1956) are collected in quantities by local entrepreneurs and exported to southern California, Arizona, New Mexico, and Texas where they are sold as curios during the jumping season, May or June. Games of chance have even been devised, with the beans used as pawns.

Just before maturing (when they reach a length of 3–5 mm), the larva cuts a circular door in the end of the pod, leaving an edge in place, like a trapdoor. The emerging moth later pushes it out of the way to escape. Beans with pupae do not jump.

The adult (fig. 10.12c) is a small (WS 20 mm), dark brown moth with broken, zigzag bluish-gray lines traversing the fore wings, the tips of which are abruptly marked with a broad dark triangular area.

The species formerly went under the name "*Laspeyresia saltitans*."

## References

BERG, C. 1891. Sobre la *Carpocapsa saltitans* Westw. y la *Grapholitha motrix* Berg, n. sp. Soc. Cient. Argentina An. 31: 97–110.

HUTCHINS, R. E. 1956. The jump in the jumping bean. Nat. Hist. 65: 102–105.

## PYRALID MOTHS

Pyralidae.

The remaining moths are all members of a large and economically important family. They are mostly small, but some are medium-sized with rather long labial palpi that project forward in front of the head. They have tympana, whose cavities open forward on the underside of the abdomen near the base. Many species fold their wing curiously, rolling them or otherwise doubling them to emulate sticks and other innocuous objects.

### Sloth Moths

Pyralidae, Chrysauginae, *Cryptoses choloepi*.
  *Spanish:* Polillas del perezozo.
  *Portuguese:* Traças da preguiça.

For a long time, an erroneous idea was perpetuated in the literature that the larvae of these moths as well as the adults lived amid the hair of sloths, feeding on algal masses that allegedly grew there. (Such masses do not exist, although a minute unicellular alga does develop in the fine hair straiae, imparting to the fibers a green color.) It has been discovered, however, that the caterpillars actually live in the dung of the sloth (Waage and Montgomery 1976). Adult female moths leave the host's fur to oviposit on the mammal's leavings when it descends once a week to the forest floor to defecate. The pupa are also found in the dung pile, and the newly emerged moths fly into the forest canopy to find a new host. Mating occurs on the body of the sloth (Greenfield 1981: 6).

Although they come to lights at night (Wolda 1985), these moths (fig. 10.12e) are normally seen only on the bodies of three-toed (*Bradypus*) and two-toed (*Choloepus*) sloths. More than one hundred individuals may be found on a single animal at a time. They are very active, slipping rapidly through the fur, aided by their small size (length when wings folded, about 12 mm), flattened body, and arrowhead shape. Their fore wings are dark brown, with three contrasting cream-colored longitudinal lines; the hind wings are dark gray-brown. The proboscis of the sloth moth is very short, but they readily drink in the laboratory; their liquid food in nature is unknown.

This species is common in Central America wherever its host occurs. There are four additional species of South American sloth moths whose relationships and biologies remain unstudied (Bradley 1982).

## References

BRADLEY, L. D. 1982. Two new species of moths (Lepidoptera, Pyralidae, Chrysauginae) associated with the three-toed sloth (*Bradypus* spp.) in South America. Acta Amazonica 12: 649–656.

GREENFIELD, M. D. 1981. Moth sex pheromones: An evolutionary perspective. Fla. Entomol. 64: 4–17.

WAAGE, J. K., AND G. G. MONTGOMERY. 1976. *Cryptoses choloepi*: A coprophagous moth that lives on a sloth. Science 193: 157–158.

WOLDA, H. 1985. Seasonal distribution of sloth moths *Crytoses choloepi* Dyar (Pyralidae; Chrysauginae) in light traps in Panama. *In* G. G. Montgomery, ed., The evolution and ecology of armadillos, sloths, and vermilinguas. Smithsonian Inst. Press, Washington, D.C. Pp. 313–318.

## Sugarcane Borer

Pyralidae, Crambinae, *Diatraea saccharalis.*
  *Spanish:* Barreno de la caña,
  barrenador de los tallos cañeros,
  taladrador de la caña de azúcar.
  *Portuguese:* Broca da cana.

This species has assumed major importance as a pest in Latin America because of the havoc it wreaks on sugarcane, one of the region's economically most important crops. The larvae (fig. 10.13b) bore into the internodes and often cause the death or significant decline of the whole plant. At harvest time, it is not uncommon to find that up to a third of the cane internodes contain larvae. They permit the entry of fungi and reduce the quantity and purity of the juice. In older canes, the tunneling of the borers causes the tops to die so that the stocks break off in strong winds. The larvae also attack rice and corn in a similar manner. The species is cosmotropical and is distributed widely through the West Indies, Central America, and South America to Buenos Aires Province.

The medium-sized (WS 18–29 mm) moth (fig. 10.13a) is pale straw colored, with black dots in a V-pattern on the also faintly black-streaked fore wings. Full-grown larvae are about 25 millimeters long and yellowish white with contrasting dark brown head and prothorax and spots at the base of the body hairs.

The females place their disk-shaped eggs on the leaves, overlapping them like roof tiles. The young larvae move into the apical funnel of leaves and feed first on the leaf surface, later moving into the stem. Prior to pupation, the mature larva makes a chamber, separated from the outside by a thin wall of plant tissue and lightly lined with silk. In this it pupates. To escape, the adult breaks through the thin wall.

The Amazon fly (*Metagonistylum mi-*

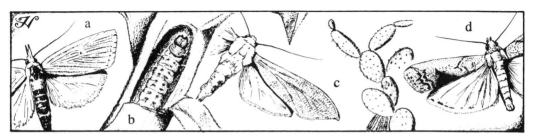

**Figure 10.13  PYRALID MOTHS (PYRALIDAE).** (a) Sugarcane borer (*Diatraea saccharalis*). (b) Sugarcane borer, larva. (c) Giant pyralid (*Myelobia smerintha*). (d) Cactus moth (*Cactoblastis cactorum*).

*nense*), a tachinid fly parasite of the larvae, and the trichogramma wasp (*Trichogramma minutum*), which destroys its eggs, are released in cane fields to help in the borer's control. Other members of the same genus (Bleszynski 1969; Box 1931, 1956) also plague this crop in Mexico, Guyana, and Trinidad. These are *D. considerata, D. magnifactella, D. grandiosella,* and *D. centrella.*

The laboratory biology of the species is presented at length by Bergamin (1949: 61f.).

## References

BERGAMIN, J. 1949. *Diatraea saccharalis. In* A. Costa Lima, Insetos do Brasil. 6: 61–81. Escuela Nac. Agron, Rio de Janeiro.

BLESZYNSKI, S. 1969. The taxonomy of crambid moth borers of sugarcane. *In* J. R. Williams, J. R. Metcalf, R. W. Montgomery, and R. Mathes, eds., Pests of sugarcane. Elsevier, Amsterdam. Pp. 1–9.

Box, H. E. 1931. The crambine genera *Diatraea* and *Xanthopherne* (Lep., Pyral.). Bull. Entomol. Res. 22: 1–50.

Box, H. E. 1956. New species and records of *Diatraea* Guilding and *Zeadiatraea* Box from Mexico, Central and South America (Lepid., Pyral.). Bull. Entomol. Res. 47: 755–776.

## Giant Pyralid

Pyralidae, Crambinae, *Myelobia*.

These pyralids are among the largest microlepidoptera. Adults have wingspans of 10 to 12 centimeters (fig. 10.13c). They have long, pale brown, pointed wings and much resemble sphinx moths with which they are often confused. The larvae (*bichos da taquara-quice*) are borers in bamboo and taquaras (bamboolike grasses). Paraguayan Indians reportedly used the larvae pharmaceutically. The head was believed to be deadly poisonous, while the intestinal tract contains substances that induce hallucinogenic trances (Schultes 1974). Adults congregate around lights in large numbers, even to plague proportions in São Paulo and Rio de Janeiro in recent history (Ihering 1917).

## References

IHERING, R. 1917. Observações sobre a mariposa *Myelobia smerintha* Hübner em São Paulo. Physis 3: 60–68.

SCHULTES, R. E. 1974. Alucinógenos tropicales americanos. Bol. Soc. Quim. Peru 40: 230–247.

## Cactus Moth

Pyralidae, Phycitinae, *Cactoblastis cactorum*.

The cactus moth (fig. 10.13d) is a native of southeastern South America, where its normal hosts are prickly pear cacti (*Opuntia*). It was introduced from Argentina into Australia in 1914 and again in 1925, after the first trial failed, for the control of this plant that had become a weed, threatening to take over vast amounts of valuable agricultural and rangeland territory (Dodd 1940). Eggs are laid at the bases of spines, and the larvae tunnel into the pads, reducing them to a rotting mass. Others in the genus may be more effective control agents under some conditions and in other areas (McFadyen 1985).

## References

DODD, A. P. 1940. The biological campaign against prickly-pear. Commonwealth Prickly Pear Board, Brisbane.

MCFADYEN, R. E. 1985. Larval characteristics of *Cactoblastis* spp. (Lepidoptera: Pyralidae) and the selection of species for biological control of prickly pears (*Opuntia* spp.). Bull. Entomol. Res. 75: 159–168.

## Pantry and Granary Moths

Pyralidae, Phycitinae; Gelechiidae.

A number of small lepidopterans (Corbet and Tams 1943), commonly called meal moths, are pests of stored products, from warehouses to the home pantry. The most damaging of these in Latin America are the pyralid flour moths (*Ephestia* spp. and *Anagasta kuehniella;* fig. 10.14a) and the Indian meal moth (*Plodia interpunctella;* fig. 10.14b). In the family Gelechiidae, there is also the Angoumois grain moth (*Sitotroga cerealella*).

**Figure 10.14.** **MOTHS.** (a) Mediterranean flour moth (*Anagasta kuehniella,* Pyralidae). (b) Indian meal moth (*Plodia interpunctella,* Pyralidae). (c) Greater wax moth (*Galleria mellonella,* Pyralidae). (d) Agave worm moth (*Comadia redtenbacheri,* Cossidae). (e) Agave worm butterfly (*Aegiale hesperiaris,* Megathymidae).

## Reference

CORBET, A. S., AND W. H. T. TAMS. 1943. Keys for the identification of the Lepidoptera infecting stored food products. Zool. Soc. London, Proc. B 113: 55–148.

## Greater Wax Moth

Pyralidae, Galleriinae, *Galleria mellonella.*
  *Spanish:* Gusano de la cera, falsa tiña
  de los colmenares.

The larva of this pyralid attacks the combs of the honeybee, feeding on the wax and spoiling the honey. It is a cosmopolitan species well known to apiculturists. The adult (fig. 10.14c) is medium-sized (WS 28–35 mm), with dull brown fore wings having a characteristic notched or falcate apex; the hind wings are cream (Koehler 1933, Whitcomb 1936).

## References

KOEHLER, P. 1933. La polilla de las colmenas (*Galleria mellonella* L.). Min. Agric. Argentina Bol. Mens. 33: 257–277.

WHITCOMB, W. 1936. The wax moth and its control. U.S. Dept. Agric. Circ. 386: 1–14.

## AGAVE WORMS

Cossidae, Cossinae, *Comadia redtenbacheri,* and Megathymidae, Aegialinae, *Aegiale hesperiaris. Spanish:* Gusanos del maguey. *Nahuatl:* Meocuilin.

In the central, dry highlands of Mexico, potent fermented beverages have been made from the sap of various species of agave plants since prehistoric times. During the harvesting of the sap, when the center leaves are cut open, the larvae of these two lepidopterans are commonly found and collected along with the fluid. Curiously, although similar, they were clearly distinguished by the Aztecs, who referred to the larvae of the moth as *chilocuiles* and of the hesperiid as *meocuili* (MacGregor 1969). Today, they are ordinarily confused by the layman, both being unicolorous, white, wormlike larvae with inconspicuous legs (a modification in keeping with their habits as borers). To the Indians and campesinos, they often become food, as they are rich in fats, and were also consumed for medicinal reasons (Bachstez and Aragón 1942), but a select number were pickled in the brew, normally mezcal, to give it a "special flavor." Even today, bottles of the better brands of this liquor contain such caterpillars (see fig. 1.14). More normally, they are fried in grease or their own fat or braised and chopped into fragments and mixed with soup or green tomato sauce or other dishes.

Unmolested, *Comedia redtenbacheri* larvae develop into moths (fig. 10.14d) with long gray, mottled wings; *Aegiale hespariaris* (fig. 10.14e) becomes a large brown skipper butterfly (Ancona 1934, Dampf 1924).

## References

ANCONA, L. 1934. Los gusanitos del maguey, *Aegiale (Acentrocneme hespariaris)* Kirby. Inst. Biol., Univ. Nac. México, An. 5: 193–200.

BACHSTEZ, M., AND A. ARAGÓN. 1942. Notes on Mexican drugs. II. Characteristics and composition of the fatty oil from "gusanos de Maguey" (caterpillars of *Acentrocneme hesperiaris*). J. Amer. Pharm. Assoc. Sci. Ed. 31: 145–146.

DAMPF, A. 1924. Estudio morfológico del gusano del maguey (*Acentrocneme hesperiaris* Wlk.) (Lepidoptera, Megathymidae). Rev. Biol. Mexicana 4: 147–161.

MACGREGOR, R. 1969. La représentation des insectes dans l'ancien Mexique. L'Entomologiste 25: 1–8.

# BUTTERFLIES

Papilionoidea (= Rhopalocera). *Spanish:* Mariposas. *Portuguese:* Borboletas. *Nahuatl:* Papalomeh, sing. papalotl (Mexico). *Quechua:* Pillpinto, ccori kente. *Tupi-Guaraní:* Panamá.

Truly, the resplendent colors and airy flight of butterflies give the Neotropical environment much of its charm and beauty, for these insects are nowhere else so abundant and diverse. The fauna is the largest for any zoogeographic region (an estimated 7,000–9,000 species or nearly 50%; Legg 1978, Robbins 1982) and contains some unique types that are among the most attractive in the world. The wings of morphos, for example, have been used to make jewelry and the famous "art deco" serving trays and montages sold as art pieces as an early tradition in the tourist centers of Brazil and other parts of South America. In fact, the demand for butterfly artwork is so great that an economically significant cottage industry has prospered for a long time (Barrett 1902) in some Latin American countries. It is especially well developed in Brazil and Peru, where organized marketing chains exist. In Peru, this begins with the jungle peasant collectors, who peddle the raw specimens to traders in Tingo María, Satipo, and Pucallpa, who in turn ship them to craftsmen and merchants in Lima where they are mounted and finally displayed in the tourist shops. Most are sold to visitors as curios; few become material for collectors or scientists.

The trade has some strange beginnings in the practices of earlier inhabitants. Collecting butterflies for sale in Europe and North America provided the only real source of survival for many French liberés from the infamous Devil's Island penal colony. These men (Norris 1955), some famous, such as René Belbenoit, expositeur of the abominable conditions in the prison (*Dry Guillotine*), and his famous fanciful counterpart Henri Charrière (*Papillon*), were condemned to spend their remaining days on the Ile Royal and eked out an existence by selling morphos and other valued species to middlemen.

The possible deleterious effects of such resource exploitation have only been partly assessed (Pyle 1976). Some entomologists (Carvalho and Mielke 1972) feel that little harm is done, since, for the most part, individuals that are harvested are males that have already mated or females that have laid their eggs. It is believed that natural enemies, such as lizards (Ehrlich and Ehrlich 1982) and birds, are more effective in controlling populations. Others, however, are of the opinion that considerable genetic damage to natural butterfly populations may ensue from commercial collecting (Owen 1974).

The attitudes of ancient Mexicans were less pecuniary. They stylized many species in their religious art (Franco 1961, de la Maza 1976) and recognized a butterfly (*Papilio multicaudata*) as Xochiquetzal, Goddess of Flowers. Numerous images of butterflies are included in the frescoes of the temple of Tlalocán in Teotihuacán near Mexico City (Beutelspacher 1976). Butterflies were thought to be the reincarnated souls of dead warriors returned to earth. The ancient Aztec name for the butterfly, *papalotl*, persists in the Mexican language of today in place-names (Papalotepec, Mexico) and in other forms; Indian children still call a kite *amapapálotl* (paper butterfly) (Hoffman 1918).

Modern scientific study of butterflies is currently very active with regard to the Neotropical fauna; the early admonition of H. W. Bates (1892: 353) seems at last to have been realized: "the study of butterflies— creatures selected as the types of airlines and frivolity—instead of being despised, will some day be valued as one of the most important branches of biological science." Indeed, active research on Neotropical butterflies is having an impact in many areas of ecology (Gilbert and Singer 1975, Young 1980*a*), evolution (Benson 1971), genetics and mimicry (Turner 1977), cytology (Wesley and Emmel 1975), physiology (Swihart 1972), and biogeography (Brown 1981) and to other aspects of general biology and science.

Nevertheless, many widespread phenomena are still poorly understood. The causes and patterns of migration of enormous numbers of many species, for example, remain unknown (Welling 1959). Clouds of butterflies, often yellow pierids, were described by early chroniclers in Amazonia and the Caribbean where they are best seen. Darwin (1962) tells of a time when the *Beagle* was some miles off the mouth of La Plata: "vast numbers of butterflies, in bands or flocks of countless myriads, extended as far as the eye could range. Even by the aid of a telescope it was not possible to see a space free from butterflies. The seamen cried out 'it was snowing butterflies,' and such in fact was the appearance." Such migrations are even recognized by a special word in the widespread Tupi language, *panapaná*.

The purpose of "puddling," the habitual practice of many butterflies (mostly yellows and swallowtails and almost always the males) to siphon liquids from moist soil, is also still a puzzle (Boggs and Jackson 1991). Dense groups of individuals thus disposed along tropical lowland riverbanks, often at spots where humans launder their clothes or bathe, are a common sight. The insects rest side by side, with the axis of their bodies aligned, and suck up with their tongues interstitial fluids from sand and mud, pumping it through their bodies until it at times fairly squirts from the anus. Some important nutritional element could be removed in the process, but there may be other reasons for the behavior (see Lepidoptera, above). Children delight in chasing through these aggregations, sending throngs of butterflies into kaleidoscopic aerial dances, from which individuals often break away into "follow the leader" games (Collenette 1928).

Many butterflies also are fond of feeding on bird droppings (Young 1984), sweet-smelling, highly odoriferous rotting fruits, and decaying fungal growths on soupy sap flows (Young 1980*b*). Some unknown special nutritive requirement is probably satisfied by these habits (ibid.). A number of species have also been observed drinking tears from the eyes of crocodilians and turtles for some unexplained reason (Lamas 1986).

Sound is perceived by many species and is probably involved with territoriality and protection. Wing sacs at the base of the wing veins on the underside react to sonic stimuli (Swihart 1967).

The larvae of many Neotropical butterflies, especially among the metalmarks (Lycaenidae, Riodininae) (Callaghan 1977), and blues (Lycaeninae) have evolved symbiotic partnerships with ants (Hinton 1951, Pierce 1987). They secrete substances that are sought by the ants, which jealously protect the larvae by aggressively biting or stinging any creature attempting to molest them. They may even protect the larvae by building shelters for them. The true nature of these associations remains largely unknown, although a considerable amount has been learned about some species (Malicky 1970). Many butterfly caterpillars are spiny (Nymphalidae), but the hairs borne on the spines are not urticating. Such larvae often have a long pair of spines arising on the head to distinguish them from similar spiny moth larvae (Saturniidae) whose heads are always hornless.

Another perplexing question about Neotropical butterfly biology is how mimetic species find their own kind for mating among the many almost identically appearing species in an area. It has been discovered that the wing patterns of many butterflies appear very different under ultraviolet illumination than under normal light (Mazokhin-Porshnyakov 1957) and constitute sexual signals. The eyes of the butterflies themselves have been found to be especially sensitive to the ultraviolet portion of the light spectrum, which explains how many mimetic species may distinguish between wing patterns that appear the same to us or to birds, their chief predators (Remington 1973). Odor clues also play an important part in sexual recognition (Pliske 1975).

A still further puzzling habit of some butterflies is their attraction to swarms of army ants (*Eciton*). Some observations of this phenomenon suggest that the butterflies drink protein and sugar-rich fluids from the droppings of ant birds that also flock to such swarms (Lamas 1983, Ray and Andrews 1980).

The richest area for butterflies is the Neotropics. There are four times as many butterfly species in Panama than in the entire Malaysian Archipelago and about twice as many skippers. The Tambopata Forest Reserve near Puerto Maldonado in southeastern Peru hosts some 1,200 species, the largest number for any comparable site in the world (Lamas pers. comm.). Butterflies are the best-known large group of insects, and it is possible to identify most to species using the many works available (Lamas 1977, 1978). Some excellent popular guides to the adults have been published (e.g., d'Abrera 1981, 1984*a*, 1984*b*, 1987*a*, 1987*b*, 1989; Barcant 1970; De Vries 1987; Riley 1975).

## References

BARCANT, M. 1970. Butterflies of Trinidad and Tobago. Collins, London.

BARRETT, O. W. 1902. Cheap tropical butterflies. Entomol. News 13: 239–240.

BATES, H. W. 1892 [1863]. The naturalist on the River Amazons. Murray, London.

BENSON, W. W. 1971. Evidence for the evolution of unpalatability through kin selection in the Heliconiinae (Lepidoptera). Amer. Nat. 105: 213–226.

BEUTELSPACHER, C. R. 1976. La diosa Xochiquetzal. Soc. Mexicana Lepidop. Bol. Inf. 2: 1–3.

BOGGS, C. L., AND L. A. JACKSON. 1991. Mud puddling by butterflies is not a simple matter. Ecol. Entomol. 16: 123–127.

BROWN, JR., K. S. 1981. Biogeography and evolution of Neotropical butterflies. *In* T. C. Whitmore and G. R. Prance, eds., Biogeography and Quaternary history in tropical America. Oxford Univ. Press, Oxford.

CALLAGHAN, C. J. 1977. Studies on restinga butterflies. I. Life cycle and immature biology of *Menander felsina* (Riodinidae), a myrmecophilous metalmark. J. Lepidop. Soc. 31: 173–182.

CARVALHO, J. C. M., AND O. H. H. MIELKE. 1972. The trade of butterfly wings in Brazil and its effects upon the survival of the species. 9th Int. Cong. Entomol. (Moscow) 1: 486–488.

COLLENETTE, C. L. 1928. Gatherings of butterflies on damp sand, with notes on the attraction of moths to human perspiration. Entomol. Soc. London Trans. 76: 400–407.

D'ABRERA, B. 1981. Butterflies of the Neotropical Region. Pt. I. Papilionidae & Pieridae. Lansdowne, East Melbourne, Australia.

D'ABRERA, B. 1984*a*. Butterflies of the Neotropical Region. Pt. II. Danaidae, Ithomiidae, Heliconidae, Morphidae. Hill House, Victoria, Australia.

D'ABRERA, B. 1984*b*. Butterflies of South America. Hill House, Victoria, Australia.

D'ABRERA, B. 1987*a*. Butterflies of the Neotropical Region. Pt. III. Brassolidae, Acraeidae and Nymphalidae—partim. Hill House, Victoria, Australia.

D'ABRERA, B. 1987*b*. Butterflies of the Neotropical Region. Pt. IV. Nymphalidae—partim. Hill House, Victoria, Australia.

D'ABRERA, B. 1989. Butterflies of the Neotropical Region. Pt. V. Nymphalidae—partim. Hill House, Victoria, Australia.

DARWIN, C. 1962. The voyage of the Beagle.

Doubleday and Amer. Mus. Nat. Hist., New York.

DE LA MAZA, R. 1976. La mariposa y sus estilizaciones en las culturas Teotihuacana (200 a 750 D.C.) y Azteca (1324 a 1521 D.C.). Rev. Soc. Mexicana Lepidop. 2(1): 39–48.

DE VRIES, P. J. 1987. Butterflies of Costa Rica and their natural history: Papilionidae, Pieridae, Nymphalidae. Princeton Univ. Press, Princeton.

EHRLICH, P. R., AND A. H. EHRLICH. 1982. Lizard predation on tropical butterflies. J. Lepidop. Soc. 36: 148–152.

FRANCO, J. L. 1961. Representaciones de la mariposa en Mesoamérica. El México Antiguo 9: 195–244.

GILBERT, L. E., AND M. C. SINGER. 1975. Butterfly ecology. Ann. Rev. Ecol. Syst. 6: 365–397.

HINTON, H. E. 1951. Myrmecophilous Lycaenidae and other Lepidoptera—A summary. Proc. Trans. So. London Entomol. Nat. Hist. Soc. (1949–50): 111–175.

HOFFMAN, C. C. 1918. Las mariposas entre los antiguos mexicanos. Cosmos 1.

LAMAS, G. 1977. Bibliografía de catálogos y listas regionales de mariposas (Rhopalocera) de América Latina. Soc. Mexicana Lepidop. Publ. Esp. Pp. 1–44.

LAMAS, G. 1978. Adiciones a la bibliografía de catálogos y listas regionales de mariposas de América Latina (Rhopalocera). Soc. Mexicana Lepidop. Bol. Inf. 4(5): 1–14.

LAMAS, G. 1983. Mariposas atraídas por hormigas legionarias en la Reserva de Tambopata, Perú. Rev. Soc. Mexicana Lepidop. 8(2): 49–51.

LAMAS, G. 1986. Drinking crocodile tears. Antenna 10(4): 162.

LEGG, G. 1978. A note on the diversity of world Lepidoptera. Biol. J. Linnean Soc. 10: 343–347.

MALICKY, H. 1970. New aspects on the association between lycaenid larvae (Lycaenidae) and ants (Formicidae, Hymenoptera). J. Lepidop. Soc. 24: 190–202.

MAZOKHIN-PORSHNYAKOV, G. A. 1957. Reflecting properties of butterfly wings and role of ultra-violet rays in the vision of insects. Biophysics 2: 352–362.

NORRIS, W. E. 1955. Dawn adventure. Chambers Journal, London.

OWEN, D. F. 1974. Trade threat to butterflies. Oryx 12: 479–483.

PIERCE, N. E. 1987. The evolution and biogeography of associations between lycaenid butterflies and ants. Oxford Surv. Evol. Biol. 4: 49–116.

PLISKE, T. E. 1975. Courtship behavior and use of chemical communication by males of certain species of ithomiine butterflies (Nymphalidae: Lepidoptera). Entomol. Soc. Amer. Ann. 68: 935–942.

PYLE, R. M. 1976. The ecogeographic basis for lepidoptera conservation. Ph.D. diss., Yale Univ., New Haven.

RAY, T. S., AND C. C. ANDREWS. 1980. Ant butterflies: Butterflies that follow army ants to feed on antbird droppings. Science 210: 1147–1148.

REMINGTON, C. L. 1973. Ultraviolet reflectance in mimicry and sexual signals in the Lepidoptera. New York Entomol. Soc. J. 81: 124.

RILEY, N. D. 1975. A field guide to the butterflies of the West Indies. Collins, London.

ROBBINS, R. K. 1982. How many butterfly species? Lepidop. Soc. News (1982): 40–41.

SWIHART, S. L. 1967. Hearing in butterflies (Nymphalidae: *Heliconius, Ageronia*). J. Ins. Physiol. 13: 469–476.

SWIHART, S. L. 1972. Modelling the butterfly visual pathway. J. Ins. Physiol. 18: 1915–1928.

TURNER, J. R. G. 1977. Butterfly mimicry: The genetical evolution of an adaptation. Evol. Biol. 10: 1636–206.

WELLING, E. C. 1959. Notes on butterfly migrations in the peninsula of Yucatán. J. Lepidop. Soc. 13: 62–64.

WESLEY, D. J., AND T. C. EMMEL. 1975. The chromosomes of Neotropical butterflies from Trinidad and Tobago. Biotropica 7: 24–31.

YOUNG, A. M. 1980a. Evolutionary responses by butterflies to patchy spatial distributions of resources in tropical environments. Acta Biotheor. 29: 37–64.

YOUNG, A. M. 1980b. The interaction of predators and "eyespot butterflies" feeding on rotting fruits and soupy fungi in tropical forests; variations on a theme developed by the Muyshondts and Arthur M. Shapiro. Entomol. Rec. J. Var. 92: 63–69.

YOUNG, A. M. 1984. Ithomiine butterflies associated with non-antbird droppings in a Costa Rican tropical rain forest. J. Lepidop. Soc. 38: 61–63.

## SWALLOWTAILS

Papilionidae.

Large (WS of most 7–11 cm), colorful, and graceful in flight, the swallowtail butterflies (so-called from the taillike extensions

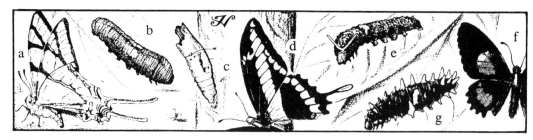

**Figure 10.15.   SWALLOWTAIL BUTTERFLIES (PAPILIONIDAE).**  (a) Kite (*Eurytides bellerophon*). (b) Kite (*Eurytides philolaus*), larva. (c) Giant swallowtail, pupa. (d) Giant swallowtail (*Papilio thaos*). (e) Giant swallowtail, larva. (f) Aristolochia swallowtail (*Parides iphidamas*). (g) Aristolochia swallowtail, larva.

of the apices of the hind wings of some genera) adorn the Neotropics. Color patterns vary, but there is frequently a red-tinted, eyelike spot at the inner notch of the hind wing. Their caterpillars are all naked, without spines or visible hairs but sometimes with tubercles, and are somewhat club shaped, with the thorax enlarged. When disturbed, the larvae arch their backs and evert an odoriferous forked organ (osmeterium) from behind the head which is a deterrent device (Eisner et al. 1971, López and Quesnel 1970, Young et al. 1986). The chrysalids mimic wood fragments with their angular form (two projecting points on the head) and rough brown or greenish integument. They rest upright, fastened to a terminal silk button and leaning back into a silken girdle.

There are three major types of swallowtails based on form and habits (Hancock 1983). The kites (pages, zebras, swordtails, Leptocircini, e.g., *Eurytides;* fig. 10.15a) are smaller than most (WS 7 cm) and have pale, thinly scaled, often white wings, with thin, transverse, black stripes; the tails are extra long and flexible. The larvae (fig. 10.15b) are usually green and smooth, and they feed on species of the custard apple family (Annonaceae). These are forest dwellers, and the males are common participants in the clouds of butterflies seen drinking from wet sand along watercourses.

The sun-loving true swallowtails (Papilionini, *Papilio;* fig. 10.15d) are the largest of the group (WS to 12 cm) and variously marked, although most are black with broad bright yellow bars through the middle of the wings and crescent-shaped spots bordering the outer margin of the hind wing. A few are "tiger marked" and mimic similarly colored heliconian and ithomiine models. Tails are lacking in the latter but are nearly always present in true swallowtails, although short. The rare *Papilio homerus* of Jamaica is the largest tailed swallowtail in the world (WS to 15 cm). The larvae (fig. 10.15e) are mottled brown and cream streaked, simulating bird droppings, and often congregate for mutual protection. Food plants are varied but often are of the pepper (Piperaceae) and citrus families (Rutaceae), the latter including orange, lemon, and lime. On citrus, those of certain species (*Papilio cresphontes, P. andraemon, P. anchisiades*) sometimes constitute pests and are known to fruit growers as "orange puppies" or "orange dogs" (Lawrence 1972). *Papilio* chrysalids often mimic broken twigs (fig. 10.15c).

Aristolochias ("poison eaters," pharmacophagous swallowtails, Troidini) are moderate-sized (WS 7–9 cm), mostly without tails, and inky black with red or magenta color fields in the center of the hind wing and green (blue or yellow) areas near the base of the fore wing (*Parides;* fig. 10.15f) or dull, greenish-black with splotches of yellow mostly on hind wings (*Battus*). These swallowtails are partial to shade and moisture and are consummate forest insects where the larvae (fig. 10.15g) feed on their pipe vine (*Aristolochia*, Aristolochiaceae) hosts. These plants contain toxic alkaloids sequestered by the caterpillars and transmitted to the adults, making them unpalatable. Many serve as models in Batesian and Müllerian mimicry complexes (Young 1971). The tuberculate larvae are also protected by these chemicals, a fact they seem to advertise with red-orange or yellow streaks, conspicuous against an otherwise completely black body. The pupae are flared out laterally along the edges of the wing cases (Young 1977).

There are more than ninety swallowtail species in Latin America (d'Abrera 1981, d'Almeida 1966); the family occurs as far south as central Chile and northern Patagonia (Slansky 1973). The biologies of most are still unknown in spite of the attention this attractive group has received from collectors and hobbyists. Some information is available on the several species, including *Papilio homerus*, above, considered possibly in danger of extinction by the International Union for the Conservation of Nature (Collins and Morris 1985).

## References

COLLINS, N. M., AND M. G. MORRIS. 1985. Threatened swallowtail butterflies of the world: The IUCN Red Data Book. IUCN, Gland, Switzerland.

D'ABRERA, B. 1981. Papilionidae. *In* B. d'Abrera, Butterflies of the Neotropical Region. Pt. I. Papilionidae and Pieridae. Lansdowne, East Melbourne, Australia.

D'ALMEIDA, R. F. 1966. Catálogo dos Papilionidae americanos. Soc. Brasil. Entomol., São Paulo.

EISNER, T., A. F. KLUGE, M. I. IKEDA, Y. C. MEINWALD, AND J. MEINWALD. 1971. Sesquiterpenes in the osmeterial secretion of a papilionid butterfly, *Battus polydamas*. J. Ins. Physiol. 17: 245–250.

HANCOCK, D. L. 1983. Classification of the Papilionidae (Lepidoptera): A phylogenetic approach. Smithsersia (Nat. Mus. Mon. Zimbabwe) 2: 1–48.

LAWRENCE, P. O. 1972. The Jamaican "orange dog," *Papilio andraemon* (Lepidoptera: Papilionidae). Fla. Entomol. 55: 243–246.

LÓPEZ, A., AND V. C. QUESNEL. 1970. Defensive secretions of some papilionid caterpillars. Carib. J. Sci. 10: 5–7.

SLANSKY, JR., F. 1973 [1972]. Latitudinal gradients in species diversity of the New World swallowtail butterflies. J. Res. Lepidop. 11: 210–217.

YOUNG, A. M. 1971. Mimetic associations in natural populations of tropical papilionid butterflies (Lepidoptera: Papilionidae). New York Entomol. Soc. J. 79: 210–224.

YOUNG, A. M. 1977. Studies on the biology of *Parides iphidamas* (Papilionidae: Troidini) in Costa Rica. J. Lepidop. Soc. 31: 100–108.

YOUNG, A. M., M. S. BLUM, H. M. FALES, AND Z. BIAN. 1986. Natural history and ecological chemistry of the Neotropical butterfly *Papilio anchisiades* (Papilionidae). J. Lepidop. Soc. 40: 36–53.

# METALMARKS

Lycaenidae, Riodininae (= Erycinidae, Nemeobiidae).

Such a large proportion of the species of this very diverse subfamily (over 90% of the world's 1,500 or so species, separated into 150 genera) live in the Neotropics that they could be considered the most characteristic butterflies of the region. Most are small (WS 20–35 mm) and delicately built, and their colors and shapes are so varied as to defy description. Many are vividly and extremely beautifully hued with metallic blues, deep scarlet, green, white, and other colors, often in complex combinations and designs (e.g., *Amarynthis menaria;* fig. 10.16d). Many have metallic gold or silver flecks on the undersides, and a few have eyespots. Some take the patterns of distasteful models among the other Lepidoptera. The wings may be rounded or angular and in some cases flamboyantly multitailed (*Helicopis;* fig. 10.16a), resembling hairstreaks, or with single, long "swallowtails" (*Chlorinea;* fig. 10.16c). Adults are fast fliers generally but often settle repeatedly on the same perch, with their resplendent wings outstretched. Some display a "false head" posteriorly on the undersides of the wings, as do the hairstreaks (see hairstreaks, below) (Robbins 1985).

The early stages of few are recorded in the scientific literature. In these, food plants are varied, and the larvae are elongate and sluglike with small lateral expansions held closely appressed to the substratum. The dorsum of the larva's prothorax is thickened and rigid and usually brown in contrast to the rest of the body and often with hornlike projections or large bristles on either side. Most are greenish or pale with many short body hairs; some have brilliantly colored protuberances marking external glands that produce exudates avidly sought by ants (De Vries 1989, Ross 1985).

Some metalmark larvae are participants in mutualistic associations with ants (as in *Juditha molpe;* fig. 10.16b), obtaining their protection in exchange for these substances (Boulard 1981, Callaghan 1977, Horvitz et al. 1987). It is suspected that the phenomenon is widespread through the subfamily. Ants actually build shelters for one species (Ross 1966). The caterpillar of one species (*Thisbe irenea*) not only feeds on leaf tissue but also drinks the extrafloral nectar of its host plant, creating a conflict between plant and herbivore for the attentions of ants (De Vries and Baker 1989).

### References

BOULARD, M. 1981. Nouveaux documents sur les chenilles de lycènes tropicaux. Alexanor 12: 135–140.

CALLAGHAN, C. J. 1977. Studies on restinga butterflies. I. Life cycle and immature biology of *Menander felsina* (Riodinidae), a myrmecophilous metalmark. J. Lepidop. Soc. 31: 173–182.

DE VRIES, P. 1989. The ant associated larval organs of *Thisbe irenea* (Riodinidae) and their effects on attending ants. Zool. J. Linnean Soc. 94: 379–393.

DE VRIES, P. J., AND I. BAKER. 1989. Butterfly exploitation of an ant-plant mutualism: Adding insult to herbivory. New York Entomol. Soc. J. 97: 332–340.

HORVITZ, C. C., C. TURNBULL, AND D. J. HARVEY. 1987. Biology of immature *Eurybia elvina* (Lepidoptera: Riodinidae), a myrmecophilous metalmark butterfly. Entomol. Soc. Amer. Ann. 80: 513–519.

ROBBINS, R. K. 1985. Independent evolution of "false head" behavior in Riodinidae. J. Lepidop. Soc. 39: 224–225.

Ross, G. N. 1966. Life-history studies on Mexican butterflies. IV. The ecology and ethology of *Anatole rossi*, a myrmecophilous metalmark (Lepidoptera: Riodinidae). Entomol. Soc. Amer. Ann. 59: 985–1004.

Ross, G. N. 1985. The case of the vanishing caterpillar. Nat. Hist. 94(11): 48–55.

## HAIRSTREAKS

Lycaenidae, Lycaeninae.

The Lycaeninae are poorly known in Latin America compared to their sister group, the metalmarks. Currently, about 1,000 species have been discovered. When all are described, it is estimated that their number will exceed the Riodininae.

In the Neotropics, well over 90 percent of this subfamily of small butterflies (WS 15–40 mm) is comprised of one tribe, the Eumaeini (Eliot 1973). These are the hairstreaks, typically with short hairlike appendages extending from a lobe at the rear of the hind wing, at the base of which are conspicuous, eyelike spots. They rest with the hind wings appressed tightly over the back which they characteristically rub together, setting these "tails" in motion so that they resemble waving antennae. The action is thought to divert the attacks of predators away from the true head to these expendable wing structures. The validity of the hypothesis has been tested on the common "false head hairstreak," *Arawacus*

**Figure 10.16    LYCAENID BUTTERFLIES (LYCAENIDAE).** (a) Multitailed metalmark (*Helicopis acis*). (b) Metalmark (*Juditha molpe*), larva being attended by ants of the genus *Hypoclinea*. (c) Long-tailed metalmark (*Chlorinea faunus*). (d) Metalmark (*Amarynthis menaria*). (e) False head hairstreak (*Arawacus aetolus*).

*aetolus* (fig. 10.16e) (Robbins 1980, 1981). On the upper sides, they are solid colored and plain, although many others display iridescent blues and green and others are vividly patterned.

The early stages of the majority of the species are unknown. The few larvae that are known are mostly sluglike with a small retracted head and are somewhat flattened, some very much so (Callaghan 1982). Most feed on a variety of dicotyledoneous plants, often the flowers, and fruit, but a few (e.g., *Chliaria*) may have specialized food preferences among other plants, such as orchids. Some are associated with ants symbiotically and have a thickened, tough cuticle, presumably to protect them from attack when entering ant nests to feed on their larvae or other guests, such as coccids. These also exude substances from special integumentary glands to attract ants for protection or to entice them to carry them to their nest. Some are agricultural pests, such as the pineapple hairstreak (*Tmolus basilides*) whose larvae eat the flowers and bore into the developing fruit of pineapple (Harris 1927), but hairstreaks are largely benign insects.

### References

CALLAGHAN, C. J. 1982. Notes on the immature biology of two myrmecophilous Lycaenidae: *Juditha molpe* (Riodininae) and *Panthiades bitias* (Lycaeninae). J. Res. Lepidop. 20: 36–42.

ELIOT, J. N. 1973. The higher classification of the Lycaenidae (Lepidoptera): A tentative arrangement. Brit. Mus. Nat. Hist. Entomol. Bull. 28: 373–505, pls. 1–6.

HARRIS, W. V. 1927. On a lycaenid butterfly attacking pineapples in Trinidad, B.W.I. Bull. Entomol. Res. 18: 183–188, pls. 7–8.

ROBBINS, R. K. 1980. The lycaenid "false head" hypothesis: Historical review and quantitative analysis. J. Lepidop. Soc. 34: 194–208.

ROBBINS, R. K. 1981. The "false head" hypothesis: Predation and wing pattern variation of lycaenid butterflies. Amer. Nat. 118: 770–775.

## WHITES AND SULFURS

Pieridae. *Spanish:* Isocas (General, larvae), pirpintos (Argentina).

In these familiar butterflies, yellow, orange, and white are the predominating colors. They frequent flowers of open glades and clearings and are among the sun-loving insect throngs that tend the blossoms of the forest canopy. They also range widely from the coastal deserts and humid swampland to well above tree line in the páramos and rocky slopes of snowcapped peaks.

Like their relatives, the swallowtails, they possess a fully functional pair of front legs, and the pupa is attached at the tip of the abdomen and held upright by a silken girdle passing around the thorax. The larvae are largely smooth except for minute papillae in the integument.

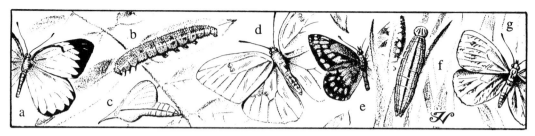

**Figure 10.17 BUTTERFLIES.** (a) Great southern white (*Ascia monuste,* Pieridae). (b) Cloudless sulfur (*Phoebis sennae,* Pieridae), larva. (c) Cloudless sulfur, pupa. (d) Cloudless sulfur, adult. (e) Alpine pierid (*Catasticta semiramis,* Pieridae). (f) Silver-winged Butterfly (*Argyrophorus argenteus,* Nymphalidae), larva. (g) Silver-winged butterfly, adult.

Several species, especially the cloudless sulfurs, *Phoebis* (d'Almeida 1940), participate in migratory swarms. The males are avid puddlers. Many mimics also are found in the family.

The larvae of a few (*gusanos de la col, isocas del repollo, lagartas da hortaliça, lagartas da couve*) are notorious pests, for example, the widespread, native great southern white (*Ascia monuste;* fig. 10.17a) and, in Chile only where it was accidentally introduced, the European cabbage butterfly (*Pieris brassicae*) (Gardiner 1974). The larvae of *Colias lesbia* (*isoca de la alfalfa, cuncuna*) destroy alfalfa in Argentina.

Adults of the genus *Catasticta* (fig. 10.17e) are atypical pierids, having a checkered wing pattern and occurring only at high elevations from Mexico to the Andes, where the greatest number of species are found. The larvae feed on Loranthaceae; they are gregarious and respond by head rearing when molested.

There are approximately 400 species of pierids in Latin America (d'Abrera 1981).

### References

D'Abrera, B. 1981. Pieridae. *In* B. d'Abrera, Butterflies of the Neotropical Region. Pt. I. Papilionidae and Pieridae. Lansdowne, East Melbourne, Australia. Pp. 80–165.

D'Almeida, R. F. 1940. Revisão do gênero *Phoebis* Hübn. (Lepidopt Pieridae). Arq. Zool. Est. São Paulo 1: 67–148.

Gardiner, B. O. C. 1974. *Pieris brassicae* L. established in Chile; another palaearctic pest crosses the Atlantic (Pieridae). J. Lepidop. Soc. 28: 269–277.

### Cloudless Sulfur

Pieridae, Pierinae, Coliadini, *Phoebis sennae.*

This is probably the most common and widespread of the Neotropical pierids. A moderately large butterfly (WS 6 cm), its intense greenish-yellow wings, punctuated on the undersides with scattered patches or lines of reddish-brown scales, immediately identify it (fig. 10.17d). The upper sides of the male wings are immaculate; those of the female are bordered by a broken black margin and an irregular black spot midway near the leading edge of the fore wing.

The species is strongly migratory. Vast clouds moving through many parts of South and Central America and even over the open sea in the Caribbean are of frequent occurrence.

The immatures are well known, the larva (fig. 10.17b) being almost a pest on leguminous ornamentals in the genus *Cassia.* It feeds on *Calliandra* and *Inga* as well. It grows to 35 to 40 millimeters and is elongate (slightly tapered at both ends), and its skin is transversely wrinkled or ridged. It is generally green to yellow-green with a lateral yellow line. The upper half of the body is speckled with small purplish dots in each of which is a minute black wart bearing fine white hairs. The pupa (fig. 10.17c) is wedge shaped, with greatly expanded wing cases that form a sharp heel where they join from each side. The very deep wing cases give it an arched-backed appearance. Although pink pupae are known, this stage is typically green with a whitish-yellow longitudinal lateral stripe and dark middorsal line.

# BRUSH-FOOTED BUTTERFLIES

Nymphalidae.

Butterflies classified into this diverse family, which is considered here in a broad sense, include all of the succeeding subfamilies, which some authors consider separate families.

## Satyrs

Nymphalidae, Satyrinae.

Satyrs are almost all drab denizens of the forest floor. Many do have bright colors on the hind wings and eyespot patterns ventrally. A few are transparent, like "glassy-wings" (Haeterini). The wings are also soft and thinly scaled and with few exceptions have the bases of the veins of the fore wing inflated to form a conspicuous swelling. The host plants of almost all species are grasses and bamboos.

### Silver-winged Butterfly

Nymphalidae, Satyrinae, *Argyrophorus argenteus. Spanish:* Mariposa plateada, cinta plateada.

This is an exceedingly beautiful, medium-sized (WS 4 cm) butterfly of the lower slopes of the Patagonian Andes, the upper wing surfaces being solid, shining silver (fig. 10.17g). It is a difficult butterfly to catch because of its erratic flight over its brushland habitat. The immatures are poorly known; the larva (fig. 10.17f) is light yellowish-brown with longitudinal stripes and apparently feeds on the grass

*Stipa (coirón)*, on which females have been observed to oviposit (Shapiro 1982).

### Reference

SHAPIRO, A. M. 1982. Notas sobre los estados inmaduros de la mariposa plateada, *Argyrophorus argenteus* Blanchard (Lepidoptera: Satyridae). Soc. Mexicana Lepidop. Rev. 7: 29–31.

## Monarch Butterflies

Nymphalidae, Danainae, *Danaus. Spanish:* Monarcas.

Like ithomiines and some other tropical Lepidoptera, male monarch butterflies seek organic compounds, called pyrrolizidine alkaloids, from withered plants, most commonly of the genera *Heliotropium, Eupatorium, Senecio,* and *Tournefortia.* These substances are imbibed with the tongue and form substrates in the butterfly's body from which are synthesized special aphrodisiacal perfumes that help attract females and ensure success in courtship and mating. The chemicals (di-hydro-pyrrolizidines) are secreted by pouched glands near the center of the hind wings from which they are first picked up and then disseminated into the air by protrusible fan-shaped hair brushes located at the tip of the abdomen (Brower and Jones 1965). The role of these chemicals in the courtship process is still not clear.

Individuals of this genus also sequester cardiac glycosides in their bodies (Boppré 1978). These compounds are acquired by the larvae when feeding on their asclepiad host plants and are toxins capable of inducing severe intestinal complaints in birds or other animals that swallow them (Roeske et al. 1976). Thus, the adults gain protection for themselves and also serve as a model for a number of other mimetic butterflies and moths. Not every milkweed (*Asclepias*) food plant provides cardiac glycosides, and butterfly individuals that eat these are not poisonous.

There are four *Danaus* species ranging

**Figure 10.18 NYMPHALID BUTTERFLIES (NYMPHALIDAE).** (a) Monarch (*Danaus plexippus*).
(b) Monarch, larva. (c) Monarch, pupa. (d) Cracker (*Hamadryas feronia*). (e) Head-for-tail (*Colobura dirce*). (f) Malachite green (*Siproeta stelenes*).

throughout the Neotropics, the most widespread being the "true" monarch (*D. plexippus;* fig. 10.18a), which exists as various subspecific and local forms (the southern *D. erippus* is sometimes considered as a separate species). *D. gilippus* and relatives are the so-called queens, recognized by dark, brownish ground color in the wings. The widespread soldier (*D. eresimus*) has found its way to several islands of the Antilles; the Jamaican monarch (*D. cleophile*) is restricted to Jamaica and Hispaniola.

All monarch butterflies (Ackery and Vane-Wright 1984) are frost-sensitive and essentially tropical butterflies (Young 1982). The northern subspecies *D. plexippus plexippus* penetrates the temperate latitudes of North America (to southern Canada) but only as a summer visitor. In the fall months, its northern populations migrate southward over two major routes, a Pacific flyway to the west of the Rocky Mountains and a continental flyway to the east (Brower 1985). The overwintering sites of most of the latter have only been recently discovered in a remote mountain range on the Mexican plateau (Calvert and Brower 1986, Urquhart and Urquhart 1976, Brower et al. 1977). This roosting phenomenon is so spectacular that the butterflies are now protected by presidential decree (López Portillo 1980, Norman 1986). A civic group, Pro Monarca, has even been formed which is dedicated to the conservation of the overwintering sites (Ogarrio 1984). The phenomenon of roosting monarchs has

been known for a long time to the local residents, who refer to them as *palomas de los Santos*. The butterflies are believed to arrive on the Day of the Dead (All Soul's Day, November 2).

Some of the North American migrants of *D. plexippus* overwinter in the West Indies where the larval host is the universal red-flowered weed, *A. curassavica*, and *Calotropis procera*, an African immigrant. On Barbados, the larvae have consumed all traces of the former, which is also the food plant of the milkweed bugs (*Oncopeltus*). The bugs, as a consequence, are no longer found there (Blakley and Dingle 1978).

The other *Danaus* species, and the southern subspecies of *D. plexippus*, are fundamentally sedentary.

Most *Danaus* are rich rusty brown in general color with black wing veins. The wing apex and margins are also black, containing small white spots. The related large tiger (*Lycorea cleobaea*, formerly *L. ceres*) and *L. ilione*, however, deviate strongly from this typical pattern, having tiger colors and a clear wing pattern, respectively, like the heliconians, ithomiines, and other Lepidoptera that they join in Müllerian mimicry complexes.

The larvae are colored in narrow multihued circular bands and bear fleshy tentacular structures on various segments (fig. 10.18b). The pupae are compact, smooth, and of various colors; that of *D. plexippus*, emerald green with golden spots, is well known for its beauty (fig. 10.18c).

## References

ACKERY, P. R., AND R. I. VANE-WRIGHT. 1984. Milkweed butterflies, their cladistics and biology. Brit. Mus. Nat. Hist., London. See esp. pp. 106–110.

BLAKLEY, N. R., AND H. DINGLE. 1978. Competition: Butterflies eliminate milkweed bugs from a Caribbean Island. Oecologia 37: 133–136.

BOPPRÉ, M. 1978. Chemical communication, plant relationships, and mimicry in the evolution of the danaid butterflies. Entomol. Exper. Appl. 24: 64–77.

BROWER, L. P. 1985. New perspectives on the migration biology of the monarch butterfly, *Danaus plexippus* L. *In* M. A. Rankin, Migration: Mechanisms and adaptive significance. Contrib. Mar. Sci. Suppl. 27:748–785.

BROWER, L. P., W. H. CALVERT, L. E. HENDRICK, AND J. CHRISTIAN. 1977. Biological observations on an overwintering colony of monarch butterflies (*Danaus plexippus*, Danaidae) in Mexico. J. Lepidop. Soc. 31: 232–242.

BROWER, L. P., AND M. A. JONES. 1965. Precourtship interaction of wing and abdominal sex glands in male *Danaus* butterflies. Royal Entomol. Soc. London Proc. 40: 147–151.

CALVERT, W. H., AND L. P. BROWER. 1986. The location of monarch butterfly (*Danaus plexippus* L.) overwintering colonies in Mexico in relation to topography and climate. J. Lepidop. Soc. 40: 164–187.

LÓPEZ PORTILLO, J. 1980. Decreto por el que por causa de utilidad pública se establece zona de reserva y refúgio silvestre los lugares donde la mariposa conocida con el nombre de "Monarca" hiberna y se reproduce. Diario Oficial (México) 9 April 1980. Pp. 7–8.

NORMAN, C. 1986. Mexico acts to protect overwintering monarchs. Science 233: 1252–1253.

OGARRIO, R. 1984 [1981]. Development of the civic group, Pro Monarca, A.C., for the protection of the monarch butterfly overwintering grounds in the Republic of México. Atala 9: 11–13.

ROESKE, C. M., J. M. SEIBER, L. P. BROWER, AND C. M. MOFFITT. 1976. Milkweed cardenolides and their comparative processing by monarch butterflies. Rec. Adv. Phytochem. 10: 93–167.

URQUHART, F. A., AND M. R. URQUHART. 1976. The overwintering site of the eastern population of the monarch butterfly (*Danaus p. plexippus:* Danaidae) in southern Mexico. J. Lepidop. Soc. 30: 153–158.

YOUNG, A. M. 1982. An evolutionary-ecological model of the evolution of migratory behavior
in the monarch butterfly and its absence in the queen butterfly. Acta Biotheor. 31: 219–237.

## Nymphalines
Nymphalidae, Nymphalinae.

Members of this subfamily are varied, small to medium size, with reduced, brushlike forelegs and strongly clubbed antennae. The central cell of the hind wing venation is always open, and the wing has a cupped depression on the inner margin to accommodate the abdomen when the wings are closed. The larvae and pupae are likewise diverse, the former usually spiny.

### Crackers
Nymphalidae, Nymphalinae, Ageroniini, *Hamadryas*. *Spanish:* Calicoes (General), tronadoras (Mexico), tabletas (Peru), cascabeles (Venezuela), soñadoras (Costa Rica), gritonas (Colombia). *Portuguese:* Assentas páu, matracas, angolinhas, etc. (Brazil).

These swift-flying and pugnacious butterflies (Perry 1964) invariably rest on tree trunks 1 to 10 meters above the ground, head down, wings spread, their colors merging completely with the mottled bark background. The edges of the wings are pressed flat to the bark so that no shadow is thrown.

On the approach of another individual, a male cracker lurches from its perch and fights off the intruder. Such encounters are usually punctuated by crackling or clicking noises that the butterflies themselves emit. The sounds are audible to the human ear, in still air, up to several meters away. Both sexes indulge in these fast aerial pursuits and place other kinds of butterflies and even humans (orig. obs.) under "attack." One author (Ross 1963) thinks that this extreme wariness may be wholly a means of escaping predators and not an instance of true territorial behavior.

The mechanism of clicking is not well understood. Sharp contact between certain thoracic sclerites during irregular wing beats appears to emit the actual sound (Hannemann 1956, Swihart 1967).

The butterflies often choose perches on wounded trees oozing fermenting sap, on which they feed. They are also fond of juices exuding from rotting food, carrion, and other moist organic matter.

The wings of crackers are marked on the upper sides with alternating, zigzag lines of brown, or blue and cream crossed by the dark wing veins, together producing an intricate, irregular, mosaic or checker-board pattern (fig. 10.18d). This is broken by a submarginal row of spots, which are smaller anteriorly on the fore wing and gradually increase to fairly conspicuous eyespots posteriorly on the hind wing. The twenty species are all medium-sized (WS 5–6 cm) and fairly common throughout the Neotropics in dry to wet forest habitats (Jenkins 1983).

Cracker caterpillars mostly feed on toxic vines in the euphorbia family, recorded species being *Dalechampia scandens* and *Tragia volubilis*. They grow to 30 to 35 millimeters (BL) and are spined all over like many other nymphalid larvae, but the main shaft of each spine process is fine and the lateral barbs long and arising near the base so that the organ appears branched in stellate fashion. The paired head spines are extra long and decidedly clubbed. Bright colors decorate the skin; cream, red, green, orange, and dark lines run along the back and sides. The chrysalids mimic dead leaves, the head manifesting a pair of very prominent flattened flange-shaped extensions with scalloped edges. They hang straight down from its terminal support and when touched are capable of violent, snapping movements (Muyshondt and Muyshondt 1975a, 1975b, 1975c; Young 1974).

## References

HANNEMANN, H. J. 1956. Über pterotarsale Stridulation und einige andere Arten der Lauterzeugung bei Lepidopteren. Deutche Entomol. Zeit., N.F., 3(1): 14–27.

JENKINS, D. W. 1983. Neotropical Nymphalidae. I. Revision of *Hamadryas*. Allyn Mus. Bull. 81: 1–146.

MUYSHONDT, A., AND A. MUYSHONDT, JR. 1975a. Notes on the life cycle and natural history of butterflies of El Salvador. IB. *Hamadryas februa* (Nymphalidae-Hamadryadine). New York Entomol. Soc. J. 83: 157–169.

MUYSHONDT, A., AND A. MUYSHONDT, JR. 1975b. Notes on the life cycle and natural history of butterflies of El Salvador. IIB. *Hamadryas guatemalena* Bates (Nymphalidae-Hamadryadinae). New York Entomol. Soc. J. 83: 170–180.

MUYSHONDT, A., AND A. MUYSHONDT, JR. 1975c. Notes on the life cycle and natural history of butterflies of El Salvador. IIIB. *Hamadryas amphinome* L. (Nymphalidae-Hamadryadinae). New York Entomol. Soc. J. 83: 181–191.

PERRY, R. 1964. Notes on the genus *Ageronia* and the butterflies of Poponté. Entomologist 97: 140–141.

ROSS, G. M. 1963. Evidence for lack of territoriality in two species of *Hamadryas* (Nymphalidae). J. Res. Lepidop. 2: 241–246.

SWIHART, S. L. 1967. Hearing in butterflies (Nymphalidae: *Heliconius*, *Ageronia*). J. Ins. Physiol. 13: 469–476.

YOUNG, A. M. 1974. On the biology of *Hamadryas februa* (Lepidoptera: Nymphalidae) in Guanacaste, Costa Rica. Zeit. Angewan. Entomol. 76: 380–393.

### Head-for-Tail Butterfly
Nymphalidae, Nymphalinae, Nymphalini, *Colobura dirce*. Zebra (Trinidad).

This butterfly is a lover of oozing, fermenting fruit and sap and spends much of its time in wooded areas, sitting on food that has fallen to the ground or perching on the trunks of the trees, head downward with wings tightly closed. In the latter position, the full confusing effect of the pattern on the undersides of the wings can be seen (fig. 10.18e). The illusion of the head and tail reversed (tergiversation) is produced by an enlargement of the tip of the hind wing with a central ocellate spot, forming a false head and eye; also, vertical bars of black at the base of the wing toward the true head simulate abdominal segments. Presumably, this pattern is a deception directing the strikes of bird beaks away from the vital head and body. The butterfly also takes flight at an instant, scurrying through the air and making a rustling noise with its wings. Thus, its behavior as well as its form and color afford it protection from vertebrate predators.

The solid dark brown of the upper surfaces is broken only by a conspicuous yellow bar crossing the middle of the fore wing obliquely. It is a medium-sized butterfly (WS 11–12 cm).

The life cycle is well known (Beebe 1952, Muyshondt and Muyshondt 1976). The mature larva (BL 36 mm), of typical nymphalid form, is velvety black or green with contrasting yellow areas around the spiracles and white to bright yellow stellately branched spines. It has an eversible fingerlike gland on the underside of the neck, whose function is unknown. The leaves of the cecropia tree are its food. The chrysalid (BL 3 cm) resembles a fragmented wood chip with its roughened exterior, jagged dorsum, and light brown color. It hangs from twigs by a fastening only at the tip of the abdomen.

## References

Beebe, W. 1952. A contribution to the life history of *Colobura* (*Gynaecia* auct.) *dirce dirce* (Linnaeus) butterfly. Zoologica 37: 199–202.

Muyshondt, Jr., A., and A. Muyshondt. 1976. Notes on the life cycle and natural history of butterflies of El Salvador. IC. *Colobura dirce* L. (Nymphalidae-Coloburinae). New York Entomol. Soc. J. 84: 23–33.

### *Malachite Green*
Nymphalidae, Nymphalinae, Nymphalini, *Siproeta stelenes*. Bamboo page (Trinidad).

This is another widespread and common butterfly associated with forest clearings throughout Central America and much of northern South America (Young and Muyshondt 1973). It is moderately large (WS 75 mm), with the upper wing surfaces black except for a broad, broken median bar of light green and small circular spots of the same color on the outer half of the hind wings (fig. 10.18f). This coloration is strikingly similar to that of the unrelated heliconian *Philaethria dido* (fig. 10.21a). The undersides are richly marbled with green and brown blotches, punctuated with black lines and ovals. Adults feed on a variety of flowers but also commonly take liquids from rotten fruit and even fresh equine or bovine dung.

The mature larva is dull black, of moderate size (BL 50–53 mm), and liberally spined; the head spines are decidedly longer than those of the body and clubbed. Food plants are diverse species of the family Acanthaceae, often small semiwoody herbs (*Justicia, Ruellia, Blechum*) growing as weeds in disturbed places such as wood lots, clearings, and coffee plantations.

The pupa is about 30 millimeters long, free-hanging from a black terminal stalk, and generally translucent light green (and often covered with a white bloom). There are short, paired head projections, short spines on the back of the anterior abdominal segments, and sparse black specks on the skin.

The species has gone under the name "*Metamorpha*" or "*Victorina*" *steneles* [sic] in the earlier literature.

## Reference

Young, A. M., and A. Muyshondt. 1973. Ecological studies of the butterfly *Victorina stelenes* (Lepidoptera: Nymphalidae) in Costa Rica and El Salvador. Stud. Neotrop. Fauna 8: 155–176.

### *Number Butterflies*
Nymphalidae, Nymphalinae, Catagrammini, *Diaethria* and relatives. *Spanish:* Ochenta y ochos, ochenta y nueves (General). *Portuguese:* Oitenta-e-oitos, oitenta-e-noves, cruzeiros do sul (Brazil). Eighty-eights, eighty-niners.

While varied in details, the patterns of the undersides on the hind wings of this famil-

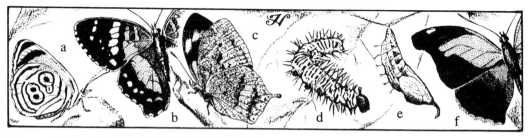

**Figure 10.19 NYMPHALID BUTTERFLIES (NYMPHALIDAE).** (a) "89" (*Diaethria clymena*). (b) Red anartia (*Anartia amathea*). (c) Leaf butterfly (*Memphis arachne*). (d) Cecropia butterfly (*Historis odius*), larva. (e) Cecropia butterfly, pupa. (f) Cecropia butterfly, adult.

iar group of medium-sized (WS 3–5 cm), entirely Neotropical nymphalids always consists of concentric dark rings on a gray to yellow background. The innermost of these are separated into two figure eights (although one pupil of the anterior figure may be fused with the other, forming more of a "9"). With a little imagination, one can read 88, 89, 80, or other numbers from the wing (fig. 10.19a), although many see letters instead ("BD"). On the upper sides, the black background color of the wings is interrupted by a diagonal bar of metallic blue or green and/or broad fields of these colors plus red or yellow.

There are some fifty species in this group (Dillon 1948), distributed widely throughout the region, mainly in forest habitats. A few species in the related and generally similar-appearing genera *Callicore, Paulogramma, Perisama, Dynamine, Callidula,* and *Catacore* have numerological hind wings as well, but more often, they sport barred, leaflike, or other patterns. In the past, many of these were lumped into the catch-all genus "*Catagramma.*"

The early stages of a few species are known (Muyshondt 1975). The mature larva of *Diaethria astala* is more or less typical; it grows to moderate size (BL 25–27 mm) and is unarmed except for the head spines, which are very long (a third of the body length) and have strong lateral branches. The body is light green and covered with tiny white warts and longitudinal rows of yellow tubercles. It exhibits an active defense behavior when disturbed, rearing its head and striking with its horns.

Food plants are vines in the family Sapindaceae (*Serjania, Cardiospermum, Urvillea,* etc.), although the euphorbs *Sapium* and *Dalechampia* are recorded for *Dynamine*. These plants have poisonous properties that they impart to the larvae feeding on them.

The chrysalid is green with dark, veinlike marks on the wing cases and lateral, light green lines. It attaches to the underside of a leaf, only by the tail, but hangs closely appressed to the surface. It may produce a faint creaking sound by wiggling sideways or contracting its abdomen accordionlike.

### References

DILLON, L. S. 1948. The tribe Catagrammini (Lepidoptera: Nymphalidae). Pt. I. The genus *Catagramma* and allies. Read. Pub. Mus. Art Gal. Sci. Pub. 8: 1–v, 1–113.

MUYSHONDT, A. 1975. Notes on the life cycle and natural history of butterflies of El Salvador. VIA. *Diaethria astala* Guérin. (Nymphalidae—Callicorinae). New York Entomol. Soc. J. 83: 10–18.

### *Peacock Butterflies*

Nymphalidae, Nymphalinae, Nymphalini, *Anartia*.

Any of the three continental species in this genus (Silberglied et al. 1979) might be found in a wayside flower patch or disturbed sunny clearing anywhere in the humid lowlands. They are among the most common and conspicuous diurnal Lepidoptera encountered in the New World tropics.

The fatima (*A. fatima*) (Young and Stein 1976), ranges from Mexico to Panama, where it is replaced by its close relative, the similar red anartia (also called the coolie or tomato, *A. amathea;* fig. 10.19b), whose distribution continues through South America to northern Argentina. Two rarer species, *A. chrysopelea* and *A. lytrea,* are of local occurrence in the larger islands of the West Indies.

The fatima and red anartia have the same basic coloration, the upper wing surface basically dark brown with transverse pale brown and red bands. The essential difference concerns the extent of a red area inside the outermost light band: here there are just a few spots in the hind wing in the fatima, but this is a broad field extending across both fore and hind wings in the red anartia. The outer bar may be white or yellow in the former species, a dimorphic characteristic apparently determined genetically and with possible (but not probable?) implications for mate selection and survival (Emmel 1972, Silberglied et al. 1979).

The white peacock (biscuit, *A. jatrophae*) has the widest distribution of the peacocks, occurring throughout the Neotropics, including the Antilles but excepting the high Andes and Chile. Its wings are off-white to gray, with complex zigzag lines crossing the wings and three dark spots just beyond the center (one on the fore wing, two on the hind wing).

All peacocks are medium-sized butterflies (WS 4 cm) and generally similar in shape, having triangular wings with scalloped borders. They are unpalatable to vertebrates and presumably rely on this and disruptive wing colors for survival, although at least one field study does not support the latter idea (Silberglied et al. 1980).

Larvae of all are similar, with coarse longitudinal stripes usually on a black background and with rows of elongate spines that are thickly bristled but not branched. The spines of the head are clubbed. Food plants are varied but are often water-loving herbs of the families Scrophulariaceae (*Bacopa monnieri* or water hyssop) and Labiaceae for the white peacock or Acanthaceae (*Blechum*) for the other species.

The pupae are suspended terminally and are smooth and of simple shape. They are usually jade green with small black spots, although occasional individuals may be black.

Adults of the red and brown species have a jaunty, erratic flight, while the white species is an inveterate glider. At rest, those of all species habitually orient with their wings open and bask in the sun (Fosdick 1973).

## References

EMMEL, T. C. 1972. Mate selection and balanced polymorphism in the tropical nymphalid butterfly, *Anartia fatima.* Evolution 26: 96–107.

FOSDICK, M. K. 1973 [1972]. A population study of the Neotropical nymphalid butterfly, *Anartia amalthea* [sic], in Ecuador. J. Res. Lepidop. 11: 65–80.

SILBERGLIED, R. E., A. AIELLO, AND G. LAMAS. 1979. Neotropical butterflies of the genus *Anartia:* Systematics, life histories and general biology (Lepidoptera: Nymphalidae). Psyche 86: 219–260.

SILBERGLIED, R. E., A. AIELLO, AND D. M. WINDSOR. 1980. Disruptive coloration in butterflies: Lack of support in *Anartia fatima.* Science 209: 617–619.

YOUNG, A. M., AND D. STEIN. 1976. Studies on the evolutionary biology of the Neotropical nymphalid butterfly *Anartia fatima* in Costa Rica. Milwaukee Pub. Mus. Contrib. Biol. Geol. 8: 1–29.

### Leaf Butterflies

Nymphalidae, Nymphalinae, Charaxini, *Memphis* (formerly *Anaea*).

These are medium-sized (WS 4–5 cm) butterflies having thin wings with angular margins, a sharp tipped fore wing and a short-tailed hind wing. *Memphis arachne* (fig. 10.19c) is an example. Above, the wings are brightly colored, but the lower surfaces are gray to brown and mottled in a way closely resembling a dead leaf, even to the details of the veins, imitated by dark streaks and mold spots and other imperfections faked by clear spots. The deception is enhanced by their habit of resting with wings tightly closed on tree trunks and among dead foliage on the ground. They are common forest dwellers and numerous in species (Comstock 1961). Their larvae feed on the leaves of various small trees and shrubs in the families Lauraceae and Flacourtiaceae. They are cylindrical and lacking spines or other elaborations. The cuticle possesses only numerous small beadlike granules, each bearing a short, white hair (Young 1981).

## References

Comstock, W. P. 1961. Butterflies of the American tropics: The genus *Anaea*, Lepidoptera, Nymphalidae. Amer. Mus. Nat. Hist., New York.

Young, A. M. 1981. Notes on the seasonal distribution of *Anaea* butterflies (Nymphalidae) in tropical dry forests. Acta Oecologica 2: 17–30.

### *Cecropia Butterfly*
Nymphalidae, Nymphalinae, Nymphalini, *Historis odius*. *Spanish:* Pescadito (Costa Rica, pupa).

This well-known butterfly (De Vries 1983) ranges widely over Latin America, including the Caribbean Islands and even isolated Coco Island, where it is the only resident butterfly. It is a moderately large (WS 10 cm) robust and powerful flier in the canopy; but its fondness for the juices of rotting fruit often bring it to the ground. The fore wings are bordered broadly in black on the upper sides around a basal median field of dull orange. The hind wings are almost entirely black (fig. 10.19f).

The large, mature larvae (BL to 75 mm) bear many branched spines and are mottled black, brown, orange, and blue (fig. 10.19d). They are often seen resting singly on the apical growing stems of the host, which is the cecropia tree. Young larvae feed in groups on the undersides of the leaves. The chrysalid has fine, fork-tipped, dorsal abdominal spines and curved head horns (fig. 10.19e).

## Reference

De Vries, P. J. 1983. *Historis odius* (Orion). *In* D. H. Janzen, ed., Costa Rican natural history. Univ. Chicago Press, Chicago. Pp. 731–732.

### Ithomiines
Nymphalidae, Ithomiinae.

Ithomiines (Fox 1956, 1960, 1967; Fox and Real 1971) superficially resemble heliconiines, with the same attenuate antennal clubs, wing shapes, flight behavior, and color patterns (except for the "transparent" type, which is lacking in that subfamily and very common in this, e.g., *Oleria* and *Hypoleria;* fig. 10.20a). They also are presumed unpalatable forest dwellers that serve as models in Müllerian mimicry complexes with heliconians and other Lepidoptera. Structural details distinguishing the family are a basally spurred anal vein in the fore wing and a weak spur only at the base of the humeral vein of the hind wing; they also lack the eversible hair tufts on the abdomen which the heliconiines possess and seem to be less complex in their behavior and life history, although this may be due to the relatively poorer extent to which they have been studied.

Like their monarch butterfly relatives, male ithomiines are attracted to dead plants of the genus *Heliotropium* (Boraginaceae) containing certain chemicals of decomposition called pyrrolizidine alkaloids (Lamas and Pérez 1983). Dried inflorescences, emitting volatile components of these alkaloids, are the most attractive part of the plant. On

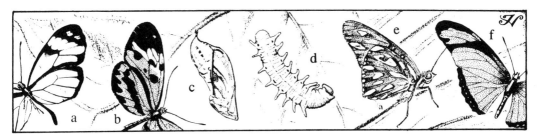

**Figure 10.20    NYMPHALID BUTTERFLIES (NYMPHALIDAE).** (a) Glassy wing ithomiine (*Hypoleria andromica*). (b) Heliconius-mimicking ithomiine (*Mechanitis polymnia*). (c) Ithomiine (*Mechanitis* sp.), pupa. (d) Ithomiine (*Mechanitis* sp.), larva. (e) Gulf fritillary (*Agraulis vanillae*). (f) Julia (*Dryas iulia*).

arriving at a moist plant, the butterflies drink surface droplets; or if the plant is dry, they regurgitate liquids and spread it over twigs and reimbibe. In the process, they obtain quantities of these alkaloids in solution. While feeding, they become docile and can even be picked off and returned without causing them to fly. Males use these substances as metabolic substrates to produce a sexual secretion that is disseminated from hair tufts on the costal margins of the hind wings (Pliske et al. 1976) and protect them from spider predators (Brown 1984). They often group together during these displays, among the few butterflies to exhibit apparent lek behavior. During normal behavior, females come to males emitting this pheromone from erect hair pencils (Pliske 1975). This male aphrodisiac of one species may pervade a territory also occupied by males of a second and lead to cross mating, thought to have significance in the family's evolution (Vasconcellos Neto and Brown 1982).

*Mechanitis* females (and those of a few other genera, e.g., *Hypothyris*) lay their eggs in clusters on thick, pilose or spiny (trichomes), poisonous solanum, and apocyanaceous (Echites Group) types and the larvae live gregariously on silk pads that they spin over the coarse leaf surfaces (Rathcke and Poole 1975; Young and Moffett 1979*a*, 1979*b*). However, they do not seem to sequester from them the distasteful or poisonous chemicals that render

them largely immune from vertebrate predation (Brown 1984). Some birds have even learned to feed selectively on the abdominal contents of a few species that may lack these chemicals (Brown and Vasconcellos Neto 1976). By contrast, most genera (*Greta*, formerly *Hymenitis*, etc.) dwell in understory habitats where they exploit nonpilose, thin, papery leaves, oviposit singly, and have solitary larvae.

The larvae of one species have been observed interacting with ants in an apparently rudimentary, mutualistic way (Young 1978). Adults may follow army ant swarms, feeding on the droppings of ant birds (Ray and Andrews 1980), but may take nourishment from any liquid bird feces (Young 1984).

Externally, the larvae (fig. 10.20d) are simple, slender, tapering at both ends, and smooth, or with sparse short hairs, (although some have elongate, fleshy lateral protuberances). They are unicolorous or banded with various colors. The silver chrysalids (fig. 10.20c) are suspended only from the tip of the abdomen and are strongly arched dorsally, the head recurved; the thorax is enlarged so that the apex of the wing covers protrude strongly.

The widespread and common members of the genera *Melinaea* and *Mechanitis* (fig. 10.20b) have been implicated as prime movers in the evolution of mimicry rings (Brown 1977).

The subfamily contains 300 species

(d'Almeida 1978, Mielke and Brown 1979, Lamas pers. comm.).

## References

BROWN, JR., K. S. 1977. Geographical patterns of evolution in Neotropical Lepidoptera: Differentiation of the species of *Melinaea* and *Mechanitis* (Nymphalidae, Ithomiinae). Syst. Entomol. 1: 161–197.

BROWN, JR., K. S. 1984. Adult-obtained pyrrolizidine alkaloids defend ithomiine butterflies against a spider predator. Nature 309: 707–709.

BROWN, JR., K. S., AND J. VASCONCELLOS NETO. 1976. Predation on aposematic ithomiine butterflies by tanagers (*Pipraeidea melanonota*). Biotropica 8: 136–141.

D'ALMEIDA, R. F. 1978. Catálogo dos Ithomiidae Americanos (Lepidoptera). Cons. Nac. Desenv. Cien. Tecn., Curitiba.

FOX, R. M. 1956. A monograph of the Ithomiidae (Lepidoptera). Pt. 1. Amer. Mus. Nat. Hist. Bull. 111: 1–75.

FOX, R. M. 1960. A monograph of the Ithomiidae (Lepidoptera). Pt. II. The Tribe Melinaeini Clark. Amer. Entomol. Soc. Trans. 86: 109–171.

FOX, R. M. 1967. A monograph of the Ithomiidae (Lepidoptera). Pt. III. The tribe Mechanitini Fox. Amer. Entomol. Soc. Mem. 22: 1–90.

FOX, R. M., AND H. G. REAL. 1971. A monograph of the Ithomiidae (Lepidoptera). Pt. IV. The tribe Napeogenini Fox. Amer. Entomol. Inst. Mem. 15: 1–368.

LAMAS, G., AND J. E. PÉREZ. 1983. Danainae e Ithomiinae (Lepidoptera, Nymphalidae) atraidos por *Heliotropium* (Boraginaceae) en Madre de Dios, Perú. Rev. Peruana Entomol. 24: 59–62.

MIELKE, O. H. H., AND K. S. BROWN, JR. 1979. Suplemento ao catalogo do Ithomiidae Americanos (Lepidoptera) de Romualdo Ferreira D'Almeida. Cons. Nac. Desenv. Cien. Tecn., Curitiba.

PLISKE, T. E. 1975. Courtship behavior and use of chemical communication by males of certain species of ithomiine butterflies (Nymphalidae: Lepidoptera). Entomol. Soc. Amer. Ann. 68: 935–942.

PLISKE, T. E., J. A. EDGAR, AND C. C. J. CULVENOR. 1976. The chemical basis of attraction of ithomiine butterflies to plants containing pyrrolizidine alkaloids. J. Chem. Ecol. 2: 255–262.

RATHCKE, B. J., AND R. W. POOLE. 1975. Coevolu-tionary race continues: Butterfly larval adaptation to plant trichomes. Science 187: 175–176.

RAY, T. S., AND C. C. ANDREWS. 1980. Ant butterflies: Butterflies that follow army ants to feed on ant bird droppings. Science 210: 1147–1148.

VASCONCELLOS NETO, J., AND K. S. BROWN, JR. 1982. Interspecific hybridization in *Mechanitis* butterflies (Ithomiidae): A novel pathway for the breakdown of isolating mechanisms. Biotropica 14: 288–294.

YOUNG, A. M. 1978. Possible evolution of mutualism between *Mechanitis* caterpillars and an ant in northeastern Costa Rica. Biotropica 10: 77–78.

YOUNG, A. M. 1984. Ithomiine butterflies associated with non-antbird droppings in Costa Rican tropical rain forest. J. Lepidop. Soc. 38: 61–63.

YOUNG, A. M., AND M. W. MOFFETT. 1979a. Behavioral regulatory mechanisms in populations of the butterfly *Mechanitis isthmia* in Costa Rica: Adaptations to host plants in secondary and agricultural habitats (Lepidoptera: Nymphalidae: Ithomiidae). Deutsche Entomol. Zeit., N.F., 26: 21–38.

YOUNG, A. M., AND M. W. MOFFETT. 1979b. Studies on the population biology of the tropical butterfly *Mechanitis isthmia* in Costa Rica. Amer. Midl. Nat. 101: 309–319.

## Passion Vine Butterflies

Nymphalidae, Heliconiinae, *Heliconius* and relatives. Heliconians.

Passion vine butterflies are a successful group, judging from their abundance in low to mid-elevation forest habitats, the fairly large number of species (65, mostly *Heliconius*), and their many highly developed behavioral and physical adaptations (Brown 1981). These uniquely Neotropical butterflies have attracted numerous biological studies since the days of their first appreciation by Henry Walter Bates (1862) in Amazonia. Of all taxa, it seems best to have fulfilled Bates's own prediction that "the study of butterflies will some day be valued as one of the most important branches of biological science."

Bates himself proposed and demonstrated the principle of simple mimicry from his Amazonian observations, mainly of ithomiines and heliconians. The phenomenon is now recognized as common among insects, fish, and even plants. Later, working in southern Brazil, Fritz Müller extended the concept by discovering that two or more distasteful species benefit by displaying the same pattern (see mimicry, chap. 2).

Substantiating conclusions from bird, reptile, plant, and other butterfly distributions, detailed investigation of *Heliconius* in Central and South America has helped reveal the existence and locations of presumed Quaternary ice age forest refugia (see life zones, chap. 2) (Brown et al. 1974, Benson 1982). Localized speciation of heliconians is a continuing dynamic process, linked to present-day ecological factors and historical fluctuations in vegetation (Brown and Benson 1977, Descimon and Mast de Maeght 1984).

Much of the heliconians' significance in evolutionary biology derives from their association with two types of plants, passion vines (*Passiflora*) and the cucurbit genus *Psiguria* (formerly *Anguria*) and close relatives (Gilbert 1975). The former is the larval food and contains noxious chemicals (cyanogenic glucosides and alkaloids) that deter most other herbivores but from which the butterflies probably derive an inedibility that determines their role as mimicry models. The vine also possesses extrafloral nectaries that maintain a defense force of pugnacious ants. The latter's presence strongly influences many aspects of the butterfly's utilization of the plant as a host.

Curiously, some plants also defend themselves against feeding by heliconians by producing small, oval, yellow growths on the leaves and tendrils, which mimic the butterfly's eggs. Females refrain from ovipositing on plants already occupied by eggs and are thus discouraged from doing so on parts with these structures (Gilbert 1982).

The bright orange flowers of *Psiguria* are produced in progressively maturing peduncles that attract regular visitations of adult heliconians. The butterflies derive from these flowers not only nectar but also pollen, which clings to their proboscises, to be later dissolved by regurgitated digestive fluids and ingested (Gilbert 1972). This rich protein food enables heliconians to live very long for butterflies, possibly up to nine months, during which time egg production is also prolonged. Other flowers may be visited at random.

Long life has enabled such sophisticated behaviors to evolve in these butterflies. They also form aggregations to pass the night, hanging from vegetation in groups of many individuals (Mallet 1986). Roosts are habitually used by the same individuals.

They have developed highly sensitive and efficient neural mechanisms that give them exceptional sight and color pattern recognition, presumably because of the role played by such in releasing courtship behavior (Swihart 1967). Females attract mates also with strong sexual pheromones. Multiple males often arrive simultaneously and elicit rapid copulation (often before the female completely escapes the chrysalid case). Interference by rival suitors is discouraged with a postmating sexual suppression pheromone (so-called olfactory chastity belt, Turner 1973) imparted to the female by the successful male (Gilbert 1976).

Heliconian adults are long winged, with attenuate antennal clubs. In the abdomen, females have yellow eversible glands with hair pencils on the tips. There is a basically simple anal vein in the fore wing (i.e., no spur) and a strongly recurved humeral vein at the base of the hind wing. Their color patterns vary greatly within the basic types (except for the "clear" type, which is not represented in this subfamily; see butterflies and moths, above) and are controlled by a simple set of genetic loci (Nijhout and Gilbert 1990). There are

**Figure 10.21 HELICONIINE BUTTERFLIES (NYMPHALIDAE).** (a) Green heliconius (*Philaethria dido*). (b) Tiger heliconius (*Heliconius melanops*). (c) Black heliconius (*Heliconius erato*), pupa. (d) Black heliconius, adult. (e) Black heliconius, larva. (f) Zebra butterfly (*Heliconius charitonius*).

some unusual patterns, as in *Philaethria dido* (fig. 10.21a), which has a light green ground color on the upper side and black bars and spots. This pattern is mimicked by an edible nymphalid (*Siproeta* [formerly *Victorina*] *stelenes;* fig. 10.18f) (Young 1974).

Two very common species often seen in gardens and weed fields are the zebra butterfly (*Heliconius charitonius;* fig. 10.21f), which is black with oblique yellow wing bars (Cook et al. 1976, Young 1976), and the julia (*Dryas iulia* [formerly spelled *julia*], fig. 10.20f), which is all dull orange except for a single subapical black wing bar (Young 1978, Muyshondt 1973). The latter species has been observed drinking tears from the eyes of caimans and turtles in Peru (Turner 1986). Because the wings of members of the genus *Agraulis* (fig. 10.20e) are silver spotted on the undersides like the Holarctic butterflies called "fritillaries," they bear that name also. A common species with "tiger" markings is *Heliconius melanops* (fig. 10.21b); a predominantly black-marked species is *H. erato* (fig. 10.21d).

The larvae (fig. 10.21e) are usually cream colored or pale with dark spots or bands and covered with fringed, single, nonurticating spines, including a pair dorsally from the oversized head. Chrysalids (fig. 10.21c) hang from the terminus of the abdomen with an irregularly shaped body and a pair of wide-set, flat head projections. This shape and their brownish or grayish color give them the appearance of a dead leaf. Silver or white streaks also are frequently present, imitating holes and imperfections in the fake leaf. A few species, such as *Laparus doris*, have gregarious, nonspiny larvae that may form enormous clusters on the leaves of the host vines; pupae of these hang in great numbers from the undersides of leaves and stems of associated trees and shrubs. Such groups arise from eggs that are grouped, even those of different females (Cook and Brower 1969).

Because of their great diversity and plastic wing patterns, the taxonomy of the group is complicated (Emsley 1965, Brown 1976, Michener 1942). It is clear that members of the same morphological series can display widely divergent mimetic patterns, and, conversely, similar species are often unrelated (Turner 1976).

## References

BATES, H. W. 1862. Contributions to an insect fauna of the Amazon valley (Lepidoptera: Heliconidae). Linnean Soc. London Trans. 23: 495–561. 6, pl. 55–56.

BENSON, W. S. 1982. Alternative models for infrageneric diversification in the humid tropics: Tests with passion vine butterflies. *In* G. T. Prance, ed., Biological diversification in the tropics. Columbia Univ. Press, New York. Pp. 608–640.

BROWN, JR., K. S. 1976. An illustrated key to the silvaniform *Heliconius* (Lepidoptera: Nymphalidae) with descriptions of new subspecies. Amer. Entomol. Soc. Trans. 102: 373–484.

BROWN, JR., K. S. 1981. The biology of *Heliconius* and related genera. Ann. Rev. Entomol. 26: 427–456.

BROWN, JR., K. S., AND W. W. BENSON. 1977. Evolution in modern Amazonian non-forest islands: *Heliconius hermathena*. Biotropica 9: 95–117.

BROWN, JR., K. S., O. H. H. MIELKE, AND H. EBERT. 1974. Quaternary refugia in tropical America: Evidence from race formation on *Heliconius* butterflies. Royal Soc. London Proc. B 187: 369–378.

COOK, L. M., AND L. P. BROWER. 1969. Observations on polymorphism in two species of heliconiine butterflies from Trinidad, West Indies. Entomologist 102: 125–128.

COOK, L. M., E. W. THOMASON, AND A. M. YOUNG. 1976. Population structure, dynamics and dispersal of the tropical butterfly *Heliconius charitonius*. J. Anim. Ecol. 45: 851–863.

DESCIMON, H., AND J. MAST DE MAEGHT. 1984. Semispecies relationships between *Heliconius erato cyrbia* Godt. and *H. himera* Hew in southwestern Ecuador. J. Res. Lepidop. 22: 229–237.

EMSLEY, M. G. 1965. Speciation in *Heliconius* (Lep., Nymphalidae): Morphology and geographic distribution. Zoologica 50: 191–254.

GILBERT, L. E. 1972. Pollen feeding and reproductive biology of *Heliconius* butterflies. Nat. Acad. Sci. Proc. 69: 1403–1407.

GILBERT, L. E. 1975. Ecological consequences of a coevolved mutualism between butterflies and plants. *In* L. E. Gilbert and P. H. Raven, ed., Coevolution of animals and plants. Univ. Texas Press, Austin. Pp. 210–240.

GILBERT, L. E. 1976. Postmating female odor in *Heliconius* butterflies: A male-contributed antiaphrodisiac? Science 193: 419–420.

GILBERT, L. E. 1982. The coevolution of a butterfly and a vine. Sci. Amer. 247: 15, 110–114, 116, 119–121.

MALLET, J. 1986. Gregarious roosting and home range in *Heliconius* butterflies. Natl. Geogr. Res. 2(2): 198–215.

MICHENER, C. D. 1942. A generic revision of the Heliconiinae (Lepidoptera, Nymphalidae). Amer. Mus. Nov. 197: 1–8.

MUYSHONDT, A. 1973. Some observations on *Dryas iulia iulia* (Heliconiidae). J. Lepidop. Soc. 27: 302–303.

NIJHOUT, H. F., AND L. E. GILBERT. 1990. An analysis of the phenotypic effects of certain colour pattern genes in *Heliconius* (Lepidoptera: Nymphalidae). Biol. J. Linnean Soc. 40: 357–372.

SWIHART, S. L. 1967. Neural adaptations in the visual pathway of certain heliconiine butterflies, and related forms, to variations in wing coloration. Zoologica 52: 1–14.

TURNER, J. R. G. 1973. Passion flower butterflies. Animals 15: 15–17, 19–21.

TURNER, J. R. G. 1976. Adaptive radiation and convergence in subdivisions of the butterfly genus *Heliconius* (Lepidoptera: Nymphalidae). Zool. J. Linnean Soc. 58: 297–308.

TURNER, J. R. G. 1986. Drinking crocodile tears: The only use for a butterfly? Antenna 10: 119–120.

YOUNG, A. M. 1974. Further observations on the natural history of *Philaethria dido dido* (Lepidoptera: Nymphalidae: Heliconiinae). New York Entomol. Soc. J. 82: 30–41.

YOUNG, A. M. 1976. Studies on the biology of *Heliconius charitonius* L. in Costa Rica (Nymphalidae: Heliconiinae). Pan-Pacific Entomol. 52: 291–303.

YOUNG, A. M. 1978. Spatial properties of niche separation among *Eueides* and *Dryas* butterflies (Lepidoptera: Nymphalidae: Heliconiinae) in Costa Rica. New York Entomol. Soc. J. 86: 2–19.

## Morphos

Nymphalidae, Morphinae. *Spanish:* Morfos. *Portuguese:* Azulonas, azulás.

Resplendent with their great, metallic blue wings, the morphos are the flying jewels of the forests. This effulgent blue is caused by the spectral reflectance of light passing through microscopic lamellae in the wing scales (Hirata and Ohsako 1966, Pillai 1968). As it is a physical color, it never fades. Its function is not understood, but it may startle and confuse attackers or function in sexual recognition (Young 1971b).

The subfamily is uniquely tropical American and contains about twenty-five species, most in the typical genus *Morpho* (LeMoult and Réal 1962). (Many more have been named by overly zealous amateur lepidopterists, but all are not considered valid entities.)

The sight of one of these magnificent iridescent butterflies lazily flapping through the verdant green is enough to dazzle the perceptions and elicit exclamations from traveler and scientist alike. Natives in Amazonian Peru tell that the morpho is one form assumed by the forest spirits, or *chullachaquis,* who lead followers into the jungle to become lost forever. Morphos are the favorites of collectors and once were prizes sought by the liberés from Devil's Island penal colony in French Guiana during its infamous history. Today, the tradition of making trays and decoupages from their wings still persists for the tourist trade in Rio de Janeiro (Carvalho and Mielke 1972). Some species are protected by federal law, but others are still caught wild without apparent damage to their populations since primarily the males are taken (Kesselring 1975). Others are reared artificially for commercial purposes (Otero 1971). Bermúdez (1966) suggested one species (*M. peleides*) as the national butterfly for Venezuela.

It is a legend that the first one of these beautiful butterflies to be taken to Europe was by Sir Walter Raleigh who presented it to Queen Elizabeth (ibid.) She is said to have adorned her hair with it one night at a ball. On seeing the horror of some naturalists also present, she gave it to the one who later used it as the type specimen of *M. peleides.*

Their fondness for the exudates of fermenting fruit makes morphos easy prey to hunters using the fruit as bait (Young and Thomason 1974). A piece of blue paper or a dead specimen also tempts their curiosity and will draw specimens into net range. They may also be taken while sleeping in loose aggregations (Young 1971*a*).

Two types of blue morphos may be recognized according to basic wing pattern: solid blue (e.g., *M. rhetenor*) and blue only in a broad medial field bordered by black (e.g., *M. achillaena;* fig. 10.22a, pl. 2f). In addition, there are all white species (e.g., *M. polyphemus*) and a single very large species that is generally orange with black markings (*M. hecuba;* fig. 10.22e). Each group represents different adaptive strategies with distinctive behavioral and ecological traits (Young 1982, Young and Muyshondt 1972). The undersides of the wings bear eyespots or are otherwise marked with dead leaf patterns to camouflage the closed-winged resting butterfly (fig. 10.22c). Eyespots tend to be more conspicuous in low-flying, ground-feeding types (Young 1980).

The larvae, which feed mainly on many species of legumes, euphorbs, and grasses (Carvalho and Mielke 1972), are very colorful, with multicolored, serial designs on the dorsum (fig. 10.22b). They are also beset with hair tufts whose color and arrangements vary according to species. All have very hairy, oversized heads. Species are generalized in their choice of hosts, and larval colors vary on different plants, a possible system of automimicry to protect palatable individuals on low-toxicity hosts (Young and Muyshondt 1973). They are crepuscular feeders. The light, leaf green chrysalids (fig. 10.22d) are obovoid and hang from hooks on a short, terminal stalk.

**Figure 10.22 MORPHO BUTTERFLIES (NYMPHALIDAE).** (a) Achilles morpho (*Morpho achillaena*). (b) Achilles morpho, larva. (c) Achilles morpho, undersides. (d) Morpho (*Morpho* sp.), pupa. (e) Hecuba (*Morpho hecuba*).

Some of them have spot patterns reminiscent of a simian face.

## References

BERMÚDEZ, P. J. 1966. El *Morpho peleides* podría ser la mariposa nacional. Natura (Caracas) 31: 20.

CARVALHO, J. C. M., AND O. H. H. MIELKE. 1972. The trade of butterfly wings in Brazil and its effects upon the survival of the species. 9th Int. Cong. Entomol. 1: 486–488.

HIRATA, K., AND N. OHSAKO. 1966. Studies on the structure of scales and hairs of insects. IV. Microstructure of scales of the butterfly *Morpho menelaus nakaharai* Le Moult. Kagoshima Univ. Sci. Rpt. 15: 49–61.

KESSELRING, J. 1975. Are morphos endangered? Atala 3(2): 31.

LEMOULT, E., AND P. RÉAL. 1962. Les *Morpho* d'Amérique du Sud et Central. Ed. LeMoult, Paris.

OTERO, L. S. 1971. Instruções para criação de borboleta "capitão-do-mato" (*Morpho achillaena*) e outras espécies do gênero *Morpho* ("azul-sêda," "bóia," "azulão-branco," "praia-grande"). Inst. Brasil. Desenv. Flor., Rio de Janeiro.

PILLAI, P. K. C. 1968. Spectral reflection characteristics of *Morpho* butterfly wing. Optical Soc. Amer. J. 58: 1019–1022.

YOUNG, A. M. 1971a. Notes on gregarious roosting in tropical butterflies of the genus *Morpho*. J. Lepidop. Soc. 25: 223–234.

YOUNG, A. M. 1971b. Wing coloration and reflectance in *Morpho* butterflies as related to reproductive behavior and escape from avian predators. Oecologia 7: 209–222.

YOUNG, A. M. 1980. The interaction of predators and "eyespot butterflies" feeding on rotting fruits and soupy fungi in tropical forests: Variations on a theme developed by the Muyshondts and Arthur M. Shapiro. Entomol. Rec. 90: 63–69.

YOUNG, A. M. 1982. Notes on the natural history of *Morpho granadensis polybaptus* Butler (Lepidoptera: Nymphalidae: Morphinae), and its relation to that of *Morpho peleides limpida* Butler. New York Entomol. Soc. J. 90: 35–54.

YOUNG, A. M., AND A. MUYSHONDT. 1972. Geographical and ecological expansion in tropical butterflies of the genus *Morpho* in evolutionary time. Rev. Biol. Trop. 20: 231–263.

YOUNG, A. M., AND A. MUYSHONDT. 1973. Notes on the biology of *Morpho peleides* in Central America. Carib. J. Sci. 13: 1–49.

YOUNG, A. M., AND J. H. THOMASON. 1974. The demography of a confined population of the butterfly *Morpho peleides* during a tropical dry season. Stud. Neotrop. Fauna 9: 1–34.

## Owl Butterflies

Nymphalidae, Brassolinae, *Caligo*.
*Spanish:* Buhitos pardos (Costa Rica), mariposas de Muzo (Colombia), vaquitas negras (Ecuador, larvae).

These are familiar members of the family, easily recognized by their great size (WS 12–15 cm) and large, conspicuous eyespot near the center of the underside of the hind wing (fig. 10.23a). The latter, in some species, is surrounded by an elongate, dark field, creating the illusion of the head profile of a lizard or frog. A secondary spot, on the leading edge of the wing, resembles the tympanum of such vertebrates. This is thought to be a false warning device that affords the butterfly protection when it is at rest with wings folded (Stradling 1976). The upper sides are dark blue, purple, or brown on the inner half or more and black on the outer half or dark bordered. The male abdomen bears conspicuous red to yellow scale-covered scent glands on each side opposite the androconia of the wings. Volatile chemicals from these glands may play a role in the territorial behavior exhibited in this sex (Blandin and Descimon 1975).

The adults are not known to take nectar from flowers but avidly suck liquids from rotting fruit. They are crepuscular but are often frightened into flight—an erratic, undulating motion close to the ground— during the daylight hours from their perches in forest glades. Individuals may be spotted resting on vegetation near the ground. They are most frequent in the vicinity of the larval food plants, which most commonly belong to the banana family and its relatives: *Heliconia* and banana (Musaceae), *Canna* (Cannaceae), and *Calathea* (Marantaceae). Some plants in the ginger (*Hedychium*), grass (*Pennisetum*), and

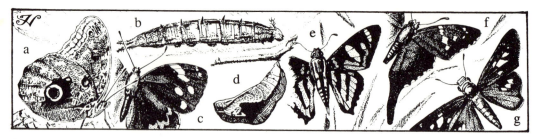

**Figure 10.23    BUTTERFLIES AND SKIPPERS.** (a) Owl butterfly (*Caligo eurilochus*, Nymphalidae), undersides. (b) Owl butterfly, larva. (c) Darius (*Dynastor darius*, Nymphalidae). (d) Darius, pupa. (e) Blue banded skipper (*Elbella polyzona*, Hesperiidae). (f) Long-tailed skipper (*Urbanus proteus*, Hesperiidae). (g) Canna skipper (*Calpodes ethlius*, Hesperiidae).

palm families (*Cocos*) are also utilized (Carvalho and Mielke 1972). At times, the larvae are even abundant enough on banana to be considered a pest (Harrison 1963, Malo and Willis 1961). Bamboo and manioc are also recorded hosts but are probably substitutes for the aforementioned. Larvae feed at all times of the day (Condie 1976).

Mature larvae are large (BL 10–15 cm), mottled brown, and fusiform (fig. 10.23b, pl. 2g). The head has a pair of well-developed clubbed "horns" projecting from its upper outer corners (and sometimes a second pair laterally) and short, sharp spines situated on the back of each of the middle segments. The terminally suspended pupae are also large (BL 4–5 cm) and have the appearance of a curled, dead leaf. Numerous minute prickly spines are present on the back of the abdomen (Young and Muyshondt 1985).

Some seventeen species are recognized in the genus.

### References

BLANDIN, P., AND H. DESCIMON. 1975. Contribution a la connaissance des Lépidoptères de l'Equateur: Les Brassolinae (Nymphalidae). Soc. Entomol. France Ann. (N.S.) 11: 3–28.

CARVALHO, J. C. M., AND O. H. H. MIELKE. 1972. The trade of butterfly wings in Brazil and its effects upon the survival of the species. 9th Int. Cong. Entomol. 1: 486–488.

CONDIE, S. 1976. Some notes on the biology and behavior of three species of Lepidoptera (Satyridae: Brassolinae) on non-economic plants in Costa Rica. Tebiwa 3: 1–28.

HARRISON, J. O. 1963. The natural enemies of some banana insect pests in Costa Rica. J. Econ. Entomol. 56: 282–285.

MALO, F., AND E. R. WILLIS. 1961. Life history and biological control of *Caligo eurilochus*, a pest of banana. J. Econ. Entomol. 54: 530–536.

STRADLING, D. J. 1976. The nature of the mimetic patterns of the brassolid genera *Caligo* and *Eryphanis*. Ecol. Entomol. 1: 135–138.

YOUNG, A. M., AND A. MUYSHONDT. 1985. Notes on *Caligo memnon* Felder and *Caligo atreus* Kollar (Lepidoptera: Nymphalidae: Brassolinae) in Costa Rica and El Salvador. J. Res. Lepidop. 24: 154–175.

### The Darius

Nymphalidae, Brassolinae, *Dynastor darius*.

This is a very large (WS 95 mm), crepuscular species (fig. 10.23c) found from Guatemala south into Brazil (Aiello and Silberglied 1978). Its larva (BL 6 cm) feeds on bromeliads and is a bizarre creature with a smooth fusiform body, a head with a corona of club-shaped horns, and a forked tail. The cuticle is longitudinally pin-striped with dark brown and dark green on a neutral ground color. A conspicuous, oval, black spot, with a reticulate green center and concentric outer rings of green and black, marks the back between abdominal segments 3 and 4 (a second, similar, smaller spot is situated behind, spanning abdominal segments 5 and 6). It is the chrysalid, however, that is extraordinary in its resemblance to the head of a snake (fig. 10.23d). It is large (4 cm long) and brown on the dorsum, beige and whitish on the undersides. Because it hangs in an incline position, its colors are reversed in position so that the pupa is darker below, this shade invading on each light upper portion, giving the impression of the constriction behind a viper's head. Reticulations in the pattern also resemble reptilian scales, and the eyes are accentuated.

## Reference

AIELLO, A., AND R. E. SILBERGLIED. 1978. Life history of *Dynastor darius* (Lepidoptera: Nymphalidae: Brassolinae) in Panama. Psyche 85: 331–345.

## SKIPPERS

Hesperioidea. *Spanish:* Gusanos cabezones (Central America, larvae).

Skippers are mothlike butterflies with robust bodies and drab colors for the most part, although some deviate from drabness with metallic blue, red, or other bright color fields on the upper wing surfaces. They are otherwise closely related to the true butterflies.

They range in size from small (WS 17 mm) to large (WS 5 cm) and possess wideset antennae with the knobs uniquely curved, some even hooked, at the tip. All six legs are completely developed. Skippers are named for their nervous, flitting flight.

Their larvae are naked, often green, cream, or yellow, and covered with a fine white powder or bloom. They have over-sized, round, brightly colored heads attached to a distinct somewhat constricted "neck" (= first thoracic segment). Their tastes are extremely varied, but many food plants are monocots, often grasses. Many have the habit of rolling themselves up in the leaves of their hosts, tying the cut edges of their shelters down with silk. Here they pupate in cocoonlike structures or on the ground.

This is a large group with some 1,800 to 2,000 Neotropical species. Representative species are the canna skipper, *Calpodes ethlius* (fig. 10.23g), and fiery skipper, *Hylephila phyleus* (Young 1982). Many in the genera *Urbanus* (fig. 10.23f) and *Chioides* are tailed; these and the larger, more brilliantly marked species belong to the subfamily Pyrginae. Black bars on a pale bluish-white background is an often-repeated pattern in several unrelated species, for example, *Elbella polyzona* (fig. 10.23e), *Phocides thermus, Tarsoctenus papias,* and *Jamadia gnetus.*

Adults are avid nectar feeders and act as important pollinators of flowering plants, some even having greatly elongated tongues for probing deep-throated blossoms of plant tubes to which they may be mutualistically associated (Emmel 1971).

## References

EMMEL, T. C. 1971. Symbiotic relationship of an Ecuadorian skipper (Hesperiidae) and *Maxillaria* orchids. J. Lepidop. Soc. 25: 20–22.

YOUNG, A. M. 1982. Notes on the interaction of the skipper butterfly *Calpodes ethlius* (Lepidoptera: Hesperiidae) with its larval host plant *Canna edulis* (Cannaceae) in Mazatlán, State of Sinaloa, Mexico. New York Entomol. Soc. J. 90: 99–114.

# 11 FLIES AND MIDGES

Diptera. *Spanish:* Moscas (General); zancudos (midges, long-legged flies, General); mosquillas (midges, General). *Portuguese:* Moscas, pernilongos (Brazil, long-legged flies); mosquinhas (Brazil, midges). *Tupi-Guaraní:* Mberú, carapanã (Brazil, long-legged flies); bironhas (= mberu-obi) (Brazil, biting flies). *Quechua:* Chuspi (fly). *Nahuatl:* Zayolmeh (sing. zayolin) (Mexico).

Except for a relatively few wingless forms, all the members of this order (Oldroyd 1964) are characterized by a single pair of wings arising from the mesothorax; the metathoracic wings have evolved into special club-shaped, sensory organs (halteres), important to the insect's flight equilibrium. Flies are very efficient aeronauts, and many are capable of powerful, sustained flight, an ability provided by an oversized thorax, accommodating hyperdeveloped flight muscles to move the wings. No other order of insects has this particular arrangement of flight organs.

All adult flies feed on liquids, which they imbibe with siphoning tubular or spongelike mouthparts. Vertebrate blood is an important food, but other natural liquors are consumed as well: blood of other insects, plant exudates, nectar, animal secretions, honeydew, decomposing organic substances, feces, and so on. The larvae feed on, and live in, a still greater variety of substances, both liquid and solid. Many live in water and are filter feeders or browsers of algae and diatoms. Pupae of many aquatic species are active swimmers.

The formal higher classification of the Diptera is controversial (Griffiths 1972, 1981; Steyskal 1974). For convenience, the order may be divided into three main groups: (1) Nematocera ("midges"), in which the adults have long, many-segmented antennae, the larvae have well-developed heads, and pupae are free; (2) Brachycera ("straight-seamed flies"), in which there are abbreviated antennae composed of several unlike segments, larval heads are reduced, and there are free pupae from which the adult escapes through a T-shaped or straight opening; and (3) Cyclorrhapha ("muscoid flies"), whose adults have very short antennae, usually with only two, very small, basal segments and an oval, large third segment with a sensory bristle, whose larvae are maggotlike, completely without a sclerotized head capsule, and whose pupae are contained in the persistent last larval skins that harden and form a capsule called a "puparium" (fig. 11.8d). The last group is further divided into small muscoid flies without calypters, enlarged lobes at the base of the wings ("acalypterate muscoids"), and larger muscoid flies with well-developed calypters ("calypterate muscoids"). (A few families do not fit into this dichotomy and are set off in a special group, the Aschiza.) The terms "lower" or "primitive" flies, in contrast to "higher" flies, are also often applied informally to refer respectively to the Nematocera versus Brachycera and muscoids.

The Neotropics support a very large dipteran fauna, mostly poorly known.

Most families have been cataloged (Papavero 1966–), and the numbers of species found in each as given in the following sections are taken from this publication.

There are many economically important dipteran groups. Adults carry human and animal diseases (see flies and disease, below), and larval feeding damages fruits, vegetables, tubers, and other crops. Flies, however, also include many parasites useful in biological control of insect pests. Certain Neotropical tachinid flies, the so-called Cuban fly (*Lixophaga diatraeae*) and the Amazon fly (*Metagonistylum minense*), have even been exported to the Oriental tropics to reduce populations of lepidopterous stem borers of rice, sugarcane, and other gramineous crops (although with only partial success; Kamran 1973).

It is curious that, along with the llama and other animals, flies (*chuspi*) were held in some reverence by the ancient Andean peoples. According to Rivero and von Tschudi (1857) flies were sacrificed to the sun by the followers of the Incas in ancient Peru and are still used as a design motif in that country (see fig. 1.8).

## References

GRIFFITHS, G. C. D. 1972. The phylogenetic classification of Diptera Cyclorrhapha, with special reference to the structure of the male postabdomen. W. Junk, The Hague.

GRIFFITHS, G. C. D. 1981. [Book review of] Manual of Nearctic Diptera. Vol. 1. J. F. McAlpine, B. V. Peterson, G. E. Shewell, H. J. Teskey, J. R. Vockeroth, D. M. Wood, eds. Minister of Supply and Services. Entomol. Soc. Can. Bull. 13: 49–55.

KAMRAN, M. A. 1973. Introduction of Neotropical tachinids into Southeast Asia for biological control of stem borers of graminaceous crops. Entomol. Soc. Amer. Bull. 19: 143–146.

OLDROYD, H. 1964. The natural history of flies. Norton, New York.

PAPAVERO, N., ed. 1966–. A catalogue of the Diptera of the Americas south of the United States. Dept. Zool., Sec. Agric. São Paulo, São Paulo. Fasc. 1–103.

RIVERO, M. E., AND J. J. VON TSCHUDI. 1857. Peruvian antiquities. Putnam, New York.

Translated from Spanish to English by F. L. Hawks.

STEYSKAL, G. C. 1974. Recent advances in the primary classification of the Diptera. Entomol. Soc. Amer. Ann. 67: 513–517.

# FLIES AND DISEASE

Exceptionally good flying and dispersal abilities, along with the bloodsucking, predaceous, unsavory habits of numerous species, have made the Diptera the single most important insect order from the medical-veterinary standpoint (Zumpt 1973). Most of the world's insect-borne diseases of humans and animals are transmitted by biting and filth flies, and several serious pathological conditions are caused by flies themselves. Very large sums of money are spent from the public health budgets of nations everywhere, especially in the tropical portions of Latin America, in efforts to control diseases effected by dipterans.

## Reference

ZUMPT, F. 1973. Diptera parasitic on vertebrates in Africa south of the Sahara and in South America, and their medical significance. *In* B. J. Meggers, E. S. Ayensu, and W. D. Duckworth, eds., Tropical forest ecosystems in Africa and South America: A comparative review. Smithsonian Inst. Press, Washington, D.C. Pp. 197–205.

## Biting Flies as Disease Vectors

Dipteran mouthparts, which are basically structured for a liquid diet, have been further modified in several families for piercing or slashing the skin of vertebrates and withdrawing blood. Blood is rich in protein and provides the nutrients necessary for egg development in these flies, which is why only the females bite. Attraction of the fly to its host is still not fully understood. Substances emanating from the host, such as carbon dioxide, perspiration, and body odors, as well as skin color and body size are some factors that stimulate biting, but others, still unknown, are probably involved. Environmental conditions are also important; man-biting activity is largely dependent on ambient temperature and water vapor pressure (Read et al. 1978).

Many disease microorganisms of birds and mammals, including humans, have evolved blood-inhabiting stages in their life cycles, taking advantage of the blood-letting habits of many insects, even developing an obligatory dependence on them for the development of critical phases in the reproduction of their species. These adaptations, and the fact that Diptera fly, promote the dissemination of these organisms and make midges and flies the most effective known biological vectors of parasites.

Among the blood feeders, the mosquitoes (Culicidae) are the best known and the most active spreaders of pathogens to both man and animals. The already lengthy list of diseases that only mosquitoes transmit promises to grow even longer as new ones, especially ones caused by viruses, are discovered (Aitken et al. 1960, Causey et al. 1961). In the New World tropics, the most serious medical and veterinary problems are mostly mosquito related and include the malarias, yellow fever, encephalitides, dengue, filariases, and a host of other poorly understood infections.

Other groups of blood-feeding Diptera with species also involved in the spread of disease are the blackflies (Simuliidae), which transmit onchocerciasis, sand flies (Psychodidae, *Lutzomyia*), which carry cutaneous leishmaniasis and oroya fever, punkies (Ceratopogonidae, *Culicoides*), and horseflies and deerflies (Tabanidae). The latter two families are both involved as vectors in several filarial, bacterial, and protozoal afflictions.

These groups are all (except Tabanidae [Brachycera]) classified in the Nematocera, but several isolated species of Cyclorrhapha, such as the stable fly (*Stomoxys calcitrans*) and horn fly (*Haematobia irritans*), also feed on blood and act as vectors (of these, both males and females bite). (The infamous tsetses [*Glossina*], carriers of sleeping sickness in Africa, are not found in the New World.)

The direct effects of the bites of flies can be as grave as the infections they bring. Mosquitoes and other biting flies are a constant nuisance in the environment of Indians, who may be forced to live in a cloud of smoke from a wood fire to survive parts of the day. Every account of travel along the great rivers tells of the merciless attacks of these insects. Explorers Alexander von Humboldt and Aimé Bonpland wrote,

> People who have not navigated the great rivers of equinoctal America . . . can scarcely conceive how, at every instant, without intermission, you may be tormented by . . . mosquitoes, zancudos, jejenes, and tempraneros, that cover the face and hands, pierce the clothes with their long needle-formed suckers, and getting into the mouth and nostrils, occasion coughing and sneezing whenever any attempt is made to speak in the open air. In the missions of the Orinoco . . . the plague of the mosquitoes affords an inexhaustible subject of conversation. When two people meet in the morning, the first questions they address to each other are: "How did you find the zancudos during the night? How are we today for the mosquitoes?" (1852: 2:273)

The last message received from the ill-fated 1925 expedition of Col. P. H. Fawcett, seeking lost cities in the Brazilian hinterland, tells how noxious insects contributed to its failure:

> The attempt to write is fraught with much difficulty owing to the legions of flies that pester one from dawn till dark—and sometimes all through the night! The worst are the tiny ones smaller than a pinhead, almost

*invisible, but stinging like a mosquito. Clouds of them are always present. . . . The stinging horrors get all over one's hands, and madden. Even the head nets won't keep them out. As for mosquito nets, the pests fly through them!* (1953: 300)

Habitation in some areas is impossible even today because of the bloodthirsty onslaughts of hordes of blackflies and sand flies. Bites often cause severe allergic reactions and may produce local sores that resist healing and become septic.

Not all species of the biting fly groups are equally offensive; those of primary importance will be noted below.

### References

AITKEN, T. H. G., W. G. DOWNS, C. R. ANDERSON, AND L. SPENCE. 1960. Mayaro virus isolated from a Trinidadian mosquito, *Mansonia venezuelensis*. Science 131: 986.

CAUSEY, O. R., C. E. CAUSEY, O. M. MAROJA, AND D. G. MACEDO. 1961. The isolation of arthropod-borne viruses, including members of two hitherto undescribed serological groups, in the Amazon region of Brazil. Amer. J. Trop. Med. Hyg. 10: 227–249.

FAWCETT, P. H. 1953. Lost trails, lost cities. Funk & Wagnalls, New York.

READ, R. G., A. J. ADAMES, AND P. GALINDO. 1978. A model of microenvironment and man-biting tropical insects. Environ. Entomol. 7: 547–552.

VON HUMBOLDT, A. AND A. BONPLAND. 1852 [1814–1825]. Personal narrative of travels to the equinoctial regions of America, during the year 1799–1804. 3 vols. Henry Bohn, London. Translated by Thomasina Ross.

### Filth Flies

Numerous species of flies live in close association with humans ("synanthropic" or "domestic" flies), their larvae breeding in wastes and adults frequenting all sorts of organic matter contaminated with human disease organisms. These are the filth flies, a name describing the foul habits of the notorious housefly (Coutinho et al. 1957) and its relatives among the muscoid flies, the green blowfly and other carrion flies, certain flesh flies, acalypterates, such as the eye gnats (*Hippelates*), and many others (Lindsay and Scudder 1956). Because of the propensity of these Diptera to pick up bacteria, viruses, amoebic cysts, and other pathogens while frequenting feces, garbage, and carrion, and the capability to mechanically transmit them to the food of people and domestic animals, or directly onto lips, eyes, and fingers, filth flies must be considered of public health and veterinary importance (Alcívar and Campos 1946, Greenberg and Bornstein 1964). The evidence for fly involvement in the perpetuation of more than sixty diseases is abundant, the most convincing being that for enteric disorders or dysenteries caused by *Shigella* and *Salmonella* bacteria. While not demonstrably involved as primary vectors of these pathogens, flies must always be considered to often have considerable potential epidemiological importance during outbreaks of polio, hepatitis, conjunctivitis, staphylococcal infections, cholera, pinta, brucellosis, and surra, especially under very unsanitary conditions such as prevail in times of warfare and famine, following disasters, or when other disruptions of social order and hygiene occur, such as the development of slums.

A comprehensive list of the filth fly species and the diseases borne by them in the vast area of Latin America is beyond the scope of this book; the subject is surveyed in some detail by Greenberg (1971, 1973).

### References

ALCÍVAR, C., AND F. CAMPOS. 1946. Las moscas, como agentes vectores de enfermedades entéricas en Guayaquil. Rev. Ecuatoriana Hig. Med. Trop. 3: 3–14.

COUTINHO, J. O., A. DE E. TAUNAY, AND L. P. DE CARVALHO LIMA. 1957. Importância de *Musca domestica* como vector de agentes patogênicos para o homen. Rev. Inst. Adolpho Lutz São Paulo 17: 5–23.

GREENBERG, B. 1971, 1973. Flies and disease. 2 vols. Princeton Univ. Press, Princeton.

GREENBERG, B., AND A. A. BORNSTEIN. 1964. Fly dispersion from a rural Mexican slaughterhouse. Amer. J. Trop. Med. Hyg. 13: 881–886.

LINDSAY, D. R., AND H. I. SCUDDER. 1956. Nonbiting flies and disease. Ann. Rev. Entomol. 1: 323–346.

## Myiasis

The larvae of many kinds of flies find their way into living animals and feed on healthy or diseased tissues, a condition called myiasis (Guimarães et al. 1983). The host acquires the infection by swallowing fly eggs or larvae in its food or from eggs (and sometimes young larvae) purposefully placed by the fly on its body or on some substratum from which it will be taken up (Dove 1937). Many fly groups may be involved (James 1947), either as (1) accidental invaders of the host, seldom causing any lasting effects and dying before reaching maturity, (2) facultative parasites usually breeding in decomposing organic matter but occasionally feeding within the body, or (3) specific myiasis producers, belonging to species having obligatory development in living flesh.

From the clinical standpoint, these infestations are classified according to the site at which the larvae feed: (1) in the skin (cutaneous myiasis), including creeping diseases such as produced by early larvae of horse bots, (2) in sores and wounds (traumatic myiasis) by calliphorids, especially screwworms, (3) in cavities of the body (sinus myiasis), (4) in internal organs or passages (organic and enteric myiasis) by almost any species that is accidentally swallowed or attracted to orifices of the digestive or urogenital tracts, and (5) in the eye (ocular myiasis) caused by various types, among them the sheep bots whose first instar larvae invade the conjunctiva. Numerous publications contain case histories of all types of myiasis in human (Isola and Osimani 1944, Donoso 1947, Tobar and Honorato 1947) and animal (Lacey and George 1981) hosts in many parts of Latin America (Mazza and Jörg 1939, Thomas 1987).

The incidence of myiasis is apparently higher in the humid tropics than in cooler temperate regions (Méndez 1981). This may be due to the poorer hygiene and faster development of larvae in the warmer climes. Myiasis lesions are seldom benign and should always be considered as potentially serious and treated promptly. The results of untreated cases in humans range from mild discomfort, with no permanent damage, to gross destruction of vital organs and dreadful external disfigurements with associated psychological disturbances. Botflies (Cuterebridae, Oestridae, and Gasterophilidae), carrion or blowflies (Calliphoridae), flesh flies (Sarcophagidae), and certain other muscoid flies are most often the cause of myiasis, but almost any species with saprophagous larvae may be involved, at least accidentally.

## References

DOVE, W. E. 1937. Myiasis of man. J. Econ. Entomol. 30: 39–49.

DONOSO, R. B. 1947. Miasis humana en Chile y consideraciones clínicas y epidemiológicas. Rev. Chilena Hig. Med. Prevent. 1947: 1–54.

GUIMARÃES, J. H., N. PAPAVERO, AND A. P. DO PRADO. 1983. As míases na região Neotropical (identificação, biologia, bibliografia). Rev. Brasil. Zool. 1: 239–416.

ISOLA, W., AND J. OSIMANI. 1944. Un nuevo caso de oftalmiasis conjuntival producida por Oestris ovis L. en el Uruguay. Arch. Uruguay Med. 25: 260–264.

JAMES, M. T. 1947. The flies that cause myiasis in man. U.S. Dept. Agric. Misc. Publ. 631: 1–175.

LACEY, L. A., AND T. K. GEORGE. 1981. Myiasis in an Amazonian porcupine. Entomol. News 92: 79–80.

MAZZA, S., AND M. E. JÖRG. 1939. Cochliomyia hominivorax americana C. y P., estudio de sus larvas y consideraciones sobre miasis. Univ. Buenos Aires, Mis. Estud. Pat. Reg. Argentina, Publ. 41: 3–46.

MÉNDEZ, E. 1981. Las miasis centroamericanas

y los dípteros que las producen. Rev. Med. Panamá 6: 146–159.

THOMAS, JR., D. B. 1987. Incidence of screwworm (Diptera: Calliphoridae) myiasis on the Yucatán Peninsula of Mexico. J. Med. Entomol. 24: 498–502.

TOBAR, G., AND A. HONORATO. 1947. Anotaciones acerca de una epidemia de miasis humana. Hospital Viña del Mar (Chile) Publ. Trimes. 3(1): 5–14.

## CRANE FLIES

Tipulidae.

Extremely long, frail, easily detached legs and elongate body and wings immediately identify these flies (fig. 11.1c). They are mistaken for mosquitoes by many people, but they never bite and have no scaly vestiture. Their size range is very great, from tiny (BL 5–7 mm) to large (BL to 50 mm).

Adults are commonplace in damp situations, usually only near water, which is a frequent larval medium, although a few species occur over open water or remain continuously submerged (Byers 1977, 1981, 1982). Bromeliad tanks, tree holes, and seashore rock crevices are specialized aquatic habitats. Many others develop in soil, in leaf mold, and in other purely terrestrial sites. The larvae are mostly cylindrical, with a leathery integument and fleshy lobes surrounding large breathing pores on the blunt terminus of the abdomen (fig. 11.1d). In the forest, crane flies often aggregate in protected, moist places like hollow rotten tree stumps or between the buttresses of large trees and "dance." They gyrate up and down and create a ghostly illusion with their diaphanous wings and widespread legs. The function of this behavior is not known. Many also vibrate rapidly on their leg tips over the substratum, another unexplained phenomenon. Mimetic relationships between crane flies and ichneumonid wasps are known (Slobodchikoff 1974).

This is the largest family of flies in the Neotropics. Practically nothing has been written regarding the biology of the well over 3,400 Latin American species, which certainly represents only a moderate fraction of the total number yet to be found (Alexander and Alexander 1970, Bruch 1939).

### References

ALEXANDER, C. P., AND M. M. ALEXANDER. 1970. Family Tipulidae. *In* N. Papavero, ed., 1967–, A catalogue of the Diptera of the Americas south of the United States. Mus. Zool., Univ. São Paulo, São Paulo. Pp. 4.1–4.259.

BRUCH, C. 1939. Contribución al conocimiento de los tipúlidos argentinos (Diptera). Physis 17: 3–28.

BYERS, G. W. 1977. Tipulidae. *In* S. H. Hurlbert, ed., Biota acuática de sudamérica austral. San Diego State Univ., San Diego. Pp. 259–265.

BYERS, G. W. 1981. Tipulidae. *In* S. H. Hurlbert, G. Rodríguez, and N. Dias dos Santos,

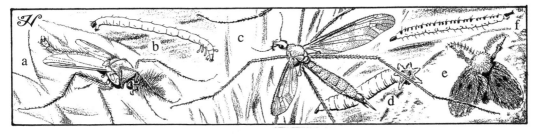

**Figure 11.1    FLIES (MIDGES).** (a) Water midge (*Chironomus* sp., Chironomidae). (b) Water midge, larva. (c) Crane fly (*Tipula* sp., Tipulidae). (d) Crane fly, larva. (e) Bathroom fly (*Clogmia albipunctata*, Psychodidae). (f) Bathroom fly, larva.

eds., Aquatic biota of tropical South America. Pt. 1. Arthropoda. San Diego State Univ., San Diego. Pp. 231–241.

BYERS, G. W. 1982. Tipulidae. *In* S. H. Hurlbert and A. Villalobos Figueroa, eds., Aquatic biota of Mexico, Central America and the West Indies. San Diego State Univ., San Diego. Pp. 407–414.

SLOBODCHIKOFF, C. N. 1974. Behavioral and morphological mimicry in a cranefly and an ichneumonid. Pan-Pacific Entomol. 50: 155–159.

## WATER MIDGES

Chironomidae. *Spanish:* Sayules (when swarming, from Nahuatl, *zayollan* = abundance of flies) (Nicaragua).

Although they resemble mosquitoes with their elongate bodies, long legs, and plumose male antennae, water midges do not have biting mouthparts and lack scales on any part of the body (fig. 11.1a) (Saether 1971).

A few larvae occur in wet decaying matter, but the majority are truly aquatic, leading submerged lives in all types of freshwater, brackish water, and marine coastal habitats. When abundant, they constitute an important food item for other aquatic animals, especially fish. Larval densities of 50,000 per square meter are not unusual. Particularly noticeable are those of the genus *Chironomus* that inhabit lakes, where their numbers build enormously during the reproductive cycle. These are often bright red, from hemoglobin in their blood ("blood worms") (fig. 11.1b), which aids in respiration on the oxygen-deficient bottom waters in the tube houses they construct in the bottom ooze. In some places during the dry season, periodic mass emergences of adults from these populations take place, filling the air with clouds of gnats and creating a public nuisance. This phenomenon is well documented at Lake Nicaragua (Bay 1964) where the offending species is *Siolimyia amazonica* (Wirth 1979).

The larvae of a few species associate as commensals with other animals in the water; some attach to other aquatic insects (Epler 1986) or even the skin of bottom-dwelling catfishes in the Amazon River (Fittkau 1974, Freihofer and Neil 1967). Except for a few carnivorous, phytophagous, and parasitic forms, chironomid larvae are microphages on organic detritus.

The taxonomy of Neotropical water midges is in an infant state. At least 1,500 species are thought to exist, many of these yet unnamed (Fittkau and Reiss 1979, Fittkau 1971). The South American fauna is divided into a warm-adapted species complex, occupying the Guiana-Brazilian area, and more or less cold-loving groups, limited to the Andean-Patagonian regions (Fittkau 1978). Of the seven subfamilies, only the Tanypodinae and Chironominae are abundant in the region. Most of the water midges living in forest streams of Central Amazonia belong to the latter. The number of species in an aquatic habitat often accounts for more than 50 percent of the total invertebrate species present. Larvae of virtually all build some sort of a case on or within the substrate in which they live.

The Telmatogetoninae are largely a marine group, found in rocky intertidal areas along the entire American coast. Their larvae are physiologically adapted to salt water (Oliver 1971).

The special contribution of certain aquatic midge groups to our understanding of the biogeography of southern South America has been referred to elsewhere (see physiographic regions, chap. 2). Literature on the Neotropical members of the family is reviewed by Reiss (1977, 1981, 1982); world bibliographies are also available (Fittkau et al. 1976, Hoffrichter and Reiss 1981).

## References

Bay, E. C. 1964. An analysis of the "sayule" (Diptera: Chironomidae) nuisance at San Carlos, Nicaragua, and recommendations for its alleviation. World Health Org./EBL20, WHO/Vector Control 86: 1–19.

Epler, J. H. 1986. A novel new Neotropical *Nanocladius* (Diptera: Chironomidae), symphoretic on *Traverella* (Ephemeroptera: Leptophlebiidae). Fla. Entomol. 69: 319–327.

Fittkau, E. J. 1971. Distribution and ecology of Amazonian chironomids (Diptera). Can. Entomol. 103: 407–413.

Fittkau, E. J. 1974. *Ichthyocladius* n. gen., eine neotropische Gattung der Orthocladiinae (Chironomidae, Diptera) deren Larven epizoische auf Welsen (Astroblepidae und Loricariidae) leben. Entomol. Tidskr. 95: 91–106.

Fittkau, E. J. 1978. Sich abzeichnende Verbreitungsmuster in die neotropische-nearktischen Chironomidenfauna. Deutsch. Ges. Allg. Angewan. Entomol. Mitt. 1: 77–81.

Fittkau, E. J., and F. Reiss. 1979. Die zoogeographische Sonderstellung der neotropischen Chironomiden. Spixiana 2: 273–280.

Fittkau, E. J., F. Reiss, and O. Hoffrichter. 1976. A bibliography of the Chironomidae. Gunnera 26: 1–177.

Freihofer, W. C., and E. H. Neil. 1967. Commensalism between midge larvae (Diptera: Chironomidae) and catfishes of the families Astroblepidae and Loricariidae. Copeia 1967: 39–45.

Hoffrichter, O., and R. Reiss. 1981. Supplement 1 to "A bibliography of the Chironomidae." Gunnera 37: 1–68.

Oliver, D. R. 1971. Life history of the Chironomidae. Ann. Rev. Entomol. 16: 211–230.

Reiss, F. 1977. Chironomidae. *In* S. H. Hurlbert, ed., Biota acuática de sudamérica austral. San Diego State Univ., San Diego. Pp. 277–280.

Reiss, F. 1981. Chironomidae. *In* S. H. Hurlbert, G. Rodríguez, and N. Dias dos Santos, eds., Aquatic biota of tropical South America. Pt. 1. Arthropoda. San Diego State Univ., San Diego. Pp. 261–268.

Reiss, F. 1982. Chironomidae. *In* S. H. Hurlbert and A. Villalobos Figueroa, eds., Aquatic biota of Mexico, Central America and the West Indies. San Diego State Univ., San Diego. Pp. 433–438.

Saether, O. A. 1971. Notes on general morphology and terminology of the Chironomidae (Diptera). Can. Entomol. 103: 1237–1260.

Wirth, W. W. 1979. *Siolimyia amazonica* Fittkau, an aquatic midge new to Florida with nuisance potential. Fla. Entomol. 62: 134–135.

## MOTH FLIES

Psychodidae. Owl midges.

So called because of a dense body covering of elongate scales, which gives them a lepidopteran appearance (fig. 11.1e), moth flies are dwellers in dampness. The larvae of most are aquatic or require considerable moisture in their habitats. These range from wet soil and rotting leaf litter to permanently moist or submerged stream and pond margins; the lance-winged midges (*Maruina*) live on smooth rocks under clean dripping or spray water and even totally submerged in torrential water in streams (Hogue 1973).

Larvae are elongate and cylindrical (fig. 11.1f), except those of *Maruina*, which are flattened and have ventral suckers. Each segment possesses distinct, transverse, dorsal sclerotized plates and well-developed setae. The head capsule is small but complete, and there is a short terminal breathing tube. Adults are all small (BL 1–4 mm) and densely covered with elongate scales. The wings are elliptical in outline (lanceolate in *Maruina*), with only longitudinal veins apparent and, when at rest, are held rooflike over the abdomen. The number of species currently known (approximately 430 in 28 genera) is probably far short of the total living in Latin America (Duckhouse 1973).

A few well-known domestic species, such as *Clogmia* (formerly *Telmatoscopus*) *albipunctata* (fig. 11.1e) (Sebastiani 1978) and *Psychoda alternata*, breed in the foul organic matter that accumulates in sink drain traps, sewage filters, septic tanks, and similar places. Adults of these are often seen indoors, especially in bathrooms and lavatories, where they are a minor nuisance.

Save for these few, little is known of Neotropical moth flies (see the literature reviews in Duckhouse 1977, 1981, 1982). Sand flies of the subfamily Phlebotominae (see below), however, are bloodsuckers and transmit several human and animal diseases and are well studied.

## References

DUCKHOUSE, D. A. 1973. Family Psychodidae. *In* N. Papavero, ed., 1967–, A catalogue of the Diptera of the Americas south of the United States. Mus. Zool., Univ. São Paulo, São Paulo. Pp. 6A.1–6A.29.

DUCKHOUSE, D. A. 1977. Psychodidae. *In* S. H. Hurlbert, ed., Biota acuática de sudamérica austral. San Diego State Univ., San Diego. Pp. 266–267.

DUCKHOUSE, D. A. 1981. Psychodidae. *In* S. H. Hurlbert, G. Rodríguez, and N. Dias dos Santos, eds., Aquatic biota of tropical South America. Pt. 1. Arthropoda. San Diego State Univ., San Diego. Pp. 241–245.

DUCKHOUSE, D. A. 1982. Psychodidae. *In* S. H. Hurlbert and A. Villalobos Figueroa, eds., Aquatic biota of Mexico, Central America and the West Indies. San Diego State Univ., San Diego. Pp. 414–416.

HOGUE, C. L. 1973. A taxonomic review of the genus *Maruina* (Diptera: Psychodidae). Los Angeles Co. Mus. Nat. Hist. Bull. 17: 1–69.

SEBASTIANI, F. L. 1978. Ciclo biológico de *Telmatoscopus albipunctatus* (Williston, 1893) (Diptera, Psychodidae). I. Comportamento sexual. Ciên. Cult. 30(6): 719–722.

## SAND FLIES

Psychodidae, Phlebotominae, *Lutzomyia.* *Spanish:* Manta-blancas (General), moscas morranos (Colombia), chitras (Panama), titiras (Peru). *Portuguese:* Arrupiados, asa-duras, mosquitos palha (Brazil). *Tupi-Guaraní:* Birigüi.

Females of several genera of Psychodidae (once combined with the Old World species as the single genus *Phlebotomus;* Lewis et al. 1977, Theodor 1965) feed on vertebrate blood and are important vectors of human and animal diseases (Lewis 1974). In Latin America, species of *Lutzomyia*

transmit several types of leishmaniases (Killick-Kendrick 1978, Ward 1977) from Mexico to Argentina (but not, strangely, in Chile and Uruguay). The most prevalent are so-called cutaneous leishmaniases (*papalomoyo, espundia, uta, pian bois, bubon de Vélez, bubon de Aleppo, úlcera de Chinácota,* etc.), resulting from infections by various species of *Leishmania, L. brasiliensis, L. peruviana,* and *L. mexicana.* In the near past, one variety of the last species (*Leishmania m. mexicana*) was common among chicle gatherers in Mexico, Guatemala, and Belize, whose ulcerated ears, acquired during forays in the forest where the disease was rife and sand flies common, became the hallmark of their profession (chiclero's ulcer). Visceral leishmaniasis (caused by *Leishmania donovani chagasi*) is also widely distributed in South America where a proven vector is *Lutzomyia longipalpis.*

Sand flies also transmit Peruvian verruga (Oroya fever, carrion's disease), which is restricted to certain Andean valleys in Peru, Colombia, and Ecuador. This is an often deadly rickettsial disease, manifested at one stage by ugly skin warts. It is transmitted through the bite of *Lutzomyia verrucarum* and *L. colombiana.*

Pizarro's troops were decimated by this disease during military expeditions in 1533–1537 in Peru, and there is evidence that verruga was the cause of death of the Inca Huayna Capac shortly before that time (Patrón 1896). In the 1870s, 7,000 deaths were reported from the affliction in the Rímac Valley of Peru, when the construction of the Trans-Andean railway was impeded by this sickness suffered by thousands of laborers (Noguchi et al. 1928). In addition to leishmanias, wild sand flies harbor trypanosomatid flagellates (Christensen and Herrer 1976) and many viruses of unknown effects on their hosts (e.g., pacuí virus in rice rats; Aitken et al. 1975).

Sand flies (Martins et al. 1971) are small, hairy psychodids (BL 1.5–4 mm), grayish-

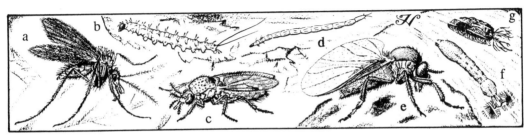

**Figure 11.2   BITING MIDGES.** (a) Sand fly (*Lutzomyia* sp., Psychodidae). (b) Sand fly, larva. (c) Punkie (*Culicoides* sp., Ceratopogonidae). (d) Punkie, larva. (e) Blackfly (*Simulium* sp., Simuliidae). (f) Blackfly, larva. (g) Blackfly, pupa in cocoon.

yellow to brown, with elongate wings that are held obliquely away from the body when at rest (fig. 11.2a). They have a slender body and a set of elongate biting mouthparts (Lewis 1975). Their flight is also characteristic: a series of short, erratic hops, in which the fly seldom moves more than a short distance.

Adults often rest in sheltered places (Chaniotis et al. 1972), soil and rock crevices, between tree buttresses, under loose tree bark, in caves (Williams 1976), tree holes, animal burrows, and termite nests, and sometimes in human dwellings. The majority of the 260 or more Neotropical species are forest dwellers (Porter and De Foliart 1981, Young 1979), and females normally suck the blood of sylvan mammals, birds, and reptiles. Both sexes also take plant sugars in the form of nectar, honeydew, and the like.

The larvae (fig. 11.2b) are separated from those of other psychodids by a double pair of long anal spines. They live in soil, feeding on organic debris such as excrement, dead insects, decaying plant matter, and fungi.

## References

AITKEN, T. H. G., J. P. WOODALL, A. H. P. DE ANDRADE, G. BENSABATH, AND R. E. SHOPE. 1975. Pacuí virus, phlebotomine flies, and small mammals in Brazil: An epidemiological study. Amer. J. Trop. Med. Hyg. 24: 358–368.

CHANIOTIS, B. N., R. B. TESH, M. A. CORREA, AND K. M. JOHNSON. 1972. Diurnal resting sites of phlebotomine sandflies in a Panamanian tropical forest. J. Med. Entomol. 9: 91–98.

CHRISTENSEN, H. A., AND A. HERRER. 1976. Neotropical sand flies (Diptera: Psychodidae), invertebrate hosts of *Endotrypanum schaudinni* (Kinetoplastida: Trypanosomatidae). J. Med. Entomol. 13: 299–303.

KILLICK-KENDRICK, R. 1978. Recent advances and outstanding problems in the biology of phlebotomine sandflies. Acta Tropica 35: 297–313.

LEWIS, D. J. 1974. The biology of Phlebotomidae in relation to Leishmaniasis. Ann. Rev. Entomol. 19: 363–384.

LEWIS, D. J. 1975. Functional morphology of the mouth parts in New World phlebotomine sandflies (Diptera: Psychodidae). Royal Entomol. Soc. London Trans. 126: 497–532.

LEWIS, D. J., D. G. YOUNG, G. B. FAIRCHILD, AND D. M. MINTER. 1977. Proposals for a stable classification of the Phlebotomine sandflies (Diptera: Psychodidae). Syst. Entomol. 2: 319–332.

MARTINS, A. V., D. J. LEWIS, AND O. THEODOR. 1971. Phlebotomine sandflies. World Health Org. Vector Biol. Contr. Unit. Pap. 71.255: 1–23.

NOGUCHI, H., R. C. SHANNON, E. B. TILDEN, AND J. R. TYPER. 1928. *Phlebotomus* and Oroya fever and verruga peruana. Science 68: 493–495.

PATRÓN, P. 1896. Apuntes históricos sobre la verruga americana. Soc. Geogr. Lima Bol. 5(4): 435–445.

PORTER, C. H., AND G. R. DE FOLIART. 1981. The man-biting activity of phlebotomine sandflies (Diptera: Psychodidae) in a tropical wet forest in Colombia. Arq. Zool. (São Paulo) 30: 81–158.

THEODOR, O. 1965. On the classification of American Phlebotominae. J. Med. Entomol. 2: 171–197.

WARD, R. D. 1977. New World leishmaniasis: A

review of the epidemiological changes in the last three decades. 15th Int. Cong. Entomol. Proc. Pp. 505–522.

WILLIAMS, P. 1976. The phlebotomine sandflies (Diptera, Psychodidae) of caves in Belize, Central America. Bull. Entomol. Res. 65: 601–614.

YOUNG, D. G. 1979. A review of the blood-sucking psychodid flies of Colombia (Diptera: Phlebotominae and Sycorocinae). Univ. Fla. Inst. Food Agric. Sci. Agric. Esp. Sta. Bull. 806: 1–266.

## PUNKIES

Ceratopogonidae. *Spanish:* Jejenes (var. ihenni), majes, plagas, mimes (General). *Portuguese:* Maruins, mosquitos pólvora, bembés (Brazil). *French:* Bigailles (Haiti). No-see-ums.

The smallest of the biting flies, punkies measure only 1 to 4 millimeters in body length. They are gnatlike, somewhat resembling blackflies, but their wings are elongate, with few veins, often dark pigmented with clear spots, and folded flat over the abdomen when at rest. The legs are slender but short.

Punkies have mouthparts adapted for liquid food like those of other biting midges, but they feed on a wider variety of animal hosts and plants as well. A few *Atrichopogon, Pterobosca,* and *Forcipomyia* are known to be ectoparasites on adult and larval lepidopterans, orthopterans, and other insects (Wirth 1956). *Culicoides, Lasiohelea,* and *Leptoconops* species suck blood from birds, mammals, including humans, and frogs. Many genera only take nectar from flowers (*Dasyhelea*). Some species of *Dasyhelea,* of *Forcipomyia,* and small, hairy members of other genera, are pollinators of Pará rubber (Warmke 1952), cacao (Bystrak and Wirth 1978, Young 1983), and other tropical crops (Winder 1978).

Many of the blood-feeding *Culicoides* (fig. 11.2c) (Forattini 1957) are notorious pests and bite intensely, seemingly well beyond the capacities of such minute crea-

tures (Aréan and Fox 1955, Kettle 1977). They often swarm in intolerable numbers in mangrove swamps and along riverbanks or marshes, making such places uninhabitable. Like blackflies and mosquitoes, they are mentioned in every account of tropical travel among the greatest hazards encountered.

*The tiny black ihenni of the Bolivian part of the Amazon watershed, has a bite like the burn of a cigarette and travels millions strong. It has an uncanny ability to get under or through any net. It belongs to no union and knows no hours—it attacks incessantly every minute of the day and night. Several travelers through this region have died, not directly from the bites of the insects, but because of the impossibility of getting a moment's rest.* (Price 1952: 143)

Their bites often induce severe allergic reactions. Skin reactions range from mild itching to blistering and open lesions often complicated by secondary infections.

In some species of temperate areas, the larger females often eat the smaller males during mating. This habit has not been observed for tropical species but is likely to occur, since many of the same or related genera, such as *Bezzia* and *Palpomyia,* inhabit both regions (Downes 1978).

*Culicoides furens,* found throughout the Caribbean and along the Middle American coasts to Ecuador and Brazil, is the most widespread biter. It and related species are vectors of filarial heart worm (*Mansonella ozzardi*) between and among humans and dogs. The genus also harbors a variety of human pathological viruses (Linley et al. 1983). In tropical America, *Culicoides* transmit onchocerciasis, blue tongue, and other sicknesses of horses, sheep, and cattle, as well as protozoan infections of birds (Kettle 1965). Curiously, the genus is virtually absent from South America south of the approximate latitude of Santiago, Chile.

Most immature punkies are aquatic or subaquatic, occupying many habitats, usually mud or wet sand on lake, pond, river, or stream margins, but they also live in water accumulations in water-holding plants (*Heliconia, Calathea* inflorescences, bromeliad leaf axils; Wirth and de J. Soria 1981). Marshes and land crab burrows are favored breeding sites as well. Other species are semiterrestrial, their larvae developing in damp situations under bark or in rotting wood, fruit, cactus stems, and banana stalks, or in decomposing leaves.

The larvae are very long and slender (BL 1–3 mm), white, and practically hairless. The eellike freely aquatic species (fig. 11.3d) are recognized by their serpentine swimming movements.

The family has about 700 Neotropical species grouped into over 30 genera (Wirth 1974). Many species certainly remain undiscovered. The literature on the Neotropical members of the group has been well indexed (Atchley et al. 1981; Wirth and Cavalieri 1977; Wirth 1981, 1982).

## References

ARÉAN, V. M., AND I. FOX. 1955. Dermal alterations in severe reaction to the bite of the sandfly, *Culicoides furens*. Amer. J. Clin. Path. 25: 1359–1366.

ATCHLEY, W. R., W. W. WIRTH, C. T. GASKINS, AND S. L. STRAUSS. 1981. A bibliography and keyword index of the biting midges (Diptera: Ceratopogonidae). U.S. Dept. Agric. Bibliog. Lit. Agric. 13: 1–544. List of all publications on the family from 1758 to 1978.

BYSTRAK, P. G., AND W. W. WIRTH. 1978. The North American species of *Forcipomyia*, subgenus *Euprojoannisia* (Diptera: Ceratopogonidae). U.S. Dept. Agric. Tech. Bull. 1591: 1–51.

DOWNES, J. A. 1978. Feeding and mating in the insectivorous Ceratopogonidae (Diptera). Entomol. Soc. Can. Mem., 104: 1–62.

FORATTINI, O. P. 1957. Culicoides da região Neotropical (Diptera, Ceratopogonidae). Fac. Hig. Saud. Púb. Univ. São Paulo Arq. 11: 161–526.

KETTLE, D. S. 1965. Biting ceratopogonids as vectors of human and animal diseases. Acta Trop. 22: 356–562.

KETTLE, D. S. 1977. Biology and bionomics of bloodsucking ceratopogonids. Ann. Rev. Entomol. 22: 33–51.

LINLEY, J. R., A. L. HOCH, AND F. P. PINHEIRO. 1983. Biting midges (Diptera: Ceratopogonidae) and human health. J. Med. Entomol. 209: 347–364.

PRICE, W. 1952. The amazing Amazon. John Day, New York.

WARMKE, H. E. 1952. Studies on natural pollination of *Hevea brasiliensis* in Brazil. Science 116: 474–475.

WINDER, J. A. 1978. Cocoa flower Diptera, their identity, pollinating activity, and breeding sites. Pest Art. News Sum. 24: 5–18.

WIRTH, W. W. 1956. New species and records of biting midges ectoparasitic on insects (Diptera, Heleidae). Entomol. Soc. Amer. Ann. 49: 356–364.

WIRTH, W. W. 1974. Family Ceratopogonidae. *In* N. Papavero, ed., 1967–, A catalogue of the Diptera of the Americas south of the United States. No. 14. Mus. Zool. Univ. São Paulo, São Paulo. Pp. 14.1–14.89.

WIRTH, W. W. 1981. Ceratopogonidae. *In* S. H. Hurlbert, G. Rodríguez, and N. Dias dos Santos, eds., Aquatic biota of tropical South America. Pt. 1. Arthropoda. San Diego State Univ., San Diego. Pp. 268–275.

WIRTH, W. W. 1982. Ceratopogonidae. *In* S. H. Hurlbert and A. Villalobos Figueroa, eds., Aquatic biota of Mexico, Central America and the West Indies. San Diego State Univ., San Diego. Pp. 438–442.

WIRTH, W. W., AND F. CAVALIERI. 1977. Ceratopogonidae. *In* S. H. Hurlbert, ed., Biota acuática de sudamérica austral. San Diego State Univ., San Diego. Pp. 280–283.

WIRTH, W. W., AND S. DE J. SORIA. 1981. Two *Culicoides* biting midges reared from inflorescences of *Calathea* in Brazil and Colombia, and a key to the species of the *discrepans* group (Diptera: Ceratopogonidae). Rev. Theobroma 11: 107–117.

YOUNG, A. M. 1983. Seasonal differences in abundance and distribution of cocoa-pollinating midges in relation to flowering and fruit set between shaded and sunny habitats of the La Lola cocoa farm in Costa Rica. J. Appl. Ecol. 10: 801–831.

## BLACKFLIES

Simuliidae. *Spanish:* Mosquitos pelones, alazanes, rodedores, polcos, tabardillos (General); majes, perrujas, moscas de café (Costa Rica); barrigones, jerjeles (Venezuela, Peru). *Portuguese:*

Borrachudos, piuns (Brazil). Kaboura flies (Guianas). Buffalo gnats, coffee flies.

These diurnal biting gnats (Darsie 1982; Vulcano 1977, 1981) are similar to punkies but are generally larger (BL 1–5 mm), often with a black body, although many are gray or yellow with silvery or golden pubescence. Their antennae and legs are short, and their wings are immaculate and broadly triangular.

The females suck blood from warm-blooded vertebrates. Near fast-flowing rocky streams and rivers where their larvae develop, they may occur in such numbers and bite with such persistence as to drive off humans and animals alike and require determined control (Laird 1981). The bite is not bothersome at first but causes a burning sensation that may last for days; the bite of some species is marked by a small, persistent blood spot. Amazon naturalist-explorer Alfred Russell Wallace (1853: 310) recorded his experience with blackflies thus: "My feet were so thickly covered with the little blood-spots produced by their bites, as to be of a dark purplish-red colour, and much swelled and inflamed. . . . The only means of taking a little rest in the day was by wrapping up hands and feet in a blanket."

The blackfly fauna of Latin America is extremely rich, with over 300 species distributed in 10 genera, occurring in practically all regions and running water habitats from high Andean melt waters and outflows of hot springs to the great cataracts of the major rivers (Coscarón 1987). Their distribution includes most of the major Caribbean islands, but they are absent from the Galápagos and Malvinas; mysteriously, they are known on the island of Margarita off the Venezuelan coast where there are no streams.

Species in the genus *Simulium* (fig. 11.2e) bear a number of serious diseases affecting their vertebrate hosts (Coscarón 1984). The most important of these is onchocerciasis (*enfermedad de robles*), caused by the filarial worm *Onchocerca volvulus*, which, before the conquest, was absent from America. These filarial worms undoubtedly were introduced with slaves from Africa and became permanently established in coffee-growing areas of Guatemala (Dalmat 1955), neighboring El Salvador, and the Mexican states of Chiapas and Oaxaca. There are now secondary foci in Venezuela, extreme northern Brazil, Colombia, and Ecuador. The affliction causes blindness in many people in these locales when the worms situate in ocular tissues. Major vector species in Guatemala are *S. ochraceum*, *S. metallicum*, and *S. callidum* (Anon. 1971, Garms 1975). The taxonomy of other vectors is still unsettled (Tidwell et al. 1981). Blackflies are also suspected of transmitting the hemorrhagic syndromes of Altamira and black fever of Lábrea, recently recognized diseases of unknown etiology discovered along the Trans-Amazonian Highway in Brazil (Anon. 1966, Pinheiro et al. 1974). The dog heart worm (*Mansonella ozzardi*) is also transmitted by *Simulium amazonicum* and has been found to affect a significant part of the human population, especially indigenes, in parts of Central America, the West Indies, and the Amazon Basin (Shelley et al. 1982). These findings haved precipitated new interest in studies on Neotropical blackflies (Lacey and Charlwood 1980). There are sixty-one species known to bite humans in the Neotropics (Travis et al. 1974).

Simuliid larvae (fig. 11.2f) are obligatory inhabitants of swift streams and river rapids where they cling to rocks and other hard substrate surfaces, head trailing in the current, in groups of a few to hundreds of individuals (Takaoka 1981). They are elongate with a swollen posterior and a well-developed head; expanded mouthbrushes projecting from the upper side of the head catch drifting plankton and detritus, closing intermittently to transfer the accumulated catch to the mouth. Around an anal sucker and on a small lobe beneath the head are fine hooks that, with the sucker, anchor larvae to a silk mat that they spin with modified salivary glands. If disturbed, larvae loop the head end into the current and may escape in unique fashion by releasing their hold and allowing themselves to be swept away into the current. Simultaneously, they let out a silk anchor line with which they reel themselves back to their feeding site when the danger has passed.

Pupation usually occurs in a silk cocoon, by whose shape species may often be identified. The respiratory organs of the pupa (fig. 11.2g) itself are also diagnostic in shape and branching pattern. The cocoon is attached to submerged rocks so that the adult must later emerge from some depth. After leaving the pupal case and cocoon, the imago rises to the surface in an air bubble, from which it "pops" into flight.

## References

ANONYMOUS. 1966. Studies on the Amazonian fever (febre de Lábrea; Febre Negra; etc.). Belem Virus Lab. Ann. Rpt. App. Pp. 139–157.

ANONYMOUS. 1971. Blackflies in the Americas. World Health Org. Vector Biol. Control Unit Pap. 71.283 4:1–24.

COSCARÓN, S. 1984. Simúlidos sudamericanos: Diptera-Insecta. 9th Cong. Latinoamer. Zool. (Arequipa) Info. Final. Pp. 165–168.

COSCARÓN, S. 1987. El género Simulium Latreille en la Región Neotropical: Análisis de los grupos sujarespecíficos, especies que los integran y distribución geográfica (Simuliidae, Diptera). Mus. Paraense Emilio Goeldi, Belem.

DALMAT, H. T. 1955. The black flies (Diptera: Simuliidae) of Guatemala and their role as vectors of onchocerciasis. Smithsonian Misc. Coll. 125: 1–425.

DARSIE, JR., R. F. 1982. Simuliidae. In S. H. Hurlbert and A. Villalobos Figueroa, eds., Aquatic biota of Mexico, Central America and the West Indies. San Diego State Univ., San Diego. Pp. 443–449.

GARMS, R. 1975. Observations on filarial infections and parous rates of anthropophilic blackflies in Guatemala, with reference to the transmission of Onchocerca volvulus. Tropenmed. Parasit. 26: 169–182.

LACEY, L. A., AND J. D. CHARLWOOD. 1980. On the biting activities of some anthropophilic Amazonian Simuliidae (Diptera). Bull. Entomol. Res. 70: 495–509.

LAIRD, M., ed. 1981. Blackflies: The future for biological methods in integrated control. Academic, London.

PINHEIRO, F. P., D. COSTA, Z. C. LINS, G. BENSABATH, O. M. MAROJA, AND A. H. P. ANDRADE. 1974. Haemorrhagic syndrome of Altamira. Lancet 1: 639–642.

SHELLEY, A. J., R. R. PINGER, AND M. A. P. MORAES. 1982. The taxonomy, biology and medical importance of Simulium amazonicum Goeldi (Diptera: Simuliidae), with a review of related species. Brit. Mus. Nat. Hist. Bull. Ser. Entomol. 44: 1–29.

TAKAOKA, H. 1981. Seasonal occurrence of Simulium ochraceum, the principal vector of Onchocerca volvulus in the southeastern endemic area of Guatemala. Amer. J. Trop. Med. Hyg. 30: 1121–1132.

TIDWELL, M. A., B. V. PETERSON, J. RAMÍREZ, M. DE TIDWELL, AND L. A. LACEY. 1981. Notas y claves preliminares de los jejenes Simulium amazonicum y S. sanguineum (Diptera: Simuliidae) incluyendo los vectores de Onchocerca volvulus y Mansonella ozzardi. Dir. Malar. San. Amb. Bol. 21: 79–89.

TRAVIS, B. V., M. VARGAS, AND J. C. SWARTZWELDER. 1974. Bionomics of black flies (Diptera: Simuliidae) in Costa Rica. I. Species biting man, with an epidemiological summary for the Western Hemisphere. Rev. Biol. Trop. 22: 187–200.

VULCANO, M. A. 1977. Simuliidae. In S. H. Hurlbert, ed., Biota acuática de sudamérica austral. San Diego State Univ., San Diego. Pp. 285–293.

VULCANO, M. A. 1981. Simuliidae. In S. H. Hurlbert, G. Rodríguez, and N. Dias dos Santos, eds., Aquatic biota of tropical South America. Pt. 1. Arthropoda. San Diego State Univ., San Diego. Pp. 275–285.

WALLACE, A. R. 1853. A narrative of travels on the Amazon and Río Negro. Reeve, London.

## MOSQUITOES

Culicidae. *Spanish:* Mosquitos, zancudos (General). *Nahuatl:* Moyomeh (sing. moyotl) (Mexico). *Portuguese:*

Mosquitos, pernilongos, muricocas (Brazil). *Tupi-Guaraní:* Carapanã.

For many, images of hordes of blood-thirsty mosquitoes arise when the word "tropics" is mentioned. Indeed, humid lowland swamps, teeming jungles, and mosquitoes are almost synonymous to those without direct experience in tropical biology. In truth, although there are more mosquito species in the lower latitudes than in the arcto-temperate zones, they seldom appear in such great numbers, and then only locally, certainly never like the vast multitudes spawned by the summer thaw on the Alaskan and Siberian tundras. Yet sometimes their attacks become sufficiently persistent to kill. Pedro Teixeira, while journeying through Mexico in 1600, wrote, "Almost all along this road [Acapulco to San Juan] is a plague of mosquitoes, so terrible and grievous that no defence avails against them, and they stung my best slave to death" (Gillett 1971).

Mosquito diversity in Latin America is phenomenal, however; almost a third of the world's more than 3,000 species, in 18 genera, are found here (Knight 1978, Knight and Stone 1977, Mattingly 1971), as many as in any other single zoogeographic region. The best known are the medically important genera, especially *Anopheles, Aedes,* and *Culex.* Some dominant groups are the subgenus *Melanoconion* of *Culex* and the sabethines, especially *Wyeomyia.* Exploitation of the innumerable available breeding sites probably has engendered their variety. Some classic, general works on the taxonomy of the family in the region are those of Dyar (1928) and Howard, Dyar, and Knab (1912–1917). Although historically important, they are outdated and are now replaced by a multitude of modern contributions on specific groups. But as yet, there is no comprehensive work on the fauna in general. This is greatly needed. Since mosquitoes are the best-known large family of organisms, much can be said about the developmental stages and other aspects of their biology (Bates 1949, Horsfall 1955, Machado-Allison 1980–1982). It is a mistake to think only of stagnant swamps as the mosquito's domain, although larvae do not survive in currents flowing much faster than a few centimeters per second. Neither is fresh water a requirement. Many species breed in coastal marshes and mangrove swamps (*Aedes taeniorhynchus, Anopheles albimanus*), limestone solution holes (*Culex opisthopus*), land crab burrows (*Deinocerites*), and even coral rock holes (*Culex bahamensis*), where brackish, salty, or sometimes even supersaline aquatic conditions prevail. Larvae that live in these situations have special physiological mechanisms for maintaining internal water balance; for example, the rectal gills, organs normally used to absorb salts, are greatly reduced.

Mosquito populations tend to build during the early part of the rainy season when the number of available breeding sites increases. Even minor rainfall fluctuations in nonseasonal areas may effect changes in numbers (Wolda and Galindo 1981).

The most common habitats of the immatures are freshwater accumulations on the ground, especially where there is abundant vegetation, such as in swamps, in marshes, or along the overgrown margins of lakes, ponds, and rivers. This is the home of the ubiquitous *Anopheles, Culex,* and *Aedes* mosquitoes, but others typically breed here, such as *Uranotaenia.* The amount of water need not be great; small puddles and pools and even small niches on the surfaces of rocks or animal footprints may hold enough water to support mosquito life. However, because the nutrient-poor groundwaters of central Amazonia support little food, the region is relatively free of mosquitoes, save those breeding in containers and epiphytes.

Larvae and pupae can thrive in incredibly small amounts of water, as is often

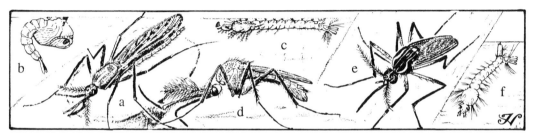

**Figure 11.3  MOSQUITOES (CULICIDAE).** (a) Malaria mosquito (*Anopheles darlingi*). (b) Malaria mosquito, pupa. (c) Malaria mosquito, larva. (d) Giant mosquito (*Toxorhynchites* sp.). (e) Yellow fever mosquito (*Aedes aegypti*). (f) Yellow fever mosquito, larva.

found in various natural and artificial containers. The former includes the hollowed or cupped parts of plants in which rainwater collects, such as rot holes in trees, limbs, and logs (*Aedes, Toxorhynchites*), on fallen leaves, especially large, broad, thick types that drop from trees like *Cecropia* and aroids (*Aedes, Wyeomyia*), and in fruit husks. *Trichoprosopon digitatum* larvae form extremely dense populations in the fetid liquid accumulating in hollow cacao pods that have had their contents eaten away by scavengers. Certain types of New World tropical plants also collect water: bromeliads in their leaf axils, bamboo in hollow stems, *Heliconia* in flower bracts, and pitcher plants (*Heliamphora*). These are the niches occupied by specialized genera with numerous species, often with highly local distributions (*Wyeomyia, Sabethes*).

Larval mosquitoes are elongate with swollen thorax and well-developed, pigmented head (fig. 11.3f). The mouth has brushlike elements (developed from the labrum), kept almost constantly in motion to sweep suspended microorganisms and organic particles into the mandibles and mouth opening. In some predaceous forms (*Toxorhynchites*), the filaments of the mouth brushes are heavy blades used to grasp and hold prey, which is then chewed and swallowed in bits.

Projecting from the last abdominal segment is a sclerotized breathing tube containing the major trachea and entry pores for

air at its apex. The larva relies on atmospheric oxygen and must come to the surface often to take in fresh air. A few types (*Mansonia, Coquillettidea*) have a breathing tube modified for piercing the air-filled vessels of aquatic plants and never need to break the surface film to breathe.

Larvae swim by rapid sideways jerking movements of the body, for which they are called "wrigglers"; pupae are also active and progress similarly, except that the action of the abdomen is forward under the large, oval thorax, and propulsion is aided by a pair of overlapping, rounded flat paddles at the terminus of the abdomen (fig. 11.3b). Rapidly moving pupae often roll over and over, a motion earning them the name "tumblers." The outer body cuticle of both of these stages is set with numerous, often complex hairs whose configurations are extensively used in classification and identification. The shape of the "trumpets," breathing organs projecting from the anterior part of the thorax of the pupa, is also of diagnostic value.

Mosquitoes also live varied and involved lives as adults. The females of nearly all but those belonging to the species with predaceous larvae (*Toxorhynchites*) take terrestrial vertebrate blood as their principal protein food source. The female bites wherever she can penetrate surface capillaries, through exposed skin almost anywhere on the bodies of mammals and amphibians, through the feet and base of the beaks of birds, and between the scales of reptiles. Blood meals may not be required by some (autogenous) strains, which explains how species may survive in areas temporarily without a normal food source.

The adult is distinguishable from Diptera in general and other biting midges by having scales on elongate wings and a very long proboscis. The legs are also very long and fragile, and the eyes are well-developed, large hemispheres. In most, males may be distinguished by densely plumose antennae; female antennae are only lightly haired.

Large swarms of males emit ultrasonic whines with their vibrating wings to attract females for mating. Most mosquitoes follow this pattern. *Deinocerites*, however, never swarm, and males seek females directly in crab burrows (see crabhole mosquitoes, below).

Mosquito females lay their eggs either directly on or near water. Eggs may be placed singly or in small floating groups of a hundred or more (egg rafts). Eggs placed out of water are drought resistant and may remain viable for months before induced to hatch by rising water in their habitat.

Adult mosquitoes are by far the most important insect vectors of diseases to humans and other vertebrates in the Neotropics. Historically, malaria, borne by *Anopheles* species, has been the greatest killer and debilitator of humankind. Its incidence was reduced to a low level as a result of the widespread control campaigns waged with DDT following World War II. However, the development of insecticide resistance by the vectors, economic restraints, drug resistance by the plasmodia, and government apathy toward control programs have allowed a dramatic resurgence of the disease in the 1960s and 1970s (Agarwal 1978, Brown et al. 1976, Chapin and Wasserstrom 1981). Although all four species of human malaria are found in the New World tropics, *Plasmodium falciparum* is dominant and the most serious.

Yellow fever is second to malaria in prevalence, but it is a more virulent disease. It is thought that the urban vector, *Aedes aegypti,* spread from Africa and was responsible for epidemics during the colonization of the Americas, although the jungle form, maintained by native *Haemagogus, Aedes,* and other vectors, may have existed here in pre-Columbian times. The disastrous effect of the disease on the construction of the Panama Canal is well known. Like malaria, yellow fever is increasing in occurrence, following a decline resulting from temporarily successful control efforts in the first half of the century.

Many other, probably indigenous, viral infections are known to have mosquito vectors in the Neotropics, including dengue, transmitted by *Aedes aegypti,* and Saint Louis, Eastern, Venezuelan, and other encephalitides transmitted by *Culex* and *Aedes* species (McLintock 1978). Bancroftian filariasis is nonindigenous but has become established in limited foci, undoubtedly by introduction from Africa during the slave days. These points of infection are scattered throughout the Antilles and northeastern coastal and subcoastal South America where the vectors are mainly house mosquitoes (*Culex quinquefasciatus*).

While the seriousness of these diseases is widely recognized, the direct effects of the bites of mosquitoes should not be discounted. Saliva injected with the bite contains proteins foreign to the human physiology and often elicits strong sensitization reactions. Allergic reactions manifest on the skin as mild to severe urticaria, formation of subdermal tubercles, or production of large, water-filled blisters; systemic symptoms are fever and nausea. Anthropophilic species often show preference for attacking different hosts and areas of the body. Studies to determine the reasons for these preferences are inconclusive, but some generalities are evident. Persons with lighter skin pigmentation seem to be more attractive than darker-skinned individuals. Exhaled carbon dioxide is a major attractant, although other odors emanating from the skin also draw mosquitoes. Favored feeding spots are the ankles, elbows, ears, and back of the hands; warm skin is preferred over cold or hot. Tightly worn clothing is not a good protection since most mosquitoes can bite through the weave. Modern chemical repellents (such as N, N-diethyl-m-toluamide, DEET) are effective if strong solutions are employed and maintained on the skin.

The effects of a few bites are usually negligible, but the persistence of mosquitoes and their abundance in some places renders them formidable pests. Tales from the colonial period of exploration tell of mosquito attacks and their curious effects on people. Seventeenth-century French naturalist Mouffet said, "The gnats in America do so slash and cut, that they will pierce through very thick clothing; so that it is excellent sport to behold how ridiculously the barbarous people, when they are bitten will skip and frisk, and slap with their hands their thighs, buttocks, shoulders, arms and sides, even as a carter doth his horses." von Humboldt and Bonpland (1852, 2: 274) found during their South American travels in 1799–1804 that "between the little harbor of Higuerote and the mouth of the Rio Unare [in coastal Venezuela near Caracas, to avoid mosquitoes], the wretched inhabitants are accustomed to . . . pass the night buried in the sand . . . leaving out the head only, which they cover with a handkerchief."

## References

AGARWAL, A. 1978. Malaria makes a comeback. New Scient. 77: 274–277.

BATES, M. 1949. The natural history of mosquitoes. Macmillan, New York.

BROWN, A. W. A., J. HAWORTH, AND A. R. ZOHAR. 1976. Malaria eradication and control from a global standpoint. J. Med. Entomol. 13: 1–25.

CHAPIN, G., AND R. WASSERSTROM. 1981. Agricultural production and malaria resurgence in central America and India. Nature 293: 181–185.

CLEMENTS, A. N. 1963. The physiology of mosquitoes. Pergamon, Oxford.

DYAR, H. G. 1928. The mosquitoes of the Americas. Carnegie Inst., Washington, D.C., Publ. 387.

FOOTE, R. H., AND D. R. COOK. 1959. Mosquitoes of medical importance. U.S. Dept. Agric. Agric. Handbk. 152: 1–158.

FORATTINI, O. P. 1962, 1965. Entomologia médica. 1. Parte geral, Diptera, Anophelini; 2. Culicini: *Culex, Aedes e Psorophora;* 3. Culicini: *Haemagogus, Mansonia, Culiseta,* Sabethini. Toxorhynchitini, Arboviroses, Filariose bancroftiana, Genetica. Fac. Hig. Saúde Pub., Univ. São Paulo, São Paulo.

GILLETT, J. D. 1971. Mosquitoes. Weidenfeld and Nicolson, London.

GILLIES, M. T. 1980. The role of carbon dioxide in host finding by mosquitoes (Diptera: Culicidae): A review. Bull. Entomol. Res. 70: 525–532.

HARBACH, R. E., AND K. L. KNIGHT. 1980. Taxonomists' glossary of mosquito anatomy. Plexus, Marlton, N.J.

HORSFALL, W. R. 1955. Mosquitoes—their bionomics and relation to disease. Ronald, New York.

HOWARD, L. O., H. G. DYAR, AND F. KNAB. 1912–1917. The mosquitoes of North and Central America and the West Indies. Carnegie Inst., Washington, D.C., Publ. 159.

JONES, J. C. 1978. The feeding behavior of mosquitoes. Sci. Amer. 238(6): 138–140, 143–144, 146, 148.

KNIGHT, K. L. 1978. Supplement to A catalog of the mosquitoes of the world (Diptera: Culicidae). Entomol. Soc. America, College Park, Md.

KNIGHT, K. L., AND A. STONE. 1977. A catalog of the mosquitoes of the world. 2 ed. Entomol. Soc. America (Thomas Say Found.), College Park, Md., 6: 1–611.

LAIRD, M. 1988. The natural history of larval mosquito habitats. Academic, New York.

MACHADO-ALLISON, C. E. 1980–1982. Ecología de los mosquitos (Culicidae). Acta. Biol. Venezuelica 10: 303–371; 11: 51–129, 133–237.

McLINTOCK, J. 1978. Mosquito-virus relationships of American encephalitides. Ann. Rev. Entomol. 23: 17–37.

MATTINGLY, P. F. 1969. The biology of mosquito-borne disease. Sci. Biol. Ser. 1: 1–184.

MATTINGLY, P. F. 1971. Illustrated key to the genera of mosquitoes. Amer. Entomol. Inst. Contrib. 7(4): 1–84.

SEIFERT, R. P. 1980. Mosquito fauna of *Heliconia aurea*. J. Anim. Ecol. 49: 687–697.

SERVICE, M. W. 1976. Mosquito ecology: Field sampling methods. Applied Science, London.

VON HUMBOLDT, A. AND A. BONPLAND. 1852 [1814–1825]. Personal narrative of travels to the equinoctial regions of America, during the years 1799–1804. 3 vols. Henry Bohn, London. Translated by Thomasina Ross.

WARD, R. A. 1977. Culicidae. *In* S. H. Hurlbert, ed., Biota acuática de sudamérica austral. San Diego State Univ., San Diego. Pp. 268–274.

WARD, R. A. 1981. Culicidae. *In* S. H. Hurlbert, G. Rodríguez, and N. Dias dos Santos, eds., Aquatic biota of tropical South America. Pt. 1. Arthropoda. San Diego State Univ., San Diego. Pp. 245–256.

WARD, R. A. 1982. Culicidae. *In* S. H. Hurlbert and A. Villalobos Figueroa, eds., Aquatic biota of Mexico, Central America and the West Indies. San Diego State Univ., San Diego. Pp. 417–429.

WOLDA, H., AND P. GALINDO. 1981. Population fluctuations of mosquitoes in the non-seasonal tropics. Ecol. Entomol. 6: 99–106.

## Malaria Mosquitoes

Culicidae, Anophelinae, *Anopheles*.

Adults in this genus are characterized by an evenly rounded, rather than trilobed, scutellum and the usual absence of scales on the abdomen. Biting specimens can be identified by their "head-standing" posture; the head is rotated forward on the thorax more than in other mosquitoes, forcing the mosquito to elevate the body when working its mouthparts into the skin. Larvae lack a respiratory tube and exhibit flat, palmate setae on the dorsum of most abdominal segments. With notched, thoracic lobes, their spiracular flaps, and these palmate setae, they anchor themselves to the underside of the surface film to feed and respire (fig. 11.3c). (Four species in the related genus *Chagasia* have a trilobed scutellum and larvae with the anterior spiracular flap produced into a long, spine-like process.) The pupa is typical of most mosquitoes (fig. 11.3b).

There are seventy-eight species, representing all the world subgenera in the Neotropical Region except *Cellia*. Most of the human malaria vectors are in the subgenera *Nysorrhynchus* (Faran 1980) and *Kerteszia* (Zavortink 1973). These are mainly the following: *A. pseudopunctipennis*, widely in South America; *A. bellator*, Trinidad and Brazil, a bromeliad breeder (Downs and Pittendrigh 1946); *A. cruzii*, Brazil, also in bromeliads; *A. albimanus*, Caribbean, Mexico, and Central America; and *A. darlingi* (fig. 11.3a), widespread and the best vector (Mendes dos Santos et al. 1981). Many others act as reservoirs for the malaria parasite (*Plasmodium*) in the wild.

Throughout all of history, malaria has rightfully been considered one of the worst scourges of mankind. Extremely debilitating and widespread, it has probably caused more severe suffering than any other single arthropod-borne disease. There is no proof, however, of its presence in America prior to the Spanish Conquest, and it was probably not common for fifty years or more after its introduction, possibly first via one of Columbus's own crews. It surely came with many subsequent trans-Atlantic expeditions. To plunder the rich West Indian cities, Sir Francis Drake left Plymouth in 1585 with twenty-nine ships carrying 1,500 seamen and 800 soldiers. He picked up malaria in the Cape Verde Islands and transported it to the Caribbean and Central America where his men became so afflicted and so many hundreds died that he was forced to abandon the mission (Keevil 1957: 92–94). Fortunately, the primary African vector species, *Anopheles gambiae*, accidently introduced to the northeast coast of Brazil about 1930, was contained and eventually eradicated (Soper and Wilson 1943).

## References

DOWNS, W. G., AND C. S. PITTENDRIGH. 1946. Bromeliad malaria in Trinidad, British West Indies. Amer. J. Trop. Med. Hyg. 26: 46–66.

FARAN, M. E. 1980. Mosquito studies (Diptera, Culicidae). XXXIV. A revision of the Albimanus Section of the subgenus *Nyssorhynchus* of *Anopheles*. Amer. Entomol. Inst. Contrib. 15(7): 1–215.

KEEVIL, J. J. 1957. Medicine and the navy 1200–1900. Vol. 1. E. and S. Livingstone, Edinburgh.

MENDES DOS SANTOS, J. M., E. P. B. CONTEL, AND W. E. KERR. 1981. Biologia de anofelinos amazônicos. 1. Ciclo biológico, postura e estádios larvais de *Anopheles darlingi* Root 1926 (Diptera: Culicidae) da Rodovia Manaus-Boa Vista. Acta Amazonica 11: 789–797.

SOPER, F. L., AND D. B. WILSON. 1943. *Anopheles gambiae* in Brazil, 1930 to 1940. Rockefeller Foundation, New York.

ZAVORTINK, T. J. 1973. Mosquito studies (Diptera, Culicidae). XXIX. A review of the subgenus *Kerteszia* of *Anopheles*. Amer. Entomol. Inst. Contrib. 9(3): 1–54.

## Giant Mosquitoes

Culicidae, Toxorhynchitinae, *Toxorhynchites*. Elephant mosquitoes.

These are very large, nonbiting mosquitoes (fig. 11.3d) seen only in deep forest (Steffan and Evenhuis 1981). The proboscis of the adults of both sexes is long and ventrally recurved, contrary to the straight or upward bend of other mosquitoes, and used to siphon nectar and plant exudates, instead of blood. The body is usually deep metallic blue or violaceous, and males often display bright red scale patches at the tip of the abdomen.

The larvae are also large and prey on active aquatic invertebrates, usually other mosquito larvae, which they search out and catch with their prehensile, flat-bladed mouthparts. Because of this habit, they are considered potential biological control agents against harmful mosquitoes (Steffan 1975). However, their aggressiveness and penchant for cannibalism diminishes their effective application in this area.

The immatures develop in natural containers, such as cut bamboo, bromeliads, and tree holes. They also develop in artificial containers discarded in or near forested areas, most commonly, metal cans, barrels, earthen pots, and old automobile tires. Adults generally rest on tree trunks or vegetation near the larval breeding sites.

Giant mosquitoes are used to detect dengue in humans suspected of having the disease (xenodiagnosis). Serum from a patient is injected into the thorax of the adult or head of the male and allowed to incubate for a prescribed number of days. Fluorescence of brain tissue under ultraviolet light is positive for the disease (Ramalingam pers. comm.).

There are sixteen regional species, all belonging to the subgenus *Lynchiella*. The most common and widespread are *T. haemorrhoidalis* and *T. theobaldi*.

## References

STEFFAN, W. A. 1975. Systematics and biological control potential of *Toxorhynchites* (Diptera: Culicidae). Mosq. Syst. 7: 59–67.

STEFFAN, W. A., AND N. L. EVENHUIS. 1981. Biology of *Toxorhynchites*. Ann. Rev. Entomol. 26: 159–181.

## Aedes Mosquitoes

Culicidae, Culicinae, Aedini, *Aedes*.

Adult *Aedes* mosquitoes are usually adorned with brilliant white, silvery, or golden scale patches on the thorax. Details identifying them are a lack of setae in the spiracular area and the presence of bristles behind the spiracle and on the anterior sides of the thorax. The larvae have a normal siphon with lateral teeth basally, an anal segment with an incomplete sclerotization, and a single pair of tufted setae placed near the middle.

This is a large and diversified genus in the Neotropics, with about 120 species. Several subgenera are represented in Latin America, some found nowhere else (Arnell 1976, Berlin 1969, Schick 1969, Zavortink 1972).

A very widespread, often troublesome, species in this genus is the salt marsh mosquito (*Aedes taeniorhynchus*), found along both of the Central and South American coasts and the inland saline areas as well as in the Antilles and Galápagos Islands. Adults may emerge in enormous numbers to make territory bordering salt marshes and swamps uninhabitable.

## References

ARNELL, J. H. 1976. Mosquito studies (Diptera, Culicidae). XXXIII. A revision of the Scapularis Group of *Aedes* (*Ochlerotatus*). Amer. Entomol. Inst. Contrib. 13(3): 12–144.

BERLIN, D. G. W. 1969. Mosquito studies (Diptera, Culicidae). XII. A revision of the Neotropical subgenus *Howardina* of *Aedes*. Amer. Entomol. Inst. Contrib. 4(2): 1–190.

SCHICK, R. X. 1969. Mosquito studies (Diptera, Culicidae). XX. The Terrens Group of *Aedes* (*Finlaya*). Amer. Entomol. Inst. Contrib. 5(3): 1–158.

ZAVORTINK, T. J. 1972. Mosquito studies (Diptera, Culicidae). XXVIII. The New World species formerly placed in *Aedes* (*Finlaya*). Amer. Entomol. Inst. Contr. 8(3): 1–206.

## Yellow Fever Mosquito

Culicidae, Culicinae, Aedini, *Aedes aegypti*.
*Portuguese:* Pernilongo rajado (Brazil).
*Tupi-Guaraní:* Carapanã pinima (Brazil).

*Aedes aegypti* (Christophers 1960) is the only member of the medically important, Old World mosquito subgenus *Stegomyia* in Latin America; it is adventitious in urban situations over much of the region and is easily recognized by the unique, white, lyre-shaped marking on the back of an otherwise black thorax (fig. 11.3e). The larva is typical of most *Aedes* (fig. 11.3f). The species has long been abhored as the most efficient vector of yellow fever and as an urban vector species most difficult to eradicate (Groot 1980).

Many times, outbreaks of this disease (known to the Spanish as "black vomit" and in English tradition as "yellow jack") may have changed the course of human lives and history. When yellow fever broke out among his troops, Sir George Clifford, Third Earl of Cumberland, failed in his siege of San Juan in 1598, and the colony remained under Spanish rule until the American conquest. A similar fate befell Napoleon's Captain-General LeClerc in his attempt to take Haiti for France in 1802, and much of America's future was probably turned from French domination (Cloudsley-Thompson 1976: 171). In 1741, Admiral Vernon sailed from Britain with 27,000 men, with the intention of conquering Mexico and Peru. But he lost 20,000 of his men to yellow fever, and his mission was aborted.

A great stride in proving the role of insects as human disease vectors and establishing the field of medical entomology resulted from the classical experiments of Walter Reed with soldier volunteers in Havana following the Spanish-American war (Reed et al. 1900, 1901). The work confirmed the suspicions announced almost twenty years earlier by the Cuban physician, Carlos Finlay (Sosa 1989).

## References

CHRISTOPHERS, S. R. 1960. *Aedes aegypti* (L.), the yellow fever mosquito: Its life history, bionomics, and structure. Cambridge Univ. Press, Cambridge.

CLOUDSLEY-THOMPSON, J. L. 1976. Insects and history. St. Martin's, New York.

GROOT, H. 1980. The reinvasion of Colombia by *Aedes aegypti:* Aspects to remember. Amer. J. Trop. Med. Hyg. 29: 330–338.

REED, W., J. CARROL, AND A. AGRAMONTE. 1901. The etiology of yellow fever: An additional note. J. Amer. Med. Assoc. 36: 431–440.

REED, W., J. CARROL, A. AGRAMONTE, AND J. W. LAZEAR. 1900. The etiology of yellow fever: A preliminary note. Philadelphia Med. J. 6: 790–796.

SOSA, JR., O. 1989. Carlos J. Finlay and yellow fever: A discovery. Entomol. Soc. Amer. Bull. 35(2): 23–25.

## Blue Devils

Culicidae, Culicinae, Aedini, *Haemagogus*.

The dorsum of the thorax of these small to medium-sized (BL 7–10 mm) forest mosquitoes is completely covered with broad, flat, reflecting scales, giving them a deep, steely blue or metallic greenish color (fig. 11.4a). The anterior pronotal lobes are

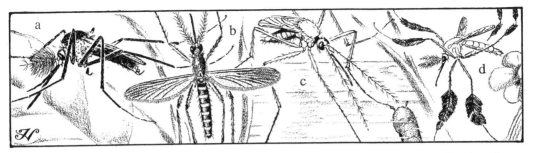

**Figure 11.4  MOSQUITOES (CULICIDAE).** (a) Blue devil (*Haemagogus* sp.). (b) Southern house mosquito (*Culex quinquefasciatus*). (c) Crabhole mosquito (*Deinocerites cancer*). (d) Sabethine mosquito (*Sabethes* sp.).

unusually large. They are particularly persistent biters and often noticed because of their diurnal habits, resplendent colors, and high-pitched whine. Several species are important in arbovirus transmission and probably are responsible for maintaining jungle yellow fever in forest mammal populations.

Structurally, the larvae are like those of *Aedes*. They often develop in containers, bamboo stumps, artificial vessels, and tree holes. This genus of twenty-seven species is exclusive to the New World tropics and subtropics (Arnell 1973).

### Reference

Arnell, J. H. 1973. Mosquito studies (Diptera, Culicidae). XXXII. A revision of the genus *Haemagogus*. Amer. Entomol. Inst. Contrib. 10(2): 1–174.

### Gallinippers

Culicidae, Culicinae, Aedini, *Psorophora*.

Gallinippers are very large, fiercely biting mosquitoes in the genus *Psorophora*, which contains over forty Neotropical species. Adults resemble *Aedes* but have spiracular bristles. The larvae, most of which develop in ground pools, also resemble those of *Aedes* but have a completely ringed sclerotization around the anal segment, which is pierced along the midventral line by tufted bristles. They are poorly studied in Latin America but familiar because of their size.

### Culex Mosquitoes

Culicidae, Culicinae, Culicini, *Culex*.

This is by far the largest genus of mosquitoes in the New World tropics. It contains over 300 species, more than half of these in one, characteristically Neotropical subgenus, *Melanoconion*. Adults are mostly drab gray with subdued thoracic coloration (except the subgenus, which has many ornamented species); the spiracular area is bare, and postspiracular bristles are absent; the tips of the tarsi possess pulvillar pads (lacking in other mosquitoes except *Deinocerites*). The larvae have a normal breathing tube with lateral teeth at the base and several hair tufts along the underside.

Species in the genus display varied habits. Larval breeding sites include all types, and adults of many species are anthropophilic. The ubiquitous southern house mosquito (*Culex quinquefasciatus*; fig. 11.4b) is a prime nuisance and is involved in the transmission of numerous human and animal pathogens. This is a domestic species whose larvae develop in all sorts of foul water in containers. Favored urban breeding grounds are rain barrels, cisterns, flooded latrines, polluted ponds, road ditches, gutters, and the like.

### References

Bram, R. A. 1967. Classification of *Culex* subgenus *Culex* in the New World (Diptera: Culicidae). U.S. Natl. Mus. Proc. 120: 1–120.

FOOTE, R. H. 1952. The larval morphology and chaetotaxy of the *Culex* subgenus *Melanoconion* (Diptera, Culicidae). Entomol. Soc. Amer. Ann. 45: 445–472.

FOOTE, R. H. 1954. The larvae and pupae of the mosquitoes belonging to the *Culex* subgenera *Melanoconion* and *Mochlostyrax*. U.S. Dept. Agric. Tech. Bull. 1091: 1–126.

SIRIVANAKARN, S. 1982. A review of the systematics and a proposed scheme of internal classification of the New World subgenus *Melanoconion* of *Culex* (Diptera, Culicidae). Mosq. Syst. 14: 265–333.

## Crabhole Mosquitoes

Culicinae, Culicini, *Deinocerites*.

This is another exclusively Neotropical mosquito group, with eighteen species (Adames 1971). The monocolorous gray-brown adults possess the general characteristics of *Culex* but have sparsely haired antennae, these organs usually much longer than the proboscis in both sexes, and the males often have enlarged fore tarsal claws. They use these claws and unusually long antennae in a highly peculiar mating strategy. They remain in the flooded crab burrow, skating on the water, while holding the tips of the antennae close to the surface (McIver and Siemicki 1976). On detecting a mature female pupa fixed to the surface film by its breathing trumpets, they hook it with an enlarged fore tarsal claw (fig. 11.4c) and remain with it until the female emerges. Copulation immediately follows or even accompanies emergence, the male sometimes grasping the female's genitalia even before she has escaped her pupal skin (Provost and Haeger 1967). The larvae also resemble those of *Culex* but possess a pair of conspicuous lateral pouches on the head and much reduced gills on the anal segment.

Breeding and much of adult life is confined to the burrows of land crabs in the genera *Cardisoma*, *Ucides*, and *Sesarma*. The genus's distribution is entirely coastal.

*Galindomyia* with a single species, *G. leei*, from Pacific coastal Colombia, is structurally similar to *Deinocerites* in the adult stage, but its larva and breeding habits are unknown.

## References

ADAMES, A. J. 1971. Mosquito studies (Diptera, Culicidae). XXIV. A revision of the crabhole mosquitoes of the genus *Deinocerites*. Amer. Entomol. Inst. Contrib. 7(2): 1–154.

McIVER, W., AND R. SIEMICKI. 1976. Fine structure of the antennal tip of the crabhole mosquito, *Deinocerites cancer* Theobald (Diptera: Culicidae). Int. J. Ins. Morph. Embryol. 5: 319–334.

PROVOST, M. W., AND F. S. HAEGER. 1967. Mating and pupal attendance in *Deinocerites cancer* and comparisons with *Opifex fuscus* (Diptera: Culicidae). Entomol. Soc. Amer. Ann. 60: 565–574.

## Water Weed Mosquitoes

Culicidae, Culicinae, Mansonini, *Mansonia* and *Coquillettidia*.

These closely related genera (previously combined as *Mansonia*) have immatures that live among dense beds of floating aquatic plants, commonly water lettuce (*Pistia stratiotes*) and water hyacinth (*Eichornia*). Both the larvae and pupae fix themselves to the stems of these plants and extract oxygen from air vessels therein, the larvae by means of a unique, attenuated, hooked siphonal tube with apical sawlike teeth and the pupae with scythelike respiratory horns. The adults are easily distinguished from other medium-sized culicines by the very broad, asymmetrical scales on the upper wing surface. Among the twenty-five Neotropical species, several are vectors of viral infections and filarial worms (Ronderos and Bachmann, 1963).

## Reference

RONDEROS, R. A., AND A. O. BACHMANN. 1963. Mansoniini Neotropicales. I (Diptera-Culicidae). Soc. Entomol. Argentina Rev. 26: 57–65.

## Sabethine Mosquitoes

Culicinae, Sabethini, *Limatus, Sabethes, Wyeomyia, Phoniomyia, Trichoprosopon,* and relatives.

These wholly Neotropical genera, containing together just under 200 species, form a varied assemblage (Lane and Cerqueira 1942, Zavortink 1979). All display a tuft of small hairs on the postnotum (rounded area beneath the scutellum) and a trilobate scutellum; postspiracular bristles are absent. Many species are brilliantly metallic or iridescent, and one or more tarsi often have conspicuous paddle-shaped areas of scales in the genus *Sabethes* (fig. 11.4d). Most use various plant containers as larval and pupal habitats. *Trichoprosopon digitatum* commonly breeds in the fetid water that accumulates in rotting cacao pods and is a major pest for plantation workers (Zavortink et al. 1983); females have been observed brooding their egg rafts in their microhabitat (Lounibos and Machado-Allison 1983). The largest genus is *Wyeomyia* with over 100 species, mostly distinguishable only by their highly complex male genitalia. Highly specialized forest mosquitoes, they choose water-filled bamboo internodes, bromeliad leaf axils, inflorescences of *Heliconia* and *Calathea,* and palm spaths as breeding sites.

### References

LANE, J., AND N. L. CERQUEIRA. 1942. Os sabetínos da América (Diptera, Culicidae). Mus. Zool., Univ. São Paulo, Arq. Zool. 3: 473–849.

LOUNIBOS, L. P., AND C. E. MACHADO-ALLISON. 1983. Oviposition and egg brooding by the mosquito *Trichoprosopon digitatum* in cacao husks. Ecol. Entomol. 8: 475–478.

ZAVORTINK, T. J. 1979. Mosquito studies (Diptera, Culicidae). XXXV. The new sabethine genus *Johnbelkinia* and a preliminary reclassification of the composite genus *Trichoprosopon.* Amer. Entomol. Inst. Contrib. 17(1): 1–61.

ZAVORTINK, T. J., D. R. ROBERTS, AND A. L. HOCH. 1983. *Trichoprosopon digitatum* morphology, biology and potential medical importance. Mosq. Syst. 15: 141–149.

## Other Mosquitoes

Culicinae, Culisetini, *Culiseta:* Orthopodomyiini, *Orthopodomyia;* Uranotaeniini, *Uranotaenia.*

*Culiseta particeps* is the only one truly Neotropical member of its genus. It occurs at higher elevations in cold mountain ponds. *Orthopodomyia* is an assemblage of seven regional species, all utilizing bamboo internodes, bromeliad leaf axils, and tree holes for development. About thirty *Uranotaenia* occur in the New World tropics; larvae are usually found in weedy ground pools (Zavortink 1968).

### Reference

ZAVORTINK, T. J. 1968. Mosquito studies (Diptera, Culicidae). VIII. A prodrome of the genus *Orthopodomyia.* Amer. Entomol. Inst. Contrib. 3(2): 1–221.

## HORSEFLIES

Tabanidae. *Spanish:* Tábanos. *Tupi-Guaraní* (adapted in *Portuguese*): Mutucas (Brazil). Greenheads, mango flies, deerflies, gadflies.

The vicious bloodletting abilities of these flies are notorious. They feed by lapping and sponging up the blood that oozes from wounds they inflict by their sidewise slashing, bladelike mandibles, a process distinctly more painful to the host than the delicate hypodermic needling of other bloodsucking Diptera. Blood often flows from wounds much in excess of the amount ingested by the fly. Most females are hematophagous, but some, as well as all males, subsist only on nectar and plant exudates. Primarily mammals but also birds and reptiles, including crocodilians (Medem 1981), are hosts. The proboscises of the subfamily Pangoniinae (*Fidena, Scione*) are very long and slender and are used to extract nectar from flowers, never to take blood. Some species and strains probably can reproduce without the females having to feed on blood (autogenous), like certain mosquitoes (Charlwood and Rafael 1980).

Aside from the hematophagous habit of most of its members, this large family (approximately 1,000 species in the Neotropics; Fairchild 1969) is recognized by their stout, nonspiny bodies, large eyes (meeting on top of the head in the males; narrowly separated in the females), which often display iridescent rainbow colors, large basal wing lobes, and a noticeable divergent fork in the wing vein (4th and 5th branches of the radius vein) that encloses the wing tip. The wings also are commonly banded or splotched with brown pigment. Their total body size ranges from small to large (BL 5–25 mm).

Females lay egg masses on objects located above the larval medium, usually standing water. This explains the abundance of the adults near aquatic habitats. Yet they are strong fliers and may be found long distances from the swamps, marshes, and weedy ponds that are their usual breeding sites. A few are even terrestrial, while others develop in bromeliad tanks.

The larvae (fig. 11.5c) are elongate, black or brownish to green, often with a translucent cuticle that has linear pigment markings. They have a well-developed head and are markedly segmented. The segments bear circlets of short lobes anteriorly, and the terminal segment has a short telescoping respiratory tube. Most lay among dead leaves or other decomposing plant matter on the bottom where they prey on other aquatic insects and invertebrates; a few are phytophagous (Goodwin and Murdoch 1974).

Because they are especially bothersome to humans and often colorfully marked, many species have attracted attention and are known by local names. Such are the widespread cabo verde or mosca congo (*Lepiselaga crassipes;* fig. 11.5a) (Fairchild 1940) in Amazonia and Central America and the colihuacho (*Scaptia* [= *Onca*] *lata*) in Chile (Berg 1881). Both are very bothersome biters. The former is an insistent pest that is often very numerous by rivers, coming aboard boats to bite around the ankles and legs. It is small (BL 8–9 mm) and all black except where the pigment abruptly terminates beyond the middle of the wing, leaving its apical third transparent. Females favor the legs and feet of its hosts when bloodletting. Development occurs in matted floating vegetation, especially water lettuce (*Pistia*) in slow or standing water (Fairchild 1940). The second species is large (BL to 2 cm) and black with tufts of crimson hairs on the thorax and abdomen and can attack insidiously and in numbers by its river habitat (orig. obs.).

These and most other tabanids are diurnal, although some are crepuscular or even nocturnal in the tropics, such as the pale green *Chlorotabanus.* Major genera contain-

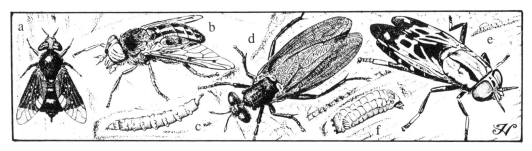

**Figure 11.5    FLIES.** (a) Mosca Congo (*Lepiselaga crassipes,* Tabanidae). (b) Horse fly (*Tabanus dorsiger,* Tabanidae). (c) Horse fly, larva. (d) Giant mydas fly (*Mydas* sp., Mydidae). (e) Timber fly (*Pantophthalmus* sp., Pantophthalmidae). (f) Timber fly, larva.

ing biting forms are *Chrysops, Tabanus* (fig. 11.5b), and *Dichaelacera.*

The significance of tabanids as disease vectors has hardly been studied in Latin America. Considering their considerable potential as such elsewhere (Krinsky 1976), they deserve attention in this respect.

The literature on Neotropical tabanids is reviewed by Coscarón (1977) and Fairchild (1981, 1982).

## References

BERG, F. W. K. 1881. Observaciones acerca de la *Osca lata* (Guér.) Lynch. Soc. Cient. Argentina An. (Bol.). Pp. 9–10.

CHARLWOOD, J. D., AND J. A. RAFAEL. 1980. Autogeny in the River Negro Horse Fly, *Lepiselaga crassipes,* and an undescribed species of *Stenotabanus* (Diptera: Tabanidae) from Amazonas, Brazil. J. Med. Entomol. 17: 519–521.

COSCARÓN, S. 1977. Tabanidae. *In* S. H. Hurlbert, ed., Biota acuática de sudamérica austral. San Diego State Univ., San Diego. Pp. 297–304.

FAIRCHILD, G. B. 1940. A note on the early stages of *Lepiselaga crassipes* Fab. (Dipt., Tabanidae). Psyche 47: 8–13.

FAIRCHILD, G. B. 1969. Notes on Neotropical Tabanidae. XII. Classification and distribution, with keys to genera and subgenera. Arq. Zool. (São Paulo) 17: 199–255.

FAIRCHILD, G. B. 1981. Tabanidae. *In* S. H. Hurlbert, G. Rodríguez, and N. Dias dos Santos, eds., Aquatic biota of tropical South America. Pt. 1. Arthropoda. San Diego State Univ., San Diego. Pp. 290–301.

FAIRCHILD, G. B. 1982. Tabanidae. *In* S. H. Hurlbert and A. Villalobos Figueroa, eds., Aquatic biota of Mexico, Central America and the West Indies. San Diego State Univ., San Diego. Pp. 452–460.

GOODWIN, J. T., AND W. P. MURDOCH. 1974. A study of some immature Neotropical Tabanidae (Diptera). Entomol. Soc. Amer. Ann. 67: 85–133.

KRINSKY, W. L. 1976. Animal disease agents transmitted by horse flies and deer flies (Diptera: Tabanidae). J. Med. Entomol. 13: 225–275.

MEDEM, F. 1981. Horse flies (Diptera: Tabanidae) as ectoparasites on caimans (Crocodylia: Alligatoridae) in eastern Colombia. Cespedesia 10: 123–191.

## GIANT MYDAS FLIES

Mydidae, *Mydas.*

The giant mydas flies are very large dipterans (BL 35–45 mm). Most species in the genus are all black, save for the tips of its antennae and its wings, which are bright orange-red; others have black wings as well (fig. 11.5d). Several are Batesian mimics of various large species of tarantula hawk wasps (*Pepsis,* Pompilidae). The genus ranges from northern Mexico to southern Brazil.

Adults feed on flower nectar, and the larvae probably are predators of other soil-dwelling insect larvae. Larvae of some Brazilian *Mydas* are known to live in leaf cutter ant (*Atta*) nests, where they are predaceous on certain larval myrmecophilous Scarabaeidae (Zikán 1944).

### Reference

ZIKÁN, J. F. 1944. Novas observações sôbre a biologia de *Mydas* (Dipt.) e sua relação com as formigueiros de saúva. Bol. Minis. Agric. Rio de Janeiro 33: 43–55. [Not seen.]

## TIMBER FLIES

Pantophthalmidae. *Portuguese:* Moscardos (Brazil).

Timber flies (Carrera and d'Andretta 1957) are the real goliaths of the Diptera. Adults reach body lengths of 40 to 50 millimeters and wingspans of 85 to 95 millimeters. They generally resemble horseflies but have small, nonbiting mouthparts and much reduced calypters. Other family characteristics are a flattened, squarish abdomen (female with a telescoping ovipositor) and no or very small tibial spurs. There are also five closed cells in the wing, as opposed to four in the horseflies. The wing membranes of all species are pigmented with large irregular blotches.

Adults (fig. 11.5e) are usually seen resting on logs or occasionally flying near tree trunks in forest habitats. They oviposit on the crevices in the bark of trees or fallen trees. After they hatch, the larvae burrow into the sound wood beneath the bark and spend approximately two years working their way through the wood, forming extensive tubular galleries. Living or dead hardwood trees of many species may be attacked. It is uncertain whether or not the larvae feed on the wood itself or on fermented sap that floods their galleries. The burrowing is forceful and creates a rasping or grating sound that may be audible some meters away (orig. obs.).

The larvae (fig. 11.5f) (Thorpe 1930) are cylindrical with a heavily sclerotized head and powerful mandibles. The last segment of the abdomen is obliquely truncate with two series of strong spines, forming a push plate used to eject the large quantities of sawdust produced by their boring activities. Ventrally, just anterior to this segment, is a unique organ (Fiebrig's body), composed of many appressed, fingerlike processes; it apparently functions as a gill that allows the larvae to respire in their liquid microhabitat.

The mobile pupae are strongly sclerotized and have a nearly spherical head structure bearing two powerful, ridged, grinding plates. The cylindrical body is abruptly truncate posteriorly like the larvae. These structures permit the pupae to plug the surface opening of their galleries when near the log's exterior just prior to adult emergence.

This is a small family with only two genera, *Opetiops*, with one species, and *Pantophthalmus* with nineteen. It is endemic to the New World tropics (Val 1976). Pantophthalmids are most common in the Amazonian and Central American rain forests.

## References

CARRERA, M., AND M. A. V. D'ANDRETTA. 1957. Sôbre a familia Pantophthalmidae. Arq. Zool. (São Paulo) 10: 253–330.

THORPE, W. H. 1930. Observations on the structure, biology and systematic position of *Pantophthalmus tabaninus* Thunb. (Diptera: Pantophthalmidae). Royal Entomol. Soc. London Trans. 82: 5–22.

VAL, F. C. 1976. Systematics and evolution of the Pantophthalmidae (Diptera, Brachycera). Arq. Zool. (São Paulo) 27: 51–164.

# FLOWER FLIES

Syrphidae.

Over 1,600 species comprise this group of common, conspicuous, generally medium-sized (BL 4–15 mm) flies with hovering flight and large eyes. Many are bright yellow or orange and banded, as in the genus *Metasyrphus*, while others are dull gray to black or iridescent green or blue, these patterns and colors of many giving them a resemblance to stinging bees and wasps. Their habit of visiting flowers makes them beneficial as pollinators. The larvae of many species are also valuable as carnivores on pests such as aphids and mealybugs (Gonçalves and Gonçalves 1976). A widespread species is *Metasyrphus americanus*, which has a yellow-spotted abdomen (fig. 11.6a).

Syrphids are not well studied in Latin America. An important taxonomic work is by Thompson (1972).

## References

GONÇALVES, C. R., AND A. J. L. GONÇALVES. 1976. Observações sobre môscas da família Syrphidae predadoras de Homópteros. Soc. Entomol. Brasil. An. 5: 3–10.

THOMPSON, F. C. 1972. A contribution to a generic revision of the Neotropical Milesinae (Diptera: Syrphidae). Arq. Zool. (São Paulo) 23: 73–215.

### Drone Fly

Syrphidae, Milesiinae, Eristalini, *Eristalis tenax. Spanish:* Mosca abeja.

This flower fly (fig. 11.6b) is so similar in coloration, size (BL 10–15 mm), and behavior to the honeybee that it is avoided by humans and animals. It is also slightly pubescent, like the bee, and has a dark,

**Figure 11.6    FLIES.** (a) Flower fly (*Metasyrphus americanus,* Syrphidae). (b) Drone fly (*Eristalis tenax,* Syrphidae). (c) Green flower fly (*Ornidia obesa,* Syrphidae). (d) Drone fly, larva. (e) Wasp fly (*Hermetia illuscens,* Stratiomyidae), larva. (f) Wasp fly, adult.

dull orange and black-banded abdomen. Although variable (Heal 1979), the basal band is usually incomplete dorsally, broken by a median dark bar, simulating the bee's narrow waist. Frequenting the same flowers, drone flies and honeybees have also evolved similar visual sensitivities (Bishop and Chung 1972).

The larva ("rat-tailed maggot") (fig. 11.6d) possesses a long, telescoping, posterior respiratory tube, consisting of three segments that can extend to several times the length of the robust, cylindrical body (BL 20–25 mm). There are eight pairs of stumpy false legs tipped with fine spines developed on the ventral side of the trunk. The larva is exceedingly hardy and lives in water contaminated by sewage, fresh excrement, moist carrion, and other foul organic matter of liquid consistency. It is sometimes involved in cases of gastrointestinal myiasis.

A cosmopolitan species, the drone fly is found in most Latin American countries, particularly in warmer parts, but is locally absent in many areas. In ancient times in Europe, its resemblance to the honeybee and its habit of breeding in decaying carcasses led to a curious popular belief that the bees were spontaneously generated from dead oxen ("bugonia") (Atkins 1948).

## References

Atkins, Jr., E. L. 1948. Mimicry between the drone-fly, *Eristalis tenex* (L.), and the honeybee, *Apis mellifera* L.: Its significance in ancient mythology and present day thought. Entomol. Soc. Amer. Ann. 41: 887–892.

Bishop, L. G., and D. W. Chung. 1972. Convergence of visual sensory capabilities in a pair of Batesian mimics. J. Ins. Physiol. 18: 1501–1508.

Heal, J. 1979. Colour patterns of Syrphidae. I. Genetic variation in the dronefly *Eristalis tenax.* Heredity 42: 223–236.

## Green Flower Fly

Syrphidae, Milesiinae, Volucellini, *Ornidia obesa. Portuguese:* Berneira (Brazil).

This species (fig. 11.6c) ranges throughout the moist lowlands of most of the Neotropical Region where it is a familiar visitor on warm days to outhouses, urinals, and rotting fruit or other foul materials. Its brilliant deep metallic color immediately identifies it even as it hovers in the air, its wings blurred by their rapid vibration. It is about the same size (BL 10 mm) as similar blowflies and orchid bees and may mimic the latter. Further identifying features are a dark spot midway on the leading edge of the wing, a much smaller similar spot near the wing tip, and a depression in the center of the scutellum. The early stages are not well studied; the larva is said to be coprophagous (Val 1972).

*Ornidia* contains three other equally beautiful but less widespread species (Thompson 1991).

## References

Thompson, F. C. 1991. The flower fly genus *Ornidia* (Diptera: Syrphidae). Proc. Entomol. Soc. Wash. 93: 248–261.

VAL, F. C. 1972. On the biometry and evolution of the genus *Ornidia* (Diptera, Syrphidae). Univ. São Paulo Mus. Zool. Pap. Avul. Zool. 26:1–28.

## WASP FLY

Stratiomyidae, Hermetiinae, *Hermetia illuscens. Spanish:* Guarero, borrachero (Central America).

This semidomestic species is found in all temperate and tropical portions of the world, owing to the broad adaptive capacities of adults and the transportation of larvae in contaminated food. Because all other members of its genus are Neotropical, it probably originated in some warm area of the New World.

The larvae develop in a wide variety of organic media, including decaying fruit and vegetables, garbage, compost, dung, carrion, and soil contaminated with these materials (Copello 1926). In Central American banana plantations, the species damages fruit by ovipositing between the fingers of young fruit and on ripe bunches after they have been picked. A strong attraction for yellow color by the female is thought to account at least partly for the latter (Stephens 1975).

When they occupy the same food, wasp fly larvae may suppress housefly populations through some indirect competitive effect, since the former do not feed on the latter (Furman et al. 1959). Larvae are also commonly found in honeybee hives where they feed on honey, wax, and waste materials. There is one record from the nest of a stingless bee (Borgmeier 1930) which may indicate a general habit in the wild for this and other members of the genus. Larvae also are occasionally the cause of human intestinal myiasis.

Identifying characteristics of the larval stage (BL 20 mm) are a thick, leathery, dark brown integument that is set with numerous short bristles, a broad, flattened, ovate general shape, and a slender well-sclerotized head (fig. 11.6e).

The moderately large (BL 15–20 mm) adults (fig. 11.6f) much resemble spider wasps in anatomy and behavior. They are elongate and dark bluish-black, with uniformly dusky wings and white tarsi (the hind tibia is also white); at the base of the abdomen are paired transparent areas separated by a dark median line, simulating a wasp's waist. The male abdomen is bronzy, in contrast to black in females. The flies comport themselves waspishly, running on the ground excitedly and flicking their wings. They also buzz and feign a sting when handled (Iide and Mileti 1976)

### References

BORGMEIER, T. 1930. Ueber das Vorkommen der Larven von *Hermetia illuscens* L. (Dipt., Stratiomyidae) in den Nestern von Meliponiden. Zool. Anz. 90: 225–235.

COPELLO, A. 1926. Biología de *Hermetia illuscens* Latr. Rev. Entomol. Argentina 1(2): 23–27. [*Sic*] Author of species is Linneaus.

FURMAN, D. P., R. D. YOUNG, AND E. P. CATTS. 1959. *Hermetia illuscens* (Linnaeus) as a factor in the natural control of *Musca domestica* Linnaeus. J. Econ. Entomol. 52: 917–921.

IIDE, P., AND D. I. C. MILETI. 1976. Estudos morfológicos sôbre *Hermetia illuscens* (Linnaeus, 1758) (Diptera, Stratiomyidae). Rev. Brasil. Biol. 36: 923–935.

STEPHENS, C. S. 1975. *Hermetia illuscens* (Diptera: Stratiomyidae) as a banana pest in Panama. Trop. Agric. 52: 173–178.

## POMACE FLIES

Drosophilidae, *Drosophilia.*

These gnatlike (BL 2–3 mm), yellowish to brownish flies (fig. 11.7a) are usually seen around decaying vegetation and ripe fruit (Shorrocks 1980, Belo and de Oliveira 1976). They are best known, however, for the roles a few species have played in the advancement of the science of heredity. Early in this century, geneticist Thomas Hunt Morgan found that one species, *Drosophilia melanogaster,* reproduced with

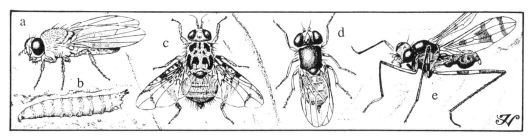

**Figure 11.7  FLIES.** (a) Pomace fly (*Drosophila melanogaster,* Drosophilidae). (b) Pomace fly, larva. (c) Mediterranean fruit fly (*Ceratitis capitata,* Tephritidae). (d) Eye gnat (*Liohippelates pusio,* Chloropidae). (e) Stilt-legged fly (*Taeniaptera* sp., Micropezidae).

great rapidity, was easily cultured in the laboratory, required but twelve days to develop from the egg to maturity, and had larval cells containing only four, large, well-marked chromosomes. In fact, the insect was so well adapted for genetics that someone once quipped that it must have been created by the Almighty solely as an object of heredity research. Since Morgan's days, many additional species have been studied in great detail and contributed more to our understanding of evolutionary mechanisms, population dynamics, phylogeny, cytology, and so on, than any other single group of higher organisms (Ashburner 1976–1981, Rubin 1988, Throckmorton 1975). As a result, a tremendous body of literature now exists concerning their taxonomy and biology. (Although also commonly called "fruit flies," they are not to be confused with the Tephritidae.)

The family is large in the Neotropics, containing almost 700 species, the majority of which belong to the genus *Drosophila* (Val et al. 1981). The vast majority of species are denizens of woodlands; a few are desert dwellers. Adults are diurnal and active for brief periods during the day. The larvae (fig. 11.7b) of these mostly feed on microorganisms, especially yeasts (da Cunha et al. 1957), associated with spoiled fruit, slime (sap) fluxes on tree trunks and roots, in rotting cacti (Benado et al. 1984), and fungi and similar fermenting vegetable matter (Spieth and Heed 1972). Some species breed and feed as adults in living flowers

(often *Calathea* and *Heliconia*) (Pipkin et al. 1966). Two remarkable species, *D. carcinophila* and *D. endobranchia*, in the Caribbean live on the bodies of land crabs, probably as commensals (Carson 1967, Carson and Wheeler 1968).

## References

ASHBURNER, M., ed. 1976–1981. The genetics and biology of *Drosophila.* 3 vols. Pergamon, Oxford.

BELO, M., AND I. J. J. DE OLIVEIRA. 1976. Espécies domésticas de *Drosophila.* V. Influências de factores ambientais no número de indivíduos capturados. Rev. Brasil. Biol. 36: 903–909.

BENADO, M., A. FONTDEVILA, H. G. CERCA, G. GARCIA, A. RUIZ, AND C. MONTERO. 1984. On the distribution and the cactiphilic niche of *Drosophila martensis* in Venezuela. Biotropica 16: 120–124.

CARSON, H. L. 1967. The association between *Drosophila carcinophila* Wheeler and its host, the land crab *Gecarcinus ruricola* (L.). Amer. Midl. Nat. 78: 324–343.

CARSON, H. L., AND M. R. WHEELER. 1968. *Drosophila endobranchia,* a new drosophilid associated with land crabs in the West Indies. Entomol. Soc. Amer. Ann. 61: 675–678.

DA CUNHA, A. B., A. M. EL-TABAY SHEHATE, AND W. DE OLIVEIRA. 1957. A study of the diets and nutritional preferences of tropical species of *Drosophila.* Ecology 38: 98–106.

PIPKIN, S. B., R. L. RODRÍGUEZ, AND J. LEÓN. 1966. Plant host specificity among flower-feeding Neotropical *Drosophila* (Diptera: Drosophilidae). Amer. Nat. 100: 135–156.

RUBIN, G. M. 1988. *Drosophila melanogaster* as an experimental organism. Science 240: 1453–1459.

SHORROCKS, B. 1980. *Drosophila.* Pergamon, Oxford.

SPIETH, H. T., AND W. B. HEED. 1972. Experimental systematics and ecology of *Drosophila*. Ann. Rev. Ecol. Syst. 3: 269–288.

THROCKMORTON, L. H. 1975. *In* R. C. King, ed., Handbook of genetics. 3. Invertebrates of genetic interest. Plenum, New York. Pp. 421–469.

VAL, F. C., C. R. VILELA, AND M. D. MARQUES. 1981. Drosophilidae of the Neotropical Region. *In* M. Ashburner, ed., The genetics and biology of *Drosophila*. 3a: 123–168. Pergamon, Oxford.

# FRUIT FLIES

Tephritidae. *Spanish:* Moscas de la fruta.
*Portuguese:* Môscas da fruta.

These are small to medium-sized flies (BL 3–7 mm) with spotted or banded wings in colors of brown and white, often forming complicated or attractive patterns. Viewed from behind, living flies may mimic jumping spiders: the wing bars simulate legs, and dark spots on the apex of the abdomen look like the spider's eyes (Eisner 1984). The apex of the subcostal vein angles sharply forward to distinguish them from other pictured-winged flies. Adults are commonly observed feeding on nectar from flowers or ovipositing on fruit. When at rest or walking, some have the odd behavior of slowly moving their wings up and down as if displaying them. The females of most species insert their eggs in living, healthy plant tissue.

The larvae, which are typical maggots, feed in stalks, leaves, flowers, and often in fruits, and thus, many species have become major agricultural pests (Bateman 1976, Cavalloro 1983). The most serious and widespread is the Mediterranean fruit fly or so-called Medfly (*Ceratitis capitata;* fig. 11.7c), which attacks citrus, papaya, mango, pineapple, and as many as 260 other hosts from Central Mexico south through most of South America (not in the Antilles) (Weems 1981). Many conspicuous, economically important species are found also in the genus *Anastrepha*, such as the South American (*A. fraterculus*), Mexican (*A. ludens*), and Caribbean fruit fly (*A. suspensa*). Larvae of the Medfly are ubiquitous in guayaba fruits and incorporated unavoidably into jelly and other products made from them. The presence of larvae of these species in fruit consumed by humans sometimes causes an intestinal form of myiasis, the principal symptom of which is diarrhea (Jirón and Zeledón 1979).

Some members of the mainly temperate genus *Rhagoletis* are also of major importance (Boller and Prokopy 1976). Unlike the others mentioned, the species have relatively narrow food preferences, for example, *R. lycopersella*, a species indigenous to cultivated valleys in the dry coastal plains of western Peru which attacks only tomatoes (Smyth 1960) and *Toxotripana curvicauda*, which attacks only papaya (Adarve 1979).

Tropical fruit flies are reproductively active throughout the year. In temperate areas, there are species with only one generation per year, the adults having a winter hibernation period. Both types tend to form local transient populations through dispersal of the strong flying females.

Except for a few economically important species (Prokopy and Roitberg 1984), the Neotropical fauna is generally poorly studied (Christenson and Foote 1960, Bateman 1972). The family is a fairly diverse one in the region, with more than 680 species and 88 genera (Foote 1980).

Adults feed on a variety of natural soupy liquids such as the juices oozing from damaged or decaying fruit, plant sap, flower nectar, and bird feces. Adults and larvae of many species harbor symbiotic microorganisms in their intestines, which most likely provide specific nutritive substances from vegetable tissues.

## References

ADARVE, R. 1979. Observaciones sobre los hábitos del *Toxotripana curvicauda* Gerst (Tephritidae) que ataca al *Carica papaya*. Ceiba 23: 63–75.

BATEMAN, M. A. 1972. The ecology of fruit flies. Ann. Rev. Entomol. 17: 493–518.

BATEMAN, M. A. 1976. Fruit flies. *In* V. L. Delucchi, ed., Studies in biological control. Int. Biol. Prog. 9: 11–49. Cambridge Univ. Press, Cambridge.

BOLLER, E. F., AND R. J. PROKOPY. 1976. Bionomics and management of *Rhagoletis*. Ann. Rev. Entomol. 21: 223–246.

CAVALLORO, R., ed. 1983. Fruit flies of economic importance. Proc. CEC/IOBC Int. Symp. 1982. Balkema, Rotterdam.

CHRISTENSON, L. O., AND R. H. FOOTE. 1960. Biology of fruit flies. Ann. Rev. Entomol. 5: 171–192.

EISNER, T. 1984. Consumer fraud. Nat. Hist. 93: 112.

FOOTE, R. H. 1980. Fruit fly genera south of the United States. U.S. Dept. Agric. Tech. Bull. 1600: 1–79.

JIRÓN, L. F., AND R. ZELEDÓN. 1979. El género *Anastrepha* (Diptera: Tephritidae) en las principales frutas de Costa Rica y su relación con pseudomiasis humana. Rev. Biol. Trop. 27: 155–161.

PROKOPY, R. J., AND B. D. ROITBERG. 1984. Foraging behavior of true fruit flies. Amer. Sci. 72: 41–49.

SMYTH, E. G. 1960. A new tephritid fly injurious to tomatoes in Peru. Dept. Agric. State Calif. Bull. 49(1): 16–22.

WEEMS, H. V. 1981. Mediterranean fruit fly, *Ceratitis capitata* (Wiedmann) (Diptera: Tephritidae). Fla. Dept. Agric. Cons. Ser., Div. Plant Indus., Entomol. Circ. 230: 1–4 (unnumbered) + 1–8.

# EYE GNATS

Chloropidae, *Liohippelates pusio* complex.
Ulcer flies (West Indies).

Although they do not bite, eye gnats are a major source of torment because of their persistent habit of entering the eyes and clustering about wounds and the exposed genitals of mammals ("dog pecker gnats"). They were implicated in the mechanical transmission of yaws in the West Indies when the disease was prevalent (Nicholls 1936). The common species here are found in the *L. pusio* complex (fig. 11.7d) in a widely distributed genus of twenty-nine Neotropical species (Legner and Bay 1965, Sabrosky 1984). All are small (BL 1.5–2.5 mm) and generally shiny black with clear wings; most were formerly included in the genus *Hippelates*.

Other small gnatlike flies have similar noxious habits. *Paraleucopis mexicana* (Chamaemyiidae), called "bobos," walk on exposed skin and hair. They do not bite, but large numbers swarm about the face and head, making it very unpleasant to be near their seaside habitat. They also cluster about wounds. These flies come every year in March to the upper Gulf of California (Smith 1981) where they make the local residents miserable with their persistent aggravations. Their normal feeding seems to be on secretions emanating from the eyes of lizards and marine birds. They have not been incriminated in the spread of any diseases.

## References

LEGNER, E. F., AND E. C. BAY. 1965. Culture of *Hippelates pusio* (Diptera: Chloropidae) in the West Indies for natural enemy exploration and some notes on behavior and distribution. Entomol. Soc. Amer. Ann. 58: 436–440.

NICHOLLS, L. 1936. Framboesia tropica: A short review of a colonial report concerning statistics and *Hippelates flavipes*. Amer. J. Trop. Med. Paras. 30: 331–335.

SABROSKY, C. W. 1984. Family Chloropidae. *In* N. Papavero, ed., A catalogue of the Diptera of the Americas south of the United States. Mus. Zool., Univ. São Paulo, São Paulo. Pp. 81.1–81.63.

SMITH, R. L. 1981. The trouble with "bobos," *Paraleucopis mexicana* Steyskal, at Kino Bay, Sonora, Mexico (Diptera: Chamaemyiidae). Entomol. Soc. Wash. Proc. 83: 406–412.

# STILT-LEGGED FLIES

Micropezidae (= Tylidae).

These are slender, medium-sized (BL 10–15 mm) flies with very long mid and hind legs, compared to the forelegs, which are about half their length. The tip of the abdomen is usually recurved ventrally. Many are dark colored with banded wings (such as members of the genus *Taeniaptera;* fig. 11.7e) and contrasting white-tipped tarsal segments; the head is also often light hued in shades of red or orange.

Adults are diurnal, slow to fly, and usually seen walking nervously about on the upper surfaces of leaves with the wings closed over the dorsum and the forelegs held erect and waving in front of the head. Their form, coloration, and this behavior gives them a strong resemblance to ants. The mimicry is enhanced by the conspicuous white-tipped forelegs, stimulating antennae, and suitably placed dark wing bands, which give the impression of a narrowed, antlike waist.

This is mainly a tropical group of about 280 species (Steyskal 1968), the adults frequenting shady, moist forest habitats. The immatures are poorly known. The maggots of some species have been found living in feces, rotting fruit, and decomposing plant parts such as the pseudo-stems of bananas (Fischer 1932). Others may bore into living plant tissue, and at least one species is a minor pest by invading ginger roots in the Old World (Steyskal 1964).

These flies are also known for their bizarre courtship habits. The antics of *Plocoscelus arthriticus* (= *Cardiacephala myrmex*) in Panama were recorded by Wheeler (1924). Males approach willing females head on and perform a peculiar dance, stepping first to one side and then to the other, swaying the abdomen. After witnessing this dance, the female bends her body in an arc by throwing back the head and turning up the tip of the abdomen. The male then mounts her and brings his proboscis in contact with hers, a drop of food on it, and simultaneously inserts his intromittent organ. Occasionally, he reaches forward with a foreleg and scratches the female's eye and places more food on the corner of it. These acts, with other detailed movements, may be repeated several times.

The female finally kicks the male free, and copulation is complete.

## References

FISCHER, C. R. 1932. Contribuição para o conhecimento da metamorphose e posição systematica da familia Tylidae (Micropezidae, Dipt.). Rev. Entomol. 2: 15–24.

STEYSKAL, G. C. 1964. Larvae of Micropezidae (Diptera), including two species that bore in ginger roots. Entomol. Soc. Amer. Ann. 57: 292–296.

STEYSKAL, G. C. 1968. Family Micropezidae. *In* N. Papavero, A catalogue of the Diptera of the Americas south of the United States. Dept. Zool., Sec. Agric., São Paulo, no. 48, 48.1–48.33.

WHEELER, W. M. 1924. Courtship of *Calobatas*, the kelep ant and the courtship of its mimic, *Cardiacephala myrmex*. J. Heredity 15: 487–494.

## KELP FLIES

Anthomyiidae, *Fucellia.*

Larval kelp flies live in old piles of the brown seaweeds (kelp, *Fucus, Macrocystis,* etc.) cast up by the waves along sea beaches. They feed on the dead plant and are important in its decomposition and nutrient cycling in the coastal habitat. The adults can be seen mostly during the warmer months of the year on heaps of such material (wrack beds) and may develop enormous populations. Sometimes during the winter or on cold days, dense masses of flies are seen packed into crevices on rocky cliffs above the shore, apparently overwintering.

Species are found throughout the tropical American region on both the Atlantic and Pacific seaboards, including offshore islands and the Caribbean and Juan Fernandez islands. They are medium-sized (BL 4–5 mm) and gray-brown, somewhat resembling the house fly, but with fine hairs beneath the scutellum, small, bulging round eyes, and with a row of 3 to 4 stout, erect bristles on the upper side of the hind tibia (fig. 11.8a) (Aldrich 1918). These are

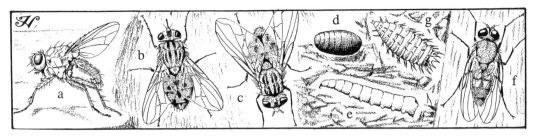

**Figure 11.8   FLIES.** (a) Kelp fly (*Fucellia* sp., Anthomyiidae). (b) Stable fly (*Stomoxys calcitrans,* Muscidae). (c) House fly (*Musca domestica,* Muscidae). (d) House fly, puparium. (e) House fly, larva. (f) Lesser house fly (*Fannia canicularis,* Muscidae). (g) Lesser house fly, larva.

widely spaced, coarse setae on the lower surface of the anteriormost wing vein. Additionally, the males have a heavily spined protuberance at the base of the hind femur on the inner side.

Very little has been written on the Neotropical *Fucellia.* The life history of the European *F. maritima* may serve as a model for the genus (Egglishaw 1960).

### References

ALDRICH, J. M. 1918. The kelp-flies of North America (Genus *Fucellia,* Family Anthomyiidae). Calif. Acad. Sci. Proc. (4 ser.) 8: 157–179.

EGGLISHAW, H. J. 1960. The life-history of *Fucellia maritima* (Haliday) (Diptera, Muscidae). Entomologist 93: 225–231.

## MUSCID FLIES

Muscidae.

This is the largest family of muscoid Diptera, with 830 species in the Neotropics (Pont 1972) having varied forms and biologies (Skidmore 1985). Adults are generally small to medium-sized, dull gray to brown flies (a few, such as *Muscina,* are metallic green or blue like calliphorids). They are moderately hairy but are unique among Calypterates (a feature shared only with the closely related Anthomyiidae) in lacking bristles on the thoracic sclerite immediately dorsal to the base of the hind leg (= hypopleuron).

Larvae are mostly typical maggots, although some are flattened and exhibit projections (*Fannia*) or other peculiar modifications of the cuticle. They are extremely varied in habits, scavenging animal feces and feeding on carrion and other insects, in many habitats, including running and standing water (*Limnophora*). The larvae of the more than thirty species of *Philornis* (= *Neomusca*) are found in birds' nests where they feed on feces and other organic matter, occasionally attacking the nestlings and causing subcutaneous myiasis (Guimarães et al. 1983). The conspicuous, noisy adults place their eggs or larvae on the chicks of orependolas and caciques. The larvae burrow into the chick's body, often killing it. Curiously, nests of these birds parasitized by the giant cowbird in Panama suffer less harm from *Philornis* because of the propensity for preening exhibited by the alien chick (Smith 1968). The reverse situation is described by Fraga (1984) for cowbirds.

Several Muscidae are cosmopolitan associates of civilization and are important as food contaminators and household invaders. Others (subfamily Stomoxyinae: *Stomoxys, Haematobia,* and *Neivamyia*) bite or are extreme nuisances to humans and domestic animals (Pinto and de Souza Lopes 1933).

### References

FRAGA, R. M. 1984. Bay-winged cowbirds (*Molothrus badius*) remove ectoparasites from their brood parasites, the screaming cowbirds (*M. rufoaxillaris*). Biotropica 16: 223–226.

GUIMARÃES, J. H., N. PAPAVERO, AND A. P. DO PRADO. 1983. As miíases na região Neotropical (Identificação, biologia, bibliografia). Rev. Brasil. Zool. 1: 239–416.

PINTO, C., AND H. DE SOUZA LOPES. 1933. Anatomia, biologia e papel patogênico da *Neivamyia lutzi*, mosca sugadora de sangue dos equídeos do Brasil. Escol. Sup. Agric. Med. Vet. Arch. 10: 77–88.

PONT, A. C. 1972. Family Muscidae. *In* N. Papavero, ed., A catalogue of the Diptera of the Americas south of the United States. Mus. Zool., Univ. São Paulo, São Paulo. Pp. 97.1–97.111.

SKIDMORE, P. 1985. The biology of the Muscidae of the World. Junk (Series Entomologica), The Hague.

SMITH, N. G. 1968. The advantage of being parasitized. Nature 219: 690–694.

ZUMPT, F. 1973. The Stomoxyine biting flies of the world. Diptera: Muscidae. Taxonomy, biology, economic importance and control measures. Fischer, Stuttgart.

## House Fly

Muscidae, Muscinae, *Musca domestica*.
*Spanish:* Mosca casera. *Portuguese:*
Môsca doméstica, môsca comun das casas. Typhoid fly.

This species is known by its ubiquity and close association with mankind. Indeed, its numbers in slums, unkempt corrals of domestic animals, dumps, and other sites of human neglect (Baumgartner 1988) may be astronomical and threaten public health through mechanical transmission of dysentery-causing bacteria and protozoa (Coutinho et al. 1957). The species was not always so abundant in the American tropics; not long after its colonization by Europe, the New World is thought to have received *Musca domestica* with human traffic from the Old World. Only in undisturbed forests and the most remote uninhabited regions is it still absent.

The adult is small (BL 6–9 mm), generally gray, with four dark lines running longitudinally on the back of the thorax; the lateral portions of the abdomen are translucent brown (fig. 11.8c). The larva (fig. 11.8e) is a typical, elongate maggot (BL to 12 mm when mature), creamy white, and may be distinguished from that of other similar domestic fly larvae by the shape of the posterior spiracles, three serpentine slits surrounded by a complete sclerotized ring. The puparium (fig. 11.8d) is outwardly indistinguishable from that of other similar-sized muscoid flies.

The species breeds in a wide range of decomposing organic matter but finds horse manure an especially favored food medium. Other materials commonly harboring larvae are cow dung, human feces, and refuse heaps of vegetables and fruit, all as long as they are fairly moist and warm. The larvae are not normally found in carcasses or commonly involved in myiasis.

Details of the life cycles are available in numerous publications (Milani 1975, West 1951, West and Peters 1973). The house fly is highly prolific, with a fantastic potential for multiplication, which, fortunately, is never realized due to a substantial mortality rate from predation, parasitism, and food and environmental limitations. However, it is difficult to control artificially (Schenone 1962).

## References

BAUMGARTNER, D. L. 1988. The housefly, *Musca domestica* (Diptera, Muscidae), in central Peru: Ecological studies of medical importance. Rev. Brasil. Entomol. 32: 455–463.

COUTINHO, J. O., A. DE E. TAUNAY, AND L. P. DE CARVALHO LIMA. 1957. Importância da *Musca domestica* como vector de agentes patogênicos para o homen. Inst. Adolfo Lutz Rev. 17: 5–23.

MILANI, R. 1975. The house fly, *Musca domestica*. *In* R. C. King, ed., Handbook of genetics. Vol. 3. Invertebrates of genetic interest. Plenum, New York. Pp. 377–399.

SCHENONE, H. 1962. Medidas de profilaxis de las moscas. Bol. Chilena Parasit. 17: 23–25.

WEST, L. S. 1951. The housefly. Comstock, Ithaca.

WEST, L. S., AND O. B. PETERS. 1973. An annotated bibliography of *Musca domestica* Linnaeus. Dawsons of Pall Mall, Folkestone and London.

## Stable Fly

Muscidae, Stomoxyinae, *Stomoxys calcitrans*. *Spanish:* Mosca de los establos. *Portuguese:* Môsca do bagaço. Dog fly.

This medium/small (BL 5–6 mm) muscoid fly also superficially resembles the common house fly with its mostly gray, black-striped thorax (fig. 11.8b). It differs significantly, however, in possessing a spotted abdomen and a rigid, elongate sucking beak with minute biting teeth at its apex. This organ is used to puncture the skin of mammals and withdraw blood. The stable fly is a major pest of horses, cattle, and humans and other large domestic animals. Both sexes suck blood.

The typical maggotlike larvae breed in a variety of rotting plant materials, commonly straw in stables, weeds washed up on lakeshores, and seacoast wrack (Kunz et al. 1977). Immense populations may develop near such larval food sources and constitute a major nuisance to equestrians and beach bathers.

### Reference

KUNZ, S. E., L. L. BERRY, AND K. W. FOERSTER. 1977. The development of the immature forms of *Stomoxys calcitrans*. Entomol. Soc. Amer. Ann. 70: 169–172.

## Lesser House Fly

Muscidae, Fanniidae, *Fannia canicularis*.

This species resembles the house fly but is somewhat smaller (BL 6–7 mm) and has a more elongate body shape and yellow spots at the base of the abdomen (fig. 11.8f). It is most easily recognized by its habit of hovering in small groups in shady places on hot days, individuals lazily zigzagging in the air, usually just a few feet off the ground, never landing. It is an abundant fly in and near human habitations and is frequent in outdoor toilets, privies, stables, pigsties, and other places where excrement accumulates. The darkly pigmented larvae have a flattened, fusiform body with series of elongate, basally fringed projections extending from the sides and dorsum (fig. 11.8g). They tolerate very moist substrata and breed in the fresh liquid feces of most animals and poultry.

## Green House Fly

Muscidae, Muscinae, *Muscina stabulans*. False stable fly.

This muscoid is unusual in being metallic blue-green in color, appearing very similar to many blowflies but generally smaller (BL 8 mm). Adults are invariably found in the near vicinity of human habitations in rural situations; they are scarce in towns. They are filth feeders but also take honeydew and tree sap. The larvae feed in human and other animal feces and a wide variety of rotting matter (rarely carrion; more likely decaying fruit, bird's nests, dead insects, etc.).

## Horn Fly

Muscidae, Stomoxyinae, *Haematobia irritans*. *Spanish:* Mosca (mosquilla) del ganado (General).

The horn fly is primarily a pasture pest of cattle (McLintock and Depner 1954, Vogelsang and de Armas 1940). It is fairly widespread but has been found in Brazil only relatively recently (Valério and Guimarães 1983). It breeds exclusively in fresh cattle dung, and the small (BL 4 mm) adults of both sexes congregate in masses at the base of the horns and suck blood. When very abundant, they are excessively bothersome to the host and contribute significantly to weight and milk losses. They are attracted to their host by olfactory, heat, and visual stimuli, the last seeming to be the most important (Hargett and Goulding 1962).

This species was formerly placed in either the genera *Lyperosia* or *Siphona*. The adult superficially resembles the lesser

**Figure 11.9 DOMESTIC FLIES.** (a) Horn fly (*Haematobia irritans*, Muscidae). (b) Black garbage fly (*Ophyra aenescens*, Muscidae). (c) Flesh fly (*Bercaea haemorrhoidalis*, Sarcophagidae), adult. (d) Flesh fly, larva. (e) Green blowfly (*Phaenicia sericata*, Calliphoridae). (f) Green blowfly, larva.

house fly (BL about 3 mm) but has a short rigid beak like the stable fly (fig. 11.9a).

### References

HARGETT, L. T., AND R. L. GOULDING. 1962. Studies on the behavior of the horn fly [*Haematobia irritans* (Linn.)]. Agric. Exper. Sta., Ore. State Univ., Tech. Bull. 61: 1–27.

MCLINTOCK, J., AND K. R. DEPNER. 1954. A review of the life-history and habits of the horn fly, *Siphona irritans* (L.) (Diptera: Muscidae). Can. Entomol. 86: 20–33.

VALÉRIO, J. R., AND J. H. GUIMARÃES. 1983. Sobre o ocorrência de uma nova praga, *Haematobia irritans* (L.) (Diptera, Muscidae), no Brasil. Rev. Brasil. Zool. 1: 417–418.

VOGELSANG, E. G., AND J. C. DE ARMAS. 1940. La mosquilla del ganado, *Lyperosia irritans* (L. 1761) en Venezuela. Rev. Med. Vet. Parasit. (Min. Agric. Cría, Venezuela) 2: 95–98.

### Black Garbage Fly

Muscidae, Muscinae, Hydrotaeini, *Ophyra aenescens*.

Adults of this small (BL 6 mm), shiny black species (fig. 11.9b) are attracted to human and animal excrement. The hairlike bristle of the antenna is very long and slightly pubescent. The maggots develop in the aforementioned materials and often produce heavy populations that become a nuisance in urban and suburban settlements. They show a fair degree of tolerance for high salinity in their food and have been found common also in carrion of marine origin and even in salted meats (Johnson and Venard 1957, de Oliveira 1941).

### References

DE OLIVEIRA, S. J. 1941. Sôbre *Ophyra aenescens* (Wiedmann, 1830) (Diptera: Anthomyidae). Arq. Zool. (São Paulo) 2: 341–355.

JOHNSON, W. T., AND C. E. VENARD. 1957. Observations on the biology and morphology of *Ophyra aenescens* (Diptera: Muscidae). Ohio J. Sci. 57: 21–26.

## FLESH FLIES

Sarcophagidae. *Spanish:* Moscas de la carne (Argentina).

Adult flesh flies of most of the common species are similar, medium-sized muscoid types, gray in general color and frequently marked with longitudinal black bands on the back of the thorax and checkerboard patterns on the abdomen that change with varied light incidence. The eyes are widely separated in both sexes and often bright green or brick red; the genitalia may also be glaring red. The body is strongly bristled.

Larvae are typical pale muscoid maggots, often fairly large (BL when mature 17 mm) and with truncate posteriors. The latter has a deep concavity rimed with prominent fleshy tubercles, in the bottom of which are located the paired rear spiracles with three straight slits each.

The family displays a great variety of biotic types (Jirón and Marín 1982) and a wider spectrum of hosts than any other, although, as a rule, the larvae are predaceous or endoparasitoids on other animals (Souza Lopes 1973). Most attack other invertebrates, such as snails, free-living grasshoppers (*Doringia acridiorum* parasitizes the South American locusts), and katydids, but some scavenge dead insects, even including provisions in wasp's nests. A minority breed in excrement and decomposing organic matter or are involved in various modes of myiasis in mammals and humans; they rarely develop in carrion. These mostly belong to various species formerly placed in *Sarcophaga,* an Old World genus, but now are dispersed among several genera, primarily *Peckia.* The genus *Dexosarcophaga* lives in termite and ant nests or in the galleries of some wood-boring insects.

The adults of many species associate with humans and may be a factor in mechanical disease transmission (Gregor 1972). *Bercaea haemorrhoidalis* (fig. 11.9c), widely cited in the genus *Sarcophaga,* is a cosmopolitan species often associated with man. Its larvae (fig. 11.9d) feed on carrion, excrement, and exposed meats and sometimes cause myiasis.

The family is large, with over 600 species in the Neotropical Region.

## References

GREGOR, F. 1972. Synanthropy of Sarcophaginae (Diptera) from Cuba. Fol. Parasit. 19: 155–163.

JIRÓN, L. F., AND R. E. MARÍN. 1982. Moscas sarcofágidas de Costa Rica (Diptera; Cyclorrhapha). Rev. Biol. Trop. 30: 105–106.

SOUZA LOPES, H. 1973. Collecting and rearing sarcophagid flies (Diptera) in Brazil during forty years. Acad. Brasil. Cien. An. 45: 279–291.

## CARRION FLIES

Calliphoridae. *Spanish:* Moscas verdes (azules) de la carne (Argentina). *Quechua:* Shinguitos (Peru, larvae). Blowflies, bluebottle flies.

Members of this family (Hall 1948, Norris 1965) are the familiar, medium to small (BL 5–10 mm), metallic green and blue muscoid flies that buzz about fresh carrion (in the "blown" stage, hence the common name). Their maggots (fig. 11.9f) thrive in dead animal tissue, and their feeding constitutes a dominant factor in the primary reduction of most untended vertebrate cadavers and thus is of considerable importance in ecological hygiene and nutrient cycling. Those of a few species are obligatory parasites, attacking healthy flesh, but most may assume facultative roles in myiasis and as veterinary and medical pests. Adults may also transmit pathogens mechanically from putrefactive or fecal material to human food. Because carrion flies breed in cadavers, their presence and developmental rates are frequently used in forensic medicine to determine the time of death of cast-off human bodies (Easton and Smith 1970).

Notable among the approximately one hundred native American tropical species (Dale 1987, James 1970) are the black blowfly (*Phormia regina*), a variable-sized, blackish-green species, and the greenbottle fly (*Lucilia illustris*), a small, bluish-green fly. Both of these occur no farther south than central Mexico. The green blowflies (sheep blowflies, fleeceworms, *Phaenicia sericata* [fig. 11.9e], *P. cuprina,* and *P. eximia*), bright, metallic green to bronze, are the usual types found in human refuse (Quattro and Wasti 1978). The bluebottles (*Calliphora* species), with shiny dark blue bodies, are normally associated with carcasses. All the preceding flies are wide-ranging and extremely common in urban as well as in rural situations. (See Greenberg and Szyska 1984 and Jirón and Marín 1984, for information on these and additional species.)

In recent years, four species of the economically important Old world genus *Chrysomya* have become established and are spreading in Latin America (Baumgartner and Greenberg 1984, Guimarães et al. 1979, Jirón 1979). These have habits similar to those of the screwworm flies and are displacing native calliphorid species. They all associate with humans and live near dwellings to a greater or lesser degree and are potential public health and veterinary nuisances (Baumgartner and Greenberg 1984).

The most serious myiasis-producing species, the screwworms, are discussed below.

### References

BAUMGARTNER, D. L., AND B. GREENBERG. 1984. The genus *Chrysomyia* (Diptera: Calliphoridae) in the New World. J. Med. Entomol. 21: 105–113.

DALE, W. E. 1987. Identidad de las moscas Calliphoridae en la costa central del Perú. Rev. Peruana Entomol. 28: 63–70.

EASTON, A. M., AND K. G. V. SMITH. 1970. The entomology of the cadaver. Med. Sci. Law 10: 208–215.

GREENBERG, B., AND M. L. SZYSKA. 1984. Immature stages and biology of fifteen species of Peruvian Calliphoridae (Diptera). Entomol. Soc. Amer. Ann. 77: 488–517.

GUIMARÃES, J. H., A. P. DO PRADO, AND G. M. BURALLI. 1979. Dispersal and distribution of three newly introduced species of *Chrysomya* Robineau-Desvoidy in Brazil (Diptera, Calliphoridae). Rev. Brasil. Entomol. 23: 245–255.

HALL, D. G. 1948. The blowflies of North America. Entomol. Soc. Amer., Washington, D.C.

JAMES, M. T. 1970. Family Calliphoridae. *In* N. Papavero, ed., A catalogue of the Diptera of the Americas south of the United States. Dept. Zool., Sec. Agric., São Paulo, no. 102.

JIRÓN, L. F. 1979. Sobre moscas califóridas de Costa Rica (Diptera: Cyclorrhapha). Brenesia 16: 221–223.

JIRÓN, L. F., AND F. J. MARÍN. 1984. Notas complementarias sobre moscas califóridas de Costa Rica (Diptera: Calliphoridae). Brenesia 22: 65–68.

NORRIS, K. R. 1965. The bionomics of blow flies. Ann. Rev. Entomol. 10: 47–68.

QUATTRO, M. J., AND S. S. WASTI. 1978. Olfactory and oviposition responses of the green bottle fly, *Phaenicia sericata* (Meig.), to a variety of natural baits (Diptera: Calliphoridae). Rev. Brasil. Biol. 38: 115–119.

### Screwworms

Calliphoridae, *Cochliomyia hominovorax* (= *americana*). *Spanish:* Larvae—gusanos tornillos (General), gusanos de las heridas (Puerto Rico), gusanos barrenadores (Mexico). Adults—mosca americana de larva-tornillo, moscas de la queresa (Panama). *Portuguese:* Larvae—vermes torneiros, mororó, coró. Adults—môscas varejeiras, beronhas (Brazil). Wounds—bicheiras. *Tupi-Guaraní:* Larvae—tapuru (Brazil).

This is one of the most serious pests of livestock in subtropical and tropical America, named the screwworm from the wriggling and twisting appearance of the larva (fig. 11.10b) when burrowing into flesh. Adults oviposit in any open wound, the umbilicus of newborns, even minor cuts. The eggs hatch into white, slender maggots that feed on healthy tissue within the wound. Infested animals become nervous and make frantic attempts to scratch and

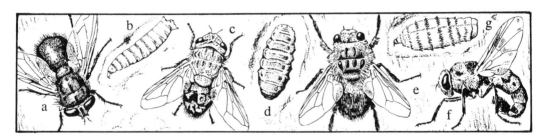

**Figure 11.10 PARASITIC FLIES.** (a) Screwworm (*Cochliomyia hominovorax,* Calliphoridae). (b) Screwworm, larva. (c) Sheep botfly (*Oestris ovis,* Oestridae). (d) Common cattle grub (*Hypoderma lineatum,* Hypodermatidae), larva. (e) Cattle grub, adult. (f) Horse botfly (*Gasterophilus intestinalis,* Gasterophilidae). (g) Horse botfly, larva.

lick the afflicted area; the majority of untreated cases end in the death of the animal, which may be any species of domestic animal, although cattle, goats, sheep, and hogs are the usual hosts; humans may also suffer infections with dire results (Aubertin and Buxton 1934, Mazza and Jörg 1939).

The adult screwworm fly is medium-sized (BL 8–10 mm) and is generally metallic blue to bluish-green (fig. 11.10a). There are three indistinct dark longitudinal bands on the back of the thorax, and the head is contrasting reddish-orange. For many years, it was confused with the secondary screwworm (*C. macellaria*) but found ultimately to be distinct (Laake et al. 1936).

The species occurs widely in the Americas and remains a major problem in spite of the availability of good control techniques. It has been eradicated for the time being from the southern parts of the United States through the use of the sterile male program. Entomologists first established a screwworm barrier zone along the 2,000-mile border between the United States and Mexico to reduce the chance of migration northward; in 1972, an agreement was signed between the two countries to move the barrier zone to the narrowest part of Mexico, the Tehuantepec Isthmus (Spencer et al. 1981). Attempts were undertaken to apply this technique in Mexico (Brenner 1984) and Central America (Snow et al. 1985) with limited success. As yet, no serious efforts have been made to apply this wide-ranging suppression measure in other parts of Latin America, partly because of its rising ineffectiveness in parts of the United States and Mexico, probably due to the genetic diversity of wild populations (Richardson et al. 1982). However, some feel that it is feasible to develop plans for eradication at least through Central America (Snow et al. 1985). It is of historical interest to note, however, that the first complete eradication of the species was accomplished experimentally on the island of Curaçao in 1954 (Bushland et al. 1955), yet the fly demonstrated its tenacity by reestablishing itself on the island in 1975 (Snow et al. 1978).

Victims of screwworms are treated by various folk remedies, usually consisting of poultices made from the leaves of medicinal plants. In Brazil, magical means involving prayers and incantations believed to force the exit of the larvae from wounds have also been applied widely. One such is performed by an exorcist, who first ties together the ends of a flexible stick to make a circle about the diameter of the wound. Holding this charm over the afflicted part of the animal, he utters the supplication: "Foge doença, De bicho mau, Da santa presença, De São Nicolau, Verme da terra, Na terra dura, São Nicolau fez tua sipurtura" (Go away sickness of the bad grub, from the presence of Saint Nicholas. Grub of the earth in the earth will Saint Nicholas make your grave) (Lenko and Papavero 1979: 423).

Two figures in the Mayan Tiro-Cortesianus Codex appear to depict larvae of these flies attacking gods. In each case, one of the maggots is proximate to the nose, a common focus of screwworms. The hosts may have been victims of human sacrifice (Tozzer and Allen 1910).

## References

AUBERTIN, D., AND P. A. BUXTON. 1934. *Cochliomyia* and myiasis in tropical America. Ann. Trop. Med. Parasit. 28: 245–254.

BRENNER, R. J. 1984. Dispersal, mating, and oviposition of the screwworm (Diptera: Calliphoridae) in southern Mexico. Entomol. Soc. Amer. Ann. 77: 779–788.

BUSHLAND, R. C., A. W. LINDQUIST, AND E. F. KNIPLING. 1955. Eradication of screw-worms through release of sterilized males. Science 122: 287–288.

LAAKE, E. W., E. C. CUSHING, AND H. E. PARISH. 1936. Biology of the primary screw worm fly, *Cochliomyia americana*, and a comparison of its stages with those of *C. macellaria*. U.S. Dept. Agric. Tech. Bull. 500: 1–24.

LENKO, K., AND N. PAPAVERO. 1979. Insetos no folclore. Sec. Cult. Cien. Tech., São Paulo.

MAZZA, S., AND M. E. JÖRG. 1939. *Cochliomyia hominivorax americana* C. y P., estudio de sus larvas y consideraciones sobre miasis. Univ. Buenos Aires Mis. Estud. Pat. Reg. Argentina Publ. 41: 3–46.

RICHARDSON, R. H., J. R. ELLISON, AND W. W. AVERHOFF. 1982. Autocidal control of screwworms in North America. Science 215: 361–370.

SNOW, J. W., J. R. COPPEDGE, AND A. H. BAUMHOVER. 1978. The screwworm *Cochliomyia hominivorax* (Diptera: Calliphoridae) reinfests the island of Curaçao, Netherlands Antilles. J. Med. Entomol. 14: 592–593.

SNOW, J. W., C. J. WHITTEN, A. SALINAS, J. FERRER, AND W. H. SUDLOW. 1985. The screwworm, *Cochliomyia hominivorax* (Diptera: Calliphoridae), in Central America and proposed plans for its eradication south to the Darien Gap in Panama. J. Med. Entomol. 22: 353–360.

SPENCER, J. P., J. W. SNOW, J. R. COPPEDGE, AND C. J. WHITTEN. 1981. Seasonal occurrence of the primary and secondary screwworm (Diptera: Calliphoridae) in the Pacific coastal area of Chiapas, Mexico, during 1978–1979. J. Med. Entomol. 18: 240–243.

TOZZER, A. M., AND G. M. ALLEN. 1910. Animal figures in the Mayan codices. Harvard Univ., Peabody Mus. Pap. Amer. Archaeol. Ethnol. 4: 273–372.

## BOTFLIES

Cuterebridae, Gasterophilidae, Hypodermatidae, and Oestridae. Warble flies, breeze flies, clegs, heel flies, gadflies (adults), grubs (larvae).

Botflies and warble flies all are somewhat large (BL 12–20 mm), robust, and usually hairy, muscoid flies, which do not feed, possessing only vestigial mouthparts. The endoparasitic larvae are likewise large, thick skinned and heavy bodied, always with rearward directed spines or denticles projecting from the cuticle, which prevent them from being easily dislodged from the host. The pupal stage is passed in a very hard, heavily pigmented puparium that develops on the ground, in surface debris, or shallowly buried in the soil.

There are various types, the larvae of all living obligatorily in mammals (Guimarães et al. 1983). Some types are not native to the New World, having been introduced with infected livestock into a few areas from elsewhere. These include the sheep warbles (*Oestrus ovis;* fig. 11.10c), which infest the sinuses and other cranial tissues of sheep (*estro del borrego, mosca de la nariz* in Peru) (Rogers and Knapp 1973) and occasionally enter the conjunctiva of the human eye as first instars (Atias et al. 1960), and two species of cattle grubs. The latter, *Hypoderma lineatum* (fig. 11.10d) and *H. bovis*, pass part of their life in the internal organs of cattle and oxen but later lodge in open boils beneath the skin on the back, from which they emerge later to pupate, doing damage to the hide. The adults resemble bumblebees (fig. 11.10e) and are much feared by livestock. The species have become pests in certain cattle areas of Latin America but are not well established.

The preceding species all belong to the families Hypodermatidae and Oestridae. The family Gasterophilidae contains three similar parasites in the genus *Gasterophilus*, *G. intestinalis*, *G. haemorrhoidalis*, and *G. nasalis*, all occupying the intestinal tract of horses. The larvae (fig. 11.10g) most commonly attach to the internal stomach wall and cause considerable irritation to the host. Like the foregoing species, horse bots (fig. 11.10f) are of local occurrence in the countries of the New World tropics, only where they have been carried in with imported animals.

Among the bots, only the family Cuterebridae, the rodent botflies, is indigenous to the Western Hemisphere. The larvae of most of the eighty-three species of *Cuterebra* and five related genera parasitize a wide variety of rodents (*Sciurus, Thomomys, Neotoma, Oryzomys, Mus*, etc.) and rabbits (*Lepus, Sylvilagus*) (Guimarães 1971) as well as marsupials and carnivores (Séguy 1948).

The life cycle of most species is similar. The female oviposits in environs frequented by the host, and the eggs hatch in response to the heat and movement of its body nearby. The newly hatched larvae are wet and sticky from egg fluids and readily adhere to the host as it contacts them. Entry to the host's body is via moist body openings (mouth, nostrils, eyes, etc.) or skin lacerations. Following entry, the larvae migrate internally to subcutaneous sites where they settle, molt, and form feeding pockets with external breathing pores. Mature larvae drop from the pore and burrow into surface soil or debris for pupation (Catts 1982).

*Alouattamyia* contains a single species, the monkey botfly, developing subcutaneously primarily in the throats of howler monkeys (Zeledón et al. 1957). Little is known of the biology of the four remaining genera, save *Dermatobia*, which contains a single, most curious species, because of its unique mode of oviposition and frequent use of humans as hosts (see below).

## References

ATIAS, A., R. DONCKASTER, H. SCHENONE, AND M. OLIVARES. 1960. Myiasis ocular producida por larvas de *Oestrus ovis*. Bol. Chilena Parasit. 15: 37–38.

CATTS, E. P. 1982. Biology of New World bot flies: Cuterebridae. Ann. Rev. Entomol. 27: 313–338.

GUIMARÃES, J. H. 1971. Notes on the hosts of Neotropical Cuterebrini (Diptera, Cuterebridae), with new records from Brazil. Univ. São Paulo, Mus. Zool., Pap. Avul. Zool. 25: 89–94.

GUIMARÃES, J. H., N. PAPAVERO, AND A. P. do PRADO. 1983. As miíases na região Neotropical (Identificação, biologia, bibliografia). Rev. Brasil. Zool. 1: 239–416.

ROGERS, C. E., AND F. W. KNAPP. 1973. Bionomics of the sheep bot fly, *Oestris ovis*. Environ. Entomol. 2: 11–23.

SÉGUY, E. 1948. Introduction à l'étude des myiases. Rev. Brasil. Biol. 8: 93–111.

ZELEDÓN, R., O. JIMÉNEZ, AND R. R. BRENES. 1957. *Cuterebra baeri* Shannon y Greene, 1926 en el mono aullador de Costa Rica. Rev. Biol. Trop. 5: 129–134.

## Human Botfly

Cuterebridae, *Dermatobia hominis*. *Spanish:* Tórsalo, torcel (Central America); tornillo (Peru, Argentina); gusano de monte, nuche (central and northern South America); barro (Bolivia); gusano peludo (Colombia). *Portuguese:* Berne (Brazil). *Nahuatl:* Colmoyotl, moyocuil (Mexico, Guatemala). *Tupi-Guaraní:* Ura. *Kaingang Indian:* Bikuru.

The method used by the female of this botfly to infect its host is one of the most devious and amazing employed by vertebrate endoparasites (Catts 1982). The female does not lay its eggs directly on its wary hosts, large mammals (cattle, horses, dogs, pigs, tapirs, deer, etc.) and some birds, but commandeers other insects to carry them, always choosing a bloodsucking or zoophilous type (mosquito—*virole zancudo*—deerfly, stable fly) that will surely be attracted to an animal in search of a meal. It attaches its eggs to the body of the vector, which then transports them to the host's skin. Stimulated by body heat and agitation, the completely mature larva within the egg capsule hatches instantly when the vector touches the animal and immediately burrows into its skin. It is interesting that the Amazonian Indian apparently had knowledge of this unique biological phenomenon, as evidenced by the existence of a word in the Tupi-Guaraní language for crane flies, thought to be carriers, *carapanã ura*. The use of the word *moyotl* (mosquito) in combinations of words for this insect by the Aztecs suggests a similiar knowledge by these people (orig. obs.).

After feeding a week and a half subcutaneously, the larvae make a breathing aperture to the outside to accommodate their increased oxygen needs, having molted to the second stage. First-stage larvae are small (BL 6 mm), fusiform, and with numerous fine to coarse surface denticles. The second stage is larger (BL 10–15 mm) and strangely shaped with a spheri-

**Figure 11.11    PARASITIC FLIES.** (a) Human botfly (*Dermatobia hominis,* Cuterebridae), larva. (b) Human botfly, adult. (c) Bat tick fly (*Basilia ferrisi,* Nycteribiidae). (d) Bat fly (*Trichobius dugesii,* Streblidae). (e) Louse fly (*Olfersia fassulata,* Hippoboscidae).

cal anterior portion with coarse denticles and elongate, smooth taillike extension. The third instar (fig. 11.11a) is large (BL to 2 cm) and grublike, with a slightly narrow, unarmed posterior section and numerous small rearward-pointing denticles arranged in transverse rows on the anterior segments. All stages remain in a feeding pocket beneath the breathing aperture; on fully maturing, they squeeze out of the pocket and drop to the ground to pupate. The whole developmental period, from egg to adult, requires a little over two months (Jobsen and Mourier 1972).

This life cycle may take place in humans (*mirunta,* Peru) as well as domestic and wild mammals, and the incidence of infestation in some parts of Latin America, particularly in cattle areas, is irritatingly high and of long standing (Blanchard 1892, Thomas 1988). The larvae are seldom able to complete their development, however, because they are not long tolerated and are removed. They frequently twist on their axis, causing pain, and occasionally cut through blood vessels while feeding, releasing copious flows of blood.

Removal of larvae is accomplished in many ways by the locals, depending on tradition. Because they are virtually impossible to remove when healthy and vigorous, the maggots' body being covered with recurved denticles that prevent its withdrawal, they are induced to relax their grip in the feeding pocket by an overnight application of a kind of poultice, often a piece of raw bacon and/or tobacco (*ampiri,* Peru). This also suffocates the larvae, which in a limp state can then be extracted with forceps or squeezed out. The modern version of this trick is the application of a tight covering of transparent adhesive tape. Usually, the larvae occur singly, but multiple infections of two, three, or more at a single site are not uncommon.

Although harboring larvae is very painful and unpleasant, the wound bleeding often and oozing noxious liquids, at least a few intrepid entomologists have allowed them to complete development in their own bodies and recorded the experience (Dunn 1930, Busck 1913). Usual sites of infection are the forearms, shoulders, and scalp. In infants, the last site is potentially dangerous because larvae have been known to bore through the soft fontanel and lodge in the brain, causing death (Rossi and Zucoloto 1973).

Adults of both sexes are heavy bodied and large (BL 12–15 mm; fig. 11.11b). (For a detailed study of the male genitalia, see Leite 1990.) The head is mainly yellow, the thorax dark bluish-grey, and the abdomen a brilliant, shiny dark blue. They are rarely seen in nature but are occasionally observed during the daytime in the forest near stagnant water where their most common mosquito egg vectors are emerging.

*Dermatobia hominis* has an equatorial distribution from 25 deg. N to 18 deg. S, in which it favors moist tropical, hilly, secondary forest. In parts of its range, cattle are severely affected, and infestations account for significant losses in meat, milk, and hide value. Insecticides are used against larvae in livestock, but possible control through the use of chemosterilants is under study.

A bibliography on the species, complete to 1966, was published by Guimarães and Papavero (1966).

## References

BLANCHARD, R. 1892. Sur les Oestrides américains dont la larve vit dans la peau de l'homme. Soc. Entomol. France Ann. 61: 109–154.

BUSCK, A. 1913. On the rearing of a *Dermatobia hominis* Linnaeus. Entomol. Soc. Wash. Proc. 14: 9–11.

CATTS, E. P. 1982. Biology of New World bot flies: Cuterebridae. Ann. Rev. Entomol. 27: 313–338.

DUNN, L. H. 1930. Rearing the larvae of *Dermatobia hominis* Linn., in man. Psyche 37: 327–342.

GUIMARÃES, J. H., AND N. PAPAVERO. 1966. A tentative annotated bibliography of *Dermatobia hominis* (Linnaeus 1781) (Diptera, Cuterebridae). Arq. Zool. (São Paulo) 14: 223–294.

JOBSEN, J. A., AND H. MOURIER. 1972. The morphology of the larval instars and pupa of *Dermatobia hominis* L. Jr. (Diptera: Cuterebridae). Entomol. Bericht. 32: 218–224.

LEITE, A. C. R. 1990. Scanning electron microscopy of male genitalia of Dermatobia hominis (Diptera: Cuterebridae). J. Med. Entomol. 27: 706–708.

ROSSI, M. A., AND S. ZUCOLOTO. 1973. Fatal cerebral myiasis caused by the tropical warble fly, *Dermatobia hominis*. Amer. J. Trop. Med. Hyg. 22: 267–269.

THOMAS, JR., D. B. 1988. The pattern of *Dermatobia* (Diptera: Cuterebridae) myiasis in cattle in tropical Mexico. J. Med. Entomol. 25: 131–135.

## ECTOPARASITIC FLIES

Streblidae, bat flies. Nycteribiidae, bat tick flies. Hippoboscidae, louse flies.

Three families of Diptera have evolved a parasitic way of life among the hairs of bats and feathers of birds. Their adaptations include flattening of the body, development of combs of flat bristles for scuttling among the host's pelage, and enlarged tarsal claws for grasping.

The blind, straw-colored or yellowish bat flies (Streblidae, fig. 11.11d) and bat tick flies (Nycteribiidae, fig. 11.11c) (Guimarães and d'Andretta 1956) are small (BL 2–4 mm) and live permanently and solely on bats. These dipterans are well studied only taxonomically and from few areas of Latin America, primarily Panama (Guimarães 1966, Wenzel et al. 1966) and Venezuela (Wenzel 1976), as a result of special surveys there. A bibliography to 1971 is provided by Maa (1971).

The streblids (Wenzel 1976), with 94 neotropical species in 23 genera, are cylindrical, have a more or less normal-appearing head and relatively short legs, and are usually winged (some have reduced or no wings); nycteribiids are flat and spiderlike, with long legs, without wings, and the head is uniquely folded back into a groove on the back of the thorax. There are 37 species in two genera (almost all in *Basilia*, on vespertilionid bats).

Louse flies (Hippoboscidae, fig. 11.11e) are larger (BL 3–11 mm) and have well-developed eyes (Bequaert 1953–1957, Maa 1963). Their integument is dark and leathery, they are usually completely winged, and they lack bristle combs; the tarsal claws are hooked and heavy for grasping. Most species live on birds; the dark-colored adults are often seen running conspicuously over the white plumage of seabirds (important pests, *moscas de gallinazos*, of guano birds in Peru belong to the genus *Olfersia;* Dale 1969). Altogether, there are 43 regional species (Maa 1969). There are also many kinds infesting mammals such as deer (*Lipoptena mazamae*) and domestic quadrupeds (e.g., the horse louse fly, *Hippobosca equina,* and sheep ked, *Melophagus ovinus;* they have never been found on bats. The last two are cosmopolitan pests, spread by man with their hosts, as is the ubiquitous pigeon louse fly (*Pseudolynchia canariensis*).

In all hippoboscids, a single larva at a time is retained within the female's abdomen where it is nourished to maturity by special "milk" glands. It is then extruded and falls to the ground where it pupates.

The blood-feeding habits of these Diptera would make them potential spreaders of disease, but they so far have not been incriminated as primary vectors for any pathogens save some bird blood protozoans transmitted by hippoboscids.

## References

BEQUAERT, J. 1953–1957. The Hippoboscidae or louse-flies (Diptera) of mammals and birds. Pt. I. Structure, physiology and natural history. Entomol. Americana (n.s.) 33: 1–209, 211–442, figs. 1–21; 34: 1–232, figs. 22–44; 35: 233–416, figs. 45–82; 36: 417–611, figs. 83–104.

DALE, W. E. A. 1969. Hippoboscidae (Diptera) del Peru. I. Nuevas identificaciones: Registros hallados en la literatura Peruana. Biota 8: 41–52.

GUIMARÃES, L. R. 1966. Nycteribiid batflies from Panama. *In* R. L. Wenzel and V. J. Tipton, eds., Ectoparasites of Panama. Field Mus. Nat. Hist., Chicago. Pp. 393–404.

GUIMARÃES, L. R., AND M. A. V. D'ANDRETTA. 1956. Sinopse dos Nycteribiidae (Diptera) do Novo Mundo. Arq. Zool. (São Paulo) 9: 1–184.

MAA, T. C. 1963. Genera and species of Hippoboscidae (Diptera): Types, synonymy, habitats and natural groupings. Pacific Ins. Monogr. 6: 1–186.

MAA, T. C. 1969. A revised checklist and concise host index of Hippoboscidae (Diptera). Pacific Ins. Monogr. 20: 262–299.

MAA, T. C. 1971. An annotated bibliography of batflies. Pacific Ins. Monogr. 28: 119–211.

WENZEL, R. L. 1976. The streblid batflies of Venezuela (Diptera: Streblidae). Brigham Young Univ. Sci. Bull. (Biol Ser.) 10(4): 1–177.

WENZEL, R. L., V. J. TIPTON, AND A. KIEWLICZ. 1966. The streblid batflies of Panama (Diptera: Streblidae). *In*. R. L. Wenzel and V. J. Tipton, eds., Ectoparasites of Panama. Field Mus. Nat. Hist., Chicago. Pp. 405–675.

# 12  SAWFLIES, WASPS, ANTS, AND BEES

Hymenoptera.

Although the primitive sawflies are free-living, phytophagous, and unarmed, outstanding characteristics of this familiar order are their social or subsocial living habits, predaceousness, and ability to sting.

Among the Hymenoptera, all the ants and thousands of species of bees and wasps (Snelling 1981) exhibit some level of sociality, from simple maternal care to the complex community organization of the honeybee hive. A progression of stages exists, recapitulating the evolution of sociality in this order (see social insects, chap. 2). The first stage is simple mass provisioning for larvae in a natural burrow such as found in certain spider wasps, followed by forms, such as the sand wasps, that construct nests that the parent provisions continually to ensure a supply of fresh food for the developing larvae. A further step beyond this is increased longevity of the female parent and a tendency for offspring to remain with her and assist in the care of subsequent broods, the pattern in polistine paper wasps. The final evolutionary step is a division of labor and a correlated formation of different body types, or "castes," and enormous numbers of colony members living in elaborate nests such as the polybiine paper wasps, leaf cutter ants, and social bees. Nest building and larval care are major concerns of the females; the males' adult lives are mainly spent in pursuit of females, often in complex ways (Alcock et al. 1978). To understand hymenopteran social structure, it is important to recall that the members of the colony are all closely related family members. All the so-called workers or sterile females are daughters of a single founding mother. (The latter is the normal case and is referred to as monogyny. In some species and groups, there are multiple queens at the head of a colony, referred to as polygyny, but this is unusual.) The queen (or queens) mates and carries the sperm in storage pouches off the reproductive tract. The fathers of colonies die soon after inseminating the queen.

The many kinds of Hymenoptera capable of stinging do so with a specially modified ovipositor, the egg-laying device of the female, which is located in the tip of the abdomen (Hermann and Blum 1981). Males, lacking this organ, are incapable of stinging. A sting is an effective way to discourage or punish enemies but also is used to subdue prey. The pain, swelling, and other adverse symptoms caused in humans is shared by other vertebrate animals, although many have a natural immunity to stings. These harmful effects are due to proteins and enzymes in the venom which are injected into the wound and are foreign to the physiology of the recipient. Direct toxicity and allergic reactions result, and either may be sufficiently severe to cause death in some cases. Some birds take advantage of the aggressiveness and stinging habits of ants, bees, and wasps by situating their own nests in close proximity (Myers 1935, Smith 1968).

There are some 105,000 known species of Hymenoptera in the world, a large percentage of these in Latin America (Willink 1982). Previously unknown to science, many are being discovered every year, especially among the small, parasitic groups. Because of the diversity of the order, it cannot be easily characterized structurally. In general, these insects have two pairs of membranous wings, the fore larger than the hind, with moderately complex venation, although many are wingless (e.g., worker ants), and others exhibit simplified venation. The mouthparts are adapted basically for biting, but bees have elongated maxillae and labia for feeding on liquids.

There are two suborders, the Symphyta (Chalastrogastra) and Apocrita (Clistogastra, Petiolata). The former is defined by the broad attachment of the abdomen to the thorax, two-segmented trochanters in the legs, at least three closed cells in the hind wing venation, and phytophagous larval habits. Members of the group also possess a well-developed cutting ovipositor that is somewhat sawlike in some families ("sawflies") or elongate for drilling into wood in others ("horntails"). The larvae of those feeding externally on plants are often caterpillarlike, with walking legs and often cryptic colors; the wood borers are wormlike, elongate, pale, and legless. Others form galls or bore in wood and are grublike. The Orussidae are parasitic on wood-boring insects.

In the Apocrita, the basal segment of the abdomen is fused to the thorax, and the next segment forms a stalk or narrow waist. The trochanters are one or two segmented, and the hind wing venation never has more than two closed cells. The grublike larvae of most are either internal or external parasitoids of other arthropods (Parasitica) or are housed in nests and fed a diet of processed plant or animal tissue, honey, or pollen by the adults (Aculeata).

Some specialized types are leaf or fruit miners or gall makers (Agaonidae, Cynipidae, etc.).

Hymenoptera are tremendously important and interesting ecologically (Rau 1933), bees especially, for the ecological roles they play as pollinators of flowering plants, many of which are human food staples. In fact, the dependence of certain plants on these insects is so great that they possess highly specialized flower structures and colors to ensure successful fertilization by specific hymenopterans. Although a few phytophagous species are pests of tropical crops, the large number of "parasites" (actually parasitoids) far offset their injurious effects by keeping other noxious insects at bay.

These insects are familiar to everyone and have a particular cultural importance to many Amerind groups. For instance, it is believed by the Kayapó Indians of Brazil that they learned how to live as social beings from an ancestral wise man (*wayanga*) who gained this knowledge from the study of ant, bee, and wasp behavior (Posey 1982).

## References

ALCOCK, J., E. M. BURROWS, G. GORDH, L. J. HUBBARD, L. KIRKENDALL, D. W. PYLE, T. L. PONDER, AND F. G. ZALOM. 1978. The ecology and evolution of male reproductive behavior in the bees and wasps. Zool. J. Linnean Soc. 64: 293–326.

HERMANN, H. R., AND M. S. BLUM. 1981. Defensive mechanisms in the social Hymenoptera. *In* H. R. Hermann, ed., Social insects. 2: 77–197. Academic, New York.

MYERS, J. G. 1935. Nesting associations of birds with social insects. Royal Entomol. Soc. London Trans. 83: 11–22.

POSEY, D. A. 1982. The importance of bees to the Kayapó Indians of the Brazilian Amazon. Fla. Entomol. 65: 452–458.

RAU, P. 1933. The jungle bees and wasps of Barro Colorado Island (with notes on other insects). Publ. by author, Kirkwood, Mo.

SMITH, N. G. 1968. The advantage of being parasitized. Nature 219: 690–694.

SNELLING, R. R. 1981. Systematics of social

Hymenoptera. *In* H. R. Hermann, ed., Social insects. 2: 369–453. Academic, New York.

WILLINK, A. 1982. Himenópteros Neotropicales, su orígen, ecología, comportamiento y distribución. *In* P. J. Salinas, ed., Zoología Neotropical. 1: 71–90. 8th Cong. Latinoamericano Zool. (Mérida) Actas.

## SAWFLIES AND HORNTAILS

Symphyta.

These two groups are very poorly studied in Latin America. They are less commonly collected in the region than in northern latitudes and are not found in numbers as they are in subarctic and temperate North America (Smith 1988).

Most Neotropical sawflies belong to the families Pergidae, Argidae, and Tenthredinidae, whose larvae feed on a large variety of plants, including ferns. Hosts include forest and ornamental trees, shrubs, and agricultural crops. Members of the native genus *Acordulecera* (Pergidae) are pests of potatoes in Peru and Bolivia (Smith 1980). Among the tenthredinid pests are many species of the genus *Waldheimia* (fig. 12.1a); *Syzygonia cyanocephala* attacks the Quaresmeira (*Tibouchina,* Melanostomaceae) in Brazil (Marques 1933).

Horntails are placed in the family Siricidae, a primarily northern group associated with conifers and arboreal angiosperms. They occur only as far south as the coniferous forests into Central America and are mostly in the genera *Sirex* and *Urocerus* (fig. 12.1b). One palearctic horntail, *Urocerus gigas gigas,* has been introduced into Chile and has become established there (Smith 1988). Another indigenous species in the same genus (*U. patagonicus*) is known as a Paleocene fossil from Argentina (Fidalgo and Smith 1987). In Mexico, horntail larvae may assume importance as pests when burrowing in pine and fir wood used for building construction or furniture.

Adults of both groups are small to large, wasplike insects with well-veined wings and many-segmented antennae. The latter are sometimes highly modified, with terminal clubs or plumose branches.

### References

FIDALGO, P., AND D. R. SMITH. 1987. A fossil Siricidae (Hymenoptera) from Argentina. Entomol. News 98: 63–66.

MARQUES, L. A. 1933. Tenthredinidae conhecida por "Mosca de Serra," cuja larva ou "falsa lagarta" é novica a várias espécies do gênero *Tibouchina.* Inst. Biol. Dept. Agric. Rio de Janeiro 1933: 1–11.

SMITH, D. R. 1980. Identification of the *Acordulecera* "potato" sawflies of Peru and Bolivia, with descriptions of these and related species from South America (Hymenoptera: Pergidae). J. Wash. Acad. Sci. 70: 89–103.

SMITH, D. R. 1988. A synopsis of the sawflies (Hymenoptera: Symphyta) of America south of the United States: Introduction, Xyelidae,

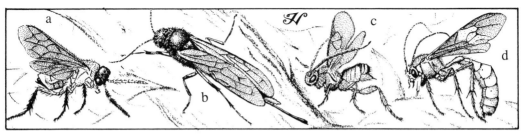

**Figure 12.1  WASPS.** (a) Sawfly (*Waldheimia ochra,* Tenthredinidae). (b) Horntail (*Urocerus californicus,* Siricidae). (c) Braconid wasp (*Apanteles congregatus,* Braconidae). (d) Ichneumon wasp (*Thyreodon* sp., Ichneumonidae).

Pamphilidae, Cimbicidae, Diprionidae, Xiphydriidae, Siricidae, Orussidae, Cephidae. Syst. Entomol. 13: 205–261.

## WASPS

Hymenoptera. *Spanish:* Avispas.
*Portuguese:* Vespas, marimbondos (larger, stinging types, Brazil), cabas (Brazil).

The term "wasp" is loosely applied to all winged adult Hymenoptera, except the bees and ants. Thus, a vast assemblage of diverse, often unrelated, forms are lumped together into one broad category (Evans and West-Eberhard 1970, Spradbery 1973). Within the wasps, however, it is useful to recognize several basic biological types: the "parasitic" (actually parasitoid) wasps, gall wasps, solitary wasps, and social wasps.

Parasitoid wasps are considered by some authors to be less numerous in species in the tropics than in the temperate latitudes. Probably, because they tend to be niche specific rather than host specific, this may be true for Ichneumonidae (Janzen 1981) but may be only an artifact of collecting for chalicidoids (Hespenheide 1979). Also, ichneumons need a humid environment and keep mainly to forests; they are comparatively rare in deserts and high mountains, while chalcidoids as a group are more widely tolerant of the environment.

### References

EVANS, H. E., AND M. J. WEST-EBERHARD. 1970. The wasps. Univ. Michigan Press, Ann Arbor.
HESPENHEIDE, H. A. 1979. Are there fewer parasitoids in the tropics? Amer. Nat. 113: 766–769.
JANZEN, D. H. 1981. The peak in North American ichneumonid species richness between 30° and 42°N. Ecology 62: 532–537.
SPRADBERY, J. P. 1973. Wasps. Univ. Washington Press, Seattle.

## PARASITOID WASPS

### Braconid and Ichneumon Wasps
Braconidae and Ichneumonidae.

Among the Neotropical parasitic Hymenoptera, two very large families are dominant, the braconid wasps (fig. 12.1c) (Matthews 1974) and the ichneumon wasps (fig. 12.1d) (Porter 1980). They are generally similar, small to medium-sized (BL 3–25 mm), somewhat frail-bodied wasps with many-segmented antennae (16 or more) and with parallel entomophagous parasitoid habits, utilizing the larvae and pupae of almost any holometabolous insect (but especially Lepidoptera) as hosts (Gauld 1988). One to many larvae develop inside the host, feeding on various tissues and eventually killing the insect. Ichneumons generally do not spin cocoons outside the host; braconids attach individual white silken cocoons on the host's exterior, and parasitized caterpillars are often seen adorned with masses of such cocoons clinging to the surface of the skin. Sometimes, the cocoons are located apart from the host either singly or in a mass on vegetation or other substratum.

The two families are distinguished from other parasitoid wasps by the absence of a costal cell in the fore wing venation, the leading margin of the fore wing being a single, heavy vein. Braconids are usually smaller (BL 14 mm maximum) and have only one recurrent vein in the fore wing (i.e., no closed cell in the outer, posterior part of the wing); the females also usually have abbreviated abdominal petioles (waists) and short ovipositors, although in *Iphiaulax*, the latter is fourteen times the length of the body. Ichneumons are all sizes but are often large (BL up to 2 cm or more) and with two recurrent veins in the fore wing (a closed cell is present in the outer posterior part of the wing); the females of many have extremely long, slender ovipositors and long waists connecting the abdomen to the thorax. The ovipositor is inserted directly into the host or used to drill through wood, leaf tissue, cocoons, soil, and so on, to oviposit in hidden insects. The abdomen is often much compressed. An additional, useful identifying characteristic is the fusion of the second and third abdominal tergites in most braconids; nearly all ichneumons have a freely movable articulation between these two segments.

Because they are parasitoids, these numerous (just under 20,000 described Neotropical Ichneumonidae and perhaps several thousand Braconidae; Townes and Townes 1966) wasps are of great value in controlling insect populations, including those of many pests, naturally or through human introduction. Hosts consist mainly of caterpillars, but aphids, beetle larvae, and other insects are used also. They subdue their prey with paralyzing venoms (Beard 1978). A few ichneumonids have a stinging apparatus sufficiently powerful to penetrate human skin. The potency of the sting of certain *Tetragonochora* is apparently sufficient to qualify them as models for Batesian mimics in the katydid genus *Aganacris,* whose nymphs resemble them remarkably (dark orange abdomen, white maculae on sides of black thorax, and black antenna with median white band). Some large species of ichneumonids in the subfamily Ophioninae, for example, *Rhynchophion,* resemble tarantula hawks (Pepsis, Pompilidae) and also undoubtedly enter into mimicry complexes with them.

## References

BEARD, R. L. 1978. Venoms of Braconidae. *In* S. Bettini, ed., Arthropod venoms. Springer, Berlin. Pp. 773–800.

GAULD, I. D. 1988. Evolutionary pattern of host utilization by ichneumonoid parasitoids (Hymenoptera: Ichneumonidae and Braconidae). Biol. J. Linnean Soc. 35: 351–377.

MATTHEWS, R. W. 1974. Biology of Braconidae. Ann. Rev. Entomol. 19: 15–32.

PORTER, C. 1980. Zoogeografía de las Ichneumonidae Latino-Americanas (Hymenoptera). Acta Zool. Lilloana 36: 1–52.

TOWNES, H., AND M. TOWNES, 1966. A catalogue and reclassification of the Neotropic Ichneumonidae. Amer. Entomol. Inst. Mem. 8: 1–367.

## Chalcidoid Wasps

Chalcidoidea.

This is an enormous assemblage of mostly parasitoid wasps, which occur worldwide and occupy almost every habitat. Despite their omnipresence, they are seldom noticed because of their small to minute size. Most have body lengths within the limits of 0.5 to 3 or 4 millimeters; in fact, the smallest insect known belongs to this group and is only 0.2 millimeters long (*Alaptus,* Mymaridae).

The adults are recognized not only by their minuteness but by the possession of elbowed antennae, a pronotum not extending to the small caps covering the anterior, outer corners of the thorax (tegulae), and wings almost without venation. Usually only the anterior, heavier wing veins are present, and sometimes all traces of nervature are absent. The wings are clear and at rest are held flat over the body. The bodies of the majority are black or dark, although many are deep metallic green or blue; some may have contrasting yellow or white patterns.

This category contains some twenty families and exhibits a wide variety of adaptations and habits. Most are endoparasitoids, attacking more host types, in different major taxonomic categories, than any other group of parasitoid insects. The spectrum includes spider eggs, ticks, aquatic beetles, ants, aculeate wasps, scale insects, lepidopterous eggs and larvae, and flies. Because of this characteristic, they are among the most useful biological control agents, and much of what has been learned about chalcidoid biology has come from studies relating to their use in this regard. The wasps' larva develops internally in the host, nearly always killing it. Hyperparasitoidism (parasitoidism of a parasitoid), superparasitoidism (host attacked by multiple parasitoids), and polyembryony (multiplication of parasitoid's eggs in host) are common phenomena exhibited by different parasitoid species.

A considerable number of chalcidoids are phytophagous, including one family, the Agaonidae, that develops in the fruit of *Ficus* trees and has evolved an intimate association with these plants, including an essential role in their pollination (see fig wasps, below). A few types produce galls or live in them as inquilines or guests of the true gall makers.

These insects are diverse (de Santis 1967) and difficult to identify, and very little is known of them in the Neotropics. De Santis (1971) recorded 1,581 species in 421 genera in continental South America. This certainly represents only a small fraction of the total fauna, literally thousands of species remaining to be discovered (some of which appear already in de Santis's later works [1979, 1981]).

Some are of economic importance, either as pests, for example, the eurytomid *Bruchophagus platyptera,* which destroys seeds of alfalfa and clover, or as beneficial, biological control agents. The number of the latter currently used within Latin America is only about thirty, all introduced from elsewhere, such as *Trichogramma minutum* (fig. 12.2d) to infect the egg stage of lepidopterous pests and *Aphelinus mali* (fig. 12.2c), which kills aphids. A great potential probably exists among the unknown forms in these applications.

As for the world generally, the largest families in the American tropics are Trichogrammatidae, Eupelmidae, Pteromalidae, and Chalcididae.

## References

DE SANTIS, L. 1967. Catálogo de los himenópteros Argentinos de la Serie Parasitica, incluyendo Bethyloidea. Prov. Buenos Aires, Comisión Investig. Cien., La Plata, Publ. Especial.

DE SANTIS, L. 1971. La fauna de chalcidoideos de América del Sur. Soc. Entomol. Peru Bol. 6: 57–63.

DE SANTIS, L. 1979. Catálogo de los himenópteros chalcidoideos de América al sur de los Estados Unidos. Prov. Buenos Aires, Comisión Investig. Cien., La Plata, Publ. Especial. Excludes Argentina and Brazil.

DE SANTIS, L. 1981. Catálogo de los himenópteros chalcidoideos de América al sur de los Estados Unidos. Primer supplemento. Rev. Peruana Entomol. 24: 1–38.

### *Fig Wasps*
Chalcidoidea, Agaonidae.

In the New World figs (*Ficus*), the female flowers are intermixed with the male flowers and are scattered over the internal

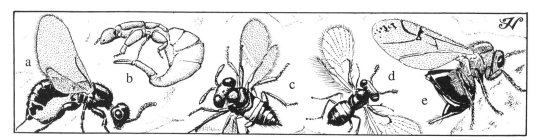

**Figure 12.2 CHALCIDOID WASPS.** (a) Fig wasp (*Blastophaga dugesi,* Agaonidae), female. (b) Fig wasp, male. (c) Parasitic chalcidoid (*Aphelinus mali,* Eulophidae). (d) Minute Egg Parasite (*Trichogramma minutum,* Trichogrammatidae). (e) Gall wasp (*Atrusca spinuli,* Cynipidae).

surface of the swollen, invaginated receptacle, called a synconium or fig fruit. The female flowers are of two types, "gall flowers" with short styles and "reproductive flowers" with long styles. The arrangement provides for a unique symbiotic relationship with many species of fig wasps, in the New World mostly with the genera *Blastophaga* and *Tetrapus* (Wiebes 1966), which are responsible for pollinating the plant as part of the process of rearing their young in the synconium. Wasps of the first genus carry the pollen on special setose areas (corbiculae, Ramírez 1978) on the body and legs; those of the second genus, in the digestive tract.

Fully winged female wasps (fig. 12.2a), after emerging from ripe figs, fly in search of new, young figs at the right stage for pollination. They enter the synconium through the aperture at its apex, forcing their way through the tight opening and in doing so lose their wings and (*Blastophaga* only) the apical six segments of their antennae. Once inside, the females use their long ovipositors to bore down through the styles of female flowers and deposit eggs among the immature ovules. As they do so, they inject a gall-producing substance as well. The ovipositor is long enough, however, only to reach the ovules of the short-styled flowers; long-styled flowers are merely probed. At the same time, both types of flowers are fertilized by the females with pollen from their bodies (Ramírez 1978), but only the long-styled flowers develop seeds. The short-styled flowers form galls in which the wasp larvae feed and mature.

Wingless male wasps (fig. 12.2b) develop first, in figs that are still unripe, thus avoiding being destroyed by fig-eating animals. They seek out mature female pupae and fertilize the females therein. Then they make tunnels through the wall of the synconium and escape. These events are followed by the emergence of impregnated females, which, also in leaving the fruit,

crawl over and pick up pollen from the male flowers that have now opened. The females escape the synconium through its natural opening with the aid of the males, which help bite through any tissues inhibiting their progress. The males then die, and the females go in search of new trees with receptive fruit to begin a new cycle.

Finally, the fig ripens, and its seeds are dispersed, usually by herbivorous animals. Thus, while accomplishing its own reproduction, the wasp ensures the life of the fig, which sacrifices only a portion of germinal tissue as food for the carriers of its pollen.

In most *Ficus*, virtually every synconium of an individual tree is pollinated the same day, or at least over a short period of up to three days. If the young figs are not pollinated during this period, even if wasps enter, they stop growing, shrink, and drop from the tree. The time of development for each species of wasp is correlated with the ripening time of the fruit, around a month. There is also considerable specificity between the species of wasp and fig (Ramírez 1970a, Wiebes 1979). Many more details of these relationships are available (Ramírez 1970b, 1976).

Other nonpollinating chalcidoids also inhabit and develop in figs, for example, the genus *Idarnes* (Torymidae) and various Eurytomidae. Their relations to agaonids and the figs have not been established (Gordh 1975). *Blastophaga* species of the New World are all placed in the subgenus *Pegoscapus*, which are characterized by having the pollen-carrying organs located on the front coxae.

### References

GORDH, G. 1975. The comparative external morphology and systematics of the Neotropical parasitic fig wasp genus *Idarnes* (Hymenoptera: Torymidae). Univ. Kansas Sci. Bull. 50: 389–455.

RAMÍREZ, W. 1970a. Host specificity of fig wasps (Agaonidae). Evolution 24: 680–691.

RAMÍREZ, W. 1970b. Taxonomic and biological

studies of Neotropical fig wasps (Hymenoptera: Agaonidae). Univ. Kansas Sci. Bull. 49: 1–44.

RAMÍREZ, W. 1976. Evolution of blastophagy. Brenesia 9: 1–13.

RAMÍREZ, W. 1978. Evolution of mechanisms to carry pollen in Agaonidae (Hymenoptera: Chalcidoidea). Tijd. Entomol. 121: 279–293.

WIEBES, J. T. 1966. Provisional check list of fig wasps (Hymenoptera, Chalcidoidea). Zool. Verh. 83: 1–44.

WIEBES, J. T. 1979. Co-evolution of figs and their insect pollinators. Ann. Rev. Ecol. Syst. 10: 1–12.

## GALL WASPS

Cynipidae, Cynipinae. *Spanish:* Avispas galígenas (General).

It is the galls, induced on plants by these wasps, that are usually noticed rather than the wasps themselves (Mani 1964). Each species of wasp causes a particular part of a particular plant to form characteristic galls. The latter are swollen masses of tissue in which the wasp's larvae feed and develop. The mechanism of gall formation is poorly understood but is thought mainly to involve redirected growth of undifferentiated plant tissues by substances in the larval saliva (Cornell 1983).

Galls are of all shapes, sizes, and locations on the host plant. Many are smooth, spherical, and singular, while others are irregular masses or swellings often with a coarse or even hairy surface. They usually grow from twigs or leaves, often along the midrib of the latter. The largest number of gall wasps use oaks as hosts and are therefore most numerous in the northern half of Latin America where these trees occur. Typical spherical galls (1.5–2 cm diameter) are the product of members of the widespread genus *Atrusca* (fig. 12.2e).

Not all gall wasps are phytophagous and produce their own galls; some live as inquilines in the galls made by other wasps, and some are parasitic. Larger galls often house a varied assemblage of other insects besides inquiline gall wasps, for example, chalcidoid wasps and parasitic Diptera (Shorthouse 1973).

Phytophagous gall wasps (Weld 1952) are small to minute (BL 1–4 mm) and black to pale reddish-brown and have filiform antennae, a pronotum with extensions laterally to touch the tegulae, and a shiny, large, oval, compressed abdomen whose second dorsal sclerite is greatly enlarged and covers over half of this body region.

Many gall wasps have a complex life cycle, with two very different generations in vastly different galls appearing at different times of the year. One generation may be parthenogenetic.

The study of galls and their makers is in a beginning phase in Latin America (Occhioni 1979). Few life cycles of indigenous species are known, and their unraveling will remain a fertile field for investigation for a long time to come. Some landmark studies were accomplished on the northern fauna by A. C. Kinsey (1930, 1936).

### References

CORNELL, B. 1983. Why and how wasps form galls: Cynipids as genetic engineers? Antenna 7(2): 53–58.

KINSEY, A. C. 1930 [1929]. The gall wasp genus *Cynips:* A study in the origin of species. Indiana Univ. Stud. 16: 1–577.

KINSEY, A. C. 1936. The origin of higher categories in *Cynips*. Indiana Univ. Publ. Sci. Ser. 4: 1–334.

MANI, M. S. 1964. Ecology of plant galls. Junk, The Hague.

OCCHIONI, P. 1979. "Galhas," "cecídeas" ou "tumores vegetais" em plantas nativas da flora do Brasil. Leandra 8–9: 5–35.

SHORTHOUSE, J. D. 1973. The insect community associated with rose galls of *Diplolepid polita* (Cynipidae, Hymenoptera). Quaest. Entomol. 9: 55–98.

WELD, L. H. 1952. Cynipoidea (Hym.) 1905–1950. L. Weld, Ann Arbor.

# WASP NESTS

Only certain solitary and social wasps construct true nests. Other wasps rear their young in a variety of plant or animal hosts.

Several authors (Richards 1978: 19–21, Jeanne 1975, Wilson 1971) have attempted to classify the different types of nests. Perhaps the most useful system is the following (simplified from Evans and West-Eberhard 1970):

I. A preexisting cavity, modified to suit the needs of the species (e.g., *Pepsis*, Pompilidae).

II. An elongate burrow dug in the ground, rotten wood, or pith (*Bembix*, Sphecidae).

III. Fabricated of foreign material and usually placed aboveground.
  A. Construction material primarily mud (some *Polybia*, Vespidae).
  B. Construction material wood pulp or other vegetable substance.
    1. Spherical cells in irregular clusters inside a ball of plant wool suspended by a filament (*Microstigmus*, Sphecidae).
    2. Tubular cells in a cluster or series on plant stems (*Parischnigaster*, Vespidae).
    3. Naked paper comb or combs suspended by a pedicel (*Polistes*, Vespidae).
    4. Paper combs surrounded by an envelope (most *Polybia*, Vespidae).

The large size, internal structure and shape of many social wasp nests make them true wonders of nature (Rau 1943). Some of the so-called ceramic types, composed of smoothly polished, colored mud (e.g., by *Polybia singularis*), are considered works of art and sold as such in curio shops in South America.

In the Neotropics, the greatest development of nest building is found in the vespid subfamily Polistinae. Generally, these wasps build rows of horizontal combs that they attach directly to a flat surface or suspend from overhanging objects with a central or lateral filament. They make the combs of relatively fragile, paperlike material, which they produce by chewing wood fibers and mixing them with salivary secretions ("wasp paper"). The combs are usually protected by an outer baglike envelope of much stronger material, either mud or a much heavier form of wasp paper. The latter may exceed the thickness and tenacity of high-grade, commercial cardboard in some nests, such as those of *Chartergus chartarius*. Both the inner combs and outer wall are made simultaneously so that the entire structure is finished at the same time. Some wasps may "add on" to a complete nest from time to time, a typical practice of *Synoeca* and *Polybia rejecta*. The same nest may be occupied year after year (perennial) or abandoned after a single reproductive season (annual). Examples of the former can persist as long as twenty-five years.

The wasps carefully form single or multiple openings in the outer envelope for passage in and out of the nest. Usually, there is only one such doorway at the bottom, but slitlike or double openings may be placed at the sides or outer corners by some species. The combs also are penetrated by passageways to allow access to all parts of the nest interior. The number of individuals per colony varies considerably, from a few to hundreds. It is not unusual for large nests of several years age to be home for several thousand adult wasps at a time. In general, the brood cells are reused for the development of several generations and must be cleaned of debris and excretions of prior larvae before receiving their new occupants (Jeanne 1980).

Most wasp nests are hung from branches high in trees to ensure protection from climbing mammalian or reptilian predators. The globular dark forms of these *nidos de avispas* are familiar sights to country folk. To youngsters in search of adventure, nests serve as attractive targets for rock throwing and other forms of general mischief. The danger involved in molesting a large colony is quickly appreciated as scores of stinging wasps soon descend on any would-be attacker and can inflict extremely painful and occasionally fatal retribution.

Certain Amazonian Indians actually invite attacks from wasps as part of rituals. Among the Kayapó of central Brazil, aggressive *Polybia* species are sought on regular occasions to take part in the reenactment of an ancient myth describing their fight with the giant rhinoceros beetle god. A scaffold is constructed which the warrior uses to reach the nests. With bare hands, the Indians strike the nests and receive the stings of wasps until they become unconscious from the pain and venom. The ceremony is important to the Kayapó as a statement of their place in the universe (Posey 1981).

Several kinds of Neotropical birds take advantage of the aggressiveness of wasps by placing their own nests in proximity to those of the insects (see above).

## References

EVANS, H. E., AND M. J. WEST-EBERHARD. 1970. The wasps. Univ. Michigan Press, Ann Arbor.
JEANNE, R. L. 1975. The adaptiveness of social wasp nest architecture. Quart. Rev. Biol. 50: 267–287.
JEANNE, R. L. 1980. Observações sobre limpeza e reutilização de células em ninhos de vespas sociais (Hymenoptera: Vespidae). Mus. Paraense Emílio Goeldi Bol. Zool. 101: 1–7.
POSEY, D. A. 1981. Wasps, warriors and fearless men: Ethnoentomology of the Kayapó Indians of central Brazil. J. Ethnobiol. 1: 165–174.
RAU, P. 1943. The nesting habits of Mexican social and solitary Vespidae. Entomol. Soc. Amer. Ann. 36: 515–536.
RICHARDS, O. W. 1978. The social wasps of the Americas, excluding the Vespinae. Brit. Mus. Nat. Hist., London.
WILSON, E. O. 1971. The insect societies. Belknap Press, Harvard Univ., Cambridge.

## SOLITARY WASPS

Solitary wasps are defined as species in which there is no cooperation involving division of labor between mother and daughters or between females of the same generation. Many are subsocial, living in nesting aggregations and building free nests of mud and plant materials. One genus of the otherwise totally solitary family Sphecidae (*Microstigmus*) is social.

Females of most form a burrow in the ground in which to rear their young, usually fed on insect prey paralyzed by her sting. Provisioning may be progressive, fresh food being brought to the developing larvae until they mature, or prey is provided in a single mass for the larvae to devour without further attention from the adults.

This assemblage is separated from the "parasitoid wasps" on the basis of their well-developed nesting habits and larvae that feed externally on the host (Evans 1966). It is comprised of generally larger wasps that are agile fliers and capable of stinging painfully. The sting is used to subdue and paralyze prey, in contrast to its purely defensive use in the social wasps. Groups of adults sleeping on plants are sometimes observed.

### Reference

EVANS, H. 1966. The behavior patterns of solitary wasps. Ann. Rev. Entomol. 11: 123–154.

### Cuckoo Wasps
Chrysididae.

The common name of this family comes from the habit many species have of entering the nests of their hosts, which are most often solitary wasps and bees, as they are being provisioned. (One anomalous subfamily—Amiseginae—of these wasps specializes on walkingsticks as prey.) In a manner analogous to that of their avian namesakes, the wasp larvae devour the

**Figure 12.3   SOLITARY WASPS.** (a) Cuckoo wasp (*Neochrysis carina*, Chrysididae). (b) Velvet ant (*Traumatomutilla indica*, Mutillidae). (c) Mammoth wasp (*Campsomeris ephippium*, Scoliidae). (d) Taruntula hawk (*Pepsis* sp., Pompilidae).

food left in the cell for the host's young, killing and eating the latter in the process.

These are among the most beautiful of insects because of their brilliantly colored bodies. The numerous species are bright, metallic purple, blue, or green and are sometimes mixtures of one or more of these colors in resplendent combinations, appreciated only with the aid of a strong magnifying lens. Most are small (BL rarely over 12 mm) and also recognized by a coarse body sculpture and ventrally concave abdomen, the latter consisting of only three or four visible segments. When the wasp is disturbed, it curls up into a ball, the head and thorax nesting snugly in the hollow of the abdomen, and remains immobile until the danger has passed.

The two largest genera in the Neotropics are *Trichrysis* and *Neochrysis* (fig. 12.3a). The family is not large here, only 111 species in 18 genera (Kimsey and Bohart 1980).

## Reference

KIMSEY, L. S., AND R. M. BOHART. 1980. A synopsis of the chrysidid genera of Neotropical America (Chrysidoidea, Hymenoptera). Psyche 87: 75–91.

## Velvet Ants

Mutillidae. *Spanish:* Perritos de Dios, hormigas terciopelas, arañas pus-pus (Argentina). *Portuguese:* Formigas feiticeiras, cachorrinhos de Nossa Senhora, formigas de onça, oncinhas (Brazil). *Quechua:* Sisi huakan ñahui (ant that makes you cry; Peru).

The females of these wasps are usually seen walking agilely on the ground or on logs and stumps, their brightly marked, hairy bodies attracting attention. The pubescence of most is dark velvety blue or black with contrasting spots of brilliant red, white, or yellow. Most are relatively small (BL 3–5 mm), but some Neotropical representatives, *Hoplomutilla, Leucospilomutilla,* and *Traumatomutilla* (fig. 12.3b), reach fairly great size, lacking wings and stinging painfully; they are sometimes mistakenly identified as large ants (e.g., *H. xanthocerata = folofilla* for *Paraponera clavata = folofa* in Panama; Méndez 1987). Species of *Hoplocrates* are also large and have outsized heads. Males are winged but less often observed. Both sexes are capable of producing a squeaking noise by moving the third abdominal segment in and out of the second, thus bringing stridulatory surfaces on each into contact.

All are solitary and develop as external parasitoids on the immatures of various wasps, bees, beetles, and flies. The excessively long stinger of the females enables them to pierce the nest cells of their hosts, into which they inject their venom and on which they place their eggs.

In some cultures, odd beliefs and practices have arisen surrounding these wasps.

Some people in Brazil employ them in love magic. A man, desiring the attentions of a woman, obtains three of her hairs, coats them with sweet syrup, and puts them in a box with a velvet ant. As the insect eats the hairs, the woman gradually falls in love with the perpetrator. Other powers of sorcery attributed to mutillids, common especially among the *caboclos* and Indians of Brazil, include the ability to cure a variety of illnesses and to enhance talents. For example, playing the violin can be improved by crossing the palm of the right hand with the insect on Friday (Lenko and Papavero 1979: 218).

The Neotropics is rich in species: just under 1,100 are now known (Fritz pers. comm.), and many more are certain to be discovered (Brothers 1975: 589–638, Schuster 1949). They are widely distributed as a group but especially abundant in warmer climates (Mickel 1952).

### References

BROTHERS, D. M. 1975. Phylogeny and classification of the Aculeate Hymenoptera, with special reference to Mutillidae. Univ. Kansas Sci. Bull. 50: 483–648.

LENKO, K., AND N. PAPAVERO. 1979. Insetos no folclor. Cons. Estad. Artes Ciên. Hum., São Paulo.

MÉNDEZ, E. 1987. Elementos de la fauna panameña. Priv. publ., Panama.

MICKEL, C. E. 1952. The Mutillidae (wasps) of British Guiana. Zoologica 37: 105–150.

SCHUSTER, R. M. 1949. Contributions toward a monograph of the Mutillidae of the Neotropical Region. III. A key to the subfamilies represented and descriptions of several new genera (Hymenoptera). Entomol. Americana 29: 59–140.

### Mammoth Wasps
Scoliidae, Campsomerinae,
  Campsomerini, *Campsomeris*.

These gigantic (BL males to 20 mm, females to 35 mm), dark wasps with black, hairy, robust bodies and shiny blue-black wings usually sport conspicuous, deep orange, transverse bands or spots at the base of the abdomen dorsally (fig. 12.3c) (Bradley 1945). The smaller males differ somewhat from the females also in having longer, straighter antennae and finer body hairs. Also, the legs of the female are shorter and adapted for digging, unlike the more delicate slender appendages of the male.

Although the several species (Bradley 1957) are fairly common over almost all of the American tropics, very little is known of their habits. There are several similar species in the genus, all of which are parasitoids of scarab beetle larvae. The latter may be so-called white grubs (*Phyllophaga*) or other large types that burrow in the ground (*Strategus, Oryctes*). The female wasp's powerful fossorial legs adapt it particularly well for digging into the ground after these hosts.

### References

BRADLEY, J. C. 1945. The Scoliidae (Hymenoptera) of northern South America, with especial reference to Venezuela. I. The genus *Campsomeris*. Bol. Entomol. Venezolana 4: 1–36.

BRADLEY, J. C. 1957. The taxa of *Campsomeris*. (Hymenoptera: Scoliidae) occurring in the New World. Amer. Entomol. Soc. Trans. 83: 65–77.

### Spider Wasps
Pompilidae.

This is a large and diverse family, mostly composed of medium-sized to very large (BL 1.5–5 cm) wasps with slender bodies and long, spiny legs (Evans 1953). Most are dark with pigmented wings, which they often flick nervously when walking on the ground.

Females seek spiders, which they paralyze with a sting and pack into subterranean cells or in existing cavities in wood to provide food for the developing young. Some construct elevated mud nests. A single prey is provided for each cell. The female uses the tip of the abdomen to tamp down the earth when closing the cell or to mold mud.

## Reference

Evans, H. E. 1953. Comparative ethology and the systematics of spider wasps. Syst. Zool. 2: 155–172.

### *Tarantula Hawks*

Pompilidae, *Pepsis. Spanish:* San Jorges, avispones, matacaballos (Argentina).

It is curious that these conspicuous spider wasps do not carry more vernacular names among the people of the region. Their gigantic size, impressive steely blue bodies and the bright orange wings of many species often attract attention and immediately identify them. They are commonly seen taking nectar from flowers (often milkweed) and can be heard making a loud buzzing sound while in flight.

Females of the largest species have body lengths of over 4.5 centimeters; males are a little smaller, 2.5 to 3.5 centimeters (fig. 12.3d). Species have one of two wing colors: bright or burnt orange or a dark, smoky hue.

The large size of these creatures adapts them to their prey, the great mygalomorph spiders, or tarantulas. Females hunt these spiders and engage them in battles that have been witnessed by naturalists and described in many photographic essays (Petrunkevitch 1926). They approach their arachnid adversary with wings raised, then grasp it by a leg with the mandibles. The spider retaliates, attempting to bite and kill the wasp, in which it is sometimes successful. But almost always, the wasp wins the foray by stinging the hapless tarantula on the underside. She then drags its paralyzed body to a burrow, often the spider's own, and buries it there after depositing a single egg on it. The larva feeds externally on the interred carcass, finally transforming into an adult within a silken cocoon spun nearby in the burrow. Wasps excited by battle or threat emit a pungent odor whose function is unknown.

Although not well known, *Pepsis* appear to be very specific with regard to prey species and are adept at discriminating them, apparently by odor clues picked up as the wasp taps the spider with its antennae.

Members of this genus are restricted to the New World, in the tropical portions of which there are several hundred species (Hurd 1952). The group seems centered in Amazonia where most of the 300 species of northern South America are found. Many species also live in Middle America, including the West Indies (Alayo 1954), and occupy all habitats from sea level to 4,000 meters in the Andes.

Several disparate insect types have evolved color patterns and behaviors mimicking tarantula hawks. The resemblances are very convincing and incredible in the way unwasplike portions of the mimic's body are molded and colored to resemble the wasp's. Such mimics are found among certain reduviid bugs (*Spiniger ater*), katydids (*Scaphura, Aganacris*), mydid flies (*Mydas rubidapex*), and arctiid moths (*Macrocneme*).

## References

Alayo, P. 1954. El género *Pepsis* Fabr. en Cuba (Hymenoptera-Pompilidae). Univ. Oriente Cuadernos 37: 1–25.

Hurd, P. 1952. Revision of the nearctic species of the pompilid genus *Pepsis* (Hymenoptera, Pompilidae). Amer. Mus. Nat. Hist. Bull. 98: 257–334.

Petrunkevitch, A. 1926. Tarantula versus tarantula-hawk: A study in instinct. J. Exper. Zool. 45: 367–397.

## Digger Wasps

Sphecidae.

Digger wasps comprise a large family (Bohart and Menke 1976) with varied form and habits. They are distinguished by the shape of the lateral portion of the pronotum (dorsal sclerite of prothorax), which is formed into a rounded lobe, well

**Figure 12.4  DIGGER WASPS (SPHECIDAE).** (a) Digger wasp (*Trypoxylon albitarsi*). (b) Mud dauber (*Sceliphron assimile*). (c) Sand wasp (*Bembix citripes*). (d) Social sphecid (*Microstigmus comes*).

separated from the base of the wing. The wings are never folded longitudinally when at rest. All are solitary, with the exception of the *Microstigmus* (see social sphecids, below), making their nest in many situations and provisioning them with various kinds of insects, such as plant bugs, spiders, grasshoppers, and caterpillars (Fritz and Genise 1980). Among the ground and mud nest makers, females apply their head to shaping and manipulating earth. The sting is used to subdue prey, not just for defense, as in the social wasps.

In Brazil, the mud nest making species of *Trypoxylon* (fig. 12.4a) are regarded with superstition by the caboclos and are known by many local names: *minguita, nhá fina, mariambola.* Their nest material is applied widely in folk therapeutics, uses ranging from aphrodisiacs to cures for constipation, spider bites, and burns. It is a very large genus (Richards 1934). An interesting feature of many species is the peculiar nest-guarding behavior of males in the absense of the females (Coville and Griswold 1984). The male sits just inside the nest with its head protruding from the entrance.

**References**

Bohart, R. M., and A. S. Menke. 1976. Sphecid wasps of the world. Univ. California Press, Berkeley and Los Angeles.

Coville, R. E., and C. Griswold. 1984. Biology of *Trypoxylon* (*Trypargilum*) *superbum* (Hymenoptera: Sphecidae), a spider-hunting wasp with extended guarding of the brood by males. Kans. Entomol. Soc. J. 57: 365–376.

Fritz, M., and J. Genise. 1980. Nido de barro Sphecidae, constructores, inquilinos, parasitoides, cleptoparásitos, detritívoros. Soc. Entomol. Argentina Rev. 39: 67–81.

Richards, O. W. 1934. The American species of the genus *Trypoxylon* (Hymenopt., Sphecoidea). Royal Entomol. Soc. London Trans. 123: 173–362.

### Mud Daubers

Sphecidae, Sphecinae, Sceliphronini, *Sceliphron. Spanish:* Celifrónes (General).

Just six of the thirty worldwide species in this genus live in the New World tropics (van der Vecht and van Breugel 1968), but these are conspicuous both for their commonness and for the nests they often build on human habitations (Shafer 1949).

The adults are readily recognized by their extra-long waist and color pattern of sharply contrasting designs of black and yellow. The sides of the rear of the thorax (propodeum) are built up also as ridges to form a U-shaped enclosure.

Nests are constructed of mud and consist of a series of parallel tubes or elongated cells. Each cell is mass provisioned with spiders. One species, *Sceliphron assimile* (fig. 12.4b), thrives on civilization and often builds its nest on houses or other man-made structures. This species occurs through the West Indies, Mexico, and Central America (Freeman 1973, Freeman and Johnston 1978); its place is assumed by *S. asiaticum* and *S. fistularium* in South America. Species occupy islands, including some in the remote Pacific (Coco Island), where they have presumably been introduced inadvertently by human action.

## References

FREEMAN, B. E. 1973. Preliminary studies on the population dynamics of *Sceliphron assimile* Dahlbom (Hymenoptera: Sphecidae) in Jamaica. J. Anim. Ecol. 42: 173–182.

FREEMAN, B. E., AND B. JOHNSTON. 1978. The biology in Jamaica of the adults of the sphecid wasp *Sceliphron assimile* Dahlbom. Ecol. Entomol. 3: 39–52.

SHAFER, G. 1949. The ways of a mud dauber. Stanford Univ., Stanford.

VAN DER VECHT, J., AND F. M. A. VAN BREUGEL. 1968. Revision of the nominate subgenus *Sceliphron* Latreille (Hymenoptera, Sphecidae) (Studies on the Sceliphronini, Part I). Tijd. Entomol. 111: 185–255.

### Sand Wasps
Sphecidae, Nyssoninae, Bembicini, *Bembix*. *Spanish:* Insectos policias (General). Cowfly tigers, horseguards.

The sand wasps are a familiar sight on beaches or sand dunes cruising and darting about in search of prey or busily excavating nests (Evans 1957). They are medium to large (BL 12–17 mm), stout-bodied wasps, basically black but usually elaborately marked with undulating white or yellow bands on the abdomen. They are also recognized by the reduced simple eyes (ocelli), the anteriormost of which may be vestigial, and elongate mouthparts. The anterior legs normally are fringed with long hairs useful to the wasp in digging in soft sand.

Nests are sloping tunnels, a few to several centimeters in depth, in sandy soil. At the terminus of the entry tunnel is a horizontal branch in which the young are reared and a final vertical spur where the females may rest or take refuge. They provision the nests progressively with various insects, most often noxious flies, including horseflies (frequently plucking them right off horses and cattle), Syrphidae, Muscidae, and other large species. Although they subdue their prey with a potent stinger, they are not prone to sting humans (Cane and Miyamoto 1979).

These are truly solitary wasps, but they occasionally engage in mass attacks against intruders of "colonies" (where many individuals are nesting in proximity). Males also take part in "sun dances," or flight rituals wherein females are met in the air and aggressively brought to bay for copulation.

Only a couple of dozen species are found in the Neotropics, either in inland areas of sandy soil or more commonly on coastal beaches. The most typical and widespread are *Bembix americana* in the Caribbean and *B. citripes* (fig. 12.4c) in South America.

## References

CANE, J. H., AND M. M. MIYAMOTO. 1979. Nest defense and foraging ethology of a Neotropical sand wasp, *Bembix multipicta* (Hymenoptera: Sphecidae). Kans. Entomol. Soc. J. 52: 667–672.

EVANS, H. E. 1957. Studies on the comparative ethology of digger wasps of the genus *Bembix*. Comstock, Ithaca.

### Social Sphecids
Sphecidae, Pemphredoninae, *Microstigmus*.

This genus of nearly fifty species (West-Eberhard 1977), known from Costa Rica to Paraguay, is considered to be the only genus of the family with some species having true social habits. Sociality was discovered recently in one species (*Microstigmus comes,* fig. 12.4d; Matthews 1968*a,* 1968*b*) and subsequently verified in many others in the genus. Other species are solitary and practice progressive provisioning.

The small (0.5–1.5 cm), unique nests

(pl. 4b) are baglike and constructed of the waxy material coating the undersides of fan-palm leaves (in *M. comes*, see Matthews and Starr 1984). This material, or wood flecks, moss, and lichens in other species, is bound together by cooperating females with silk produced from abdominal glands. Nests are usually suspended from the undersides of leaves but sometimes hang from other inclined objects (rocks, logs, eaves of dwellings) by a slender coiled or straight pedicel. The upper portion of the nest is hollow and acts as a chamber for the adult male and female wasp occupants. The lower portion consists of a few to several cells in which the brood is reared. These cells are mass provisioned with very small insects, normally springtails, thrips, or leafhoppers (Cicadellidae) by females working in cooperation. The prey type and nest shape are species specific. One female may show reproductive dominance and, although structurally indistinct from the others, may be considered a kind of "queen." Foraging and other activities are diurnal.

The wasps themselves are minute (BL 2–5 mm) and usually reddish or pale yellowish-brown and narrow waisted. In life, the yellow species have bright green eyes. The lower tooth of the mandible is often very long.

## References

MATTHEWS, R. W. 1968a. *Microstigmus comes:* Sociality in a sphecid wasp. Science 160: 787–788.

MATTHEWS, R. W. 1968b. Nesting biology of the social wasp *Microstigmus comes* (Hymenoptera: Sphecidae, Pemphredoninae). Psyche 75: 23–45.

MATTHEWS, R. W., AND C. K. STARR. 1984. *Microstigmus comes* wasps have a method of nest construction unique among social insects. Biotropica 16: 55–58.

WEST-EBERHARD, M. J. 1977. Morphology and behavior in the taxonomy of *Microstigmus* wasps. 8th Int. Cong. Int. Union Stud. Soc. Ins. Proc. Wageningen, the Netherlands. Pp. 123–125.

## Caterpillar Hunters
Eumenidae.

These are builders of a greater variety of nests than any other group of solitary wasps, but they almost always employ mud in some way. Females may choose a simple hollowed twig or burrow into clay or compact soil banks. More familiar are the so-called mason wasps, which construct free mud nests. Those of the "potter wasps" may take the shape of a well-formed jug (spherical with flared spout) placed atop a twig. A few actually camouflage their mud cells with chewed wood fibers or build with macerated leaves, plant fibers, and resins, a practice thought to represent a step in the direction of paper nest building, as performed by their higher social relatives.

The nests are usually supplied with caterpillars, although beetle larvae are sometimes used. The female may provision singly or return repeatedly to provide her developing larvae with a continuous supply of food (progressive provisioning).

Caterpillar hunters are like their close relatives, the social wasps, in general appearance, all having the wings folded longitudinally at rest and some even with narrowed waists. They differ, however, in possessing long mandibles, which often cross when closed, one apical spur on the middle tibia, bifurcate tarsal claws, and, of course, lacking the social habit.

### Potter Wasps
Eumenidae, Eumeninae, *Eumenes*.
*Portuguese:* Mariambolas (Brazil).

The quaint little (1–2 cm diameter) mud jug nests of these wasps are unique. They are spherical, with the edges of the opening (in the center or at one end) on a slight projection and flared like the spout of an old-fashioned, clay water pot. They are constructed by all the hundred or so species in this genus, the most common and widespread being *E. consobrinus* (fig. 12.5a).

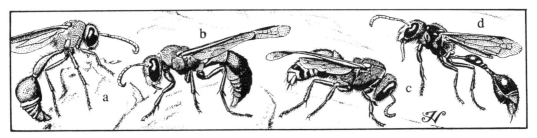

**Figure 12.5    CATERPILLAR HUNTING WASPS (EUMENIDAE).** (a) Potter wasp (*Eumenes consobrinus*). (b) Montezuma wasp (*Montezumia azurescens*). (c) Wanderer (*Pachodynerus nasidens*). (d) Zethus wasp (*Zethus matzicatzin*).

Their taxonomy is complex and unsettled (Soika 1978).

These are medium-sized (BL 10–15 mm), solitary wasps with long mandibles and elongate waists like zethus wasps (see below). They are distinguished from the latter, however, by the strongly convex thorax and by the swollen part of the waist segment near the joint with the abdomen (rather than near the middle). Neither is the basal portion of the abdomen extended as it is in their near relatives.

The nests are placed, often evenly spaced in linear series, sometimes in clusters, along the upper sides of slender twigs. They are provisioned with caterpillars, and each is the nursery for a single larva. The latter hatches from an egg suspended from the roof of the nest on a threadlike stalk.

### Reference

Soika, A. G. 1978. Revisione degli Eumenidi Neotropicali appartenenti ai generi *Eumenes* Latr., *Omicron* (Sauss.), *Pararaphidoglossa* Schulth. et affini. Mus. Civ. Storia Nat. Venezia Bol. 29: 1–420. (The several new genera proposed are considered synonymous with *Eumenes* by most other authorities.)

### *Montezuma Wasps*

Eumenidae, Eumeninae, *Montezumia*.

Members of this genus (fig. 12.5b) of solitary mason wasps superficially resemble common *Polistes* paper wasps in their black and yellow and other maculate color patterns and fairly large size (BL 2 cm). They are distinguished by family characteristics and by the combination of a completely sessile abdomen, a thickening in the veins immediately before the spot (pterostigma) on the outer leading edge of the fore wing, and the peculiar shape of the small cap (tegula) overlying the base of the wing, this having a posterior lobe or extension that curves inward.

Not much is known of their habits, save that nests seem to be made both in the ground and banks of soil as well as in the form of free mud structures affixed to hard substrates, depending on the species. Free nests contain only a few cells, generally arranged in parallel fashion in a single layer, much as in the nests of mud daubers but much more irregularly and roughly shaped. Females apparently provision these cells progressively (Evans 1973).

The genus is entirely American and contains fifty-two species mostly living in warm lowland climes (Willink 1982).

### References

Evans, H. E. 1973. Notes on the nests of *Montezumia* (Hymenoptera, Eumenidae). Entomol. News 84: 285–290.

Willink, A. 1982. Revisión de los géneros *Montezumia* Saussure y *Monobia* Saussure (Hymenoptera: Eumenidae). Acad. Nac. Cien. (Argentina) Bol. 55: 1–321.

## Wanderer

Eumenidae, Eumeninae, *Pachodynerus nasidens.*

This mason wasp is named for its penchant for turning up in out-of-the-way places. It is almost continuously distributed over Mexico, Central America, and northwestern South America but also is found on the Antilles, on Coco Island, and across the Pacific in Hawaii and Micronesia. The very similar *Pachyodynerus galapagensis* occurs in the Galápagos Archipelago. The wanderer so closely resembles the honey wasp (*Brachygastra lecheguana*) that they are believed to be Müllerian mimics (Snelling pers. comm.).

It is a small species (BL 8–14 mm) and recognizable by its abdominal color pattern, black with yellow bands completely ringing the hind borders of each segment, except the first (fig. 12.5c). The thorax is black with a thick, rear, marginal yellow border and line of yellow arching across the front. The thorax also bears a dense vestiture of short golden hairs. A unique trait identifying the males in the genus to which it belongs is the greatly reduced terminal antennal segments, which are tiny lobes residing in a pit on the tip of the organ.

Some has been written about the species' biology (Freeman and Jayasingh 1975, Jayasingh and Taffe 1982). Its mud nests of a few cells are mostly found on man-made objects, cracks in windows, keyholes, electrical sockets, wall recesses, and the like. The wasps must seek natural holes and cavities such as hollow twigs in the wild, but such nests have not yet been observed. The only well-known natural sites are cells in abandoned mud dauber (*Sceliphron*) and mason wasp (*Zeta*) nests.

The female wasps stock their brood cells with caterpillars of various species, showing little specificity as to type. They lay their eggs in the cells before provisioning them with four to eighty prey pieces. Each egg is suspended from the food or side of the cell by a fine thread.

## References

FREEMAN, B. E., AND D. B. JAYASINGH. 1975. Population dynamics of *Pachodynerus nasidens* (Hymenoptera) in Jamaica. Oikos 26: 86–91.

JAYASINGH, D. B., AND C. A. TAFFE. 1982. The biology of the eumenid mud-wasp *Pachodynerus nasidens* in trap nests. Ecol. Entomol. 7: 283–289.

## Zethus Wasps

Eumenidae, Discoeliinae, *Zethus.*

The 195 species of solitary wasps in this genus (fig. 12.5d) are typically Neotropical, occurring throughout almost all of the region with the exception of the Lesser Antilles, western Andes, and coastal deserts (one species is known from the Chilean desert) (Bohart and Stange 1965).

The genus is of great interest ecologically, as it appears to bridge the gap between social and solitary wasps. Two distinct types of nesting behavior are found: utilization of old insect burrows in twigs, wood, or the ground by the less specialized species, and construction of elevated nests from masticated and salivated plant material by the more advanced species. Females practice both progressive and mass provisioning, with caterpillars always forming the preferred food source.

Adults are usually seen frequenting flowering shrubs, often *Acacia* and *Mimosa*. They are quick, nervous fliers, darting about in search of prey. They are most easily recognized by their rather long waists (the swollen portion of which is near the middle rather than the apex, as in potter wasps). The abdomen often has an additional constricted segment next to the waist, making the latter appear even longer.

The terminal abdomen segments are frequently telescoped into the basal segments. The mandibles are somewhat short for eumenids but are still significantly longer than those of their vespid relatives. Color patterns are typically vespoid, black and yellow or other light-colored bands and thoracic patches on a dark background.

### Reference

BOHART, R. M., AND L. A. STANGE. 1965. A revision of the genus *Zethus* Fabricius in the Western Hemisphere (Hymenoptera: Eumenidae). Univ. Calif. Publ. Entomol. 40: 1–208.

## SOCIAL WASPS

Vespidae.

True social wasps, those that exhibit cooperative brood care by specialized castes and overlap of generations (Jeanne 1980), belong only to the family Vespidae, also called paper wasps (Bequaert 1944; Dias Filho 1975; Richards 1978; Richards and Richards 1951; West-Eberhard 1975). (The genus *Microstigmus* of the family Sphecidae is also social; see social sphecids, above.)

In most social wasps, one egg-laying female is dominant, and she is dubbed the "queen"; all other females usually have reduced ovarian function. The nonreproductive females forage, care for the brood, and build the nest. These are the "workers," and they may exhibit some anatomical differences from the queen, usually smaller overall size. There may be multiple functional reproductives (polygyny), however, sometimes very many, as in the honey wasps, with nothing to distinguish any dominant individual.

All make nests, which are often large and elaborate (Jeanne 1975). They are fabricated from plant fibers, usually chewed wood, but a few form mud houses or use hairs, floss, or vegetable wool of one kind or another for construction material.

Colonies are founded in three ways: in the Polistinae, independently by one or several queens without the aid of workers from the parent colony or by one (usually) or several queens accompanied by swarms of workers, and in the Vespinae, by single queens alone (Jeanne 1980).

Although ants and wasps are usually mortal enemies, certain wasps benefit by nesting in trees within ant territory, even on ant plants (Herre et al. 1986). They manage to evade predation directly by the ants and find themselves freer of raids by their other enemies through the ant's presence. Associations of nests of one (Jeanne 1978) or more (Windsor 1972) wasp species are also known, an arrangement that probably enhances the defense effectiveness of any one colony against vertebrate predators (ibid.). Birds nest also, but not always highly successfully, in the near vicinity of social wasps and likewise share in their natural protection (Windsor 1976). Social wasps protect their nests from ant attack in a variety of ways: by direct attack, application of deterring sticky substances, temporary nest abandonment, and even by effacing their own trail pheromones (West-Eberhard 1989).

Social wasps take a variety of insect prey. They do not sting the prey but pounce on it, cut it up with their mandibles, and chew it into a paste on which they and their larvae feed. The sting is used solely as a protective defense against vertebrate predators, and for that reason, its venom contains substances to make it especially painful (Edery et al. 1978).

Because of the social structure and nest-building abilities of these wasps, they hold religious significance for some Indian groups. The Kayapó of Brazil give them status as totem animals and respect them as fellow inhabitants of their lands. They also play a role in the tribe's pharmacology. Stings are considered to be beneficial, even if painful, and are recommended in the treatment of bone diseases (Overal 1984).

### References

BEQUAERT, J. C. 1944. The social Vespidae of the Guianas, particularly of British Guiana.

Mus. Comp. Zool. (Harvard Univ.) Bull. 94: 249–304.

DIAS FILHO, M. M. 1975. Contribuição à morfologia de larvas de vespídeos sociais do Brasil (Hymenoptera, Vespidae). Rev. Brasil. Entomol. 19: 1–36.

EDERY, H., J. ISHAY, S. GITTER, AND H. JOSHUA. 1978. Venoms of Vespidae. In S. Bettini, ed., Arthropod venoms. Springer, Berlin. Pp. 691–771.

HERRE, E. A., D. M. WINDSOR, AND R. B. FOSTER. 1986. Nesting associations of wasps and ants on lowland Peruvian ant-plants. Psyche 93: 321–330.

JEANNE, R. L. 1975. The adaptiveness of social wasp nest architecture. Quart. Rev. Biol. 50: 267–287.

JEANNE, R. L. 1978. Intraspecific nesting associations in the Neotropical social wasp *Polybia rejecta* (Hymenoptera: Vespidae). Biotropica 19: 234–235.

JEANNE, R. L. 1980. Evolution of social behavior in the Vespidae. Ann. Rev. Entomol. 25: 371–396.

OVERAL, W. L. 1984. Wasp studies among the Kayapó Indians of Brazil. Sphecos 8: 19–22.

RICHARDS, O. W. 1978. The social wasps of the Americas excluding Vespinae. Brit. Mus. Nat. Hist., London.

RICHARDS, O. W., AND M. J. RICHARDS. 1951. Observations of the social wasps of South America (Hymenoptera: Vespidae). Royal Entomol. Soc. London Trans. 102: 1–170.

WEST-EBERHARD, M. J. 1975. Estudios de las avispas sociales (Himenoptera, Vespidae) del Valle del Cauca. Cespedesia 4: 245–267.

WEST-EBERHARD, M. J. 1989. Scent-trail diversion, a novel defense against ants by tropical social wasps. Biotropica 21: 280–281.

WINDSOR, D. M. 1972. Nesting association between two Neotropical polybiine wasps (Hymenoptera: Vespidae). Biotropica 4: 1–3.

WINDSOR, D. M. 1976. Birds as predators on the brood of *Polybia* wasps (Hymenoptera: Vespidae: Polistinae) in a Costa Rican deciduous forest. Biotropica 8: 111–116.

## Polistes Paper Wasps

Vespidae, Polistinae, Polistini, *Polistes*.

This genus (and tribe) is immediately separated from the similar Polybiini by the shape of the slit (propodeal orifice, muscle slit) at the rear of the thorax (actually the first abdominal segment fused to the abdomen) through which the major muscle passes to connect with the first waist segment. In the Polybiini, the orifice is broadly rounded, never more than twice as long as broad; in this group, it is narrow, much longer than wide. These wasps otherwise resemble most social vespids in their yellow and black color patterns, moderately long waist, and plaited wings. The cheeks by the eyes tend to be fairly wider and more discretely margined than in other vespids, however.

The nest is a single exposed comb without an envelope and supported by a single (sometimes multiple or branched) stalk. Colonies are small, rarely consisting of more than two hundred adults. They are usually founded by a single female, but often several potentially fecund females may act as co-founders. In the latter case, after the new nest is established, a dominant female assumes the queen's function by a behavioral competition with the other females. Several may lay eggs, but these are eaten by competitors until one consumes all the others at a faster rate; this female then becomes the queen. This method of queen determination is the most common in the genus, but in some species (such as the varied paper wasp, see below), open direct struggle decides the outcome (Eberhard 1969).

The genus is fairly large and cosmopolitan in distribution. In the New World tropics, there are some 80 species arranged in 5 subgenera. They range widely in all habitats throughout the region, including the West Indies and some of the more isolated islands (*P. dorsalis clarionensis* in the Revillagigedo Islands; *P. fuscatus* on Ascensión, Bermuda, and Cape Verde islands).

*P. carnifex* is the largest member of the genus (BL to 3 cm) (Corn 1972), and *P. canadensis* (see below) is probably the most common. Both are found throughout the Neotropical Region.

A general review of the group is given by West-Eberhard (1983). *P. erythrocephalus* biology is discussed by Nelson (1971).

## References

Corn, M. L. 1972. Notes on the biology of *Polistes carnifex* (Hymenoptera, Vespidae) in Costa Rica and Colombia. Psyche 79: 150–157.

Eberhard, M. J. 1969. The social biology of polistine wasps. Mus. Zool., Univ. Michigan, Misc. Publ. 140: 1–101.

Nelson, J. M. 1971. Nesting habits and nest symbionts of *Polistes erythrocephalus* Latreille (Hymenoptera: Vespidae) in Costa Rica. Rev. Biol. Trop. 18: 89–98.

West-Eberhard, M. J. 1983. *Polistes* (quita calzón, lengua de vaca [name of nest], paper wasp). *In* D. H. Janzen, ed., Costa Rican natural history. Univ. Chicago Press, Chicago. Pp. 758–760.

## Varied Paper Wasp

Vespidae, Polistinae, Polistini, *Polistes canadensis. Spanish:* Chia (Costa Rica), panelera, pulate (Peru). *Portuguese:* Cavapita, caboclo (Brazil). Jack Spaniard (Trinidad).

Populations of this wasp vary greatly in color pattern from place to place, and accordingly, many varieties and subspecies have been named (West-Eberhard 1969). Most are of solid color, from nearly black to reddish-brown (described by Bequaert [1943] as the "dirty color of dried blood"). Different amounts of yellow and light brown intrude on these solid backgrounds in other forms producing blotched patterns on the thorax or abdominal bands. The status of these entities is still in question, although some are now thought to represent separate species. All these pattern variations occur on structurally consistent wasps of medium size (BL 17–25 mm), with a long, slender abdomen with a pointed tip and constriction about the sixth segment (fig. 12.6a).

This is the most widely distributed *Polistes* species among the approximately fifty in the New World, found from Mexico (not Canada, despite its name) to Patagonia (although the genus is present on only a few of the Lesser Antilles). It lives in all habitats, except the driest.

Its nest is a single, exposed comb, usually situated on a vertical or strongly inclined surface so that it extends downward from the pedicel asymmetrically and parallel to the substratum (Rau 1943). Sites include man-made structures, if available, but in the wild are normally large limbs, hollows in trees, and the undersides of palm fronds. In outline, the nest is an elongate, oval structure with numerous cells, reaching a length of 20 to 30 centimeters and a width of 10 to 11 centimeters. The larvae line their cells with silk, and the adults cover the pedicel and roof of the nest with a black tarry substance that waterproofs and strengthens it.

There is a single queen who initiates the nest but who is joined by other females soon after the building begins. Queen dominance is determined by fighting, a rare phenomenon in the genus. Males tend

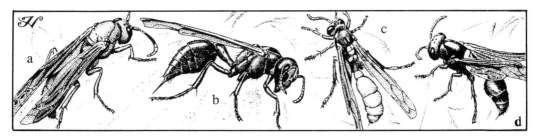

**Figure 12.6  SOCIAL WASPS (VESPIDAE).** (a) Varied paper wasp (*Polistes canadensis*). (b) Drumming wasp (*Synoeca surinama*). (c) Parasol wasp (*Apoica pallens*). (d) Polybia wasp (*Polybia scutellaris*).

to stay on the comb after emergence much longer than in most paper wasps. The workers are very protective of the colony and quick to attack anyone or anything coming too close and sting forcefully (Meneses 1969). They spend most of their time, however, seeking nectar, fruit juices, and caterpillars to feed to the larvae.

In Amazonia, an unnamed tineid moth infests the combs. Its larvae scavenge the liquid body wastes that accumulate in the intestines of developing adults and are ejected when they emerge (meconia) and occasionally burrow into cells with wasp pupae and destroy them (Jeanne 1979).

### References

BEQUAERT, J. 1943. Color variation in the American social wasp *Polistes canadensis* (Linnaeus), with descriptions of two new forms (Hymenoptera, Vespidae). Bol. Entomol. Venezolana 2: 107–124.

JEANNE, R. L. 1979. Construction and utilization of multiple combs in *Polistes canadensis* in relation to the biology of a predaceous moth. Behav. Ecol. Sociobiol. 4: 293–310.

MENESES, B. 1969. Aspectos farmacológicos de la ponzoña de *Polistes canadensis*. II. Simp. Foro Biol. Trop. Amazônica. Pp. 399–413.

RAU, P. 1943. Nesting habits of Mexican social and solitary Vespidae. Entomol. Soc. Amer. Ann. 36: 515–536.

WEST-EBERHARD, M. J. 1969. The social biology of polistine wasps. Mus. Zool. Univ. Mich. Misc. Publ. 140: 1–101.

### Drumming Wasps

Vespidae, Polistinae, Polybiini, *Synoeca*.
   *Spanish:* Carachupa avispas (Peru), guitarrera (*S. septentrionalis,* Costa Rica). *Portuguese:* Marimbondo tatu, caba tatu (*S. cyanea*).

The nests of these species are familiar objects and have given the wasps their common name in Brazil, *marimbondo tatu,* meaning "armadillo" wasp, from the fancied resemblance of the envelope to the armature of that mammal. The combs are attached directly to the substratum, usually an inclined tree trunk, and are covered with an elongate, half oval outer envelope. This is ribbed transversely and has a small access hole at the upper end. New combs and envelopes are added at the end from time to time, the boundaries of which remain plainly visible.

The wasps themselves are formidable protectors of their domiciles. They have a potent venom dispatched with a barbed stinger that remains imbedded as the wasp pulls away. They are not prone to immediate attack, however. Instead, they display a warning ritual, which may be related to the fact that their sting is barbed and only used under extreme duress since it must be sacrificed along with the worker's life in a consummate attack. When disturbed, the wasps produce a rhythmic drumming sound by vibrating the envelope from within. The sound may be heard several meters away. If further aroused, hundreds of individuals rush onto the nest's outer surface, where they continue to produce thumping sounds and raise and lower their wings in synchrony. The threat is sufficient to scare off all but the most insistent or naive assailants.

These are large (BL 20 mm), shiny blue wasps (fig. 12.6b), with a short waist and a constricted segment at the base of the abdomen. The latter is heart-shaped, with a sharp apex; the apical segments are also somewhat compressed.

Drumming wasps belong to that group of social wasps that found new colonies by swarming. A mass of workers accompanies one queen (sometimes several) to the new site. One egg-laying female always dominates, repressing the others with inhibitory pheromones. Queens are replaced in tandem as they expire in their perennial, long-lived colonies (one known to be 16 years of age).

The genus *Synoeca* is restricted to the New World. The five similar species are found only in tropical, forested areas.

## Parasol Wasps

Vespidae, Polistinae, Polybiini, *Apoica*.
  *Portuguese:* Beiju-cabas, marimbondos
  de chapéu, cabas de ladrão (Brazil).

The shape of its nest gives this genus its common name. The structure is an umbrella-shaped, single, open paper comb with a rudimentary envelope partially covering the upper surface. It is circular in outline and measures up to 29 centimeters in diameter. It is constructed of tightly packed branched plant hairs, giving it a feltlike texture.

During the day, the colony may be seen sleeping on the nest, each wasp side by side, forming a tightly packed mass and covering the entire lower surface (pl. 4f). Peripheral females align themselves, facing outward, eyes and antennae forming a defense perimeter. At night, the workers are active in foraging and nest building and are frequently attracted to artificial light. Although nocturnal, *Apoica* will vehemently defend their nest in the daytime.

Nocturnal habits are correlated with the pale colors of these wasps. In the most common species, *A. pallens* (fig. 12.6c), the abdomen is usually a light cream or yellow hue, the thorax buff with pale spots on the shoulders and hind portion. The body color in other species is light red to brown. The genus stands apart from other vespids in having enlarged ocelli and compound eyes that nearly meet on top of the head. The entire body is long and slender and the abdomen roughly sausage-shaped with an abruptly narrowed, short waist.

This genus of just six species belongs to that group of social wasps in which colonies are founded by a swarm of workers accompanying multiple queens to the new nest site. Males seem to fly in the swarm as well, indicating an unusual and not yet understood reproductive pattern.

The general biology of the genus is discussed by van der Vecht (1973) and Schremmer (1972).

### References

SCHREMMER, F. 1972. Beobachtungen zur Biologie von *Apoica pallida* (Oliver, 1791), einer neotropischen sozialen Faltenwespe (Hymenoptera, Vespidae). Ins. Soc. 19: 343–357.

VAN DER VECHT, J. 1973. The social wasps (Vespidae) collected in French Guyana by the Mission du Muséum National d'Histoire Naturelle with notes on the genus *Apoica*. Soc. Entomol. France Ann. (n.s.) 8: 735–743.

## Polybia Paper Wasps

Vespidae, Polistinae, Polybiini, *Polybia*.
  *Tupi-Guaraní:* Lamborinas (Brazil).
  Boraseries (Guyana).

These are the most common and dominant social wasps of the Neotropical Region (Richards 1978, Windsor 1983). The genus is large (54 species) and composed of wasps that are variable in color pattern and other features (fig. 12.6d). For this reason, they are identified with difficulty, and common names must be applied with considerable caution. The vulgates of many are derived from the shapes of the nest, which are fairly diagnostic.

The tendency for color variation is marked in most species. Ground color tends to be dark (bluish to black) or shades of brown and yellow. Often the abdomen is banded with thin, yellow borders to the segments and the rear portion of the thoracic dorsum spotted with the same color. This is especially conspicuous in dark species, in which the yellow contrasts sharply with the background. The wings may be clear to smoky brown.

Structural characteristics that distinguish this genus are the more or less petiolate abdomen and the waist segment that is about as long as the hind femur. The thorax and attached basal segment form a smooth, rounded curve in profile, and the entire body is often long and slender in general form.

Nest shapes, though varied through the genus, are consistent and diagnostic within species. Shape, size, type of construction material, placement on the substratum, comb structure, and other criteria are all particular to each species of wasp (pl. 4c). Most nests are made of paper or carton, as is usual with social wasps, but a few species (*P. singularis, P. emaciata*) form the outer envelope of mud. This may be highly polished and is referred to as the "ceramic" type. In these, the combs are still made of paper, however. Nest shapes and engineering are a source of admiration. Oval or bell-shaped forms are common, but one also finds pear-shaped, bun-shaped, sausage-shaped, stellate, cylindrical, and other configurations. The outer surface is usually smooth but may also exhibit spiny processes, projections, flanges, and textures of many types. Colors range widely, too, from buff to gray and brown to rusts and yellows.

The nest is normally suspended from tree branches by an apical fastening, but that of the largest species (*P. dimidiata*) is situated close to the ground and may be pierced by saplings that support it internally. Many suspended nests are small, only a few centimeters long, but others may attain lengths of up to 100 centimeters or more (*P. scutellaris,* pl. 4d). Nests of different species are sometimes found in close proximity (Jeanne 1978).

Prey consists of a wide range of soft-bodied arthropods, mostly katydids, caterpillars, and beetle grubs. A few species store nectar in select cells, and this takes on a similarity to honey when abundant and concentrated. The nectar of *P. jurinei* is said to be "perfumed."

*Polybia* are notoriously aggressive and pugnacious. Most wasp attacks in tropical America are from these feisty defenders of their nest and territory. Anyone molesting a colony will suffer the stings of a maddened throng of relentless polybiine pursuers, who dive into the hair and clothing and attack any spot of skin they can reach.

Human sweat, however, has been found to suppress the stinging urge at least in some individuals (Young 1978).

Colonies are founded by swarming. They attain enormous populations in older, larger perennial nests, up to several thousand individuals. Colonies of *P. scutellaris* may persist as long as twenty-five to thirty years and, at any one time, may have 4,000 to 5,000 inhabitants. More commonly, nests are relatively small with only a few dozen or one hundred occupants. Females are similar to queens, although generally larger. In one species, *P. dimidiata,* the reverse is true.

## References

Jeanne, R. 1978. Intraspecific nesting associations in the Neotropical social wasp *Polybia rejecta* (Hymenoptera: Vespidae). Biotropica 10: 234–235.

Richards, O. W. 1978. The social wasps of the Americas excluding Vespinae. Brit. Mus. Nat. Hist., London.

Windsor, D. M. 1983. *Polybia occidentalis* (Cojones de toro [name of nest], paper wasp). *In* D. H. Janzen, ed., Costa Rican natural history. Univ. Chicago Press, Chicago. Pp. 760–762.

Young, A. M. 1978. A human sweat-mediated defense against multiple attacks by the wasp *Polybia diguetana* in northeastern Costa Rica. Biotropica 10: 73–74.

## Bell Wasp

Vespidae, Polistinae, Polybiini, *Chartergus chartarius. Spanish:* Campana-avispa.

It is the enormous, bell-shaped carton nests of this species that are most often seen, hanging from trees in a variety of forest habitats. They are especially common in the Amazonian hyalea on trees near river courses. Their beauty and peculiarity have made them well-known tourist trophies that are sold in curio shops in South American cities.

Elaborate and beautifully engineered structures (pl. 4e), they may reach a length of nearly a meter, weigh 10 to 15 kilograms, and harbor a colony of over 5,000 individu-

**Figure 12.7    SOCIAL WASPS (VESPIDAE).** (a) Bell wasp (*Chartergus chartarius*). (b) Honey wasp (*Brachygastra lecheguana*). (c) Long-waisted paper wasp (*Mischocyttarus drewseni*).

als. The nest envelope is made of a durable, cardboardlike substance, tough, thick (6–7 mm), smooth, and dirty white. The interior is intricately formed into several horizontal tiers of vertical cells. The fastening envelope at the top is wide and completely encompasses the branch supporting the nest. There is a single entrance in the center of the slightly conical bottom, leading to a single passageway running opposite the entrance through the middle of each of the stories of combs.

The wasp (fig. 12.7a) is a fairly typical polybiine, recognizable by its moderate size (BL 8–10 mm) and color pattern, which is mostly black but with narrow, transverse, yellow bands running across the front and hind edges of the thorax and bordering the posterior margin of each abdominal segment; the face is yellow spotted between the antennal base and eye and on the lower part of the frons; and the wing has a clear membrane with dark veins. Definitive structural features are not conspicuous: rounded scutellum, short ridge separating the side of the head from its back, and broad, first abdominal segment.

This and two other similar species in the genus range throughout most of midtropical America (Bequaert 1938).

### Reference

BEQUAERT, J. 1938. A new *Charterginus* from Costa Rica, with notes on *Charterginus, Pseudochartergus, Chartergus, Pseudopolybia, Epipona* and *Tatua* (Hymenoptera: Vespidae). Rev. Entomol. 9: 99–117.

### Honey Wasps

Vespidae, Polistinae, Polybiini, *Brachygastra* (= *Nectarina*). *Tupi-Guaraní:* Eixy, chiguana, capij (Brazil). *Nahuatl:* Yzaxalasmitl (Mexico). Mexican bee.

The nectar-storing habits of wasps have been perfected by this group of twelve species (Naumann 1968), the best known of which is *B. lecheguana*, to which most of the following discussion applies (Schwarz 1929). The cells of the nest of this species often contain large stores of a very palatable honey that is widely exploited by humans (Bequaert 1933). Numerous early accounts speak of the use of the substance by the Indians of Mexico and Brazil and leave no doubt about the identity of the insect because of its characteristic nest, which has a paper envelope and cells that are distinct from the quite different enclaves of stingless bees, the only other common source of native honey in early days. In some areas (e.g., Jalisco), the honey is gathered regularly and even sold on the market. It is very limpid and strongly scented and said to crystallize more rapidly than bees' honey. Some Indians keep the nests, cutting them when they are small and carrying them to their gardens. The source of the honey is flower nectar gathered by the wasps in bee fashion. When *Datura* or other toxic blooms are available, the honey may be tainted, and cases of poisoning are not rare.

Mature nests are large (40–50 cm long) and ovoid. When filled with honey, they may weigh several kilograms. There are usually only a few large combs (less than 10 but as many as 20) arranged concentrically within the thin envelope. The entrance is circular or slitlike and located toward the lower end. New colonies are founded by swarming, and fertile females (polygynous "queens") are numerous, forming up to 17 percent of the total population. The number of nest inhabitants can become very large, as high as 15,000 wasps. The wasps survive off-seasons by feeding on their stores of honey, the nest being perennial and persistent for several years. It appears that the larvae are fed exclusively on the honey and pollen, an unusual diet for wasps.

*Brachygastra* is a typically Neotropical genus, occurring in all but the driest habitats, from extreme southern Texas to northern Argentina. Its members are small (BL 7–9 mm) and appear stubby because of the shape of the abdomen (fig. 12.7b). The latter is nonpetiolate, wider than long, and its second segment is greatly enlarged, often concealing the succeeding segments. The elevated scutellum, which often projects over the rest of the posterior of the thorax, is also highly characteristic; the rear portion of this forms a flat, vertical surface. Color patterns range from almost all black to solid yellow with many intermediate patterns of both.

Workers are mild mannered by polybiine standards but sting hard when sufficiently provoked. The sting is barbed and stays in the wound if the victim is a human or other large animal.

## References

BEQUAERT, J. 1933. The nearctic social wasps of the subfamily Polybiinae (Hymenoptera; Vespidae). Entomol. Americana 13: 87–150.

NAUMANN, M. 1968. A revision of the genus *Brachygastra* (Hymenoptera: Vespidae). Univ. Kans. Sci. Bull. 47: 929–1003.

SCHWARZ, H. F. 1929. Honey wasps. Nat. Hist. 39: 421–426.

## Long-Waisted Paper Wasps

Vespidae, Polistinae, Polybiini, *Mischocyttarus. Tupi-Guaraní:* Capuxu (Brazil, *M. ater*).

*Mischocyttarus* is the largest genus of social wasps (186 species) and has achieved its diversity entirely in the New World, primarily the tropical portions (Richards 1945, Zikán 1949). It is also a morphologically distinct genus, characterized by a rather long waist and asymmetrical terminal segments of the mid and hind tarsi, the inner lobes much more strongly projecting and pointed than the outer. The latter is possibly an adaptation to assist the wasp in searching for prey caught in spider webs. The dorsum of the thorax is evenly curved and, in profile, slopes directly onto the attached basal abdominal segment. There is also a characteristic conspicuous yellow spot preceding a dark patch at the apex of the hind tibia on the inside, which may serve as a recognition signal when the wasps fly near the nest with hind legs dangling. These wasps otherwise generally resemble *Polistes* in body coloring and shape but are consistently smaller (BL 10–15 mm).

All members of the genus make small (rarely more than 100 cells), stalked, uncovered combs somewhat like those of *Polistes,* although in many, the combs may be turned to face the substrate and may be irregularly layered in series rather than horizontally. There also may be more than one supporting stalk. The latter is varnished with a secretion from a gland on the abdomen that has been shown in some species to be an ant repellent. Nest sites of some species are carefully situated amid spiny plants or near other wasps for maximum protection (Gorton 1978).

Colonies are usually founded by a single female (haplometrosis) or in company with a few female nestmates (pleometrosis) (Poltronieri and Rodrígues 1976). Working progeny soon develop and a dominance hierarchy ensues. Subordinate females may eventually oust the queen and take over the colony (Little 1981). A strong trophallactic bond exists between workers and queens in one species (Machado and Wiendl 1976).

The peculiar biology of *M. drewseni* (fig. 12.7c) has been given some direct attention (Jeanne 1972, Jeanne and Castellón Bermúdez 1980).

## References

GORTON, R. E. 1978. Observations on the nesting behavior of *Mischocyttarus immarginatus* (Rich.) (Vespidae: Hymenoptera) in a dry forest in Costa Rica. Ins. Soc. 25: 197–204.

JEANNE, R. L. 1972. Social biology of the Neotropical wasp *Mischocyttarus drewseni*. Mus. Comp. Zool. (Harvard Univ.) Bull. 144: 63–150.

JEANNE, R. L., AND F. G. CASTELLÓN BERMÚDEZ. 1980. Reproductive behavior of a male Neotropical social wasp, *Mischocyttarus drewseni* (Hymenoptera: Vespidae). Kans. Entomol. Soc. J. 53: 271–276.

LITTE, M. 1981. Social biology of the Polistine wasp *Mischocyttarus labiatus:* Survival in a Colombian rain forest. Smithsonian Contrib. Zool. 327: 1–27.

MACHADO, V. L. L., AND F. M. WIENDL. 1976. Aspectos do comportamento de colônias de *Mischocyttarus cassununga* von Ihering, tratadas com alimento marcado por radiofósforo. Soc. Entomol. Brasil Ann. 5: 79–85.

POLTRONIERI, H. S., AND V. M. RODRÍGUES. 1976. Vespídeos sociais: Estudo de algumas espécies de *Mischocyttarus* Saussure, 1853 (Hymenoptera-Vespidae-Polistinae). Dusenia 9: 99–105.

RICHARDS, O. W. 1945. A revision of the genus *Mischocyttarus* de Saussure (Hymen., Vespidae). Royal Entomol. Soc. London Trans. 95: 295–462.

ZIKÁN, J. F. 1949. O gênero *Mischocyttarus* Saussure (Hym.-Vespidae), com a descrição de 82 novas espécies. Par. Nac. Itatiaia Bol. 1: 1–251.

# ANTS

Formicidae. *Spanish:* Hormigas. *Portuguese:* Formigas. *Tupi-Guaraní:* Quenquem. *Nahuatl:* Azcameh (sing. azcatl). *Quechua:* Sisi, cuki.

If anyone doubts who is in possession of the Earth, let them consider the ant (Hölldobler and Wilson 1990, Wheeler 1910, Wilson 1971: 27–74). It is hardly possible to find a place, particularly in the American tropics, where ants are not visible, often abundantly and obnoxiously so (although they are conspicuously absent at high elevations). Because of their small size, social structure, and adaptiveness, they have evolved into a fairly large number of species in Latin America, where well over a third of the world's 7,600 known species are found (Kempf 1972). These are currently divided into 116 genera (Kusnezov 1978), representing nearly 50 percent of the world's total. Many of these are known only to ant specialists, but many are familiar pests. Yet, destructive though some may be, most ants deserve our appreciation for the roles they play in controlling other injurious insects, regulating vegetative growth, and increasing the fertility of the soil (Branner 1900).

Resemblance of the social organization of ants to human society has earned them special significance to the Indian. Indeed, ants hold central, even revered, positions in the mythologies of many tribes. In earlier days, such was the dominance of these insects in their forest environment that the Kaingang of southeastern Brazil believed the glory of their people to be derived from being transformed into ants after death (Lenko and Papavero, 1979: 225). Among the Kayapó, even today, ants are spoken of in terms of their "power," or ability to inflict pain, which is used by shamans to manipulate spirits to cause harm. Stinging ants are collected by the men, pounded into a paste with red urucu dye and painted on hunting dogs to make them keep their noses to the ground and hunt with determination as the ants do (Posey 1979: 143).

Lest we be amused by such beliefs among simpler cultures, consider the friars of the Convent of San Antonio at Maranhão, Brazil, who prosecuted some local red ants under ecclesiastical law. The insects had devoured the altar cloths and brought up into the chapel pieces of shrouds from the churchyard graves (Cowan 1865: 168).

Ants are easily distinguished from other Hymenoptera by the unique narrowed "waist" (petiole) connecting the abdomen and thorax. The waist may consist of one or two segments, one or both of which often have a strong, erect projection, or "node," arising from it. This arrangement prevails whether the ant is a wingless, sterile female worker or a winged, reproductively active adult male or female. Ants are also recognized by their elbowed antennae (the basal segment of which is very long), usually well-developed mandibles, and large mobile prothorax.

Equally characteristic of the family is their habit of living in colonies in a highly organized social state. In most species, this has led to a division of labor correlated with differing morphotypes called "castes." The reproductive castes are the males, who die after mating, and the queens, who establish new colonies after mating and shedding their wings. The bulk of the population consists of sterile females who form a series of worker types from the smallest "minors" to the largest "majors." The latter, as a result of disproportionate (allometric) body growth, may develop outsized heads and jaws used in defense of the colony and are functional "soldiers."

Colonies live in nests constructed of a variety of materials and located in diverse sites, both terrestrial and arboreal (Wilson 1987). Most are excavations in the soil, sometimes simple chambers at the end of a direct entrance; in other instances, they are enormous, complex, underground labyrinths. Nests are also made of soil particles or chewed wood fragments glued to vegetation (carton nests) or are intricately woven of silk (*Camponotus senex*). A special category of the former are those attached to leaves (Black 1987). Many species live in hollowed-out plant structures and some even have developed symbiotic relationships with their hosts (see ants and plants, following).

Ants make use of elaborate communication systems, based on chemical messenger chemicals produced by various glands, to organize their foraging and other communal activities. Such are the trail-making substances laid down from anal glands, alarm odors produced by the mandibular gland in the head and abdominal Dufour's gland, noxious odors or antifungal compounds secreted by the metapleural glands in the thorax, and venoms for subduing prey and defense against enemies which are generated by venom glands associated with the sting.

Food often consists of special types of prey or plant products, the exploitation of which requires cooperation and initiative among colony members (Sudd 1967). Almost any edible material may be taken: living or dead arthropods and small vertebrate animals, fungi grown on leaf fragments, seeds, flower pollen and nectar (including that from extrafloral nectaries), and honeydew, which they take from scale insects, aphids, and treehoppers.

Food is shared with the whole colony by foraging workers. This includes the helpless larvae and the queen, which delivers pheromonal secretions in exchange. The movement of such substances through the colony (trophallaxis) acts as a bond, keeping the colony members together and regulating the physiology and behavior of its members. A variety of diverse insects often share in the food exchange and form symbiotic bonds with the colony. There are the so-called ant guests, or "myrmeco-philes," such as silverfish, crickets, cock-roaches, beetles, flies, and even mites, spiders, and millipedes (see symbiosis, chap. 2). Cooperative or mutualistic associations are also common between ants and honeydew-producing aphids, scale insects, and other Homoptera (Way 1963).

Secondary only to their ubiquity and sociality, ants attract attention with their ability to inflict wounds. Many have stings, some, especially species among the primitive tribes, very potent ones. Contrary to popular belief, formic acid is only one, minor, constituent of ant venom, except in the subfamily Formicinae where it predominates. In most ants, various alkaloids and protein derivatives are much more important in producing lasting and debilitating effects. These are only now being identified and understood by insect pharmacologists (Schultz and Arnold 1977).

Because they sting or are malodorous, ants serve as models for Batesian mimics. They are closely resembled in form and behavior by stilt flies (Micropezidae), immature assassin bugs, adult grass bugs, weevils, and other insects, as well as spiders (Jackson and Drummond 1974).

The phenomenom of "ant mosaics," that is, colonies of abundant, static species that compete to establish foraging territories in vegetation dominated by trees (first discovered in West Africa), is thought to occur in some tropical American habitats (Jackson 1984; Leston 1978; Winder 1978; Young 1983). Mosaics may be important in the cultivation of tropical tree crops, for many pests and plant diseases are associated with one or more of the dominant ant species.

Most of the subfamilies of the Formicidae are represented in the American tropics (Alayo 1974; Kusnezov 1963; Smith 1936; Snelling and Hunt 1976). These are the Ponerinae (primitive hunting ants), Ecitoninae (New World army ants), Pseudo-myrmecinae (fever ants), Myrmicinae (myrmicine ants), Dolichoderinae (odiferous ants), and Formicinae (formicine ants).

## References

ALAYO, D. P. 1974. Introducción al estudio de los himenópteros de Cuba, Superfamilia Formicoidea. Acad. Cien., Cuba, Inst. Zool. Ser. Biol. 53: 1–58.

BLACK III, R. W. 1987. The biology of leaf nesting ants in a tropical wet forest. Biotropica 19: 319–325.

BRANNER, J. C. 1900. Ants as geologic agents in the tropics. J. Geology 8: 151–153.

COWAN, F. 1865. Curious facts in the history of insects. Lippincott, Philadelphia.

HÖLLDOBLER, B., AND E. O. WILSON. 1990. The ants. Harvard Univ. Press, Cambridge.

JACKSON, D. 1984. Competition in the tropics: Ants on trees. Antenna 8(1): 19–25.

JACKSON, J. F., AND B. A. DRUMMOND III. 1974. A Batesian ant mimicry complex from the Mountain Pine Ridge of British Honduras, with an example of transformational mimicry. Amer. Midl. Nat. 91: 248–251.

KEMPF, W. W. 1972. Catálogo abreviado das formigas da Região Neotropical (Hym. Formicidae). Studia Entomol. 15: 3–344.

KUSNEZOV, N. 1963. Zoogeografía de las hormigas en Sudamerica. Acta Zool. Lilloana 19: 25–186.

KUSNEZOV, N. 1978. Hormigas argentinas-clave para su identificación. Min. Cult. Educ., Fundación Miguel Lillo, Misc. 61: 1–147.

LENKO, K., AND N. PAPAVERO. 1979. Insetos no folclore. Cons. Estad. Artes Ciên. Hum., São Paulo.

LESTON, D. 1978. A Neotropical ant mosaic. Entomol. Soc. Amer. Ann. 71: 649–653.

POSEY, D. A. 1979. Ethnoentomology of the Gorotire Kayapó of central Brazil. Ph.D. diss., Univ. Georgia.

SCHULTZ, D. R., AND P. I. ARNOLD. 1977. Venom of the ant *Pseudomyrmex* sp.: Further characterization of two factors that affect human complement proteins. J. Immunol. 119: 1690–1699.

SMITH, M. R. 1936. The ants of Puerto Rico. Univ. Puerto Rico J. Agric. 20: 819–875.

SNELLING, R., AND J. HUNT. 1976. The ants of Chile. Rev. Chilena Entomol. 9: 63–129.

SUDD, J. 1967. An introduction to the behavior of ants. Arnold, London.

WAY, M. J. 1963. Mutualism between ants and

honeydew producing Homoptera. Ann. Rev. Entomol. 8: 307–344.

WHEELER, W. M. 1910. Ants, their structure, development and behavior. Columbia Univ., New York.

WILSON, E. O. 1971. Insect societies. Belknap Press, Harvard Univ., Cambridge. See esp. pp. 27–74.

WILSON, E. O. 1987. The arboreal ant fauna of Peruvian Amazon forests: A first assessment. Biotropica 19: 245–251.

WINDER, J. A. 1978. The role of non-dipterous insects in the pollination of cocoa in Brazil. Bull. Entomol. Res. 68: 559–574.

YOUNG, A. M. 1983. Patterns of distribution and abundance of ants (Hymenoptera; Formicidae) in three Costa Rican cocoa farm localities. Sociobiology 8: 52–76.

## ANTS AND PLANTS

Quite a few unrelated kinds of woody plants in the American tropics have evolved structures adapting them for habitation by ants (Beattie 1985, Wheeler 1942). Some have existed so long in close association with their formicid guests that an obligatory mutualism between plant and insect exists ("myrmecophytes"). Other associations are merely casual or facultative, although they may represent incipient symbioses demonstrating how the most highly evolved cases may have originated (Jolivet 1986).

Myrmecophytes are found among several plant families and genera. All are similar in having at least some part of the stem, trunk, or leaf enlarged and hollowed to provide living quarters for ant colonies (formicarium, myrmecodomatium). Some also have special tissues or glands that provide food (pearl bodies, Beltian bodies, etc.) or nectar sources (extrafloral nectaries) for the ants and no doubt assure the ants' dependence on, and jealous protection of, the host plant. On this level, the best example is the swollen thorn (bull's horn) acacias (Mimosaceae, *Acacia cornigera,* and many other species known by many names, e.g., *cornizuelas, palines,* and *guisaches corteños* in Mexico and *cachitos* in Panama) (Janzen 1969).

These are small trees or shrubs with delicately pinnate compound leaves and large bifurcate thorns that the ants enter by gnawing an opening just below the tip of one of the pair and hollowing them out. Extrafloral nectaries situated on the upper side of the petioles and at the base of the pinnae where they join the rachis provide sweet attractions for the ants, and there are elongate whitish food bodies (Beltian bodies) produced at the tips of the young leaflets, which the ants collect, store, and eat.

Swollen thorn acacias are most abundant in Mexico and Central America; they do not extend south beyond Venezuela and Colombia. Throughout their range, they live in mutualistic associations with several species of ants of the genus *Pseudomyrmex* (Janzen 1966, 1967). Workers of these ants begin to patrol the limbs and leaves about nine months following colonization by new queens and effectively protect it from the ravages of browsing mammals, vegetarian insects, and even vines whose tendrils attempt to entwine it. They reward human intruders with a sting whose potency rivals that of some ponerine hunting ants.

A similar relationship exists between several *Azteca* ants and the cecropia tree (Moraceae, *Cecropia adenopus* and other species), also called the trumpet tree, imbaúba (Brazil), cetico (Peru), guarumo (Central America), and yagram (Antilles). Here the jointed trunks of small trees are the ant dwellings. The internodes are hollow and separated from each other by partitions. Externally, these joints are marked by a node from which arises a leaf, and below each leaf base, there is a groove enclosing a deep pit. The founding queen perforates the trunk by gnawing through this pit and gaining entrance to the chamber inside, there establishing a colony that grows to occupy adjoining chambers when the separating partitions are drilled through by progeny workers. In the interior, the ants build a kind of carton nest for their larvae out of a brown, waxlike substance.

Like the acacias, the cecropias provide nutritive bodies for the ants. At the base of each leaf petiole, there is a velvety brown cushion in which egg-shaped, food-rich structures (Müllerian bodies, "pearl bodies"; O'Dowd 1982) are produced. The ants also coax honeydew from the white scale insects or mealybugs that invariably also cohabit in the trunks and stems of this tree.

When a cecropia is disturbed, its ant guardians boil out of their chambers in attack. Bereft of a sting, they rub noxious anal secretions in wounds made by their mandibles and just as effectively discourage the plant's enemies as their armed cousins. This partnership is not ubiquitous; it breaks down under some circumstances (de Andrade and Carauta 1982) at high elevations and on some islands where large plant browsers are absent (Janzen 1973).

Other ant plants have simpler relationships with ants. Lacking nutritive bodies and extrafloral nectaries is the South American tangarana, a name applied in Peru by natives to both the tree and its protector ants. The reputation of the ants has earned the tree (Polygonaceae, *Triplaris americana* and other species in the genus) a host of telling names: *pau-santo, pau de novato, pau formigueiro* (Brazil); *tachizeiro* (Pará); *itassi,* Long John (Guyana); *jacuna* (Arawak); *hormigo, tabaco* (Costa Rica); *árvore de tachi* (Amazonia); *palo María, barabas* (Venezuela).

During its flowering period in the dry season, *Triplaris* is a large and decorative tree that grows along watercourses in the wet forests of the northern half of the South American continent. Its flowers are small and insignificant in themselves, but the conspicuous calyxes form dense white masses that turn bright red, transforming the slender-trunked trees into masses of flame.

Flame would also describe the character of the ants associated with the tangarana tree. These mostly belong to the genus *Pseudomyrmex* (Latinodus Group), but some *Azteca* are also known from this plant. Local names for the former are fever ants (general), *tachi* (Amazonia), *novato* (Pará), and *jacuna sae* (Arawak).

The small terminal branches and twigs are hollow and occupied by the ants, which obtain entry by cutting through the thin tissue in the base of slitlike scars located toward the far end of each internode. Tales of the attacks of these ants are legend among the Indians and were known even to the earliest jungle explorers. One of the earliest accounts is that of P. Bernabé Cobo in 1653, who wrote in his *History of the New World* (1890–1895), "Since these ants are concealed within the tree, they are not seen and this is the reason why those who do not know the secret, are not on their guard; but if a single leaf be touched, so many of the ants swarm forth from all parts of the tree as to excite wonder, and they assail the person who touches the tree and, if he does not withdraw in time, martyr him with their stings."

These same ant genera and others (*Solenopsis, Crematogaster*) and still other plants have formed intricate partnerships, with different parts of the plant's anatomy modified as formicaries. In *Tachigalia* (Caesalpinaceae), the bases of the petioles of the large pinnate leaves are swollen and hollow; access to the interior is through an elongate slit on one side (Wheeler 1921a). Bulbous swellings immediately below and surrounding the nodes on the stems of certain *Cordia* species (Boraginaceae) are also the abodes of noxious ants (the genera already named plus *Allomerus* and *Azteca*; pl. 4g), which move in and out through a single opening between the opposite leaves at one end of the node. One side of the bases of the large velvety leaves of *Tococa* (and other genera of Melostomaceae) are inflated, double, globular chambers harboring these formicids (pl. 3c). A pair of tiny holes at the bases of the forks of the three strong main veins on the underside of the leaf provide access to the chamber.

Another possibly coevolved mutualistic relationship exists between *Pheidole* ants and plants of the genus *Piper* (Piperaceae) (Letourneau 1983, Risch et al. 1977). The ants live in petiolar cavities and in the stems, which they hollow out. The plant provides lipid-rich food bodies inside the petiolar cavities, and the ants appear to increase the survival of the plants by protecting them from encroaching vines. The plant may also receive nutritional benefit by absorbing decay products from the nest.

Still other plants are known with similar but ill-defined symbiotic relationships to the Formicidae. For example, extrafloral nectaries on many species (e.g., *Bixa, Costus, Inga, Byttneria*) attract ants whose presence may discourage the attacks of herbivores (Bentley 1977; Hespenheide 1985; Koptur 1984; Schemske 1982). However, Hespenheide (1984) thinks that the protection of the plant by parasitoid wasps attracted to the nectaries may be more important. These wasps might then significantly reduce the incidence of phytophagous caterpillar attacks on the plant hosts, but this has not yet been amply demonstrated.

Ants may inhabit the stems of almost any hollow-stemmed plant and nest in all kinds of cavities in roots, bark, dead twigs, or parts of practically any woody type. But only those plants that have adapted to the presence of ants by evolving formicaries, and some also by providing nutritive bodies, can be considered true myrmecophytes.

Labyrinthine formicaries in the pseudobulbs of orchids and growing at the base of other epiphytes have been described from the forests of Asia but have not been found with certainty in tropical America. Their importance in the nutrition of air plants on nutrient-poor substrates has been demonstrated and ought to have evolved in the New World where it may yet be found.

An additional curious intimacy between ants and plants are the so-called ant gardens (pl. 4h). Species of *Azteca, Crematogaster, Dolichoderus*, and *Camponotus* in the Amazonian and other lowland forests form spongelike nests on tree branches from particles of soil. Sometimes pairs of species live together in such structures ("parabiosis"; Wheeler 1921b). Epiphytic plants of several types take root in these nests, which forms a kind of nutritional "potting mix," anchoring and fostering their growth. Species of *Codonanthe* (Gesneriaceae) are the best-known benefactors of this arrangement, even tempting the ants to disperse their seeds by their shape and/or odor that mimics those of ant larvae. Other plants in such associations are *Peperomia, Anthurium, Philodendron, Epiphyllum, Markea*, and various bromeliads. The ant benefits from a constant fruit and nectar supply as well as structural support of their nest by the plant's roots. The ants do not appear to pollinate the plants but may protect them from herbivores and possibly even actively plant their seeds in the nest matrix (Kleinfeldt 1978).

## References

BEATTIE, A. J. 1985. The evolutionary ecology of ant-plant mutualisms. Cambridge Univ. Press, New York.

BENTLEY, B. L. 1977. Extrafloral nectaries and protection by pugnacious bodyguards. Ann. Rev. Ecol. Syst. 8: 407–427.

COBO, B. 1890–1895. Historia del Nuevo Mundo. 4 vols. Marcos Jiménez de la Espada, Sevilla.

DE ANDRADE, J. C., AND J. P. P. CARAUTA. 1982. The *Cecropia-Azteca* association: A case of mutualism. Biotropica 14: 15.

HESPENHEIDE, H. A. 1984. *Agrilus xanthonotus* (yellow-spotted *Byttneria* borer). *In* D. H. Janzen, ed., Costa Rican natural history. Univ. Chicago Press, Chicago. Pp. 681–682.

HESPENHEIDE, H. A. 1985. Insect visitors to extrafloral nectaries of *Byttneria aculeata* (Sterculiaceae): Relative importance and roles. Ecol. Entomol. 10: 191–204.

Janzen, D. H. 1966. Coevolution of mutualism between ants and acacias in Central America. Evolution 20: 249–275.

Janzen, D. H. 1967. Interactions of the bull's horn acacia (*Acacia cornigera* L.) with an ant inhabitant (*Pseudomyrmex ferruginea* F. Smith) in eastern Mexico. Univ. Kans. Sci. Bull. 47: 315–558.

Janzen, D. H. 1969. Allelopathy by myrmecophytes: The ant *Azteca* as an allelopathic agent of *Cecropia*. Ecology 50: 147–153.

Janzen, D. H. 1973. Dissolution of mutualism between *Cecropia* and its *Azteca* plants. Biotropica 5: 15–28.

Jolivet, P. 1986. Les fourmis et les plants—un example de coevolution. Soc. Nouv. Ed., Boubée, Paris.

Kleinfeldt, S. E. 1978. Ant-gardens: The interaction of *Codonanthe crassifolia* (Gesneriaceae) and *Crematogaster longispina* (Formicidae). Ecology 59: 449–456.

Koptur, S. 1984. Experimental evidence for defense of *Inga* (Mimosoideae) saplings by ants. Ecology 65: 1787–1793.

Letourneau, D. K. 1983. Passive aggression: An alternative hypothesis for the *Piper-Pheidole* association. Oecologia 60: 122–126.

O'Dowd, D. J. 1982. Pearl bodies as ant food: An ecological role for some leaf emergences of tropical plants. Biotropica 14: 40–49.

Risch, S., M. McClure, J. Vandermeer, and S. Waltz. 1977. Mutualism between three species of tropical *Piper* (Piperaceae) and their ant inhabitants. Amer. Midl. Nat. 90: 433–444.

Schemske, D. W. 1982. Ecological correlates of a Neotropical mutualism: Ant assemblages at *Costus* extrafloral nectaries. Ecology 63: 932–941.

Wheeler, W. M. 1921a. The tachygalia ants. Zoologica 3: 137–168.

Wheeler, W. M. 1921b. A new case of parabiosis and the "ant gardens" of British Guiana. Ecology 2: 89–103.

Wheeler, W. M. 1942. Studies of Neotropical ant-plants and their ants. Mus. Comp. Zool. (Harvard Univ.) Bull. 90: 1–262.

## GIANT HUNTING ANTS

Formicidae, Ponerinae, *Dinoponera* and *Paraponera*. *Spanish:* Calenturas, perros negros, jaulas (Peru). *Portuguese:* Tocandiras (vars. tucanderas, tacanduiras, toucandeiras, etc.), vinte-e-quatros, formigas de febre, formigas de quatro picadas (Brazil).

The following two best-known and largest of the gigantic forest ants are often confused as one and with other large, black species under the name *tocandira*. According to Sampaio (Lenko and Papavero 1979), this appellation is a contraction of *tucaba-ndy*, or *tucana-ngia*, in Tupi for "the one that hurts much with its underparts," an obvious reference to the extremely severe sting they can inflict. The smaller *Paraponera* are the more aggressive and potent and can subdue a healthy adult with a venomous stab that is often described as "like a blow from a hammer or bullet." The serious symptoms that may (but not invariably) follow a sting and last a day or longer are prolonged aching pain that rapidly spreads from the site of the wound, sometimes labored breathing, cardiac palpitations, and fever (Weber 1939, orig. obs.). It is alleged that death is an outcome in rare instances, although no such cases are documented in the literature. The larger *Dinoponera* are gentler ants, less apt to sting and with less effect than the fiercer *Paraponera* (Hermann et al. 1984).

Because of their ferocity and ability to inflict pain, these ants are employed in male initiation and virility rites among some Amazonian Indian groups (Liebrecht 1886). Large numbers of ants are caught and tied to specially contrived wickerwork panels and applied to the initiate's bare skin (see fig. 1.12). The maddened mass of ants need no inducement to sting and do so with impatience and repeatedly. Only youths who endure the excruciating experience without complaint (and survive) are deemed worthy of passage to manhood.

The tocandira are also the object of a curious belief through much of Amazonia. The usual story has the ant allegedly eating the tiny seed of a specific liana or aerial root, called *tamshi* (*Carludovica divergens*), used by the Indians for binding together the framework of huts. The seed germinates in the ant's body, even while it lives, and grows from it into a new plant. In another version, the ant simply metamorphoses directly into the plant. The basis for these myths undoubtedly is found in the attacks of a common parasitic fungus of the genus *Cordyceps* on these ants. *Cordyceps australis,* in fact, is a highly specialized pathogen infecting the ant tribe Ponerini in tropical forest habitats. Before dying from an infection, the ant attaches itself to a leaf or tree and becomes transfixed by the fungus while the fruiting bodies of the fungus grow out from it, in long filaments resembling the early stages of a developing vine (pl. 3a).

These tocandira ants are representative of several large ants in the tribe Ponerini, all of which live in small colonies of a hundred or so workers and a queen (Zahl 1972). Their nest is normally a burrow situated in the ground at the base of trees; the nest entrance is a gap in the soil, usually contiguous with the tree trunk. Workers forage alone for arthropod prey and plant materials and do not exhibit coordinated group behavior, a trait considered primitive in this insect family. They are also virtually casteless, the workers not being well differentiated from the sexual females.

The giant hunting ants may be encountered anywhere in the moister portions of tropical America but are absent from the Antilles and isolated islands.

## References

HERMANN, H. R., M. S. BLUM, J. W. WHEELER, W. L. OVERAL, J. O. SCHMIDT, AND J. T. CHAO. 1984. Comparative anatomy and chemistry of the venom apparatus and mandibular glands in *Dinoponera grandis* (Guérin) and *Paraponera clavata* (F.) (Hymenoptera: Formicidae: Ponerinae). Entomol. Soc. Amer. Ann. 77: 272–279.

LENKO, K., AND N. PAPAVERO. 1979. Insetos no folclore. Cons. Estad. Artes Ciên. Hum., São Paulo.

LIEBRECHT, F. 1886. Tocandyrafestes. Zeit. Ethnol. 18: 350–352.

WEBER, N. A. 1939. The sting of the ant, *Paraponera clavata.* Science 89: 127–128.

ZAHL, P. A. 1972. Giant ants of the Amazon. Nat. Geogr. Soc. Res. Repts. 1955–1960: 193–201.

## Greater Giant Hunting Ant

Formicidae, Ponerinae, Ponerini, *Dinoponera gigantea. Spanish:* Isula (Peru).

This ant and its few congeners (Kempf 1971) are primarily Amazonian but extend westward into the Andean foothills, southward to Paraguay, and across the Guiana and Brazilian highlands to the Atlantic coast. The worker is decidedly larger than all other large black ponerines of the American tropics; specimens with body lengths of over 3 centimeters are known, making them the largest ants in the world (Hermann et al. 1984).

The females are shiny black ants with a smooth cuticle, although it is often clothed in a sparse, close-set, golden pile (fig. 12.8a) (Haskins and Zahl 1971). The node of the petiole is narrow and rectangular and higher than long in outline, as viewed from the side. There are no overt projections or spines on the thorax. Males are considerably smaller than females and chestnut brown, with long silky pubescence.

Colonies number only a hundred or so individuals and live in shallow burrows at the bases of trees; Fowler (1986) records only thirty workers per nest of *D. australis.* Habitat is typically forest, but colonies survive in cut-over or second growth scrubby areas as well. The workers emerge during the daytime to forage for food, normally on the ground, ascending vegetation only to take liquids from extrafloral nectaries.

In Amazonian Peru, this ant is mimicked by an unidentified species of cerambycid beetle in the genus *Tillomorpha*

**Figure 12.8   ANTS (FORMICIDAE).** (a) Greater giant hunting ant (*Dinoponera gigantea*). (b) Lesser giant hunting ant (*Paraponera clavata*). (c) Kelep (*Ectatomma tuberculatum*). (d) Trap jaw ant (*Odontomachus* sp.). (e) Cobra ant (*Pachycondyla villosa*).

(Cerambycinae, tribe Tillomorphini) (orig. obs.).

## References

FOWLER, H. G. 1986. Populations, foraging and territoriality in *Dinoponera australis* (Hymenoptera, Formicidae). Rev. Brasil. Entomol. 29: 443–447.

HASKINS, C. P., AND P. A. ZAHL. 1971. The reproductive pattern of *Dinoponera grandis* Roger (Hymenoptera, Ponerinae) with notes on the ethology of the species. Psyche 78: 1–11.

HERMANN, H. R., M. S. BLUM, J. W. WHEELER, W. L. OVERAL, J. O. SCHMIDT, AND J. T. CHAO. 1984. Comparative anatomy and chemistry of the venom apparatus and mandibular glands in *Dinoponera grandis* (Guérin) and *Paraponera clavata* (F.) (Hymenoptera: Formicidae: Ponerinae). Entomol. Soc. Amer. Ann. 77: 272–279.

KEMPF, W. W. 1971. A preliminary review of the ponerine ant genus *Dinoponera* Roger (Hymenoptera: Formicidae). Studia Entomol. 14: 369–391.

## Lesser Giant Hunting Ant

Formicidae, Ponerinae, Ectatommini, *Paraponera clavata*. *Spanish:* Chacha, folofa (Panama); isulilla (Peru); bala (Costa Rica). *Portuguese:* Tapiaí (Brazil).

This species (Janzen and Carroll 1983) is much more widespread than *Dinoponera gigantea,* living in forest habitats from Nicaragua through Central America to Amazonia. It is the sole member of its genus and distinguished from the other hunting ants primarily by its large size (BL 20 mm), but it is smaller (fig. 12.8b) than *Dinoponera*. It tends to be hairier than the others (stiff, bristly hairs generally) and reddish-brown, with a coarsely corrugate cuticle. The anterior corners of the thorax also bear large, conical projections. The node of the petiole is large, compressed and rectangular in outline. The structure of the species mouthparts (Whiting et al. 1989) and stinging apparatus has been described in detail (Hermann and Douglas 1976, Hermann and Blum 1966), as has the effects of the sting on humans (Weber 1939).

There is some evidence of worker types with different tasks associated with allometric growth (Breed and Harrison 1988). Foraging workers actively pursue prey, plant exudates (Young 1981), and sap, usually at night, searching for insects from dusk to dawn on low vegetation (although daytime foraging occurs on overcast days) (McCluskey and Brown 1972, Young and Hermann 1980). They are commonly seen on tree trunks and lianas near the forest floor. Nest burrows are usually placed between tree buttresses and have multiple openings, the latter sometimes with a thin cone or weak chimney. Some arboreal nests have been observed (Breed and Harrison 1989). There appears to be some degree of association between the ant and the gavilán tree (*Pentaclethra macroloba*) (Bennett and Breed 1985), but other trees are also selected, possibly because they possess extrafloral nectaries (Belk et al. 1989). Workers guard the nest entrance and defend the nest against intrusion by vertebrates, usually stinging them, and against other insects, including members of other colonies of their own kind. There may also be aggression between ants from different colonies in foraging areas, indicating the possible existence of territoriality (Hermann and Young 1980). Injured workers may be attacked by the parasitoid phorid fly *Apocephalus paraponerae*, which apparently locates its hosts by odors given off from their wounds (Brown and Feener 1991). The species serves as a model for Batesian mimicry by the cerambycid beetles *Acyphoderes sexualis* (Silberglied and Aiello 1976) and possibly also *Stenygra contracta* (orig. obs.).

## References

BELK, M. C., H. L. BLACK, AND C. D. JORGENSEN. 1989. Nest tree selectivity by the tropical ant, *Paraponera clavata*. Biotropica 21: 173–177.

BENNETT, B., AND M. D. BREED. 1985. On the association between *Pentaclethra macroloba* (Mimosaceae) and *Paraponera clavata* (Hymenoptera: Formicidae) colonies. Biotropica 17: 253–255.

BREED, M. D., AND J. M. HARRISON. 1988. Worker size, ovary development and division of labor in the giant tropical ant, *Paraponera clavata* (Hymenoptera: Formicidae). Kans. Entomol. Soc. J. 61: 285–291.

BREED, M. D., AND J. M. HARRISON. 1989. Arboreal nesting in the giant tropical ant, *Paraponera clavata* (Hymenoptera: Formicidae). Kans. Entomol. Soc. J. 62: 133–135.

BROWN, B. V., AND D. H. FEENER. 1991. Behavior and host location cues of *Apocephalus paraponerae* (Diptera: Phoridae), a parasitoid of the giant tropical ant, *Paraponera clavata* (Hymenoptera: Formicidae). Biotropica 23: 182–187.

HERMANN, JR., H. R., AND M. S. BLUM. 1966. The morphology and histology of the hymenopterous poison apparatus. I. *Paraponera clavata* (Formicidae). Entomol. Soc. Amer. Ann. 59: 397–409.

HERMANN, JR., H. R., AND M. DOUGLAS. 1976. Sensory structures on the venom apparatus of a primitive ant species. Entomol. Soc. Amer. Ann. 69: 681–686.

HERMANN, JR., H. R., AND A. M. YOUNG. 1980. Artificially elicited defensive behavior and reciprocal aggression in *Paraponera clavata* (Hymenoptera: Formicidae: Ponerinae). Georgia Entomol. Soc. J. 15: 8–10.

JANZEN, D. H., AND C. R. CARROLL. 1983. *Paraponera clavata* (bala, giant tropical ant). *In* D. H. Janzen, ed., Costa Rican natural history. Univ. Chicago Press, Chicago. Pp. 752–753.

MCCLUSKEY, E. S., AND W. L. BROWN, JR. 1972. Rhythms and other biology of the giant tropical ant *Paraponera*. Psyche 79: 335–347.

SILBERGLIED, R. E., AND A. AIELLO. 1976. Defensive adaptations of some Neotropical longhorned beetles (Coleoptera, Cerambycidae): Antennal spines, tergiversation, and double mimicry. Psyche 83: 256–262.

WEBER, N. A. 1939. The sting of the ant, *Paraponera clavata*. Science 89: 127–128.

WHITING, JR., J. H., H. L. BLACK, AND C. D. JORGENSEN. 1989. A scanning electron microscopy study of the mouthparts of *Paraponera clavata* (Hymenoptera: Formicidae). Pan-Pacific Entomol. 65: 302–309.

YOUNG, A. M. 1981. Giant Neotropical ant *Paraponera clavata* visits *Heliconia pogonantha* flower bracts in premontane tropical rain forest. Biotropica 13: 223.

YOUNG, A. M., AND H. R. HERMANN, JR. 1980. Notes on the foraging of the giant tropical ant *Paraponera clavata* (Hymenoptera: Formicidae: Ponerinae). Kans. Entomol. Soc. J. 53: 35–55.

## TRAP JAW ANTS

Formicidae, Ponerinae, Ponerini, *Odontomachus. Spanish:* Tingoteros (Peru). *Portuguese:* Bate-bicos, taracutingas (var. saracutingas), formigas porta pinças (Brazil).

Workers of this and related (Moffett 1989) genera are among the most easily recognized of ants (fig. 12.8d). Their head shape is uniquely widened in the anterior part, with the eyes situated at the broadest part on prominences. The straight mandibles are elongate with a hooked tip and lie nearly parallel at full closure but can be held open to a full 180 degrees. The node of the petiole (waist) is also distinctive: a tall cone with a sharp, single point at its apex. These ants are slender bodied, long legged, and variously colored in dull reddish-black hues to bright reds, blacks, and yellows.

Like other ponerines, *Odontomachus* are active hunters. The peculiar mandibles play an essential part in this activity. While searching for prey, the jaws are carried at an open ready position, under tension by a locking mechanism at their bases. Special sensory trigger hairs pointing forward from the bases of the mandibles initiate a sudden convulsive snap of the jaws when something touches them. Insects and other small arthropods are thus caught and then subdued with a sting.

Empty mandibles snapping together emit an audible clicking sound. If by chance when an ant closes its jaws, they impinge on some immovable large object like a steel instrument, the ant may be flipped backward several centimeters. Such acrobatics have been described by a number of observers and are testimony to the viciousness of the ant's ability to attack. Jaw snapping is also used as a defense (Carlin and Gladstein 1989).

Like other fairly large ponerines (BL 6–20 mm), many *Odontomachus* sting quickly and painfully.

Trap jaw ants live in all parts of the American tropics, including the West Indies and Galápagos, and other Pacific islands. Here they may be encountered usually during the day walking to and fro on vegetation, engaged in foraging activities. Their nests are situated in the soil, most commonly under rotting wood, an old stump, or a log; nest tunnels may even extend into the wood. Another frequent nesting site is in humus and leaf litter at the base of large trees.

Throughout the Neotropics, there are some twenty-four species, a number that may be increased when some taxonomic complexes are better known, for example, that going under the name of "*Odontomachus haematodus*," a ubiquitous entity that some authors consider a superspecies (Brown 1976).

## References

BROWN, JR., W. L. 1976. Contributions toward a reclassification of the Formicidae. Pt. VI. Ponerinae, tribe Ponerini, subtribe Odontomachiti. Sect. A. Introduction, subtribal characters. Genus *Odontomachus*. Studia Entomol. 19: 67–171.

CARLIN, N. F., AND D. S. GLADSTEIN. 1989. The "bouncer" defense of *Odontomachus ruginodis* and other odontomachine ants (Hymenoptera: Formicidae). Psyche 96: 1–19.

MOFFETT, M. W. 1989. Trap-jaw ants: Set for prey. Natl. Geogr. 175(3): 395–400.

## COBRA ANT

Formicidae, Ponerinae, Ponerini,
*Pachycondyla villosa. Spanish:* Pungara (Peru). *Portuguese:* Saracutinga, beijo de moça, cachorro magro (Brazil).

This large, black, fiercely stinging species (fig. 12.8e) resembles the giant hunting ants. Workers are always smaller (BL 12–13 mm), however, and have a distinctively shaped petiole. Viewed from above, it is very wide and appears like another abdominal segment; in side view, it has straight anterior and rounded posterior margins. The cuticle is densely punctate and set with golden yellow and white pubescence. The abdomen is also more or less straight sided, without a strong constriction as in the giants.

The cobra ant has a very wide range, occurring from southern Texas to northern Argentina, throughout continental, tropical America. Small colonies nest underground at the bases of trees or in rotting logs and stumps. Workers run rapidly on the soil, often in the bright sun, in search of insect prey.

## KELEP

Formicidae, Ponerinae, Ectatommini,
*Ectatomma tuberculatum*. Weevil ant.

The kelep (fig. 12.8c) is another fairly large solitary hunting ant (BL worker 11–

12 mm) but with a rather milder sting than its larger relatives. It is recognized by a coarsely sculptured, reddish-brown cuticle and the structure of the thorax, the dorsum of which has a central convexity bounded on all sides by distinct grooves and three protuberances dorsally on its anterior portion. Another anatomical characteristic is the large, compressed petiole node that, in side view, shows concave anterior and convex posterior margins. There are also strong bulbs at the bases of the antennae. Viewed from above, the abdominal segments appear swollen, the intersegmental joints constricted.

Widespread geographically (southern Mexico to Paraguay and southern Brazil), this ant is also common in a wide variety of habitats, from arid lands to rain forest. Its nests are ground burrows with multiple openings at the base of trees. From these emerge prominent tubular runways made of thatch or mud appressed against the side of the tree a few centimeters to a meter or more in height. Workers forage arboreally for small insects (especially social insects), nectar, and other food items. They are often seen tending extrafloral nectaries (pl. 3b). In Panama, acrobat ants (*Crematogaster limata*) have been observed to file into kelep nests, perhaps to steal food (Wheeler 1986).

The species was the subject of an early, abortive biological control scheme (Weber 1946). On a trip to Guatemala in 1904, O. F. Cook, an entomologist from the U.S. Department of Agriculture, found the ant associated with wild cotton. The natives of the area he visited, who called the ant "kelep," knew of the association and convinced Cook that the ant preyed on adult cotton boll weevils, keeping the plant free of that pest. The ants were reputed to carry off the weevils and feed them to their larvae, which have long flexible necks specially adapting them to reach inside and consume the soft tissues in the body of the prey (Cook 1905). In the same year, Cook

gathered keleps and introduced them into cotton fields in Texas, but they showed little inclination to colonize their new home and died out after a short time, failing to control the weevils that had recently invaded the fields in that state. Unfortunately, Cook's observations on the kelep were superficial, and many of his conclusions regarding its biology fanciful, as pointed out by William Morton Wheeler (1905, 1906), his more sagacious contemporary.

## References

Cook, O. F. 1905. The social organization and breeding habits of the cotton-protecting kelep of Guatemala. U.S. Dept. Agric. Tech. Ser. 10: 1–55.
Weber, N. A. 1946. Two common ants of possible economic significance, *Ectatomma tuberculatum* (Olivier) and *E. ruidum* Roger. Entomol. Soc. Wash. Proc. 48: 1–16.
Wheeler, D. E. 1986. *Ectatomma tuberculatum:* Foraging biology and association with *Crematogaster* (Hymenoptera: Formicidae). Entomol. Soc. Amer. Ann. 79: 300–303.
Wheeler, W. M. 1905. Dr. O. F. Cook's "Social organization and breeding habits of the cotton-protecting Kelep of Guatemala." Science (n.s.) 22: 706–710.
Wheeler, W. M. 1906. The kelep excused. Science (n.s.) 23: 348–350.

## ARMY ANTS

Formicidae, Ecitoninae, *Eciton* et al.
  *Spanish:* Hormigas militares, guerreras, soldados (General); arrieras (Panama); ronchadores (Costa Rica); tepeguas, cazadoras, carniceras (Peru). *Portuguese:* Formigas correções, f. taiocas, f. morupetecas, etc. (Brazil).

All the members of the subfamily Ecitoninae are loosely referred to as "army ants," a category also including the tropical driver or legionary ants (Dorylinae, *Dorylus*) of Africa (Schneirla 1971). The name is more appropriately applied to the 130 or so New World species, especially those in the genus *Eciton* (Borgmeier 1955, Franks

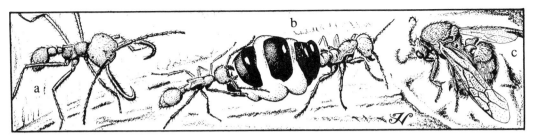

**Figure 12.9 ARMY ANTS (ECITON: FORMICIDAE).** (a) Major worker. (b) Physogastric queen, aided by minor worker. (c) Male.

1989). These are found throughout the lowland moist and wet forests of the American tropics from coastal northern Mexico to northern Argentina and southeastern Brazil. The range extends to the larger islands of the West Indies as well. Other, common, lesser army ants belong to the genera *Labidus* and *Neivamyrmex*. Several species are primarily subterranean, emerging to the soil's surface only at intervals to swarm and seek food. The larvae of a few species of these have been described (Wheeler and Wheeler 1984).

The best-known and most widespread species of army ants are *Eciton hamatum* and *E. burchelli*. Major workers of the former species are light brown to yellow, with shiny yellowish heads; the latter are dark brown to blackish, with reddish abdomens and dull-surfaced whitish heads.

A colony of either consists of the queen (fig. 12.9b), males (fig. 12.9c), workers, and immatures. The males are winged and live long enough only to exit the colony to mate with wingless queens. Workers are graded in size and proportions from the very large majors, or "soldiers," with outsized heads and gigantic hooked mandibles (fig. 12.9a), to tiny minors, one quarter the size of the majors. The nest is not a fixed structure but an enclave, called the bivouac, formed around the queen and brood by the interlocking bodies and limbs of the workers (pl. 2h).

The life of *Eciton* colonies is organized into a biphasic behavior cycle by the repro-ductive rhythm of the queen. A stationary ("statary") phase, lasting around 20 days, alternates with a slightly shorter migratory ("nomadic") phase, typically of 17 days.

In the stationary phase, the colony locates the bivouac in a well-protected place, usually within a hollow log or stump or under a partially buried fallen log. Within the bivouac, eggs are being laid by the queen, and pupae from the previous brood are developing. The presence of workers freshly emerged from their cocoons (so-called callow workers) releases the nomadic instinct, and the entire colony emigrates to a new bivouac site each day for the period of this phase. During these translocations, a retinue of workers closely guards the queen (Rettenmeyer et al. 1978). The bivouacs are placed in relatively exposed places, such as the space beneath low overhanging branches or between the buttresses of a tree. During this phase, the eggs hatch, and the larvae are brought to maturity. Return to the stationary phase ensues when the larvae have spun cocoons and pupated.

Virtually without fail, a raiding party issues from the bivouac each day to obtain animal prey, which is the only food of these ants. This habit represents the most complex instance of organized mass behavior occurring regularly outside the nest in any social insect or subhuman animal.

Raids of nomadic colonies of *E. hamatum* almost always start shortly after dawn; stationary colonies may raid in the morning or afternoon. Raids consist of thousands of workers reaching out over the forest floor and lower portions of the vegetation in search of any insects or small animals that may be found and captured. With some exceptions (Young 1979), *E. hamatum* shows a preference for paper wasp larvae and pupae; *E. burchelli* is less specialized, taking cockroaches, spiders, and katydids. The advancing front of the raid also takes two forms depending on the species. *E. hamatum* is a "column raider," in which the files of workers remain more or less distinct. This contrasts with the "swarm raid" of *E. burchelli*, in which the columns anastamose at the front into a solid, fan-shaped mass. Other species of *Eciton* demonstrate habits differing only in detail (Burton and Franks 1985).

Numerous imaginative and sensational accounts of the ferocity of ants in these swarms are to be found in popular literature. The model of these is the classic short story, "Leiningen versus the Ants," by Carl Stephenson. The author created an image of an ant capable of total devastation and destruction of anything in its path (including vegetation, an apparent confusion with the leaf cutter ants). This myth is now widespread, except among natives of regions where these ants abound. These people may actually welcome invasions by swarms for their value in exterminating household cockroaches and other vermin.

Raids are accompanied by much commotion among the other creatures of the forest. The whining wings of several species of tachinid (*Calodexia* and *Androeuryops*) and conopid flies (*Stylogaster*) mingle with the calls of ant birds and the rustle of fleeing insects, frogs, lizards, and other small animals (Rettenmeyer 1961). The flies take advantage of the panic to locate and parasitize certain of the insects roused by the ants. The insectivorous ant birds and other birds (Willis and Oniki 1978, Gochfeld and Tudor 1978) also find that hunting is made easier within the melee.

These regular "camp followers," and occasionally other insects, may be attracted to the swarm. Butterflies have been observed mingling with army ant swarms, possibly attracted by the latter's odor, which is said to be unpleasant and detectable by the sensitive human nose. The odor may simulate the butterflies' sexual pheromones (Drummond 1976) or resemble decay smells associated with proteinaceous food sources (Young 1977), including the droppings of ant birds (Lamas 1983, Ray and Andrews 1980) on which they commonly have been seen to feed. Many myrmecophilous members of the beetle families Staphylinidae, Histeridae, and Limulodidae, as well as silverfish and mites, have also been found intimately associated with army ants.

Amerinds have capitalized on the tenacity of the powerful "ice-tong" jaws of *Eciton* to aid in the healing of flesh wounds. They pinch the edges of the injury together and hold the ant close so that it bites across the cut. Then they wrench off the body, leaving the head intact and jaws fixed like sutures (Majno 1975). Large leaf cutter ant workers may also be used this way (Gudger 1925).

## References

BORGMEIER, T. 1955. Die Wanderameisen der neotropischen region. Studia Entomol. 3: 1–716, pls. 1–87.

BURTON, J. L, AND N. R. FRANKS. 1985. The foraging ecology of the army ant *Eciton rapax*: An ergonomic enigma? Ecol. Entomol. 10: 131–141.

DRUMMOND, B. A. 1976. Butterflies associated with an army ant swarm raid in Honduras. J. Lepidop. Soc. 30: 237–238.

FRANKS, N. R. 1989. Army ants: A collective intelligence. Amer. Sci. 77: 139–145.

GOCHFELD, M., AND G. TUDOR. 1978. Ant-following birds in South American subtropical forests. Wilson Bull. 90: 139–141.

GUDGER, E. 1925. Stitching wounds with the mandibles of ants and beetles. J. Amer. Med. Assoc. 84(24): 1861–1864.

LAMAS, G. 1983. Mariposas atraídas por hormigas legionarias en la Reserva de Tambopata,

Perú. Rev. Soc. Mexicana Lepidop. 8(2): 49–51.

MAJNO, G. 1975. Ant saga. *In* G. Majno, The healing hand: Man and wound in the ancient world. Harvard Univ. Press, Cambridge. Pp. 304–309.

RAY, T. S., AND C. C. ANDREWS. 1980. Ant butterflies: Butterflies that follow army ants to feed on ant bird droppings. Science 210: 1147–1148.

RETTENMEYER, C. W. 1961. Observations on the biology and taxonomy of flies found over swarm raids of army ants (Diptera: Tachinidae, Conopidae). Univ. Kans. Sci. Bull. 42: 993–1066.

RETTENMEYER, C. W., H. TOPOFF, AND J. MIRANDA. 1978. Queen retinues of army ants. Entomol. Soc. Amer. Ann. 71: 519–528.

SCHNEIRLA, T. 1971. Army ants, a study in social organization. Edited by H. Topoff. Freeman, San Francisco.

WHEELER, G. C., AND J. WHEELER. 1984. The larvae of the army ants (Hymenoptera: Formicidae). Kans. Entomol. Soc. J. 57: 263–275.

WILLIS, E. O., AND Y. ONIKI. 1978. Birds and army ants. Ann. Rev. Ecol. Syst. 9: 243–263.

YOUNG, A. M. 1977. Butterflies associated with an army ant swarm raid in Honduras: The "feeding hypothesis" as an alternative explanation. J. Lepidop. Soc. 31: 190.

YOUNG, A. M. 1979. Attacks by the army ant *Eciton burchelli* on nests of the social paper wasps *Polistes erythrocephalus* in northeastern Costa Rica. Kans. Entomol. Soc. J. 52: 759–768.

# FIRE ANTS

Formicidae, Myrmicinae, Solenopsidini, *Solenopsis*. *Spanish:* Hormigas de fuego, hormigas bravas (Cuba). *Quechua:* Pucacuro-kuna. *Portuguese:* Formigas de fogo, f. brasas, lavapés, cagafogos, doceiras, queima-queima, etc. (Brazil). *Indian:* Aracaras (Amazonia).

This is a large, ubiquitous genus with over one hundred species in the American tropics, including the Caribbean and Pacific oceanic islands (Creighton 1930, Snelling pers. comm.). A couple of dozen species (especially *S. saevissima* and *S. geminata*) are particularly noxious pests, which, because of their potent sting, are referred to as "fire ants." In spite of the widespread occurrence of these nuisance species, little is known of their biology, save with the two South American species that were introduced into the southern United States and created a furor that eventually reached political significance in that nation's capitol (Lofgren et al. 1975, Williams and Whitcomb 1973). These are the red imported fire ant (*S. invicta*) and the black imported fire ant (*S. richteri*). The former (fig. 12.10a) is the more important of the two, having spread over most of seven southeastern states since its accidental introduction in Mobile, Alabama, probably in the late 1930s.

It is not appropriate to review the fire ant saga here. From intensive studies on these invaders, however, some generalizations can be made about the genus's vital characteristics. All species nest in the ground, many forming large mounds around the entrance from soil castings. Tunnels extend belowground to a meter or more. Colonies are large (up to 230,000 workers) and often have two to many queens (polygyny). They are generally adapted to warm wet climates, although some species are desert loving. Their food habits are catholic, but they are primarily predators on other insects. They are highly competitive and often drive colonies of other ant species from their area. They suffer from a variety of parasites, which keep populations under control in South America (Williams and Whitcomb 1973).

They are also fierce defenders of their own nests, using highly toxic stings as effective weapons, for which they are best known and to which most of their popular names refer. The pharmacology of the venom has been analyzed and its composition possibly better known than that of any other ant. Major active components are piperidine alkaloids; proteins are noticeably lacking (MacConnell et al. 1971). The effects of venom injected into humans by

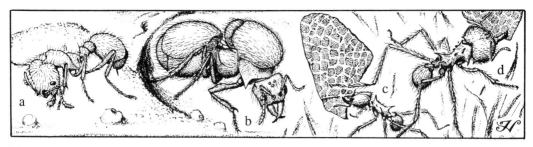

**Figure 12.10    ANTS (FORMICIDAE).** (a) Fire ant (*Solenopsis invicta*). (b) Leaf cutter ant (*Atta cephalotes*), dealate queen. (c) Leaf cutter ant, minor worker with leaf fragment. (d) Leaf cutter ant, major worker.

the ant's sting can be severe. Pustules form, and necrotic effects often follow. Allergic sensitivity frequently develops in response to random stings. Death has been known to occur in rare cases in highly sensitive individuals.

Some authors tell of fire ants in some areas so numerous and omnipresent as to be a scourge. Such was how the residents found Aveyros on the Tapajós River in Bates's time. He states (1892) that the village, "was deserted a few years before [his] visit on account of this little tormentor, and the inhabitants had only recently returned, thinking its numbers had decreased." Carvalho explains that the oldest hunting dogs on the Río Negro were blind in former days from corneas made opaque by continual fire ant stings. A local saying went, "the dog with white eyes is always a good hunter" (Lenko and Papavero 1979).

Workers of *Solenopsis* are recognizable by the structure of the antennae, which is unique among ants. They have ten segments with a two-segmented club. The body, whether dark or light, has a reddish tint and is smooth and shiny without spines but with sparse erect hairs. The petiole is two segmented, with distinct rounded nodes. The clypeus has a pair of elevated ridges that diverge toward the anterior end and (usually) distinct teeth that project beyond the anterior margin of this structure.

The little fire ant, belonging to a related genus (*Wasmannia auropunctata*), is a well-known, tropical mainland tramp species. It has become established in Puerto Rico (where it is called *abayalde*) and in the Galápagos Archipelago where it is now a serious pest, displacing native species and preying on the indigenous fauna and stinging humans (Silberglied 1972).

### References

BATES, H. W. 1892. The naturalist on the River Amazons. John Murray, London.

CREIGHTON, W. S. 1930. The New World species of the genus *Solenopsis* (Hymenop. Formicidae). Amer. Acad. Arts Sci. Proc. 66: 39–151.

LENKO, K., AND N. PAPAVERO. 1979. Insetos no folclore. Cons. Estad. Artes Ciên. Hum., São Paulo.

LOFGREN, C. S., W. A. BANKS, AND B. M. CLANCEY. 1975. Biology and control of imported fire ants. Ann. Rev. Entomol. 20: 1–30.

MACCONNELL, J. G., M. S. BLUM, AND H. M. FALES. 1971. The chemistry of fire ant venom. Tetrahedron 27: 1129–1139.

SILBERGLIED, R. 1972. The "little fire ant," *Wasmannia auropunctata*, a serious pest in the Galápagos Islands. Noticias Galápagos 19–20: 13–15.

WILLIAMS, R. N., AND W. H. WHITCOMB. 1973. Parasites of fire ants in South America. Tall Timbers Conf. Ecol. Anim. Hab. Mgmt. Proc. 5: 49–59.

## LEAF CUTTER ANTS

Formicidae, Myrmicinae, Attini, *Atta, Acromyrmex,* and other genera. *Spanish:* Zampopos, hormigas arrieras (Central

America, Colombia); mochomas (Mexico); bachacos (Venezuela); curuhuinsi (Peruvian Amazon); coqui (Peru); bibijaguas (Cuba). *Tupi-Guaraní:* Saúva, içá (Brazil). Wee wee (Belize), parasol ants, town ants.

These are such conspicuous ants that they have inspired a large number of fanciful indigenous names (Weber 1982). Those given above are only a sample of the more than fifty listed (Weber 1971: 6–7). Only two genera of about two hundred species of "gardening ants" (ants that cultivate fungi for food), or Attini, actually cut leaves and comprise the true leaf cutters, *Atta* and *Acromyrmex* (Cherrett and Cherrett 1989). These genera are distributed widely in the Americas, from the southernmost portions of the United States to northern Argentina and Uruguay, although they are absent from the Caribbean Islands except Cuba. In this range, these ants prefer relatively humid habitats and are therefore also completely absent from the southern half of the Andes and South America's coastal deserts.

Leaf cutters are fairly large, reddish-brown ants with strongly polymorphic workers. Major workers (fig. 12.10d) are often several times larger than other worker castes (fig. 12.10c) and have disproportionately enlarged, heart-shaped heads. The body is spiny and the appendages long and gangly. The base of the antenna is hidden by an overlapping frontal lobe. In the major genus, *Atta,* the males and females are much larger than even the major workers. In this genus, the female, especially, is an enormous ant (30–40 mm long, head to wing tips) with a huge, spherical abdomen (fig. 12.10b). In other genera, such as *Acromyrmex,* the reproductives are not the largest castes by much.

Because of the abundance of queens at emergence times and their size, they are exploited as food by aboriginals almost everywhere. They have even found their way into trade as traditional and novelty foods. The dried, packaged *hormigas colonas* (or *hormigas santandereanas*) of Bucaramanga, Colombia, are famous.

Nests of leaf cutters are highly complex and varied in architecture (Weber 1969, 1971). The surface mounds of common, well-known species of *Atta* (*cephalotes, sexdens*) may cover an area of 40 to 50 square meters. Nest density seems to be greater in disturbed habitats than in primary forest (Jaffe and Vilela 1989). Mature nests have many craters composed of soil fragments, broken down gardens, and other ejected debris (mostly from spent leaf substrate) and are conspicuous features of the landscape in many parts of tropical America, especially in open country like the pampas. Colonies may be fairly readily maintained in artificial nests on which detailed study is possible (Weber 1976).

These ants are best known for their ability to cultivate specific types of fungi on which they feed (Quinlan and Cherrett 1978). These fungi are cultured on snips of leaves and other parts of plants carried to special nest chambers by the workers. The ants tend these "gardens" continually, removing all alien bacterial and unwanted fungal growths, and they maintain a highly pure strain of preferred fungi (Boyd and Martin 1975). The tips of the hyphae produce peculiar round swellings (gongylidia) that are plucked and eaten or fed to larvae. Few of the latter have been identified to species, but *Leucocoprinus gongylophora* and *Lepiota* sp. may be the most important of those known. The identities of the fungal species involved are still controversial and certainly differ among their ant "host" species.

The source of the cuttings for the gardens may be any tree or shrub growing up to several hundred meters from the nest (Rockwood 1977). The long trails of workers carrying oval fragments over their heads (fig. 12.10c), winding their way over well-worked paths on the forest floor, are a familiar sight to tropical travelers (Fowler 1978). Workers maintain a trail by laying down a marking pheromone from rectal glands, although ants may be repelled or attracted by other odors along their foraging paths (Littledyke and Cherrett 1978).

Occasionally, a large worker is seen in these processions with a very tiny worker riding on its leaf cargo. Observations in Trinidad (Eibl-Eibesfeldt and Eibesfeldt 1967) have revealed the hitchhiker as protector of the larger ant that is otherwise preoccupied with its load and unable to avoid endoparasitic phorid flies seeking to oviposit on its neck. Milichiid flies of the genus *Pholeomyia* ride leaves similarly but for unknown reasons (Waller 1980).

Beneath the ground, the nest consists of numerous oval chambers connected by anastamosing tunnels and passageways. The latter may penetrate to considerable depths, up to 4 or 5 meters, and connect dozens of chambers. The latter are mostly used for fungus gardens, lesser numbers for brood rearing, and a single royal chamber is reserved for the queen. Depending on the species, there may be as many as 200,000 to 300,000 in such colonies.

Nests of *Acromyrmex* generally differ from those of *Atta* in that they are simpler and less extensive. In part of the seasonally flooded parts of the Amazon Basin, this genus is often found nesting in trees two or more meters above the ground.

Because leaf-cutting ants often strip foliage from cultivated vegetation and their mounds spoil otherwise level farmland, they have long been considered pests (Blanton and Ewel 1985). Even as early as the sixteenth century, they earned the title "King of Brazil." French naturalist St. Hilaire, impressed by the problem during his visit to that country in 1816–1822, said, "Either Brazil kills the saúva or the saúva kills Brazil!" All sorts of control measures have been applied against them, most with little success, although poison-laced citrus pulp, which the ants relish (Mudd et al. 1978), or other baits (Robinson et al. 1982), show promise. The improvement of tropical soils by their burrowing and mixing activities, as well as the multitudes of microorganisms they introduce, perhaps should earn them more appreciation than scorn (Jonkman 1978).

## References

BLANTON, C. M., AND J. J. EWEL. 1985. Leaf-cutting ant herbivory in successional and agricultural tropical ecosystems. Ecology 66: 861–869.

BOYD, N. D., AND M. M. MARTIN. 1975. Faecal proteinases of the fungus-growing ant, *Atta texana:* Properties, significance and possible origin. Ins. Biochem. 5: 619–635.

CHERRETT, J. M., AND F. J. CHERRETT. 1989. A bibliography of the leaf-cutting ants, *Atta* spp. up to 1975. Overseas Devel. Adm., Nat. Res. Inst. Bull. 14: 1–58. Includes *Acromyrmex*.

EIBL-EIBESFELDT, V. I., AND E. EIBESFELDT. 1967. Parasitenabwehren Minima-Arbeiterinen Blattschneide-Ameisen. Zeit. Tierpsychol. 24: 278–281.

FOWLER, H. G. 1978. Foraging trails of leaf-cutting ants. New York Entomol. Soc. J. 86: 132–136.

JAFFE, K., AND E. VILELA. 1989. On nest densities of the leaf-cutting ant *Atta cephalotes* in tropical primary forest. Biotropica 21: 234–236.

JONKMAN, J. C. M. 1978. Nest of the leaf-cutting ant *Atta vollenweideri* as accelerators of succession in pastures. Zeit. Angewan. Entomol. 86: 25–34.

LITTLEDYKE, M., AND J. M. CHERRETT. 1978. Olfactory responses of the leaf-cutting ants *Atta cephalotes* (L.) and *Acromyrmex octospinosus* (Reich) (Hymenoptera: Formicidae) in the laboratory. Bull. Entomol. Res. 68: 273–282.

MUDD, A., D. J. PEREGRINE, AND J. M. CHERRETT. 1978. The chemical basis for the use of citrus pulp as a fungus garden substrate by the leaf-cutting ants *Atta cephalotes* (L.) and *Acromyrmex octospinosus* (Reich) (Hymenoptera: Formicidae). Bull. Entomol. Res. 68: 673–685.

QUINLAN, R. J., AND J. M. CHERRETT. 1978. Aspects of the symbiosis of the leaf-cutting ant *Acromyrmex octospinosus* (Reich) and its food fungus. Ecol. Entomol. 3: 221–230.

ROBINSON, S. W., A. R. JUTSUM, J. M. CHERRETT, AND R. J. QUINLAN. 1982. Field evaluation of methyl 4-methylpyrrole-2-carboxylate, and ant trail pheromone, as a component of baits

for leaf-cutting ant (Hymenoptera: Formicidae) control. Bull. Entomol. Res. 72:
345–356.

ROCKWOOD, L. L. 1977. Foraging patterns and
plant selection in Costa Rican leaf cutting
ants. New York Entomol. Soc. J. 85: 222–233.

WALLER, D. A. 1980. Leaf-cutting ants and leaf-
riding flies. Ecol. Entomol. 5: 305–306.

WEBER, N. A. 1969. A comparative study of the
nests, gardens and fungi of the fungus growing ants, Attini. 6th Congr. Int. Union Study
Soc. Ins. (Bern 1969) Proc. Pp. 299–307.

WEBER, N. A. 1971. Gardening ants, the Attines.
American Philos. Soc., Philadelphia.

WEBER, N. A. 1976. A ten-year laboratory colony of *Atta cephalotes*. Entomol. Soc. Amer.
Ann. 69: 825–829.

WEBER, N. A. 1982. Fungus ants. *In* H. R.
Hermann, Jr., ed., Social insects. 4: 255–363.
Academic, New York.

## CORK-HEAD ANTS

Formicidae, Myrmecinae, Cephalotini,
*Zacryptocerus. Portuguese:* Cascudas,
chiazinhas (Brazil). Turtle ants.

The remarkable thing about *Zacryptocerus*
(Wilson 1976) is the large, disk-shaped
head and widely expanded prothorax of
the major workers. Colonies nest in abandoned burrows of wood-boring beetles and
other tubular hollows in dead twigs and
branches of trees and large grasses. In
some species, the head is convex and
lowered, so its long axis is about 120
degrees to the midline of the body and the
prothoracic projections thrust above it.
With the anterior part of the body in this
position at the entrance, the nest is effectively plugged, and unwanted intrusions
are thwarted. In other species, the head is
cup shaped and very round, held at a 90-
degree angle, and neatly fitted into the
circular entrance hole of the nest without
involvement of the thorax.

Cork-head ants may even move against
opponents, "bulldozing" them out of the
nest passageways. Some species emit a
volatile chemical secretion from the tip of
the abdomen, which can be tipped forward
over the thorax; others rely on their hardened, rigid body, heavy spines, and quick
movements for defense.

This head plugging ability (called
"phragmosis") was described early in this
century by the famous myrmecologist,
William Morton Wheeler, for the Caribbean species *Z. varians* and since confirmed by observations on *Z. texanus*
(Creighton and Gregg 1954, Creighton
1963) and other species (Wheeler and
Hölldobler 1985). These were all once
placed in *Cryptocerus* (and in *Paracryp-
tocerus*), as were others, but are now properly assigned to *Zacryptocerus*. The most
extremely modified heads are found in
the subgenus *Cyathomyrmex*. The surface
of the head is also often covered with a
fine granular encrustment that probably
functions as camouflage.

*Zacryptocerus* (fig. 12.11a) are small ants
(BL 4–6 mm) and hide in their nests most
of the time, although they occasionally are
seen crawling on the outsides of twigs and
grasses. Under a strong lens, the cephalic
and prothoracic disks of the majors of
some *Zacryptocerus* exhibit numerous flattened scalelike setae whose function is
unknown but probably is associated with
phragmosis.

The feeding habits of *Zacryptocerus* are
unknown, although they have been frequently seen taking honeydew. One species
(*Z. maculatus*) is a social parasite of *Azteca*
ants (Adams 1990). Their foraging activities apparently are diurnal, when they are
most often seen. A number of species in
Central America are believed to serve as
Batesian mimicry models for a wide variety
of other insects, primarily buprestid beetles of the genus *Agrilus* (Hespenheide
1986). The primary defense of these ants
seems to be distastefulness, their sting and
bite being of little consequence.

Cork-head ants of the related genus
*Cephalotes* (fig. 4.2d) are encountered in
small groups on tree trunks where their
large size (BL 10–15 mm), deep black

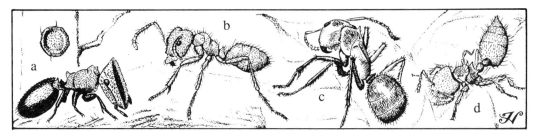

**Figure 12.11    ANTS (FORMICIDAE).** (a) Cork-head Ant (*Zacryptocerus varians*). (b) Aztec ant (*Azteca instabilis*). (c) Carpenter ant (*Camponotus sericeiventris*). (d) Acrobat ant (*Crematogaster stolli*).

color (often with a silvery sheen), odd shape, and lethargic movements attract notice.

### References

ADAMS, E. S. 1990. Interaction between the ants *Zacryptocerus maculatus* and *Azteca trigona:* Interspecific parasitization of information. Biotropica 22: 200–206.

CREIGHTON, W. S. 1963. Further studies on the habits of *Cryptocerus texanus* Santschi (Hymenoptera: Formicidae). Psyche 70: 133–143.

CREIGHTON, W. S., AND R. E. GREGG. 1954. Studies on the habits and distribution of *Cryptocerus texanus* Santschi (Hymenoptera: Formicidae). Psyche 61: 41–57.

HESPENHEIDE, H. A. 1986. Mimicry of ants of the genus *Zacryptocerus* (Hymenoptera: Formicidae). New York Entomol. Soc. J. 94: 394–408.

WHEELER, D. E., AND B. HÖLLDOBLER. 1985. Cryptic phragmosis: The structural modifications. Psyche 92: 337–353.

WILSON, E. O. 1976. A social ethogram of the Neotropical arboreal ant *Zacryptocerus varians* (Fr. Smith). J. Anim. Behav. 24: 354–363.

## AZTEC ANTS

Formicidae, Dolichoderinae, Tapinomini, *Azteca.*

All members of this genus are arboreal. They place their nests in shrubs and trees, sometimes in exposed situations but more usually in natural cavities, hollow stems, rot holes, insect burrows, and the like (Eidmann 1948). There are four groups, depending on the type of nest constructed.

Several species (among them the well-known *A. muelleri, A. alfari;* Longino 1991*b*) form spongy nests of a waxy material within the hollowed, jointed stems of trumpet trees (*Cecropia*) (Barnwell 1967, Harada and Benson 1988, Longino 1991*a*). Their association with the plant is mutualistic: the ant protects the tree from predation by herbivorous animals, and the plant provides the ant with an abode and nutritive substances (see ants and plants, above). Each tree supports a single colony of ants.

A second group makes carton nests, usually on the underside of a large inclined or horizontal tree branch. These nests, composed of masticated wood mixed with saliva, have a paperlike exterior; the interior is a labyrinth of cells and tunnels whose walls are made of a substance resembling cardboard. Old nests of *A. chartifex* (called *caçarema* in Brazil) have external stringy excrescences, giving them a shaggy appearance.

Some nests are very large and may attain a length of 2 to 3 meters and stand away from the supporting branch 30 to 40 centimeters. The nests of one Amazonian species (*A. trigona; barba* or *casicero* in the Peruvian Amazon) hang down in conical shapes that remind the natives of a man's beard (pl. 4*g*). These nests resemble those of arboreal termites and are often confused with them. However, the latter usually have runways leading from them and are of a more granular material than the structures made by the ants.

The third assemblage (*A. olitris, A. ulei, A. traili, A. delpini*) make "ant gardens," masses of soil and sprouting young plants placed in the crotches of limbs in the forest (pl. 4h). A fourth category consists of generalized cavity nesters.

*Azteca* workers are all small to very small ants (BL 1–1.5 mm) and mostly dull brown (fig. 12.11b). They are highly aggressive and, although lacking a sting, are capable of making themselves exceedingly offensive by descending on their enemies in droves, biting and emitting a repugnant odor (resembling butyric acid). This odor is quite noticeable even from a crushed individual. The abdomen is held erect during attacks, "cocktail" fashion like that of *Crematogaster,* but its apex is blunt rather than acute as in that genus and less capable of total flexion. There are no obvious structural characteristics to distinguish the genus. The workers are polymorphic, the majors have a large, heart-shaped head. The integument is sometimes soft, giving the body unusual flexibility.

About 150 species and subspecies can be listed in the genus from all parts of the American tropics, to which region the genus is restricted; certainly many more live there and will be found when this poorly known genus is properly studied. Because the application of many names is uncertain and different species are often confused, literature records must be used with caution (especially with *A. alfari*).

## References

BARNWELL, F. H. 1967. Daily patterns in the activity of the arboreal ant *Azteca alfari.* Ecology 48: 991–993.

EIDMANN, H. 1948. Zur Kenntnis der Ökologie von *Azteca muelleri* Em. (Hym. Formicidae), ein Beiträg zum Problem der Myrmecophyten. Abt. Syst. Okol. Geog. Tiere 77: 1–48.

HARADA, A. Y., AND W. W. BENSON. 1988. Espécies de *Azteca* (Hymenoptera, Formicidae) especializadas em *Cecropia* spp. (Moraceae): Distribuição geográfica e considerações ecológicas. Rev. Brasil. Entomol. 32: 423–435.

LONGINO, J. T. 1991a. *Azteca* ants in *Cecropia* trees: Taxonomy, colony structure, and behaviour. *In* C. R. Huxley and D. F. Cutler, eds., Ant-plant interactions. Oxford, Oxford.

LONGINO, J. T. 1991b. Taxonomy of the *Cecropia*-inhabiting *Azteca* ants. J. Nat. Hist. 25: 1571–1602.

## CARPENTER ANTS

Formicidae, Formicinae, Camponotini, *Camponotus. Spanish:* Hormigas agrias (Costa Rica). *Portuguese:* Sará sará (esp. *C. rufipes*), boca azedas, formigas de cupim, jejá (*C. abdominalis*), tracuá (*C. femoratus*) (Brazil).

*Camponotus* is the largest ant genus in the world, with over 500 species and subspecies listed for the Neotropics. It is also the most widespread and ecologically tolerant group of species, reaching all parts of the area including the most remote islands. Unfortunately, and undoubtedly because of their great number and confusing polymorphism, the species are in a very sad state taxonomically, and little is known of their biology. Various subgeneric categories have been devised to break this enormous generic taxon into workable divisions, but these have largely proved ineffectual in dealing with its complexity (Wheeler 1921).

These ants nest in almost every situation, commonly in sound or rotting wood, even that used in construction (hence the name carpenter ants). Many live under rocks, logs, and other objects on the ground and in ant gardens, bromeliads, and other epiphytes. Members of the subgenus *Colobopsis* nest in hollow twigs or branches in trees, in insect galls and nuts, and have soldiers with peculiar flat-fronted heads (like *Zacryptocerus*) for blocking the single entrance hole (phragmosis). A species in the subgenus *Myrmobrachys, Camponotus senex,* constructs large bag nests in trees, binding together leaves with silk produced by the larvae in a fashion similar to that of the Old World weaver ants, *Oecophylla* (Schremmer 1979).

*Camponotus* are not equipped with stings but are pugnacious and capable of inflicting considerable pain with their mandibles and secretions of formic acid and other chemicals that they spray into the bite. Those living in silk nests rush out onto the surface of the nest at the slightest provocation, setting up a rattle that resonates loudly in the dry bag and scares away any potential attacker.

Species whose food habits are known are generally omnivorous but prey to a large extent on other insects and arachnids (Matthiesen 1980). Few actually burrow into sound wood and do structural damage. *C. abdominalis* is one such pest, which also attacks commercial beehives, whose populations it may decimate.

*C. sericeiventris* (fig. 12.11c) is a common, well-known, polymorphic species that attracts attention because of its handsome silver or gold pubescence (Busher et al. 1985). Workers are unique in the genus in the possession of a sharp median crest running along the dorsum of the meso- and metathorax and sharp spines projecting obliquely from the anterolateral angles of the prothorax. The species nests in decayed portions of standing tree trunks some distance above the ground and is often seen during the day scrambling up and down tree trunks in search of prey. Major workers are very aggressive and able to bite severely. *C. sericeiventris* is mimicked by a cerambycid beetle, *Eplophorus velutinus* (W. M. Wheeler 1931), and a velvet ant, *Pappognatha myrmiciformis* (G. C. Wheeler 1983). The various subgenera previously recognized (W. M. Wheeler 1921) have now mostly been found invalid.

*Camponotus* are so diverse structurally that they defy diagnosis. One constant feature is the isolation of the antennal sockets from the clypeus, but this is sometimes difficult to observe. The workers are polymorphic, and those of many species are rather large (BL 10–18 mm).

## References

BUSHER, C. E., P. CALABI, AND J. F. A. TRANIELLO. 1985. Polymorphism and division of labor in the Neotropical ant *Camponotus sericeiventris* Guerin (Hymenoptera: Formicidae). Entomol. Soc. Amer. Ann. 78: 221–228.

MATTHIESEN, F. A. 1980. Sará-sará, formiga predadora de escorpiões e opiliões. Rev. Agric. (Piracicaba, São Paulo) 55: 239–241.

SCHREMMER, F. 1979. Das nest der neotropischen Weberameise *Camponotus (Myrmobrachys) senex* Smith (Hymenoptera, Formicidae). Zool. Anz. 203: 273–282.

WHEELER, G. C. 1983. A mutillid mimic of an ant (Hymenoptera: Mutillidae and Formicidae). Entomol. News 94: 143–144.

WHEELER, W. M. 1921. Professor Emery's subgenera of the genus *Camponotus* Mayr. Psyche 28: 16–19.

WHEELER, W. M. 1931. The ant *Camponotus (Myrmepomis) sericeiventris* Guérin and its mimic. Psyche 38: 86–98.

## OTHER ANTS

The acrobat ants form a large, nearly cosmopolitan genus (Myrmicinae, Crematogastrini, *Crematogaster*) and are well represented in the Neotropics (Buren 1958). They occupy many habitats and are common mutualistic associates of myrmecophytes. Small, monomorphic, and somewhat resembling aztec ants in the habit of elevating their abdomen when aroused, they have a sharply pointed, heart-shaped abdomen that is flat on top, convex below (fig. 12.11d). They lack any odor when disturbed or crushed. Nests often are of the carton type, placed in trees or shrubs, but are also situated in the ground, in termite nests, in natural plant cavities, in rotting logs, in dead twigs and branches and tree trunks, and in epiphytes.

The structure and use of the sting by acrobat ants is unique. The organ is flattened and blunt (spatulate) at the tip and quite unable to pierce the integument of prey or an enemy. When a droplet of venom is secreted, it clings to the tip of the sting from which it is applied topically to its target. To complete this process, the abdomen is elevated and arched all the way forward over the thorax and head in order to bring its apex in contact with the substratum ahead of the ant's body.

Members of the genus seem to have a particular proclivity for honeydew and are notorious protectors of the homopterous producers of this substance. Some construct succursal nests or "tents" around colonies of aphids and coccids which they keep in good repair and which hide and protect their charges (W. M. Wheeler 1906). One species raids the nests of the kelep (*Ectatomma tuberculatum*) (D. E. Wheeler 1986).

Big-headed ants (Myrmicinae, Pheidolini, *Pheidole; hormigas cabezones*) are also small but by contrast form a large assemblage of several hundred species in Latin America (Kusnezov 1951). They are ubiquitous and varied in their nesting and feeding habits, which are generally practiced along two lines, graminivory and predation. Workers of both are decidedly dimorphic, consisting of small minors with normal form and majors with enormous heads and powerful, ridged mandibles (fig. 12.12a). These formidable jaws are used for fighting in the predaceous group; for cracking seeds among the graminivores. In all workers, the head tends to be longer than wide and the antennae rather long. Nests of both types are usually situated in the ground but are sometimes made in decaying logs, termite nests, natural cavities in twigs, and even in the formicaries of myrmecophytes.

The stinging habits of fever ants (Pseudomyrmecinae, *Pseudomyrmex; tachí, formigas-de-novato* in Brazil, *hormiga del cornizuelo* in Costa Rica) have already been described in connection with ant plants. Not all, however, have close and specific associations with myrmecophytes, although they always are arboreal to some degree, usually nesting in preformed cavities, hollow stems, and the like (Kempf 1960). Colonies of some species have been found ensconced in termite nests. Colonies in acacias ("bull's-horn acacia ants") may have one queen or multiple queens. In the latter case, colonies may spread to many plants in an area and expand to enormous size, the largest of any ant in the world (as many as 1.8–3.6 million workers) (Janzen 1973).

These are medium to large (5 to 11 mm BL), rather elongate, slender ants, with slightly down curved abdomens and a two-segmented waist, usually reddish or yellowish brown in color (fig. 12.12b). Each of the two segments of the waist bears a low rounded node. Workers have a good-sized head, with large eyes. They are active and agile and quick to sting. The sting and venom apparatus is exceptionally well developed, and the workers can inflict pain to a degree second only to the giant solitary hunting ants (Ponerini). Sensitive persons

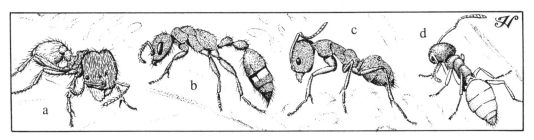

**Figure 12.12    ANTS (FORMICIDAE).** (a) Big-headed ant (*Pheidole fallax*). (b) Bull's-horn acacia ant (*Pseudomyrmex ferrugineus*). (c) Argentine ant (*Iridomyrmex humilis*). (d) Stink ant (*Tapinoma melanocephalum*).

react severely after the sting, the skin swelling and often blistering, occasionally with fever following.

A polysaccharide recently isolated from this ant's venom is known to be remedial in the treatment of rheumatoid arthritis (Schultz and Arnold 1977). The genus is found only in the New World, where it has evolved over one hundred fairly uniform species.

A very well-known Neotropical ant is the Argentine ant, *Iridomyrmex humilis* (Dolichoderinae, Tapinomini) (fig. 12.12c) (Mallis 1964). The species' fame derives from its particularly noxious nature in the southern half of the United States, where it was introduced via New Orleans very late in the nineteenth century. It is native and common to most of tropical South America (described originally from near Buenos Aires) and probably was transported north in shipments of coffee.

Colonies of this ant may number in the many thousands of individuals and often merge and have numerous queens (polygyny). Called the "Genghis Khan of the emmet world," it is aggressive and often displaces other ground-nesting species in its territory of invasion.

Workers are small (BL 2.25–2.75 mm) and dull brown and have an abdomen capable of considerable distension when they are engorging with food, especially honeydew, of which they are fond. The single waist segment has a well-developed node.

One of the most objectionable traits of the species is its habit of fostering aphids, mealybugs, and other homopterous pests in orchards and horticultural plots. It is the most common of the several house-infecting types in South and North America, hence its common name, "sugar ants" (*ciganas açucareiras* in Brazil). While its presence in the north has evoked a severe antipathy, it seems to be of little concern in its native countries.

Stink ants (Dolichoderinae, Tapino-mini, *Tapinoma*) (*hormigas hediondas*) rather resemble *Iridomyrmex* in their thin flexible integument, small size, and single segment in the waist. The node of the latter, however, is undeveloped. Only a dozen or so species live in the Neotropics. Some of these are easily recognized by the type of nest they build, a loose encrustment of earth laid as a linear tunnel along twigs and as a flat pancake on the undersides of leaves. Excited workers become aggressive and emit a characteristic odor like that of butyric acid. A widely known pest species is *Tapinoma melanocephalum*, the "crazy ant" (fig. 12.12d), whose wild gyrations on the table at mealtime always attract notice.

Other ant genera characteristic of the American tropics are *Hypoclinea* (formerly *Monacis;* fig. 10.16b), a dweller of the rain forest canopy; *Daceton*, a large myrmicine with a grotesque head and spined thorax; and *Pogonomyrmex*, the familiar harvester ants that are restricted to drier areas (Kugler 1984, González-Espinoza 1984).

## References

BUREN, W. F. 1958. A review of the species of *Crematogaster,* sensu stricto, in North America (Hymenoptera: Formicidae). Pt. I. New York Entomol Soc. J. 66: 119–134.

GONZÁLEZ-ESPINOZA, M. 1984. Patrones de comportamiento de forrajeo de hormigas recolectoras *Pogonomyrmex* spp. en ambientes fluctuantes (Hymenoptera Formicidae). Fol. Entomol. Mexicana 61: 147–158.

JANZEN, D. H. 1973. Evolution of polygynous obligate acacia-ants in western Mexico. J. Anim. Ecol. 42: 727–750.

KEMPF, W. W. 1960. Estudos sobre *Pseudomyrmex* I–III (Hymenoptera: Formicidae). (I) Rev. Brasil. Entomol. 9: 5–32. (II) Studia Entomol. 1: 433–462. (III) Studia Entomol. 4: 369–408.

KUGLER, C. 1984. Ecology of the ant *Pogonomyrmex mayri:* Foraging and competition. Biotropica 16: 227–234.

KUSNEZOV, N. 1951. El género *Pheidole* en la Argentina (Hymenoptera, Formicidae). Acta Zool. Argentina 12: 5–88.

MALLIS, A. 1964. Handbook of pest control. 4th ed. MacNair-Dorland, New York. Pp. 536–554.

SCHULTZ, D. R., AND P. I. ARNOLD. 1977. Venom of the ant *Pseudomyrmex* sp.: Further characterization of two factors that affect human complement proteins. J. Immunol. 119: 1690–1699.

WHEELER, D. E. 1986. *Ectatomma tuberculatum:* Foraging biology and association with *Crematogaster* (Hymenoptera: Formicidae). Entomol. Soc. Amer. Ann. 79: 300–303.

WHEELER, W. M. 1906. The habits of the tent-building ant (*Crematogaster lineolata* Say). Amer. Mus. Nat. Hist. Bull. 22(1): 1–18.

# BEES

Apoidea. *Spanish:* Abejas. *Portuguese:* Abelhas. *Quechua:* Amo, urunku. *Tupi-Guaraní:* Eira. *Nahuatl:* Pipiyolmeh, sing. pipiyolin. *Mayan:* Cab.

Bees (Stephen et al. 1969) comprise a major group of evolutionarily advanced Hymenoptera, derived from sphecid wasps through specialization for close association with flowers, from which they acquire pollen and nectar, their primary food sources; in contrast, wasps mostly take food of animal origin. In the process of collecting these substances, bees pollinate plants and thus are of major importance ecologically and economically to both natural plant communities and crops (Frankie and Coville 1979). This process is aided by the most distinctive feature of bees, a thick body vesture of finely branched (plumose) hairs, to which the pollen grains cling tenaciously and on which they are dispersed among flowers.

Bees usually collect pollen and nectar directly from the anthers and petal bases. Some however, have developed special techniques for forcing these substances from difficult flowers. Educated bees hover over blossoms with tubular anthers and buzz their wings violently, blowing the pollen onto their bodies ("buzz pollination"; Buchmann 1974), or bite holes near the base of the corolla of deep-throated flowers, taking a shortcut to steal the nectar from the side ("nectar robbing"; Wille 1963). The latter technique circumvents the anthers and fails to provide the service of pollination (Barrows 1976). Different species may compete directly for nectar sources (Hedström 1984).

Most bees are solitary and make tubular burrows for their nests in the soil or in other locations (hollow twigs, termite nests, rotting or sound wood, etc.) (Janvier 1955). Here they form cells in which to rear their larvae. An unidentified species of *Anthophora* (Anthophoridae) is doing considerable damage to Incan and recent adobe structures in the Urubamba Valley of Peru by boring into them in large numbers to make their nests (orig. obs.). Many types are parasitic and develop in the nests of other bees. These tend to be less hairy than nonparasitic forms and to more closely resemble wasps. Still others are social (Michener 1974).

Like wasps, female bees usually can sting but only do so in defense. The venoms of some species, particularly those in the subfamily Apinae (honeybees and others), may cause serious consequences when injected into humans (O'Connor and Peck 1978). A curious habit of the males of many bees is the formation of sleeping aggregations. Masses of dozens or even hundreds of individuals lock mandibles to leaves or twigs to pass the later afternoon and night.

Hundreds of species of the several families of bees live in Latin America (Michener 1954). Bees generally are most diverse in the arid and semiarid portions, although more highly evolved groups have apparently arisen in lowland rain forests (Michener 1979, Moldenke 1976).

## References

BARROWS, E. M. 1976. Nectar robbing and pollination of *Lantana camara* (Verbenaceae). Biotropica 8: 132–135.

BUCHMANN, S. L. 1974. Buzz pollination of *Cassia quiedondilla* (Leguminosae) of bees of the genera *Centris* and *Melipona*. So. Calif. Acad. Sci. Bull. 73: 171–173.

FRANKIE, G. W., AND R. COVILLE. 1979. An experimental study on the foraging behavior of selected solitary bee species in the Costa Rican dry forest (Hymenoptera: Apoidea). Kans. Entomol. Soc. J. 52: 591–602.

HEDSTRÖM, I. 1984. Interference competition between two species of *Ptiloglossa* bees (Hymenoptera, Colletidae) in the central valley of Costa Rica. Brenesia 22: 219–231.

JANVIER, H. 1955. Le nid et la nidification chez quelques abeilles des Andes tropicales. Ann. Sci. Natur. Zool. Biol. Anim. 17: 311–349.

MICHENER, C. D. 1954. Bees of Panamá. Amer. Mus. Nat. Hist. Bull. 104: 1–175.

MICHENER, C. D. 1974. The social behavior of the bees: A comparative study. Belknap Press, Harvard Univ., Cambridge.

MICHENER, C. D. 1979. Biogeography of the bees. Missouri Bot. Garden Ann. 66: 277–347.

MOLDENKE, A. R. 1976. Evolutionary history and diversity of the bee faunas of Chile and Pacific North America. Wasmann J. Biol. 34: 147–178.

O'CONNOR, R., AND M. L. PECK. 1978. Venoms of Apidae. *In* S. Bettini, ed., Venoms of arthropods. Springer, Berlin. Pp. 613–659.

STEPHEN, W. P., G. E. BOHART, AND P. F. TORCHIO. 1969. The biology and external morphology of bees. Agric. Exper. Sta., Ore. State Univ., Corvallis.

WILLE, A. 1963. Behavioral adaptations of bees for pollen collecting from *Cassia* flowers. Rev. Biol. Trop. 11: 205–210.

## SOLITARY BEES

The majority of bees are solitary, conducting their lives independent of others of their own kind and quite apart from hives or communal nests. Among these, a few forms have developed subsocial habits, however, and live in aggregations wherein females may actually cooperate with nest-building chores or form primitive colonies with a few offspring who remain with her and make up a kind of primitive worker caste.

### Leaf Cutter Bees

Megachilidae, Megachilinae, Megachilini.
*Spanish:* Ronsapitas (Peru).

The females of many species of this tribe, the most numerous of which belong to the genus *Megachile* (fig. 12.13a), cut circular pieces out of leaves which they use to line their nests. The nests are almost always placed in preexisting cavities, in pockets or narrow cracks and crevices in rocks and hard soil, and in hollow stems. The nests are elongate and the cells laid in tandem. Other species form brood cells from sawdust, plant down, leaf pulp, and other materials cemented together with resin collected from plants. These do not secrete wax or cementlike substances as do other bees.

Leaf cutter bees are widespread in the Neotropics in all habitats, from rain forest to coastal desert to mountains. They are always recognized by the presence of only two submarginal cells in the fore wing, extra-long tongues, and a large labrum, covered by the mandibles when they are closed. They are often dark, solid, black or

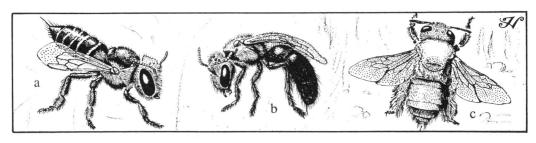

**Figure 12.13   BEES.** (a) Leaf cutter bee (*Megachile leucographa,* Megachilidae). (b) Sweat bee (*Lasioglossum* sp., Halictidae). (c) Centris bee (*Centris inermis,* Anthophoridae).

brown colors. Many others are buff with banded abdomens, and males of many species sport "flowerlike" structural developments on the forelegs which are used in mating. Leg segments are dilated and excavated on the inner surface and fringed with elongate hairs, a unique arrangement among bees. Leaf cutters also differ from other bees by carrying their pollen loads on the underside of the abdomen instead of on the legs; resting individuals often elevate the abdomen characteristically.

*Megachile* do not extend their range south farther than Panama, but other genera, like *Cressoniella, Pseudocentron,* and *Chrysosarus,* occupy most all of temperate to tropical Latin America. The taxonomy of leaf cutter bees has been recently updated by Mitchell (1980).

## Reference

MITCHELL, T. B. 1980. A generic revision of the megachiline bees of the Western Hemisphere (Hymenoptera: Megachilidae). Dept. Entomol. N.C. State Univ. Contrib.

## Sweat Bees

Halictidae, Halictinae, Augochlorini, *Augochlora* et al., Halictini, *Lasioglossum* and other genera. *Spanish:* Lameojos, morrocujes (Peru); chupadores (Costa Rica).

Small bees of the subfamily Halictinae are called "sweat bees" because so many have a fondness for human perspiration. An overheated person acts as an irresistible attraction for these bees, which become an extreme nuisance because of their insistence on drinking sweat, in so doing entering the eyes, nose, and ears and generally pestering one's body on hot, humid days. There are two basic kinds of sweat bees (Eickwort 1969a): those belonging to the tribe Halictini, principally the genus *Lasioglossum* (fig. 12.13b), and members of the tribe Augochlorini, primarily *Agopostemon.*

The latter also attract attention because of their brilliant metallic green or blue colors ("green sweat bees") (Eickwort 1969b). (Small stingless bees in the genus *Trigona* also display the sweat-drinking habit and likewise are called "sweat bees.")

Sweat bees are medium-sized (BL 5–10 mm), common, flower-visiting bees that nest in the soil. The nests are dug in compact soil (a few hollow out wood; Eickwort and Eickwort 1973a, 1973b) and consist of an oblique tube with brood cells issuing directly from its walls (most Augochlorini) or at the ends of long, lateral tubes (*Agopostemon* and allies).

Halictine bees are also remarkable for the primitively eusocial behavior exhibited by many species: colonies of the most highly evolved forms have legitimate queens and worker castes, although the latter are scarcely distinguishable from the former. Colonies are very small, never numbering more than several individuals. Other species of *Augochlorella* and *Augochlora* show habits ranging from completely solitary to subsocial or semisocial.

## References

EICKWORT, G. C. 1969a. Tribal positions of Western Hemisphere green sweat bees, with comments on their nest architecture (Hymenoptera: Halictidae). Entomol. Soc. Amer. Ann. 62: 652–660.

EICKWORT, G. C. 1969b. A comparative morphological study and generic revision of the Augochlorine bees (Hymenoptera: Halictidae). Univ. Kans. Sci. Bull. 48: 325–524.

EICKWORT, G. C., AND K. R. EICKWORT. 1973a. Aspects of the biology of Costa Rican halictine bees, V. *Augochlorella edentata* (Hymenoptera: Halictidae). Kans. Entomol. Soc. J. 46: 3–16.

EICKWORT, G. C., AND K. R. EICKWORT. 1973b. Notes on the nests of three wood-dwelling species of *Augochlora* from Costa Rica (Hymenoptera: Halictidae). Kans. Entomol. Soc. J. 46: 17–22.

SAKAGAMI, S. F., AND C. D. MICHENER. 1962. The nest architecture of the sweat bees. Univ. Kansas, Lawrence.

## Centris Bees

Anthophoridae, Anthophorinae,
  Centridini, *Centris.*

Bees of the genus *Centris* form a ubiquitous group in the Neotropics where some 150 species occur widely. They are fairly large, robust bees (BL 1–2.5 cm) and are conspicuously marked with contrasting black, yellow, and buff colors (fig. 12.13c). A common pattern is an all-red or all-black abdomen and light brown thorax, although several species groups display banded, bumblebeelike pelages. Others (subgenus *Melanocentris*) are very large, are mostly black (buff thorax), and much resemble carpenter bees. Structurally, centris bees are characterized by mandibles with pointed teeth, the basal tarsal segment of the hind leg which is shorter than the tibia, and a small marginal cell in the fore wing.

All are solitary, although nesting colonies and aggregations of sleeping males are often observed. Nesting habits vary greatly: nests are constructed in diverse substrates, such as clay banks, rotten wood, tree holes (Frankie et al. 1988), and even in arboreal termite nests, but in all cases, they are tubular burrows with one to a few branches, each containing several brood cells in series (Vinson and Frankie 1977). The young are reared entirely on pollen. Some species use empty cells in abandoned bees' nests. Preexisting burrows are used, or a new one is excavated (Coville et al. 1983).

Males of many *Centris* species are known to be highly territorial, patrolling specific areas, fighting off rival bees, and attacking other intruders. They even mark their areas with a citrallike substance produced by the pharyngeal glands (Raw 1975). Males also display two diverse mating strategies correlated with dimorphism in body size. Much larger males, called "metanders," fly over nesting areas and locate emerging females by odor. On locating the odor, they dig down and mate with females as they emerge. Smaller, "satellite" males hover within territories at the periphery of nesting areas, waiting for females that escape the larger forms unmated (Alcock et al. 1977, Frankie et al. 1980).

## References

ALCOCK, J., C. E. JONES, AND S. L. BUCHMANN. 1977. Male mating strategies in the bee *Centris pallida* Fox (Anthophoridae: Hymenoptera). Amer. Nat. 111: 145–155.

COVILLE, R. E., G. W. FRANKIE, AND S. B. VINSON. 1983. Nest of *Centris segregata* (Hymenoptera: Anthophoridae) with a review of the nesting habits of the genus. Kans. Entomol. Soc. J. 56: 109–122.

FRANKIE, G. W., S. B. VINSON, AND R. E. COVILLE. 1980. Territorial behavior of *Centris adani* and its reproductive function in the Costa Rican dry forest (Hymenoptera; Anthophoridae). Kans. Entomol. Soc. J. 53: 837–857.

FRANKIE, G. W., S. B. VINSON, L. E. NEWSTROM, AND J. F. BARTHELL. 1988. Nest site and habitat preferences of *Centris* bees in the Costa Rican dry forest. Biotropica 20: 301–310.

RAW, A. 1975. Territoriality and scent marking by *Centris* males (Hymenoptera, Anthophoridae) in Jamaica. Behavior 54: 311–321.

VINSON, S. B., AND G. W. FRANKIE. 1977. Nest of *Centris aethyctera* (Hymenoptera: Apoidea: Anthophoridae) in the dry forest of Costa Rica. Kans. Entomol. Soc. J. 50: 301–311.

## Carpenter Bees

Anthophoridae, Xylocopinae, Xylocopini,
  *Xylocopa. Spanish:* Ronsapas (Peru).
  *Nahuatl:* Xicotes (Mexico).

These are the largest bees in the Neotropics, the body length of some species (*Xylocopa fimbriata* or *X. frontalis*) measuring up to 26 millimeters or more. Thickly pubescent, with dark wings and ponderous flight, they somewhat resemble bumblebees but lack their well-defined yellow color bands on the abdomen or thorax (fig. 12.14a). The back of the thorax is flattened, and they possess powerful, blunt mandibles, characteristics not present in that other group of large bees. Another special feature identifying the genus is a triangular cell formed by the veins in the center of the

**Figure 12.14    BEES.** (a) Carpenter bee (*Xylocopa fimbriata,* Anthophoridae). (b) Orchid bee (*Euglossa purpurea,* Apidae). (c) Orchid bee (*Eulaema meriana,* Apidae).

fore wings toward the leading edge. Usually all black, some species possess green, orange, red, white, yellow, or other colored pelage; some species are sexually dimorphic, the males being shaded entirely with tan, the females dark. The wings are often dark tinted and display violaceous, even resplendent metallic reflections (*X. ornata*).

The biology of only a few species is known (Gerling et al. 1989, Janzen 1966, Sage 1968). Their habit of burrowing into wood gives these nonsocial bees their common name. The burrows ramify and anastomose through the wood to form complex tubular galleries in which the larvae are reared. During their developmental period, the larvae are provisioned with pollen and sealed into linear rows of cells with disk-shaped partitions. On emerging, the neotenic bees must wait their turn to escape, since less-developed siblings from eggs laid late block their exit paths.

All sorts of dead wood serve as substrata for the nests, including tree trunks, stumps, logs, large branches, and stems. Rarely do the bees select living plant tissues. A few species nest in the hollow, jointed stems of bamboo (subgenus *Stenoxylocopa;* Hurd 1978) and some even in the ground, rock crevices, or earthen tunnels. Occasionally, structural timbers and construction wood are attacked and weakened, so that carpenter bees are sometimes destructive. The Galápagos carpenter bee (*X. darwini*) is the principal pollen vector among plants in the Galápagos Islands (Linsley et al. 1966).

There are about 150 New World tropical species of *Xylocopa,* grouped into 17 subgenera (Hurd and Moure 1963). The group is considered poorly developed in comparison to the Old World, where most species also occur in warmer climes. Carpenter bees have been successful in reaching distant oceanic islands, probably through their wood-nesting habits. Infested logs, rafting on ocean currents, likely have carried colonizing females to the Galápagos, Revillagigedo, and West Indian islands (Hurd 1958).

### References

Gerling, D., H. H. W. Velthuis, and A. Hefetz. 1989. Bionomics of the large carpenter bees of the genus *Xylocopa.* Ann. Rev. Entomol. 34: 163–190.

Hurd, Jr., P. D. 1958. The carpenter bees of the eastern Pacific oceanic islands (Hymenoptera: Apoidea). Kans. Entomol. Soc. J. 31: 249–255.

Hurd, Jr., P. D. 1978. Bamboo-nesting carpenter bees (Genus *Xylocopa* Latreille) of the subgenus *Stenoxylocopa* Hurd and Moure (Hymenoptera: Anthophoridae). Kans. Entomol. Soc. J. 51: 746–764.

Hurd, Jr., P. D., and J. S. Moure. 1963. A classification of the large carpenter bees (Xylocopini) (Hymenoptera: Apoidea). Univ. Calif. Publ. Entomol. 29: 1–365.

Janzen, D. H. 1966. Notes on the behavior of the carpenter bee *Xylocopa fimbriata* in Mexico (Hymenoptera: Apoidea). Kans. Entomol. Soc. J. 39: 633–641.

LINSLEY, E. G., C. M. RICK, AND S. G. STEPHENS. 1966. Observations on the floral relationships of the Galápagos carpenter bee. Pan-Pacific Entomol. 42: 1–18.

SAGE, R. D. 1968. Observations on feeding, nesting and territorial behavior of carpenter bees genus *Xylocopa* in Costa Rica. Entomol. Soc. Amer. Ann. 61: 884–889.

## Orchid Bees

Apidae, Bombinae, Euglossini, *Euglossa*, *Eulaema* et al. *Spanish:* Chiquizás (Costa Rica, *Eulaema meriana*). Gold Bees, emerald bees.

Orchid bees (Dressler 1982) comprise five genera (Kimsey 1987) of more than two hundred species restricted to the forested, tropical portions of the New World (Moure 1967). The males of all are attracted to orchid flowers only by their odors, no food in the form of nectar or pollen being taken. On arriving at the flower, the bee scrapes the petals with its front tarsi and sponges up the exudate from the wound with masses of hairs on the tarsal segments. It then transfers the substance to storage tissues inside the enlarged hind tibia (Dodson 1966). In the process, the flower is pollinated, usually in a very specific manner. Males also collect fragrances from the blooms (and other parts) of plants in other families, timing their visits to coincide with the time of flowering (Armbruster and McCormick 1990).

The value of the substance to the bee is not known. Theories suggest a possible internal metabolic role or use as an acquired pheromone. If the latter, the substance volatilizes and may call other males. A swarm of such males, with their brilliant colors and loud buzzing, could lure females and effect the meeting of the sexes much like the lek aggregations of brightly colored birds (Kimsey 1980). It has also been suggested that males may also utilize the chemical as a metabolite (like a vitamin) or as a scent to mark a mating territory.

Many of the floral fragrances have been identified chemically and may be used in their pure state to attract male bees in the wild. Several are common drugstore compounds or perfume ingredients such as cineole (oil of eucalyptus), methyl salicylate (oil of wintergreen), benzyl acetate ("bubblegum flavor"), and even vanillin and are used by entomologists to survey and study these bees (Dodson et al. 1969, Janzen et al. 1982).

Female euglossines do not visit orchids but do visit other kinds of flowers for nectar and pollen, like most bees. Their nests vary considerably in size, construction, and location (Young 1985). Nests contain only a relatively few ovoid cells and are of three types: (1) subterranean or in cavities and constructed of chewed or shredded wood fragments, cemented together with wax and resin; (2) attached to exposed leaves or twigs and made entirely of resin (Young 1985); and (3) in hollow stems. Although not truly social, several females may form cells in a single globular mass. Those of *Eulaema* may even cooperate to some degree in nest building and may be considered subsocial (Zucchi et al. 1969).

Euglossine bees are recognized morphologically by their extremely long tongues, greatly swollen hind tibiae (males only), and small secondary lobe at the base of the main hind wing lobe, all features absent in their near bumblebee relatives. They also possess hind tibial spurs and are commonly bright metallic blue, green, or copper in color. The latter is true of *Euglossa* (fig. 12.14b), but *Eulaema* (fig. 12.14c) and *Eufriesea* (Kimsey 1982) have metallic green only on portions of the integument and are large (BL 2.5 mm), hairier, and with abdomens colored with black, yellow, and orange (often in bands), much resembling bumblebees. Females of *Eufriesea* sting fiercely and have evolved Müllerian mimetic complexes in South America (Dressler 1979) and have Batesian mimics among the Diptera. The much smaller (BL

12 mm), metallic *Euglossa* may be mimicked by the common green "drone flies," *Ornidia* spp.

*Exaerete* and *Aglae* generally resemble *Euglossa* with their shiny, bright colors but are generally twice as large (BL 25 mm) and have a more streamlined body shape with a narrower head and a tapered abdomen. They also lack the contrasting white facial markings of that genus. *Exaerete* has two tubercles on the scutellum; in *Aglae,* this is a large, flat, platelike structure. Biologically, these two genera are still more distinct, both being kleptoparasites in the nests of their *Eulaema* cousins.

A bibliography of orchid bees has been published (Williams 1978).

## References

ARMBRUSTER, W. S., AND K. D. McCORMICK. 1990. Diel foraging patterns of male euglossine bees: Ecological causes and evolutionary response by plants. Biotropica 22: 160–171.

DODSON, C. H. 1966. Ethology of some bees of the tribe Euglossini. Kans. Entomol. Soc. J. 39: 607–629.

DODSON, C. H., R. L. DRESSLER, H. G. HILLS, R. M. ADAMS, AND N. H. WILLIAMS. 1969. Biologically active compounds in orchid fragrances. Science 164: 1243–1249.

DRESSLER, R. L. 1979. *Eulaema bombiformis, E. meriana* and Müllerian mimicry in related species (Hymenoptera: Apidae). Biotropica 11: 144–151.

DRESSLER, R. L. 1982. Biology of the orchid bees (Euglossini). Ann. Rev. Ecol. Syst. 13: 373–394.

JANZEN, D. H., P. J. DeVRIES, M. L. HIGGINS, AND L. S. KIMSEY. 1982. Seasonal and site variation in Costa Rican euglossine bees at chemical baits in lowland deciduous and evergreen forests. Ecology 63: 66–74.

KIMSEY, L. W. 1980. The behavior of male orchid bees (Apidae, Hymenoptera, Insecta) and the question of leks. Anim. Behav. 28: 996–1004.

KIMSEY, L. S. 1982. Systematics of bees of the genus *Eufriesea* (Hymenoptera, Apidae). Univ. Calif. Publ. Entomol. 95: 1–125.

KIMSEY, L. S. 1987. Generic relationships within the Euglossini (Hymenoptera: Apidae). Syst. Entomol. 12: 63–72.

MOURE, J. S. 1967. A check list of the known euglossine bees (Hymenoptera, Apidae). Atas Simposio Biota Amazonica 5 (Zool.): 395–415.

WILLIAMS, N. H. 1978. A preliminary bibliography on euglossine bees and their relationships with orchids and other plants. Selbyana 2: 345–355.

YOUNG, A. M. 1985. Notes on the nest structure and emergence of *Euglossa turbinifex* Dressler (Hymenoptera: Apidae: Bombinae: Euglossini) in Costa Rica. Kans. Entomol. Soc. J. 58: 538–543.

ZUCCHI, R., S. F. SAKAGAMI, AND J. M. F. DE CAMARGO. 1969. Biological observations on a Neotropical parasocial bee, *Eulaema nigrita,* with a review of the biology of Euglossinae: A comparative study. Hokkaido Univ. J. Fac. Sci., Ser. 6, Zool. 17: 271–380.

## SOCIAL BEES

Bees with large numbers of worker offspring that cooperate in building and maintenance of complex colonies in nests or hives are considered true social insects and are represented in the American tropics by the bumblebees and stingless bees. The social honeybee (*Apis mellifera*), although well established in this part of the world, is an introduced species living in domestic hives or as a feral insect in a foreign environment.

### Reference

MICHENER, C. D. 1974. The social behavior of the bees: A comparative study. Belknap Press, Harvard Univ., Cambridge.

### Bumblebees

Apidae, Bombinae, Bombini, *Bombus.*
  *Spanish:* Abejorros, abejones (General).
  *Portuguese:* Mamangabas (var.
  mangangás). *Quechua:* Huayronco-
  kuna. *Nahuatl:* Xicohtin, sing. xicohtli.

Bumblebees (Alford 1975, Morse 1982) are large bees (BL 20–30 mm), distinguished from other large bees by their woolly hair that is colored in black and yellow (to orange) bands. The wings are tinged with dark pigment, and the hind pair lack a small second lobe at the extreme base of the usual posterior lobe. They also have very long tongues, a characteristic not found in other bees of great size.

They are widely distributed over the Americas, even to Tierra del Fuego, but are decidedly less abundant in warm lowland wet and moist forests (Dias 1958) than in more temperate, drier, deciduous forests and mountain habitats (Moure and Sakagami 1962). They are among the few insects active on cold days high in the mountains, occurring up to 4,000 meters in the Andes and Central American cordillera, where they are important pollinators. It is postulated by Janzen (1971) that nest sites are fewer and predation higher in the former habitats, contributing to their relative scarceness. This determines a tendency to nest on, rather than in, the ground and even in trees, also noted in the more tropical areas (Laroca 1976).

In mountainous terrain, it is typical for queens emerging from hibernation after the retreat of the cold season to establish new nests in natural hollows in the ground, including those made by rodents and birds. The larvae are reared in saclike, waxen cells ("honey pots") and fed on honey and pollen. Eventually, the colonies, which are true insect societies, may come to harbor several hundred queens, sterile worker females, and males in annual resting species. Colonies of 2,000 to 3,000 are

known in some perennial tropical species. The females sting readily and forcefully in defense of the nests. Males of many species of bumblebees fly along established routes, hovering momentarily at certain places (Stiles 1976) that they mark with scent to attract females (Blum 1981).

An outdated general taxonomic study indicates forty-five Neotropical species (Franklin 1913). The list has been modified some in more recent works (Milliron 1971), but the overall number is still about the same. Some are very widespread, as the ubiquitous, typically marked *Bombus tucumanus* (fig. 12.15a) and the all-yellow *Bombus dahlbomi*, which ranges over most of southern South America. A useful bibliography is available (Milliron 1970).

## References

ALFORD, D. V. 1975. Bumblebees. Davis-Paynter, London.

BLUM, M. S. 1981. Sex pheromones in social insects: Chemotaxonomic potential. *In* P. E. Howse and L. L. Clément, eds., Biosystematics of social insects. 19: 163–174. Academic, New York.

DIAS, D. 1958. Contribuição para o conhecimento da bionomia de *Bombus incarum* Franklin da Amazônia (Hymenoptera: Bombidae). Rev. Brasil. Entomol. 8: 1–20.

FRANKLIN, H. J. 1913. The Bombidae of the New World. Amer. Entomol. Soc. Trans. 39: 73–200.

JANZEN, D. H. 1971. The ecological significance of an arboreal nest of *Bombus pullatus* in Costa Rica. Kans. Entomol. Soc. J. 44: 210–216.

**Figure 12.15   SOCIAL BEES (APIDAE).** (a) Bumblebee (*Bombus tucumanus*). (b) Stingless bee (*Trigona fuscipennis*). (c) Stingless bee (*Trigona fulviventris*). (d) Honeybee (*Apis mellifera*).

LAROCA, S. 1976. Sôbre a bionomia de *Bombus morio* (Hymenoptera, Apoidea). Acta Biol. Paranaense 5(1–2): 107–127.

MILLIRON, H. E. 1970. A monograph of the Western Hemisphere bumblebees (Hymenoptera: Apidae; Bombinae). Entomol. Soc. Canada Mem. 65: 1–52. Bibliography only.

MILLIRON, H. E. 1971. A monograph of the Western Hemisphere bumblebees (Hymenoptera: Apidae; Bombinae). I. The genera *Bombus* and *Megabombus* subgenus *Bombias*. Entomol. Soc. Canada Mem. 82: 1–80.

MORSE, D. H. 1982. Behavior and ecology of bumble bees. *In* H. R. Hermann, ed., Social insects. 3: 245–322. Academic, New York.

MOURE, J. S., AND S. F. SAKAGAMI. 1962. As mamangabas sociais do Brasil (*Bombus* Latr.) (Hym. Apoidea). Studia Entomol. 5: 65–194.

STILES, E. W. 1976. Comparison of male bumblebee flight paths: Temperate and tropical (Hymenoptera: Apoidea). Kans. Entomol. Soc. J. 49: 266–274.

## Stingless Bees

Apidae, Meliponinae, Meliponini, *Melipona, Trigona,* and *Lestrimelitta.*
*Spanish:* Abejas sin aguijón (General); zeganas (Panama); abejas bobos, angelitos (Colombia); arambasas (Amazonian Peru); pegones (Venezuela); culos de vaca, abejas jicotes, abejas atarrá (Costa Rica). *Portuguese:* Abelhas sem ferrão, torcecabelos, irapua, abelhas de cupim, cupira, jataí, xupé, abelhas bravas, cagafogos, etc. (Brazil). Juanats (*Melipona*), pegones (*Trigona*) (Trinidad). *Nahuatl:* For specific types only, e.g., necutli, pipiolin.

Certainly more can be said of the natural history of this tribe than any other Neotropical bee group (Sakagami 1982, Schwarz 1948). That they have been profoundly important in the region's human culture also is evidenced by the many common names applied to them, not only as a group but to individual species (see above and Lenko and Papavero 1979: 171).

Before the introduction of sugarcane and the European honeybee (*Apis mellifera*) to the New World, the chief source of sweets was stingless bee honey (Crane 1983). To this day, the sweet product of these bees is preferred widely among Indians and country people, who make from it many condiments, beverages, and medicinals. The honey of specific bees was even supposed to have value for particular ailments, for example, honey from *Trigona jaty,* a widely cultivated species, in southern Brazil, is a folk remedy for a sore throat. Balche, made from stingless bee honey, was a prime ceremonial drink of the Maya and drunkenness from imbibing it was compulsory in religious rituals. Honey from some stingless bees, notably, the lemon bee (*Lestrimelitta limao*), is poisonous, a quality curiously contributing to, rather than detracting from, its therapeutic use. The Guarayo Indians of Bolivia were reputed to use honey from this species for the cure of paralysis.

Because of their long association with stingless bees, Amerinds succeeded in developing a form of apiculture crudely parallel to that of Africa with the honeybee which is still practiced today ("meliponiculture"; Nogueira-Neto 1953, Weaver and Weaver 1981). Native hives commonly consist of hollow logs that are seeded with portions of comb from a wild nest. After the new colony develops to a healthy size, its honey pots are harvested by removing the end plugs. Although a number of species are occasionally kept in this manner, *Melipona beecheii* is most frequently domesticated and even referred to in early accounts as *Melipona* "domestica" (Weaver and Weaver 1981).

No less important than its honey, the wax of the stingless bees found numerous applications among Pre-Columbian people, many of which continue among rurals even today. Miscellaneous applications include making candles, waterproofing articles, and forming religious icons. It is also a common adhesive, calking material, filler, lubricant, and occasional medicinal in minor therapeutics such as the removal of corns and warts. Its greatest historical significance, however, surely derives from its use in metallurgy, which all the classic Pre-Columbian civilizations discovered without influence from the East. Ancient goldsmiths molded gold jewelry and other items of the finest quality with a "lost wax technique" identical to that practiced by Old World artesans but employing wax from native meliponines in place of that of the honeybee (Bird 1979).

Such was the prominence of stingless bees to primitives that they inevitably became entwined in the culture (Posey 1983). Certain Paraguayan tribes recognized property rights in wild honey. Tributes were often paid in honey and wax; the Codex Mendoza text specifies quantities that were to be delivered to Moctezuma by lowland Aztec communities. Even among the Yucatán Maya today, a major ceremony known as the *u hanli cab* is celebrated in which the ancient Maya bee gods are beseeched to bless the cultivated bees (Weaver and Weaver 1981).

These bees are common and conspicuous throughout the central Neotropics (Wille 1961), especially in moist lowland forest environments. They are absent from the high Andean valleys, coastal deserts, and Antilles except for the large islands and those close to the mainland. They are most directly recognized by their bothersome habits when aroused and their associations with their nests, which are always densely populated, aggressively defended, and uniquely constructed. Most are located in natural cavities, usually in the ground or tree trunks but sometimes in odd sites such as dry mammal carcasses and bird, termite, or ant nests. The particular arrangement of structural elements vary (Roubik 1979, Wille and Michener 1973), but the nest always contains brood cells in a cluster (*Trigona*) or layered in horizontal combs (*Melipona*), these surrounded by a layered envelope, storage pots for honey and pollen located outside the envelope, and waxen entrance canal that often extends outside the nest as a freely projecting tube. The whole complex is walled off from the exterior by hard end plates or an outer shell called the batumen. Several kinds of building materials go into nest construction, primarily wax, but this is usually mixed with other matter, such as propolis, plant resin, and gum collected by the bees (Ramírez and Gómez 1978). This habit may have led to the entombment of individuals, several of which have been found as fossils in copal (Moure and Camargo 1978). Mixed with wax (cerumen), this is the substance of the brood cells. Mud, feces, plant fibers, and leaf fragments also are used in nest formation. The larval provisions of some species are known to support a rich bacterial flora that may play a fundamental role in the preservation and metabolic conversion (Gilliam et al. 1985) of these substances.

This type of nest is the most elaborate of all native social bees in the New World and identifies the stingless bees at once, as does their method of defense in the absence of a sting. The sting organ is vestigial and of no use in inflicting wounds on large enemies, but these bees are by no means impotent. In numbers, they hurl themselves on those who threaten the nest, crawling into nostrils, ears, hair, and eyes. Although most employ only mandibles to pinch, a few also deposit a caustic fluid from glands at the bases of the jaws. Many of the native names of these refer to their belligerence and potency (e.g., *cagafogos*, "spit fires"; *torcecabellos*, "hair twisters").

Structurally, stingless bees resemble other apids in having broadly expanded hind tibia fringed with hairs, which form the pollen basket (but these are absent in *Lestrimelitta*). They are recognized readily from other social bees by their usual smaller size (BL at most 15 mm), relative hairlessness (no pelage on legs), and blunt tip to the abdomen. The wing venation is also unique, the marginal cell of the fore wing being open to the wing margin at the latter's apex.

Their biology (de Camargo 1972, Johnson and Hubbell 1974, Kempf 1962) also resembles that of the honeybee in many respects, both being social and producing large broods. The honeybee is mass provisioned, however. Foragers also communicate distance and direction to nestmates like honeybees but with only sound signals, a complete symbolic dance not having evolved, although returning foragers do move through the hive performing "buzzing runs" similar to the acoustic portion of the waggle run of honeybees (Esch et al. 1965). Other differences are found in the composition of reproductive swarms, which include only young, never original, queens, and production of wax from dorsal rather than ventral glands on the abdomen. Some stingless bees forage for various forms or organic matter, including dead animals (Baumgartner and Roubik 1989).

The tribe includes three genera, each containing many common species, except *Lestrimelitta,* which has but two species. One, *L. limao,* is known for its lemon odor when crushed (from citral; Blum et al. 1970). These lack a pollen basket on the hind tibia, are shiny black with round heads, and are medium sized (BL 8 mm). Obligate robbers of honey from the storage chambers of other stingless bees, the workers are not adapted morphologically or behaviorally for normal foraging (Sakagami and Laroca 1963). When scouts locate a suitable nest, they are probably killed at the entrance, and the process releases large amounts of citral, the trail-marking pheromone for this and other stingless bees. The odor pervades the nest and diffuses into the air, attracting more *Lestrimelitta* and confusing the victim bees (Blum et al. 1970), but this may not always be highly effective (Johnson 1987). *Trigona*

also produce a volatile alarm pheromone, composed of aliphatic alcohols, ketones, and benzaldehyde (Luby et al. 1973).

Although there are exceptions, *Melipona* species are generally the largest of the three genera (BL 6–15 mm), are relatively hairy, and have wings that do not extend beyond the tip of the abdomen when folded (fig. 12.15b) (Schwarz 1932). The integument surface is dull and nonreflective. Some of the larger species are all black with white-tipped wings. The very widespread *Melipona beechii* is pale brown and superficially looks much like the honeybee. *Trigona* are mostly smaller (BL 2–8 mm; *Trigona duckei* is the smallest bee known), are sparsely hairy, have wings that extend well beyond the abdomen when at rest, and are often shiny. *T. fulviventris* (fig. 12.15c) is perhaps the most common; it is recognized by its largeness for the genus (BL 5–6.5 mm) and contrasting black thorax and orange abdomen (Johnson 1983).

Stingless bees are generally beneficial through their pollination activities. However, they occasionally harm fruit crops (especially citrus) by cutting the flowers into pieces, which they use in nest construction.

## References

BAUMGARTNER, D. L., AND D. W. ROUBIK. 1989. Ecology of necrophilous and filth-gathering stingless bees (Apidae: Meliponinae) of Peru. Kans. Entomol. Soc. J. 62: 11–22.

BIRD, J. 1979. Legacy of the stingless bee. Nat. Hist. 88(5): 48–51.

BLUM, M. S., R. M. CREWE, W. E. KERR, L. H. KEITH, A. W. GARRISON, AND M. M. WALKER. 1970. Citral in stingless bees: Isolation and functions in trail-laying and robbing. J. Ins. Physiol. 16: 1637–1648.

CRANE, E. 1983. The archaeology of beekeeping. Cornell Univ. Press, Ithaca.

DE CAMARGO, C. A. 1972. Mating of the social bee *Melipona quadrifasciata* under controlled conditions (Hymenoptera, Apidae). Kans. Entomol. Soc. J. 45: 520–523.

ESCH, H., I. ESCH, AND W. E. KERR. 1965. Sound: An element common to communication of stingless bees and to dances of honey bees. Science 149: 320–321.

GILLIAM, M., S. L. BUCHMANN, AND B. J. LORENZ. 1985. Microbiology of the larval próvisions of the stingless bee, *Trigona hypogea*, an obligate necrophage. Biotropica 17: 28–31.

JOHNSON, L. K. 1983. *Trigona fulviventris* (Abeja atarrá, abeja jicote, culo de vaca, trigona, stingless bee). *In* D. H. Janzen, ed., Costa Rican natural history. Univ. Chicago Press, Chicago. Pp. 770–772.

JOHNSON, L. K. 1987. The pyrrhic victory of nest-robbing bees: Did they use the wrong pheromone? Biotropica 19: 188–189.

JOHNSON, L. K., AND S. P. HUBBELL. 1974. Aggression and competition among stingless bees: Field studies. Ecology 55: 120–127.

KEMPF, N. 1962. Mutualism between *Trigona compressa* Latr. and *Crematogaster stolli* Forel (Hymenoptera: Apidae). New York Entomol. Soc. J. 70: 215–217.

LENKO, K., AND N. PAPAVERO. 1979. Insetos no folclore. Cons. Estad. Artes Ciên. Hum., São Paulo.

LUBY, J. M., R. E. REGNIER, E. T. CLARKE, E. C. WEAVER, AND N. WEAVER. 1973. Volatile cephalic substances of the stingless bees, *Trigona mexicana* and *Trigona pectoralis*. J. Ins. Physiol. 19: 1111–1127.

MOURE, J. S., AND J. M. F. CAMARGO. 1978. A fossil stingless bee from copal (Hymenoptera: Apidae). Kans. Entomol. Soc. J. 51: 560–566.

NOGUEIRA-NETO, P. 1953. A criação de abelhas indigenas sem ferrão. Ed. Chácaras e Quintais, São Paulo.

POSEY, D. A. 1983. Keeping of stingless bees by the Kayapó Indians of Brazil. J. Ethnobiol. 3: 63–73.

RAMÍREZ, W., AND L. D. GÓMEZ. 1978. Production of nectar and gums by flowers of *Monstera deliciosa* (Araceae) and of some species of *Clusia* (Guttiferae) collected by New World *Trigona* bees. Brenesia 14–15: 407–412.

ROUBIK, D. W. 1979. Nest and colony characteristics of stingless bees from French Guiana (Hymenoptera: Apidae). Kans. Entomol. Soc. J. 52: 443–470.

SAKAGAMI, S. F. 1982. Stingless bees. *In* H. R. Hermann, ed., Social insects. 3: 361–423. Academic Press, New York.

SAKAGAMI, S. F., AND S. LAROCA. 1963. Additional observation on the habits of the cleptobiotic stingless bees, the genus *Lestrimelitta* Friese (Hymenoptera, Apoidea). Hokkaido Univ. Fac. Sci., Ser. 6, Zool. J. 15: 319–339.

SCHWARZ, H. F. 1932. The genus *Melipona*: The type genus of the Meliponidae or stingless bees. Amer. Mus. Nat. Hist. Bull. 63: 231–460.

SCHWARZ, H. F. 1948. Stingless bees of the Western Hemisphere. Amer. Mus. Nat. Hist. Bull. 90: 1–546.

WEAVER, N., AND E. C. WEAVER. 1981. Beekeeping with the stingless bee *Melipona beecheii*, by the Yucatecan Maya. Bee World 62: 7–19.

WILLE, A., AND C. D. MICHENER. 1973. The nest architecture of stingless bees with special reference to those of Costa Rica (Hymenoptera, Apidae). Rev. Biol. Trop. 21(suppl. 1): 1–278.

WILLE, A. 1961. Las abejas jicotes de Costa Rica. Univ. Costa Rica Rev. 22: 1–30.

## Honeybee

Apidae, Apinae, Apini, *Apis mellifera*. *Spanish:* Abeja de miel. *Portuguese:* Abelha doméstica.

Among all of the dubious gifts bestowed by the Old World on the New World following its discovery, at least one was equal to the riches returned. This was the honeybee (fig. 12.15d), from which a quantity and quality of honey (Crane 1979) is obtained much superior to that of native meliponine (stingless) bees. Unfortunately, early records of the establishment of *Apis mellifera* in tropical America are scanty. Apparently, its introduction came much later than into North America, where it arrived with the earliest colonists in the early 1600s. It was not until 1839 that honeybees were brought to Brazil (Nogueira-Neto 1962) and 1857 that they first reached Chile and Peru. Other dates quoted for the introduction and spread of the species into various other countries of Central and South America are at variance, and the history of apiculture in tropical America still needs to be documented.

The bee industry today is well established and an important factor in rural economics in Latin America (Crane 1978; Ordetx 1952; Ordetx and Pérez 1966; Smith 1960). Until a short time ago, the biggest exporter and producer was certainly Mexico, followed closely by Argentina. Other countries with major commitments to beekeeping are Brazil, Chile, Colombia, Costa Rica, and Venezuela.

This picture changed in the 1960s and 1970s as a result of the unfortunate, accidental introduction to the New World of the African subspecies, *Apis mellifera scutellata* (cited until most recently as *A. m. adansonii*), which is hardly distinguishable from other subspecies (Daly and Balling 1978, Ruttner 1976).

This unexpected chapter in the story of the honeybee in America began in 1956 with the intentional introduction of forty-six queens from South Africa which were shipped to Rio Claro, São Paulo State, Brazil, where they were to be carefully interbred with bees of European origin already established in the country to improve the latters' productivity. The following year, a misfortune permitted the escape of twenty-six of these queens to the wild, where they started a series of events that became exaggerated into a horror story (Michener 1975, Taylor 1977, Taylor and Levin 1978). As a result of their intense defense of their nests, there were mass attacks on humans and even a few deaths. These happenings received great notoriety, which led to an image of a horde of so-called killer bees heading across the land, threatening to cause great loss of life and wreak havoc on established bee colonies in its path. The new entity was dubbed also variously as the "Mau Mau bee," "Brazilian bee," "African bee," and "kamikaze bee" and even inspired incredible novels (*The Swarm*, by Arthur Herzog, 1974) and motion pictures (*Savage Bees*).

Although not the monster it was first purported to be, the new mixed strain has persisted, nevertheless, and retains many of its undesirable traits as it slowly expands its range outward over South and Central America. As of 1991, it had reached northern Mexico and spread to most northern South American countries (Venezuela, Trinidad) and touched much of Colombia,

Peru, Ecuador, and even points on the northern coast of Chile. Some entomologists predict that it will eventually occupy most of Latin America and extend to the southern fourth of the United States where it will displace the gentler European bees now tended there by beekeepers, seriously hindering honey production and pollination management. There is evidence also that Africanized bees displace important native bee pollinators (Roubik 1979). In 1983, a program was initiated to quarantine the bee to a control zone in Costa Rica, along the Panama border, in an effort to halt its northward spread (Stibick 1984). Apiculturists will either have to adapt to the new bee (Roubik and Boreham 1990), as they have in southern Brazil (Erickson et al. 1986, Gonçalves 1982, Wiese 1977), or be forced out of business, as has been the case in many parts of South America.

The biology of the honeybee is too well known to repeat in detail here (Butler 1955, Dade 1962, Dietz 1982, Winston 1987). Some basic facts may be reviewed, however, especially as they differ in the Africanized strain and relate to the survival of a species primarily adapted to a temperate climate in its adopted, largely tropical, home (Crane 1980, Ordetx 1952, Smith 1960).

The life of the colony follows a similar pattern whether free in the wild or pampered in a beekeeper's box. It is largely regulated by the workers whose pheromones help control the internal reproductive instincts of the queen, which, in turn, determines the number of other workers, timing of drone production, new queen development, and other biological phenomena such as swarming. The latter is the method of founding new colonies and is of two basic types: "reproductive swarming," in which a portion of the parental hive leaves with a new queen, and "absconding," which is the total removal of the entire colony to a new site. The tendency to abscond appears to be more common in

Africanized than in European bees (Winston et al. 1979). Under human management, swarming is unwanted and suppressed but still occurs from commercial hives and results in feral populations. These take up residence in hollow trees, in rock crevices, and in other protected places. Here, workers quickly construct a waxen comb with rows of hexagonal cells placed back to back in hanging layers. The wax is secreted from glands on the underside of the abdomen and molded into shape by the mandibles and legs.

A portion of the cells are used as rearing chambers for the larvae of new workers. These will produce females similar in structure to the queen but smaller, sterile, and burdened with a multitude of duties, including the gathering of pollen and nectar from flowers. The pollen is fed to the young unchanged; the nectar is converted to honey through enzymatic action in the worker's gut. Excesses are stored in other cells, apart from the brood cells, and are mainly for use during hard times. Worker larvae are fed almost exclusively on pollen and honey. Africanized bees are considered more industrious than European, one of the desirable traits for which it was imported originally, though in reality, it is not a better honey producer. Workers of all honeybees communicate the location of nectar sources with a "dance language," scent marking not being of primary importance (Gould 1976).

From time to time, the colony produces drones and/or virgin queens. The former develop from unfertilized eggs and therefore have half the complement of chromosomes in their tissues. The latter grow from normal eggs, but their larvae are fed a special diet of "royal jelly," a protein-rich substance secreted by glands emptying into the workers' pharynx. Queens also develop in oversized, irregularly shaped cells.

Emergence of virgin queens, with fully functional reproductive organs, follows swarming by a few days. The pattern varies considerably but most commonly goes as follows. After providing for a new generation of queens, the old queen leaves the hive with a retinue of workers bound for a new home. From the queen cells left behind, one potential new queen hatches soon, ahead of the others, and destroys her rivals, often with the help of workers. She then issues from the hive, mates with drones from other hives (who immediately die), and returns to assume the role of her departed parent.

Insemination lasts the queen's lifetime; sperm stored in a special pouch off the oviduct are released as needed to fertilize eggs passing down the tract. Queens may survive a year or longer, laying up to 1,500 or more eggs a day in favorable seasons, and produce mature colonies of up to 60,000 workers.

Both queens and workers have stings. The sting of the latter is barbed so that it usually becomes anchored in the wound when used on large animals, and the attached poison sac and other internal organs are wrenched out when the bee pulls free. The venom is potent, a complex mixture of proteins that elicit varied, often severe, immune reactions in humans. These are sometimes found to remedy certain complaints, including arthritis and some neural disorders, and controlled venom injections are occasionally prescribed by physicians in the treatment of these ailments. The venom of Africanized bees is no more potent than that of other honeybees, but a victim is likelier to be attacked in mass as a result of an irresistible alarm pheromone given off by enraged workers.

The species, in general, is adapted to semidry, temperate climates and lives best at higher elevations and latitudes in Latin America. For this reason, it has never been successfully kept in wet lowland tropical zones; it is largely absent from Amazonia in spite of attempts to establish it there; the Africanized strain, however, is hardier and seems to be making inroads even into this warm and humid environment.

The honeybee is susceptible to a number of harmful parasites, including several parasitic mites (Camazine 1986, DeJong et al. 1982), the bee louse, (*Braula coeca*) (Weems 1983), and phorid flies of the genus *Melaloncha* (Ramírez 1984).

## References

BUTLER, C. G. 1955. The world of the honeybee. Macmillan, New York.

CAMAZINE, S. 1986. Differential reproduction of the mite, *Varroa jacobsoni* (Mesostigmata: Varroidae), on Africanized and European honey bees (Hymenoptera: Apidae). Entomol. Soc. Amer. Ann. 79: 801–803.

CRANE, E. 1978. Bibliography of tropical apiculture. Int. Bee Res. Assoc., London.

CRANE, E. 1979 [1975]. Honey: A comprehensive survey. Heinemann, London.

CRANE, E. 1980. The scope of tropical apiculture. Bee World 61: 19–28.

DADE, H. A. 1962. Anatomy and dissection of the honeybee. Int. Bee Res. Assoc., London.

DALY, H. V., AND S. S. BALLING. 1978. Identification of Africanized honeybees in the Western Hemisphere by discriminant analysis. Kans. Entomol. Soc. J. 51: 857–869.

DEJONG, D., R. A. MORSE, AND G. C. EICKWORT. 1982. Mite pests of honey bees. Ann. Rev. Entomol. 27: 229–252.

DIETZ, A. 1982. Honey bees. *In* H. R. Hermann, Jr., ed., Social insects. I: 333–360. Academic, New York.

ERICKSON, JR., E. H., B. J. ERICKSON, AND A. M. YOUNG. 1986. Management strategies for "Africanized" honey bees: Concepts strengthened by our experiences in Costa Rica. Pts. I, II. Glean. Bee Cult. 1986(Oct.): 456–459, 506–507.

GONÇALVES, L. S. 1982. The economic impact of the Africanized honey bee in South America. 9th Int. Cong. Int. Union Study Social Ins. (Boulder, Colo., 1982), Proc. Pp. 134–137.

GOULD, J. L. 1976. The dance-language controversy. Quart. Rev. Biol. 51: 211–244.

MICHENER, C. D. 1975. The Brazilian bee problem. Ann. Rev. Entomol. 20: 399–416.

NOGUEIRA-NETO, P. 1962. O início da apicultura no Brasil. Biol. Agric. São Paulo 49: 5–14.

ORDETX, G. S. 1952. Flora apícola de la América tropical. Ed. Lux, La Habana. [Not seen.]

ORDETX, G. S., AND D. E. PÉREZ. 1966. La apicultura en los trópicos. Priv. publ., México.

RAMÍREZ, W. 1984. Biología del género *Melaloncha* (Phoridae), moscas parasitoides de la abeja doméstica (*Apis mellifera* L.) en Costa Rica. Rev. Biol. Trop. 32: 25–28.

ROUBIK, D. W. 1979. Africanized honeybees, stingless bees and the structure of tropical plant-pollinator communities. Proc. 4th Int. Symp. Pollination, Maryland Agric. Exper. Sta., Spec. Misc. Publ. 2: 403–417.

ROUBIK, D. W., AND M. M. BOREHAM. 1990. Learning to live with Africanized honeybees. Interciencia 15: 146–153.

RUTTNER, F. 1976. African races of honeybees. 25th Int. Apicul. Cong. (Grenoble, 1976). Pp. 325–347.

SMITH, F. G. 1960. Beekeeping in the tropics. Longmans, Green, London.

STIBICK, J. N. L. 1984. Animal and plant health inspection service strategy and the African honey bee. Entomol. Soc. Amer. Bull. 30(4): 22–26.

TAYLOR, JR., O. R. 1977. The past and possible future spread of Africanized honeybees in the Americas. Bee World 58: 19–30.

TAYLOR, JR., O. R., AND M. D. LEVIN. 1978. Observations on Africanized honeybees reported to South and Central American government agencies. Entomol. Soc. Amer. Bull. 24: 412–414.

WEEMS, JR., H. V. 1983. Bee louse, *Braula coeca* Nitzsch (Diptera: Braulidae). Fla. Dept. Agric. Consum. Serv. Entomol. Circ. 252: 1–2.

WIESE, H. 1977. Apiculture with Africanized bees in Brazil. Amer. Bee J. 117: 166–168, 170.

WINSTON, M. L. 1987. The biology of the honey bee. Harvard Univ. Press, Cambridge.

WINSTON, M. L., G. W. OTIS, AND O. R. TAYLOR, JR. 1979. Absconding behaviour of the Africanized honeybee in South America. J. Apicul. Res. 18: 85–94.

# 13 INSECT STUDY

The study of insect life in Latin America has a long history and has advanced today to a high level of excellence. The subject is taught in academic, agricultural, and medical curricula in all countries, and researchers and natural historians are progressing rapidly on all fronts. For this reason, information is readily available on most aspects, at least in the major cities. This chapter is included to aid the student who may not have ready access to entomological facilities.

## INFORMATION SOURCES

Information on entomology is available from a variety of sources (Gilbert and Hamilton 1983). These are basically of two kinds: (1) personal consultation with authorities, directly or through their institutions, and (2) reference to the records of authorities by reading the literature, both printed and in its other forms.

### Reference

GILBERT, P., AND C. J. HAMILTON. 1983. Entomology: A guide to information sources. Mansell, London.

### Institutions

Listed below are selected institutions in Latin America that employ entomologists; they are segregated according to administrative or organizational purpose; some are not primarily entomological. The list is not complete; entries are intended to reflect the variety of types and those that are well established or especially active currently. (See Anonymous [1983, 1987] for more complete lists.)

### References

ANONYMOUS. 1983. The world of learning 1983–84. Europa, London.
ANONYMOUS. 1987. Resources in entomology. Entomological Society of America, College Park, Md. (See pp. 199–232 for Latin America.)

### *Teaching: Universities, Colleges, and Other Schools*

*Argentina*
División de Entomología, Facultad de Agronomía; Instituto de Limnología, Facultad de Ciencias Naturales y Museo Universidad Nacional de La Plata, La Plata.
Departamento de Ciencias Biológicas, Facultad de Ciencias Exactas, Físicas y Naturales; Cátedra de Zoología Agrícola, Facultad de Agronomía: Universidad Nacional de Buenos Aires, Buenos Aires.

*Bolivia*
Instituto de Biología: Universidad Boliviana Mayor, Real y Pontificia de San Francisco Xavier, Le Paz.

*Brazil*
Departamento de Zoología: Universidade de Brasília, Brasília.
Instituto de Biologia—Parasitologia: Universidade Federal Rural do Rio de Janeiro, Rio de Janeiro.

Departamento de Entomologia, Departamento de Zoologia: Escola Superior de Agricultura "Luiz de Queiroz," Piracicaba.

Departamento de Biologia Geral, Faculdade de Higiene e Saúde Pública, Instituto de Biociências: Universidade de São Paulo, São Paulo.

Departamento de Zoologia, Instituto de Biologia: Universidade Estadual de Campinas, Campinas.

Departamento de Zoologia: Universidade Federal do Paraná, Curitiba.

Chile

Departamento de Zoología, Instituto de Biología: Universidad de Concepción, Concepción.

Departamento de Biología, Facultad de Ciencias: Universidad de Chile, Santiago.

Colombia

Departamento de Biología: Universidad de Antioquia, Medellín.

Facultad de Agronomía: Universidad de Caldas, Manizales.

Departamento de Biología, Departamento de Microbiología: Universidad de Los Andes, Bogotá.

Facultad de Agronomía, Instituto de Ciencias Naturales: Universidad Nacional de Colombia, Bogotá.

Departamento de Microbiología, Departamento de Biología: Universidad del Valle, Cali.

Costa Rica

Departamento de Biología, Departamento de Microbiología, Facultad de Agronomía: Universidad de Costa Rica, San José.

Organization for Tropical Studies, San José.

Dominican Republic

Departamento de Biología: Universidad Autónoma de Santo Domingo, Santo Domingo.

Ecuador

Instituto de Ciencias: Pontíficia Universidad Católica del Ecuador, Quito.

El Salvador

Departamento de Ciencias Biológicas: Universidad de El Salvador, San Salvador.

Guatemala

Escuela de Biología: Universidad de San Carlos, Guatemala City.

Facultad de Biología: Universidad del Valle, Guatemala City.

Guyana

Faculty of Natural Sciences: University of Guyana, Georgetown.

Haiti

Faculté de Science: Université d'état d'Haiti, Port-au-Prince.

Honduras

Department of Plant Protection: Escuela Agrícola Panamericana, Tegucigalpa.

Departamento de Biología: Universidad Nacional Autónoma de Honduras, Tegucigalpa.

Jamaica

Zoology Department: University of the West Indies, Mona Campus, Kingston.

Mexico

Laboratorio de Acarología, Facultad de Ciencias; Departamento de Parasitología, Facultad de Medicina Veterinaria; Instituto de Biología, Departamento de Zoología: Universidad Nacional Autónoma de México, Mexico City.

Escuela Nacional de Ciencias Biológicas: Instituto Politécnico Nacional, Mexico City.

Centro de Entomología y Acarología: Universidad Autónoma Chapingo, Chapingo.

Universidad Autónoma Agraria "Antonio Narro," Saltillo, Coahuila.

Escuela Superior de Agricultura "Hermanos Escobar," Ciudad Juárez, Chihuahua.

Facultad de Ciencias: Universidad Autónoma de Nuevo León, Monterrey.

Nicaragua

Facultad de Ciencias: Universidad Nacional Autónoma de Nicaragua, León.

## Panama

Facultades de Biología, Agronomía y Medicina; Departamento de Ciencias Naturales y Farmacia: Universidad de Panamá, Panama City.

## Paraguay

Facultad de Ciencias y Tecnología: Universidad Católica "Nuestra Señora de La Asunción," Asunción.

## Peru

Departamento de Zoología, Facultad de Ciencias Biológicas: Universidad Nacional Mayor de San Marcos, Lima.
Departamento de Biología: Universidad Nacional de Trujillo, Trujillo.
Departamento de Entomología: Universidad Nacional Agraria, Lima.
Departamento de Zoología: Universidad Nacional "San Antonio Abad," Cuzco.

## Puerto Rico

Departamento de Biología: Universidad de Puerto Rico, Río Piedras and Mayagüez.

## Trinidad and Tobago

Department of Biological Sciences: University of the West Indies, St. Augustine.

## Uruguay

Facultad de Agronomía; Facultad de Humanidades y Ciencias, Departamento de Artrópodos: Universidad de la República, Montevideo.

## Venezuela

Instituto de Zoología Agrícola, Facultad de Agronomía: Universidad Central de Venezuela, Maracay.
Departamento de Entomología, Facultad de Agronomía: Universidad del Zulia, Maracaibo.
Departamento de Entomología, Facultad de Agronomía: Universidad Centro-Occidental Lisandro Alvarado, Barquisimeto.
Departamento de Entomología, Fundación Museo de Ciencias, Caracas.
Instituto de Zoología Tropical: Universidad Central de Venezuela, Caracas.

## *Government Agencies*

In every country, these exist primarily to serve the public and for the protection of natural resources, crops, and public health. A sampling follows.

## Argentina

Instituto Nacional de Tecnología Agropecuaria (INTA), various experimental stations, Buenos Aires and other locations.

## Barbados

Entomology Division, Ministry of Agriculture, Food and Consumer Affairs, Bridgetown.

## Belize

Ministry of Agriculture and Lands, Belmopan.

## Bolivia

Departamento de Sanidad Vegetal, Instituto Boliviano de Tecnología Agropecuaria (IBTA), La Paz.

## Brazil

Empresa Brasileira de Pesquisa Agropecuária (EMBRAPA), Brasília and several branch locations.
Secção de Entomologia, Instituto Agronômico do Estado de São Paulo, Campinas.

## Chile

Instituto Nacional de Investigaciones Agropecuarias, Santiago.

## Colombia

Instituto Nacional de Salud, Bogotá.
Servicio Nacional de Erradicación de la Malaria (SEM), Bogotá.
Centro Internacional de Agricultura Tropical (CIAT), Cali.

## Costa Rica

Departamento de Agricultura, Departamento de Ganadería, Departamento de Ciencias Forestales, Centro Agronómico Tropical de Investigación y Enseñanza (CATIE), formerly Institución Internacional de Ciencias Agrícolas (IICA), Turrialba.

Departamento de Entomología, Ministerio de Agricultura y Ganadería, San José.

*Ecuador*

Instituto Nacional de Investigaciones Agropecuarias (INIAP), Quito and several branch locations.

Departamento de Entomología, Instituto Nacional de Higiene "Leopoldo Izquieta Pérez," Guayaquil.

*El Salvador*

Centro Nacional de Tecnología Agropecuaria, Ministerio de Agricultura y Ganadería, San Andrés.

*Guadeloupe*

Institut Nacional de la Rescherche Agronomique, Station de Zoologie de Lutte Biologique, Domaine Duclos, Petit-Bourg.

*Guyana*

Central Experimental Station, Ministry of Agriculture, East Coast Demerara.

*Honduras*

Ministerio de Recursos Naturales, Tegucigalpa.

*Mexico*

Centro de Investigaciones Biológicas de Baja California Sur, La Paz, Baja California Sur, CONOCYT.

Comisión México Americana para la erradicación del gusano barrendador del ganado, SARH-USDA.

Dirección General de Sanidad Vegetal, Secretaria de Agricultura y Recursos Hidraulicos (SARH), Programa Moscamed, Chiapas.

Instituto de Ecología, Asociación Civil, Mexico City.

Instituto Nacional de Investigaciones Agrícolas (INIA), Mexico City.

Instituto de Salubridad y Enfermadades Tropicales, Mexico City.

*Nicaragua*

Instituto de Recursos Naturales y del Ambiente (IRENA).

*Panama*

Entomology Section, Servicio Nacional de la Erradicación de la Malaria.

Instituto de Investigaciones Agrícolas de Panamá.

*Paraguay*

Dirección de Investigación y Extención Agropecuaria y Forestal, Ministerio de Agricultura y Ganadería, Asunción.

*Peru*

Programa de Erradicación de la Malaria y Mal de Chagas, Ministerio de Salud, Lima.

Instituto Nacional de Investigaciones Agrarias, Ministerio de Agricultura, Lima and other locations.

*Puerto Rico*

College of Agriculture and Mechanical Arts, University of Puerto Rico, Mayagüez.

*Surinam*

Centre for Agricultural Research in Surinam, Paramaribo.

*Trinidad and Tobago*

Department of Biological Sciences, University of the West Indies, St. Augustine.

*Venezuela*

División de Sanidad Vegetal, Ministerio de Agricultura y Cría, Caracas.

Fondo Nacional de Investigaciones Agropecuarias (FONAIAP), Maracay.

Centro Nacional de Investigaciones Agropecuarias (CENIAP), Maracay.

Several international governmental organizations also operate agencies in many areas which are concerned with insects and employ entomologists. Some of these are the following:

1. U. S. Peace Corps. Volunteer entomologists often participate in public health, agricultural, and academic programs.

2. Caribbean Research and Development Institute (CARDI). Main office in Trinidad, subsidiary offices on Grenada, St. Vincent, St. Lucia, and other West Indian islands.

3. United Nations, World Health Organization, Pan American Health Organization.

### Societies
Only the following few societies in Latin America are expressly dedicated to insect study (Sabrosky 1956).

*Argentina*
> Asociación Argentina de Artropodología, Buenos Aires, 1944–.
> Sociedad Entomológica Argentina, Buenos Aires, 1925–.

*Brazil*
> Sociedade Brasileira de Entomologia, São Paulo, 1937–.
> Sociedade Entomológica do Brasil, Rio de Janeiro, 1922–1945.
> Sociedade Entomológica do Brasil, Rio Grande do Sul, 1972–.

*Chile*
> Sociedad Chilena de Entomología, Santiago, 1933–.
> Sociedad Entomologíca de Chile, Santiago, 1922–1929.

*Colombia*
> Sociedad Colombiana de Entomología, Bogotá, 1973–.

*Mexico*
> Sociedad Mexicana de Entomología, Mexico City, 1952–.
> Sociedad Mexicana de Lepidopterología, Mexico City, 1974–.

*Peru*
> Sociedad Entomológica del Perú (formerly Sociedad Entomológica Agrícola del Perú) Lima, 1956– (Aguilar 1987).

*Uruguay*
> Sociedad Uruguaya de Entomología, Montevideo, 1956–.

*Venezuela*
> Sociedad Venezolana de Entomología, Maracay, 1964–.

### References
AGUILAR, P. G. 1987. Algunos apuntes sobre la Sociedad Entomológica del Perú, a los treinta años de su fundación. Rev. Peruana Entomol. 29: 127–140.

SABROSKY, C. W. 1956. Entomological societies. Entomol. Soc. Amer. Bull. 2(4): 1–22.

### Directories
Entomologists may be contacted personally to obtain information on insects and other related arthropods. Their addresses may be found in lists of members of societies and various directories (e.g., Yantko and Golley 1977, Arnett and Arnett 1985).

### References
ARNETT, JR., R. H., AND M. E. ARNETT. 1985. The naturalists' directory and almanac (international). 44th ed. Flora and Fauna, Gainesville.

YANTKO, J. A., AND F. B. GOLLEY. 1977. A world census of tropical ecologists. Institute of Ecology, Univ. of Georgia, Athens.

### Museums (Insect Collections)
Included here are all sizable or important collections of insects, arachnids, and so on, regardless of status, that is, independent museums as well as those supported by teaching institutions, government agencies, or privately. Those in Brazil are discussed by Papavero (1985), in Honduras by O'Brien and Ward (1987), and in Mexico by Anaya et al. (1991); a world listing has been compiled by Arnett and Samuelson (1986). Some collection managers have published lists of their holdings, especially of type material (see below). Abbreviations for most collections are available (Heppner and Lamas 1982).

*Argentina*
> Sección de Entomología, Museo Argentino de Ciencias Naturales "Bernardino Rivadavia," Buenos Aires.
> Museo de Ciencias de La Plata, La Plata.
> Insect collections, Fundación e Instituto "Miguel Lillo" e Instituto Superior de Entomología, Facultad de Ciencias Naturales, Tucumán.
> Centro de Entomología, Facultad de Ciencias Exactas, Físicas y Naturales, Córdoba.

Museo Territorial, Ushuaia, Tierra del Fuego.

Instituto Patagónico de Ciencias Naturales, San Martín de Los Andes.

*Bolivia*

Entomología, Museo Nacional de Historia Natural, La Paz.

*Brazil*

Departamento de Entomologia, Museu Nacional, Rio de Janeiro.

Entomologia, Museu de Zoologia, Universidade de São Paulo, São Paulo.

Departamento de Entomologia, Museu Paraense "Emílio Goeldi," Belém, Pará.

Coleção sistematica de Entomologia, Instituto Nacional de Pesquisas da Amazônia, Manaus.

Entomology collection, Departamento de Zoología, Universidade Federal do Paraná, Curitiba, Paraná.

Museu Rio-Grandense de Ciências Naturais, Porto Alegre, Rio Grande do Sul.

Departamento de Entomologia, Instituto Oswaldo Cruz, Rio de Janeiro.

*Chile*

Insect collections, Departamento de Zoología, Facultad de Ciencias Biológicas y Reacursos Naturales, Universidad de Concepción, Concepción.

Sección Entomología, Museo Nacional de Historia Natural, Santiago (Camousseight 1980).

Museo Entomológico, Facultad de Agronomía, Universidad de Chile, Santiago.

*Colombia*

Entomología, Museo de Historia Natural, Instituto de Ciencias Naturales, Universidad Nacional de Colombia, Bogotá.

*Costa Rica*

Museo Nacional de Costa Rica, San José.

Museo de Insectos, Facultad de Agronomía, Universidad de Costa Rica, Ciudad Universitaria, San José.

Instituto Nacional de Biodiversidad de Costa Rica (INBio), Ciudad Universitaria, San José. (Janzen 1991).

*Cuba*

Museo Poey, Facultad de Ciencias, Universidad de La Habana, La Habana.

Instituto de Zoología, Instituto de Ecología y Sistemática, Museo Nacional de Historia Natural: Academia de Ciencias de Cuba, La Habana.

*Dominican Republic*

Museo Nacional de Historia Natural, Santo Domingo.

*Ecuador*

Entomología, Museo Ecuatoriano de Ciencias Naturales, Quito.

*El Salvador*

Entomología, Museo de Historia Natural, San Salvador.

*Guadeloupe*

Institut de Recherches Entomologique de la Caribe, Pointe-a-Pitre.

*Guatemala*

Colección Nacional Guatemalteca de Arthrópoda, Universidad del Valle, Guatemala City.

*Guyana*

Guarana Museum, Georgetown.

*Haiti*

Musée National, Port-au-Prince.

*Jamaica*

Institute of Jamaica, Kingston.

*Mexico*

Sección de Entomología, Instituto de Biología, Universidad Nacional Autónoma de Mexico, Mexico City (Vázquez 1981, Vázquez and Zaragoza 1979).

Departamento de Entomología, Museo de Historia Natural de la Cuidad de México, Mexico City (Barrera 1966, Barrera and Martín 1968).

Laboratorio de Acarología, Museo de Zoología "Alfonso L. Herrera": Facultad de Ciencias, Universidad Nacional Autónoma de México, Mexico City (Llorente 1984, Muñiz et al. 1981).

Instituto Nacional de Investigaciones Agrícolas, Servicio de Agricultura y Ganadería, Chapingo (Carrillo et al. 1966).

Invertebrados Terrestres, Departamento de Biología Terrestre, Centro de Investigaciones de Baja California Sur, La Paz, Baja California Sur.

*Nicaragua*

Zoología, Museo Nacional de Nicaragua, Managua.

Departamento de Entomología, Ministerio de Desarrollo Agropecuario y Reforma Agraria (MIDINERA), Managua.

Museo Entomológico, Facultad de Ciencias, Universidad Nacional Autónoma de Nicaragua, León.

*Panama*

Insect collections, Laboratorio Conmemorativo Gorgas, Panama City. (Arthropods of medical importance.)

Smithsonian Tropical Research Institute, Panama City.

Museo de Invertebrados "G. B. Fairchild," Facultad de Ciencias Naturales y Exactas, Universidad de Panamá, Panama City.

*Paraguay*

Museo de Historia Natural, Sociedad Científica del Paraguay, Asunción.

*Peru*

Departamento de Entomología, Museo de Historia Natural, Universidad Nacional Mayor de San Marcos, Lima.

Museo de Entomología, Universidad Nacional Agraria la Molina, Lima (Ortiz and Raven 1972).

*Puerto Rico*

Entomología, Museo de Historia Natural, San Juan.

Laboratorio de Entomología, Facultad de Ciencias y Artes, Universidad de Puerto Rico, Mayagüez.

*Uruguay*

Departamento de Biología, Museo Nacional de Historia Natural, Montevideo.

Ministerio de Agricultura y Pesca, Dirección de Sanidad Vegetal, Montevideo (agricultural pests).

Facultad de Agronomía, Facultad de Humanidades y Ciencias, Departamento de

Artrópodos, Universidad de la República, Montevideo.

*Venezuela*

Departamento de Entomología, Instituto de Zoología Agrícola, Facultad de Agronomía, Universidad Central de Venezuela, Maracay.

Centro Nacional de Investigaciones Agropecuarios (CENIAP), Universidad Central de Venezuela, Maracay.

Entomología, Sociedad de Ciencias Naturales "La Salle," Caracas.

Departamento de Entomología, Facultad de Agronomía, Universidad del Zulia, Maracaibo.

Departamento de Entomología, Facultad de Agronomía, Universidad Centro-Occidental "Lisandro Alvarado," Barquisimeto.

Museo de Biología, Universidad Central de Venezuela Caracas.

Collection of Sr. Carlos Bordón, Maracay, Coleoptera.

Departamento de Entomología, Fundación Museo de Ciencias, Caracas.

## References

Anaya, S., F. Cervantes, R. Peña, N. Bautista, and R. Campos, eds. 1991. Colecciones entomológicas de México: Objetivos y estado actual. Mem. Prim. Simp. Nac. Col. Entomol. Veracruz.

Arnett, Jr., R. H., and G. A. Samuelson. 1986. The insect and spider collections of the world. E. J. Brill/Flora and Fauna, Gainesville.

Barrera, A. 1966. Primera lista de tipos depositados en el Museo de Historia Natural de la Ciudad de México. Acta Zool. Mexicano 8(4): 1–3.

Barrera, A. and E. Martín. 1968. Segunda lista de tipos depositados en el Museo de Historia Natural de la Ciudad de México. Acta Zool. Mexicano 9(4): 1–5.

Carrillo S., J. L., A. Ortega C., and W. C. Gibson. 1966. Lista de insectos en la colección entomológica del Instituto Nacional de Investigaciones Agrícolas. Inst. Nac. Inves. Agric. (Mexico) Fol. Misc. 14: 1–133.

Camousseight M., A. 1980. Catálogo de los tipos de insecta depositados en la colección del Museo Nacional de Historia Natural (Santiago, Chile). Mus. Nac. Hist. Nat. Publ. Occ. 32: 1–45.

HEPPNER, J. B., AND G. LAMAS. 1982. Acronyms for world museum collections of insects, with an emphasis on Neotropical Lepidoptera. Entomol. Soc. Amer. Bull. 28: 305–315.

JANZEN, D. H. 1991. How to save tropical biodiversity. Amer. Entomol. 37: 158–171.

LLORENTE, J. E. 1984. Las colecciones zoológicas de la Facultad de Ciencias, Acervo del Museo de Zoología "Alfonso L. Herrera." Univ. Nac. Autón. México, Mexico City.

MUÑIZ, A. M., J. C. MORALES, R. A. BARAJAS, AND J. L. BOUSQUETS. 1981. Primera lista de tipos depositados en el Museo de Zoología "Alfonso L. Herrera" de la Facultad de Ciencias de la Universidad Nacional Autónoma de México: Colección de insectos ectoparásitos "Alfredo Barrera." Fol. Entomol. Méxicana 49: 155–168.

O'BRIEN, C. W., AND C. R. WARD. 1987. Current state of insect collections in Honduras. Fol. Entomol. Mexicana 71: 87–101.

ORTIZ, M., AND K. RAVEN. 1972. Catálogo preliminar del Museo de Entomología de la Universidad Nacional Agraría. Univ. Nac. Agrar. La Molina, Mus. Entomol., Lima.

PAPAVERO, N. 1985. Entomological collections and human resources in Brazil. Assoc. Syst. Coll., Newsl. 13(3): 21–24.

VÁZQUEZ, L. 1981. Los tipos existentes en la colección entomológica del Instituto de Biología, de la Universidad Nacional Autónoma de México. Inst. Biol. Univ. Nac. Autón. México, Ser. Zool. 1, An. 52: 493–505.

VÁZQUEZ, L., AND S. ZARAGOZA. 1979. Tipos existentes en la colección entomológica del Instituto de Biología de la Universidad Nacional Autónoma de México. Inst. Biol. Univ. Nac. Autón. México, Ser. Zool. 1, An. 50: 575–632.

### Miscellaneous Institutions

Other institutions with entomological departments exist for private industry or commercial purposes (mainly control of crop pests), purely for experimental and research aims, for special projects, or to serve other ends. Some major institutions appear in the following list.

*Barbados*
The Barbados Sugar Producers Association, St. Michael.

Bellair's Research Institute, McGill University, St. James (Peck and Peck 1980).

*Bolivia*
Centro de Investigaciones de Mejoramiento de Caña de Azúcar, Santa Cruz.

*Brazil*
Instituto Nacional de Pesquisas de Amazônia (INPA), Manaus.
Fundação Instituto Oswaldo Cruz, Rio de Janeiro.
Instituto Butantan, São Paulo.
Instituto Agronómico de Campinas, Campinas.

*Costa Rica*
Tropical Science Center, San José.

*Dominican Republic*
Jardin Botánico y Parque Zoológico Nacional, Santo Domingo.

*Ecuador*
Asociación Nacional de Cultivadores de Palma Africana, Quito.

*French Guiana*
Institut Pasteur, Cayenne.
Office de la Recherche Scientifique et Technique d'Outre-Mer (ORSTOM), Cayenne.

*Guyana*
Guyana Rice Board, Georgetown.

*Honduras*
Division of Tropical Research, United Fruit Company, Cortés.

*Jamaica*
The Institute of Jamaica, Kingston.

*Panama*
Smithsonian Tropical Research Institute (STRI), Panama City. Maintains field research station on Barro Colorado Island.
Gorgas Memorial Laboratory of Tropical and Preventive Medicine, Panama City.
Entomology Unit, Panama Canal Commission.

*Peru*
Fundación para el Desarrollo Algodonero (FUNDEAL), Lima.

## Surinam

Foundation for Scientific Research in Surinam and the Netherlands Antilles, Paramaribo.

*Trinidad and Tobago*

Commonwealth Institute of Biological Control, Curepe.

*Venezuela*

Instituto Pedagógico de Caracas (Dr. González-Sponga), Caracas. Arachnids.

*Virgin Islands*

Caribbean Research Institute, College of the Virgin Islands, St. John.

### Reference

Peck, S. B., and J. Peck. 1980, Insect field work opportunities in Barbados, Lesser Antilles. Entomol. News 91: 63–64.

### Literature

The entomological literature is vast and complex. No general text other than the present treats the subject of Latin American insects and entomology in a comprehensive way, although portions of some general natural history books refer to a significant number of regional species (Cendrero 1971, von Ihering 1968, Shelford 1926). Access to pertinent references requires a knowledge of information sources (Blanchard and Farrell 1981, Trauger et al. 1974) and library sources (Davis 1989).

### References

Blanchard, J. R., and L. Farrell. 1981. Guide to sources for agricultural and biological research. Univ. California, Berkeley.

Cendrero, L., ed. 1971. Zoología hispanoamericana. Vol. 2. Ed. Porrúa, Mexico City.

Davis, T. J., compiler. 1989. Latin American research libraries in natural history: a survey. American Ornithol. Union, Washington, D.C.

Shelford, V. E. 1926. Naturalist's guide to the Americas. Williams and Wilkins, Baltimore.

Trauger, S. C., R. D. Shenefelt, and R. H. Foote. 1974. Searching entomological literature. Entomol. Soc. Amer. Bull. 20: 303–315.

von Ihering, R. 1968. Dicionário dos animais do Brasil. (Ed. Univ. Brasília, São Paulo.

## Monographs and Serials

According to their frequency and finality of issue, publications may be classified into two categories, monographs and serials. Monographs are individual and self-contained treatments of a particular subject. They may be composed of a single volume or several volumes, all published at one time. An example of a single volume monograph is: Price, P. W., 1984 (2 ed.), *Insect ecology* (Wiley, New York). A multivolume monograph is: Kerkut, G. A., and L. I. Gilbert, eds., 1985, *Comprehensive insect physiology, biochemistry and pharmacology* (Pergamon, Oxford), vols. 1–13.

Serial publications are characterized by issue over an interval of time. Periodicity may be regular or occasional, continuing (meant to be open-ended, although some have ceased publication after a period) or temporal (complete after a preset number of issues). Many are organs of learned societies (journals), but others are independent outlets for original reports of scientific investigation. Major continuing serials published in Latin America which are strictly entomological or more general but with significant space dedicated to regional entomology are the following (see Anonymous 1983, Hammack 1970, and King 1986 for more complete lists).

*Argentina*

*Acta Científica:* Institutos de Investigación de San Miguel, Instituto de Ciencias Naturales, San Miguel, 1955–.

*Agro:* Ministerio de Asuntos Agrarios, La Plata. 1959–.

*Arthropoda:* Asociación Argentina de Artropodología, Buenos Aires, 1947–.

*Bibliografía Entomológica Argentina,* Suplemento (Augusto A. Pirán), 1961–.

*Boletin* de la Sociedad Entomológica Argentina, Buenos Aires, 1925, 1931.

*Boletin Técnico* del Instituto Científico de Medicina Veterinaria, Buenos Aires, 1957–.

*Comunicaciones, Entomología* del Museo

Argentino de Ciencias Naturales "Bernardino Rivadavia," Buenos Aires, 1964–.

*Physis:* Asociacíon Argentina de Ciencias Naturales, Buenos Aires, 1915–.

*Revista Argentina de Entomología:* Museo Argentino de Ciencias Naturales, Buenos, Aires, 1935–.

*Revista* de la Sociedad Entomológica Argentina, Buenos Aires, 1926–.

*Brazil*

*Acta Amazonica:* Instituto Nacional de Pesquisas da Amazônia, Manaus, 1971–.

*Anais* da Sociedade de Biologia de Pernambuco, Instituto de Antibióticos, Universidade de Recife, Recife, 1941–.

*Arquivos de Entomologia,* Series A and B, Escola de Agronomia "Eliseu Maciel," Instituto Agronômico do Sul, Pelotas, Rio Grande do Sul, 1958–.

*Arquivos* do Instituto Biológico, Departamento de Defensa Sanitária da Agricultura, São Paulo, 1928–.

*Boletim* do Instituto Biológico da Bahia, Bahia, 1954–.

*Boletim* do Museu Nacional de Rio de Janeiro (Zoologia), Brazil, 1942–.

*Boletim,* Nova Serie, Zoologia, Museu Paraense "Emílio Goeldi," Instituto Nacional de Pesquisas da Amazônia, 1956–.

*Boletim* do Serviço de Entomologia, Secretaria da Agricultura, Industria e Comercio, Rio Grande do Sul, 1956–.

*Boletim* da Sociedade Brasileira de Entomologia, 1948–.

*Dipan:* Directoria da Producão Animal, Secretaria de Agricultura, Rio Grande do Sul, Brasil, 1948–.

*Entomologista Brasileiro,* São Paulo, 1908–09.

*Iheringia,* Zoologia, Museu Rio-Grandense de Ciencias Naturais, Porto Alegre, Rio Grande do Sul, 1957–.

*Memorias* do Instituto Butantan, São Paulo, 1921–.

*Revista Brasileira de Entomologia:* Sociedade Brasileira de Entomologia, São Paulo, 1954–.

*Revista Brasileira de malariologia e doenças tropicais:* Departamento Nacional de Endemias Rurais, Divisão de Cooperação e Divulgação, Rio de Janeiro, 1951–.

*Revista de Entomologia,* Rio de Janeiro, 1931–1951.

*Studia Entomologica* (Revista Internacional de Entomologia), Petropolis, Rio de Janeiro, 1952–.

*Revista Brasileira de Biologia:* Academia Brasileira de Ciências, Rio de Janeiro, 1941–.

*Chile*

*Boletin Chileno de Parasitología:* Departamento de Parasitología, Universidad de Chile, Santiago, 1954–.

*Acta Entomologica Chilena:* Instituto de Entomología, Universidad Metropolitana de Ciencias de la Educación, Santiago, 1975–.

*Investigaciones Zoológicas Chilenas:* Centro de Investigaciones Zoológicas, Universidad de Chile, Santiago, 1950–.

*Publicaciones* del Centro de Estudios Entomológicos, Facultad de Filosofía y Educación, Universidad de Chile, Santiago, 1958–.

*Revista Chilena de Entomología:* Sociedad Chilena de Entomología, Santiago, 1951–.

*Colombia*

*Anales* de la Sociedad de Biología de Bogotá, Bogotá, 1945–.

*Caldasia:* Instituto de Ciencias Naturales, Universidad Nacional de Colombia, Bogotá, 1940–.

*Revista* de la Facultad de Medicina Veterinaria, Universidad Nacional de Colombia, Bogotá, 1928–.

*Agricultura Tropical:* Asociación Colombiana de Ingenieros Agronómicos, Bogotá, 1945–.

*Costa Rica*

*Revista de Biologia Tropical,* San José, 1953–.

*Brenesia:* Museo Nacional de Costa Rica, San José, 1972–.

*Cuba*

*Boletin* and other series: Secretaría de Agricultura, Industria y Trabajo, Sección de Sanidad Vegetal, La Habana, 1916–.

*Poeyana,* Series A and B, Comisión Nacional de la Academia de Ciencias de la República de Cuba, Instituto de Biología, La Habana, 1964–.

*Ecuador*

*Revista Ecuatoriana de Entomología y Parasitología:* Centro Ecuatoriano de Investigaciones Entomológicas, Quito, 1953–.

*El Salvador*

*Boletin Técnico* de la Dirección General de Investigaciones agronómicas, Sección de Entomología, San Salvador, 1960–.

*Guyana*

*Entomological Bulletin:* Department of Agriculture, British Guiana (Guyana), Georgetown, 1930–.

*Jamaica*

*Entomological Bulletin:* Department of Agriculture of Jamaica, Kingston, 1921–1932.

*Entomology Circular:* Department of Agriculture of Jamaica, Kingston, 1921–1934.

*Mexico*

*Boletin de Divulgación:* Instituto para el Mejoramiento de la Producción de Azúcar, Balderas, 1956–.

*Folia Entomológica Mexicana:* Sociedad Mexicana de Entomología, Mexico City, 1961–.

*Revista* de la Sociedad Mexicana de Entomología, Mexico City, 1955–.

*Revista* de la Sociedad Mexicana de Lepidopterología, Mexico City, 1975–.

*Nicaragua*

*Circular Entomológica:* Departamento de Entomología, Servicio Técnico Agrícola, Managua, 1952–.

*Revista Nicaraguense de Entomología,* privately published, León, 1987–.

*Peru*

*Boletín* de la Sociedad Entomológica Agrícola del Peru, Lima, 1959–.

*Informe mensual* de la Estación Experimental Agrícola "La Molina," Lima, 1927–.

*Revista Peruana de Entomología* (formerly Revista Peruana de Entomología Agrícola), Sociedad Entomológica del Peru, Lima, 1958–.

*Puerto Rico*

*Journal of Agriculture:* University of Puerto Rico, Río Piedras, 1917–.

*Uruguay*

*Revista* de la Sociedad Uruguaya de Entomología, Montevideo, 1956–.

*Venezuela*

*Acta Biológica Venezuelica:* Caracas, 1951–.

*Boletín de Entomología Venezolana:* Departamento de Entomología, Instituto de Higiene, Caracas, 1941–.

*Boletín Técnico* del Instituto Nacional de Agricultura de Venezuela, Maracay, 1951–1956.

*Revista* de la Facultad de Agronomía, Universidad Central de Venezuela, Maracay, 1968–.

*Revista de Medicina Veterinaria y Parasitología:* Facultad de Medicina Veterinaria, Universidad Central de Venezuela, Maracay, 1939–.

In addition, there are numerous international or foreign serials, purely entomological and more inclusive, that contain significant numbers of articles on Latin American insects and related arthropods:

*Acta Tropica,* Basel, 1944–.

*Amazoniana,* Instituto Nacional de Pesquisas da Amazônia, Manaus, 1965–.

*Beiträge zur Neotropischen Fauna:* Humboldt-Hauses in Miraflores und des Instituto Colombo-Alemán in Santa Marta, Stuttgart, Germany, 1956–.

*Biotropica,* Association for Tropical Biology, Florida, 1963–.

*Boletín* de la Oficina Sanitaria Panamerica, Pan American Sanitary Bureau, Washington, D.C., 1923–.

*Boletín de Patología Vegetal y Entomología Agrícola,* Laboratory of Entomology, Estación Nacional Agronómica, Madrid, Spain, 1927–.

*Ceiba,* Escuela Agrícola Panamerica, Tegucigalpa, 1950–.

*International Journal of Tropical Insect Science,* Pergamon, Oxford, 1980–.

*Neotropica,* Notas Zoológicas Sudamericanas, Buenos Aires, 1954–.

*Revista Sudamericana de Entomología Aplicada,* Serie A, Entomología Agrícola, 1946–.

*Studies on Neotropical Fauna and Environment,* 1956–. Continues Beiträge zur Neotropischen Fauna and Studies on the Neotropical Fauna.

*Tropical Ecology* International Society for Tropical Ecology, Allahabad. 1960–. Continues Bulletin of the International Society for Tropical Ecology.

*Tropical Titles and Pest Management,* Centre for Overseas Pest Research, London. 1955–. Continues Pest Articles News Summary (PANS).

Temporal serials are too numerous to list.

## References

ANONYMOUS. 1983. Serial sources for the Biosis data base. Biosciences Information Service, Philadelphia.

HAMMACK, G. M. 1970. The serial literature of entomology: A descriptive study. Entomol. Soc. America, College Park, Md.

KING, A. H. 1986. Latin American entomological serials. Fla. Entomol. 69: 30–45.

The entomological literature can also be classified into a number of categories according to its purpose. Only a few of the more important areas can be mentioned (see Trauger et al. 1974 for an extensive review).

## Reference

TRAUGER, S. C., R. D. SHENEFELT, AND R. H. FOOTE. 1974. Searching entomological literature. Entomol. Soc. Amer. Bull. 20: 303–315.

### Reference Works

These are primary guides to information sources, citing publications and availability of all other kinds of literature. They also may provide access to other reference works. An example is: Arnett, R. H., Jr., 1970, Entomological information and retrieval (Bio-Rand Fd., Baltimore).

### Catalogs

These are listings of taxa, usually species, with complete or partial citations of applicable publications and other information. An example is: *Trichopterorum catalogus.* A simple file of taxonomic names, or "checklist," is often used by collectors and museum curators to arrange their collections and is a vital entry point into the technical literature.

### Bibliographies and Literature Indexes

Compendia of citations of literature concerning a particular subject are often published for the convenience of investigators, reviewers, and writers. They range from very limited, individual treatments of a single species to very elaborate series covering a large taxonomic and/or geographic area. They may be simple listings of papers, with or without subject analyses and annotations. Numerous limited bibliographies to particular taxa are cited in the text references. An example is: Atchley, W. R., et al., 1981, *A bibliography and keyword index of the biting midges (Diptera: Ceratopogonidae),* U.S. Dept. Agric., Bibl. Lit. Agric. 13: 1–544. Geographically oriented general entomological bibliographies are the following:

*Argentina*
    Pirán A. A. 1946. Bibliografía entomológica Argentina. Min. Agr. Dir. Gen. Invest. Inst. San. Veg., Ser. B 2(5): 1–144.

## Costa Rica

Jirón, L. F., and M. E. Sancho de Barquero. 1983. Indice de publicaciones entomológicas de Costa Rica. Consej. Nac. Inves. Cien. Téc. and Org. Trop. Stud., San José.

## Mexico

Trujillo, P. 1967. Bibliografía entomológica de Baja California. Ed. Californidad, Tijuana.

Several comprehensive, world bibliographies are available which pertain to Latin American entomology. The single most important continuing work is the *Zoological Record* (BioSciences Information Service, Philadelphia, and Zoological Society of London, London), which attempts to index all the entomological literature from 1849 to date (see Trauger et al. 1974 for others).

### Reference

TRAUGER, S. C., R. D. SHENEFELT, AND R. H. FOOTE. 1974. Searching entomological literature. Entomol. Soc. Amer. Bull. 20: 303–315.

### *General Works and Textbooks*

The most significant information on a subject is ultimately digested and organized into works primarily for teaching and primary reference. An example is: Coronado, R. A., and A. Márquez, 1976, *Introducción a la entomología* (Ed. Limusa, Mexico City).

### *Natural History and Travel Books*

Valuable information on insects is often contained in books of a general nature, especially those on natural history and travel. It may be necessary to read the entire book to find pertinent passages, especially older works, because they commonly have inadequate or missing indexes. An example is: Hogue, C. L., 1972, *Armies of the ant* (World, New York) 234p.

### *Dictionaries*

Semantics is the subject of a variety of glossaries, lexicons, and dictionaries compiled especially for entomology. Useful are: (1) *The Torre-Bueno Glossary of Entomology,* S. W. Nichols, compiler, and R. T. Schuh, editor. Revised edition of a glossary of entomology by J. R. de la Torre-Bueno, including Supplement A by G. S. Tulloch (New York Entomol. Soc. and American Mus. Nat Hist., New York, 1989); (2) *Dictionnaire des termes techniques d'Entomologie élémentaire* by E. Seguy (Lechevalier, Paris, 1967); (3) *Entomologisches Wörterbuch*, by S. von Kéler (Akademie, Berlin, 1963); (4) *Glossário de Entomología* by M. B. de Carvalho, E. Carvalho de Arruda, and G. Pereira de Arruda (Univ. Fed. Rur. Pernambuco, Recife, 1977). (5) *Spanish-English-Spanish Lexicon of Entomological and Related Terms with Indexes of Spanish Common Names of Arthropods and Their Latin and English Equivalents*, by M. Grieff (Commonwealth Inst. Entomol., Slouth, England, 1985). Three parts, pages separately.

Key words for entomological topics and names are listed and classified in *Thesaurus of Entomology*, by R. H. Foote (Entomol. Soc. America, College Park, Md., 1977).

### *Compendia of Research Papers*

Research papers on related topics are frequently collected under one cover and the editorship of one or more specialists. They differ from monographs in lacking an integrated coverage of the topic. An example is: Nault, L. R., and J. G. Rodríguez, eds., 1985, *The leafhoppers and planthoppers* (Wiley, Chichester).

### *Research Papers*

The majority of scientific literature consists of the original reports of research. They are usually narrowly specialized and cover only a small part of a subject and number in the hundreds of thousands. An example is: Bullock, S. H., and A. Pescador, 1983, Wing and proboscis dimensions in a sphingid fauna from western Mexico, *Biotropica* 15: 292–294.

### Review Papers

These are summaries of a subject, usually to discuss current thought and bring up to date an analysis of the literature. Some journals are dedicated entirely to this type of paper, for example, *Annual Review of Entomology.*

### Popular Articles

Many scientific subjects are of interest to the general public and lay readers. They are discursive, often accompanied by numerous illustrations, and published in popular magazines. An example is: Hogue, C. L., 1982, La Viboruga: El Extraño insecto que se parece a una víbora, *Geomundo* 6(10): 308–309.

### Field and Identification Guides

Very few of these kinds of publications, which are very useful for both the amateur and professional alike, are available for the Latin American entomofauna, and these tend to cover only the more popular and better-known groups, such as butterflies. They are usually profusely illustrated. One such work is D'Abrera, B., 1984, *Butterflies of South America* (Hill House, Victoria, Australia).

### Faunal Surveys and Species Lists

This category includes publications on the kinds of insects or other terrestrial arthropods found in a particular geographic area or reports of faunistic studies (see faunistics, chap. 2). They serve for identification purposes and often include annotations and other data useful for identifying potential economic pests, for ecology, or for biogeographic studies. They are seldom even close to complete and generally extremely limited in coverage. Some examples follow for whole insect faunas and some locally important, smaller areas.

*Argentina*
Brewer, M. M., and N. V. de Argüelo. 1980. Guía ilustrada de insectos comunes de la Argentina. Min. Cult. Educ. Fund. Miguel Lillo. Misc. 67: 1–131.
Havrylenko, D. 1949. Insectos del Parque Nacional Nahuel Huapi. Adm. Gen. Parq. Nac. Tur., Buenos Aires.

*Brazil*
da Costa Lima, A. 1939–1962. Insetos do Brasil. Escuela Nac. Agron., Rio de Janeiro, Ser. Didac. Vols. 2–5, 7–10, 12–14.
Zikáan, J. F., and W. Zikán. 1967. Insetofauna do Itatiaia e da Mantiqueira. Rev. Brasil. Entomol. 12: 117–154.
Zikán, J. F., and W. Zikán. 1968. Insetofauna do Itatiaia e da Mantiqueira. 3. Lepidoptera. Pesq. Agropec. Brasil. 3: 45–109.

*Central America*
Godman, F. D., and O. Salvin, eds. 1879–1915. Biologia Centrali-Americana. 41 vols. Dulau, London.
Selander, R. B., and P. Vaurie. 1962. A gazetteer to accompany the "Insecta" volumes of the "Biologia Centrali-Americana." American Mus. Nov. 2099: 1–70.

*Chile*
Irwin, M. E., and E. I. Schlinger. 1986. A gazetteer for the 1966–67 University of California-Universidad de Chile arthropod expedition to Chile and parts of Argentina. Calif. Acad. Sci. Occ. Pap. 144: 1–11.

*Cuba*
Alayo, P. various dates. Catálogo de la fauna de Cuba. Trab. Divulg. Mus. "Felipe Poey," Acad. Cien. Cuba, La Habana. Several sections, mostly on Hemiptera.
de Zayas, F. 1974. Entomofauna Cubana. Vol 3. Ed. Cien.-Tech., Insto. Cubano Libro, La Habana. Polyneoptera.

*El Salvador*
Berry, P. A., and M. S. Vaquera. 1957. Lista de insectos clasificados de El Salvador. Min. Agric. Ganad. El Salvador Bol. Téc. 21: 1–134.

*Galápagos Islands*

Linsley, E. G. 1977. Insects of the Galápagos (Supplement). Calif. Acad. Sci. Occ. Pap. 125: 1–50.

Linsley, E. G., and R. L. Usinger. 1966. Insects of the Galápagos Islands. Calif. Acad. Sci., Ser. 4., Proc. 33: 113–196.

Peck, S. B. 1990. Eyeless arthropods of the Galápagos Islands, Ecuador: Composition and origin of the cryptozoic fauna of a young, tropical, oceanic archipelago. Biotropica 22: 366–381.

Roth, V. D., and P. R. Craig. 1970. VII. Arachnida of the Galápagos Islands. Miss. Zool. Belgique Galápagos, Ecuador. N. and J. Leleup. 1964–65. Paris. Vol. 2.

*Haiti*

Wolcott, G. N. 1927. Entomologie d'Haiti. Service technique du Département de l'Agriculture et de l'enseignement professionnel, Port-au-Prince.

*Antilles*

Beatty, H. A. 1944. The insects of St. Croix, V.I. J. Agric. Univ. Puerto Rico 28: 114–172.

Bonfils, J. 1969. Catalogue raisonné des insects des Antilles françaises. 2. Dictyoptera: Blattaria et Mantida. Ann. Zool. Ecol. Anim. 1: 107–120.

Gruner, L., and J. Riom. 1977. Insectes et papillons des Antilles. Ed. Pacifique, Papeete, Tahiti.

Miskimin, G. W., and R. M. Bond. 1975. The insect fauna of St. Croix, United States Virgin Islands. Science Surv. Puerto Rico 13(1): 1–114.

Stiling, P. D. 1986. Butterflies and other insects of the eastern Caribbean. Macmillan, London.

Tucker, R. W. E. 1952. The insects of Barbados. J. Agric. Univ. Puerto Rico 36: 330–363.

*Jamaica*

Gowdey, C. C. 1928. Catalogus insectorum Jamaicensis. Dept. Agric. Jamaica Entomol. Bull. 4: 1–47.

*Panama*

Weber, N. A. 1972. The entomology of Panamá. Biol. Soc. Wash. Bull. 2: 187–197.

*Puerto Rico*

Drewry, G. E. 1970. A list of insects from El Verde, Puerto Rico. *In* H. T. Odum, ed., A tropical rain forest: A study of irradiation and ecology at El Verde, Puerto Rico. U.S. AEC, Washington, D.C.

Maldonado, J., and C. A. Navarro. 1967. Additions and corrections to Wolcott's Insects of Puerto Rico. Carib. J. Sci. 7: 45–64.1.

Various authors. 1919–. The insects of Porto Rico and the Virgin Islands. New York Acad. Sci., Scientific survey of Puerto Rico and the Virgin Islands. Several volumes published in this incomplete series.

Wolcott, G. N. 1936. "Insectae Borinquenses." A revised annotated check-list of the insects of Puerto Rico with a host-plant index by José I. Otero. J. Agric. Univ. Puerto Rico, 20: 1–627.

Wolcott, G. N. 1941. Supplement to "Insectae Borinquenses," J. Agric. Univ. Puerto Rico 25: 33–158.

Wolcott, G. N. 1951 [1948]. The insects of Puerto Rico. J. Agric. Univ. Puerto Rico 32: 1–975.

*Surinam*

Geijskes, D. C. 1967. De insektenfauna van Suriname, ook vergeleken met die van de Antillen. Speciaal wat betreft de Odonata. Entomol. Ber. Amsterdam 27: 69–72.

*Various islands*

Alvarenga, M. 1962. A entomofauna do Arquipélago de Fernando de Noronha, Brasil. Mus. Nac. Rio de Janeiro Arq. 52: 21–25.

Campos, L., and L. E. Peña. 1973. Los insectos de la Isla de Pascua. Rev. Chilena Entomol. 7: 217–229.

Duffy, E. 1964. The terrestrial ecology of Ascension Island. J. Appl. Ecol. 1: 219–251.

Clarke, J. F. G. 1965. Microlepidoptera of the Juan Fernandez Islands. U.S. Natl. Mus. Proc. 117: 1–106.

Graham, J. B., ed. 1975. The biological investigation of Malpelo Island, Colombia. Smithsonian Contrib. Zool. 176: 1–98.

Hogue, C. L., and S. E. Miller. 1981. Entomofauna of Cocos Island, Costa Rica. Atoll Res. Bull. 250: 1–29.

Palacios-Vargas, J. G., J. Llampallas, and C. L. Hogue. 1982. Preliminary list of the insects and related terrestrial arthropods of Socorro Island, Islas Revillagigedo, Mexico. So. Calif. Acad. Sci. Bull. 81: 138–147.

Ramos, J. A. 1946. The insects of Mona Island (West Indies). J. Agric. Univ. Puerto Rico 30: 1–74.

Robinson, G. S. 1984. Insects of the Falkland Islands: A check list and bibliography. Brit. Mus. Nat. Hist., London.

Schiapelli, R. D., and B. S. Gerschman de Pikelin. 1974. Arañas de las Islas Malvinas. Rev. Mus. Argentino Cien. Nat. Bernardino Rivadavia, Insto. Nac. Invest. Cien. Nat., Entomol. 4: 79–93.

Skottsberg, C., ed. 1920–1956. The natural history of Juan Fernandez and Easter Island. Vols. 1–3. Almquist and Wiksells, Uppsala. A few articles on insects in vol. 3 (Zoology).

Smith, D. S., S. J. Ramos, F. McKenzie, E. Munroe, and L. D. Miller. 1988. Biogeographical affinities of the butterflies of a "forgotten" island: Mona (Puerto Rico). Allyn Mus. Bull. 121: 1–35.

*Venezuela*

Martorell, L. F. 1939. Insects observed in the state of Aragua, Venezuela, South America. J. Agric. Univ. Puerto Rico 23: 177–264.

### Expedition Reports

A special type of research report covers the results of expeditions. These are often taxonomic in format and may include the description of new species. Localities, dates, and routes of travel are also usually covered and become a useful resource for the preparation of faunal surveys. An example is: Vaurie, C., and P. Vaurie, 1949, Insect collecting in Guatemala 65 years after Champion, New York Entomol. Soc. J. 57: 1–18. Some are very elaborate, multivolume series such as the famous Biologia Centrali-Americana (see faunal surveys, above). Much insect and arachnid material was collected by entomologist H. W. Foote on the Yale Peruvian Expedition of 1911 when the Inca citadel of Machu Picchu was discovered by Hiram Bingham (several reports by taxonomists on various groups were published).

### Computerized Data Banks

A new and rapidly expanding method of storing information, offering quick access to many specialized fields, is by capture on electromagnetic tapes and disks. These data banks are available commercially to subscribers (on-line via telephone modem), either directly or through libraries. In the field of entomology only a few are currently available and are of limited usefulness for Latin America. Some of these are (1) Agricultural On-Line Access (AGRICOLA), U.S. Department of Agriculture (National Agricultural Library), Science and Education Administration, Technical Information Systems, Washington, D. C. (= electronic version of Bibliography of Agriculture); (2) Biosciences Information Service (BIOSIS = Biological Abstracts); and (3) CAB abstracts, Commonwealth Agricultural Bureau, Royal Slough, England. Recent volumes of the *Zoological Record* are also available on computer. All index many entomological journals, internationally. A listing of these and other such data banks is available (Kruzas and Schmittroth 1981).

### Reference

Kruzas, A. T., and J. Schmittroth, Jr., eds. 1981. Encyclopedia of information systems and services. 4th ed. Gale Res. Co., Detroit.

### Government Publications

The nature of the publisher is yet another way the entomological literature can be classified. Of special importance in this category are government documents. These are issued for all purposes and normally for the benefit of a wide audience. Many are often very current and of practical utility. Their availability tends to be limited and short-lived, however, and it is often hard to find older issues. Those of more active departments, commonly those concerned with agriculture and public health, may be numerous and complex bibliographically. Agencies, for example, are often the authors and series can be broken by long periods of inactivity or changes in authority. The special assistance of librarians in the different countries may have to help researchers in these areas.

### Access to Literature

The primary source of entomological literature is libraries. Municipal public libraries seldom include extensive technical holdings, and it is usually necessary to consult university, national, or even private libraries to find all but the most popular works. Individual articles from serials, if recently published, often are available as reprints (also called separates or offprints) from the authors. Items may also be purchased from their publishers if recent or from used or antiquarian book dealers if out of print. Quite a few of the latter are in business around the world which include entomological items in their inventories. The best known of these are the following:

Antiquariaat Junk, Van Eeghenstraat 129, 1071 GA Amsterdam, The Netherlands.

A. Asher and Company, Kaisergracht 489, 1017 DM, Amsterdam, The Netherlands.

Australian Entomological Press, 14 Chisolm Street, Greenwich, 2065 New South Wales, Australia.

Bioquip Products, 17803 La Salle Ave., Gardena, CA 90248, U.S.A.

E. W. Classey Ltd., P.O. Box 93, Park Road, Faringdon, Oxon, SN7 7DR, England.

Entomological Reprint Specialists, P.O. Box 77224, Dockweiler Station, Los Angeles, CA 90007, U.S.A.

Dieter Schierenberg BV, Prinsengracht 485–487, 1016 HP Amsterdam, The Netherlands.

Sciences Nat, 2 rue André Méllenne, Venette, F-60200 Compiègne, France.

Wheldon and Wesley, Ltd., Lytton Lodge, Codicote, Hitchin, Herts, SG4 8TE, England.

## RESEARCH

Research is investigative activity leading to the discovery and recording of new knowledge. A tremendous amount of such activity is taking place in the field of entomology. Hundreds of insect scientists in academic as well as applied areas of the subject are rapidly unveiling new facts and interpretations to be added to humanity's store of information. The process is well ordered and follows the logical steps of the scientific method, which works as well on those problems subject to "proof" by repeatable results (experimental method) and those demonstrable only by application of logical principles (deductive method). Both, however, require first the collection of facts (data, specimens, in laboratory and field), their subsequent analysis, and finally, the making of conclusions that are published.

### Fields of Study

Entomologists, arachnologists, myriapodologists, and acarologists pursue research in many different areas. Basically, they work in either the academic or applied (economic) realms, although some cross over between both. Particular approaches are numerous and increasing in number as the tendency for greater and greater specialization progresses. The major specialties at present, and references to their practices, are:

*Taxonomy (systematics):* classification and nomenclature of species and other categories (Papavero 1983).

*Evolution:* reconstruction of phylogeny and analysis of the evolutionary processes.

*Morphology:* structure, cellular to organismal levels.

*Physiology:* function.

*Genetics:* cytology and heredity.

*Toxicology:* properties and use of pesticides.

*Behavior:* insect comportment.

*Ecology:* insects in relation to their environment.

*Agriculture:* pests of crops.

*Medicine:* vectors and agents of human diseases.

*Veterinary medicine:* agents and vectors of diseases of domestic animals.

Sometimes, specialists combine these in various ways, for example, functional anatomy or physiological ecology.

## Reference

Papavero, N. 1983. Fundamentos práticos de taxonomia zoológica: coleções, bibliografia, nomenclatura. Mus. Paraense Emilio Goeldi, Belém.

## Fieldwork

The collection of facts to apply to studies of insect ecology and many aspects of behavior, taxonomy, functional morphology, and other areas, takes place in the field. The entomologist must go to the arena of natural occurrence of the species under investigation and observe uninhibited activity or manipulate the environment and/or organisms experimentally. Often considerable planning and logistical preparation are necessary before the work can commence, especially if it involves travel and extended stay in remote localities. Considerable benefit is to be gained from the use of established field stations, where technical facilities are already available and where lodging necessities are handled by resident staff, freeing the scientist to concentrate on the research. In Latin America, there are a number of such field stations, run by various agencies, private and governmental, for academic and applied work. Many are associated with forest reserves, wildlife preserves, nature centers, or are part of national parks. A list of the better known of these that have provided service to entomologists in the recent past follows (some may not be in operation at present). See Castner (1990) for an exhaustive list.

*Argentina*
 Museo Territorial, Ushuaia, Tierra del Fuego.

*Belize*
 Desmond Slattery Research Station, Blue Creek River, near Dangriga, sea level, rain forest.

*Brazil*
 Estação Biológica de Boracéia, Museu de Zoologia, Universidade de São Paulo. Near Mogi das Cruces, São Paulo, 850–900 meters, subtropical wet forest (Travassos and de Camargo 1958).
 Estação de Campo, Parque Nacional de Itatiaia, 20 kilometers north of Itatiaia, Paraná, 300–1,000 meters, mixed montane forest.
 Reserva Campinas, Instituto Nacional de Pesquisas Amazônia, 60 kilometers north of Manaus on Boa Vista Highway, 25 meters, mixed lowland wet forest (white sand area).
 Ducke Forest Reserve, Instituto Nacional de Pesquisas da Amazônia, 26 kilometers northeast of Manaus on Manaus-Ita-

coatiara Highway, 25 meters, mixed lowland wet forest.

Reserva Humboldt, University of Matto Grosso, Matto Grosso, at Aripuanã, 200 meters, lowland wet forest.

*Colombia*

Los Llanos Tropical Research Station, Centro Internacional de Agricultura Tropical, near Villavicencio, 400 meters, dry forest.

El Refugio, private consortium, Canyon of Río Claro, 20 kilometers west of Dorodal, Antioquia Department, 1000 meters, wet forest.

Merenberg Preserve, near La Plata (road between Popayán and Neiva), 230 meters, wet forest (Buch n.d.).

Reserva La Planada, near Ricaurte, Nariño, 2,000 meters, cloud forest.

El Rufugio Biological Station, Chocó Department, 23 kilometers west of Cali on road to Buenaventura, 1,600–1,900 meters, wet forest (Calderón 1989).

*Costa Rica* (Anonymous 1972)

La Selva Field Station, Organization for Tropical Studies, near confluence of Ríos Puerto Viejo and Sarapiquí, Heredia Province, 37–150 meters, premontane tropical wet forest (Clark 1988).

Palo Verde Field Station, Organization for Tropical Studies, head of Golfo de Nicoya, Guanacaste Province, 3–183 meters, lowland tropical dry forest.

Las Cruces Field Station, Organization for Tropical Studies, 7 kilometers south of San Vito de Javá, Cartago Province, 1,200 meters, premontane rain forest.

Monteverde Cloud Forest Reserve Field Station, Tropical Science Center, 1,500 meters, cloud forest.

*Ecuador*

La Chiquita, Ecuadorian Department of Forest Resources, approximately 11 kilometers southeast and inland from coastal city of San Lorenzo, sea level, rain forest (Peck and Kulakova-Peck 1980).

Centro Científico de Río Palenque, Uni-

versity of Miami and Universidad Católica de Quito, 47 kilometers south of Santo Domingo de Los Colorados on the road to Quevedo, 220 meters, rain forest.

Tinalandia, private, 5 kilometers south of Santo Domingo de Los Colorados, 700 meters, cloud forest.

Jatun Sacha Biological Station, private, 8 kilometers east of Puerto Misahualli, upper Río Napo, 400 meters, tropical wet forest (Neill and Neill 1988).

Charles Darwin Research Station, Charles Darwin Foundation, Santa Cruz Island, Galápagos Islands, sea level, tropical scrub.

*Jamaica*

Blue Mountain Field Station, Irish Town, Foothills of Blue Mountains, 700 meters, montane dry forest (Freeman 1986).

*Mexico*

Estación de Biología Chamela, Instituto de Biología, Universidad Nacional Autónoma de México, 125 kilometers northwest of Manzanillo (5 km north of Puerto Careyes), near sea level, deciduous dry forest.

Estación de Biología Tropical Los Tuxtlas, Instituto de Biología, Universidad Nacional Autónoma de México, Sierra de San Martin, Veracruz state, 33 kilometers northeast of Catemaco, near coast, humid seasonal forest (Anonymous n.d.).

Vermillion Sea Field Station, San Diego Museum of Natural History, Bahia de Los Angeles, Baja California, sea level, desert.

Reserva Ecológica "El Morro de la Mancha," Instituto Nacional de Investigaciones y Recursos Biológicos. Near Jalapa, Veracruz, sea level, tropical forest.

*Panama*

Barro Colorado Island Biological Research Station, Smithsonian Tropical Research Institute (STRI n.d., Ingles 1954, Leigh et al. 1982), Gatún Lake, Panama Canal, 100 meters, tropical rain forest.

*Peru*

Tambopata Reserved Zone, private, at the junction of the Ríos La Torre and Tambopata, Madre de Dios, 290 meters, mixed lowland wet forest (Erwin 1985).

*Puerto Rico*

El Verde Station, U.S. Department of Energy, Luquillo Experimental Forest, 510 meters, montane forest (Odum 1970).

Biological Field Station, University of Puerto Rico (Río Piedras Campus), Biology Department, Sierra Luquillo, Route 191, 600 meters, montane forest.

Toro Negro Station, University of Puerto Rico (Mayaguez Campus), Biology Department, Cordillera Central, Villalba, 800 meters, montane forest.

*Trinidad*

Asa Wright Nature Center, private, Springhill Estate, 4.2 kilometers north of Arima, 365 meters, wet forest.

*Venezuela*

Estación Biológica Rancho Grande, Parque Nacional Henri Pittier, 15 kilometers north of Maracay, 1,100 meters, cloud forest.

Hato Masagural, private, 40 kilometers south of Calabozo, Anzoátegui, 200 meters, llanos.

Estación Culebra, Fundación Terramar, Caracas, Parque Nacional Duida-Marawaka, Territorio Federal Amazonas, 250 meters, lowland forest.

*Virgin Islands*

Virgin Islands Ecological Research Station, Caribbean Research Institute, College of the Virgin Islands, Lameshur Bay, St. John, sea level, coastal scrub.

## References

ANONYMOUS. n.d. Estación de biología tropical Los Tuxtlas. Cent. Univ. Comun. Cien., Univ. Nac. Auton. México, Mexico City. Pamphlet.

ANONYMOUS. 1972. Field stations of the Organization for Tropical Studies in Costa Rica. OTS, So. Miami.

BUCH, G. n.d. "Merenberg," un santuario selvático en Los Andes de Colombia. López, Popayán. Pamphlet.

CALDERÓN, E. 1989. Announcement: "El Refugio" biological station (Chocó biogeographical region, Colombia): An invitation to tropical biologists. Biotropica 21: 177.

CASTNER, J. L. 1990. Rainforests: A guide to research and tourist facilities at selected tropical forest sites in Central and South America. Feline, Gainesville.

CLARK, D. A. 1988. Research on tropical plant biology at the La Selva Biological Station, Costa Rica. Evol. Trends Plants 2: 75–78.

ERWIN, T. L. 1985. Tambopata Reserved Zone, Madre de Dios, Perú: History and description of the reserve. Rev. Peruana Entomol. 27: 1–8.

FREEMAN, B. 1986. Blue Mt. Field Sta., Irishtown, Jamaica, West Indies. Entomol. News 97: 197.

INGLES, L. G. 1954. Barro Colorado—tropical island laboratory. Smithsonian Ann. Rep. 1953: 361–366, pls. 1–6.

LEIGH, JR., E. G., A. S. RAND, AND D. M. WINDSOR, eds. 1982. The ecology of a tropical forest: Seasonal rhythms and long-term changes. Smithsonian Inst., Washington, D.C. Several articles on the environment of Barro Colorado Island.

NEILL, D., AND D. NEILL. 1988. Jatun Sacha Biological Station, Amazonian Ecuador: An invitation to tropical biologists. Biotropica 20: 59.

ODUM, H. T. 1970. The El Verde study area and the rain forest systems of Puerto Rico. *In* H. T. Odum, ed., A tropical rain forest: A study of irradiation and ecology at El Verde, Puerto Rico. U.S. AEC, Washington, D.C. Pp. B3–B32.

PECK, S. B., AND J. KULAKOVA-PECK. 1980. A guide to some natural history field localities in Ecuador. Stud. Neotrop. Fauna Environ. 15: 35–55.

STRI. n.d. Smithsonian Tropical Research Institute, Information for visitors. Typewritten brochure.

TRAVASSOS, L., AND H. F. A. DE CAMARGO. 1958. A Estação Biológica de Boracéia. Arq. Zool. 11: 1–21.

## Collection and Preservation

The collection and preservation of specimens and data are basic to all areas of entomological research (Kim 1978). It is a truism that living or dead insects must be acquired before any investigations can begin, either for experimental or deductive studies. Extensive collections of specimens are the very foundation of taxonomy, and vouchers should be kept for any ecological, physiological, or other studies to permit future verification of identifications. Therefore, considerable attention must be paid to the processes of acquiring and proper handling, preserving, and processing of material and data (Gibson 1960; Oldroyd 1958; Pastrana 1985; Peterson 1955; Southwood 1966; Steyskal et al. 1986; Valenzuela n.d.).

## References

Gibson, W. W. 1960. Cómo manejar y usar la colección de insectos. Secr. Agric. Ganad. Of. Estud. Espec., México.

Kim, K. C., ed. 1978. The changing nature of entomological collections: Uses, functions, growth and management. Entomol. Scand. 9: 145–177.

Oldroyd, H. 1958. Collecting, preserving and studying insects. Macmillan, New York.

Pastrana, J. A. 1985. Caza, preparación y conservación de insectos. 2d ed. Libr. "El Ateneo," Buenos Aires.

Peterson, A. 1955. A manual of entomological technique. 8th ed. Edwards Bros., Ann Arbor.

Southwood, T. R. E. 1966. Ecological methods with particular reference to the study of insect populations. 2d ed. Methuen, London.

Steyskal, G. C., W. L. Murphy, and E. M. Hoover, eds. 1986. Insects and mites: Techniques for collection and preservation. U.S. Dept. Agric. Misc. Publ. 1443: 1–103.

Valenzuela, G. O. n.d. Recolección, montaje y classificación de insectos. Agric. Trop., Bogotá.

### Acquiring Material

Although there are commercial insectaries from which living specimens may be purchased and insect dealers who sell preserved specimens (some listed below), it is more often necessary for the entomologist to find his or her own material in the field. Of course, if a specific type is sought, it is necessary to look for it in its proper locale and habitat. After these have been determined, various methods for location of specimens and their capture may then be employed.

The success of the search will depend on the abilities of the collector, who should go forth armed with as much knowledge of the insect's microhabitat and habits as possible. It may take much time and detective work to locate rarities; in this pursuit, locals familiar with their natural surroundings are often a great help. The quarry may be located in its home and forced or enticed from it in different ways. A useful procedure for finding those forms that live hidden among shrubbery is to knock them off onto a clean piece of cloth or paper by bludgeoning the main stems with a stick. Grass- and herb-dwelling types may be swept into an insect net. The latter is the standard implement of the collector and comes in a variety of types for special purposes (sweeping nets, aerial nets, aquatic nets, etc.).

Many kinds of traps for catching insects have been developed. One of the most useful is the Malaise trap (Malaise 1937). There are different designs (e.g., Townes 1962), but all basically combine a vertical baffle to stop flying insects and a tentlike umbrella to funnel them into a killing chamber. The trap works passively or in combination with lures. It can be placed on the ground or in vegetation, even suspended from trees.

A device for extracting soil- and litter-dwelling microarthropods is the Berlese (or Tullgren) funnel (Allison 1983, Merchant and Crossley 1970). It consists simply of a metal or plastic funnel suspended over a collecting chamber (killing bottle, alcohol reservoir) over which is placed a strong light or heater. The sample is put into the funnel (prevented from falling through by a screen over the mouth of the stem), and the arthropods, seeking refuge from the light and heat, travel downward, eventually dropping into the chamber.

Lures also work and include chemicals as well as light for nocturnally active forms. Volatile or aromatic substances of many kinds (eugenol, feces, rotting fruit and meat, etc.) attract insects; they may be specific in their effect, especially the pheromones that have been identified chemically and are available in a bottle. Such is Medlure, used to catch and monitor Mediterranean fruit fly infestations. Others are more general, as eucalyptol or oil of wintergreen, which will draw in the males of many species of orchid bees, or phenylacetaldehyde, which attracts various kinds of moths. Naturally occurring scents from blossoms or dried plants can be used also, a prime example being heliotrope, which is irresistible to ithomiid butterflies and other Lepidoptera. Lures are used in traps (colored pans with liquid to drown the insects, enclosures, or sticky surfaces), or to attract specimens directly to the collector.

Light, especially that in the near-ultraviolet portion of the spectrum, draws many night insects to its source. Specimens may be collected as they congregate around street lamps or the outside of windows or even by a gas lantern placed on a reflecting background. However, sophisticated emitters and traps of many kinds have been invented to take direct advantage of this phenomenon and provide special convenience for the entomologist. Some are made to operate underwater. Currently, most employ mercury vapor or fluorescent bulbs that operate on electricity. Portable battery-powered units are available commercially or can be easily built.

## References

ALLISON, A. 1983. An inexpensive, portable Tullgren extractor suitable for the tropics. Int. J. Entomol. 25: 321–323.

MALAISE, R. 1937. A new insect-trap. Entomol. Tidsskr. 58: 148–160.

MERCHANT, V. A., AND D. A. CROSSLEY, JR. 1970. An inexpensive, high-efficiency Tullgren extractor for soil microarthropods. Georgia Entomol. Soc. J. 5: 83–87.

TOWNES, H. 1962. Design for a Malaise trap. Entomol. Soc. Wash. Proc. 64: 253–262.

### Commercial Dealers

There are numerous dealers in business to sell living and preserved insect specimens and/or equipment and supplies for collectors, researchers, and teachers. Those currently active in the different countries can be ascertained by reference to the other entomological sources listed in this chapter.

### Protocol

It is imperative that the collector comply not only with the laws of the country or district where the collecting is being done but also with common courtesies regarding entry of private property. Permissions may have to be obtained in advance of fieldwork. Collecting and export permits are required by most countries and special documents especially needed to work in nature preserves, national parks, and otherwise protected or militarily sensitive areas. International laws, such as those administered by the Convention on International Trade in Endangered Species of Flora and Fauna (CITES), also apply to the taking of certain species of insects and spiders and their transport around the world. Many species are considered threatened or endangered and may be acquired only under the most stringent restrictions. Often, different regulations apply to living versus preserved specimens (Fuller and Swift 1985).

### Reference

FULLER, K. S., AND B. SWIFT. 1985. Latin American wildlife trade laws. World Wildlife Fund, Washington, D.C.

### Handling Material

Specimens to be kept alive will be treated very differently from those to be immediately preserved. They will have to be provided with sustenance and proper environmental conditions to ensure their survival both on their way from the field and in the laboratory. Special climate-controlled insectaries or vivaria may have to be constructed for types with narrow requirements. Others are more easily maintained with a minimum of care, although all should be treated with the utmost concern due any living thing. Instructions for rearing and culturing many kinds of insects, spiders, and like creatures are available in the materials and methods sections of research papers and in special treatments (Siverly 1962; Smith 1966; Singh and Moore 1985; Singh 1977).

Those specimens to be dispatched immediately are best killed quickly in a tight gas chamber or, if large, by injection. The most convenient chambers are bottles that use dry cyanide crystals, held in place in the bottom by a plug of plaster, cardboard, or other material. Volatile toxic liquids, such as ethyl acetate, benzene, and so on, may be employed likewise. A fraction of a cubic centimeter of ethyl acetate or isopropyl alcohol from a hypodermic syringe, injected into the thorax of big beetles, orthopterans, or moths, will kill them instantly. Soft-bodied and many small types, and those desired for morphological study, are collected directly into fluids (commonly 75–90% ethyl or isopropyl alcohol or various fixatives) where they will die and remain for transport.

Dead specimens quickly become brittle or will deteriorate, especially in humid climes, unless cared for properly. They should be placed in an airy, rigid container (e.g., cardboard box) for transport to the laboratory or field base. Plastic bags are appropriate only for very temporary storage of living or dead material. If time is available for immediate mounting of specimens, this should be done while they are still flaccid and body parts can be moved to desired positions. If mounting is to be delayed, the specimens should be dried to prevent decomposition. This can be done

by placing them in delicate paper (glassine, tissue) envelopes or layering them between sheets of similar material. They should not be placed in contact with cotton because they will become entangled in the fibers and will be difficult to extract later.

Once back in the laboratory, the specimens can be mounted or permanently preserved in one of three ways, depending on their body structure and later use. The soft-bodied and many small types that were taken originally in fluids will be transferred to the same or other, but clean, fluids. Most dry, hard-bodied insects will be pinned. Lepidoptera may also have their wings spread. Specific instructions for pinning and spreading are available in the references listed above. It is important to point out here that only pins made especially for insect mounting must be used. These are high-quality steel and coated with shellac or made of stainless steel, so that they do not easily bend or corrode. They also are extra-long (approximately 30 mm) to accommodate the bodies of the specimens and allow labels to be fixed beneath them.

## References

SINGH, P. 1977. Artificial diets for insects, mites, and spiders. Plenum, New York.

SINGH, P., AND R. F. MOORE, eds. 1985. Handbook of insect rearing. 2 vols. Elsevier, Amsterdam.

SIVERLY, R. E. 1962. Rearing insects in schools. W. C. Brown, Dubuque.

SMITH, C. N. 1966. Insect colonization and mass production. Academic, New York.

### Data Collection and Labeling

Information associated with specimens is as important as the specimens themselves, and considerable care must be given to collecting and recording it. At the very least, for taxomonic purposes, the precise location and date of capture, plus the collector's name, should be recorded and attached. Very useful also will be notes on observations of ecological variables, behavior, or other pertinent details that were manifest at the time of the insect's capture. These are best recorded in some standardized written format or printed form. The latter can be specialized for particular insect types (e.g., aquatic) or general for any taxon (Hogue 1966). Experimental studies or well-defined research projects may require extensive and highly organized data capture methods. Thought should be given to making forms and data computer compatible (Erwin 1976, Hodges and Foote 1982); devices are even available for directly reading data in the field onto electronic storage tapes or disks. The information also must be encoded to ensure correlation with the specimens, various alphanumeric systems being most useful.

## References

Erwin, L. J. M. 1976. Application of a computerized general purpose information management system (SELGEM) to a natural history research data bank (Coleoptera: Carabidae). Coleop. Bull. 30: 1–32.

Hodges, R. W., and R. H. Foote. 1982. Computer-based information system for insect and arachnid systematists. Entomol. Soc. Amer. Bull. 28(1): 31–38.

Hogue, C. L. 1966. A field note form for special collecting. Entomol. Soc. Amer. Ann. 59: 230–233.

### Permanent Preservation

Final handling of specimens involves their placement into some sort of permanent storage facility. Pinned, dry insects are normally kept in tight-closing wooden containers of various sorts. Most modern collections employ small cardboard "unit trays" with soft material in the bottom to receive the pin which, in turn, fit into shallow drawers with glass lids. This method permits the maximum convenience for adding or rearranging specimens with a mimimum of danger of breakage. The drawers are stored in cabinets or racks, which, in some fortunate institutions, are mechanized (movable module storage facilities, "compactors").

Placing specimens under glass in picture frames (Riker mounts) is not recommended for scientific collections. They are often broken by such treatment and are impossible to manipulate for close examination.

Fumigation is often necessary to prevent the ravages of museum pests such as dermestid beetles (*Anthrenus* and *Thylodrius*), psocids ("book lice"), silverfish, and the like (Edwards et al. 1981, Story 1985). This can be done on a continuous basis with napthalene, thymol, or paradichlorobenze, which act mainly as repellents, or intermittently by use of lethal gases such as dimethyl bromide in controlled chambers or under tents. The atmosphere must also be kept dry, to prevent the growth of molds that also destroy insect specimens. Light should be kept off specimens as well, because it fades colors and contributes to specimen deterioration.

Liquid ("wet") collections should be maintained in glass, not plastic, vials. The practice of using individually stoppered and separately stored vials should be avoided for large collections. No truly perfect seal that permits easy removal has yet been devised, and such vials dry up without fail, ruining the specimens they contain. Curators are warned against rubber stoppers, especially, because chemicals they contain dissolve in the fluid and damage the specimens. The recommended method is to use simple shell vials, stoppered with cotton plugs and immersed in a large reservoir (200–300 ml jar). Vigilance is still necessary to keep such reservoirs filled, but considerable time must pass before the vials themselves completely dessicate.

## References

Edwards, S. R., B. M. Belland, and M. E. King, eds. 1981. Pest control in museums: A status report (1980). Assoc. System. Coll., Lawrence, Kans. Paginated by sections.

STORY, K. O. 1985. Approaches to pest management in museums. Conserv. Anal. Lab., Smithsonian Inst., Washington, D.C.

## Illustration and Photography

As an integral part of both the recording of data and for presentation of published results, graphics are an important part of research. Pictures or illustrations can be made in a variety of ways, including hand renderings (drawings, paintings, sketches), photography, and with computer imaging.

Illustrations prepared by hand have long been and will continue to be essential to research. A variety of media are used. The simplest and most direct is pencil and paper, but this suffers from impermanence. Ink and paint are more durable but also require more time and care to produce. All types may be reproduced readily by copy machines and printing processes.

Some basic principles apply to the preparation of biological illustrations (Hodges 1989, Wood 1979). Unless one is an accomplished artist, for the sake of accuracy and efficiency, mechanical aids are usually necessary to obtain proper size, proportions, and shape. Such are the camera lucida and other mirror and prism devices that attach to microscopes. Grids may also be superimposed on microscope fields or on objects directly and drawing made by matching line by line on a corresponding grid on the paper. To allow changes, pencil should be used for preliminary figures and these either inked or painted over directly or transferred to a second, more durable, final surface. Size of the subject should always be indicated on technical drawings, the best way by a scale line to one side. The size of the illustration should fit its purpose: very large for exhibit, two to three times the page printed length for publication, smaller for record.

Photography for research purposes (Blaker 1989, Lefkowitz 1979) generally falls into one of three types: general photography, macrophotography (close-up), and photomicrography. The first can be accomplished with almost any camera and elementary knowledge of proper focusing and lighting. The others require some special equipment and specialized training.

Macrophotography is practiced fairly close to the subject and with the intent of producing an image usually in the range of 0.3 to 2.0 times its actual size. An absolute essential is a single-lens-reflex (SLR) camera, with through-the-lens viewing, to permit accurate focusing and avoid the parallax effect of cameras with view finders. Lenses with focal lengths from 30 to 200 millimeters are used; 50- or 55-millimeters zoom or "macro" (long throw) types serve for the most common magnifications and are the most versatile. Extension tubes, "tele-extenders," or bellows give further choices of magnification.

Pictures are usually taken at small apertures (f16 to f32) to maximize depth of focus. These require extra light for proper exposure, which only stroboscobic lights provide.

Inanimate objects or dead organisms are subjects easily manipulated and lighted. Photography of living insects, spiders and the like, is much more difficult because of their usual uncooperative behavior. Many kinds are very timid and seldom remain quiet or in a proper attitude for the photographer. They may be stunned or stilled with cold or chemicals, but this destroys their normal appearance and may even give erroneous information to viewers of the photograph. Skill, patience, and experience with insect behavior are all necessary prerequisites to this kind of photography. Very good pictures are possible using electronic flash units connected to SLR cameras with a bracket to hold a constant distance from light to subject. The extremely short flash duration stills all motion and allows small apertures to be used which give sharp images over a wide range of magnifications.

Photomicrography is photography through a microscope, which in effect takes the place of the camera's lens. Microscope lenses are much shorter in focal length than those of cameras and are used for magnification much higher than two or three times actual size. Almost any microscope can be fitted with a camera body, but lighting must be carefully controlled to give good color or contrast. Detail is often poor because of the very shallow focusing range of microscope lenses. Often, only a portion of a subject can be shown clearly, and multiple photographs are necessary to tell the whole story. Special types of photomicrographs may be taken with modified microscopes such as the transmission electron (EM) and scanning electron (SEM) microscopes, phase contrast, X-ray, diffraction microscopes, and others. They all have their own applications in entomological research, particularly with work in morphology, histology, and physiology. SEM is now used extensively to depict surface details of very small insect structure, because of its great clarity and depth of focus.

Cinematography is particularly useful in many insect studies, especially the behavior of living, active specimens. The same basic principles that govern still photography apply, with the added restriction of a fixed shutter speed. Special effects are available, however, such as stop action and time lapse exposures, of much utility to behavioral and other analyses. Cameras using film are being rapidly replaced by videotape cameras because of the latter's utility at lower ambient light levels and reusable recording surfaces. Images are also instantly viewable (not needing chemical developing) on a common television monitor.

Computers are available which also have graphics capabilities. Charts, graphs, tables, and even high resolution pictures can be done rapidly with programs designed for this purpose.

## References

BLAKER, A. A. 1989. Handbook for scientific photography. 2d ed. Focal, Boston.

HODGES, E. R. S., ed. 1989. The Guild handbook of scientific illustration. Van Nostrand Reinhold, New York.

LEFKOWITZ, L. 1979. Manual of close-up photography. Amphoto, Garden City, N.Y.

WOOD, P. 1979. Scientific illustration. Van Nostrand Reinhold, New York.

## Identification

Inherent to taxonomic research and essential to all other areas of entomology is the correct identification of specimens under study. Many errors have been committed in both academic studies and in the application of information to control as a result of incorrect species recognition. The process of identifying specimens is difficult, owing to the vast numbers of insect species and the special knowlege needed to work out their identities, and falls properly in the purview of the taxonomist. Even they trust themselves only within their own specialty areas.

Workers are assisted by well-written and illustrated taxonomic papers. These contain various kinds of summaries of identification characteristics, ordered in some way to permit step-by-step analysis of the diverse features often used. The most universal system for this is the taxonomic key, a series of mutually exclusive statements about the organisms, with one of which the specimen must agree. The statements are progressively more and more exclusive and refined until a terminus is reached, which is the name of the taxon. The same result may be had with pictorial keys, diagnostic tables, matrices, and other forms of systematizing characters and their states. Lately, computer programs have been developed for identification as well.

Nontaxonomists are urged to submit properly prepared and preserved specimens to taxonomic authorities. To do this, they must determine who and where the best person is, for which there are few directories, unfortunately. Otherwise, they must inquire of their taxonomist colleagues, who can make recommendations. Sometimes names and addresses can be found by reference to papers on the general group to which the specimens belong. To obtain help with identification, one should also follow accepted protocol, first contacting a prospective collaborator and asking permission to submit specimens. Taxonomists have the prerogative of retaining some material in exchange for determinations and may properly charge a fee for the service in some cases.

Many technical guides exist primarily for identification. Only a few are generally useful in Latin America (Bachmann 1966, Brues et al. 1954, Hollis 1980).

### References

Bachmann, A. O. 1966 [1967]. Nueva clave para determinación de los órdenes de insectos sudamericanos. Soc. Entomol. Argentina Rev. 29: 11–16.

Brues, C. T., A. L. Melander, and F. M. Carpenter. 1954. Classification of insects. Mus. Compar. Zool. Harvard Univ. Bull. 108: 1–917.

Hollis, D., ed. 1980. Animal identification. 3. Insects. Brit. Mus. Nat. Hist., London.

### Publication

The final step in the research process is publication of results (Alley 1987, Trelease 1982). This requires the utmost care and ability and represents the goal of all the foregoing activities. Writing and graphic talents as well as knowledge of the writings of other authorities are called for. Research papers have different formats, depending on purpose and methodology. The results of experimental studies are usually reported under a series of headings: introduction (outlining history, significance, and aims or hypotheses of the study), material and methods (telling precisely how the study was carried out, so that it may be repeated by others), results (which data resulted from the experiment), discussion (presenting varied or contradictory aspects of the results), and conclusions (what the results mean toward proving or disproving the initial hypothesis or goal of the study). The literature cited section at the end gives the references used to guide and substantiate the research.

Taxonomic papers follow other outlines (Mayr and Ashlock 1990), generally with headings such as introduction (as in experimental papers), synonymies (listings of varied names used for the taxa covered), material examined (inventory of specimens used for the study), systematics (presentation of new groupings and descriptions of existing species or newly discovered species). Keys and tables for identification of new and old species are also usually included. Authors of good taxonomic papers have the responsibility of making clear the identity of the taxa included; this is accomplished with precisely written descriptions and keys as well as good illustrations and the placing of voucher specimens in multiple public museums.

Other deductive studies are written up under whatever headings best explain the hypothesis and application of logic to support or disclaim them.

### References

Alley, M. 1987. The craft of scientific writing. Prentice-Hall, Englewood Cliffs, N.J.

Mayr, E., and P. D. Ashlock. 1990. Principles of systematic zoology. 2d ed. McGraw Hill, New York.

Trelease, S. F. 1982. How to write scientific and technical papers. MIT, Cambridge.

### Entomological education

Research on insect material forms a fund of knowledge that ultimately will become available to everyone through the education process. This is formally achieved by teaching in schools, universities, and colleges but is also accomplished through more popular and informal media, such as museums, television, newspapers, and even in amusement parks. Effective in the process of conveying information, especially to unmotivated people, are displays that incorporate actual specimens, photographs, and data into an integrated exposition on some topic or phenomenon. Particularly well received are so-called insect zoos where the public can see at close hand examples of living specimens of spectacular, colorful, or important species. Many such zoos are open in various parts of the world, although none to date in Latin America. Education of the lay public on insect life is immensely important to the future of agriculture, health, and conservation of natural resources.

# Included Insect and Arthropod Taxa

The following table will aid the reader in locating taxa by their classification. All names or orders, families, genera, and species that appear in the book are included plus a few intermediate or higher categories as needed. The arrangement of orders is the same as given in chapter 1; lower taxa within orders are in the approximate sequence of their appearance in the text throughout the book or group according to general evolutionary relationships. Common names are provided only when in general usage or if given in the book. Synonyms are excluded. Asterisk (*) indicates that taxon is figured.

TERRESTRIAL ARTHROPODS OTHER
THAN INSECTS

Phylum Onychophora—onychophorans

Order Onychophora
   Peripatopsidae
      *Metaperipatus*
   Peripatidae
      *Macroperipatus torquatus**
      *Peripatus heloisae*
      *Speleoperipatus speloeus*

Phylum Arthropoda—arthropods
   Subphylum Biramia
   Class Crustacea—crustaceans
   Subclass Percarida

Order Isopoda—isopods
   Tylidae
   Ligiidae—sea roaches
      *Ligia exotica**
      *Ligidium*
   Trichoniscidae

Porcellionidae—woodlice
      *Porcellio laevis**
Oniscidae
      *Trichorhina*
Armadillidiidae—pillbugs
      *Armadillidium vulgare**
Armadillidae—pillbugs

Order Amphipoda—amphipods
   Talitridae—beach hoppers
      *Hyale*
      *Orchestia platensis*—sandflea*

Subphylum Chelicerata

Class Arachnida

Order Araneae—spiders
  Suborder Orthognatha
    Theraphosidae—tarantulas
      *Acanthoscurria*
      *Brachypelma smithi*—Mexican red-legged tarantula
      *Hapalopus*
      *Theraphosa lablondi**
      *Trechona*
  Suborder Labidognatha
    Salticidae—jumping spiders
      *Aphantochilus**
    Ctenidae
      *Ctenus*
      *Phoneutria fera**
      *P. nigriventer*
    Lycosidae—wolf spiders
      *Lycosa raptoria**
    Araneidae—orb web spiders
      *Araneus*
      *Argiope argentata*—silver orb weaver*
      *A. aurantia*—golden orb weaver

*A. trifasciata*—banded orb weaver
*Eustala anistera**
*Mastophora*—bolas spiders
*M. dizzydeani**
*M. gasteracanthoides*
*Nephila clavipes*—golden silk spider*
Gasteracanthinae—spiny orb weavers
    *Gasteracantha cancriformis**
    *G. tetracantha*
    *Micrathena*
Heteropodidae—giant crab spiders
    *Heteropoda venatoria*—huntsman spider*
Selenopidae—giant crab spiders
Theridiidae—comb-footed spiders
    *Anelosimus eximius*
    *Argyrodes*
    *Conopistha*
    *Latrodectus*—widow spiders
    *L. curacaviensis*
    *L. geometricus*
    *L. mactans*—black widow*
    *Mallos gregalis*
Loxoscelidae—loxoscelid spiders
    *Loxosceles*—violin spiders
    *L. laeta**
Clubionidae
    *Castianeira rica*

Order Opiliones—harvestmen
  Suborder Cyphopalpitores
    Gagrellidae
      *Prionostemma**
  Suborder Laniatores
    Cosmetidae
      *Vonones sayi*
    Gonyleptidae
      *Gonyleptus janthinus**
      *Zygopachylus albomarginis*

Superorder Acari—mites and ticks
Order Astigmata
  Analgidae
  Chirodiscidae
  Chirorhynchobiidae
  Dermoglyphidae
  Pyroglyphidae

*Dermatophagoides farinae*—American house dust mite*
*D. neotropicalis*
*D. pteronyssinus*—European house dust mite
  Acaridae
    *Tyrophagus putrescentiae*—mold mite*
  Carpoglyphidae
    *Carpoglyphus lactis*—dried fruit mite
  Sarcoptidae
    *Sarcoptes scabiei*—scabies mite*

Order Prostigmata
  Cheyletidae
  Halacaridae
  Trombiculidae
    *Eutrombicula*—chiggers
    *E. alfreddugesi* group
    *E. batatas*—sweet potato chigger*
    *Iguanacarus*
    *Leptotrombidium*
    *Parascoschoengastia nunezi*
    *Pseudoschoengastia*
  Myobiidae
    *Archemyobia latipilis*
  Eriophyidae—gall mites
    *Eriophyes guerreronis*—coconut mite
    *E. sheldoni*—citrus bud mite*
  Tetranychidae—spider mites
    *Eotetranychus sexmaculatus*—six-spotted mite
    *Metatetranychus citri*
    *Mononychellus*
    *Tetranychus bimaculatus*—two-spotted mite
    *T. cinnabarinus*
    *T. telarius**
  Pyemotidae
    *Pyemotes ventricosus*
  Demodicidae—follicle mites
    *Demodex bovis*
    *D. canis*
    *D. caprae*
    *D. cati*
    *D. equi*

*D. folliculorum*—human follicle mite*
*D. ovis*
*D. phylloides*

Order Mesostigmata
  Laelaptidae
    *Hypoaspis dasypus*
    *Varroa jacobsonii*—varroa mite
  Macrochelidae
    *Macrocheles*
  Arrhenuridae
    *Arrhenurus*
  Antennophoridae
    *Ophiomegistus*
  Halarachnidae
    *Pneumonyssus*
  Dermanyssidae
    *Dermanyssus*
  Spelaeorhynchidae
  Spinturnicidae
  Ascidae
    *Proctolaelaps*
    *Rhinoseius*

Order Cryptostigmata
  Oribatulidae
    *Oribatula minuta**

Order Metastigmata—ticks
  Ixodidae—hard ticks
    *Amblyomma cajennense*—Cayenne tick*
    *A. variegatum*—tropical bont tick
    *Aponomma*
    *Boophilus microplus*—southern cattle tick
    *Dermacentor nitens*—tropical horse tick*
    *Haemaphysalis*
    *Ixodes pararicinus*
    *Rhipicephalus*
  Argasidae—soft ticks
    *Antricola*
    *Argas miniatus**
    *A. moreli*
    *A. persicus*—fowl tick
    *A. transversus*
    *Nothoaspis*
    *Ornithodorus darwini*

*O. galapagensis*
*O. talaje*
*O. rudis*
*Otobius*

Order Uropygi—whip scorpions
  Elyphonidae
    *Mastigoproctus giganteus*—vinegarroon*
    *Thelyphronellus*
  Hypoctonidae
    *Arnauromastigon*

Order Amblypygi—whipless whip scorpions
  Suborder Apulvillata
    Phrynidae
      *Acanthophrynus*
      *Heterophrynus longicornus**
      *Paraphrynus*
      *Phrynus*
    Damonidae
      *Trichodamon*
  Suborder Pulvillata
    Charontidae
      *Charinides*
      *Chirinus*
      *Paracharon*
      *Tricharinus*

Order Pseudoscorpionida—pseudoscorpions
  Chernetidae
    *Cordylochernes scorpioides*
    *Lustrochernes*
  Cheliferidae
    *Chelifer cancroides**
  Cheiridiidae
    *Cheiridium muesorum*
  Withiidae
    *Withius piger*

Order Scorpionida—scorpions
  Buthidae
    *Centruroides limpidus*
    *C. suffusus*—Durango scorpion*
    *Tityus serrulatus**

Order Solpugida—sunspiders
  Daesiidae
    *Amacata penai*

*Syndaesia mastix*
Ammotrechidae
Eremobatidae
    *Eremobates**

Subphylum Uniramia
Class Myriapoda—myriapods
Subclass Chilopoda—centipedes
Order Scolopendromorpha
Scolopendridae
    *Scolopendra gigantea*—giant
    centipede*
Order Geophilomorpha
Order Lithobiomorpha
Order Scutigeromorpha
Scutigeridae
    *Scutigera coleoptrata*—house
    centipede*
Subclass Diplopoda—millipedes
Superorder Helminthomorpha
Order Spirostreptida
Spirostreptidae
    *Orthoporus**
    *Vilcastreptus hoguei*
Order Polydesmida
Platyrhacidae
    *Amplinus*
    *Barydesmus**
    *Nyssodesmus python*
    *Polylepiscus*
    *Pycnotropis*
    *Psammodesmus*
Chelodesmidae
    *Chondrodesmus*
Order Spirobolida
Rhinocricidae
    *Eurhinocricis*
    *Rhinocricus lethifer*

INSECTS

Class Hexapoda—insects
Subclass Parainsecta—subinsects
Order Collembola—springtails
Sminthuridae
    *Temeritas surinamensis**
Entomobryidae

*Ctenocyrtinus prodigus**
Coenaletidae
Subclass Insecta—true insects
Infraclass Apterygota—primitive wingless
    insects
Order Thysanura—thysanurans
Suborder Zygentoma—silverfish
Lepismatidae
    *Ctenolepisma longicaudata*—long-
    tailed house silverfish*
    *Lepisma saccharina*
    *L. wasmanni*
    *Stylifera gigantea*
Maindroniidae
    *Maindronia*
Nicoletiidae
Suborder Microcoryphia—bristletails
Meinertellidae
    *Meinertellus*
    *Neomachillelus scandens**
Machilidae
    *Machilinus*
    *Machiloides*

Infraclass Pterygota—winged insects
Superorder Paleoptera—ancient-winged
    insects
Order Ephemeroptera—mayflies
Baetidae
    *Callibaetis*
Heptageniidae
    *Epearus*
Tricorythidae
    *Tricorythodes*
Leptophlebiidae
    *Nousia*
    *Thraulodes**
Polymitarcyidae
    *Campsurus albicans**
    *Tortopus*
Siphlonuridae
    *Chaquihua*
    *Chiloporter*
    *Metamonius*
    *Siphlonella*

Order Odonata—dragonflies and
damselflies

Suborder Anisoptera—dragonflies
  Petaluridae
    *Phenes raptor*
  Corduliidae
    *Gomphomacromia chilensis*
  Petaliidae
  Libellulidae
    *Diastatops*—black wings
    *D. dimidiata**
    *Libellula herculea*—ruby tail
    *Orthemis ferruginea*—ferruginous skimmer
    *Pantala flavescens*—globetrotter*
    *Perithemis*—amber wings
    *P. indensa**
    *Zenithoptera*—butterfly dragonflies
Suborder Zygoptera—damselflies
  Calopterygidae
    *Hetaerina americana*—ruby spot*
  Coenagrionidae
    *Acanthagrion*
    *Argia vivida**
    *Telchasis*
  Heliocharitidae
  Perilestidae
  Polythoridae
  Pseudostigmatidae
    *Mecistogaster*
    *Megaloprepus coerulatus**

Superorder Neoptera—modern-winged insects

Orthopteroids

Order Plecoptera—stoneflies
  Austroperlidae
  Eustheniidae
  Gripopterygidae
    *Araucanioperla*
    *Pelurgoperla*
  Notonemouridae
    *Neonemura illiesi*
  Perlidae
    *Anacroneuria**
  Nemouridae
    *Amphinemura*
  Diamphipnoidae

Order Grylloptera—katydids and crickets
  Tettigoniidae—katydids
    Decticinae—shield-backed katydids
      *Eremopedes colonialis*
    Pseudophyllinae—broad-winged katydids
      *Ancistrocercus*
      *Celidophylla albimacula*
      *Cocconotus**
      *Cycloptera speculata**
      *Mimetica*
      *Panoploscelus*
      *Pterochroza ocellata**
      *Tanusia*
      *Thliboscelus hypericifolius*—Tananá
      *Typophyllum*
    Phaneropterinae—narrow-winged katydids
      *Aganacris**
      *Championica*
      *Dysonia fuscifrons**
      *Scaphura*
      *Steirodon**
      *Vellea*
    Conocephalinae—cone-headed and meadow katydids
      *Coniungoptera*
      *Conocephalus**
      *Copiphora**
      *Neoconocephalus**
      *Panacanthus**
  Gryllidae—crickets
    Phalangopsinae
      *Amphiacusta annulipes**
      *A. maya*
    Eneopterinae
      *Eneoptera surinamensis**
    Oecanthinae
      *Neoxabea*
      *Oecanthus**
    Gryllinae
      *Acheta domesticus*—house cricket
      *Grylloides supplicans*—Indian house cricket
      *Gryllus assimilis*
  Gryllotalpidae—mole crickets
    *Neocurtilla*
    *Scapteriscus**

*S. abbreviatus*
*S. didactylus*
*S. imitatus*
*S. oxydactylus*

Order Orthoptera—grasshoppers and allies
    Pauliniidae
        *Cornops aquaticum*
        *Marellia remipes*
        *Paulinia acuminata**
    Eumastacidae
        *Eumastax**
    Acrididae—grasshoppers
        *Achurum sumichrasti**
        *Melanoplus*
        *Schistocerca americana*
        *S. cancellata*
        *S. piceifrons*—American locust*
        *Sphenarium*
        *Trimerotropis pallidipennis**
    Romaleidae—lubber grasshoppers
        *Brachystola magna*
        *Chromacris speciosa*—independence grasshopper*
        *Taenopoda eques*
        *T. varipennis**
        *Titanacris gloriosa**
        *T. velazquezii*
        *Tropidacris cristata**
    Proscopiidae—jumping sticks
        *Apioscelis**

Order Blattodea—cockroaches
    Euthyrrhaphidae
        *Holocampsa*
    Nyctiboridae
        *Paratropes*
        *Plectoptera*
    Blattidae
        *Blatta orientalis*—Oriental cockroach*
        *Litopeltis*
        *Neostylopyga rhombifolia*—harlequin cockroach*
        *Periplaneta americana*—American cockroach*
        *P. australasiae*—Australian cockroach*

Blatellidae
    *Blatella germanica*—German cockroach*
    *Megaloblatta*
    *Pseudomops**
    *Supella longipalpa*—brown-banded cockroach*
Oxyhaloidae
    *Leucophaea maderae*—Madeira cockroach*
    *Nauphaeta cinerea*—lobster cockroach*
Atticolidae
    *Achroblatta luteola**
    *Attiphila*
    *Myrmecoblatta**
Epilampridae
    *Epilampra**
Blaberidae
    *Blaberus colosseus*
    *B. craniifer*
    *B. giganteus*—death's-head cockroach*
    *B. parabolicus*
Pycnoscelididae
    *Pycnoscelus surinamensis*—Surinam cockroach*
Panchloridae—green cockroaches
    *Panchlora nivea*—Cuban cockroach*

Order Mantodea—mantids
    Mantidae
        *Acanthops falcataria*—dead leaf mantid*
        *Angela*
        *Choeradodis rhombicollis*—leaf mantid*
        *Liturgusa*—bark mantids*
        *Mantoida maya*
        *Stagmomantis*
        *Stagmotoptera**
        *Vates**

Order Phasmatodea—walkingsticks
    Bacteriidae
        *Bactridium grande*
        *Bostra scabrinota*
        *Otocrania aurita*

Bacunculidae
*Libethra minuscula*
Phibalosomatidae
*Phibalosoma phyllinum**
Anisomorphidae
*Paradoxomorpha crassa*—
chinchemoyo*
Pseudophasmatidae
*Pseudophasma**
*Pterinoxylus spinulosus*

Order Dermaptera—earwigs
Anisolabiidae
*Anisolabis maritima*—maritime
earwig
*Carcinophora americana**
*Euborellia annulipes*—ring-legged
earwig
*Metresura ruficeps**
Forficulidae
*Doru lineare*—lined earwig*
*Forficula auricularia*—European
earwig
Labiduridae
*Labidura riparia*—shore earwig*
Sparattidae
*Sparatta pelvimetra**
Labiidae
*Marava*
Pygidicranidae

Order Isoptera—termites
Termitidae
*Amitermes*
*Cornitermes cumulans*
*Mimeutermes*
*Nasutitermes corniger*
*N. costalis**
*N. fulviceps*
*Neocapritermes braziliensis**
*Syntermes dirus**
Rhinotermitidae
*Coptotermes havilandi*
*C. niger*
Kalotermitidae
*Cryptotermes brevis*
*Incisitermes snyderi*

Order Embiidina—web spinners
Clothodidae

*Clothoda urichi**
Anisembiidae
*Chelicera*
Oligotomidae
*Oligotoma saundersii*

Hemipteroids

Order Psocoptera—psocids
Liposcelidae
*Belaphapsocus*
*Liposcelis bostrychophila*—
booklouse*
Asiopsocidae
*Notiopsocus*
Psocidae
*Ceratipsocus*
*Graphocaecilius*
*Poecilopsocus iridescens**
*Thrysophorus*
Elipsocidae
*Drymopsocus*
Trogiidae

Order Mallophaga—chewing lice
Suborder Amblycera
Abrocomophagidae
Trochiliphagidae
Gyropidae—guinea pig lice
*Gliricola porcelli*—slender guinea
pig louse
*Gyropus ovalis*—oval guinea pig
louse*
Menoponidae
*Bovicola*
*Columbicola columbae*—pigeon
louse
*Menacanthus stramineus*—chicken
louse*
*Menopon gallinae*—shaft louse
*Piagetiella bursaepelecani*—pelican
louse
Ricinidae
Trimenoponidae
Laemobothriidae
*Laemobothrion opisthocomi*
Suborder Ischnocera
Philopteridae
*Paragoniocotes mirabilis**

Trichodectidae—mammal chewing lice
- *Cebidicola*
- *Felicola felis*—cat louse*
- *Geomydoecus*
- *Lymeon*
- *Neotrichodectes*
- *Trichodectes*

Order Anoplura—sucking lice
Haematopinidae
- *Haematopinus suis*—hog louse*
- *Pecaroecus javalii**
- *Solenopotes*

Linognathidae
- *Linognathus peddalis*
- *Microthoracicus mazzai**
- *M. minor*
- *M. praelongiceps*

Hoplopleuridae
- *Hoplopleura*
- *Polyplax*

Pediculidae—primate lice
- *Pediculus humanus capitis*—head louse
- *P. h. corporis*—body louse*
- *Phthirus pubis*—crab louse*

Order Hemiptera—true bugs
Suborder Heteroptera-heteropterans
Coreidae—big-legged bugs
- *Anasa*
- *Diactor bilineatus**
- *Pachylis pharaonis**
- *Paryphes blandus*
- *Thasus acutangulus*

Pentatomidae—stinkbugs
- *Antiteuchus tripterus*
- *Chlorochroa ligata*—conchuela*
- *Edessa**
- *Euschistus*
- *Mormidea*
- *Oebalus poecilus*—rice stinkbug*

Lygaeidae—seed bugs
- *Blissus leucopterus*—chinch bug*
- *Geocoris punctipes*
- *Lygaeus*
- *Oncopeltus*—milkweed bugs

*O. fasciatus*—large milkweed bug*

Miridae—plant bugs
- *Barberiella*
- *Engytatus*
- *Lygus lineolaris*—tarnished plant bug*
- *Monalonion*
- *Paracarnus*

Aradidae—flat bugs
- *Dysodius lunatus*—lunate flat bug*

Tingidae—lace bugs
- *Corythucha gossypii*—cotton lace bug*

Reduviidae—assassin bugs
- *Apiomerus lanipes**
- *A. pictipes*
- *Arilus carinatus*—cogwheel bug*
- *Empicoris rubromaculatus**
- *Graptocleptes*
- *Hiranetix*
- *Notocyrtus vesiculosus*
- *Salyavata variegata**
- *Spiniger**
- *S. ater*

Triatominae—kissing bugs
- *Panstrongylus megistus**
- *Rhodnius pallescens*
- *R. prolixus**
- *Triatoma dimidiata*
- *T. infestens*

Polyctenidae—bat bugs
- *Hesperoctenes**

Cimicidae—bedbugs
- *Cimex hemipterus*—tropical bedbug
- *C. lectularius*—bedbug*
- *Haematosiphon inodorus*—Mexican chicken bug
- *Ornithocoris toledoi*—Brazilian chicken bug
- *Psitticimex*

Pyrrhocoridae—red bugs
- *Thaumastaneis montandoni*

Alydidae
- *Hyalymenus**

Largidae
- *Arhaphe*

Cydnidae—burrowing bugs
  *Amnestus*—pepper flies
Nabidae—damsel bugs
  *Arachnocoris albomaculatus*
Belostomatidae—giant water bugs
  *Lethocerus grandis*
  *L. maximus**
Naucoridae—creeping water bugs
  *Cryphocricos*
Notonectidae—backswimmers
  *Buenoa pallens**
  *Martarega*
  *Notonecta unifasciata*
Corixidae—water boatmen
  *Corisella*
  *Trichocorixa reticulata*—salt marsh
  water boatman*
Gerridae—water striders
  *Gerris remigus*—common water
  strider*
  *Halobates micans**
  *H. robustus*
  *Rheumatobates*
Hydrometridae—water measurers
  *Bacillometra woytkowskii*
Saldidae
Suborder Homoptera—homopterans,
wax bugs
  Cicadidae—cicadas
    *Fidicina chlorogena*
    *F. manifera*
    *Quesada gigas**
    *Zammara smaragdina**
  Membracidae—treehoppers
    *Bocydium**
    *Combophora**
    *Cyphonia*
    *Eucyphonia*
    *Hemikyptha*
    *Heteronotus flavomaculatus**
    *Membracis**
    *Metcalfiella monogramma*—
    periquito del aquacate
    *Polyglypta*
    *Spongophorus**
    *Umbonia spinosa**
  Aetalionidae
    *Aetalion reticulatum*

Cicadellidae—leafhoppers
  *Amblyscartidia albofasciata**
  *Baleja flavoguttata**
  *Dalbulus*
  *Empoasca kraemeri**
  *Saccharosydne saccharivora*—cane
  leafhopper
Cercopidae—spittlebugs,
froghoppers
  *Aeneolamia varia saccharina*—
  sugarcane froghopper*
  *Tomapsis inca**
Aleyrodidae—whiteflies
  *Dialeurodes*
  *Trialeurodes*
Aphididae—aphids
  *Acyrthosiphon pisum*—pea aphid*
  *Aphis fabae*—bean aphid
  *A. gossypii*—cotton aphid*
  *A. sacchari*
  *Sipha flava*—yellow sugarcane
  aphid*
  *Therioaphis maculata*—spotted
  alfalfa aphid*
  *Toxoptera aurantii*—black citrus
  aphid
Superfamily Coccoidea—scale insects
and mealybugs
  Coccidae—tortoise scales
    *Icerya purchasi*—cottony cushion
    scale*
    *Neolecanium sallei*
  Margarodidae—giant coccids
    *Llaveia axin*—axin*
    *Margarodes formicarum*—ground
    pearls*
    *M. vitium*
    *Termitococcus*
  Diaspididae—armored scales
    *Aonidiella aurantii*—California
    red scale*
    *Aspidiotus destructor*—coconut
    scale
    *Lepidosaphes beckii*—purple scale
    *Quadraspidiotus perniciosus*—San
    Jose scale
    *Sclenaspidus articulatus*—West
    Indian red scale

Pseudococcidae—mealybugs
  *Dysmicoccus brevipes*—pineapple
  mealybug
  *Planococcus citri*
  *Pseudococcus adonidum*
  *P. comstocki*
  *P. longispinus*—longtailed
  mealybug*
Eriococcidae—mealy scales
Dactylopidae—cochineal scales
  *Dactylopius coccus**
  *D. opuntiae*
  *D. tomentosus*
Superfamily Fulgoroidea
  Fulgoridae—planthoppers
  *Cathedra serrata**
  *Cerogenes auricoma*—flying mouse*
  *Fulgora laternaria*—dragon-
  headed bug*
  *Lystra strigata**
  *Phenax variegata**
  *Phrictus diadema**
  *Pterodictya reticularis*—reticulate
  planthopper*
Acanaloniidae
Derbidae
Cixiidae
  *Myndus crudus*
Flatidae

Order Thysanoptera—thrips
  Phlaeothripidae
  *Leptothrips mali*—black hunter*
  Thripidae
  *Arachisothrips*
  *Chaetanaphothrips*—banana thrips
  *Dasythrips regalis*
  *Frankliniella parvula*—banana
  flower thrips
  *F. tritici*—wheat thrips
  *Heliothrips haemorrhoidalis*—
  greenhouse thrips*
  *Hercinothrips bicinctus*—banana
  thrips
  *Scirtothrips*
  *Selemnothrips rubrocinctus*—cacao
  thrips
  *Taeniothrips simplex*—gladiolus
  thrips

*Thrips*
Uzelothripidae
  *Franklinothrips vespiformis*—
  vespiform thrips
    Neuropteroids
Order Megaloptera—alderflies and
dobsonflies
  Sialidae—alderflies
  Corydalidae-dobsonflies
  *Archichaulides*
  *Chloronia*
  *Corydalus armatus*
  *C. cornutus**
  *Platyneuromus*
  *Protochauliodes*

Order Neuroptera—nerve-winged insects
  Coniopterygidae—dustywings
  Myrmeliontidae—antlions
  *Brachynemurus*
  *Dimares*
  *Glenurus peculiaris**
  *Maracandula*
  *Morocordula apicalis*
  *Myrmeleon**
  *Navasoleon*
  *Vella*
  Mantispidae—mantispids
  *Anchieta*
  *Climaciella**
  *Drepanicus gayi*
  Chrysopidae—lacewings
  *Ceraeochrysa*
  *Chrysopa slossonae*
  *Chrysoperla**
  *Leucochrysa*
  Hemerobiidae—brown lacewings
  Ascalaphidae—owlflies
  *Albardia furcata*
  *Ameropterus*
  *Corduleceris maclachlani**
  *Ululodes*
    Panorpoids
Order Diptera—flies and midges
  Tipulidae—crane flies
  *Tipula**
  Blephariceridae
  Chironomidae—water midges
  *Chironomus**

*Siolimyia amazonica*
Psychodidae—moth flies
    *Clogmia albipunctata*—bathroom fly\*
    *Maruina*
    *Psychoda alternata*
Phlebotominae—sand flies
    *Lutzomyia*\*
    *L. colombiana*
    *L. longipalpis*
    *L. verrucarum*
Ceratopogonidae—punkies
    *Atrichopogon*
    *Bezzia*
    *Culicoides*\*
    *C. furens*
    *Dasyhelea*
    *Forcipomyia*
    *Lasiohelea*
    *Leptoconops*
    *Microhelea*
    *Palpomyia*
    *Pterobosca*
Simuliidae—blackflies
    *Simulium*\*
    *S. amazonicum*
    *S. callidum*
    *S. metallicum*
    *S. ochraceum*
Culicidae—mosquitoes
  Anophelinae
    *Anopheles*—malaria mosquitoes
    *A. albimanus*
    *A. bellator*
    *A. cruzii*
    *A. darlingi*\*
    *A. gambiae*
    *A. pseudopunctipennis*
    *Chagasia*
  Toxorhynchitinae
    *Toxorhynchites*—giant mosquitoes\*
    *T. haemorrhoidalis*
    *T. theobaldi*
  Culicinae
    *Aedes*—Aedes mosquitoes
    *A. aegypti*—yellow fever mosquito\*
    *A. taeniorhynchus*—salt marsh mosquito

*Coquillettidea*
*Culex*—Culex mosquitoes
    *C. bahamensis*
    *C. opisthopus*
    *C. quinquefasciatus*—southern house mosquito\*
*Culiseta particeps*
*Deinocerites*—crabhole mosquitoes
    *D. cancer*\*
*Galindomyia leei*
*Haemagogus*—blue devils\*
*Limatus*
*Mansonia*
*Orthopodomyia*
*Phoniomyia*
*Psorophora*—gallinippers
*Sabethes*\*
*Trichoprosopon digitatum*
*Uranotaenia*
*Wyeomyia*
Bibionidae
Cecidiomyiidae—gall midges
    *Latrophobia brasiliensis*—manioc gall midge
Tabanidae—horseflies
    *Chlorotabanus*
    *Chrysops*
    *Dichaelacera*
    *Fidena*
    *Lepiselaga crassipes*—mosca congo\*
    *Scaptia lata*—colihuacho
    *Scione*
    *Tabanus dorsiger*\*
Conopidae—conopid flies
    *Stylogaster*
Mydidae—mydas flies
    *Mydas*\*
    *M. rubidapex*
Asilidae
Bombyliidae
Pantophthalmidae—timber flies
    *Opetiops*
    *Pantophthalmus*\*
Phoridae
    *Apocephalus paraponerae*
    *Melaloncha*
Syrphidae—flower flies
    *Copestylum*
    *Eristalis tenax*—drone fly\*

*Metasyrphus americanus\**
*Ornidia obesa*—green flower fly*
Stratiomyidae—soldier flies
    *Hermetia illuscens*—wasp fly*
    *Merosargus*
Ephydridae—shore flies
    *Dimecoenia*
Drosophilidae—pomace flies
    *Drosophila carcinophila*
    *D. endobranchia*
    *D. melanogaster\**
Lonchaeidae—lonchaeid flies
    *Neosilva perezi*—cassava shoot fly
Tephritidae—fruit flies
    *Anastrepha fraterculus*—South
    American fruit fly
    *A. ludens*—Mexican fruit fly
    *A. suspensa*—Caribbean fruit fly
    *Ceratitis capitata*—Mediterranean
    fruit fly*
    *Euxesta*
    *Rhagoletis lycopersella*
    *Toxotripana curvicauda*—papaya
    fruit fly
Milichiidae
    *Pholeomyia*
    *Phyllomyza*
Braulidae
    *Braula coeca*
Chloropidae—fruit flies
    *Hippelates*
    *Liohippelates pusio* complex—eye
    gnats*
    *Pseudogaurax*
Coelopidae
Chamaemyiidae
    *Paraleucopis mexicana*
Micropezidae—stilt-legged flies
    *Plocoscelus arthriticus*
    *Taeniaptera\**
Agromyzidae
Anthomyiidae
    *Fucellia*—kelp flies*
    *F. maritima*
Muscidae—muscid flies
    *Fannia canicularis*—lesser house
    fly*
    *Haematobia irritans*—horn fly*
    *Limnophora*

    *Musca domestica*—house fly*
    *Muscina stabulans*—green house
    fly
    *Neivamyia*
    *Ophyra aenescens*—black garbage
    fly*
    *Philornis*
    *Stomoxys calcitrans*—stable fly*
Sarcophagidae—flesh flies
    *Bercaea haemorrhoidalis\**
    *Dexosarcophaga*
    *Doringia acridiorum*
    *Peckia*
    *Sarcophaga*
Calliphoridae—carrion flies
    *Calliphora*
    *Chrysomya*
    *Cochliomyia hominovorax*—
    screwworm*
    *C. macellaria*—secondary
    screwworm
    *Lucilia illustris*—greenbottle fly
    *Phaenicia cuprina*
    *P. eximia*
    *P. sericata*—green blowfly*
    *Phormia regina*—black blowfly
Tachinidae—tachinid flies
    *Androeuryops*
    *Calodexia*
    *Lixophaga diatraeae*—Cuban fly
    *Metagonistylum minense*—Amazon
    fly
Cuterebridae—robust botflies
    *Alouattamyia*
    *Cuterebra*
    *Dermatobia hominis*—human
    botfly*
Oestridae—bot flies
    *Oestrus ovis*—sheep botfly*
Hypodermatidae—cattle grubs
    *Hypoderma bovis*—northern cattle
    grub
    *H. lineatum*—common cattle
    grub*
Gasterophilidae—horse botflies
    *Gasterophilus haemorrhoidalis*—
    nose botfly
    *G. intestinalis*—horse botfly*
    *G. nasalis*—throat botfly

Hippoboscidae—louse flies
  *Hippobosca equina*—horse louse fly
  *Lipoptena mazamae*
  *Melophagus ovinus*—sheep ked
  *Olfersia fassulata*\*
  *Pseudolynchia canariensis*—pigeon louse fly
Streblidae—bat flies
  *Trichobius dugesii*\*
Nycteribiidae—bat tick flies
  *Basilia ferrisi*\*

Order Siphonaptera—fleas
  Pulicidae
    *Ctenocephalides canis*—dog flea
    *C. felis*—cat flea\*
    *Leptopsylla segnis*—mouse flea
    *Nosopsyllus fasciatus*—northern rat flea
    *Pulex irritans*—human flea
    *Xenopsylla cheopis*—Oriental rat flea\*
  Tungidae
    *Echidnophaga gallinacea*—sticktight flea
    *Tunga penetrans*—burrowing flea\*
  Ceratophyllidae
    *Ceratophyllus*
  Dolichopsyllidae
    *Dasypsyllus lasius*\*
  Ischnopsyllidae
  Malacopsyllidae
  Rhopalopsyllidae
  Pygiopsyllidae
  Stephanocircidae—helmeted fleas

Order Trichoptera—caddisflies
  Anamolopsychidae
  Helicophidae
  Helicopsychidae
    *Helicopsyche*
  Hydrobiosidae
    *Atopsyche callosa*\*
  Hydropsychidae
    *Leptonema albovirens*\*
  Hydroptilidae
  Kokiriidae

Lepidostomatidae
Leptoceridae
  *Atanatolica*
  *Grumichella*
  *Hudsonema*
  *Nectopsyche punctata*\*
  *Notalina*
  *Triplectides*
Limnephilidae
Philorheithridae
Sericostomatidae
  *Grumicha*
  *Phylloicus*
Stenopsychidae
Tasimiidae

Order Lepidoptera—butterflies and moths

Moths
Saturniidae—wild silk moths
  Ceratocampinae
    *Citheronia*—regal moths
    *C. laocoon*\*
    *Eacles*—imperial moths
    *E. imperialis decoris*\*
  Arsenurinae
    *Arsenura*
    *A. ponderosa*
    *Copiopteryx semiramis*\*
    *Dysdaemonia*
    *Loxolomia*
    *Paradaemonia*
    *Rhescyntis*
  Saturniinae
    *Copaxa*—copaxas
    *C. cydippe*
    *C. decrescens*
    *C. lavendera*\*
    *C. moinieri*
    *Lonomia achelous*
    *Rothschildia*—window-winged saturnians
    *R. aurota*
    *R. erycina*\*
    *R. orizaba*\*
  Hemileucinae
    *Automerella*
    *Automeris*—eyed saturnians

A. illustris*
Dirphia—dirphias
D. avia*
Gamelia
Hylesia—hylesias
H. canitia
H. lineata*
H. metabus
Hyperchiria
Leucanella
Paradirphia
Pseudautomeris
Sphingidae—sphinx moths
  Macroglossinae
    Erynnis—ashy sphingids
    E. ello—ashy sphinx*
    Eumorpha—harlequin sphingids
    E. fasciata*
    E. labruscae
    Hemeroplanes—viper worms
    H. ornatus*
    Isognathus
    Pachylia ficus—fig sphinx*
    Pseudosphinx tetrio—frangipani sphinx*
  Sphinginae
    Agrius cingulata
    Amphimoeca walkeri
    Cocytius antaeus—giant sphinx*
    Manduca quinquemaculata—tomato hornworm
    M. sexta—tobacco hornworm*
Sematuridae—eyetails
  Nothus luna*
Uraniidae—rainbow moths
  Urania fulgens
  U. leilus*
Arctiidae—arctiids
  Arctiinae—tiger moths
    Bertholdia
    Cratoplastis diluta
    Eucereon
    Hypercompe decora*
    Idalus herois*
    Leuconopsis
    Opharus bimaculatus
    Paranerita
    Viviennea moma*

  Ctenuchinae—wasp moths
    Antichloris viridis
    Correbia*
    C. lycoides
    Correbida assimilis
    Macrocneme chrysitis*
    Pseudopompilia
    Pseudosphex
  Pericopinae—flag moths
    Chetone angulosa*
    Daritis howardi*
    Dysschema jansoni
    D. leucophaea*
Lithosiidae
  Ptychoglene coccinea
  P. phrada
Zygaenidae—smoky moths
  Harrisina tergina*
  Seryda constans
Dioptidae—dioptid moths
  Dioptis restricta*
  Josia
Castniidae—giant day-flying moths
  Castnia cyparissias
  C. licoides*
  Microcastnia
Agaristidae—forester moths
Noctuidae—owlet moths
  Erastrinae
    Cydosia
  Agrotinae
    Agrotis ipsilon—black cutworm
    A. malefida—palesided cutworm
    A. subterranea—granulate cutworm*
    Euxoa
    Mamestra
    Mocis
    Peridroma saucia—variegated cutworm*
    Polia
    Prodenia
    Spodoptera exigua—beet armyworm*
    S. frugiperda—fall armyworm
    S. latifascia—lateral lined armyworm

S. *ornithogalli*—yellow-striped
armyworm
Heliothidinae
    *Helicoverpa zea*—corn earworm*
Catocalinae
    *Alabama argillacea*—cotton leaf
    worm*
Plusiinae
    *Pseudoplusia includens*
    *Rachiplusia ou*—upsilon looper*
    *Trichoplusia ni*—cabbage looper*
Ophiderinae
    *Ascalapha odorata*—black witch*
    *Dipthera festiva* —hieroglyphic
    moth*
    *Thysania aggrippina*—birdwing
    moth*
    *T. zenobia*—Zenobia's birdwing
    moth*
Hadeninae
    *Pseudaletia adultera*
    *P. unipuncta*—armyworm*
    *Xanthopastis timais*—Spanish
    moth*
Notodontidae—prominents
    *Cliara croesus**
Lasiocampidae—lappet moths
    *Euglyphis cribraria**
    *Gloveria psidii*
Geometridae—measuring worm
moths
    *Atyria dicroides*
    *Pantherodes pardalaria*—polka dot
    moth*
Megalopygidae—flannel moths
    *Endobrachus revocans*
    *Megalopyge lanata**
    *Podalia*
    *Trosia*
Limacodidae—shag moths
    *Acharia nesea**
    *Phobetron hipparchia*—monkey
    slug*
    *Stenoma cecropia*
Dalceridae
    *Dalcerina tijucana*
Lymantriidae—tussock moths
    *Elnoria noyesi*

Bombycidae—silk moths
    *Bombyx mori*—domestic silk moth
Tineidae
    *Tinea pellionella*—case-bearing
    clothes moth
    *Tineola bisselliella*—webbing
    clothes moth
Tischeriidae
Nepticulidae
Psychidae—bagworm moths
    *Oiketicus kirbyi**
    *O. platensis*
Tortricidae—tortricids
    *Cydia deshaisiana*—Mexican
    jumping bean moth*
Lyonetiidae—lyonetiids
    *Leucoptera coffeella*—coffee leaf
    miner
Pyralidae—pyralid moths
  Epipaschiinae
    *Chilozela*
    *Macalla thrysisalis*—mahogony
    webworm
  Chrysauginae
    *Cryptoses choloepi*—sloth moth*
  Crambinae
    *Diatraea centrella*
    *D. considerata*
    *D. grandiosella*
    *D. magnifactella*
    *D. saccharalis*—sugarcane borer*
    *Myelobia smerintha**
  Phycitinae
    *Anagasta kuehniella*—
    Mediterranean flour moth*
    *Cactoblastis cactorum*—cactus
    moth*
    *Ephestie*
    *Plodia interpunctella*—Indian
    meal moth*
  Galleriinae
    *Galleria mellonella*—greater wax
    moth*
Gelechiidae
    *Pectinophora gossypiella*
    *Phthorimaea operculella*
    *Sitotroga cerealella*—Angoumois
    grain moth

Cossidae—cossid moths
>    *Comadia redtenbacheri*—agave
>    worm moth*

Butterflies
Papilionidae—swallowtail butterflies
>    *Battus*
>    *Eurytides*—kites
>    *E. bellerophon**
>    *E. philolaus**
>    *Papilio*—true swallowtails
>    *P. anchisiades*
>    *P. andraemon*
>    *P. cresphontes*—giant swallowtail
>    *P. homerus*
>    *P. multicaudata*
>    *P. thaos*—giant swallowtail*
>    *P. zagreus*
>    *Parides*—aristolochias
>    *P. ascanius*
>    *P. iphidamas**
Lycaenidae—blues, hairstreaks, and
metalmarks
>    Riodininae—metalmarks
>    *Amarynthis menaria**
>    *Chlorinea faunus**
>    *Helicopis acis**
>    *Juditha molpe**
>    *Thisbe irenea*
>    Lycaeninae—blues and
>    hairstreaks
>    *Arawacus aetolus**
>    *Chliaria*
>    *Tmolus basilides*—pineapple
>    hairstreak
Pieridae—whites and sulfurs
>    *Ascia monuste*—great southern
>    white*
>    *Catasticta semiramis**
>    *Colias lesbia*
>    *Dismorphia amphiona**
>    *Eucheira socialis*—madrone
>    butterfly
>    *Phoebis sennae*—cloudless sulfur*
>    *Pieris brassicae*—European
>    cabbage butterfly
Nymphalidae—brush-footed
butterflies

Satyrinae—satyrs
>    *Argyrophorus argenteus*—silver-
>    winged butterfly*
Danainae—monarch butterflies
>    *Danaus cleophile*—Jamaican
>    monarch
>    *D. eresimus*—soldier
>    *D. erippus*—southern monarch
>    *D. gilippus*—queen
>    *D. plexippus*—monarch*
>    *Lycorea cleobaea*—large tiger
>    *L. halia**
>    *L. ilione*
Nymphalinae—nymphalines
>    *Anaea*
>    *Anartia amathea*—red anartia*
>    *A. chrysopelea*
>    *A. fatima*—fatima
>    *A. jatrophae*—white peacock
>    *A. lytrea*
>    *Callicore*
>    *Callidula*
>    *Catacore*
>    *Colobura dirce*—head-for-tail
>    butterfly*
>    *Consul fabius**
>    *Diaethria astala*
>    *D. clymena**
>    *Dynamine*
>    *Eresia phillyra**
>    *Hamadryas*—crackers
>    *H. feronia**
>    *Historis odius*—cecropia
>    butterfly*
>    *Memphis*
>    *M. arachne*
>    *Paulogramma*
>    *Perisama*
>    *Siproeta stelenes*—malachite
>    green*
Acraeinae
Ithomiinae—ithomiines
>    *Greta*
>    *Hypoleria andromica**
>    *Hypothyris*
>    *Mechanitis polymnia**
>    *Melinaea ethra**
>    *Oleria*

Heliconiiae—passion vine
butterflies
    *Agraulis vanillae*—gulf fritillary*
    *Dryas iulia*—Julia*
    *Heliconius charitonius*—zebra
    butterfly
    *H. erato*\*
    *H. ismenius*\*
    *H. melanops*\*
    *Laparus doris*
    *Philaethria dido*\*—green
    heliconius
Morphinae—morphos
    *Morpho achillaena*—Achilles
    morpho*
    *M. hecuba*\*—hecuba
    *M. peleides*
    *M. polyphemus*
    *M. rhetenor*
Brassolinae
    *Caligo*—owl butterflies
    *Dynastor darius*—Darius*
Hesperiidae—skippers
    *Calpodes ethlius*—canna skipper*
    *Hylephila phyleus*—fiery skipper
    *Urbanus proteus*—bean leafroller*
    *Chioides*
    *C. eurilochus*\*
    *Elbella polyzona*\*
    *Phocides thermus*
    *Tarsoctenus papias*
    *Jamadia gnetus*
Megathymidae
    *Aegiale hesperiaris*—agave worm
    butterfly*

    Orders of Uncertain Affinities

Order Coleoptera—beetles
Carabidae—ground beetles
    *Agra*\*
    *Calosoma alternans*\*
    *Enceladus gigas*
    *Eurycoleus*
    *Lebia*
    *Notiobia peruviana*\*
    *Selenophorus*\*
Cincindelinae—tiger beetles
    *Cincindela carthagena*\*

    *Ctenostoma*
    *Megacephala*\*
    *Odontocheila*\*
    *Pseudoxychila bipustulata*\*
Cucujidae—flat bark beetles
    *Oryzaephilus*
Nitiduidae
    *Haptoncus*
    *Mystrops*
Dermestidae—dermestid beetles
    *Anthrenus*
    *Dermestes*
    *Thylodrius*
Bostrichidae—branch borers
    *Apate*
Anobiidae—death-watch beetles
    *Lasioderma serricorne*—cigarette
    beetle
    *Stegobium paniceum*—drugstore
    beetle
Lyctidae—powderpost beetles
    *Lyctus*
Tenebrionidae—darkling beetles
    *Cuphotes*\*
    *C. immaculipes*
    *Gyriosomus*\*
    *Mylaris*\*
    *Nyctelia*\*
    *Poecilesthus*
    *Proacis bicarinatus*\*
    *Scotobius gayi*\*
    *Strongylium*\*
    *Tauroceras*\*
    *Tenebrio*
    *Tribolium*
    *Trogoderma granarium*
    *Zophobas*—attic beetles*
Zopherinae
    *Zopherus chilensis*—ma'kech*
Erotylidae—giant fungus beetles
    *Cypherotylus dromedarius*\*
    *Erotylus*\*
    *Priotelus*
    *Pselaphicus giganteus*
Dytiscidae—predaceous diving
beetles
    *Cybister*\*
    *Megadytes giganteus*\*

Hydrophilidae—water scavenger beetles
*Berosus**
*Gillisius*
*Hydrophilus insularis**
*Tropisternus lateralis**
Helodidae—marsh beetles
Gyrinidae—whirligig beetles
*Dineutus*
*Gyretes**
*Gyrinus**
Histeridae—hister beetles
Ptiliidae—feather-winged beetles
*Nanosella fungi*
Silphidae
Staphylinidae—rove beetles
*Amblyopinus**
*Bledius**
*Cryptomimus*
*Dioploeciton*
*Ecitophya**
*Odontolinus*
*Paederus irritans*—whiplash beetle*
*Quichuana*
*Spirachtha*
*Termitogaster**
*Termitonannus*
Limulodidae—horseshoe crab beetles
Psephenidae—water-penny beetles
Dryopidae
Elmidae
Elateridae—click beetles
*Aeolus*
*Chalcolepidius bonplanni**
*Conoderus**
*Pyrophorus*—headlight beetles
*P. nyctophanus**
*Semiotus**
Lampyridae—fireflies
*Aspisoma*
*Cratomorphus*
*Lucidota**
*Photinus**
*Photuris**
Cantharidae—soldier beetles
Phengodidae—glowworms

*Phrixothrix**
Passalidae—passalid beetles
*Passalus**
*Ptichopus*
Lucanidae—stag beetles
*Chiasognathus granti*—Chilean stag beetle*
Scarabaeidae—scarab beetles
Aphodiinae
*Ataenius*
Scarabaeinae—dung scarabs
*Canthidium*
*Canthon*—dung rollers
*C. smaragdulum**
*Coprophanaeus lancifer**
*Dichotomius carolinus*—black dung beetle*
*Eurysternus deplanatus**
*Glaphyrocanthon*
*Liatongus*
*Onthophagus*
*Phanaeus*—dung diggers
*P. demon**
*Scarabaeus*
*Trichillum*
*Uroxys*
*U. gorgon*
Dynastinae—horned scarabs
*Cyclocephala**
*Dynastes hercules*—Hercules beetle*
*D. hyllus*
*D. neptunus*
*D. satanas*
*Enema pan*—pan beetle*
*Golofa aegeon*
*G. eacus*
*G. porteri*—caliper beetle*
*Megaceras jasoni*—great horned scarab
*Megasoma actaeon*
*M. elephas*—elephant beetle*
*M. mars*
*Oryctes*
*Strategus*—ox beetles
*S. aloeus**
*S. oblongus*—coconut rhinoceros beetle

S. talpa—sugarcane rhinoceros beetle
Cetoniiae—flower scarabs
  Cotinus mutabilis—green fruit beetle*
  Gymnetis holocericae circumdata*
Rutelinae—shiny scarabs
  Chrysina*
  Chrysophora chrysochlora—green-gold beetle*
  Heterosternus
  Macropoidelimus
  Macropoides
  Paraheterosternus
  Pelidnota sumptuosa*
  P. virescens
  Plusiotus batesi—gold beetle*
  P. chrysargyrea—silver beetle
Melolonthinae—June beetles
  Macrodactylus—cockchafers*
  Phyllophaga portiricensis*
Buprestidae—metallic wood borers
  Agrilus
  Euchroma gigantea—giant metallic ceiba borer*
Coccinellidae—ladybird beetles
  Cycloneda sanguinea*
  Epilachna paenulata—melon beetle
  E. tredecimnotata—southern squash beetle*
  E. varivestis—Mexican bean beetle
  Rodolia cardinalis—vedalia beetle
Lycidae—net-winged beetles
  Calopteron brasiliense*
  Lycus arizonensis
  L. fernandezi
Oedemeridae—false blister beetles
Meloidae
Platypodidae—ambrosia beetles
Scolytidae—bark beetles
  Hypothenemus hampei—coffee borer*
  Platypus parallelus
  Xyleborus ferrugineus—cacao borer*
Cerambycidae—long-horned beetles

Acrocinus longimanus—harlequin beetle*
Acyphoderes sexualis*
Ancistrotus cummingi
Callichroma velutinum
Callipogon—imperious sawyers
C. armillatum—giant imperious sawyer*
C. barbatum—bearded imperious sawyer*
C. senex
Elytropeltus apicalis
Eplophorus velutinus
Hypocephalus armatus—mole beetle*
Lycoplasma
Macrodontia cervicornis—giant jawed sawyer*
M. dejeani—big jawed sawyer*
M. flavipennis
Neoptychodes trilineatus—three-lined fig tree borer*
Oncideres sara*
Plinthocoelium*
Psalidognathus friendi*
P. modestus
Schwarzerion
Stenygra
Taeniotes scalaris*
Thelgetra*
Tillomorpha
Titanus giganteus—titanic longhorn*
Chrysomelidae—leaf beetles
Alticinae
Donaciinae
  Eumolpinae
    Colaspis hypochlora
  Cassidinae—tortoise beetles
    Acromis spinifex*
    Charidotis circumducta*
    Coptocycla arcuata
    Cyclosoma mirabilis*
    Omaspides pallidipennis
    Omocerus eximius*
    Polychalca
    Stolas cyanea*
    Tauroma

Hispinae—mining leaf beetles
  *Cephaloleia*
  *Chelobasis*—rolled-leaf hispine
  beetles
  *C. bicolor**
  *Pseudocalaspidea cassidea**
  *Xenarescus monocerus*
Galerucinae
  *Diabrotica undecempunctata*—
  spotted cucumber beetle*
Bruchidae—seed beetles
  *Acanthoscelides**
  *Callosobruchus*
  *Pachymerius nucleorum*—bicho de
  coco
Curculionidae—weevils
  *Anthonomus grandis*—cotton boll
  weevil
  *Cosmopolites sordidus*—banana
  weevil
  *Entimus*—jeweled weevils
  *E. imperialis*—jeweled weevil*
  *E. nobilis*
  *Rhinostomus barbirostris*—bearded
  weevil*
  *Rhynchophorus cruentatus*
  *R. palmarum*—palm weevil*
  *Sitophilus*
Brentidae—brentids
  *Brentus anchorago**

Order Hymenoptera—wasps, ants, and
bees
  Suborder Symphyta
    Pergidae—pergid sawflies
      *Acordulecera*
    Argidae—argid sawflies
    Orussidae
    Tenthredinidae—common sawflies
      *Syzgonia cyanocephala*
      *Waldheimia ochra**
    Siricidae—horntails
      *Sirex*
      *Urocerus californicus**
      *U. gigas gigas*
      *U. patagonicus*
  Suborder Apocrita
    Ichneumonidae—ichneumon wasps

  *Rhynchophion*
  *Tetragonochora*
  *Thyreodon**
Braconidae—braconid wasps
  *Apanteles congregatus**
  *Iphiaulax*
Superfamily Chalcidoidea—chalcids
Torymidae
  *Idarnes*
Myrmaridae
  *Alaptus*
Eurytomidae
  *Bruchophagus platyptera*
  *Desantisca*
Trichogrammatidae
  *Trichogramma minutum*—minute
  egg parasite*
Eulophidae
  *Aphelinus mali**
Eupelmidae
Encyrtidae
  *Tetracnemus peregrinus*
Pteromalidae
Chalcididae
Agaonidae—fig wasps
  *Blastophaga dugesi**
  *Tetrapus*

Other Superfamiles
  Cynipidae—gall wasps
    *Atrusca spinuli**
  Scelionidae
  Chrysididae—cuckoo wasps
    *Neochrysis carina**
    *Trichrysis*
  Mutillidae—velvet ants
    *Hoplocrates*
    *Hoplomutilla xanthocerata*
    *Leucospilomutilla*
    *Pappognatha myrmiciformis*
    *Traumatomutilla indica**
  Scoliidae—mammoth wasps
    *Campsomeris ephippium**
  Pompilidae—spider wasps
    *Pepsis**
  Sphecidae—digger wasps
    *Ammophila*
    *Bembix*—sand wasps

B. americana
B. citripes*
Larra
Microstigmus comes*
Sceliphron—mud daubers
S. asiaticum
S. assimile*
S. fistularium
Trypoxylon albitarsi*
Eumenidae—caterpillar hunters
Eumenes—potter wasps
E. consobrinus*
Montezumia azurescens*
Pachodynerus galapagensis
P. nasidens—wanderer*
Zethus matzicatzin*
Zeta
Vespidae—social wasps
Polistinae
Apoica—parasol wasps
A. pallens*
Brachygastra—honey wasps
B. lecheguana*
Chartergus chartarius—bell wasp*
Mischocyttarus—long-waisted
paper wasps
M. ater
M. drewseni*
Parischnigaster
Polistes—Polistes paper wasps
P. canadensis—varied paper
wasp*
P. carnifex
P. dorsalis clarionensis
P. erythrocephalus
P. fuscatus
Polybia—polybia paper wasps
P. dimidiata
P. emaciata
P. jurinei
P. rejecta
P. scutellaris*
P. singularis
Synoeca—drumming wasps
S. cyanea
S. septentrionalis
S. surinama*
Vespinae

Formicidae—ants
Ponerinae—giant hunting ants
Dinoponera australis
D. gigantea*
Ectatomma tuberculatum—kelep*
Odontomachus—trap jaw ants*
Pachycondyla villosa—cobra ant*
Paraponera clavata*
Ecitoninae—army ants
Eciton
E. burchelli
E. hamatum*
Labidus
Neivamyrmex
Myrmicinae
Acromyrmex
Allomerns
Atta—leaf cutter ants
A. cephalotes
A. sexdens
Cephalotes*
Crematogaster—acrobat ants
C. limata
C. stolli*
Daceton
Pheidole—big-headed ants
P. fallax*
Pogonomyrmex
Solenopsis—fire ants
S. geminata
S. invicta*
S. richteri
S. saevissima
Wasmannia auropunctata
Zacryptocerus—cork-head ants
Z. maculatus
Z. texanus
Z. varians*
Dolichoderinae
Azteca—aztec ants
A. alfari
A. chartifex
A. delpini
A. instabilis*
A. muelleri
A. olitris
A. traili
A. trigona

*A. ulei*
*Dolichoderus*
*Hypoclinea*
*Iridomyrmex humilis*—Argentine ant*
*Tapinoma*—stink ants
*T. melanocephalum*
  Formicinae
    *Camponotus*—carpenter ants
    *C. abdominalis*
    *C. femoratus*
    *C. planatus*
    *C. rufipes*
    *C. senex*
    *C. sericeiventris**
  Pseudomyrmecinae
    *Pseudomyrmex*—fever ants
    *P. ferrugineus**

Superfamily Apoidea—bees
  Megachilidae—leaf cutter bees
    *Chrysosarus*
    *Cressoniella*
    *Megachile leucographa**
    *Pseudocentron*
  Halictidae
    *Agopostemon*
    *Augochlora*

*Augochlorella*
*Lasioglossum**
Anthophoridae
  *Anthophora*
  *Centris*—centris bees
  *C. inermis**
  *Xylocopa*—carpenter bees
  *X. darwini*
  *X. fimbriata**
  *X. frontalis*
  *X. ornata*
Apidae
  *Aglae*
  *Apis mellifera mellifera*—honeybee*
  *A. m. scutellata*—African bee
  *Bombus*—bumblebees
  *B. dahlbomi*
  *B. tucumanus**
  *Eufriesea*
  *Euglossa purpurea**
  *Eulaema meriana**
  *Exeaerete*
  *Lestrimelitta limao*
  *Melipona beecheii*
  *Trigona duckei*
  *T. fulviventris**
  *T. fuscipennis**
  *T. jaty*

# Index

All taxa have been indexed at the generic or species level, and all arthropods and plants are indexed by family. Common names for arthropod species are indexed, but vernacular names and common names at the family level or higher are generally not indexed. References to illustrations are printed in boldface.

Designer:     Linda Robertson
Compositor:   Huron Valley Graphics, Inc.
Text:         10/12 Baskerville
Display:      Helvetica
Printer:      Malloy Lithographing, Inc.
Binder:       Malloy Lithographing, Inc.